Sustainable Co
Structures

Woodhead Publishing Series in Civil and Structural Engineering

Sustainable Concrete Materials and Structures

Edited by

Ashraf Ashour

Xinyue Wang

Baoguo Han

Woodhead Publishing is an imprint of Elsevier
50 Hampshire Street, 5th Floor, Cambridge, MA 02139, United States
125 London Wall, London EC2Y 5AS, United Kingdom

ISBN: 978-0-443-15672-4 (print)

ISBN: 978-0-443-15673-1 (online)

For information on all Woodhead Publishing publications
visit our website at https://www.elsevier.com/books-and-journals

Publisher: Mathew Deans
Acquisitions Editor: Gwen Jones
Editorial Project Manager: Fernanda Oliveira
Production Project Manager: Surya Narayanan Jayachandran
Cover Designer: Christian Bilbow

Typeset by MPS Limited, Chennai, India

Contents

List of contributors

Musab Alhawat Faculty of Engineering and Digital Technologies, University of Bradford, Bradford, United Kingdom; Departement of Civil Engineering, Elmergib University, Al Khums, Libya

Mazen Al-Kheetan Civil and Environmental Engineering Department, College of Engineering, Mutah University, Mutah, Jordan

Yazeed Al-Noaimat Department of Civil and Environmental Engineering, Brunel University, London, Uxbridge, Middlesex, United Kingdom

Ali Alsalman Department of Civil Engineering, University of Basrah, Basrah, Iraq

Ashraf Ashour Faculty of Engineering and Digital Technologies, University of Bradford, Bradford, United Kingdom

Farhad Aslani Materials and Structures Innovation Group, School of Engineering, The University of Western Australia, WA, Australia

Joseph Assaad Department of Civil and Environmental Engineering, University of Balamand, Al Koura, Lebanon

Lateef N. Assi Terracon Consultants Inc., Tulsa, OK, United States

Waqas Latif Baloch Department of Civil Engineering, Toronto Metropolitan University, Toronto, ON, Canada

Jad Bawab Department of Civil and Environmental Engineering, United Arab Emirates University, Al Ain, United Arab Emirates

Shin Hau Bong Department of Civil and Environmental Engineering, National University of Singapore, Singapore

Alexander S. Brand Charles Edward Via, Jr. Department of Civil and Environmental Engineering, Virginia Polytechnic Institute and State University, Blacksburg, VA, United States

Kealy Carter Darla Moore School of Business, University of South Carolina, Columbia, SC, United States

Beng Wei Chong Ingram School of Engineering, Texas State University, San Marcos, TX, United States

Mehdi Chougan Department of Civil and Environmental Engineering, Brunel University, London, Uxbridge, Middlesex, United Kingdom

Solomon Debbarma Department of Civil Engineering, Indian Institute of Technology Bombay, Mumbai, Maharashtra, India

Siqi Ding Department of Civil and Environmental Engineering, The Hong Kong Polytechnic University, Hong Kong, P.R. China

Sufen Dong School of Infrastructure Engineering, Dalian University of Technology, Dalian, P.R. China

Hongjian Du Department of Civil and Environmental Engineering, National University of Singapore, Singapore

Antonella D'Alessandro Department of Civil and Environmental Engineering, University of Perugia, Perugia, Italy

Hilal El-Hassan Department of Civil and Environmental Engineering, United Arab Emirates University, Al Ain, United Arab Emirates

Seyed Hamidreza Ghaffar Department of Civil Engineering, University of Birmingham, Dubai International Academic City, Dubai, United Arab Emirates

Baoguo Han School of Civil Engineering, Dalian University of Technology, Dalian, Liaoning, P.R. China; School of Infrastructure Engineering, Dalian University of Technology, Dalian, P.R. China

Payam Hosseini Genex Systems, Turner-Fairbank Highway Research Center, McLean, VA, United States

Rahman S. Kareem Department of Structure, Shatrah Technical Institute, Southern Technical University, Shatrah, Dhi Qar, Iraq

Jamal Khatib Faculty of Engineering, Beirut Arab University, Beirut, Lebanon; Faculty of Science and Engineering, University of Wolverhampton, Wolverhampton, United Kingdom

Anil Kul Institute of Science, Hacettepe University, Beytepe, Ankara, Turkey; Department of Civil Engineering, Hacettepe University, Ankara, Turkey

Mohamed Lachemi Department of Civil Engineering, Toronto Metropolitan University, Toronto, ON, Canada

Yushi Liu School of Civil Engineering, Harbin Institute of Technology, Harbin, P.R. China; Key Lab of Structures Dynamic Behavior and Control of the Ministry of Education, Harbin Institute of Technology, Harbin, P.R. China; Key Lab of Smart Prevention and Mitigation of Civil Engineering Disasters of the Ministry of Industry and Information Technology, Harbin Institute of Technology, Harbin, P.R. China

Dong Lu School of Civil Engineering, Harbin Institute of Technology, Harbin, P.R. China; Key Lab of Structures Dynamic Behavior and Control of the Ministry of Education (Harbin Institute of Technology), Harbin, P.R. China

Emircan Ozcelikci Institute of Science, Hacettepe University, Beytepe, Ankara, Turkey; Department of Civil Engineering, Hacettepe University, Ankara, Turkey

Yunshi Pan School of Civil Engineering, Harbin Institute of Technology, Harbin, P.R. China

Mustafa Sahmaran Department of Civil Engineering, Hacettepe University, Ankara, Çankaya, Turkey

Mostafa Seifan The University of Waikato, School of Engineering, Hamilton, New Zealand

Xijun Shi Ingram School of Engineering, Texas State University, San Marcos, TX, United States

Hocine Siad Department of Civil Engineering, Toronto Metropolitan University, Toronto, ON, Canada

Rafat Siddique Civil Engineering Department, Thapar Institute of Engineering and Technology, Patiala, India

Amandeep Singh Sidhu Civil Engineering Department, Thapar Institute of Engineering and Technology, Patiala, India

Surender Singh Department of Civil Engineering, Indian Institute of Technology Madras, Chennai, Tamil Nadu, India

Tong Sun School of Civil Engineering, Dalian University of Technology, Dalian, Liaoning, P.R. China

Meltem Tanguler-Bayramtan Department of Civil Engineering, Faculty of Engineering, Cukurova University, Adana, Turkey

Xinyue Wang School of Civil Engineering, Tianjin University, Tianjin, P.R. China

Ismail Ozgur Yaman Department of Civil Engineering, Faculty of Engineering, Middle East Technical University, Ankara, Turkey

Yingzi Yang School of Civil Engineering, Harbin Institute of Technology, Harbin, P.R. China; Key Lab of Structures Dynamic Behavior and Control of the Ministry of Education, Harbin Institute of Technology, Harbin, P.R. China; Key Lab of Smart Prevention and Mitigation of Civil Engineering Disasters of the Ministry of Industry and Information Technology, Harbin Institute of Technology, Harbin, P.R. China

Gurkan Yildirim Faculty of Engineering and Digital Technologies, University of Bradford, Bradford, United Kingdom; Department of Civil Engineering, Hacettepe University, Ankara, Turkey

Feng Yu School of Infrastructure Engineering, Dalian University of Technology, Dalian, P.R. China

Kunyang Yu School of Civil Engineering, Harbin Institute of Technology, Harbin, P.R. China

Maziar Zareechian Department of Civil Engineering, Toronto Metropolitan University, Toronto, ON, Canada

Mingzhong Zhang Department of Civil, Environmental and Geomatic Engineering, University College London, London, United Kingdom

Yifan Zhang Materials and Structures Innovation Group, School of Engineering, The University of Western Australia, WA, Australia

Hui Zhong Department of Civil, Environmental and Geomatic Engineering, University College London, London, United Kingdom

Jing Zhong School of Civil Engineering, Harbin Institute of Technology, Harbin, P.R. China; Key Lab of Structures Dynamic Behavior and Control of the Ministry of Education (Harbin Institute of Technology), Harbin, P.R. China

Wei Zhou School of Civil Engineering, Harbin Institute of Technology, Harbin, P.R. China

Introduction

1

Ashraf Ashour[1], *Xinyue Wang*[2] *and Baoguo Han*[3]
[1]Faculty of Engineering and Digital Technologies, University of Bradford, Bradford, United Kingdom, [2]School of Civil Engineering, Tianjin University, Tianjin, P.R. China, [3]School of Civil Engineering, Dalian University of Technology, Dalian, Liaoning, P.R. China

1.1 Background and context

Concrete is the most widely used construction material and the most used substance in the world, after water owing to its unique combination of properties and versatility, providing the foundation of modern development. In 2020, it was estimated that concrete accounts for 40% of human-made mass that accumulates to approximately 30 Gt of mass per year (Elhacham et al., 2020), as shown in Fig. 1.1. It can be seen from Fig. 1.2 that concrete possesses excellent mechanical properties, especially high compressive strength and resistance to cyclic loading, ensuring the structural integrity and longevity of concrete buildings and infrastructure.

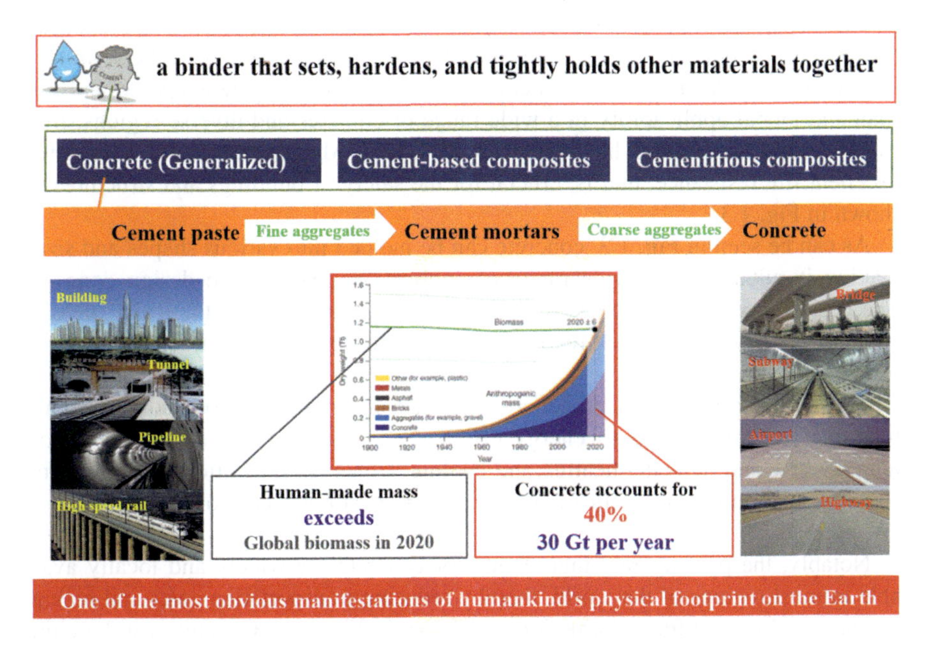

Figure 1.1 Widespread utilization of concrete in modern infrastructure.

Sustainable Concrete Materials and Structures. DOI: https://doi.org/10.1016/B978-0-443-15672-4.00001-2

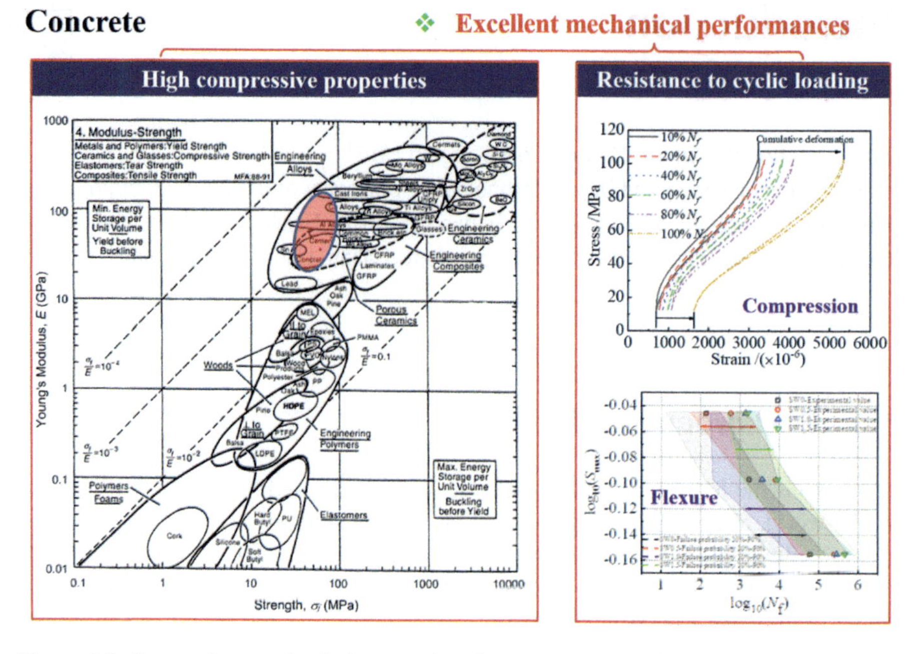

Figure 1.2 Outstanding mechanical properties of concrete.

Concrete, further, exhibits durability in harsh environments, including exposure to moisture, chemicals, and weathering. This resilience guarantees the long-term structural integrity of concrete structures, thereby prolonging their operational lifespan and global applicability in a wide range of climates and diverse regions, from extreme cold to extreme heat (Han et al., 2017). Moreover, it inherently offers excellent fire resistance, that is, crucial for the safety of buildings and structures, as shown in Fig. 1.3.

As can be seen in Fig. 1.3, concrete can be molded into various shapes and sizes, making it suitable for a wide range of applications and diverse design concepts. The design flexibility of concrete also allows for creative and esthetically pleasing structures. Its historical track record in construction provides confidence in its performance and reliability.

The high thermal mass of concrete enables it to absorb, store, and gradually release heat, contributing to energy efficiency in buildings by stabilizing indoor temperatures and reducing heating and cooling demands. In addition, its remarkable sound insulation characteristics play a crucial role in noise control within buildings (Wang et al., 2021).

Notably, the primary raw ingredients for concrete are widely and locally available in most regions of the world, thereby reducing transportation costs and making it a practical choice for construction (as shown in Fig. 1.4). Concrete can be either cast on-site or prefabricated off-site, allowing for relatively quick construction to

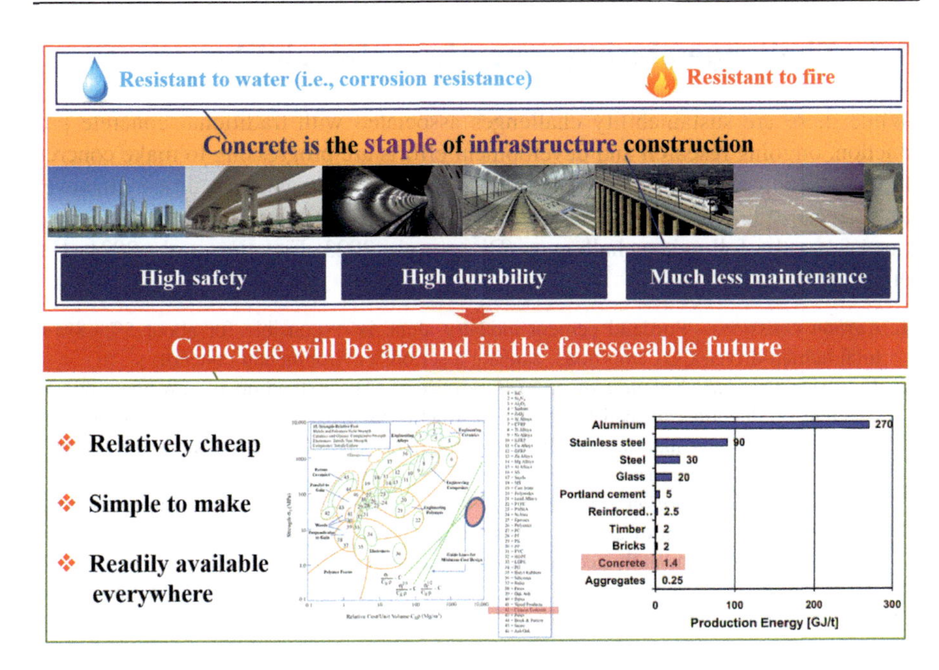

Figure 1.3 Other unique properties and versatility of concrete.

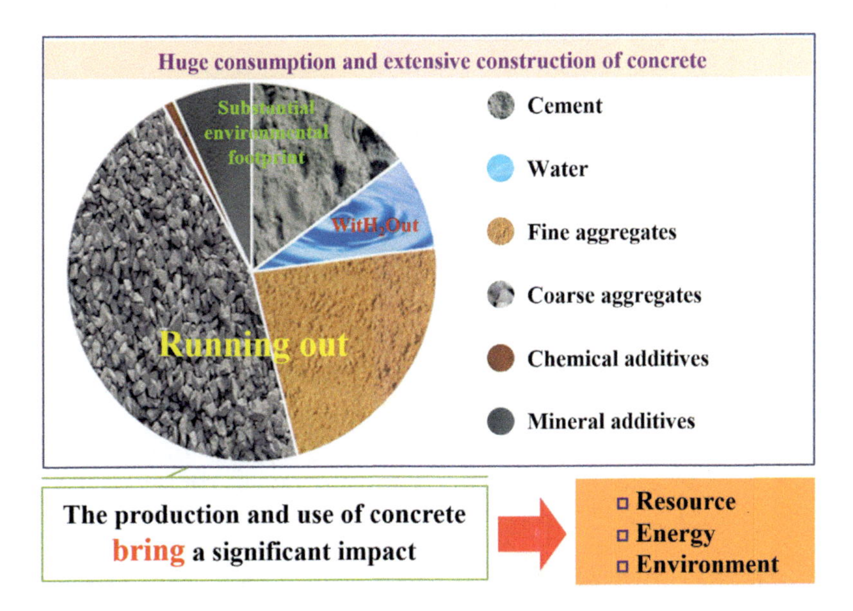

Figure 1.4 Concrete impact on resources, energy consumption, and environment.

meet tight project schedules. Nevertheless, given its extensive utilization and substantial construction demands, concrete production has significant impact on resources, energy consumption, and environment (as shown in Fig. 1.4).

1.2 Challenges facing concrete sustainability

While there are sustainability challenges associated with traditional concrete production, ongoing research and development efforts are progressing to make concrete more environmentally responsible while preserving its essential qualities. However, as shown in Fig. 1.5, the sustainable development of concrete materials and structures faces the following challenges (Ding et al., 2023; Figueira et al., 2021; Han et al., 2014, 2019):

1. *Carbon emissions*—Portland cement industry alone is responsible for about 5%–8% of total human-driven CO_2 emissions, causing a serious environmental concern.
2. *Excessive use and depletion of natural resources*—The extraction of raw materials for concrete production can lead to habitat destruction and resource depletion. Furthermore, some concrete raw materials, such as natural river sand and fresh water, are scarce in some countries or regions.
3. *High energy consumption and pollution*—The process of fabricating cement clinkers requires a great amount of energy. Moreover, the process of aggregate extraction and construction process are usually accompanied by dust and noise pollution.
4. *Waste generation*—Construction and demolition waste, which often includes concrete, is a significant contributor to landfills.
5. *Low tensile strength*—The tensile strength of traditional cementitious materials is quite low, leading to formation of cracks in concrete structures. These cracks compromise the ability of concrete to prevent corrosion in steel reinforcements, thereby reducing the extended durability of concrete structures.

Figure 1.5 Challenges of sustainable development for concrete materials and structures.

6. *Lack of multifunctionality and smart performances*—Traditional concrete only used as a structural material, lacking additional functionality (e.g., self-sensing, self-healing, self-cleaning, energy-harvesting properties).

Overall, collaborative initiatives among academia, industry, and government bodies are essential to address the challenges facing concrete sustainability through a multifaceted approach that combines research, innovation, collaboration, and applications. The potential pathways for the sustainable advancement of concrete materials and structures are expected to include (Alhawat et al., 2022; Figueira et al., 2021; Han et al., 2017; Ozcelikci et al., 2023; Wang et al., 2021), as shown in Fig. 1.6.

1. Developing alternative cementitious materials that are less energy intensive and produce fewer carbon emissions during production, including low-carbon and carbon-neutral cement.
2. Exploring methods to capture and utilize CO_2 emissions from cement production to reduce its environmental impact.
3. Enhancing cement performance aiming to improve the durability of concrete structures through innovations in materials science, for example, self-healing concrete.
4. Sustainable construction practices, including optimizing mix designs, reducing over-design, reduction of cement/concrete consumption, and minimizing waste during construction.
5. Utilization of supplementary cementitious materials and other industrial wastes as well as regional natural materials, reducing the transportation of concrete ingredients.

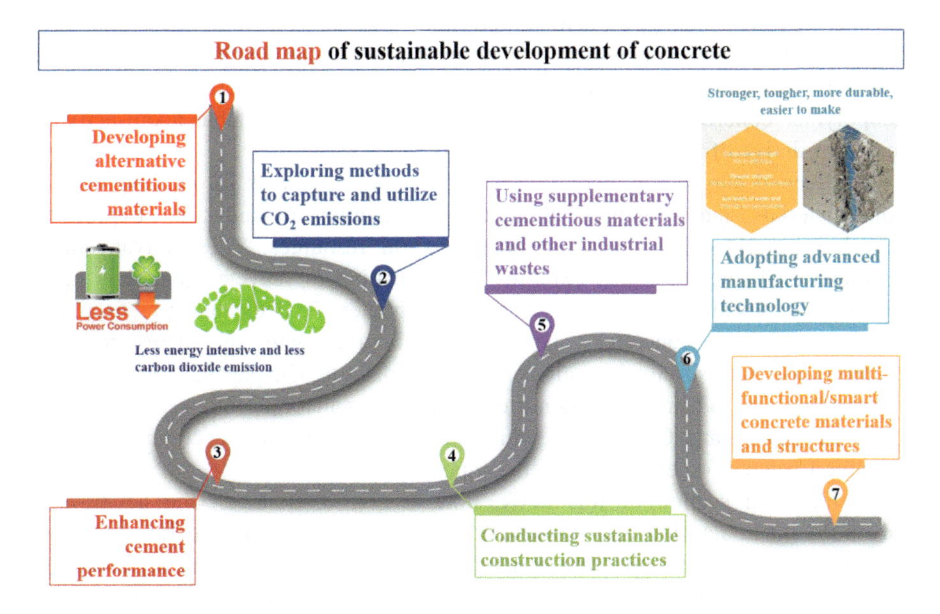

Figure 1.6 Potential pathways for the sustainable advancement of concrete materials and structures.

6. Adoption of advanced manufacturing/construction technology of concrete raw materials, concrete materials, and concrete structures. Furthermore, the use of digital technologies to develop advanced modeling and simulation tools that are used to optimize concrete structures and reduce material use and energy consumption.
7. Developing multifunctional or smart concrete materials and structures—Multifunctional concrete materials incorporate properties beyond traditional strength and durability, for example, self-healing, self-sensing, and even energy harvesting capabilities. The development of multifunctional and smart concrete materials and structures not only contributes to the advancement of construction technology but also offers immense potential for enhancing the performance and sustainability of infrastructure.

To date, substantial efforts have been dedicated to advancing sustainable concrete materials and structures, resulting in significant achievements in both innovative approaches and practical implementations.

1.3 Scope and objectives

The main objective of this book is to provide an overview of the current state of fundamental concepts, cutting-edge methodologies and techniques, and promising applications, aimed at extending the longevity and sustainability of concrete materials and structures. In the first part, this book presents novel varieties of high-performance cement and binders that are not only less energy intensive but also result in reduced CO_2 emissions. These developments are discussed in detail within Chapter 2, which focuses on cement and innovative sustainable binders, Chapter 3, exploring sustainable concrete incorporating supplementary cementitious materials, Chapter 4, which explores the mechanical and durability properties of sustainable geopolymer concrete, and Chapter 5, centered around sustainable alkali-activated materials derived from construction and demolition waste.

The second part of this book outlines recent developments for creating sustainable materials and structures through the utilization of industrial byproducts, recycled materials, and locally sourced natural resources. These advancements encompass Chapter 6, which explores the utilization of recycled materials in the creation of sustainable pervious concrete, Chapter 7, dedicated to sustainable recycled aggregate concrete, Chapter 8, which investigates the incorporation of crumb rubber into sustainable self-compacting concrete, Chapter 9, focusing on the development of sustainable cementitious composites incorporating recycled aggregates and fibers, and Chapter 10, which investigates into the development of sustainable fiber-reinforced geopolymer composites.

In the third section of this book, notable contributions on the adoption of advanced manufacturing and construction technology for concrete raw materials, concrete materials, and concrete structures are covered. This includes Chapter 11, which examines sustainable additive manufacturing of concrete with low-carbon materials, Chapter 12, dedicated to the advancements, challenges, and future prospects of sustainable 3D printing concrete, Chapter 13, exploring emerging resources for the development of low-carbon cementitious composites for 3D printing applications, and Chapter 14, which discusses sustainable 3D-printed concrete structures using high-quality secondary raw materials.

Moving to the fourth part of this book, Chapter 15 explores sustainable seawater sea-sand concrete materials and structures, Chapter 16 focuses on sustainable ultra-high-performance concrete materials and structures, and Chapter 17 presents sustainable nanoconcrete materials and structures. These chapters elucidate the latest techniques aimed at enhancing the sustainability of concrete materials and structures by modifying their performance characteristics.

In the final section, this book highlights sustainable concrete materials and structures with functional or smart capabilities. This includes Chapter 18, which explores sustainable concrete incorporating phase change materials for thermal energy storage, Chapter 19, dedicated to smart sustainable concrete materials and structures, Chapter 20, which discusses various demountable connections for future reuse of structural concrete elements, Chapter 21, focusing on the integration of new technology with sustainable concrete materials and structures, and Chapter 22, which examines CO_2 capture and storage methods for sustainable concrete production.

In summary, this book offers a comprehensive compilation of the latest developments in sustainable concrete materials and structures, contributed by experts in this field. Its primary objective is to inspire readers and industry professionals to adopt and invest in cutting-edge sustainable advancements in concrete materials and structures, ultimately leading to substantial environmental and economic benefits. This book can serve as a valuable resource and point of reference for scientists, engineers, industrial professionals, as well as postgraduate and undergraduate researchers with interests in civil engineering, materials science, and chemical engineering. This is particularly relevant for individuals whose research and focus revolve around concrete materials and structures.

References

Alhawat, M., Ashour, A., Yildirim, G., Aldemir, A., & Sahmaran, M. (2022). Properties of geopolymers sourced from construction and demolition waste: A review. *Journal of Building Engineering*, *50*, 104104. Available from https://doi.org/10.1016/j.jobe.2022.104104.

Ding, S., Wang, X., & Han, B. (2023). *New-generation cement-based nanocomposites*. Springer. Available from https://doi.org/10.1007/978-981-99-2306-9.

Elhacham, E., Ben-Uri, L., Grozovski, J., Bar-On, Y. M., & Milo, R. (2020). Global human-made mass exceeds all living biomass. *Nature*, *588*, 442–444. Available from https://doi.org/10.1038/s41586-020-3010-5.

Figueira, D., Ashour, A., Yıldırım, G., Aldemir, A., & Şahmaran, M. (2021). *Demountable connections of reinforced concrete structures: Review and future developments*, *Structures* (34, pp. 3028–3039). Available from https://doi.org/10.1016/j.istruc.2021.09.053.

Han, B., Ding, S., Wang, J., & Ou, J. (2019). *Nano-engineered cementitious composites: Principles and practices*. Springer. Available from https://doi.org/10.1007/978-981-13-7078-6.

Han, B., Yu, X., & Ou, J. (2014). *Self-sensing concrete in smart structures*. Elsevier. Available from https://doi.org/10.1016/C2013-0-14456-X.

Han, B., Zhang, L., & Ou, J. (2017). *Smart and multifunctional concrete toward sustainable infrastructures*. Springer. Available from https://doi.org/10.1007/978-981-10-4349-9.

Ozcelikci, E., Kul, A., Gunal, M. F., Ozel, B. F., Yildirim, G., Ashour, A., & Sahmaran, M. (2023). A comprehensive study on the compressive strength, durability-related parameters and microstructure of geopolymer mortars based on mixed construction and demolition waste. *Journal of Cleaner Production, 396*. Available from https://doi.org/10.1016/j.jclepro.2023.136522.

Wang, X., Dong, S., Ashour, A., & Han, B. (2021). Energy-harvesting concrete for smart and sustainable infrastructures. *Journal of Materials Science, 56*(29), 16243–16277. Available from https://doi.org/10.1007/s10853-021-06322-1.

Cement and innovative sustainable binders

2

Meltem Tanguler-Bayramtan[1] *and Ismail Ozgur Yaman*[2]
[1]Department of Civil Engineering, Faculty of Engineering, Cukurova University, Adana, Turkey, [2]Department of Civil Engineering, Faculty of Engineering, Middle East Technical University, Ankara, Turkey

2.1 Introduction to Portland cement and concrete

For more than centuries, concrete has been utilized in making resilient structures, including buildings, roads, dams, and other critical infrastructure and is renowned as the material that forms the backbone of our modern society. Among the many advantages like versatility and longevity, concrete is also known as a cost-effective and energy-efficient material, with the production of 1 m^3 of concrete requiring approximately 2–3 GJ of energy, equivalent to 0.9–1.3 MJ/kg in terms of mass (Fig. 2.1). As a result, it has earned its reputation as the most extensively utilized construction material, which is estimated to be around 20–25 billion cubic meters (m^3) per year. However, due to its widespread use, the cumulative energy demand for concrete

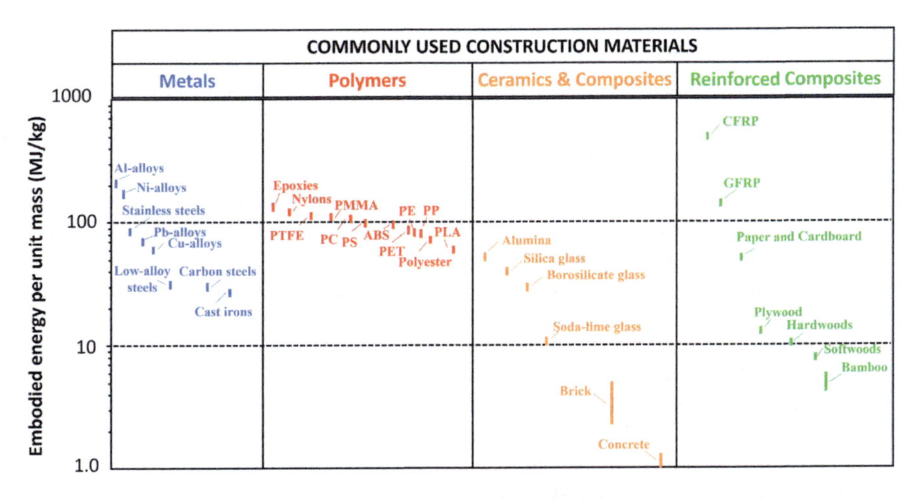

Figure 2.1 Embodied energies of construction materials. Concrete stands out among commonly used construction materials with its energy efficiency, requiring only 2–3 GJ/m^3, equivalent to 0.9–1.3 MJ/kg in mass terms.
Source: Modified from Ashby, M. F. (2016). *Materials and sustainable development*. Butterworth-Heinemann.

Sustainable Concrete Materials and Structures. DOI: https://doi.org/10.1016/B978-0-443-15672-4.00002-4

production is substantial. Consequently, there is an urgent need to minimize the energy consumption and the associated carbon footprint of concrete to ensure its sustainability. Thus, the sustainability of the construction industry can be enhanced upon reducing energy consumption and adopting low-carbon practices in concrete production.

On the other hand, when considering the main activities and ingredients involved in concrete production, it becomes evident that the majority of energy consumption and carbon footprint is associated with the production of Portland cement (PC). PC serves as the essential binding material or "glue" of concrete, undergoing chemical reactions upon mixing with water to bind the inert aggregates together. In simple terms, PC is produced by pulverizing a small amount of naturally or artificially obtained gypsum with an industrial product known as PC clinker.

The production process of PC clinker primarily involves grinding of the raw materials, mainly limestone and clay, and mixing them in appropriate proportions. The mixture is then burned in a kiln at temperatures ranging from 1400°C to 1500°C until it sinters and partially fuses into small clinker balls. As a result, cement production is listed as one of the energy-intensive industries, contributing to approximately 6%–10% of man-made carbon dioxide emissions (UN Environment, 2017). These emissions originate not only from the fossil fuels required to achieve the necessary temperatures in the process but also from the calcination process of limestone within the cement kiln.

Since its first industrial production in the 1890s, PC production has been increasing rapidly. Over the past decade, the average global cement production was estimated at 4.17 billion tons (Cembureau, 2021). Given the substantial production volume and its consequent contribution to global man-made emissions, the cement sector is compelled to prioritize sustainable development and actively pursue innovative approaches to reduce its carbon dioxide emissions. Numerous studies have focused on potential emissions reductions in the cement industry (IEA & WBCSD, 2018; UN Environment, 2017). Through various scenarios, baseline emissions, and future demand forecasts, these studies reach similar conclusions, highlighting the impacts of the following five key levers for reducing carbon emissions in cement production:

- **Improving thermal and electric efficiency:** Implementing state-of-the-art technologies in new cement plants and retrofitting energy-efficient equipment in the existing facilities to improve thermal and electric efficiency.
- **Use of alternative fuels and raw materials:** Utilizing less carbon-intensive fossil fuels and increasing the use of alternative fuels, including biomass, in the cement production process. This approach allows for the utilization of waste materials that would otherwise be incinerated, landfilled, or improperly disposed of.
- **Clinker substitution:** Substituting carbon-intensive clinker, which is an intermediate-product in cement manufacturing, with other materials that have lower carbon content but still possess cementitious properties.
- **Carbon capture and storage (CCS):** Capturing carbon dioxide (CO_2) emissions before they are released into the atmosphere and securely storing them to prevent future emissions.
- **Development of innovative sustainable binders:** Researching and developing innovative low-carbon binders that exhibit similar or better mechanical and durability performance when compared to traditional PC.

These five key levers encompass a range of strategies and technologies that can contribute to the reduction of carbon emissions in the cement industry while promoting more sustainable practices. This chapter will look into the last lever, that is, the innovative sustainable binders.

The widespread use of PC can be attributed, in part, to its composition which consists of elements readily available in the earth's crust. As depicted in Fig. 2.2, PC production primarily relies on the five most abundant elements: oxygen, silicon, aluminum, iron, and calcium (Haynes et al., 2017). Therefore, when developing low-carbon binders, besides the wastes of other industrial processes, it is advantageous to leverage from these abundant resources.

In this regard, we will first focus on calcium (Ca)-based hydraulic cements, such as calcium aluminate cements (CACs), calcium sulfoaluminate (CSA) cements, belite-rich cements, alinite cements, and limestone calcined clay cements (LC3). These binders utilize the abundant availability of calcium as the primary element. We will then explore magnesium (Mg)-based cements which utilize the seventh element depicted in the figure. Finally, we will also look into some other innovative methods, either in the production of the binder itself or in the production of the concrete, some of which will be covered in more detail in the subsequent chapters of the book.

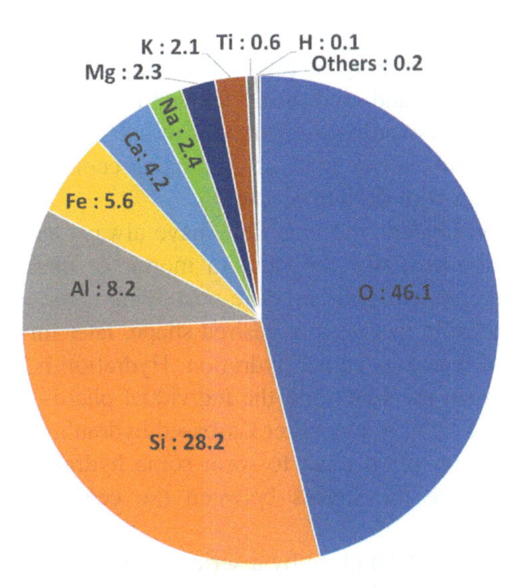

Figure 2.2 The estimated abundance of the elements in the continental crust, in percent. The extensive utilization of Portland cement is due to its composition, relying on the five most abundant elements in the Earth's crust: oxygen, silicon, aluminum, iron, and calcium. *Source*: Data from Haynes, W. M., Lide, D. R., & Bruno, T. J. (Eds.). (2017). *CRC handbook of chemistry and physics* (97th ed.). CRC Press – Taylor & Francis Group.

2.2 Calcium-based hydraulic cements

The use of Ca-based binders in building construction dates back to ancient times, including the Neolithic ages. For instance, at the archeological site of Çatalhöyük in central Anatolia, Türkiye, it was reported that "Marl," a natural material rich in carbonate minerals, clays, and silt, was used as a plastering material to cover the internal surfaces of mudbrick walls (Çamurcuoğlu & Siddall, 2016). Additionally, since ancient times Ca-based binders have been employed in various applications such as masonry mortars between bricks or stones, foundations for flooring, infillings for walls (rubble mortars), casings for water conduits, jointing compounds for terracotta pipes, and even decoration mortars (Elsen, 2006).

Over time, Ca-based binders gradually evolved into hydraulic PC, which was patented with the same name, by Joseph Aspdin in 1824. However, it is generally agreed that today's PCs, that will be called ordinary Portland cement (OPC), bear closer resemblance to those produced by Isaac Charles Johnson in 1850 (Erdoğan & Erdoğan, 2007).

In OPC, the main constituents or compounds responsible for the hydraulic binding properties are calcium silicates, specifically alite (C_3S)[1] and belite (C_2S) phases (Table 2.1). These two compounds, upon hydration, contribute to the development of cement's strength and binding properties. In addition to calcium silicates, OPC also contains aluminate (C_3A) and ferrite (C_4AF) compounds, which not only influence cement properties such as setting time and strength but are also valuable from a manufacturing perspective. They facilitate the cement manufacturing process by aiding in the fusion and sintering of raw materials, promoting particle agglomeration for efficient grinding, and controlling the setting characteristics through reactions with gypsum. Understanding the behavior of calcium silicates, aluminate, and ferrite-bearing compounds is essential for optimizing cement production and ensuring high-quality cement with desired properties.

Ca-based hydraulic binders and thus OPC have always been used with water to bind other inert substances, like the sand in mortar or plaster, and the fine and coarse aggregates in concrete. This mixture occupying a certain volume is initially plastic, thus it is workable to give any desired shape, later turning into a solid volume through a set of reactions called hydration. Hydration is the chemical process that takes place between the water and the individual phases of a hydraulic binder. Therefore whether it is OPC or any other Ca-based hydraulic cement, the anhydrous cement phases and water often react to form some hydrate phases, that form the solid volume and bridge the spaces between the cement grains and the inert substances.

The hydration reactions take place by a series of processes called dissolution and precipitation. At first, the anhydrous phases present in the grains of hydraulic cement dissolve in water, releasing ions into the solution. This phenomenon is

[1] Note that since cement is made up of oxides, a special notation is used to shorten the chemical formulae of the oxides and thus the compounds present in cement. Table 2.1 presents this notation which will be used throughout the chapter.

Table 2.1 Notation of cement chemistry.

Oxide name	Chemical formula	Notation	Some cement compounds	Chemical formula	Notation
Calcium oxide	CaO	C	Tricalcium silicate	Ca_3SiO_5	C_3S
Silicon dioxide	SiO_2	S	Dicalcium silicate	Ca_2SiO_4	C_2S
Aluminum oxide	Al_2O_3	A	Tricalcium aluminate	$Ca_3Al_2O_6$	C_3A
Iron oxide	Fe_2O_3	F	Tetracalcium alumina ferrite	$Ca_4Al_2Fe_2O_{10}$	C_4AF
Sulfate	SO_3	$\overline{S}$	Calcium aluminate	$CaAl_2O_4$	CA
Carbon dioxide	CO_2	$\overline{C}$	Mayenite	$Ca_{12}Al_{14}O_{39}$	$C_{12}A_7$
Water	H_2O	H	Ye'elimite	$Ca_4Al_6\overline{S}O_{16}$	$C_4A_3\overline{S}$
Magnesium oxide	MgO	M	Gehlenite	$Ca_2Al_2SiO_7$	C_2AS

similar to the dissolution of sugar in water. However, unlike sugar, the cement ions can combine with water ions in various proportions. These new combinations, having lower solubility, subsequently precipitate in the solution, forming the hydrate phases. The newly formed solid is now stable, not soluble in water, and these sets of binders are called as hydraulic binders.

Today, in addition to OPC, there are other Ca-based binders being produced at an industrial scale, with some already at the pilot scale and others still in the development stage. Most of these cements exhibit properties that are similar to, yet slightly different from, those of PC. It is important to note that these alternative novel clinkers and cements, which utilize new production processes and often make use of waste-like raw materials, offer exciting possibilities.

2.2.1 Calcium aluminate cements

CACs, also known as high-alumina cements or simply aluminate cements, encompass a range of cements in which calcium aluminates serve as the primary constituents, in contrast to the calcium silicates found in PC. CACs were first patented by Jules Bied in 1908 and began to be produced on an industrial scale during World War I (Scrivener, 2008). As a result, despite not being produced in the same quantities as OPC, CACs stand out as the only type of cement that has maintained continuous production over the past century.

In earlier applications, CACs containing 32%–45% of alumina (Al_2O_3) were developed to provide enhanced sulfate resistance. Nowadays, however, they are not commonly used in the structural systems of buildings. Instead, they find application in niche and specialized areas that involve harsh chemical or abrasive environments. For example, they are utilized in sewage pipes that are susceptible to acid corrosion caused by bacteriological actions, as well as in areas of hydraulic dams where high abrasive wear is prevalent. In these specific applications, CACs effectively broaden the performance range of cementitious materials and compete with alternative materials such as metals or plastics. On the other hand, the majority of CACs used today are employed in the production of refractory concretes which contain higher amounts ($>50\%$) of alumina. Another area where CACs are increasingly being utilized is the dry-mortar applications also known as building chemistry applications. In these applications CACs are blended with OPC, calcium sulfates and lime to produce various products such as rapid hardening repair mortar, self levelling flooring compounds and ceramic tile adhesives (Scrivener, 2001).

The production of low alumina grades of CAC relies on limestone and bauxite as the main raw materials. On the other hand, the higher alumina content grades are produced using relatively pure raw materials, which makes them more expensive. The iron and alumina contents also play a significant role in determining the production method. The low alumina grades are produced in reverberatory furnaces, where the limestone and bauxite undergo complete melting. In contrast, the high-alumina grades are produced in small rotary kilns, similar to those used in the production of PC. Table 2.2 presents the typical composition ranges of the main oxides corresponding to each CAC grade (Scrivener & Capmas, 1998).

Table 2.2 Composition ranges for different CACs.

CAC grade	Color	Al_2O_3	CaO	SiO_2	$Fe_2O_3 + FeO$
High alumina	White	≥ 80	<20	<0.2	<0.2
Medium alumina	White	65–75	25–35	<0.5	<0.5
Low alumina–low iron	Light buff or gray to white	48–60	36–42	3–8	1–3
Standard low alumina	Gray or buff to black	36–42	36–42	3–8	12–20

Source: Modified from Scrivener, K. L., & Capmas, A. (1998). Calcium aluminate cements. In P. C. Hewlett (Ed.), *Lea's chemistry of cement and concrete* (4th ed., pp. 713–782). Butterworth-Heinemann. https://doi.org/10.1016/B978-075066256-7/50025-4

When considering the compound phases of CACs, it is well-known that all CACs contain monocalcium aluminate (CA) as the principal hydraulic phase, which is responsible for the main properties of the binders. The other reactive phase is mayenite ($C_{12}A_7$), which is strictly controlled during the production as it helps initiate the setting process. Unlike OPC, sulfates are not added to CACs to control the setting process. In low alumina CACs, silica is typically present in the form of C_2S and C_2AS, while iron oxide usually exists as a ferrite phase referred to as C_4AF. The combined quantities of silica and iron-bearing phases have significantly lower reactivity and play a negligible role in the initial hydration reactions. However, they may undergo partial reaction at later stages or higher temperatures. It should be noted that these phases can constitute over half of the CACs' composition, making them important from a manufacturing standpoint, and needs to be monitored for control of cement compositions (Scrivener, 2003; Scrivener & Capmas, 1998).

When the CAC phase, CA, comes into contact with water, calcium ions, Ca^{2+}, and aluminate ions, $(Al(OH)_4)^-$, dissolve in the water, forming a solution. These ions can combine to form various types of hydrates such as CAH_{10}, C_2AH_8, C_3AH_6, AH_3 and poorly crystallized gel-like phases. However, unlike the hydration of calcium silicate compounds of OPC, where the formed hydrates remain broadly similar at temperatures up to about 100°C, the hydrates of CACs exhibit a strong temperature dependence. This leads to the formation of both stable (C_3AH_6 and AH_3) and metastable (CAH_{10} and C_2AH_8) or temporary phases during hydration. Therefore, over time and under the influence of temperature, these metastable phases gradually convert into stable phases of hydration. This process is referred to as conversion reactions of CACs. These sets of hydration and conversion reactions are depicted in Table 2.3 (Scrivener, 2003).

As a result of the above-mentioned conversion reactions, some of the water bound within the crystal structure of CAH_{10} and C_2AH_8 is released depending on the w/c of the mixture. This liberation of water leads to an increase in the porosity of CAC matrix, subsequently causing a reduction in strength. Fig. 2.3 shows the effects of those reactions on CAC mixtures cured initially at 20°C, later cured at elevated temperatures (Kırca et al., 2013). As seen from that figure the rate of conversion reactions, thus the loss in strength is more rapid at elevated temperatures.

Table 2.3 Hydration and conversion reactions of CAC.

Chemical reactions	Temperature range	Explanation
$3CA + 12H \rightarrow C_3AH_6 + 2AH_3$	At all temperatures	Formation of stable phases
$CA + 10H \rightarrow CAH_{10}$	<27°C–35°C	Formation of metastable phases
$2CA + 11H \rightarrow C_2AH_8 + AH_3$	Between 35°C–65°C	
$2CAH_{10} \rightarrow C_2AH_8 + AH_3 + 9H$ $3C_2AH_8 \rightarrow 2C_3AH_6 + AH_3 + 9H$	At all temperatures	Conversion of metastable phases to stable phases

Source: Data from Scrivener, K. (2003). Calcium aluminate cements. In J. Newman & B. S. Choo (Eds.), *Advanced concrete technology* (Vol. 3, pp. 1–31). Butterworth-Heinemann. https://doi.org/10.1016/b978-075065686-3/50278-0

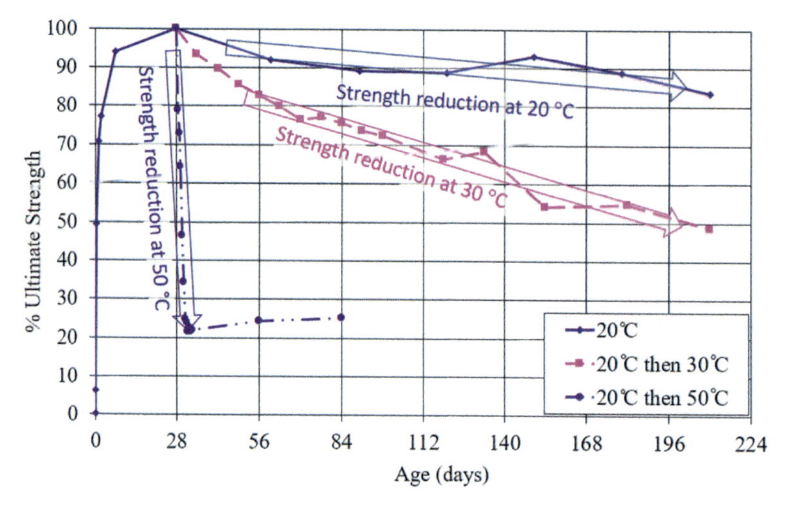

Figure 2.3 Effect of temperature on the conversion reactions. When the impact of conversion reactions on CAC mixtures initially cured at 20°C and later subjected to elevated temperatures is observed, it becomes evident that the rate of conversion reactions accelerates at higher temperatures, leading to a more rapid loss of strength.
Source: Data from Kırca, Ö., Yaman, İ. Ö., & Tokyay, M. (2013). Compressive strength development of calcium aluminate cement–GGBFS blends. *Cement and Concrete Composites, 35*(1), 163–170. https://doi.org/10.1016/j.cemconcomp.2012.08.016.

On the other hand, the same study also concluded that addition of supplementary cementitious materials (SCM) like ground granulated blast furnace slag (GGBFS) also affects the hydration reactions. In CAC–GGBFS blends, particularly where GGBFS ratio is higher than 40%, the formation of stable strätlingite (C_2ASH_8) instead of calcium aluminate hydrates hindered the probable conversion reactions, without causing a strength reduction.

Besides the GGBFS addition, the effects of other mineral admixtures such as calcium sulfate, limestone, and OPC on the hydration reactions of CAC blends are also studied by various researchers (Bizzozero & Scrivener, 2014; Martin et al., 2014; Torrens-Martin & Fernández-Carrasco, 2014). Today, one of the most widely utilized applications of CAC is in dry-mortar preparation, where blends of CAC, OPC, and calcium sulfate are used as ternary blends. In such blends, the ingredients are so-proportioned that they have rapid setting and hardening features so that they can be used in developing repair type of dry-mortars. Additionally, these blends can also be proportioned to exhibit rapid humidity reduction and shrinkage compensation in addition to strength development, making them suitable for formulating self-leveling floor screeds (Scrivener, 2001).

In terms of a future outlook, the high cost of alumina sources suggests that the production of CAC may become limited, due to its relatively high price. As a result, it appears that CAC will not be a direct substitute for OPC, but instead it is likely to find its path in niche and specific applications as discussed in this chapter.

2.2.2 Calcium sulfoaluminate cements

Ye'elimite ($C_4A_3\overline{S}$), the main phase in CSA cement, was classified and patented as a cementitious phase by Alexander Klein in the 1960s, therefore is also known as Klein's compound. The China Building Materials Academy introduced CSA cements in the mid-1970s, marketing them as the "third cement series" following OPC and CAC. The third series encompass sulfoaluminate cements (SAC) and ferroaluminate cements, with their specific properties dependent on the quantity of pulverized calcium sulfate and other mineral additives (Zhang et al., 1999). While CSA cements find widespread applications in China, including precast concrete, cold weather applications, and waterproof construction, their utilization beyond China has predominantly been confined to specialized cements and nonstructural applications (Juenger et al., 2011). Aside from China, CSA cements are currently manufactured commercially in Europe and the United States. They are especially utilized for paving in the United States due to their fast strength development and low shrinking properties (Thomas et al., 2018).

Based on the stoichiometry of the processes involved, ye'elimite production results in lower CO_2 emissions compared to the production of C_3S and C_2S, with emissions of 0.216 tons versus 0.578 and 0.511 tons, respectively. This makes CSA cement a potentially more carbon-efficient alternative to OPC. Furthermore, ye'elimite formation requires significantly less enthalpy, approximately 800 kJ/kg, compared to the 1848 kJ/kg needed for C_3S formation, allowing for production at 200°C–300°C of lower temperatures (Sharp et al., 1999). Additionally, the grinding of CSA clinker requires less energy compared to OPC owing to its more friable nature (Beretka et al., 1996). The combination of lower CO_2 emissions, reduced energy requirements during heating and grinding, and spesific technical advantages, has renewed the interest of both the industry and the scientific community in CSA cements in recent years.

CSA clinkers can be produced in OPC plants by making minor adjustments to the raw material proportions. This involves reducing limestone and increasing aluminum sources. The basic raw materials for CSA clinker production are bauxite, limestone, and calcium sulfate. Depending on the SO_3 level, gypsum or anhydrite can be added to the clinker to produce CSA cement. A stoichiometric-based formula has been developed in China to determine the optimal sulfate content for various types of CSA cements (Winnefeld & Barlag, 2010; Zhang, 2000):

$$C_T = 0.13MA\overline{S} \quad (2.1)$$

The formula involves the molar ratio of gypsum to clinker (C_T), the weight percentage of ye'elimite in the clinker (A), the weight percentage of SO_3 in the gypsum ($\overline{S}$), the molar ratio of gypsum to ye'elimite (M), and a conversion coefficient of 0.13 for mass and molar values. M value dictates the characteristics of the cement.

- $M = 0-1.5$ for rapid hardening or high strength,
- $M = 1.5-2.5$ for expansion,
- $M = 2.5-6$ for self-stressing.

As also observed from the above formulations, the CSA cement family exhibits a considerable degree of diversity (Scrivener, 2014). Therefore, these cements have been designated with multiple names and abbreviations, resulting in distinct classifications (Odler, 2000). For instance, Aranda and De la Torre (2013) have categorized CSA cements on the basis of their primary compounds, while Bescher et al. (2018) have classified CSA binders based on their properties. Unlike OPC, CSA cements lack a universally recognized standard, with Chinese standards being the primary reference. The European Technical Evaluation Organization has also published two evaluation documents on "CSA-based cement" (EOTA, 2017a) and "fast-setting sulfate-resistant CSA-based cement" (EOTA, 2017b). Moreover, the characteristics of CSA cements are influenced by numerous factors, making it challenging to establish a definitive framework for their classification.

China has been a long-standing producer and user of CSA cements with significant ye'elimite content since the 1970s. Limited commercial availability of CSA clinkers/cements can also be found in Europe and the United States. The oxide ranges and phase compositions of CSA clinkers currently available on the market, as reported in relevant literature, are summarized in Tables 2.4 and 2.5. Notably, most CSA clinkers available comprise primarily of ye'elimite and belite (C_2S), except for Rockfast CSA clinker, which does not include any belite. Gehlenite (C_2AS) and calcium aluminate (CA) phases predominate instead. Additional mineral phases observed include fluorellestadite ($C_{10}S_3\overline{S}_3F_2$), merwinite ($C_3MS_2$), bredigite ($C_7MS_4$), akermanite ($C_2MS_2$), periclase (M), and perovskite (CT) and various calcium aluminate phases such as CA, CA_2, C_3A, and $C_{12}A_7$.

Just like the CAC cements, CSA cements also have economic issues due to availability and high cost of bauxite. Therefore there are numerous laboratory-scale production studies using locally available ingredients and by-products/wastes that can be sources of CaO, SiO_2, Al_2O_3, SO_3. In this regard, various industrial by-products/wastes, such as red mud, aluminum slag, alumina powder, fly ash, granulated blast furnace slag, electric arc furnace slag, desulfurization gypsum, and phosphogypsum,

Table 2.4 Oxide composition (%) of commercially available CSA clinkers.

CaO	**Al_2O_3**	**SiO_2**	**SO_3**	**Fe_2O_3**	**Na_2O**	**K_2O**	**MgO**	**TiO_2**
36.2–45.3	27.3–47.4	3.6–11	6.5–13.9	0.9–8.6	0.04–1.4	0.1–0.5	0.3–4.1	0.4–2.2

Source: From Tangüler-Bayramtan, M. (2022). *Synthesis of green calcium sulfoaluminate cements using an industrial symbiosis approach* [PhD thesis]. Middle East Technical University.

Table 2.5 Phase composition of commercially available CSA clinkers.

Brand/producer	Phases	References
CS1 BELITH[a]	72.3% $C_4A_3\bar{S}$; 14.5% C_2S; 6.8% CT; 2.5% C_4AF; 1.6% M; 1.4% C_2MS_2; 0.9% $C\bar{S}$	García-Maté et al. (2012)
ALIPRE (2009)[b]	69.5% $C_4A_3\bar{S}$; 17.1% C_2S; 9% $C\bar{S}$; 3.5% CT; 0.52% M	Álvarez-Pinazo et al. (2012)
Mirae C&C Corp., South Korea	69.4% $C_4A_3\bar{S}$; 20% C_2S; 4.2% $C_{12}A_7$; 3.8% CT; 1.5% C_2AS; 1% M	Jeong et al. (2018)
KTS 100- Belitex	68.5% $C_4A_3\bar{S}$; 15.9% C_2S; 9.5% $C_{12}A_7$; 2.9% CT; 1.5% M; 1.2% Fe_2O_3; 0.5% $C\bar{S}$; 0.5% quartz	Berger et al. (2011)
ALI PRE GREEN[c]	68.4% $C_4A_3\bar{S}$; 16.9% C_2S; 7.40% C_7MS_4 3.67% M; 1.90% C_2AS; 1.75% C_3A	Padilla-Encinas et al. (2020)
No brand specified	68.1% $C_4A_3\bar{S}$; 14.8% C_2AS; 7.8% CA; 3.6% CT; 3.4% C_3A; 1.2% CA_2; 1% M	Pelletier et al. (2010)
ALIPRE[d]	65% $C_4A_3\bar{S}$; 9% C_2S; 11% $C_{10}S_3\bar{S}_3F_2$; 5% M; 3% $C\bar{S}$	Trauchessec et al. (2015)
No brand specified	64.3% $C_4A_3\bar{S}$; 10.6% C_2S; 7.4% $C_{10}S_3\bar{S}_3F_2$; 3.8% C_7MS_4; 3.3% M, 2.8% C_3MS_2; 2.3% $C_{12}A_7$; 2% $C\bar{S}$; 1.5% C_3S; 0.9% CT; 0.9% C_2AS	Martin et al. (2017)
42.5R CSAC[e]	58.68% $C_4A_3\bar{S}$; 28% C_2S; 13.35% C_4AF	Gao et al. (2021)
R.SAC 42.5[f]	57.37% $C_4A_3\bar{S}$; 25.55% C_2S; 6.56% C_4AF; 1.92% f-SO_3	Chang et al. (2009)
Rockfast 450[g]	57% $C_4A_3\bar{S}$; 16% C_2AS; 15% CA; 4% C_4AF; 4% CT; 1% $C_{12}A_7$; 0.5% $C\bar{S}$; 0.3% f-CaO	Zhou et al. (2006)
Chinese CSAC[h]	57% $C_4A_3\bar{S}$; 17% C_2S; 7% C_2AS; 7% CA; 6% CT; 3% $C_{12}A_7$; 2% C_3A; 1% $C\bar{S}$	Galan et al. (2016)
S.A.Cement[i]	56.2% $C_4A_3\bar{S}$; 31.1% C_2S; 6.3% $C\bar{S}$; 3.5% CT; 1.9% CA; 1.1% M	Álvarez-Pinazo et al. (2012)
Grade 72.5 belite-CSAC[j]	53.6% $C_4A_3\bar{S}$; 20.6% C_2S; 11.7% C_2AS; 4.8% CT; 4.2% C_3MS_2; 3.3% $C_{12}A_7$	Jen et al. (2017)
TS-Belitex[k]	53.5% $C_4A_3\bar{S}$; 21.2% C_2S; 16.5% C_4AF; 9% CT	Cau Dit Coumes et al. (2009)
Italian CSAC	52.1% $C_4A_3\bar{S}$; 23.8% C_2S; 9.4% C_3A; 4.9% $C\bar{S}$; 4.7% C_4AF; 1.6% C_2AS; 1.4% M; 1.2% $C_{12}A_7$; 0.9% $C_5S_2\bar{S}$	Telesca et al. (2014)

[a]Produced in China and marketed in Europe by BELITH S.P.R.L.
[b]Produced by Italcementi, Italy.
[c]Supplied by HeidelbergCement Hispania, Spain.
[d]Produced by Italcementi with recycled materials, Italy.
[e]Produced by Dengfeng Electric Power Group Cement Co. Ltd., Henan, China.
[f]Obtained from Zibo Jinhu Highwater Material Co. Ltd., China.
[g]Provided by Lafarge Cement, UK.
[h]Obtained from Shenzhen Chenggong Building Materials Co. Ltd., China.
[i]Produced by Buzzi Unicem, Italy.
[j]Obtained from China.
[k]Produced by Carrie'res du Boulonnais, France (Péra & Ambroise, 2004).
Source: From Tangüler-Bayramtan, M. (2022). *Synthesis of green calcium sulfoaluminate cements using an industrial symbiosis approach* [PhD thesis]. Middle East Technical University.

have been utilized in the production of CSA cement (Bullerjahn et al., 2015; Canbek et al., 2020; Tangüler-Bayramtan, 2022).

Unlike PC, whose binding properties originate from calcium silicate hydrate (C–S–H) formation, the binding abilities of CSA cement is due to ettringite production (Aïtcin, 2008). Several significant factors, including the composition of the clinker, the reactivity and quantity of added calcium sulfate, and the production parameters influence the hydration process of CSA cements (Zajac et al., 2016). CSA cements exhibit accelerated reactivity compared to OPC, with a notable amount of heat of hydration released during the initial 12-hour period. The early-age hydration products of CSA cement include ettringite, monosulfate, and amorphous aluminum hydroxide. Depending on the cement composition, additional hydrates such as strätlingite, C–S–H, monocarboaluminate, or hydrogarnet may form at later stages (Winnefeld & Barlag, 2009; Zajac et al., 2016). The basic hydration reactions of CSA cement are as follows:

$$C_4A_3\overline{S} + 18H \rightarrow C_4A\overline{S}H_{12} + 2AH_3 \quad (2.2)$$

$$C_4A_3\overline{S} + 2C\overline{S}H_2 + 34H \rightarrow C_6A\overline{S}_3H_{32} + 2AH_3 \quad (2.3)$$

$$2C_4A_3\overline{S} + 2C\overline{S}H_2 + 52H \rightarrow C_6A\overline{S}_3H_{32} + C_4A\overline{S}H_{12} + 4AH_3 \quad (2.4)$$

$$C_4A_3\overline{S} + 8C\overline{S}H_2 + 6CH + 74H \rightarrow 3C_6A\overline{S}_3H_{32} \quad (2.5)$$

In fact, ye'elimite hydrates slowly, and as shown in Eq. (2.2), produces monosulfate and aluminum hydroxide after a dormant period of a few hours. However, the addition of calcium sulfate (Eqs. 2.3 and 2.4) and/or calcium hydroxide (Eq. 2.5) in the media changes the rate of reaction, the products formed, and also water requirement. Complete hydration of ye'elimite with two moles of anhydrite requires a greater water-to-cement ratio (0.78) than that required by OPC. However, the overall water requirement can vary based on the content of belite or other phases present in the CSA cement (Thomas et al., 2018).

In CSA cements, particularly those with a high C_2S content, the presence of both C_2S and the amorphous AH_3 hydrates facilitates the creation of strätlingite (C_2ASH_8), which contributes to enhanced strength development (Eq. 2.6). When AH_3 is fully depleted, the direct hydration of C_2S, similar to that in PC, can take place at later stages (Eq. 2.7) (Aranda & De la Torre, 2013).

$$C_2S + AH_3 + 5H \rightarrow C_2ASH_8 \quad (2.6)$$

$$C_2S + (x + 2 - y)H \rightarrow C_ySH_x + (2 - y)CH \quad (2.7)$$

Due to the rapid crystallization of ettringite, CSA cements typically exhibit rapid setting. Factors such as the content of ye'elimite, other minor compounds, calcium sulfate and their reactivity affect the setting time, resulting in setting times ranging from 30 minutes to 4 hours (Juenger et al., 2011). Notably, especially at low water/cement ratios, CSA cements can also achieve setting times of 10 minutes or less, but the potential workability issues can be overcome by the use of retarders (Thomas et al., 2018).

Even though the strength development of CSA cement is affected by the composition and amount of cement phases, when compared to OPC, the compressive strengths of CSA cements during early-age and later-age periods are generally greater (Juenger et al., 2011; Quillin, 2001). The early strength development is attributed to the precipitation of ettringite crystals formed as a result of the chemical reactions between ye'elimite and calcium sulfate. When the calcium sulfate levels are optimal, increasing the ye'elimite phase of the cement enhances its early strength. On the other hand, the belite phase predominantly affects the strength of the cement at later stages (Odler, 2000).

As the hydration products of CSA cements have a dense microstructure with low porosity and permeability, they are highly resistant to freezing-thawing and chemical attacks. The pore solution alkalinity of CSA cements is lower compared to OPC, but it still forms a passive layer on embedded steel reinforcement, protecting steel from corrosion. The absence of lime and the reduced alkalinity offer advantages over alkali–silica reaction (Juenger et al., 2011; Thomas et al., 2018). Moreover, CSA cements have demonstrated high resistance against sulfate attack (Glasser & Zhang, 2001). However, some studies have shown that the carbonation resistance of mortars and concretes made by CSA cement is lower than that of those made by PC (Hargis et al., 2017; Ioannou et al., 2010). Therefore although the field and laboratory studies have shown that materials made from CSA-based cements generally perform well compared to PC-based materials, further research is needed to draw firm conclusions about their long-term behavior.

2.2.3 Belite-rich cements

Dicalcium silicate (C_2S) or belite in cement nomenclature, as shown in Fig. 2.4, exists in five different polymorphs, designated as α, α'_H, α'_L, β, and γ. When pure compounds of C and S are used in a reaction, the γC_2S is formed, which is the only

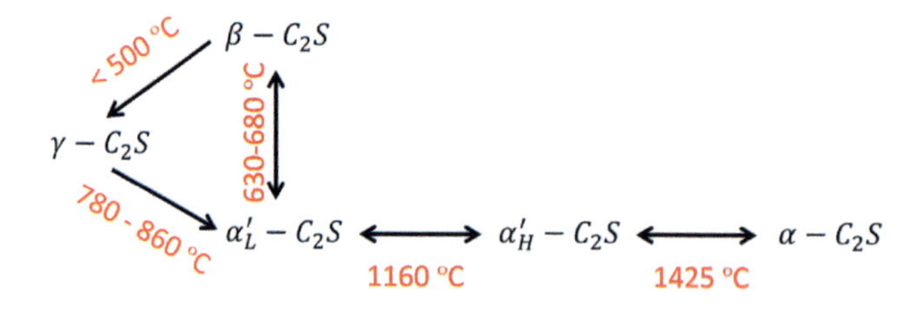

Figure 2.4 Polymorphs of dicalcium silicate. At ordinary pressure, dicalcium silicate has five recognized polymorphs, each with distinct transformation temperatures.
Source: Modified from Odler, I. (2000). In A. Bentur & S. Mindess (Eds.), *Special inorganic cements*. Modern Concrete Technology Series. E & FN Spon.

thermodynamically stable form at ordinary temperatures, but which is also nonreactive with water. The most reactive and therefore the most important modification is the βC_2S, which is thermodynamically unstable at any temperature. Upon cooling, it tends to transform into γC_2S, unless it is doped with some stabilizing ions. In cement nomenclature, its doped form, βC_2S, is called belite or reactive belite (Odler, 2000).

In the existing literature, there is a lack of consistency in the terminology used to describe cements containing a significant amount of belite phase, resulting in confusion. Various terms such as belite cements, activated belite cements, belite–aluminate cements, sulfoaluminate–belite cements, sulfoferroaluminate–belite cements, belite-SACs, belite–sulfoferroaluminate cements, and belite–sulfoaluminate–ferrite cements have been used (Lawrence, 1998; Odler, 2000; UN Environment, 2017). However, in this section, we will group all these into two categories: reactive belite-rich Portland cement clinkers (RBPC) and belite–ye'elimite–ferrite (BYF) clinkers.

RBPC are produced with the same process as OPC. Compared to alite-rich OPC, the production of RBPC results in a lower $CaCO_3$ requirement in the raw meal and therefore a lower lime saturation factor. This allows a lower calcination temperature to be used, resulting in a reduction in fuel consumption for heating and a consequent reduction in CO_2 emissions (Lawrence, 1998). The expected reduction in direct CO_2 emissions from belite cements is estimated to be around 10% (Gartner & Sui, 2018). On the other hand, it should be noted that there are certain challenges associated with the production of RBPC. These include the need for more rapid cooling and increased grinding efforts. The process of achieving a very rapid cooling rate leads to increased heat losses, which reduces the energy savings obtained from a lower lime saturation factor and lower burning temperature (Odler, 2000). In addition, energy requirements during the grinding process are increased due to the harder grindability of belite and the increased fineness required to achieve the required strength. Consequently, these two factors have the potential to offset some of the energy gains achieved during industrial production.

Today, RBPCs can be produced according to the existing internationally accepted standards such as Type IV of ASTM C150 standard (ASTM C150/C150M, 2022). However, some countries, such as China and Türkiye, have specific standards for RPBC to achieve higher reactivity, and thus higher 28-day standard strength levels (GB 200., 2003; TS 13353, 2008). For example, TS 13353 specifies the use of boron (B_2O_3) as a dopant, and thus the cement is called "boron-modified active belite cement," which can reach 28-day strengths on the order of 42.5 MPa. As a pilot study, this cement was first produced at an industrial scale in 2008 and has been used in roads and irrigation canals.

RBPCs have been reported to display comparable setting times to OPC, along with several advantageous properties. These include lower water demands, reduced drying shrinkage, lower heat generation, and early strength development. Furthermore, RBPCs have shown improved compatibility with water reducers, primarily due to the lower C_3A content. In terms of durability, RBPCs exhibit improved resistance to sulfates and chlorides, primarily due to the reduced amount of portlandite formed during hydration (Gartner & Sui, 2018).

Despite these favorable properties, their slower strength development compared to most OPCs is the main reason why RBPCs are not widely used at present. However, they are particularly well suited for applications where immediate strength gain is not critical. In addition, their low heat of hydration makes them suitable for the construction of large concrete dams and foundations. Although fly ash-blended OPCs have been used to reduce concrete temperature rise since the 1960s, one of the most effective methods of reducing thermal stresses and cracking in very large structures, including dam concrete, particularly in China, is to use RPBCs. Most industrial RPBCs are reported to contain 40–50 wt.% belite and 25–30 wt.% alite to ensure satisfactory early mechanical performance. While their early-age strength may be lower than that of OPC, their later-age strength and durability performance is reported to be quite impressive (Cuesta et al., 2021).

In addition to RBPCs, belite can also be used in the development of BYF clinkers. In such clinkers, the amount of belite is typically higher than the ye'elimite, which is also higher than the ferrite phase. The primary difference between BYF clinkers and existing commercial CSA cements mentioned earlier is in their target markets. CSA cements are currently used for their specific properties such as rapid strength development and shrinkage compensation, both of which are mainly derived from the ye'elimite phase. As a result, CSA cements are limited to specialty applications. In contrast, BYF clinkers offer a cost advantage by using less expensive aluminum-rich raw materials. In addition, BYF clinkers produce products that are more suitable for the production of ordinary concretes with a significantly lower carbon footprint compared to conventional OPC-based cements, while maintaining similar performance levels (Gartner & Sui, 2018).

There are no specific standards for BYF cements, except for the small range of BYF compositions that fall under existing Chinese standards for CSA-based cements.

2.2.4 Alinite cements

Alinite cement, originally developed in the former Soviet Union and patented by Noudelman in 1977, is characterized by the presence of an alinite phase rather than an alite phase in its clinker composition. Its industrial production has been reported in India, Russia, and Japan (Gür et al., 2010). Alinite is known as a calcium oxychloro-aluminosilicate compound, and its composition has been proposed in different ways by various researchers (Ilyukhin et al., 1977; Lampe et al., 1986; Noudelman et al., 1980). The following equation is one of the most widely accepted expressions of its composition, where the symbol "□" denotes a vacancy in the lattice structure (Neubauer & Pöllmann, 1994).

$$Ca_{10}Mg_{1-x/2}\square_{x/2}\left[(SiO_4)_{3+x}(AlO_4)_{1-x}/O_2/Cl\right] \quad 0.35 < x < 0.45 \tag{2.8}$$

Alinite clinker can be produced by burning a raw meal composed of limestone, clay, MgO, and $CaCl_2$ (6–18 wt.%) at temperatures of 1000°C–1300°C. Due to such low temperatures, it has been reported to be a special type of low-energy cement. Many types of industrial wastes such as fly ash, magnesite dust, steel mill wastes, municipal wastes, and soda sludge with a high chlorine concentration can be used in the preparation of the raw meal (Pradip et al., 1990). The presence of chloride in the raw meal enhances the burnability of the mixture and facilitates the formation of alinite and belite at lower temperatures, about 1000°C, leading to a decrease in the amount of free lime (Singh et al., 2008). The alinite phase remains stable within a temperature range of 900°C–1150°C, whereas the actual clinkering process occurs at temperatures between 1150°C and 1200°C, accompanied by a heat consumption of about 535 kcal/kg of clinker (Chatterjee, 2002). The presence of MgO is also a critical factor for the formation and stabilization of alinite. A minimum of 1%–2% MgO is reported to be required for alinite synthesis. While increased levels of MgO and chloride content will promote alinite formation, it is important to note that high MgO content is undesirable as it can lead to the formation of belinite instead of alinite (Singh et al., 2008). The typical phase composition range of alinite clinkers is as follows: 50%–80% alinite, 10%–40% belite, 5%–10% calcium aluminochloride, and 2%–10% calcium aluminoferrite. Belinite, calcium chloride orthosilicate, calcium ferrite, periclase (MgO), alite, and belite may also be present in small amounts in the clinker (Odler, 2000).

Energy consumption during the burning process is approximately 30% less than that of OPC (Locher, 1986). Because of the weak Ca–Cl bonds, alinite clinkers have a soft and friable nature. Consequently, the grinding energy required for these cements is also reduced. The resulting increase in specific surface area contributes to improved workability in cement pastes and ultimately leads to higher strengths (Pradip & Kapur, 2004).

In the production of alinite cement, there is no need to intergrind the clinker with calcium sulfate to regulate setting; nonetheless, intergrinding the clinker with calcium silicate increases both short-term and long-term strengths (Pradip et al., 1990; Uçal et al., 2018).

Alinite cement is reported to hydrate faster than OPC. The hydration of alinite and belite produces C–S–H and CH phases. The hydration of calcium aluminochloride and ferrite phases results in the formation of Friedel's salt ($3CaO \cdot Al_2O_3 \cdot CaCl_2 \cdot 10H_2O$) or a phase with some substitution of $CaCl_2$ by $Ca(OH)_2$ or $CaCO_3$. Additionally, other phases such as CAH_{10} and C_3AH_6 may also be present in trace amounts. During hydration, much of the chlorine present in the alinite clinker is incorporated into the $3CaO \cdot Al_2O_3 \cdot CaCl_2 \cdot 10H_2O$ phase. The generated C–S–H phase can also adsorb up to 3.5% of chloride. On the other hand, notable quantities of chloride ions migrate into the liquid phase, which poses a possible risk for reinforcement corrosion in steel-reinforced concrete (Odler, 2000).

The low-energy requirement and the potential usage of chlorine-rich wastes as suitable raw material, rarely utilized in various applications, make alinite cement noteworthy in terms of its positive environmental impacts. The alinite-based "eco-cement" in Japan, which is produced by recycling of incinerated municipal wastes with a chlorine content of up to 10%, serves as a successful example in this context (Chatterjee, 2018). Recently, Kesim et al. (2013) and Uçal et al. (2018) also showed that soda sludge, the waste of the Solvay process of the soda industry, can be a suitable raw material for alinite cement. Nevertheless, alinite cement is unlikely to be employed in high-performance applications due to concerns associated with corrosion of the reinforcement.

2.2.5 Limestone calcined clay cements

As mentioned earlier, substituting the carbon-intensive clinker with SCM is one of the five key levers for reducing carbon emissions in cement production, and thus widely utilized. However, when the three most widely utilized SCMs, that is, slag, fly ash, and limestone, that are used in the cement and concrete technology are considered, it will be realized that all three have some drawbacks limiting their widespread use. For example, it is reported that the amount of slag available on a global basis is approximately 5%–10% of the amount of cement produced. Since the demand for steel is growing less rapidly than the demand for cement and more steel is being recycled for environmental reasons, this proportion is unlikely to increase. In addition, availability in developing countries where cement demand is growing the fastest is even more limited because iron (and therefore slag) production is concentrated in a relatively small number of developed countries. On the other hand, even though the amount of fly ash available is somewhat higher (about 30% compared to cement), its quality is very variable, with less than a third suitable for blending into cement. Furthermore, with increasing pressure to reduce environmental emissions, the burning of coal for power generation is being questioned in many countries, so the long-term availability of fly ash is also in question. Finally, limestone is abundantly available, adding more than 10% of limestone alone to cement tends to increase porosity and degrade properties. Therefore the limited availability of SCMs (limestone, fly ash, and slag) in developing countries is an obstacle to their more widespread use (Scrivener et al., 2018).

Clays, however, are abundant materials worldwide, and clay calcination has been known and used for decades. For example, "metakaolin," which is based on calcining a high purity kaolinitic clay, has been used in the cement and concrete sector for improved properties, especially in white PC applications. However, its high price poses an obstacle for its widespread use in the construction sector.

Although the individual use of limestone and calcined clay in cement technology is not new, the combination of the two with cement clinker is a relatively new technology. It has been developed through a series of projects funded by the Swiss National Science Foundation and the Swiss Agency for Development and Cooperation (LC3-Project, 2014). The cement mixture of calcined clay, limestone, cement clinker, and gypsum is defined as "limestone calcined clay cement." It is commonly referred to as LCCC or LC3 blend or simply LC3. The amount of clinker in an LC3 mix is often referred to as LC3-X, where X is the percentage of clinker used in the mix. For example, most LC3 blends referenced contain 50% clinker and are therefore referred to as LC3–50. The original LC3 cement incorporated a clay that has a kaolinite content of only about 40% in a mixture of LC3–50 (50% ground clinker, 30% calcined clay, 15% limestone, 5% gypsum), and this was sufficient to give mechanical properties comparable to the reference OPC from about 7 days. Moreover, a study of 46 clays from around the world has shown that there does not seem to be much impact of the secondary materials present, which typically include, quartz, other clay minerals and iron oxides (Fig. 2.5).

The first industrial scale production of LC3 was performed in a cement plant in Cuba (Vizcaíno-Andrés et al., 2015), and later two others in India (Bishnoi et al., 2014; Emmanuel et al., 2016). In these three productions, different kilns and grinding mills were used ranging from nonsophisticated to modern equipment and technologies. The produced LC3−50 was used in the manufacturing of various building materials, precast concrete elements such as paving blocks, hollow concrete blocks, kerbstones, autoclaved aerated concrete blocks, door and window frames, and joists, a model house was also built in Cuba. All building materials are also produced with OPC and PPC to compare the results, and no admixtures were used. It was stated that in almost all cases, LC3 products performed better than PPC and performed similarly to OPC in terms of strength. Moreover, to further compare LC3−50 with two other cement types, OPC and pozzolanic Portland cement (PPC), a study (Sánchez Berriel et al., 2015) was conducted to analyze the associated costs. It was found out that compared to OPC production, PPC production led to 13% cost

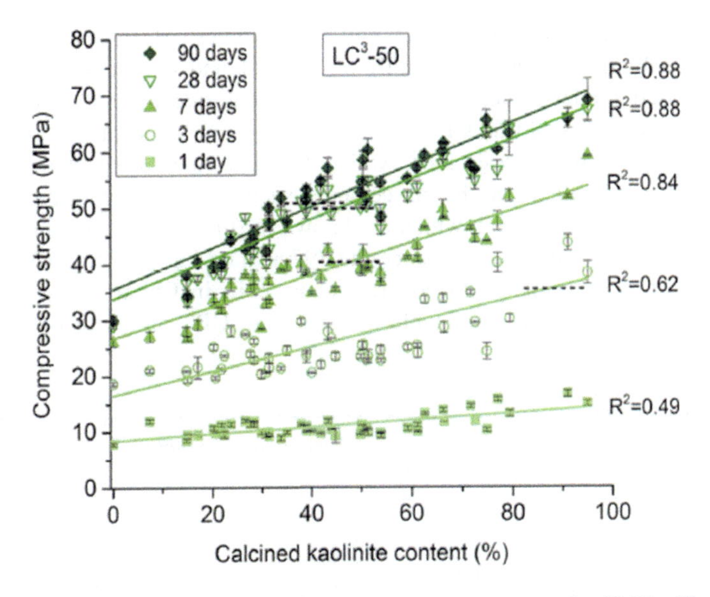

Figure 2.5 The effect of calcined kaolinite amount on mortar strength of LC3−50 cements. The study has shown that the reactivity of calcined clay is mainly dependent on the kaolinite content of the clay. When used in LC3−50, clays with a kaolinite content of around 40% or more give a strength comparable to that of OPC (indicated by the dashed line). In addition, the secondary minerals present in the clays do not have a significant effect.
Source: From Scrivener, K. L., Martirena, F., Bishnoi, S., & Maity, S. (2018). Calcined clay limestone cements (LC3). *Cement and Concrete Research, 114*, 49−56. https://doi.org/10.1016/j.cemconres.2017.08.017

savings while the savings are between 16% and 49% for LC3, depending on the method of transportation and distance from the clay deposit. The study also states that the LC3 production is 4%–40% more cost-effective compared to PPC, depending on the transportation factors. Therefore the results of all industrial productions in Cuba and India show that the LC3 cement technology has the potential to improve the sustainability of a cement plant while possibly reducing certain costs without compromising performance, durability, or rheological properties.

2.3 Magnesium-based cements

Magnesium is the seventh most abundant element in the Earth's crust (Fig. 2.2), at approximately 2.3% by weight, and occurs naturally in a variety of rock formations in the form of carbonates, silicates, sulfates, and chlorites, and after sodium and chlorine is the third most abundant element in solution in seawater (Shand, 2006). Magnesia (magnesium oxide, MgO) is mainly produced from the calcination of magnesite ($MgCO_3$) in a process similar to the production of lime from limestone and depending on the temperature to which it is subjected to, it can have different forms with different properties (Table 2.6). A smaller proportion of the world's MgO production comes from seawater and brine sources (Al-Tabbaa, 2013).

Mg-based binders are a broad family of cements, and some commercially available Mg-based binders, such as magnesium oxychloride cement, also known as Sorel cements, are as old as OPC. Despite their many technical advantages, they suffer from poor water resistance, which has prevented their widespread use. Therefore various other Mg-based cements such as the magnesium phosphate cements have been developed, which are known for their fast setting and high early strength properties. Additionally, since the 1970s, hard burned MgO has been used to compensate for shrinkage in concrete dams built in China (Mo et al., 2014). More recently, as a more sustainable alternative to OPC and with the promise of

Table 2.6 Different MgO products.

Name	Production temperature (°C)	Reactivity	Note
Dead-burned MgO, periclase	1400–2000	Least	Problematic in Portland cement.
Hard-burned MgO	1000–1400	Intermediate	Can be used as an expansive additive in concrete production.
Light-burned MgO, reactive magnesia	700–1000	Highest	Main ingredient in Mg-based binders.

Source: Data from Al-Tabbaa, A. (2013). Reactive magnesia cement. In F. Pacheco-Torgal, S. Jalali, J. Labrincha, & V. M. John (Eds.), *Eco-efficient concrete* (pp. 523–543). Woodhead Publishing. Woodhead Publishing Series in Civil and Structural Engineering. https://doi.org/10.1533/9780857098993.4.523

superior technical performance, reactive MgO cements (Harrison, 2008), which are blends of light-burned MgO and OPC, have emerged (Al-Tabbaa, 2013). Despite nearly 150 years of research for some types, critical questions remain regarding the durability and cost-effectiveness of these binders, particularly for new members of the family. Use in reinforced concrete or large-scale cast-in-place construction appears to present significant technical difficulties, but their use in the production of smaller-scale unreinforced elements under controlled conditions seems to be promising. In addition, before labeling Mg-based cements as environmental saviors in the construction field, it is essential to conduct further research and development to ensure their sustainability in large-scale applications, as emphasized by (Walling & Provis, 2016).

In recent decades, the ability of MgO to absorb CO_2 and form carbonates has led to the discussion of some Mg-based cements being called as "carbon-neutral" or even "carbon-negative." However, although Mg-based binders is considered as a low-CO_2-emission option, most research was conducted by producing MgO through the calcination of magnesium carbonates, resulting in high CO_2 emissions, and making this approach unsustainable. However, the utilization of Mg-based clinkers derived from abundant ultramafic rocks, specifically basic magnesium silicates, could make this approach feasible. These rocks, rich in basic MgO and devoid of CO_2, have the ability to effectively sequester CO_2 as stable magnesium carbonates, potentially enabling a carbon-negative binder (Gartner et al., 2014; UN Environment, 2017).

In this section, the following Mg-based binders will be briefly explained; magnesium oxychloride cement (MOC), magnesium oxysulfate cement (MOSC), magnesium phosphate cements (MPC), reactive magnesia cement (RMC), magnesium silicate hydrate cements (MSHCs), and magnesium oxalate cements (MOxCs).

2.3.1 Magnesium oxychloride cement

MOC, also known as Sorel cement, was produced by Stanislas Sorel in 1867. This nonhydraulic cement is produced through the mixture of pulverized MgO, obtained through the calcination of magnesite at around 750°C, with a concentrated solution of magnesium chloride ($MgCl_2$). The dissolution of MgO in the $MgCl_2$ solution results in the formation of a gel, which is responsible for the setting of the cement, occurring before the crystallization of the other hydrates (Odler, 2000). The quality of MgO plays a critical role in determining the effectiveness of chemical reactions, as it must be consistently and adequately calcined. Insufficient calcination leads to an excessively reactive product, while excessive calcination yields a less reactive material (Bensted, 1998). Besides MgO reactivity, molar ratios between raw components and curing conditions directly impact the hydration phase and microstructure, thus affecting the strength of the cement. MOC has gained significant interest due to its numerous advantages over OPC, including fast setting, air hardening, low density, high mechanical strength, good fire and abrasion resistance, low thermal conductivity, and resistance to oils, greases, and paints. It exhibits great bonding capabilities with various fillers like gravel, sand, marble flour, and wood particles. This versatility in bonding allows MOC to be used effectively in a wide range of applications where different filler materials are needed (Guo et al., 2018). MOC are extensively used for industrial flooring. Some other applications for this cement include polishing bricks, fireproof boards, wood-cement boards, and lightweight interior partition walls (Odler, 2000). However, the drawbacks of MOC, such as inadequate water resistance, dimensional instability, and the potential for efflorescence and corrosion, have severely restricted its use over the past 50 years (Guo et al., 2018; Walling & Provis, 2016).

2.3.2 Magnesium oxysulfate cement

The history of MOSC, a variant of Sorel cement, dates back to 1981. MOSC is produced by mixing calcined magnesia with a solution of magnesium sulfate ($MgSO_4$). The less hygroscopic nature of magnesium sulfate compared to magnesium chloride has aroused interest in MOSC. However, these cements also seem to have a similar susceptibility to water as MOC binders, which makes them unsuitable for use in structural applications. During the hardening process, a variety of magnesium oxysulfate phases can occur (Odler, 2000). Steam curing is needed for the formation of many phases due to the limited solubility of $MgSO_4$, which restrict the use of MOSC to rapidly produced interior boards or faux-wood panels, as well as spray-on fireproof coatings (Bensted, 1998; Walling & Provis, 2016).

2.3.3 Magnesium phosphate cements

MPCs are produced through an acid-based reaction between dead burnt magnesia and a soluble acid phosphate, which is commonly an ammonium or potassium phosphate. This reaction results in the formation of a magnesium phosphate salt that exhibits cementitious characteristics (Walling & Provis, 2016). MPC was initially developed as a dental cement during the late 19th century, but nowadays it is being used as a repair material for concrete structures. MPC is thought to be a promising and eco-friendly alternative to OPC based on its characteristics. It exhibits properties such as rapid setting and hardening, high early strength, minimal drying shrinkage and thermal expansion, good bond strength, resistance to abrasion and frost, and fireproof behavior. The acid-based reactions in MPCs are extremely quick and highly exothermic, which makes their application challenging. The reaction rate can be controlled by incorporation of admixtures such as borax and fly ash (Bilginer & Erdoğan, 2021). MPC has been used in diverse applications such as rapid repair of concrete structures, stabilization of toxic waste, design of biomedical materials, three-dimensional powder printing materials, and wastewater treatment. The low internal pH and high cost of these cements limit their use in structural steel-reinforced concrete, limiting their applicability to lower volume niches (Haque & Chen, 2019; Walling & Provis, 2016).

2.3.4 Reactive magnesia cements

RMCs were developed and patented by John Harrison. Generally, RMC is made by incorporating fine reactive magnesia into OPC. RMC can permanently bind CO_2 through the formation of magnesium carbonates, resulting in increased strength depending on the degree of carbonation. The maximum carbon sequestration capacity of RMC is theoretically 110%. This is significantly higher than that of OPC, which is 46%–58% (Meng et al., 2023). Because of this fact, it has been claimed that RMC can greatly reduce the net CO_2 emissions, even leading to carbon negativity by using a carbon recycling cement kiln during the production of reactive MgO and considering the subsequent CO_2 absorption by the cement during its service life (Walling & Provis, 2016). Although they have the potential to replace OPC in the production of unreinforced blocks and precast units, their use in large-scale in-situ concrete will be limited as their performance under environmental conditions, that is, their durability, remains unconfirmed. Moreover, as will be mentioned in the next section, the commercial applications utilizing carbonation curing have confirmed that cost-effective and sustainable production methods are not yet possible, and carbonate formation is currently limited to porous products exposed to elevated CO_2 curing conditions. Therefore the potential for these cements to be carbon-neutral or carbon-negative may only be feasible through the use of natural brucite deposits and renewable fuels for calcination (Ardoğa, 2020; Walling & Provis, 2016).

2.3.5 Magnesium silicate hydrate cements

For over a century, the idea of using a magnesium silicate bond to create a cementitious material has been known. The patent in 1889 described a magnesium silicate cement by mixing and calcining magnesium carbonate and finely pulverized silica. After the first patent, a few additional patents were issued, but the interest in magnesium silicate cements diminished for a long time. In the second half of the 20th century, the possibility of magnesium silicate cements as a lower carbon alternative to OPC aroused renewed interest. However, research on magnesium silicate cements is limited, making them less understood compared to other binders. Therefore magnesium silicate cements are unlikely to emerge as a significant alternative to OPC, and their usage is expected to remain confined to niche applications (Walling & Provis, 2016).

2.3.6 Magnesium oxalate cements

Recently, the production of MOxC, a novel substitute for OPC has been demonstrated, which involves the reaction of dead-burned magnesia and oxalic acid salts at ambient temperature. Oxalic acid has the potential to be produced through the capture of carbon dioxide, resulting in oxalate cements that could potentially be carbon negative. However, the emissions associated with magnesite decarbonation at high temperatures pose a challenge to achieving the desired result. To further reduce the carbon footprint of oxalate cements, the dead-burned MgO must be replaced with a less carbon-intensive waste or natural material. MOxC cements exhibit rapid setting characteristics, have the ability to achieve medium to high ultimate strength, and exhibit water resistance properties (Bilginer & Erdoğan, 2023; Bilmez, 2023; İçinsel, 2020).

2.4 Development of binders by different production methods

The above-mentioned binders based on Ca or Mg are usually produced after the raw materials are pulverized and mixed in accordance with their prescribed formulations to obtain the raw meal. The raw meal is then subjected to a heating process in kilns, possibly followed by a cooling process. The clinkers produced are then mixed or interground with other inorganic materials. Finally, these binders are mixed with water and with other inert fillers, which are then water-cured so that hydration reactions will take place and the material solidifies to form concrete. In this section, we will look at some other innovative methods, either in the production of the binder itself or in the production of the concrete. These methods offer potential advances in the binder or the concrete production techniques that differ from the conventional processes described above.

2.4.1 *Hydrothermally processed binders*

The use of hydrothermal processing is reported to be a promising technique for obtaining a new type of hydraulic binder, often referred to as "prehydrated calcium silicates." In fact, this processing technique was first used by Jiang and Roy (1992) to obtain a fly ash cement, which can also be used for SCMs. During the hydrothermal process, the pozzolanic reaction of a lime–silica mixture is accelerated through an autoclave to form a form of calcium silicate hydrate (C–S–H) and/or calcium aluminosilicate hydrate (C–A–S–H). The hydration products are then partially dehydrated by calcination at temperatures less than 1000°C, resulting in a mixture of reactive C_2S polymorphs, which can be further mechanically activated by intergrinding with hard fillers. As briefly explained, the overall manufacturing process is quite different than the conventional OPC production and requires the need for more processing steps. As this processing technique is still under development at the laboratory level, reliable estimates of their overall energy and CO_2 efficiency in a real industrial context cannot yet be made (Garbev et al., 2014; Rungchet et al., 2016; UN Environment, 2017).

2.4.2 *Binders subjected to carbonation-hardening*

Another aspect in developing low-carbon binders is the use of the mineralization process, by accelerated carbonation curing (ACC). In ACC the cementitious system is deliberately subjected to a CO_2 supply during the early ages of strength development. This causes the mineral carbonation of metal oxides present in cement, and thus allows sequestering of CO_2 in cement in the form of stable silica gel and carbonates (Daehn et al., 2022; Goyal & Sharma, 2020). There has been considerable research on the manufacture of concrete products by carbonation instead of hydration, and one of them was proposed by Novacem, which proposed to use accelerated carbonation of magnesium silicates under elevated levels of temperature and pressure (Walling & Provis, 2016). Another one was proposed by Calera, which developed an innovative method for carbon sequestration through carbonate mineralization of Ca- and Mg-containing oxides by aqueous precipitation method (Monteiro et al., 2013). However, both of these techniques were not utilized by the industry and therefore not commercialized yet. Another novel binder which utilizes AAC, has been developed by Solidia Technologies, and is also known as Solidia Cement (Solidia, 2023) This binder comprises of low-lime calcium silicate minerals such as wollastonite (CS), and it can be produced in conventional cement kilns using common raw materials (limestone and silica). Solidia cement technology is still being funded not only by some cement manufacturers but also by US Federal Government agencies (Majcher, 2015).

2.4.3 Geopolymers/alkali-activated binders

Another alternative binder technology is the use of geopolymers, which are a subset of alkali-activated binders (AABs). In this section, AABs will be mentioned only briefly, as this group of binders will be covered in later chapters of the book. AABs are formed by the reaction between aluminosilicate precursors and alkaline solutions. The most popular precursors for alkali activation are ground granulated blast furnace slag, fly ash, metakaolin, and natural pozzolans (Ulugöl, 2021). There has been a tremendous amount of research on AABs, and they have been promoted as a popular alternative to OPC. However, since most of these precursors are already utilized in the cement industry and there are problems with the availability of slag and fly ash, these facts limit the use of AABs. As a result, they have found only small amount of commercial applications, especially where slag is relatively abundant. Recently, however, construction and demolition waste, which poses significant problems for the construction industry, has been shown to be another precursor, and may even have self-healing properties when used (Ulugöl et al., 2021). Therefore AABs are likely to form an important component of a broader toolkit of cements to be used globally in the development of a sustainable future construction materials industry (Provis, 2018).

2.5 Concluding remarks

Compared to other structural building materials, concrete offers various technical advantages and therefore holds the title of "the most widely used construction material." In addition, the need for concrete is expected to grow, especially in developing countries, as the demand for infrastructure increases with economic growth. Although concrete has a very low-carbon footprint, it faces challenges as the construction industry transitions to reduce its environmental impact due to its enormous consumption. On the other hand, it is the PC—the primary binder of concrete—that contributes significantly to the environmental footprint of concrete. Therefore the cement industry is pursuing several strategies to reduce its carbon footprint, including the development of innovative low-carbon binders.

The aim of this chapter was to summarize those innovative binders, including Ca-based binders that use the abundance of calcium as the primary element, followed by Mg-based binders. In addition, nonconventional innovative binders or innovative concrete production techniques were also mentioned. It should be mentioned that among the large number of existing and even commercially available binders, PC clinker-based cements are the most widely used and are likely to continue to dominate the market in the near future. The availability of raw materials, economies of scale, lack of existing standards, and market confidence and acceptance of these novel binders limit their widespread use. In the longer term, these novel technologies may have a role to play in reducing emissions. They are therefore worthy of further investigation.

As a concluding remark, it should also be noted that all these efforts of cement industry are taking place in the upstream of the construction industry value chain. On the downstream we need to consider how we can make concrete structures with longer life spans (i.e., durable); make use of its thermal mass and operational efficiency; use less concrete in those concrete structures; use less cement in that concrete; and use less clinker in that cement. This comprehensive outlook will only lead us to what is required for a more sustainable future.

References

Aïtcin, P.-C. (2008). Binders for durable and sustainable concrete. In A. Bentur, & S. Mindess (Eds.), *Modern concrete technology series, Book 16*. Taylor & Francis Group.

Al-Tabbaa, A. (2013). Reactive magnesia cement. In F. Pacheco-Torgal, S. Jalali, J. Labrincha, & V. M. John (Eds.), *Eco-efficient concrete* (pp. 523–543). Woodhead Publishing, Woodhead Publishing Series in Civil and Structural Engineering. Available from https://doi.org/10.1533/9780857098993.4.523.

Álvarez-Pinazo, G., Cuesta, A., García-Maté, M., Santacruz, I., Losilla, E. R., De la Torre, A. G., León-Reina, L., & Aranda, M. A. G. (2012). Rietveld quantitative phase analysis of Yeelimite-containing cements. *Cement and Concrete Research*, *42*(7), 960–971. Available from https://doi.org/10.1016/j.cemconres.2012.03.018.

Aranda, M. A. G., & De la Torre, A. G. (2013). Sulfoaluminate cement. In F. Pacheco-Torgal, S. Jalali, J. Labrincha, & V. M. John (Eds.), *Eco-efficient concrete* (pp. 488–522). Woodhead Publishing.

Ardoğa, M. K. (2020). *Properties of reactive magnesia-incorporated cements* [PhD thesis]. Middle East Technical University, Ankara, Turkey.

ASTM C150/C150M. (2022). *Standard specification for Portland cement*. ASTM International. Unpublished content.

Bensted, J. (1998). Special cements. In P. C. Hewlett (Ed.), *Lea's chemistry of cement and concrete* (4th ed., pp. 783–840). Butterworth-Heinemann. Available from https://doi.org/10.1016/B978-075066256-7/50026-6.

Beretka, J., Marroccoli, M., Sherman, N., & Valenti, G. L. (1996). The influence of C4A3S̄ content and WS ratio on the performance of calcium sulfoaluminate-based cements. *Cement and Concrete Research, 26*(11), 1673–1681. Available from https://doi.org/10.1016/s0008-8846(96)00164-0.

Berger, S., Cau Dit Coumes, C., Le Bescop, P., & Damidot, D. (2011). Stabilization of ZnCl2-containing wastes using calcium sulfoaluminate cement: Cement hydration, strength development and volume stability. *Journal of Hazardous Materials, 194*, 256–267. Available from https://doi.org/10.1016/j.jhazmat.2011.07.095.

Bescher E. P., Kim J., Ramseyer C., Vallens J. K. (June 2018). Low carbon footprint pavement: History of use, performance and new opportunities for belitic calcium sulfoaluminate. *Proceedings of the 13th international symposium on concrete roads*, Berlin.

Bilginer, B. A., & Erdoğan, S. T. (2021). Effect of mixture proportioning on the strength and mineralogy of magnesium phosphate cements. *Construction and Building Materials, 277*. Available from https://doi.org/10.1016/j.conbuildmat.2021.122264.

Bilginer, B. A., & Erdoğan, S. T. (2023). Development of magnesium/calcium oxalate cements. *Materiales de Construcción., 73*(350). Available from https://doi.org/10.3989/mc.2023.298122.

Bilmez, S.B. (2023). *Development of oxalate cements with dead-burned magnesia and calcium sulfoaluminate clinker* [MSc thesis]. Middle East Technical University.

Bishnoi, S., Maity, S., Mallik, A., Joseph, S., & Krishnan, S. (2014). Pilot scale manufacture of limestone calcined clay cement: The Indian experience. *The Indian Concrete Journal, 88*(7), 22–28.

Bizzozero, J., & Scrivener, K.L. (May 18–21, 2014). Hydration of calcium aluminate cement based systems with calcium sulfate and limestone. *Proceedings of the* international conference on calcium aluminates, Avignon, France, pp. 189–196.

Bullerjahn, F., Zajac, M., & Ben Haha, M. (2015). CSA raw mix design: Effect on clinker formation and reactivity. *Materials and Structures., 48*, 3895–3911. Available from https://doi.org/10.1617/s11527-014-0451-z.

Çamurcuoğlu D., Siddall R. (2016). Plastering the prehistory: Marl as a unique material to cover, maintain and decorate the Neolithic walls of Çatalhöyük. In: Papayianni, I., Stefanidou, M., & Pachta, V. (Eds.) (10–12 October, 2016) *(Proceedings) 4th historic mortars conference HMC2016*, Santorini, Greece, pp. 482–489.

Canbek, O., Shakouri, S., & Erdoğan, S. T. (2020). Laboratory production of calcium sulfoaluminate cements with high industrial waste content. *Cement and Concrete Composites, 106*. Available from https://doi.org/10.1016/j.cemconcomp.2019.103475.

Cau Dit Coumes, C., Courtois, S., Peysson, S., Ambroise, J., & Pera, J. (2009). Calcium sulfoaluminate cement blended with OPC: A potential binder to encapsulate low-level radioactive slurries of complex chemistry. *Cement and Concrete Research, 39*(9), 740–747. Available from https://doi.org/10.1016/j.cemconres.2009.05.016.

Cembureau. (2021). *2021 activity report*. The European Cement Association. Unpublished content.

Chang, W., Li, H., Wei, M., Zhu, Z., Zhang, J., & Pei, M. (2009). Effects of polycarboxylic acid based superplasticiser on properties of sulphoaluminate cement. *Materials Research Innovations, 13*(1), 7–10. Available from https://doi.org/10.1179/143307509X402101China.

Chatterjee, A. K. (2018). *Cement production technology principles and practice*. CRC Press – Taylor & Francis Group.

Chatterjee, A. K. (2002). Special cements. In J. Bensted, & P. Barnes (Eds.), *Structure and performance of cements* (2nd ed., pp. 186–236). Taylor & Francis Group.

Cuesta, A., Ayuela, A., & Aranda, M. A. G. (2021). Belite cements and their activation. *Cement and Concrete Research, 140.* Available from https://doi.org/10.1016/j.cemconres.2020.106319.

Daehn, K., Basuhi, R., Gregory, J., Berlinger, M., Somjit, V., & Olivetti, E. A. (2022). Innovations to decarbonize materials industries. *Nature Reviews Materials, 7*, 275–294. Available from https://doi.org/10.1038/s41578-021-00376-y.

Elsen, J. (2006). Microscopy of historic mortars—a review. *Cement and Concrete Research, 36*(8), 1416–1424. Available from https://doi.org/10.1016/j.cemconres.2005.12.006.

Emmanuel, A. C., Haldar, P., Maity, S., & Bishnoi, S. (2016). Second pilot production of limestone calcined clay cement in India: The experience. *The Indian Concrete Journal, 90*(5), 57–63.

EOTA. (2017a). *European assesment document EAD 150001-00-0301: Calcium sulphoaluminate based cement.*

EOTA. (2017b). *European assesment document EAD 150004-00-0301: Rapid hardening sulfate resistant calcium sulphoaluminate based cement*

Erdoğan, S. T., & Erdoğan, T. Y. (2007). (in Turkish) *Ten thousand years of history of the binding materials and the concrete*. METU Press.

Galan, I., Beltagui, H., García-Maté, M., Glasser, F. P., & Imbabi, M. S. (2016). Impact of drying on pore structures in ettringite-rich cements. *Cement and Concrete Research, 84*, 85–94. Available from https://doi.org/10.1016/j.cemconres.2016.03.003.

Gao, D., Zhang, Z., Meng, Y., Tang, J., & Yang, L. (2021). Effect of flue gas desulfurization gypsum on the properties of calcium sulfoaluminate cement blended with ground granulated blast furnace slag. *Materials, 14*(2), 1–15. Available from https://doi.org/10.3390/ma14020382.

Garbev, K., Beuchle, G., Schweike, U., Merz, D., Dregert, O., Stemmermann, P., & Viehland, D. (2014). Preparation of a novel cementitious material from hydrothermally synthesized C-S-H phases. *Journal of the American Ceramic Society, 97*(7), 2298–2307. Available from https://doi.org/10.1111/jace.12920.

García-Maté, M., Santacruz, I., De La Torre, A. G., León-Reina, L., & Aranda, M. A. G. (2012). Rheological and hydration characterization of calcium sulfoaluminate cement pastes. *Cement and Concrete Composites, 34*(5), 684–691. Available from https://doi.org/10.1016/j.cemconcomp.2012.01.008.

Gartner, E., Gimenez, M., Meyer, V., Pisch. (April 28–May 1, 2014). A novel atmospheric pressure approach to the mineral capture of CO2 from industrial point sources. In: *13th annual conference on carbon capture, utilization and storage*, Pittsburgh, PA.

Gartner, E., & Sui, T. (2018). Alternative cement clinkers. *Cement and Concrete Research, 114*, 27–39. Available from https://doi.org/10.1016/j.cemconres.2017.02.002.

GB 200. (2003). *Moderate heat Portland cement, low heat Portland cement, low heat Portland slag cement*. Chinese Standard Unpublished content.

Glasser, F. P., & Zhang, L. (2001). High-performance cement matrices based on calcium sulfoaluminate–belite compositions. *Cement and Concrete Research, 31*(12), 1881–1886. Available from https://doi.org/10.1016/s0008-8846(01)00649-4.

Goyal., & Sharma, D. (2020). CO_2 sequestration on cement. In F. Pacheco-Torgal, E. Rasmussen, C.-G. Granqvist, V. Ivanov, A. Kaklauskas, & S. Makonin (Eds.), *Start-up creation* (2nd ed.). Woodhead Publishing, Woodhead Publishing series in Civil and Structural Engineering.

Guo, Y., Zhang, Y., Soe, K., & Pulham, M. (2018). Recent development in magnesium oxychloride cement. *Structural Concrete*, *19*(5), 1290–1300. Available from https://doi.org/10.1002/suco.201800077.

Gür, N., Aktaş, Y., & Civaş, A. (2010). Utilization of solid waste of soda ash plant as a mineral additive in cement. *Cement and Concrete World*, *15*, 57–69.

Haque, M. A., & Chen, B. (2019). Research progresses on magnesium phosphate cement: A review. *Construction and Building Materials*, *211*, 885–898. Available from https://doi.org/10.1016/j.conbuildmat.2019.03.304.

Hargis, C. W., Lothenbach, B., Müller, C. J., & Winnefeld, F. (2017). Carbonation of calcium sulfoaluminate mortars. *Cement and Concrete Composites*, *80*, 123–134. Available from https://doi.org/10.1016/j.cemconcomp.2017.03.003.

Harrison, A.J. W. (2008). United States Patent 7,347,896. Reactive magnesium oxide cements. Unpublished.

Haynes, W. M., Lide, D. R., & Bruno, T. J. (2017). *CRC handbook of chemistry and physics* (97th ed.). CRC Press – Taylor & Francis Group.

İçinsel, N. (2020). *Development of magnesium oxalate cements with recycled portland cement paste* [MSc thesis]. Middle East Technical University.

IEA WBCSD. (2018). *Technology roadmap: Low-carbon transition in the cement industry*. OECD/IEA/WBCSD.

Ilyukhin, V. V., Nevsky, N. N., Bickbau, M. J., & Howie, R. A. (1977). Crystal structure of alinite. *Nature*, *269*(5627), 397–398. Available from https://doi.org/10.1038/269397a0.

Ioannou, S., Paine, K., & Quillin, K. (21–27 September, 2010). Strength and durability of calcium sulfoaluminate based concretes. IC-NOCMAT 2010 international conference on non-conventional materials and technologies: Ecological materials and technologies for sustainable building, Cairo, Egypt.

Jen, G., Stompinis, N., & Jones, R. (2017). Chloride ingress in a belite-calcium sulfoaluminate cement matrix. *Cement and Concrete Research*, *98*, 130–135. Available from https://doi.org/10.1016/j.cemconres.2017.02.013.

Jeong, Y., Hargis, C. W., Chun, S.-C., & Moon, J. (2018). The effect of water and gypsum content on strätlingite formation in calcium sulfoaluminate-belite cement pastes. *Construction and Building Materials.*, *166*, 712–722. Available from https://doi.org/10.1016/j.conbuildmat.2018.01.153.

Jiang, W., & Roy, D. M. (1992). Hydrothermal processing of new fly ash cement. *American Ceramic Society Bulletin*, *71*(4), 642–647.

Juenger, M. C. G., Winnefeld, F., Provis, J. L., & Ideker, J. H. (2011). Advances in alternative cementitious binders. *Cement and Concrete Research*, *41*(12), 1232–1243. Available from https://doi.org/10.1016/j.cemconres.2010.11.012.

Kesim, A. G., Tokyay, M., Yaman, I. O., & Ozturk, A. (2013). Properties of alinite cement produced by using soda sludge. *Advances in Cement Research*, *25*(2), 104–111. Available from https://doi.org/10.1680/adcr.11.00040.

Kırca, Ö., Yaman, İ. Ö., & Tokyay, M. (2013). Compressive strength development of calcium aluminate cement–GGBFS blends. *Cement and Concrete Composites*, *35*(1), 163–170. Available from https://doi.org/10.1016/j.cemconcomp.2012.08.016.

Lampe, Fv, Hilmer, W., Jost, K. H., Reck, G., & Boikova, A. I. (1986). Synthesis, structure and thermal decomposition of alinite. *Cement and Concrete Research*, *16*(4), 505–510. Available from https://doi.org/10.1016/0008-8846(86)90088-8.

Lawrence, C. D. (1998). The production of low-energy cements. In P. C. Hewlett (Ed.), *Lea's chemistry of cement and concrete* (4th ed., pp. 421–470). Butterworth-Heinemann. Available from https://doi.org/10.1016/B978-075066256-7/50021-7.

LC3-Project. (2014). *LC3 – Limestone calcined clay cement.* https://lc3.ch/.

Locher, F. W. (1986). Low energy clinker. *8th international congress on the chemistry of cement.* Proceedings, Vol. I, Rio de Janiero, Brazil, pp. 57–67.

Majcher, K. (2015). *MIT technology review – What happened to green concrete?* Unpublished content. https://www.technologyreview.com/2015/03/19/73210/.

Martin, I., Patapy, C., & Cyr, M. (18–21 May, 2014). Parametric study of binary and ternary Ettringite-based systems. C. H. Fenttiman, R. J. Managbhai, & K. L. Scrivener, Eds., *Proceedings of the international conference on calcium aluminates*, Avignon, France, pp. 210–221.

Martin, L. H. J., Winnefeld, F., Tschopp, E., Müller, C. J., & Lothenbach, B. (2017). *Influence of fly ash on the hydration of calcium sulfoaluminate cement, . Switzerland cement and concrete research.* (95, pp. 152–163). . Available from https://doi.org/10.1016/j.cemconres.2017.02.030.

Meng, D., Unluer, C., Yang, E.-H., & Qian, S. (2023). Recent advances in magnesium-based materials: CO_2 sequestration and utilization, mechanical properties and environmental impact. *Cement and Concrete Composites, 138*. Available from https://doi.org/10.1016/j.cemconcomp.2023.104983.

Mo, L., Deng, M., Tang, M., & Al-Tabbaa, A. (2014). MgO expansive cement and concrete in China: Past, present and future. *Cement and Concrete Research, 57*, 1–12. Available from https://doi.org/10.1016/j.cemconres.2013.12.007.

Monteiro, P. J. M., Clodic, L., Battocchio, F., Kanitpanyacharoen, W., Chae, S. R., Ha, J., & Wenk, H.-R. (2013). Incorporating carbon sequestration materials in civil infrastructure: A micro and nano-structural analysis. *Cement and Concrete Composites, 40*, 14–20. Available from https://doi.org/10.1016/j.cemconcomp.2013.03.013.

Neubauer, J., & Pöllmann, H. (1994). Alinite—Chemical composition, solid solution and hydration behaviour. *Cement and Concrete Research, 24*(8), 1413–1422. Available from https://doi.org/10.1016/0008-8846(94)90154-6.

Odler, I. (2000). Special inorganic cements. In: Bentur, A. & Mindess, S. (Eds.), *Modern concrete technology series, Book 8*. Taylor & Francis Group.

Padilla-Encinas, P., Palomo, A., Blanco-Varela, M. T., & Fernández-Jiménez, A. (2020). Calcium sulfoaluminate clinker hydration at different alkali concentrations. *Cement and Concrete Research, 138*. Available from https://doi.org/10.1016/j.cemconres.2020.106251.

Pelletier, L., Winnefeld, F., & Lothenbach, B. (2010). The ternary system Portland cement–calcium sulphoaluminate clinker–anhydrite: Hydration mechanism and mortar properties. *Cement and Concrete Composites, 32*(7), 497–507. Available from https://doi.org/10.1016/j.cemconcomp.2010.03.010.

Péra, J., & Ambroise, J. (2004). New applications of calcium sulfoaluminate cement. *Cement and Concrete Research, 34*(4), 671–676. Available from https://doi.org/10.1016/j.cemconres.2003.10.019.

Pradip., & Kapur, P. C. (2004). Manufacture of eco-friendly and energy-efficient alinite cements from fly ashes and other bulk wastes. *Resources Processing, 51*(1), 8–13. Available from https://doi.org/10.4144/rpsj.51.8.

Pradip., Vaidyanathan, D., Kapur, P. C., & Singh, B. N. (1990). Production and properties of alinite cements from steel plant wastes. *Cement and Concrete Research, 20*(1), 15–24. Available from https://doi.org/10.1016/0008-8846(90)90112-b.

Provis, J. L. (2018). Alkali-activated materials. *Cement and Concrete Research, 114*, 40–48. Available from https://doi.org/10.1016/j.cemconres.2017.02.009.

Quillin, K. (2001). Performance of belite–sulfoaluminate cements. *Cement and Concrete Research*, *31*(9), 1341–1349. Available from https://doi.org/10.1016/s0008-8846(01)00543-9.

Rungchet, A., Chindaprasirt, P., Wansom, S., & Pimraksa, K. (2016). Hydrothermal synthesis of calcium sulfoaluminate–belite cement from industrial waste materials. *Journal of Cleaner Production*, *115*, 273–283. Available from https://doi.org/10.1016/j.jclepro.2015.12.068.

Sánchez Berriel, S., Cancio Díaz, Y., Martirena Hernández, J. F., & Habert, G. (2015). Assessment of sustainability of low carbon cement in Cuba. Cement pilot production and prospective case. *Calcined Clays for Sustainable Concrete. RILEM Bookseries*, *10*. Available from https://doi.org/10.1007/978-94-017-9939-3_23.

Noudelman, B., Bikbaou, M., Sventsitski, A., & Ilyukhin, V. (1980). Structure and properties of alinite and alinite cement. *7th international congress on the chemistry of cement. Paris*, *4*, 702–706.

Scrivener, K.L. (2001). Historical and present day applications of calcium aluminate cements. *Proceedings of the international conference on calcium aluminate cements*, Edinburgh, UK, pp. 3–23.

Scrivener, K. L. (2008). 100 years of calcium aluminate cements. In C. H. Fentiman, R. J. Mangabhai, & K. L. Scrivener (Eds.), *Proceedings of the centenary conference on calcium aluminate cements* (pp. 3–6). IHS BRE Press.

Scrivener, K. L., & Capmas, A. (1998). Calcium aluminate cements. In P. C. Hewlett (Ed.), *Lea's chemistry of cement and concrete* (4th ed., pp. 713–782). Butterworth-Heinemann. Available from https://doi.org/10.1016/B978-075066256-7/50025-4.

Scrivener, K. L., Martirena, F., Bishnoi, S., & Maity, S. (2018). Calcined clay limestone cements (LC3). *Cement and Concrete Research*, *114*, 49–56. Available from https://doi.org/10.1016/j.cemconres.2017.08.017.

Scrivener, K. L. (2003). Calcium aluminate cements. In J. Newman, & B. S. Choo (Eds.), *Advanced concrete technology* (3, pp. 1–31). Butterworth-Heinemann. Available from https://doi.org/10.1016/b978-075065686-3/50278-0.

Scrivener, K. L. (2014). Options for the future of cement. *Indian Concrete Journal*, *88*(7), 11–21.

Shand, M. A. (2006). *The chemistry and technology of magnesia* (pp. 1–266). John Wiley & Sons, Inc.. Available from https://doi.org/10.1002/0471980579.

Sharp, J. H., Lawrence, C. D., & Yang, R. (1999). Calcium sulfoaluminate cements—Low-energy cements, special cements or what? *Advances in Cement Research*, *11*(1), 3–13. Available from https://doi.org/10.1680/adcr.1999.11.1.3.

Singh, M., Kapur, P. C., & Pradip. (2008). Preparation of alinite based cement from incinerator ash. *Waste Management*, *28*(8), 1310–1316. Available from https://doi.org/10.1016/j.wasman.2007.08.025.

Solidia. (2023). *Patents, investors & awards*. https://www.solidiatech.com/.

Tangüler-Bayramtan, M. (2022). *Synthesis of green calcium sulfoaluminate cements using an industrial symbiosis approach* [PhD thesis]. Middle East Technical University.

Telesca, A., Marroccoli, M., Pace, M. L., Tomasulo, M., Valenti, G. L., & Monteiro, P. J. M. (2014). A hydration study of various calcium sulfoaluminate cements. *Cement and Concrete Composites*, *53*, 224–232. Available from https://doi.org/10.1016/j.cemconcomp.2014.07.002.

Thomas, R. J., Maguire, M., Sorensen, A. D., & Quezada, I. (2018). Calcium sulfoaluminate cement: Benefits and applications. *Concrete International*, *40*, 65–69.

Torrens-Martin, D., & Fernández-Carrasco, L. (18–21 May, 2014). Long-term hydration and mechanical behaviour of Portland cement, calcium aluminate cement and calcium sulfate blends. *Proceedings of the* international conference on calcium aluminates, Avignon, France, pp. 242–249.

Trauchessec, R., Mechling, J.-M., Lecomte, A., Roux, A., & Le Rolland, B. (2015). Hydration of ordinary Portland cement and calcium sulfoaluminate cement blends. *Cement and Concrete Composites*, *56*, 106–114. Available from https://doi.org/10.1016/j.cemconcomp.2014.11.005.

TS 13353. (2008). *Boron modified active belite cement – Definitions, composition, specifications and conformity criteria*. Turkish Standards Institute. Unpublished content.

Uçal, G. O., Mahyar, M., & Tokyay, M. (2018). Hydration of alinite cement produced from soda waste sludge. *Construction and Building Materials*, *164*, 178–184. Available from https://doi.org/10.1016/j.conbuildmat.2017.12.196.

Ulugöl, H. (2021). *Development of construction and demolition waste-based engineered geopolymer composites with self-healing capability* [PhD thesis]. Middle East Technical University.

Ulugöl, H., Günal, M. F., Yaman, İ. Ö., Yıldırım, G., & Şahmaran, M. (2021). Effects of self-healing on the microstructure, transport, and electrical properties of 100% construction- and demolition-waste-based geopolymer composites. *Cement and Concrete Composites*, *121*. Available from https://doi.org/10.1016/j.cemconcomp.2021.104081.

Vizcaíno-Andrés, L. M., Sánchez-Berriel, S., Damas-Carrera, S., Pérez-Hernández, A., Scrivener, K. L., & Martirena-Hernández, J. F. (2015). Industrial trial to produce a low clinker, low carbon cement. *Materiales de Construcción*, *65*(317), e045. Available from https://doi.org/10.3989/mc.2015.00614.

Walling, S. A., & Provis, J. L. (2016). Magnesia-based cements: A journey of 150 years, and cements for the future? *Chemical Reviews*, *116*(7), 4170–4204. Available from https://doi.org/10.1021/acs.chemrev.5b00463.

Winnefeld, F., & Barlag, S. (2009). Influence of calcium sulfate and calcium hydroxide on the hydration of calcium sulfoaluminate clinker. Einfluss von Calciumsulfat und Calciumhydroxid auf die Hydratation von Calciumsulfoaluminat-Klinker. *ZKG International.*, *62*(12), 42–53.

Winnefeld, F., & Barlag, S. (2010). Calorimetric and thermogravimetric study on the influence of calcium sulfate on the hydration of ye'elimite. *Journal of Thermal Analysis and Calorimetry*, *101*(3), 949–957. Available from https://doi.org/10.1007/s10973-009-0582-6.

Zajac, M., Skocek, J., Bullerjahn, F., & Ben Haha, M. (2016). Effect of retarders on the early hydration of calcium-sulpho-aluminate (CSA) type cements. *Cement and Concrete Research*, *84*, 62–75. Available from https://doi.org/10.1016/j.cemconres.2016.02.014.

UN Environment. (2017). *Eco-efficient cements: Potential economically viable solutions for a low-CO_2 cement based industry*.

Zhang, L. (2000). *Microstructure and performance of calcium sulfoaluminate cements* [PhD thesis]. University of Aberdeen.

Zhang, L., Su, M., & Wang, Y. (1999). Development of the use of sulfo- and ferroaluminate cements in China. *Advances in Cement Research*, *11*(1), 15–21. Available from https://doi.org/10.1680/adcr.1999.11.1.15.

Zhou, Q., Milestone, N. B., & Hayes, M. (2006). An alternative to Portland cement for waste encapsulation — The calcium sulfoaluminate cement system. *Journal of Hazardous Materials*, *136*(1), 120–129. Available from https://doi.org/10.1016/j.jhazmat.2005.11.038.

Sustainable concrete containing supplementary cementitious materials

3

Jad Bawab[1], *Jamal Khatib*[2,3] *and Hilal El-Hassan*[1]
[1]Department of Civil and Environmental Engineering, United Arab Emirates University, Al Ain, United Arab Emirates, [2]Faculty of Engineering, Beirut Arab University, Beirut, Lebanon, [3]Faculty of Science and Engineering, University of Wolverhampton, Wolverhampton, United Kingdom

3.1 Introduction

Concrete is indispensable to our built environment owing to its remarkable features. Its versatility, strength, and affordability made it the most widely used man-made material on Earth (Bawab, Khatib, El-Hassan, et al., 2021). Concrete can be cast into complex forms, allowing the creation of esthetically pleasing architectural designs that are not feasible with other materials (Neville, 1995). It also possesses excellent compressive strength, making it a well-established structural element (Bawab, Khatib, Jahami, et al., 2021). In addition, its availability and relatively low cost make it a cost-effective choice for large-scale construction projects (Alzard et al., 2021). Indeed, the use of concrete dates back to ancient civilizations, with the earliest known concrete structures built over 2000 years ago (Jackson et al., 2013). Nowadays, over 30 billion tons of concrete are produced annually worldwide (Bawab, Khatib, Kenai, et al., 2021; Yang et al., 2015). The applications of concrete span a wide spectrum, from building construction to infrastructure development (Anwar et al., 2022; El-Mir et al., 2023). Although occupying only 10%–15% of the volume of a typical concrete mixture, the cement is considered the most significant ingredient in concrete (Neville, 1995). Correspondingly, the annual production of cement has reached more than 4.1 billion tons (USGS, 2021).

Cement production exerts a substantial environmental impact for a number of reasons. The primary concern is the emission of carbon dioxide (CO_2). Cement manufacturing is the second-largest CO_2-emitting industry in the world (Leeson et al., 2017). In fact, CO_2 discharges from cement production account for around 8% of all human-made CO_2 emissions (Korol et al., 2020). This is attributed to the calcination process of limestone at extremely high temperatures and the combustion of huge amounts of fossil fuels (Hwalla, El-Hassan, et al., 2023; Korol et al., 2020; Miller et al., 2018). On average, the calcination process alone contributes to about 50% of the total CO_2 emissions associated with cement manufacturing (Sousa & Bogas, 2021). These environmental ramifications would result in a sharp increase

Sustainable Concrete Materials and Structures. DOI: https://doi.org/10.1016/B978-0-443-15672-4.00003-6

in CO_2 emissions in the coming years, possibly reaching 2.34 billion tons in 2050 in case no mitigation strategies are used (International Energy Agency, 2009).

Alleviating the environmental impacts of cement production is crucial for achieving sustainable construction practices. Currently, several strategies are being pursued to address these environmental challenges. Carbon capture and storage (CCS) is one of the proposed strategies that would effectively target the reduction of CO_2 emissions. However, the technology is dependent on finding suitable reservoirs, especially in developing countries where the demand for cement is increasing sharply (Leeson et al., 2017). The use of less carbon-intensive fossil fuels and biomass fuels in the cement production process was also suggested as a possible carbon emissions reduction method (International Energy Agency, 2009). Further, supplementary cementitious materials (SCMs) are another straightforward technique to mitigate CO_2 emissions from the cement industry. Being typically waste materials or industrial by-products, SCMs partially replace cement in concrete to lower its demand and consequently reduce CO_2 emissions (Hwalla et al., 2020; Lothenbach et al., 2011; Najm et al., 2022). Meanwhile, SCMs can maintain or improve the mechanical and durability properties of concrete. Fly ash, ground granulated blast furnace slag, and silica fume are some of the well-established SCMs in the cement industry. As these materials were incorporated into concrete mixtures for decades, their influence on the properties of the produced concrete is renowned (Juenger et al., 2019; Kachouh et al., 2022; Lothenbach et al., 2011). Nevertheless, as the production of the aforementioned SCMs is projected to decrease in the coming decades (Dwivedi & Jain, 2014), the pressure on using sustainable and emerging SCMs will certainly rise.

The main aim of this chapter is to discuss the properties of cementitious composites incorporating selected SCMs. The chapter starts with statistical information about the published work on SCM. This is then followed by the properties of concrete containing selected SCMs. The referred SCMs are metakaolin (MK), flue gas desulfurization (FGD) gypsum, municipal solid waste incineration bottom ash (IBA), ceramic waste powder (CWP), calcium carbide residue (CCR), and natural pozzolana (NP). The studied features of blended cement-based materials include fresh, mechanical, and durability properties. The main findings are summarized at the end of the chapter.

3.2 Statistics on supplementary cementitious materials

The publications on SCMs have significantly grown in recent years. Fig. 3.1 shows the number of publications with time from 1990 to 2023. The general trend of the number of yearly publications is increasing with time. The number of publications per year surpassed 100 for the first time in 2009 and reached its peak (695) in 2022. The increase in publication is attributed to the growing research interest in the use of SCMs that can contribute toward the sustainability of our environment. Another reason for the increasing research is the need for sustainable alternatives to cement coupled with the decrease in the supply of the most commonly used SCMs, specifically fly ash and ground granulated blast furnace slag (GGBS).

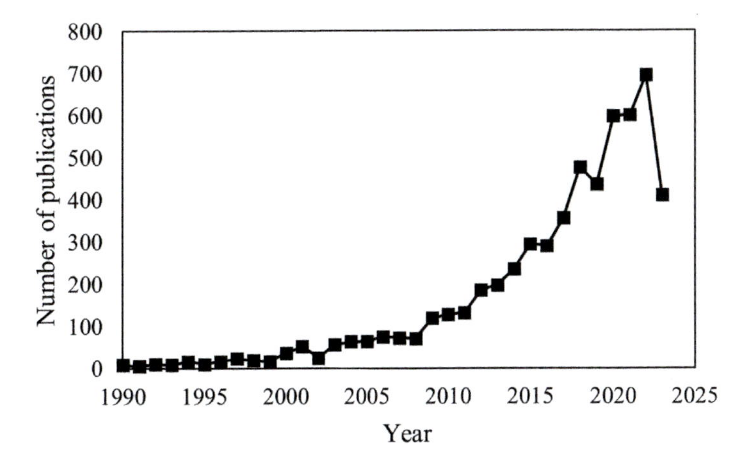

Figure 3.1 Number of supplementary cementitious material publications with time.

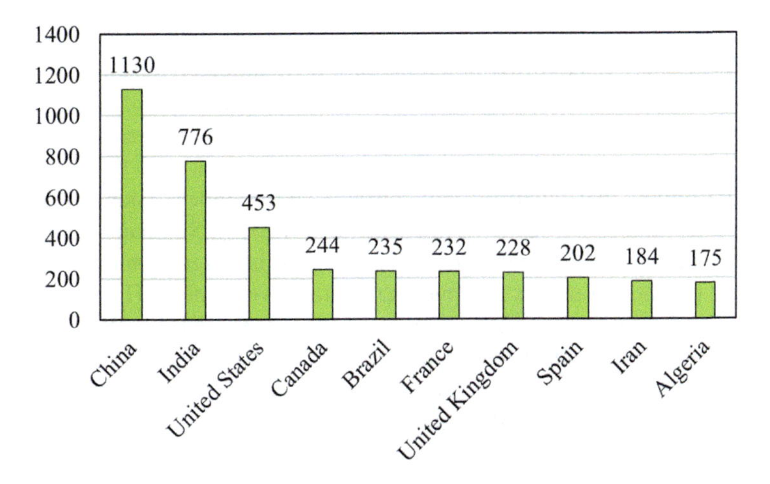

Figure 3.2 Top 10 countries publishing on supplementary cementitious materials.

The top 10 countries by the number of publications are shown in Fig. 3.2. China has the highest number of publications at 1130, followed by India at 776 and the United States at 453. The remaining countries had a very close number of publications ranging between 175 and 244. Several reasons would contribute to the diversity in the number of publications between countries. The presence of developed academic institutions and research facilities would support research projects and consequently a higher number of publications. Another reason would be the availability of local material resources and waste management strategies, where countries possessing materials suitable for use as SCMs would have an advantage in terms of research and development. For example, India, China, and the United States are the top three countries globally in the production of fly ash (Dwivedi & Jain, 2014).

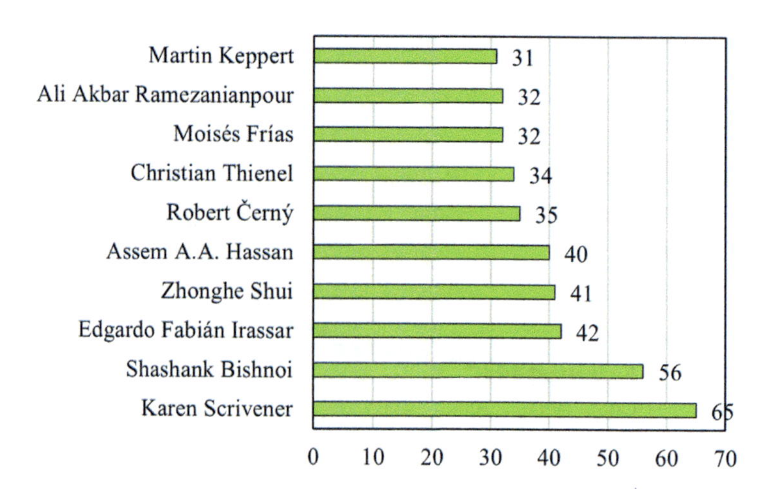

Figure 3.3 Top 10 authors published on supplementary cementitious materials by the number of publications.

Karen Scrivener ranked first in the list of top 10 authors publishing on SCMs by the number of publications (Fig. 3.3). As an expert in the area of cement and concrete technology, the authors' 65 publications on the subject present breakthrough findings and significant advancement in the area. In particular, the work includes a literature review that identifies key concepts and provides a standardized framework for later studies on SCMs (Lothenbach et al., 2011). Shashank Bishnoi ranked second on the list with 56 publications. The author has significant contributions in the area of hydration kinetics and microstructural development. The number of publications for the remaining authors ranged between 31 and 42. While targeting a wide variety of SCMs in different parts of the world, these publications provide valuable insights into the efficient use of these materials, allowing policymakers to make informed decisions based on scientific evidence.

Fig. 3.4 shows the classification of publications by type. The most common type of publication is journal articles (4247), followed by conference papers (1067), book chapters (232), reviews (138), and others (104). While publishing journal articles for original research findings is a standard practice in academia, it is evident that academic conferences played an important role in the dissemination of the topic. In addition, the high number of book chapters and reviews indicates a well-established research field with solid outcomes (Alzard et al., 2022; Hwalla, Bawab, et al., 2023).

The top 10 sources published on SCMs are listed in Fig. 3.5. By far, the *Construction and Building Materials* journal tops the list with 704 publications. As the scope of the journal includes the assessment of novel materials for different elements of civil engineering and infrastructure, it is normal for authors to target the journal for showcasing their research outcomes. The number of publications in the remaining sources ranged between 93 and 254 each. It is worth noting that the variety in the types of documents reported earlier showing the remarkable number

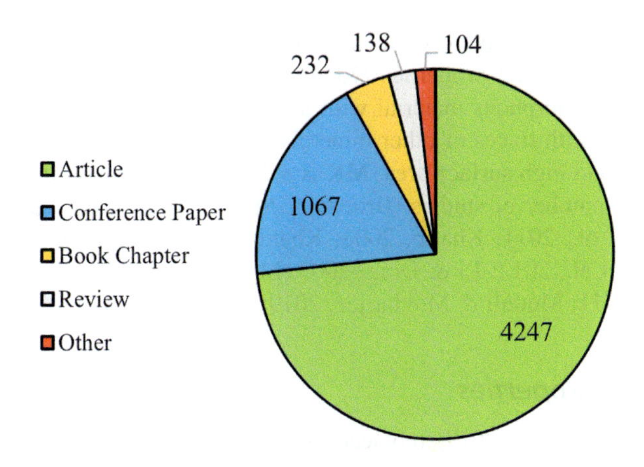

Figure 3.4 Classification of publications by type.

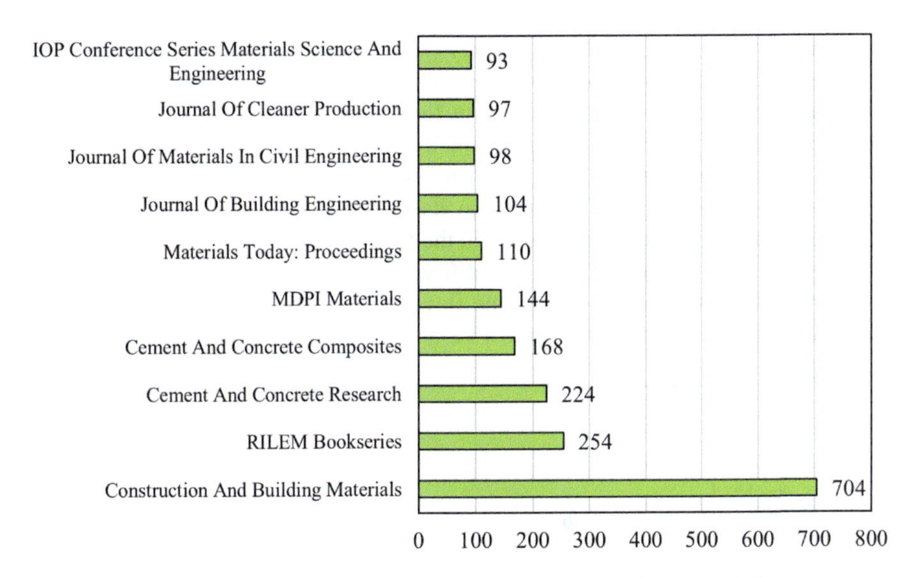

Figure 3.5 Top 10 sources publishing on supplementary cementitious materials.

of conference papers is in line with the detected sources, namely "RILEM Bookseries," "Materials Today: Proceedings," and "IOP Conference Series Materials Science and Engineering."

3.3 Metakaolin

MK originates from the calcination of kaolin clay at a temperature between 650°C and 800°C (Wild et al., 1996). When heated, chemically combined water is removed, and changes in the crystal structure take place, where the produced material possesses pozzolanic properties (Khatib & Wild, 1996). MK is typically produced as a fine, amorphous material with high reactivity. It is mainly composed of SiO_2 and Al_2O_3 with traces of other minerals (Wild & Khatib, 1997). With its pozzolanic nature and high surface area, MK is a desirable cement substitution and was addressed in a number of studies (Brooks & Megat Johari, 2001; Gill & Siddique, 2018; Kadri et al., 2011; Khatib, 2008; Khatib & Clay, 2004; Khatib & Hibbert, 2005; Khatib et al., 2012; Li & Ding, 2003; Madandoust & Mousavi, 2012; Megat Johari et al., 2011; Muduli & Mukharjee, 2019; Shekarchi et al., 2010).

3.3.1 Fresh properties

Muduli and Mukharjee (2019) replaced cement with MK by up to 20% by mass and observed a gradual decrease in the flowability of concrete with the increase in MK replacement level. This was attributed to the higher surface area of MK used in comparison with cement. However, the reduction was insignificant (reaching a maximum of 12% drop). A different study (Madandoust & Mousavi, 2012) reported a gradual drop in the flowability of concrete when using MK, where using a chemical admixture alleviated the reduction in flow. Similarly, Khatib (2008) observed that the flowability of concrete was gradually reduced as the MK content increased (Fig. 3.6). Kadri et al. (2011) reported that using chemical admixtures is necessary when using MK in concrete to maintain its flowability. Gill and Siddique (2018) observed that MK caused a slight drop in the flowability of concrete due to its higher water demand. On the contrary, a different investigation (Megat Johari et al., 2011) reported a significant drop in the flowability of concrete when incorporating an increasing amount of MK of up to 15% as partial cement replacement.

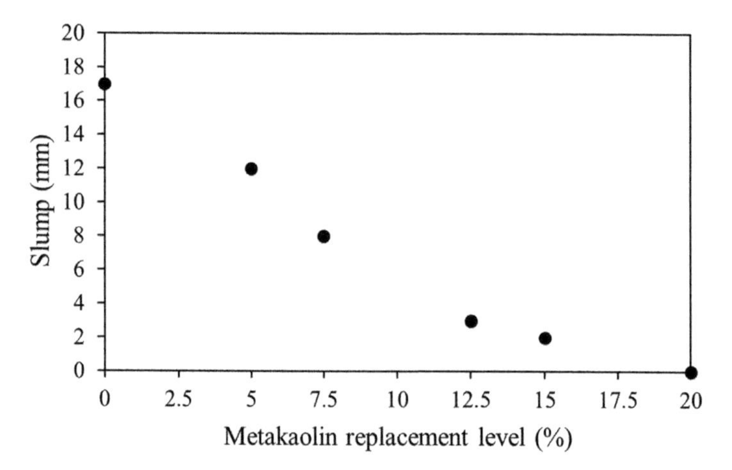

Figure 3.6 Effect of metakaolin on the flowability of concrete.
Source: Data from Khatib, J. M. (2008). Metakaolin concrete at a low water to binder ratio. *Construction and Building Materials, 22*(8), 1691–1700. https://doi.org/10.1016/j.conbuildmat.2007.06.003.

Brooks and Megat Johari (2001) studied the initial and final setting times of concrete containing up to 15% MK by cement mass and reported an extension of both durations with the inclusion of MK. The setting times increased with the increase in MK replacement level. On the contrary, Li and Ding (2003) reported that replacing cement with 10% MK by mass resulted in a shortening of both initial and final setting times. In another different study (Shekarchi et al., 2010) it was reported that the initial setting time dropped with the inclusion of MK in concrete as cement replacement, where the reduction was more evident at higher replacement levels. The authors also observed that the final setting time did not change significantly with the incorporation of MK.

3.3.2 Mechanical properties

Muduli and Mukharjee (2019) reported that the compressive strength of concrete increased with higher MK content, reaching its peak at 15% replacement by mass. The improvement in strength was observed at all curing periods. The better particles' packing and pozzolanic reaction were the main contributors to the increase in strength. Gill and Siddique (2018) stated that using MK at 5%, 10%, or 15% replacement of cement by mass would improve the compressive strength of concrete. The highest increase was at 15% replacement at 7 days and 10% replacement at 28 days and beyond. Madandoust and Mousavi (2012) reported that the most significant improvement in the strength of concrete containing MK was detected at replacement levels of 10%–15%. The authors also realized that the highest strength development was during the first 14 days of curing. Megat Johari et al. (2011) reported an increase in strength when using MK as a cement replacement in concrete. The optimum replacement level was determined to be 5% by mass at 3 and 7 days and 10%–15% by mass at 28 days and beyond. Khatib et al. (2012) replaced cement with MK by up to 50% by mass and observed an increase in strength with the increase in MK content, reaching its peak at a 20% replacement level. Beyond this replacement, the strength decreased. Similar results were reported in a different study (Khatib & Hibbert, 2005), where the compressive strength of concrete was significantly improved when replacing cement with MK at a 20% replacement level by mass, in particular at late curing periods.

Madandoust and Mousavi (2012) found that the tensile strength of concrete improved when using MK, specifically at replacement levels of 10%–15% by cement mass. Further inclusion of MK caused the opposite effect. Muduli and Mukharjee (2019) studied the flexural and tensile strengths of concrete containing MK as a partial replacement of cement for up to 20% by mass. It was found that MK improved both properties by up to 15% replacement level; however, beyond this amount, the flexural and tensile strengths were reduced. Khatib and Hibbert (2005) observed an increase in the flexural strength of concrete containing MK as a partial substitution of cement at a 20% replacement level.

3.3.3 Durability

Gill and Siddique (2018) reported that the incorporation of MK in concrete lowered its water absorption, with the maximum improvement being at a 5% replacement level.

This was attributed to the pozzolanic reaction products that helped densify the microstructure. Muduli and Mukharjee (2019) observed that incorporating MK in concrete gradually decreased its water absorption until reaching a replacement level of 15% by mass. The authors stated that MK could be effectively used to offset the higher water absorption caused by using some types of aggregates, such as recycled concrete aggregates. Madandoust and Mousavi (2012) stated that the water absorption of concrete decreased due to the incorporation of MK. The improvement was comparable at all replacement levels up to 20% by mass. On the contrary, Khatib and Clay (2004) studied the water absorption of concrete containing MK and reported an increase in water absorption when using MK as a cement replacement. The difference in water absorption became more prominent at higher replacement levels. The authors attributed this effect to the higher total pore volume of pastes containing MK compared with the control paste.

3.4 Flue gas desulfurization gypsum

FGD gypsum is a by-product generated from the process of removing sulfur dioxide (SO_2) emissions from the flue gases of coal-fired power plants (Khatib et al., 2016). Consequently, the production of FGD gypsum has been increasing due to tighter environmental regulations (Mangat et al., 2006; Wright & Khatib, 2016). The chemical composition of FGD gypsum varies, but it often contains calcium sulfate compounds, along with other trace minerals (Wan et al., 2022). These compounds are similar to those found in natural gypsum; therefore FGD gypsum was addressed by a few studies as a possible component in the production of cement-based materials (Khatib et al., 2016; Lei et al., 2017; Mangat et al., 2006; Wan et al., 2022; Wansom et al., 2019; Yao et al., 2019; Zhang et al., 2016).

3.4.1 *Fresh properties*

Khatib et al. (2016) investigated the flowability of concrete containing FGD gypsum as a partial replacement of cement by up to 90% by mass and reported that the flowability decreased with the increase in FGD gypsum especially when the FGD gypsum content exceeded 40%. Zhang et al. (2016) studied the flowability of mortar including dried FGD gypsum or thermally activated FGD gypsum by up to 33% by mass and observed that cement replacement with either material had a marginal impact on the flowability. Nevertheless, the flowability of mortar made with FGD gypsum was higher than that containing thermally activated FGD gypsum, possibly due to its lower surface area.

Zhang et al. (2016) reported that the initial and final setting times of mortar containing dried FGD gypsum as cement replacement increased, whereas that comprising thermally activated FGD gypsum was close to the reference mortar. Nonetheless, the reported initial and final setting times were within acceptable limits. In a different study (Wansom et al., 2019) a gradual increase in the initial setting time of concrete

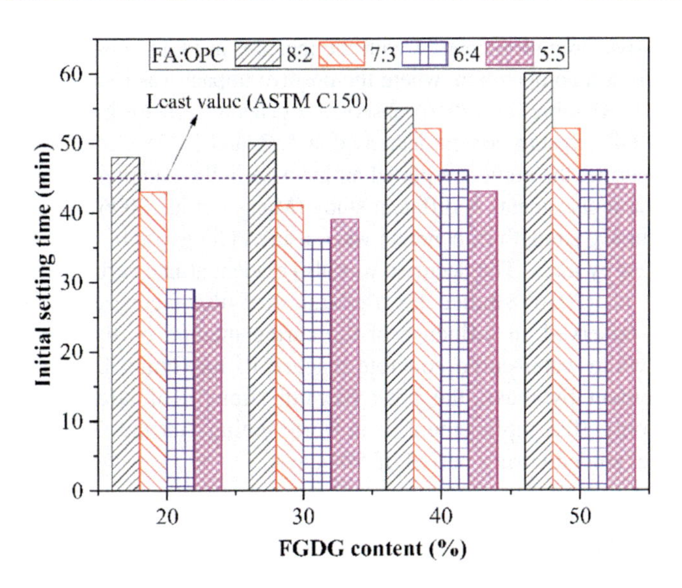

Figure 3.7 Effect of flue gas desulfurization gypsum on the initial setting time of cement-fly ash binder.
Source: Reprinted with permission from Wan, Y., Hui, X., He, X., Li, J., Xue, J., Feng, D., Liu, X., & Wang, S. (2022). Performance of green binder developed from flue gas desulfurization gypsum incorporating Portland cement and large-volume fly ash. *Construction and Building Materials, 348*, 128679. https://doi.org/10.1016/j.conbuildmat.2022.128679.

was noted with the increase in FGD gypsum content. Wan et al. (2022) recommended using FDG at high volumes for retarding the initial setting time of cement and fly ash binder beyond 45 minutes (Fig. 3.7).

3.4.2 Mechanical properties

Zhang et al. (2016) reported that replacing cement with dried FGD gypsum by up to 33% by mass reduced the compressive strength of mortar, in particular at early curing periods. However, at lower replacement levels and late curing periods (i.e., after 120 days), the mortar mixes containing dried FGD gypsum slightly surpassed the reference mortar due to the continued strength development. The authors realized that thermally activating the FGD gypsum rendered an improvement in strength starting at 28 days of curing, whereas, after 1 year, the mortar mixes containing thermally activated FGD gypsum had a strength substantially higher than that of the reference mortar. Lei et al. (2017) produced high-strength concrete containing 60% untreated FDG gypsum, where its strength reached around 42 MPa at 28 days. FDG gypsum was also incorporated in lightweight porous concrete along with other industrial by-products, where it positively impacted the compressive strength, especially at 28 days of curing (Yao et al., 2019). Wan et al. (2022) reported that the effect of FGD gypsum on the compressive strength of cement-fly ash binder was significantly related to the cement and fly ash content.

The compressive strength increased with the increase in FGD gypsum content in mixes with higher fly ash composition, where the positive impact was more prominent at early curing periods. Khatib et al. (2016) observed a general decrease in the strength of concrete as the FGD gypsum content increased at 1, 7, and 28 days of curing. However, at 365 days of curing, the authors reported an increase in the compressive strength of concrete at a 10% replacement. A different study (Mangat et al., 2006) reported a decrease in the compressive strength of concrete when using FGD gypsum at a 25% replacement level of cement by mass. The decrease was also evident at late curing periods.

Khatib et al. (2016) studied the flexural strength of concrete containing FGD gypsum and observed an influence of the curing method on the optimum replacement level; while water-cured concrete reached its maximum flexural strength at a 30% replacement by mass, the peak flexural strength of air-cured concrete was achieved at only 10% replacement. It is worth noting that the flexural strength was tested at an extended curing period of 570 days.

3.4.3 Durability

Khatib et al. (2016) studied the effect of including FGD gypsum in concrete as a partial cement replacement on its water absorption at 28 and 365 days. The authors reported an increase in water absorption with the increase in FGD gypsum content. At 28 days, using up to 10% replacement by mass kept the concrete within low water absorption levels; however, mixes with 20%–70% replacement by mass had medium water absorption, whereas up to 70% replacement by mass rendered low water absorption at 365 days.

3.5 Municipal solid waste incineration bottom ash

Municipal solid waste IBA is generated as a residue from the combustion of municipal solid wastes in waste-to-energy facilities worldwide (Bawab, Khatib, Kenai, et al., 2021). It consists of noncombustible materials, such as glass, ceramics, and metals (Tang et al., 2016). The chemical composition of IBA varies but typically includes silicates, alumina, iron oxides, and other minerals in addition to some heavy metals, which make proper management and treatment crucial to ensure environmental safety (Alderete et al., 2021). Several studies considered using IBA as a partial cement replacement, which would contribute to both waste reduction and resource conservation (Alderete et al., 2021; Li et al., 2012; Lin et al., 2003; Matos & Sousa-Coutinho, 2022; Simões et al., 2021; Tang et al., 2016; Vaičienė & Simanavičius, 2022; Zhang et al., 2021).

3.5.1 Fresh properties

Tang et al. (2016) considered replacing cement with thermally treated IBA at a 30% replacement level by mass in mortar and reported that its flowability decreased as a result of this replacement. Nevertheless, untreated IBA had the opposite effect on flowability, where it was slightly increased. In another study (Alderete et al., 2021) it was indicated that concrete containing IBA as a partial cement replacement had a lower flowability than the reference concrete, mainly due to the increased fineness and high water demand. Another investigation (Vaičienė & Simanavičius, 2022) reported a gradual decrease in the flowability of concrete with an increase in IBA content and this reduction reached 20% at 12% IBA replacement level. Simões et al. (2021) investigated the effect of including up to 50% IBA by mass as cement replacement and reported a decrease in the flowability of concrete as the IBA content increased. Nevertheless, the authors stated that the reduction was not substantial, even at high replacement levels.

Matos and Sousa-Coutinho (2022) reported that including only 10% IBA by cement mass resulted in an elongation of the initial and final setting times of cement and IBA blended paste. The increase in the initial and final setting times reached 40% and 60%, respectively. The authors explained that the heavy metals present in the IBA would probably retard the setting times. Li et al. (2012) also observed an increase in the initial and final setting times of blended cement paste containing up to 50% IBA by mass.

3.5.2 Mechanical properties

Tang et al. (2016) stated that the compressive strength of mortar decreased when 30% of cement was replaced with thermally treated IBA. The reduction in strength was more apparent at early curing ages. The finer the particle size of IBA the least reduction in strength. Zhang et al. (2021) observed an increase in the compressive strength of concrete when replacing cement with IBA at a 20% replacement. This was explained by the improvement in pozzolanic reactivity of IBA that reduced the $Ca(OH)_2$ content and contributed to strength development. Alderete et al. (2021) reported a reduction in strength when using IBA as a partial substitution for cement. The reduction was more pronounced at early curing periods. Vaičienė and Simanavičius (2022) observed an increase in strength as the replacement level of cement with IBA increased up to 6% by mass. Beyond this substitution amount, the strength started to drop. The improvement in strength was more prominent at 28 days than at 7 days of curing (Fig. 3.8). A different study (Simões et al., 2021) reported that the compressive strength of concrete containing IBA decreased as the IBA content increased. They also observed that the strength gain between 7 and 91 days was higher for the reference concrete in comparison with mixes containing IBA. Lin et al. (2003) replaced cement with IBA by up to 40% (by mass) and observed a slight increase in strength at a 10% replacement level at late curing periods. Beyond 10% replacement by mass, the strength dropped, where the reduction was more prominent at early curing ages.

Tang et al. (2016) observed that the flexural strength of mortar was slightly decreased after cement replacement with thermally treated IBA at a 30% replacement level by mass. The authors highlighted the irregular particle size of IBA as a main contributor to alleviating the reduction, where they had a positive effect on

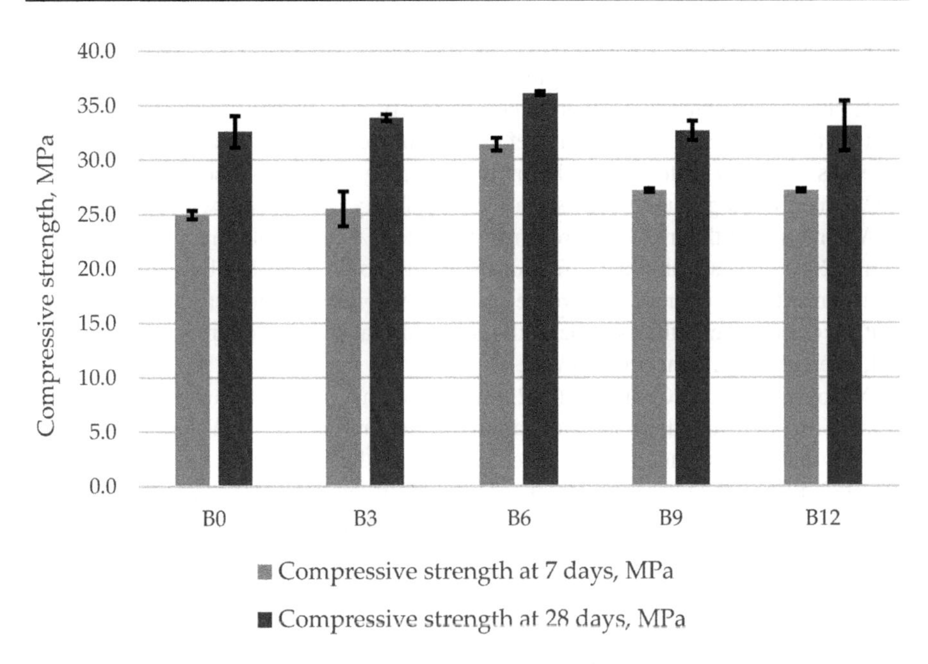

Figure 3.8 Compressive strength of concrete containing incineration bottom ash, where the number in the mix designation refers to the replacement level by mass.
Source: From Vaičienė, M., & Simanavičius, E. (2022). The effect of municipal solid waste incineration ash on the properties and durability of cement concrete. *Materials, 15*(13), 4486. https://doi.org/10.3390/ma15134486.

flexural strength. Alderete et al. (2021) studied the flexural and tensile strength of concrete containing IBA as a partial cement replacement and observed comparable results for both properties in comparison with the reference concrete. Simões et al. (2021) reported that the tensile strength of concrete containing IBA as a partial cement replacement decreased with the inclusion of IBA. The decrease was gradual and associated with the IBA replacement level.

3.5.3 Durability

Vaičienė and Simanavičius (2022) studied the water absorption of concrete containing IBA as a partial cement substitution by up to 12% by mass and observed that the water absorption decreased when including IBA by up to 6%, then increased beyond this substitution level. Simões et al. (2021) studied the water absorption of concrete containing IBA at 28 and 91 days of curing. They reported that including IBA as a cement replacement by up to 40% by mass resulted in comparable water absorption with the reference concrete.

3.6 Ceramic waste powder

CWP originates from discarded ceramic products, such as broken tiles and pottery. The global production of CWP exceeds 22 billion tons, and it is usually landfilled causing soil and groundwater pollution (El-Dieb et al., 2019). It is mainly composed of SiO_2 and Al_2O_3, making it a suitable component of cementitious systems due to its pozzolanic properties (El-Dieb & Kanaan, 2018). A number of studies considered partially replacing cement with CWP and in cement-based materials (Aly et al., 2019; El-Dieb & Kanaan, 2018; Heidari & Tavakoli, 2013; Kannan et al., 2017; Mohit & Sharifi, 2019; Pacheco-Torgal & Jalali, 2010; Subaşı et al., 2017; Vejmelková et al., 2012).

3.6.1 Fresh properties

Mohit and Sharifi (2019) reported that CWP had a marginal effect on the flowability of mortar, where it remained nearly constant even after replacing cement with up to 25% CWP by mass. Aly et al. (2019) reported that replacing cement with by up to 60% CWP by mass caused a limited decrease in the flowability of concrete. This was mainly attributed to the higher surface area of CWP in comparison with cement. In another investigation (Heidari & Tavakoli, 2013) it was found that the flowability of concrete decreased with the increase in CWP content, reaching a 30% reduction at a 40% replacement level by mass. Kannan et al. (2017) observed that the flowability of concrete containing CWP as a partial replacement of cement increased with the increase in CWP until the replacement level reached 20% by mass; beyond this amount, the flowability was reduced. Similar results were reported elsewhere (Vejmelková et al., 2012).

Mohit and Sharifi (2019) reported that cement replacement with CWP had a minimal and disordered effect on the initial and final setting times of mortar. Also, the higher water demand induced by the CWP would affect the hydration kinetics. Comparable results were reported in a different study (Kannan et al., 2017), where replacing cement with WCP at 10% and 20% replacement levels by mass had a marginal effect on initial and final setting times while they were shortened at 30% replacement level and then elongated at 40% replacement level. This could be caused by accelerated hydration at a 30% replacement level that was offset by the large replacement volume at a 40% replacement level causing retardation of the hydration reaction.

3.6.2 Mechanical properties

Mohit and Sharifi (2019) studied the compressive strength of mortar containing WCP and observed a reduction at all replacement levels of WCP at 7 and 14 days. However, an improvement in strength was possible at 5% and 10% replacement of cement with WCP by mass at 28 days and up to 15% by mass at 56 days. El-Dieb and Kanaan (2018) stated that there was no possibility of achieving an improvement in the compressive strength of concrete at 7 days, regardless of the replacement level of WCP (Fig. 3.9). However, at 28 and 90 days, the compressive strength increased at 10% replacement level by mass. A different study (Vejmelková et al., 2012) reported that the compressive strength of concrete slightly improved when WCP replaced cement by 10% by mass, whereas it decreased at higher replacement levels up to 60% by mass. In this study the improvement was observed at both 7 and 28 days of curing. Heidari and Tavakoli (2013) observed a reduction in compressive strength when replacing cement with WCP by up to 40% by mass. The authors stated that the difference in strength diminished with longer curing periods. Aly et al. (2019) reported that as the CWP percentage replacement increased, the compressive strength of concrete gradually decreased and the reduction was more noticeable at 7 days in comparison with that at 28 days. Moreover, Chokkalingam et al. (2022a) reported that up to 60% ceramic waste replacement in alkali-activated slag concrete could provide a 40-MPa compressive strength after 28 days.

Mohit and Sharifi (2019) reported that the flexural strength of mortar improved when cement was replaced with WCP by up to 15% by mass. Beyond this replacement level, the flexural strength dropped. A different study (Subaşı et al., 2017) investigated the tensile strength of concrete containing WCP as a partial replacement of cement for up to 20% by mass at 7 and 28 days of curing and reported a reduction in tensile strength at both curing periods and all replacement levels. Nevertheless, the drop in tensile strength was more evident at 7 days.

3.6.3 *Durability*

Heidari and Tavakoli (2013) reported that water absorption decreased when utilizing WCP in mortar as a partial replacement for cement. While the optimum reduction of around 13.5% was achieved at a replacement level of 20% by mass, the authors explained this improvement by the denser microstructure triggered by the pozzolanic reaction. Pacheco-Torgal and Jalali (2010) studied the water absorption of concrete containing WCP of different types as a partial replacement of cement by 20% by mass and observed that the water absorption decreased with the use of WCP. Meanwhile, in slag-based geopolymer concrete the replacement of 40% slag with ceramic waste resulted in an improvement in water absorption, sorptivity, abrasion resistance, and bulk resistivity. Higher replacement percentages led to a decrease in said properties (Chokkalingam et al., 2022b).

3.7 Calcium carbide residue

CCR is an industrial by-product generated by the production of acetylene gas (Kelechi, Uche, et al., 2022). It is produced as a slurry and disposed of in landfills, causing pollution due to its high alkalinity (Goswami et al., 2021). In addition, it has very limited recycling options. Calcium oxide (CaO) is the main component of CCR, followed by traces of silicon dioxide (SiO_2) and some impurities (Yang et al., 2021). Very recently, CCR was implemented in carbonation-cured concrete as a partial

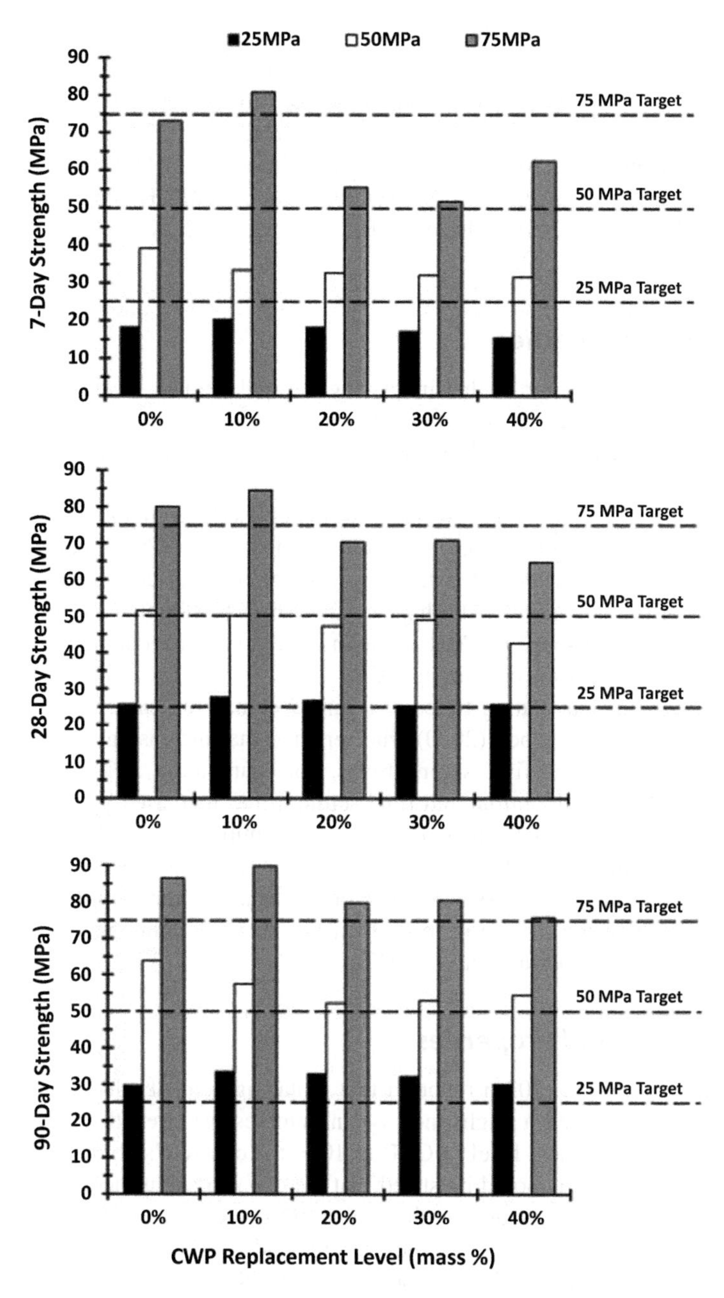

Figure 3.9 Compressive strength of concrete containing ceramic waste powder.
Source: Reprinted with permission from El-Dieb, A. S., & Kanaan, D. M. (2018). Ceramic waste powder an alternative cement replacement – Characterization and evaluation. *Sustainable Materials and Technologies, 17*. https://doi.org/10.1016/j.susmat.2018.e00063.

cement substitution, where it promoted CO_2 sequestration and contributed to the strength and durability of the developed concrete (El-Hassan et al., 2023). Considering its chemical composition and the need for finding recycling routes, several studies investigated using CCR as a partial replacement of cement in mortar and concrete (Adamu et al., 2020, 2021, 2022; Al-Khaja et al., 1992; Babako & Apeh, 2020; Goswami et al., 2021, 2022; Haruna et al., 2021; Karthiga et al., 2020; Kelechi, Adamu, et al., 2022; Kelechi, Uche, et al., 2022; Kou et al., 2021; Uche et al., 2022; Yang et al., 2021).

3.7.1 Fresh properties

Adamu et al. (2021) reported that increasing the replacement level of CCR in concrete rendered a decrease in its flowability, reaching a substantial reduction of 47% when replacing cement with CCR by 30% by mass. This drop in flowability was explained by the higher surface area of CCR used in comparison with the partially replaced cement. Another study (Kelechi, Uche, et al., 2022) reported similar observations concerning the influence of replacing cement with CCR on concrete's flowability; however, it was reported that using other SCMs, such as fly ash, would effectively offset this effect. Al-Khaja et al. (1992) highlighted the high fineness and irregular particle shape of CCR as the main reasons for reducing the flowability of cement-based materials containing CCR.

The initial and final setting times of a binder containing cement and CCR were studied by Babako and Apeh (2020) who reported that increasing the CCR content prolonged the initial and final setting times. For example, at 20% cement replacement with CCR, both the initial and final setting times increased by 50%. Goswami et al. (2021) compared the effect of CCR on the initial and final setting times of cement paste with that of fly ash and reported that although CCR elongated these durations, its influence is less than that of fly ash. In other words the presence of fly ash retarded the hydration more than that of CCR. This agrees with results reported elsewhere (Goswami et al., 2022).

3.7.2 Mechanical properties

Kelechi, Adamu, et al. (2022) reported that replacing cement with CCR by 5% by mass in concrete caused a slight increase in compressive strength. However, when increasing the replacement level of CCR to 10%, the concrete's compressive strength decreased at 90 days. It should be stated that there was a slight increase in compressive strength at early ages due to the high CaO content in the CCR. Uche et al. (2022) observed similar results and they recommended using only 5% CCR replacement by mass. Yang et al. (2021) processed the CCR by using wet grinding before using it as a partial replacement of cement in concrete and reported that replacing cement by up to 12% CCR influenced an increase in the compressive strength with an optimum at 8% replacement. A different study (Kelechi, Uche, et al., 2022) stated that using CCR as a partial replacement of cement by up to 10% by mass would render acceptable compressive strength of concrete. Karthiga et al. (2020) recommended

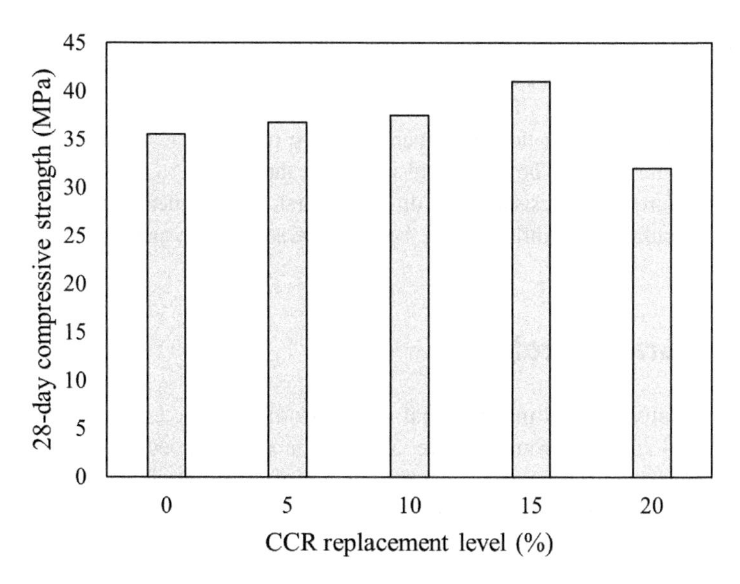

Figure 3.10 Effect of calcium carbide residue on compressive strength of concrete.
Source: Data from Karthiga, S., Devi, C. H. R., Ramasamy, N. G., Pavithra, C., Sudarsan, J. S., & Nithiyanantham, S. (2020). Analysis on strength properties by replacement of cement with calcium carbide remains and ground granulated blast furnace slag. *Chemistry Africa, 3* (4), 1133–1139. https://doi.org/10.1007/s42250-020-00169-w.

replacing cement with CCR by up to 15% by mass for strength improvement, where an increase in strength at different curing ages at this replacement level was obtained (Fig. 3.10). Another study (Adamu et al., 2021) also stated that using CCR at a replacement level of 15% slightly increased the compressive strength of concrete. In general, limiting the CCR content to 15% by cement mass would at least maintain the compressive strength of concrete. This is also dependent on the processing of the CCR before use in concrete (Amnadnua et al., 2013), and the use of other SCMs in the concrete mixture (Kou et al., 2021).

Adamu et al. (2020) studied the flexural and tensile strengths of concrete containing CCR as a partial replacement of cement for up to 15% by mass. The authors reported a positive impact of the CCR on both properties at all replacement levels. A different study (Adamu et al., 2022) reported similar results, where including CCR at up to 15% by cement mass rendered an improvement in both tensile and flexural strength. On the contrary, Haruna et al. (2021) reported that increasing the tensile and flexural strength of concrete was possible only at a CCR content of 5%, where higher inclusion of up to 20% caused a decrease in these properties.

3.7.3 Durability

Adamu et al. (2021) studied the water absorption of concrete containing CCR as a partial replacement of cement for up to 30% by mass. They reported a decrease in water absorption as the CCR content increased, reaching its minimum value at 15%.

Nevertheless, the water absorption increased beyond this replacement level. This effect was attributed to the finer particle size and pore-filling ability of the CCR. Uche et al. (2022) reported that using CCR at a 5% cement replacement would render an improvement in water absorption; however, increasing the replacement level to 10% by mass had the opposite effect. The authors also studied the impact of including CCR in the concrete mixture on its resistance to different harsh environments and concluded that CCR had an unfavorable influence on the concrete's durability performance.

3.8 Natural pozzolana

NP is a naturally existent material that results from volcanic eruptions. Albeit abundant, NP is localized around active and dormant volcanoes globally. While the materials generated from volcanic eruptions are of different sizes, NP generally has a particle size diameter of less than 2 mm (Ghrici et al., 2007). NP is mainly composed of SiO_2 while containing considerable amounts of Al_2O_3 and Fe_2O_3 (Hossain & Lachemi, 2006a). As such, NP is a pozzolanic material, where its involvement in cementitious systems triggers a pozzolanic reaction (Celik et al., 2019; Lothenbach et al., 2011). Several studies investigated the use of NP as a partial replacement for cement for this reason (Bawab et al., 2023; Belaidi et al., 2012; Celik et al., 2019; Hammat et al., 2021; Hossain, 2003; Hossain & Lachemi, 2006a, 2006b, 2007; Karolina & Simanjuntak, 2018; Khan et al., 2022; Khan & Alhozaimy, 2011; Mohamad et al., 2020; Mouli & Khelafi, 2008; Targan et al., 2003).

3.8.1 *Fresh properties*

Hossain and Lachemi (2006a) reported that NP significantly influenced the flowability of concrete. The authors observed that the flowability increased with the increase of the NP content for up to 20% by mass but then decreased beyond this replacement level. Correspondingly, Khan and Alhozaimy (2011) demonstrated that replacing cement with NP by 15% by mass increased the flowability of concrete from 120 to 130 mm. However, a further increase of NP to 20% and 25% by mass triggered a reduction in slump to 125 and 120 mm, respectively. A similar trend was observed elsewhere (Belaidi et al., 2012), where the slump of concrete slightly increased after the inclusion of 5% NP by mass but then noticeably decreased as the NP content increased to 25% by mass. Another study (Hammat et al., 2021) revealed a reduction in the flowability of concrete at both 15% and 30% NP replacement by cement mass. The authors also reported that the reduction was exacerbated as the fineness of the NP increased.

Khan and Alhozaimy (2011) studied the initial and final setting times of concrete containing NP as a partial replacement of cement for up to 25% and concluded that the initial setting time was gradually shortened with the inclusion of the NP but the final setting time decreased at a replacement level of 15% by mass then increased at an NP content of 20% and 25% by mass. Conversely, a different study (Hossain, 2003) observed that both initial and final setting times of concrete were stretched due to the inclusion of NP, as increasing the NP content from 0% to 50% by mass triggered an increase of 90% and 58% for the initial and final setting times, respectively. The author explained the increase in the setting time by the reduction in cement content and consequently retarding the hydration process.

3.8.2 Mechanical properties

Belaidi et al. (2012) replaced cement with NP by up to 25% by mass and concluded that including NP triggered a decline in strength at all replacement levels and all curing ages. This was attributed to the slower strength development when using NP as cement replacement. Hossain and Lachemi (2006a) reported a reduction in the compressive strength of concrete with the inclusion of NP as a partial replacement for cement. The drop in strength was more evident at an early age. Nevertheless, a slight improvement in strength was achieved for minimal amounts of NP (i.e., 2%–4% by mass) and at a curing duration of 28 days and beyond. A similar observation was presented elsewhere (Khan & Alhozaimy, 2011), as the compressive strength of concrete dropped with the replacement of cement with NP at 15%, 20%, and 25% by mass. However, it was reported that a comparable strength could be reached for an NP replacement level of 15% by mass and a curing age of 90 days and beyond. Similarly, Celik et al. (2019) found that cement substitution with NP could produce mortar of analogous compressive strength to that of the control mixture at 91 days of curing. A different study (Hammat et al., 2021) also reported similar findings but revealed that NP with higher fineness (500 m^2/kg) would reflect a better improvement in the strength of concrete at late curing periods in comparison with NP having lower fineness (350 m^2/kg). On the other hand, Mohamad et al. (2020) reported that the inclusion of NP in mortar, especially at a 15% replacement level of cement by mass, influenced a slight increase in compressive strength at all curing ages. Elsewhere, concrete with a high compressive strength exceeding 60 MPa was also possible even at an NP replacement level of 20% by cement mass (Hossain & Lachemi, 2007).

Mohamad et al. (2020) studied replacing cement with NP in mortar for up to 25% by mass and observed a reduction in both flexural and tensile strength at all replacement levels of NP. Nevertheless, comparable tensile and flexural strengths were obtained when the NP content reached 15% by mass. Targan et al. (2003) studied the flexural strength of concrete containing NP as a cement replacement by up to 30% by mass and found that limiting the replacement level to 10% improved the flexural strength, in particular at longer curing durations. A different study (Mouli & Khelafi, 2008) investigated the tensile and flexural strength of concrete containing NP as a partial replacement of cement for up to 50% by mass (Fig. 3.11). At curing ages of 28 days and beyond, an evident increase at both properties at a replacement level of 20% by mass. Nevertheless, using up to 40% NP yielded tensile and flexural strengths comparable to that of the control mixture, where the flexural strength reached around 7 MPa. Bawab et al. (2023) optimized the flexural strength of mortar incorporating NP and reported that using up to 40%

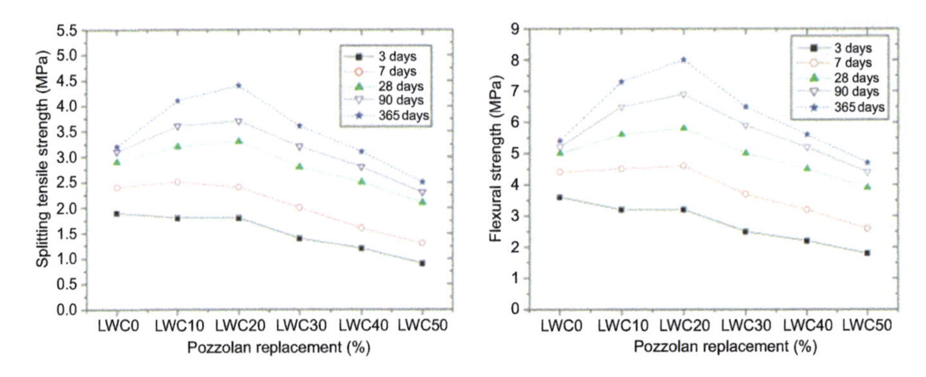

Figure 3.11 Effect of incorporating natural pozzolana on the flexural and tensile strengths of concrete.

Source: Reprinted with permission from Mouli, M., & Khelafi, H. (2008). Performance characteristics of lightweight aggregate concrete containing natural pozzolan. *Building and Environment, 43*(1), 31–36. https://doi.org/10.1016/j.buildenv.2006.110.038.

NP by cement mass would still render high flexural strength while altering other mix design parameters.

3.8.3 Durability

Khan et al. (2022) investigated the water absorption of mortar containing 20% NP as a replacement for cement. The authors reported that NP had a marginal influence on water absorption. A different study (Karolina & Simanjuntak, 2018) indicated that the inclusion of NP rendered an increase in the water absorption of concrete. Concrete incorporating NP in partial cement replacement was also observed to have poor performance against sulfate attack (Hossain & Lachemi, 2006b). In addition, NP inclusion in concrete resulted in an increase in its drying shrinkage, in particular in the first few weeks, with less effect in the long term (Hossain & Lachemi, 2007).

3.9 Conclusion

In this chapter, the effect of selected SCMs on various engineering properties of cement-based materials was discussed. The growing research interest, international collaboration, and varied publication sources collectively feature the importance of SCMs in shaping the future of sustainable construction practices. The rising trajectory in publications implies the dedication to innovation and sustainability of SCMs.

A summary of the major findings is presented in Table 3.1. MK generally reduces the flowability of concrete. The impact of MK on setting times was inconsistent across studies; some reported shortening of both initial and final setting times with increasing MK content, while others observed no significant change or even extended setting times at certain replacement levels. MK had a positive influence on the

Table 3.1 Summary of the major findings.

Supplementary cementitious material	Flow	Setting times	Strength	Water absorption
MK	↓[a]	↑↓[b]	↑[c] up to 15%	↓ up to 20%
FGD gypsum	←→[d]	↑	↓	↑
IBA	↓	↑	↑↓	↓ up to 6%
CWP	↑↓	←→	↑↓	↓ up to 20%
CCR	↓	↑	↑ up to 10%	↓ up to 5%
NP	↓	↑↓	↑↓	↑

Notes: *CCR*, calcium carbide residue; *CWP*, ceramic waste powder; *FGD*, flue gas desulfurization; *IBA*, incineration bottom ash; *MK*, metakaolin; *NP*, natural pozzolana.
[a]Decrease.
[b]Increase or decrease.
[c]Increase.
[d]Marginal effect.

mechanical properties of concrete, as various studies noticed an improvement in strength, reaching peak enhancements at around 15%–20% by mass. In addition, incorporating MK generally leads to reduced water absorption of concrete.

The substitution of cement with FGD gypsum exhibited limited influence on the flowability of concrete in most cases. In addition, while both initial and final setting times increased in some instances, they generally remained within acceptable limits. The impact of FGD gypsum on the mechanical properties was influenced by replacement levels, curing periods, and the treatment of FGD gypsum. While certain studies reported reduced early strengths with higher FGD gypsum content, the strength often improved with longer curing periods or when using thermally activated FGD gypsum. In general, higher FGD gypsum content correlated with increased water absorption.

The incorporation of municipal solid waste IBA as a cement replacement impacted the flowability, where trends generally indicated a decrease in flowability. It also often extended both initial and final setting times. The mechanical properties of concrete containing IBA showed varying responses. While some studies reductions in strength, especially at early curing ages, others observed strength improvements. Water absorption often decreased with the inclusion of IBA up to a certain level, after which it could increase.

CWP had varying effects on flowability. Some studies reported marginal changes, while others observed decreased flowability. The impact of CWP on setting times was generally limited. Strength responses to CWP replacement exhibited variation. In some instances, strength was reduced, particularly at early curing ages. However, improvements in strength were observed at certain replacement levels and curing durations. In addition, studies indicated that CWP incorporation led to reduced water absorption.

Replacing cement with CCR generally led to decreased flowability in concrete. Both initial and final setting times were extended due to the incorporation of CCR in concrete. The effect of CCR replacement on strength properties varied with replacement levels and curing durations. Positive effects on strength were observed at lower replacement levels. The impact on water absorption was influenced by factors like particle size and replacement level, with some studies reporting reductions in water absorption and others observing an increase beyond certain replacement levels.

NP generally influenced a decrease in the flowability of concrete. Some studies found that the initial and final setting times were extended due to the presence of NP, with other contradictory observations in different studies. The influence of NP on mechanical properties, particularly compressive strength, was generally positive. Many studies reported that incorporating NP led to improvements in strength, particularly at later curing ages. The extent of strength improvement depended on the replacement level and the curing duration. In addition, NP was often found to decrease the water absorption of concrete.

References

Adamu, M., Haruna, S. I., Ibrahim, Y. E., & Alanazi, H. (2022). Evaluation of the mechanical performance of concrete containing calcium carbide residue and nano silica using response surface methodology. *Environmental Science and Pollution Research*, *29*(44), 67076–67102. Available from https://doi.org/10.1007/s11356-022-20546-x.

Adamu, M., Ibrahim, Y. E., Al-Atroush, M. E., & Alanazi, H. (2021). Mechanical properties and durability performance of concrete containing calcium carbide residue and nano silica. *Materials*, *14*(22). Available from https://doi.org/10.3390/ma14226960.

Adamu, M., Olalekan, S. S., & Aliyu, M. M. (2020). Optimizing the mechanical properties of pervious concrete containing calcium carbide and rice husk ash using response surface methodology. *Nigeria Journal of Soft Computing in Civil Engineering*, *4*(3), 106–123. Available from https://doi.org/10.22115/SCCE.2020.229019.1216.

Alderete, N. M., Joseph, A. M., Van den Heede, P., Matthys, S., & De Belie, N. (2021). Effective and sustainable use of municipal solid waste incineration bottom ash in concrete regarding strength and durability. *Resources, Conservation and Recycling*, *167*, 105356. Available from https://doi.org/10.1016/j.resconrec.2020.105356.

Al-Khaja, W. A., Madany, I. M., Al-Sayed, M. H., & Darwish, A. A. (1992). The mechanical and drying shrinkage properties of cement mortars containing carbide lime waste. *Resources, Conservation and Recycling*, *6*(3), 179–190. Available from https://doi.org/10.1016/0921-3449(92)90029-2.

Aly, S. T., El-Dieb, A. S., & Taha, M. R. (2019). Effect of high-volume ceramic waste powder as partial cement replacement on fresh and compressive strength of self-compacting concrete. *Journal of Materials in Civil Engineering*, *31*(2). Available from https://doi.org/10.1061/(ASCE)MT.1943-5533.0002588.

Alzard, M. H., El-Hassan, H., & El-Maaddawy, T. (2021). Environmental and economic life cycle assessment of recycled aggregates concrete in the United Arab Emirates. *Sustainability*, *13*(18), 10348. Available from https://doi.org/10.3390/su131810348.

Alzard, M. H., El-Hassan, H., El-Maaddawy, T., Alsalami, M., Abdulrahman, F., & Aly Hassan, A. (2022). A bibliometric analysis of the studies on self-healing concrete published between 1974 and 2021. *Sustainability*, *14*(18), 11646. Available from https://doi.org/10.3390/su141811646.

Amnadnua, K., Tangchirapat, W., & Jaturapitakkul, C. (2013). Strength, water permeability, and heat evolution of high strength concrete made from the mixture of calcium carbide residue and fly ash. *Materials and Design*, *51*, 894–901. Available from https://doi.org/10.1016/j.matdes.2013.040.099.

Anwar, F. H., El-Hassan, H., Hamouda, M., El-Mir, A., Mohammed, S., & Hung Mo, K. (2022). Optimization of pervious geopolymer concrete using TOPSIS-based Taguchi method. *Sustainability*, *14*(14), 8767. Available from https://doi.org/10.3390/su14148767.

Babako, M., & Apeh, J. A. (2020). Setting time and standard consistency of Portland cement binders blended with rice husk ash, calcium carbide and metakaolin admixtures. *IOP Conference Series: Materials Science and Engineering*, *805*(1), 012031. Available from https://doi.org/10.1088/1757-899x/805/1/012031.

Bawab, J., El-Hassan, H., El-Dieb, A., & Khatib, J. (2023). Effect of Mix design parameters on the properties of cementitious composites incorporating volcanic ash and dune sand. *Developments in the Built Environment*, *16*, 100258. Available from https://doi.org/10.1016/j.dibe.2023.100258.

Bawab, J., Khatib, J., El-Hassan, H., Assi, L., & Kırgız, M. S. (2021). Properties of cement-based materials containing cathode-ray tube (CRT) glass waste as fine aggregates—A review. *Sustainability*, *13*(20), 11529. Available from https://doi.org/10.3390/su132011529.

Bawab, J., Khatib, J., Jahami, A., Elkordi, A., & Ghorbel, E. (2021). Structural performance of reinforced concrete beams incorporating cathode-ray tube (CRT) glass waste. *Buildings*, *11*(2), 67. Available from https://doi.org/10.3390/buildings11020067.

Bawab, J., Khatib, J., Kenai, S., & Sonebi, M. (2021). A review on cementitious materials including municipal solid waste incineration bottom ash (MSWI-BA) as aggregates. *Buildings*, *11*(5), 179. Available from https://doi.org/10.3390/buildings11050179.

Belaidi, A. S. E., Azzouz, L., Kadri, E., & Kenai, S. (2012). Effect of natural pozzolana and marble powder on the properties of self-compacting concrete. *Construction and Building Materials*, *31*, 251–257. Available from https://doi.org/10.1016/j.conbuildmat.2011.120.109.

Brooks, J. J., & Megat Johari, M. A. (2001). Effect of metakaolin on creep and shrinkage of concrete. *Cement and Concrete Composites*, *23*(6), 495–502. Available from https://doi.org/10.1016/S0958-9465(00)00095-0.

Celik, K., Hay, R., Hargis, C. W., & Moon, J. (2019). Effect of volcanic ash pozzolan or limestone replacement on hydration of Portland cement. *Construction and Building Materials*, *197*, 803–812. Available from https://doi.org/10.1016/j.conbuildmat.2018.110.193.

Chokkalingam, P., El-Hassan, H., El-Dieb, A., & El-Mir, A. (2022a). Development and characterization of ceramic waste powder-slag blended geopolymer concrete designed using Taguchi method. *Construction and Building Materials*, *349*, 128744. Available from https://doi.org/10.1016/j.conbuildmat.2022.128744.

Chokkalingam, P., El-Hassan, H., El-Dieb, A., & El-Mir, A. (2022b). Multi-response optimization of ceramic waste geopolymer concrete using BWM and TOPSIS based Taguchi methods. *Journal of Materials Research and Technology*, *21*, 4824–4845. Available from https://doi.org/10.1016/j.jmrt.2022.11.089.

Dwivedi, A., & Jain, M. K. (2014). Fly ash—waste management and overview: A review. *Recent Research in Science and Technology*, *6*, 30–35.

El-Dieb, A. S., & Kanaan, D. M. (2018). Ceramic waste powder an alternative cement replacement – Characterization and evaluation. *Sustainable Materials and Technologies*, *17*. Available from https://doi.org/10.1016/j.susmat.2018.e00063.

El-Dieb, A. S., Taha, M. R., & Abu-Eishah, S. I. (2019). *The use of ceramic waste powder (CWP) in making eco-friendly concretes. Ceramic Materials - Synthesis, Characterization, Applications and Recycling* (pp. 1–35.). IntechOpen.

El-Hassan, H., Bawab, J., Khatib, J., El-Dieb, A., & Hassan, A. A. (2023). Decarbonization of concrete through cement replacement of calcium carbide residue and accelerated carbonation curing. *US Patent US11713279B1*.

El-Mir, A., Hwalla, J., El-Hassan, H., Assaad, J. J., El-Dieb, A., & Shehab, E. (2023). Valorization of waste perlite powder in geopolymer composites. *Construction and Building Materials*, *368*, 130491. Available from https://doi.org/10.1016/j.conbuildmat.2023.130491.

Ghrici, M., Kenai, S., & Said-Mansour, M. (2007). Mechanical properties and durability of mortar and concrete containing natural pozzolana and limestone blended cements. *Cement and Concrete Composites*, *29*(7), 542–549. Available from https://doi.org/10.1016/j.cemconcomp.2007.040.009.

Gill, A. S., & Siddique, R. (2018). Durability properties of self-compacting concrete incorporating metakaolin and rice husk ash. *Construction and Building Materials*, *176*, 323–332. Available from https://doi.org/10.1016/j.conbuildmat.2018.050.054.

Goswami, S., Kumar Shukla, D., & Kumar Singh, P. (2021). Technical evaluation of sustainable cement containing fly ash and carbide lime sludge. *IOP Conference Series: Earth and Environmental Science*, *795*(1), 012040. Available from https://doi.org/10.1088/1755-1315/795/1/012040.

Goswami, S., Kumar Shukla, D., & Kumar Singh, P. (2022). Sustainable transformation of sewage sludge ash and waste industrial additive into green cement blend. *Structural Concrete*, *23*(4), 2337–2351. Available from https://doi.org/10.1002/suco.202100120.

Hammat, S., Menadi, B., Kenai, S., Thomas, C., Kirgiz, M. S., & Sousa Galdino, A. G. D. (2021). *The effect of content and fineness of natural pozzolana on the rheological, mechanical, and durability properties of self-compacting mortar. Journal of Building Engineering* (44). Available from 10.1016/j.jobe.2021.103276.

Haruna, S. I., Malami, S. I., Adamu, M., Usman, A. G., Farouk, A., Ali, S. I. A., & Abba, S. I. (2021). Compressive strength of self-compacting concrete modified with rice husk ash and calcium carbide waste modeling: A feasibility of emerging emotional intelligent model (EANN) versus traditional FFNN. *Arabian Journal for Science and Engineering*, *46*(11), 11207–11222. Available from https://doi.org/10.1007/s13369-021-05715-3.

Heidari, A., & Tavakoli, D. (2013). A study of the mechanical properties of ground ceramic powder concrete incorporating nano-SiO_2 particles. *Construction and Building Materials*, *38*, 255–264. Available from https://doi.org/10.1016/j.conbuildmat.2012.070.110.

Hossain, K. M. A. (2003). Blended cement using volcanic ash and pumice. *Cement and Concrete Research*, *33*(10), 1601–1605. Available from https://doi.org/10.1016/S0008-8846(03)00127-3.

Hossain, K. M. A., & Lachemi, M. (2006a). Development of volcanic ash concrete: Strength, durability, and microstructural investigations. *ACI Materials Journal*, *103*(1), 11–17.

Hossain, K. M. A., & Lachemi, M. (2006b). Performance of volcanic ash and pumice based blended cement concrete in mixed sulfate environment. *Cement and Concrete Research*, *36*(6), 1123–1133. Available from https://doi.org/10.1016/j.cemconres.2006.030.010.

Hossain, K. M. A., & Lachemi, M. (2007). Strength, durability and micro-structural aspects of high performance volcanic ash concrete. *Cement and Concrete Research*, *37*(5), 759–766. Available from https://doi.org/10.1016/j.cemconres.2007.020.014.

Hwalla, J., Bawab, J., El-Hassan, H., Abu Obaida, F., & El-Maaddawy, T. (2023). Scientometric analysis of global research on the utilization of geopolymer composites in construction applications. *Sustainability*, *15*(14), 11340. Available from https://doi.org/10.3390/su151411340.

Hwalla, J., El-Hassan, H., Assaad, J. J., & El-Maaddawy, T. (2023). Performance of cementitious and slag-fly ash blended geopolymer screed composites: A comparative study.

Case Studies in Construction Materials, *18*, e02037. Available from https://doi.org/10.1016/j.cscm.2023.e02037.

Hwalla, J., Saba, M., & Assaad, J. J. (2020). Suitability of metakaolin-based geopolymers for underwater applications. *Materials and Structures/Materiaux et Constructions*, *53*(5), 13595997. Available from https://doi.org/10.1617/s11527-020-01546-0.

International Energy Agency. (2009). *Cement technology road-map 2009 carbon emissions reductions up to 2050*. IEA.

Jackson, M. D., Moon, J., Gotti, E., Taylor, R., Chae, S. R., Kunz, M., Emwas, A. H., Meral, C., Guttmann, P., Levitz, P., Wenk, H. R., & Monteiro, P. J. M. (2013). Material and elastic properties of Al-tobermorite in ancient roman seawater concrete. *Journal of the American Ceramic Society*, *96*(8), 2598–2606. Available from https://doi.org/10.1111/jace.12407.

Juenger, M. C. G., Snellings, R., & Bernal, S. A. (2019). Supplementary cementitious materials: New sources, characterization, and performance insights. *Cement and Concrete Research*, *122*, 257–273. Available from https://doi.org/10.1016/j.cemconres.2019.05.008.

Kachouh, N., El-Maaddawy, T., El-Hassan, H., & El-Ariss, B. (2022). Shear response of recycled aggregates concrete deep beams containing steel fibers and web openings. *Sustainability*, *14*(2), 945. Available from https://doi.org/10.3390/su14020945.

Kadri, E. H., Kenai, S., Ezziane, K., Siddique, R., & De Schutter, G. (2011). Influence of metakaolin and silica fume on the heat of hydration and compressive strength development of mortar. *Applied Clay Science*, *53*(4), 704–708. Available from https://doi.org/10.1016/j.clay.2011.060.008.

Kannan, D. M., Aboubakr, S. H., EL-Dieb, A. S., & Reda Taha, M. M. (2017). High performance concrete incorporating ceramic waste powder as large partial replacement of Portland cement. *Construction and Building Materials*, *144*, 35–41. Available from https://doi.org/10.1016/j.conbuildmat.2017.030.115.

Karolina, R., & Simanjuntak, M. P. (2018). The influence of using volcanic ash and lime ash as filler on compressive strength in self compacting concrete. *IOP Conference Series: Earth and Environmental Science*, *126*(1), 012038. Available from https://doi.org/10.1088/1755-1315/126/1/012038.

Karthiga, S., Devi, C. H. R., Ramasamy, N. G., Pavithra, C., Sudarsan, J. S., & Nithiyanantham, S. (2020). Analysis on strength properties by replacement of cement with calcium carbide remains and ground granulated blast furnace slag. *Chemistry Africa*, *3*(4), 1133–1139. Available from https://doi.org/10.1007/s42250-020-00169-w.

Kelechi, S. E., Adamu, M., Mohammed, A., Obianyo, I. I., Ibrahim, Y. E., & Alanazi, H. (2022). Equivalent CO_2 emission and cost analysis of green self-compacting rubberized concrete. *Sustainability (Switzerland)*, *14*(1). Available from https://doi.org/10.3390/su14010137.

Kelechi, S. E., Uche, O. A. U., Adamu, M., Alanazi, H., Okokpujie, I. P., Ibrahim, Y. E., & Obianyo, I. I. (2022). Modeling and optimization of high-volume fly ash self-compacting concrete containing crumb rubber and calcium carbide residue using response surface methodology. *Arabian Journal for Science and Engineering*, *47*(10), 13467–13486. Available from https://doi.org/10.1007/s13369-022-06850-1.

Khan, K., Amin, M. N., Usman, M., Imran, M., Al-Faiad, M. A., & Shalabi, F. I. (2022). Effect of fineness and heat treatment on the pozzolanic activity of natural volcanic ash for its utilization as supplementary cementitious materials. *Crystals*, *12*(2). Available from https://doi.org/10.3390/cryst12020302.

Khan, M. I., & Alhozaimy, A. M. (2011). Properties of natural pozzolan and its potential utilization in environmental friendly concrete. *Canadian Journal of Civil Engineering*, *38*(1), 71–78. Available from https://doi.org/10.1139/L10-112.

Khatib, J. M. (2008). Metakaolin concrete at a low water to binder ratio. *Construction and Building Materials*, *22*(8), 1691–1700. Available from https://doi.org/10.1016/j.conbuildmat.2007.060.003.

Khatib, J. M., & Clay, R. M. (2004). Absorption characteristics of metakaolin concrete. *Cement and Concrete Research*, *34*(1), 19–29. Available from https://doi.org/10.1016/S0008-8846(03)00188-1.

Khatib, J. M., & Hibbert, J. J. (2005). Selected engineering properties of concrete incorporating slag and metakaolin. *Construction and Building Materials*, *19*(6), 460–472. Available from https://doi.org/10.1016/j.conbuildmat.2004.070.017.

Khatib, J. M., Negim, E. M., & Gjonbalaj, E. (2012). High volume metakaolin as cement replacement in mortar. *World Journal of Chemistry*, *7*, 7–10.

Khatib, J. M., & Wild, S. (1996). Pore size distribution of metakaolin paste. *Cement and Concrete Research*, *26*(10), 1545–1553. Available from https://doi.org/10.1016/0008-8846(96)00147-0.

Khatib, J. M., Wright, L., & Mangat, P. S. (2016). Mechanical and physical properties of concrete containing FGD waste. *Magazine of Concrete Research*, *68*(11), 550–560. Available from https://doi.org/10.1680/macr.15.00092.

Korol, J., Hejna, A., Burchart-Korol, D., & Wachowicz, J. (2020). Comparative analysis of carbon, ecological, and water footprints of polypropylene-based composites filled with cotton, jute and kenaf fibers. *Materials*, *13*(16). Available from https://doi.org/10.3390/MA13163541.

Kou, R., Guo, M. Z., Han, L., Li, J. S., Li, B., Chu, H., Jiang, L., Wang, L., Jin, W., & Sun Poon, C. (2021). Recycling sediment, calcium carbide slag and ground granulated blast-furnace slag into novel and sustainable cementitious binder for production of eco-friendly mortar. *Construction and Building Materials*, *305*. Available from https://doi.org/10.1016/j.conbuildmat.2021.124772.

Leeson, D., Mac Dowell, N., Shah, N., Petit, C., & Fennell, P. S. (2017). A techno-economic analysis and systematic review of carbon capture and storage (CCS) applied to the iron and steel, cement, oil refining and pulp and paper industries, as well as other high purity sources. *International Journal of Greenhouse Gas Control*, *61*, 71–84. Available from https://doi.org/10.1016/j.ijggc.2017.03.020.

Lei, D. Y., Guo, L. P., Sun, W., Liu, J. P., & Miao, C. W. (2017). Study on properties of untreated FGD gypsum-based high-strength building materials. *Construction and Building Materials*, *153*, 765–773. Available from https://doi.org/10.1016/j.conbuildmat.2017.070.166.

Li, X. G., Lv, Y., Ma, B. G., Chen, Q. B., Yin, X. B., & Jian, S. W. (2012). Utilization of municipal solid waste incineration bottom ash in blended cement. *Journal of Cleaner Production*, *32*, 96–100. Available from https://doi.org/10.1016/j.jclepro.2012.030.038.

Li, Z., & Ding, Z. (2003). Property improvement of Portland cement by incorporating with metakaolin and slag. *Cement and Concrete Research*, *33*(4), 579–584. Available from https://doi.org/10.1016/S0008-8846(02)01025-6.

Lin, K. L., Wang, K. S., Tzeng, B. Y., & Lin, C. Y. (2003). The reuse of municipal solid waste incinerator fly ash slag as a cement substitute. *Resources, Conservation and Recycling*, *39*(4), 315–324. Available from https://doi.org/10.1016/S0921-3449(02)00172-6.

Lothenbach, B., Scrivener, K., & Hooton, R. D. (2011). Supplementary cementitious materials. *Cement and Concrete Research*, *41*(12), 1244–1256. Available from https://doi.org/10.1016/j.cemconres.2010.120.001.

Madandoust, R., & Mousavi, S. Y. (2012). Fresh and hardened properties of self-compacting concrete containing metakaolin. *Construction and Building Materials*, *35*, 752–760. Available from https://doi.org/10.1016/j.conbuildmat.2012.040.109.

Mafalda Matos, Ana, & Sousa-Coutinho, Joana. (2022). Municipal solid waste incineration bottom ash recycling in concrete: Preliminary approach with Oporto wastes. *Construction and Building Materials*, *323*, 126548. Available from https://doi.org/10.1016/j.conbuildmat.2022.126548.

Mangat, P. S., Khatib, J. M., & Wright, L. (2006). Optimum utilisation of FGD waste in blended binders. *Proceedings of Institution of Civil Engineers: Construction Materials*, *159*(3), 119–127. Available from https://doi.org/10.1680/coma.2006.159.30.119.

Megat Johari, M. A., Brooks, J. J., Kabir, S., & Rivard, P. (2011). Influence of supplementary cementitious materials on engineering properties of high strength concrete. *Construction and Building Materials*, *25*(5), 2639–2648. Available from https://doi.org/10.1016/j.conbuildmat.2010.120.013.

Miller, S. A., John, V. M., Pacca, S. A., & Horvath, A. (2018). Carbon dioxide reduction potential in the global cement industry by 2050. *Cement and Concrete Research*, *114*, 115–124. Available from https://doi.org/10.1016/j.cemconres.2017.08.026.

Mineral Commodity Summaries - Cement. USGS. (2021).

Mohamad, S. A., Al-Hamd, R. K. S., & Khaled, T. T. (2020). Investigating the effect of elevated temperatures on the properties of mortar produced with volcanic ash. *Innovative Infrastructure Solutions*, *5*(1). Available from https://doi.org/10.1007/s41062-020-0274-4.

Mohit, M., & Sharifi, Y. (2019). Ceramic waste powder as alternative mortar-based cementitious materials. *ACI Materials Journal*, *116*(6). Available from https://doi.org/10.14359/51716819.

Mouli, M., & Khelafi, H. (2008). Performance characteristics of lightweight aggregate concrete containing natural pozzolan. *Building and Environment*, *43*(1), 31–36. Available from https://doi.org/10.1016/j.buildenv.2006.110.038.

Muduli, R., & Mukharjee, B. B. (2019). Effect of incorporation of metakaolin and recycled coarse aggregate on properties of concrete. *Journal of Cleaner Production*, *209*, 398–414. Available from https://doi.org/10.1016/j.jclepro.2018.100.221.

Najm, O., El-Hassan, H., & El-Dieb, A. (2022). Optimization of alkali-activated ladle slag composites mix design using taguchi-based TOPSIS method. *Construction and Building Materials*, *327*, 126946. Available from https://doi.org/10.1016/j.conbuildmat.2022.126946.

Neville, A. M. (1995). *Properties of concrete*. London: Longman.

Pacheco-Torgal, F., & Jalali, S. (2010). Reusing ceramic wastes in concrete. *Construction and Building Materials*, *24*(5), 832–838. Available from https://doi.org/10.1016/j.conbuildmat.2009.100.023.

Shekarchi, M., Bonakdar, A., Bakhshi, M., Mirdamadi, A., & Mobasher, B. (2010). Transport properties in metakaolin blended concrete. *Construction and Building Materials*, *24*(11), 2217–2223. Available from https://doi.org/10.1016/j.conbuildmat.2010.040.035.

Simões, J. R., da Silva, P. R., & Silva, R. V. (2021). Binary mixes of self-compacting concrete with municipal solid waste incinerator bottom ash. *Applied Sciences (Switzerland)*, *11*(14). Available from https://doi.org/10.3390/app11146396.

Sousa, V., & Bogas, J. A. (2021). Comparison of energy consumption and carbon emissions from clinker and recycled cement production. *Journal of Cleaner Production*, *306*, 127277. Available from https://doi.org/10.1016/j.jclepro.2021.127277.

Subaşı, S., Öztürk, H., & Emiroğlu, M. (2017). Utilizing of waste ceramic powders as filler material in self-consolidating concrete. *Construction and Building Materials*, *149*, 567–574. Available from https://doi.org/10.1016/j.conbuildmat.2017.050.180.

Tang, P., Florea, M. V. A., Spiesz, P., & Brouwers, H. J. H. (2016). Application of thermally activated municipal solid waste incineration (MSWI) bottom ash fines as binder substitute. *Cement and Concrete Composites*, *70*, 194–205. Available from https://doi.org/10.1016/j.cemconcomp.2016.03.015.

Targan, S., Olgun, A., Erdogan, Y., & Sevinc, V. (2003). Influence of natural pozzolan, colemanite ore waste, bottom ash, and fly ash on the properties of Portland cement. *Cement and Concrete Research*, *33*(8), 1175–1182. Available from https://doi.org/10.1016/S0008-8846(03)00025-5.

Uche, O. A., Kelechi, S. E., Adamu, M., Ibrahim, Y. E., Alanazi, H., & Okokpujie, I. P. (2022). Modelling and optimizing the durability performance of self consolidating concrete incorporating crumb rubber and calcium carbide residue using response surface methodology. *Buildings*, *12*(4). Available from https://doi.org/10.3390/buildings12040398.

Vaičienė, M., & Simanavičius, E. (2022). The effect of municipal solid waste incineration ash on the properties and durability of cement concrete. *Materials*, *15*(13), 4486. Available from https://doi.org/10.3390/ma15134486.

Vejmelková, E., Keppert, M., Rovnaníková, P., Ondráček, M., Keršner, Z., & Černý, R. (2012). Properties of high performance concrete containing fine-ground ceramics as supplementary cementitious material. *Cement and Concrete Composites*, *34*(1), 55–61. Available from https://doi.org/10.1016/j.cemconcomp.2011.090.018.

Wan, Y., Hui, X., He, X., Li, J., Xue, J., Feng, D., Liu, X., & Wang, S. (2022). Performance of green binder developed from flue gas desulfurization gypsum incorporating Portland cement and large-volume fly ash. *Construction and Building Materials*, *348*, 128679. Available from https://doi.org/10.1016/j.conbuildmat.2022.128679.

Wansom, S., Chintasongkro, P., & Srijampan, W. (2019). Water resistant blended cements containing flue-gas desulfurization gypsum, Portland cement and fly ash for structural applications. *Cement and Concrete Composites*, *103*, 134–148. Available from https://doi.org/10.1016/j.cemconcomp.2019.04.033.

Wild, S., & Khatib, J. M. (1997). Portlandite consumption in metakaolin cement pastes and mortars. *Cement and Concrete Research*, *27*(1), 137–146. Available from https://doi.org/10.1016/S0008-8846(96)00187-1.

Wild, S., Khatib, J. M., & Jones, A. (1996). Relative strength, pozzolanic activity and cement hydration in superplasticised metakaolin concrete. *Cement and Concrete Research*, *26* (10), 1537–1544. Available from https://doi.org/10.1016/0008-8846(96)00148-2.

Wright, L., & Khatib, J. M. (2016). *Sustainability of desulphurised (FGD) waste in construction. Sustainability of construction materials* (2nd ed., pp. 683–715). Elsevier BV. Available from 10.1016/b978-0-08-100370-1.00026-3.

Yang, J., Zhang, Y., He, X., Su, Y., Tan, H., Ma, M., & Strnadel, B. (2021). Heat-cured cement-based composites with wet-grinded fly ash and carbide slag slurry: Hydration, compressive strength and carbonation. *Construction and Building Materials*, *307*, 124916. Available from https://doi.org/10.1016/j.conbuildmat.2021.124916.

Yang, K. H., Jung, Y. B., Cho, M. S., & Tae, S. H. (2015). Effect of supplementary cementitious materials on reduction of CO_2 emissions from concrete. *Journal of Cleaner Production*, *103*, 774–783. Available from https://doi.org/10.1016/j.jclepro.2014.030.018.

Yao, X., Wang, W., Liu, M., Yao, Y., & Wu, S. (2019). Synergistic use of industrial solid waste mixtures to prepare ready-to-use lightweight porous concrete. *Journal of Cleaner Production*, *211*, 1034–1043. Available from https://doi.org/10.1016/j.jclepro.2018.11.252.

Zhang, S., Ghouleh, Z., He, Z., Hu, L., & Shao, Y. (2021). Use of municipal solid waste incineration bottom ash as a supplementary cementitious material in dry-cast concrete. *Construction and Building Materials*, *266*, 120890. Available from https://doi.org/10.1016/j.conbuildmat.2020.120890.

Zhang, Y., Pan, F., & Wu, R. (2016). Study on the performance of FGD gypsum-metakaolin-cement composite cementitious system. *Construction and Building Materials*, *128*, 1–11. Available from https://doi.org/10.1016/j.conbuildmat.2016.090.134.

Mechanical and durability properties of sustainable geopolymer concrete

Lateef N. Assi[1], *Ali Alsalman*[2], *Rahman S. Kareem*[3], *Kealy Carter*[4] *and Jamal Khatib*[5]

[1]Terracon Consultants Inc., Tulsa, OK, United States, [2]Department of Civil Engineering, University of Basrah, Basrah, Iraq, [3]Department of Structure, Shatrah Technical Institute, Southern Technical University, Shatrah, Dhi Qar, Iraq, [4]Darla Moore School of Business, University of South Carolina, Columbia, SC, United States, [5]Faculty of Engineering, Beirut Arab University, Beirut, Lebanon

4.1 Introduction

The annual global production of ordinary Portland cement (OPC) has reached 4100 million metric tons, making it the second most used material after water (Statista, 2023). Due to the global population increase and concrete building demolition and construction, the production volume of OPC is likely to continue increasing (Alsalman et al., 2022; Kareem, Dang, et al., 2021). Several studies have reported that OPC manufacturing is considered one of the most significant contributors to our global carbon footprint, with emissions from OPC representing between 7% and 10% of total CO_2 emissions (Kareem, Assi, et al., 2021). Hence, finding a way, through chemical or structural modifications, to reduce OPC production is an urgent matter (McLellan et al., 2011). Extant research has shown that geopolymer concrete is a more sustainable alternative to OPC concrete, as it reduces CO_2 emissions and utilizes supplementary cementitious materials such as fly ash (FA), metakaolin (MK), and ground granulated furnace slag that are waste materials.

4.2 Mechanical properties of alkaline-activated concrete

A number of studies have been carried out to investigate the mechanical properties of alkaline-activated concrete (AAC). Several base materials have been used to develop this type of ecofriendly concrete. These base materials include FA, ground granulated blast-furnace slag (GGBFS), MK, and rice husk (RH). The following section discusses the factors affecting the mechanical properties of AAC.

Sustainable Concrete Materials and Structures. DOI: https://doi.org/10.1016/B978-0-443-15672-4.00004-8

4.2.1 Effect of the alkaline solution to total binder content on compressive strength

Hadi et al. (2017) developed an AAC of GGBFS based on using three different binder contents, as shown in Fig. 4.1. Three alkaline to binder (*A/B*) ratios were used. They used sodium hydroxide (SH) and sodium silicate (SS) as the activating solution.

Fig. 4.1 shows the relationship between the compressive strength (7 and 28 days of age) and the *A/B* ratio. For a given binder content, the increase in *A/B* from 0.35 to 0.45 decreased the 7- and 28-day compressive strength by an average of 19% and 25%, respectively. The increase in *A/B* from 0.45 to 0.50 decreased the 7-day strength by an average of 13%. A similar trend was found regarding the 28-day compressive strength; however, the increase in *A/B* from 0.45 to 0.50 enhanced the strength 9% for binder content of 400 kg/m^3 only. The reduction in concrete compressive strength as the *A/B* increases is related to the fact that the increase in the alkaline activator resulted in an increase in the amount of water in the mixture, which delayed polymerization (Ruiz-Santaquiteria et al., 2012).

Mallikarjuna Rao and Gunneswara Rao (2018) investigated the compressive strength of AAC using a mixture of FA and GGBFS with SH and SS as the activating solution. They used four ratios of *A/B*. The considered ratios were 0.45, 0.50, 0.55, and 0.60. Also, for every binder content and *A/B*, three FA to GGBFS ratios were considered. Fig. 4.1 shows the relationship between the *A/B* and compressive strength. For a given binder content, *A/B* of 0.50 resulted in the highest compressive strength. The average increase in strength for binder content of 360, 420, and 450 kg/m^3 (regardless of the FA and GGBFS content) was 4%, 8%, and 19%, respectively, when the *A/B* increased from 0.45 to 0.50. On the other hand, the average reduction in strength for binder content of 360, 420, and, 450 kg/m^3

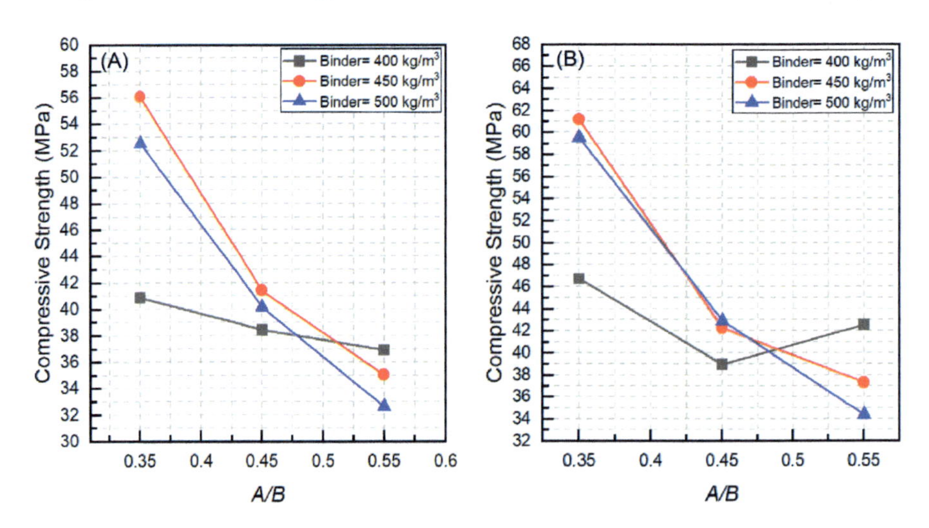

Figure 4.1 Effect of *A/B* ratio on the (A) 7-day compressive strength and (B) 28-day compressive strength alkaline-activated GGBFS-based geopolymer (Hadi et al., 2017).
Notes: *A/B*, alkaline to binder ratio; GGBFS, ground granulated blast-furnace slag.

(regardless of the FA and GGBFS content) was 154%, 18%, and 15%, respectively, when the *A/B* increased from 0.50 to 0.55. Furthermore, the average reduction in strength was 21%, 23%, and 22% for binder content of 360, 420, and 450 kg/m^3, respectively, when the *A/B* increased from 0.50 to 0.60.

Fig. 4.2 shows the effect of *A/B* on the 3-day compressive strength of AAC of different percentages of SS to SH ratios (Patankar et al., 2013).

Several data have been collected from several studies in the literature to examine the optimum *A/B* ratios (Ahmed et al., 2011; Deb, 2012; Gunasekara et al., 2018; Hadi et al., 2017; Hardjito & Rangan, 2005; Kumar et al., 2018; Kusbiantoro et al., 2012; Nath & Sarker, 2015, 2017; Nazari & Sanjayan, 2015; Nuruddin et al., 2011; Okoye et al., 2015; Olivia & Nikraz, 2012; Parthiban & Mohan, 2017; Pouhet & Cyr, 2016; Sarker et al., 2013; Shi et al., 2012; Sujatha et al., 2012; Sumajouw & Rangan, 2006; Vora & Dave, 2013; Xie & Ozbakkaloglu, 2015). The collected data represent various AAC mixture designs, such as different base materials, binder to aggregate (*B/A*) ratio, and different curing regimens. Compressive strength ranges from 0.6 to 85.66 MPa, as shown in Table 4.1. The *A/B* ratios range from 0.25 to 0.92 (does not include any additional water or superplasticizer). Fig. 4.3 shows the relationship between the *A/B* and compressive strength subjected to different curing regimens. It can be seen that the majority of the data (91%) fall between *A/B* range from 0.25 to 0.55 (shaded area of Fig. 4.2). In addition, there is no correlation between compressive strength and *A/B* as in conventional concrete (CC) (correlation between water to binder ratio and compressive strength).

4.2.2 Effect of other factors on compressive strength of alkaline-activated concrete

Other parameters can influence the compressive strength of AAC. These factors include binder content, aggregate to binder (B/A) ratio, curing regimen (duration and temperature), the particle size distribution of the base materials, and CaO content in base materials (Farhan et al., 2020). Typically, CC is cured under the ambient temperature of moist cured at a temperature of 21°C for 28 days (Mehta & Monteiro, 2006; Yashwanth Reddy & Harihanandh, 2023). On the other hand, AAC needs a higher temperature to enhance the polymerization process and increase the bond of the materials' matrix (Nagral et al., 2014). Fig. 4.4A shows the effect of heat curing con compressive strength reported by Hardjito & Rangan (2005). It can be seen that for a given mixture proportioning (binder, aggregate, SS, SH contents, and curing period) as the curing temperature increased, the compressive strength enhanced. The increase in strength using 90°C for 24 hours was 61%, 29%, 20%, and 9% compared with 30°C 45°C, 60°C, and 75°C, respectively. On the other hand, Ahmed et al. (2011) and Vidyadhara and Ranganath (2023) found that the curing temperature of 70% resulted in the highest strength; the enhancement in strength was 14%, 5%, and 6% compared with 60°C, 80°C, and 90°C, respectively, for the same mixture proportions and curing period. Xie et al. (2020) stated that the compressive strength of AAC was not necessarily increased as the curing

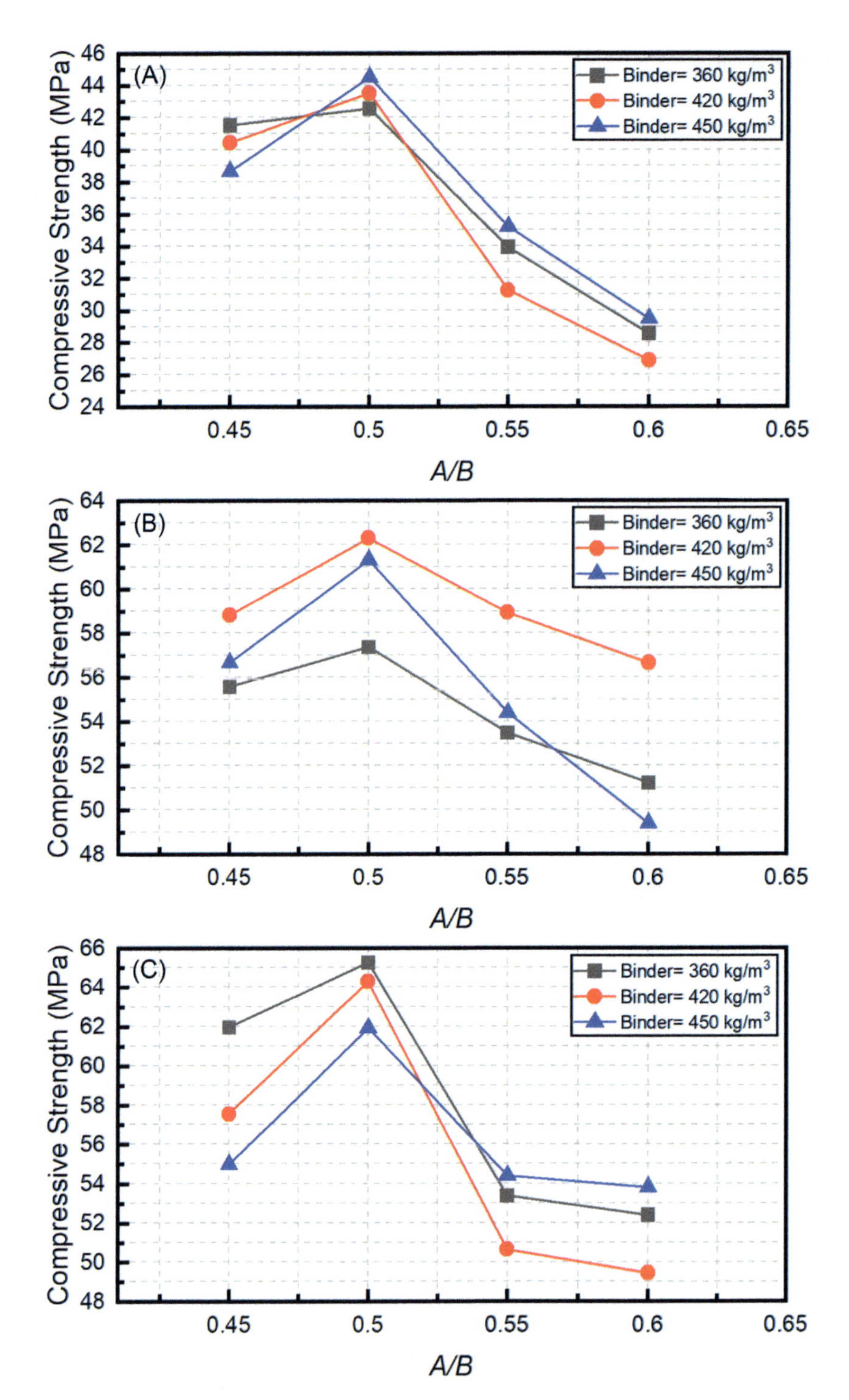

Figure 4.2 Effect of alkaline solution to binder content on compressive strength alkaline-activated concrete (A) 70% FA and GGBFS = 30%, (B) 60% FA and GGBFS = 40%, and (C) 50% FA and GGBFS = 50% of the total binder (Mallikarjuna Rao, & Gunneswara Rao, 2018). *Notes*: *A/B*, alkaline to binder ratio; FA, fly ash; GGBFS, ground granulated blast-furnace.

Table 4.1 Correlations between compressive strength and other mechanical properties.

References	Equation	Notes
Modulus of elasticity		
ACI Committee 318 (2019)	$E_c = 4{,}700(f'_c)^{0.5}$	Normal weight concrete and no limits for f'_c
ACI Committee 363 (2010)	$E_c = 3{,}320(f'_c)^{0.5} + 6{,}900$	High strength concrete 21 MPa $\leq f'_c \leq$ 83 MPa
Hardjito et al. (2004)	$E_c = 2{,}707(f'_c)^{0.5} + 5{,}300$	AAC of FA-based
Xie et al. (2020)	$E_c = 3{,}650(f'_c)^{0.5}$	AAC
Flexural strength		
ACI Committee 318 (2019)	$E_c = 0.62(f'_c)^{0.5}$	Normal weight concrete
Tensile strength		
ACI Committee 318 (2019)	$E_c = 0.56(f'_c)^{0.5}$	Normal weight concrete
Xie et al. (2020)	$E_c = 0.35(f'_c)^{0.65}$	AAC

AAC, alkaline-activated concrete; *FA*, fly ash.

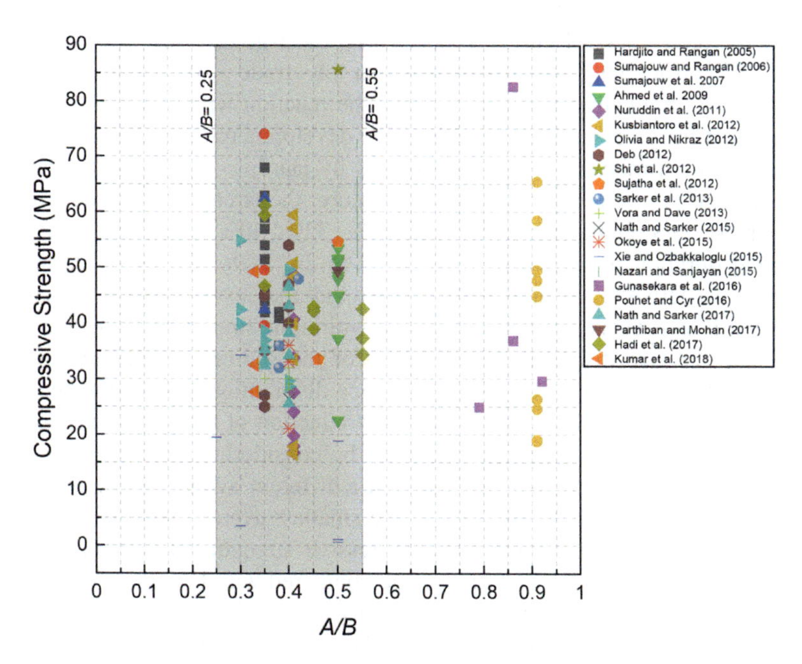

Figure 4.3 Relationship between compressive strength and *A*/*B* ratios collected from the literature. *Note*: *A*/*B*, alkaline to binder ratio.

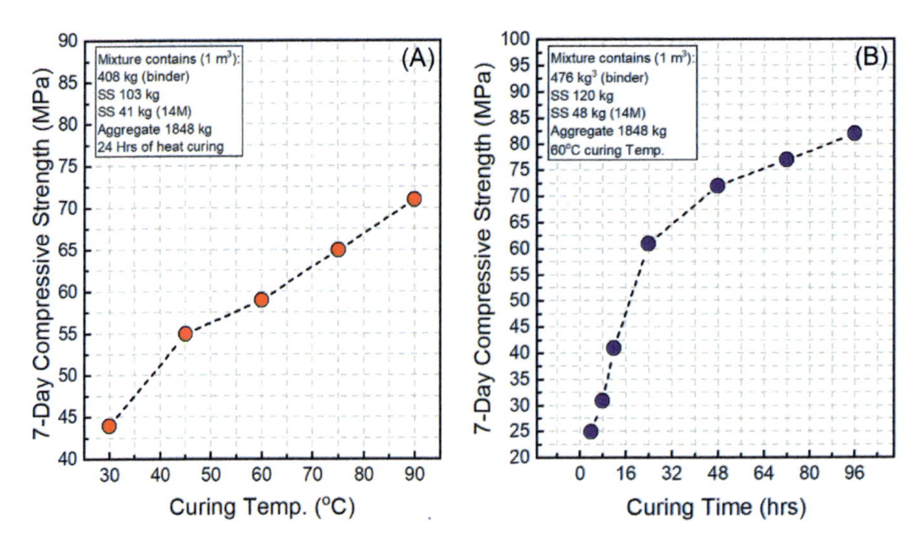

Figure 4.4 Effect of curing (A) temperature and (B) duration on compressive strength reported by (Hardjito and Rangan, 2005).

temperature increased. The high temperature (≥90°C) can negatively influence the strength due to the continuous loss of moisture, which eventually leads to micro-cracks and reduces the strength. They also found that the grade of concrete influences the effectiveness of curing temperature; lower-grade concrete seemed not to be affected considerably. The reason was that low-grade AAC had adequate source materials and an alkaline activator to assist the reaction with elevated temperature. The duration for given curing also affects the development of compressive strength. Fig. 4.4B demonstrates the effect as reported by Hardjito & Rangan (2005). For a given mixture proportion and curing temperature, the strength increased over time. The increase in strength was 24% from 24%, 32%, 49%, 18%, 7%, and 7%, for the curing duration from 4 to 8, 8 to 12, 12 to 24, 24 to 48, 48 to 72, and 72 to 96 hours, respectively. Xie et al. (2020) stated that the development of the compressive strength of AAC is typically completed within the first 24 to 24 hours.

The particle size distribution or fineness of the source material can affect the compressive strength of AAC; the particle size distribution is linked to the surface area, which may affect the reaction and the production of the polymerization. Assi, Deaver, et al. (2018) investigated this effect by considering three types of FA that had the same chemical composition and had a different average particle size distribution; 38.8, 17.9, and 4.78 μm. FA that the smallest average particle size distribution (4.78 μm) achieved the highest compressive strength. The obtained strength was 222% and 124% higher at 1 day of age compared with FA which had a particle size distribution of 38.8 and 17.9 μm, respectively. For 7-day strength, the increase in strength was 212% and 40%. On the other hand, the strength enhancement at 28 days of age was 201% and 38%. The substantial-high enhancement in strength was attributed to the increase in the surface area, which enhanced the reaction.

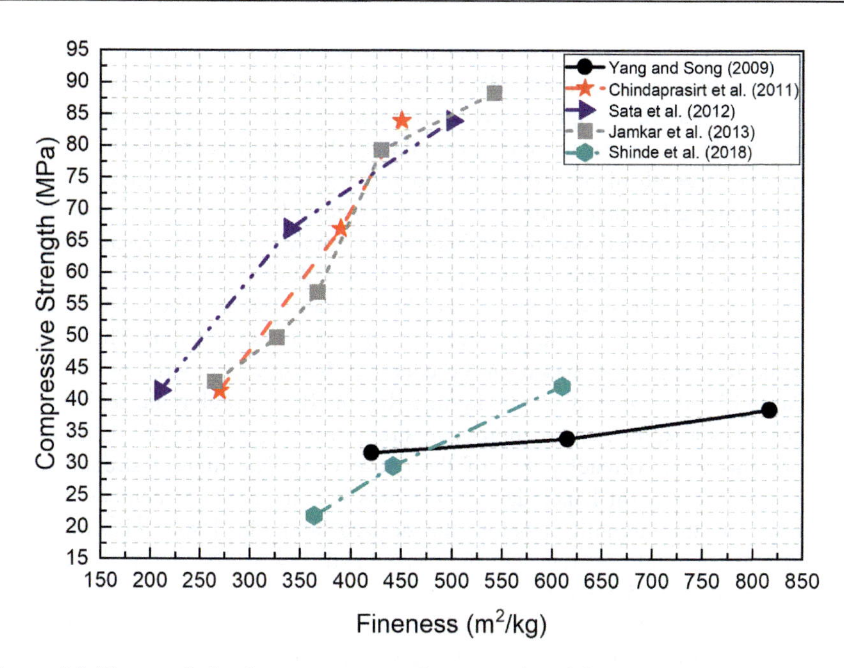

Figure 4.5 The correlation between compressive strength and fineness.

Also, the extremely fine FA particles filled the voids and achieved a denser and stronger matrix. Similar findings regarding the effect of the particle size distribution of the base materials were stated by Detphan and Chindaprasirt (2009), Firdaus et al. (2017), Jegan et al. (2023), Khale and Chaudhary (2007), and Nazari et al. (2011). It is well known that the fineness of OPC affects the strength of CC (e.g., the difference in fineness between Type I, Type II, and Type III OPC). Similarly, the fineness of the base materials in AAC influences the compressive strength. Fig. 4.5 shows the compressive strength with the fineness of the base materials. The data were collected from a number of studies in the literature (Chindaprasirt et al., 2011; Isa & Awang, 2023; Jamkar et al., 2013; Sata et al., 2012; Shinde et al., 2018; Yang & Song, 2009; Zhang, Cheng, et al., 2023). The compressive strength ranges from 21.85 and 88.37 MPa. The figure clearly shows that compressive strength increases as the fineness increases per each study. The highest compressive strength (88.37 MPa) corresponds to a fineness of 542 m^2/kg, which is approximately similar to the fineness of OPC Type III; fineness of 547 m^2/kg (Mehta & Monteiro, 2006). Other factors also have been discussed and presented in previous research, such as the incorporation of Portland cement to overcome the curing challenge, the effect of binder to aggregate, and the effect of CaO on the base materials (Assi, Deaver, et al., 2018; Farhan et al., 2020; Xie et al., 2020).

4.2.3 Alkaline-activated concrete and conventional concrete

When choosing a concrete mixture by the design engineer, several parameters are to be considered, such as cost, workability, strength, and durability. Strength, especially compressive strength, is the measurement to specify the concrete for a specific type of work commonly. To make a reasonable comparison between AAC and CC, compressive strength and cost are major parameters; the amount of binder content can play a significant role in calculating the cost. Concrete with a specified compressive strength of 35 MPa is considered very common in structural applications. This strength typically requires approximately 400 kg/m^3 of cement depending on the recommended slump, nominal maximum aggregate size, water to cement ratio, and coarse aggregate volume per unit of volume (ACI Committee, 1991). Fig. 4.6 shows compressive strength and their binder content. The graph is divided into four regions based on compressive strength and binder content. The classification of the four regions is based on the concrete that has a compressive strength of 35 MPa and binder content of 400 kg/m^3. Region II is a preferable region; binder content is $\leq$400 kg/m^3, and compressive strength is $\geq$35 MPa. The data in region IV represents approximately 28% of the entire data. The highest strength in this region is 85.66. Region I includes all mixtures that have binder content <400 kg/m^3. However, the strength is below the desired strength. These mixtures can be used in nonstructural concrete. On the other hand, region III concretes have compressive strength >35 MPa; however, the binder content is higher than 400 kg/m^3. Region II is the least desired region when it compares the binder content with CC. All mixtures shown in Fig. 4.6 have 0% cement as a binder; however, most of them

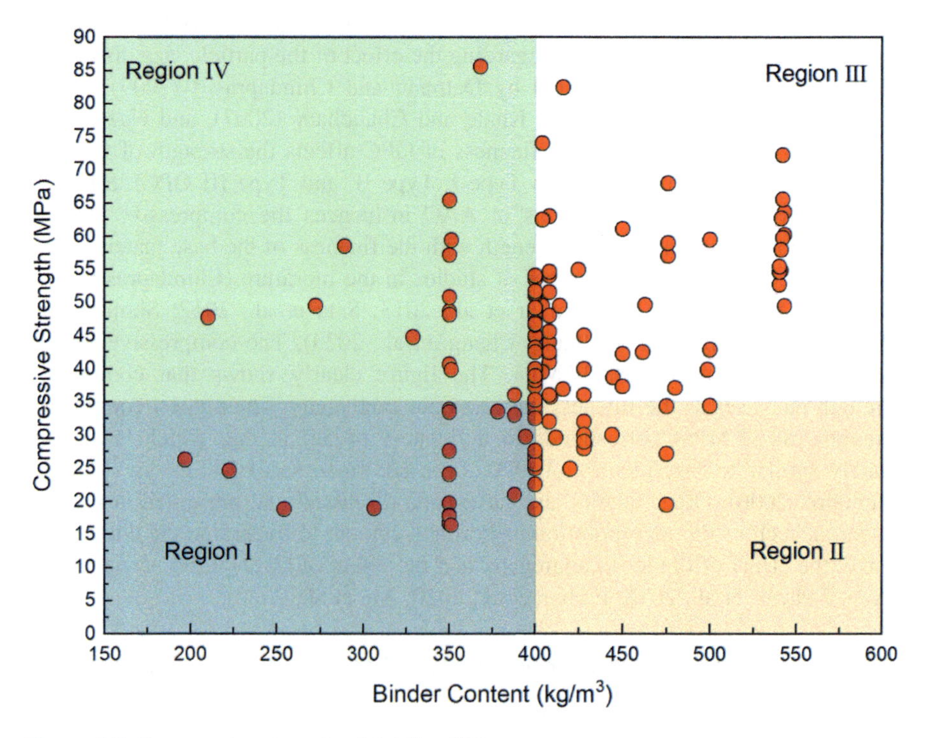

Figure 4.6 Compressive strength of AAC to CC.

satisfy regular strength concrete requirements. Therefore, in terms of strength requirements, AAC can replace CC.

4.2.4 Other mechanical properties of alkaline-activated concrete

Typically, design codes propose a correlation between the compressive strength of CC and other mechanical properties (e.g., modulus of elasticity, tensile strength, flexural strength [modulus of rupture]). Table 4.1 presents the most common correlations (ACI Committee 363, 2010). For CC, factors that affect the compressive strength can influence the other mechanical properties and durability. Several studies have stated that the factors that affect compressive strength can influence the other mechanical properties of AAC. To make a reasonable comparison between the mechanical properties of CC and AA, it is convenient to examine the equation developed for CC on AAC. The compressive strength, modulus of elasticity, tensile strength, and flexural strength range from 0.8 to 89, 8200 to 38,570, 1 to 7.47, and 0.67 to 9.85 MPa, respectively. The data represent the mechanical properties at different ages. The lower limits of the data represent the mechanical properties at an early age.

Fig. 4.7 examines these properties of AAC with the most common correlations in Table 4.1. It can be seen that ACI 318-19 and ACI 363R-10 equations overestimate the modulus of elasticity of AAC; the overestimation is 90% and 81% of the entire data, respectively. This reduction can be attributed to the aggregate type and not necessarily because of the base materials used in AAC. This finding was also confirmed by Hardjito & Rangan (2005) and Nath and Sarker (2017). The average ratio of the tested modulus of elasticity to the predicted using ACI 318 and ACI 363R equations is 0.8 and 0.9, respectively. The equation proposed by Hardjito et al. (2004), which was developed for AAC (FA-based), gives a decent prediction for the modulus of elasticity of the collected data, as shown in Fig. 4.7. It overestimates 35% of the data (it can be considered as a conservative prediction). The average of tested to predicted data is 1.0.

On the other hand, the equation (Xie et al., 2020) overestimates and underestimates 50% of the data. By giving an error of $\pm 25\%$ (shaded area in Fig. 4.7), the equation by Xie et al. (2020) fits approximately 78% of the literature data. Australian Standard (2009) allows predicting the modulus of elasticity with an error of $\pm 20\%$. Fig. 4.7B shows the correlation between compressive strength and tensile strength (indirect tensile strength). It is very evident that the equation provides by the ACI-318 overestimates and underestimates the tensile strength of AAC 60% and 40%, respectively. The average of testing tensile strength to the predicted is 0.97 of the ACI 318 equation. On the other hand, the equation (Xie et al., 2020) overestimates most of the data. As an average the ratio of tensile strength to compressive is equal to 8.4%. For CC, this ratio ranges from 7% to 11% (Mehta & Monteiro, 2006). Fig. 4.7C presents the relationship between compressive strength and flexural strength (modulus of rupture). The ACI 318 correlation underestimates the modulus of rupture of AAC; approximately 91% of the entire data. This equation can be used as a conservative estimation of the modulus of rupture for AAC. The equation by

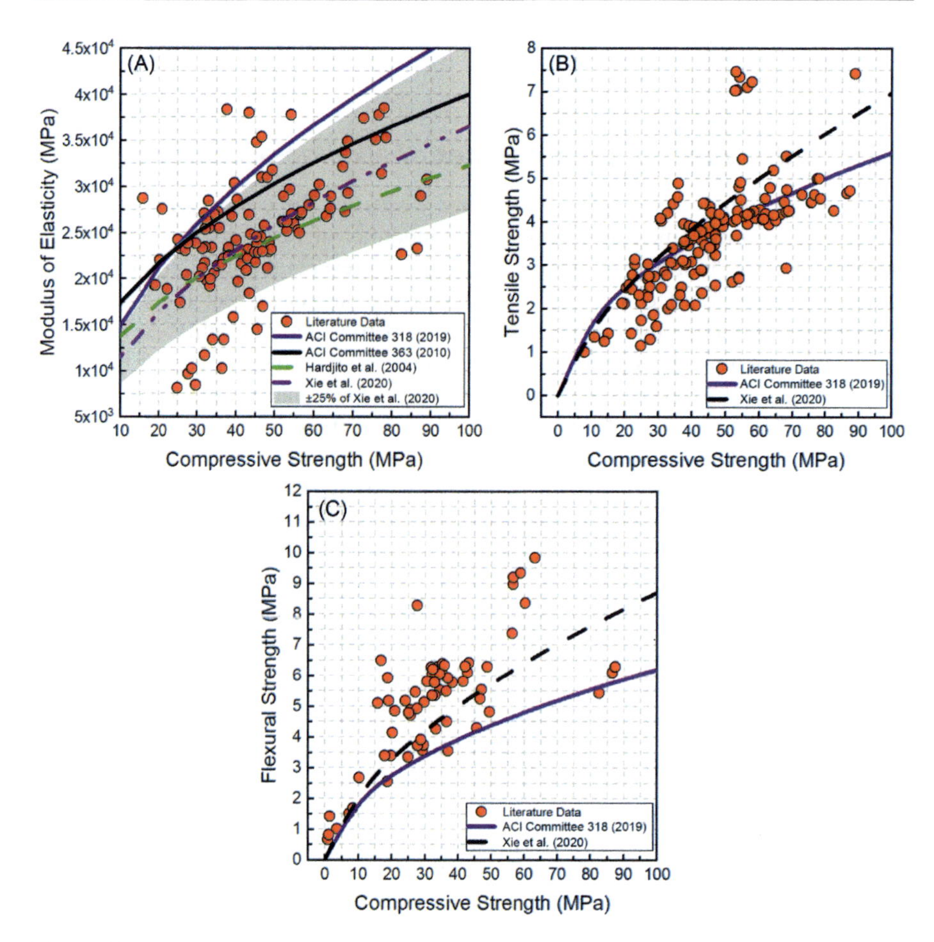

Figure 4.7 Correlation between compressive strength and (A) modulus of elasticity, (B) flexural strength, and (C) tensile strength.

Xie et al. (2020) also underestimates the majority of the collected data. Overall average, the ratio of flexural strength to compressive is equal to 20%. For CC, this ratio ranges from 11% to 23% (Mehta & Monteiro, 2006). As a conclusion the currently available correlations as shown in Table 4.1 can give an acceptable predication of mechanical properties of AAC in terms of compressive strength.

4.3 Durability properties

Geopolymer concrete has superior durability relative to OPC. In general, the resistance of geopolymer concrete under high elevated heat is greater than OPC concrete (Assi et al., 2020; Cao et al., 2017; Kong & Sanjayan, 2010). At high temperatures, the strength of OPC concrete is reduced as a result of chemical and physical

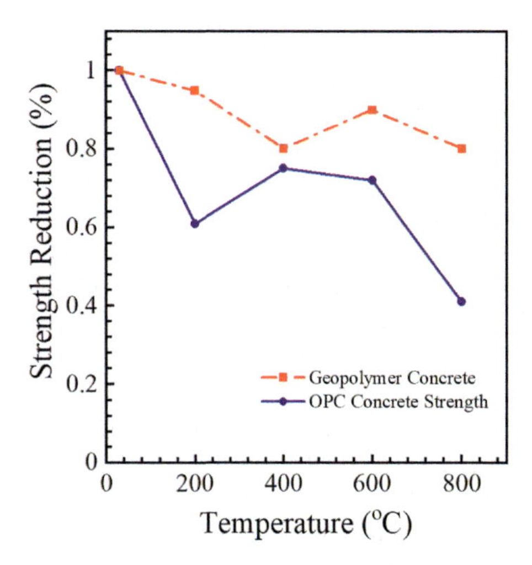

Figure 4.8 Fire resistance of geopolymer concrete and OPC concrete (Cao et al., 2017).

changes (Kong & Sanjayan, 2010). On the other hand, geopolymer concrete generally provides considerable strength at high temperatures and fire resistance due to its ceramic-like properties (Kong & Sanjayan, 2010; Türkmen et al., 2016; Zhang, Yan, et al., 2023). Fig. 4.8 shows the fire effect on the strength reduction of geopolymer concrete and OPC concrete (Cao et al., 2017). The strength reduction is the ratio of the compressive strength at elevated temperature to the initial compressive strength (at ambient temperature); where the initial strengths of geopolymer concrete and OPC concrete are 55.8 and 57.7 MPa, respectively. In general, the strength decreases with increasing the temperature. The reduction in strength with increasing temperature up to 400°C is accounted for an increase in porosity caused by evaporation. At the temperature of 800°C, OPC concrete suffered a severe decrease in the mechanical properties which accounted for the chemical decomposition of its matrix. On the other hand, the strength of geopolymer dropped slowly as a result of the ceramic-like property of matrix.

Kong and Sanjayan (2010) investigated the effect of elevated temperature on geopolymer paste, mortar, and concrete made using FA as a precursor. The authors showed that aggregate size and specimen size are considered as the two main factors that govern geopolymer behavior at elevated temperatures up to 800°C. Aggregate sizes larger than 10 mm resulted in good strength performances in both ambient and elevated temperatures. Aggregate sizes smaller than 10 mm promote spalling and extensive cracking in the geopolymer concrete. For the size of the geopolymer paste specimens, thermal cracking is developed due to the significant temperature variation between the surface and core of the specimen. Consequently, the specimen size affects the elevated temperature strength due to the thermal incompatibility arising from the thermal gradient.

The strength of geopolymer concrete under elevated temperatures is affected by the expansion rate of the aggregate with temperature. Strength loss in geopolymer concrete at elevated temperatures is attributed to the thermal mismatch between the geopolymer matrix and the aggregates (Kong & Sanjayan, 2010; Türkmen et al., 2016). The strength of geopolymer concrete specimens at elevated temperatures is reduced due to the thermal incompatibility between the geopolymer matrix and aggregate. The strength reduction of geopolymer concrete is decreased by using aggregate having less incompatibility (Türkmen et al., 2016). Moreover, conventional superplasticizer affects the strength of the geopolymer matrix (Cao et al., 2017; Kong & Sanjayan, 2010). The use of superplasticizers in geopolymer concrete deteriorates the concrete strength at elevated temperatures (Kong & Sanjayan, 2010).

The durability of geopolymer concrete is affected by the percentage content of OPC and the material used, such as FA, slag, nanosilica, RH ash, and MK. The strength of geopolymer concrete-based FA is decreased by reducing the OPC content (Mehta & Siddique, 2017). Mehta and Siddique investigated the OPC content as replacement 0%, 10%, 20%, and 30% of FA. It was concluded that the compressive strength of geopolymer concrete increases with increasing the OPC content by up to 20%. However, the porosity, water absorption, chloride permeability, and sorptivity are decreased. This reduction is due to the presence of both N−A−S−H and C−A−S−H, where the microstructure geopolymer matrix became dense and compacted by adding OPC (Garg et al., 2020; Mehta & Siddique, 2017).

The workability of geopolymer concrete is lower than OPC concrete due to the high viscous silicate compound (Abdulkareem et al., 2014; Jindal et al., 2017). In general, the workability of slag-based geopolymer concrete is reduced by increasing the slag content in the concrete matrix (Fang et al., 2018). It is confirmed that the mechanical properties of geopolymer concrete can be improved by adding ground-granulated blast-furnace slag, depending on the slag content (Fang et al., 2018; Kumar et al., 2010). This improvement in geopolymer concrete is due to the formed C−(A)−S−H gel bond water content and fills the space between pores (Chi & Huang, 2013). Therefore adding GGBFS can reduce porosity volume and enhance the pore size of geopolymer concrete (Li & Liu, 2007).

Moreover, the percentage content of RH ash, used as a based material for geopolymer concrete, affects durability (Hwang & Huynh, 2015; Kim et al., 2014). RH ash-based geopolymer concrete has considerable durability properties. Jindal et al. (2020) investigated the properties of geopolymer concrete by using RH ash up to 400 kg/m^3 and an alkaline mixture of SH and SS as the activator. The durability properties are improved by increasing the percentage content of RH ash. Also, the authors showed that the addition of ultrafine slag to RH ash-based geopolymer concrete substantially enhances the durability properties. Using a higher percentage of ultrafine slag up to 10% in RH ash geopolymer concrete has a substantial effect in improving the strength properties, such as concrete strength and water absorption.

The geopolymer concrete-based MK is more reactive in comparison to FA-based concrete in terms of the polymerization process (Görhan et al., 2016). For workability, the MK-based geopolymer concrete requires more water than FA-based geopolymer concrete because of its plate-like structure and fineness of MK particles (Chi & Huang, 2013). As a result, MK-based geopolymer concrete is more cohesive, so the filling of microstructure and durability are improved (Alanazi et al., 2016;

Duan et al., 2016). Moreover, the microstructure matrix of geopolymer concrete is improved by using nanosilica up to 6% (Adak et al., 2014; Aggarwal et al., 2015). Adak et al. (2014) showed by X-ray diffraction analysis that there are new crystalline phases formed in nanosilica (12 M6)-based geopolymer mortar. Consequently, the use of nanosilica up to 6% can improve concrete durability as a result of reducing porosity and improving the microstructure matrix of geopolymer concrete (Adak et al., 2014; Aggarwal et al., 2015; Zhang et al., 2012).

Permeability is considered one of the most significant factors in concrete durability. When the matrix permeability is high, the chances of penetration of harmful ions are increased. In the presence of chloride and sulfate salts the alkalinity of the matrix is reduced due to dissolving and leaching out CH (Reddy et al., 2013). On the other hand, geopolymer concrete does not include calcium hydroxide in the matrix. Apparent volume of permeable voids is mostly used for testing the permeability (Reddy et al., 2013). For FA-based geopolymer concrete, the percentage of water absorption varies about 3%–5% (Olivia & Nikraz, 2008; Rendell et al., 2002). The permeability of geopolymer concrete reduces by decreasing the ratio of alkaline to FA (Olivia & Nikraz, 2011; Shaikh, 2016; Xie & Ozbakkaloglu, 2015). Alternatively, increasing the compressive strength increases the permeability coefficient (Olivia & Nikraz, 2011).

Moreover, RH ash-based geopolymer concrete has a significant permeability (Jindal et al., 2020; Hwang & Huynh, 2015; Kim et al., 2014). RH ash is an industrial by-product, and the disposal of this material is a major concern. In addition to improving the concrete strength and reducing the release of CO_2, the use of RH ash in geopolymer concrete can address the waste material disposal and develop geopolymer concrete with considerable durability properties, such as permeability and water absorption (Jindal et al., 2020; Kim et al., 2014). The inclusion of ultrafine slag in RH ash-based geopolymer concrete develops better durability properties (Jindal et al., 2020). As a result of the coexistence of sodium aluminosilicate hydrate, calcium silicate hydrate, and aluminum calcium silicate hydrate gels develop ultrafine particle size (He et al., 2013; Kim et al., 2014; Munshi & Sharma, 2016). Reducing the permeability and improving the strength are the results of developing these ultrafine particles which block the micropores and fill the permeable voids (Hwang & Huynh, 2015; Venkatesan & Pazhani, 2016).

The permeability of geopolymer concrete determines by water absorption. Both permeability and water absorption are reduced with a decrease in void content (Munshi & Sharma, 2016; Venkatesan & Pazhani, 2016). Fig. 4.9 shows the initial and final water absorption for RH ash-based geopolymer concrete (Jindal et al., 2020). The initial and final water abruption is measured after 30 minutes and 72 hours, respectively. Nine geopolymer concrete mixtures having a different amount of RH ash and including various percentages of ultrafine slag were tested. The content of RH ash in Mix1–Mix3, Mix4–Mix6, and Mix7–Mix9 is 350, 370, and 400 kg/m^3, respectively. The percentage of ultrafine slag from RH ash is 0% in Mix1, Mix3, and Mix7, 5% in Mix2, Mix4, and Mix8, and 10% in Mix3, Mix6, and Mix9. It can be concluded that both initial and final water absorption decreases with an increase in binder content (RH ash). Moreover, the addition of ultrafine

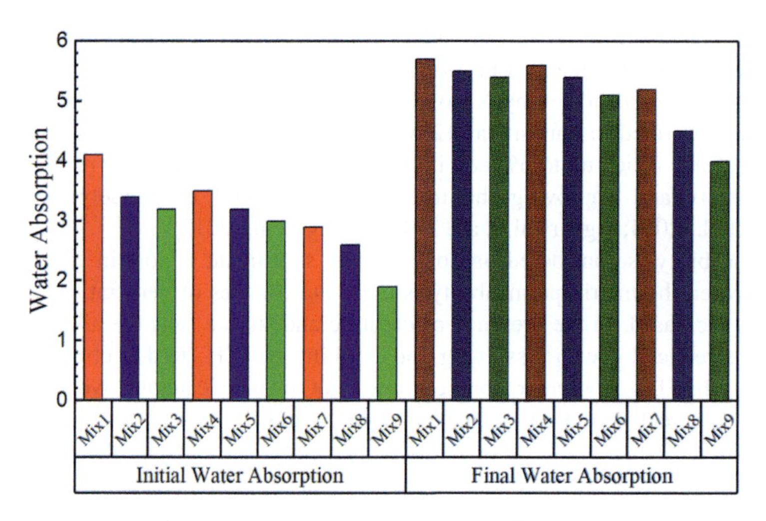

Figure 4.9 Water absorption of rice husk-based geopolymer concrete (Jindal et al., 2020).

slag improves the enhanced water absorption of RH ash-based geopolymer concrete up to 10%.

The resistance to chemical attacks such as acids and sulfates is a durable concern of concrete. As a result of the alkaline nature of OPC, CC is susceptible to acid attack. Due to the dissolution of calcium hydroxide, the components of the cement paste are broken down during contact with acids (Pacheco-Torgal et al., 2008). Acid attack has not traditionally attracted much attention although cement composites are severely damaged by acids (Kim et al., 2014; Pacheco-Torgal et al., 2008). Calcium hydroxide is dissolved in acid, so the hydrated silicate and aluminum phases are decomposed. In addition, magnesium sulfate is considered to be the most harmful for the reason that results in loss of cementitious properties (Garg et al., 2020). Consequentially, the concrete strength is reduced, and the concrete deteriorates immediately (Duxson et al., 2007; Pacheco-Torgal et al., 2008).

On the other hand, geopolymer concrete exhibits much excellent resistance against acid and sulfate attack in addition to its environmental friendliness (Hardjito et al., 2015; Pacheco-Torgal et al., 2008). The resistance of geopolymer concrete to acid attack is higher in comparison with OPC concrete when exposed to varied acid concentrations (Ariffin et al., 2013). Geopolymer concrete remains structurally undamaged even though the surface turned a little softer in comparison to OPC concrete, which shows severe deterioration after 18 months of sulfuric acid exposure. The mass losses for geopolymer concrete and OPC concrete specimens are 8% and 20%, respectively, after 18 months of sulfuric acid exposure. For concrete strength, the reduction in compressive strength of geopolymer concrete is up to 35% (Ariffin et al., 2013; Okoye et al., 2017). However, the OPC concrete starts deteriorating in the first month of sulfuric acid exposure. The compressive strength of OPC concrete is reduced up to 68% and the concrete is severely deteriorated after 18 months of sulfuric acid exposure (Ariffin et al., 2013).

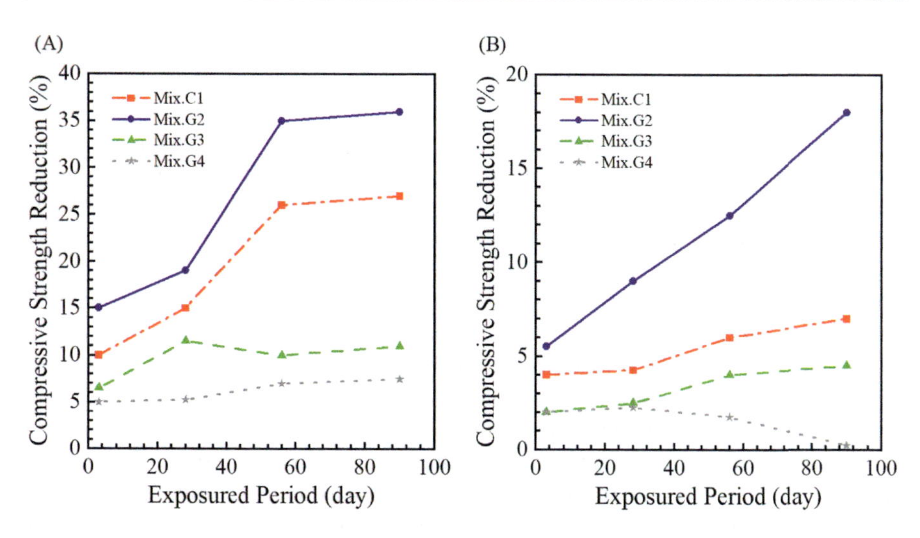

Figure 4.10 The effect of chemical attack (Okoye et al., 2017).

The comparison in the durability of geopolymer concrete and OPC concrete of chemical attack is shown in Fig. 4.10 (Okoye et al., 2017). Four concrete mixtures were investigated. Mix.C1 is OPC concrete, the control mixture. Mix.G2, MixG3, and Mix.G4 are geopolymer concrete mixtures. Mix.G3 and Mix.G4 include silica fume 10% and 20%, respectively. The variation of percent losses in compressive strengths of specimens after exposure to 2% of H_2SO_4 and 5% of NaCl for different duration are shown in Fig. 4.10A and B, respectively. Fig. 4.10A shows there is a rapid increase in the compressive strength reduction with exposure time Mix.C1. In the case of geopolymer concrete the compressive strength reduction also increased with exposure time but much lower than that of Mix.C1. Fig. 4.10B shows the compressive strength reduction of Mix.C1 increases continuously with exposure time, while there is a slow increase in the compressive strength reduction with exposure time in Mix.G2 and MixG3. Furthermore, there was no loss in compressive strength in Mix.G4 including 20% silica fume. The results indicated that geopolymer concrete containing silica fume performs better durability than OPC concrete when exposed to H_2SO_4 and NaCl. In the presence of silica fume the porosity is decreased, so the geopolymer concrete develops better resistance to chemical attack (Okoye et al., 2017; Zhuang et al., 2016).

Moreover, geopolymer concrete has great resistance to the aggressive environment. Geopolymer mortar exposed to chloride environment for 90 days exhibits no considerable variation in compressive strength and mass (Davidovits, 1994; Junaid et al., 2015; Olivia & Nikraz, 2011; Pavithra et al., 2016). Zhuang et al. (2017) investigated the mechanical properties and mass of geopolymer mortar exposed to an aggressive environment for up to 360 days. The authors studied the geopolymer mortar resistance when the specimens were soaked for up to 360 days in water, sodium chloride, and sulfuric acid solutions, respectively. In comparison to the samples soaked in tap water, it was concluded that the maximum degradation in flexural strength, compressive strength, and tensile strength of geopolymer mortar is 6%, 15%, and 11% when the samples were soaked in sodium chloride and sulfuric acid solutions for 360 days.

Salt scaling and freeze–thaw tests indicate that geopolymer concrete resistance against freeze–thaw cycles and salt scaling is generally higher than OPC concrete (Fu et al., 2011). Geopolymer concrete prepared by using slag and Na_2SiO_3 has excellent freeze–thaw resistance with a frost-resisting grade of F300 at the lowest (Fu et al., 2011). The effect of freeze–thaw cycles can be investigated by scanning electron microscope and energy dispersive spectroscopy tests. Geopolymer concrete utilized industrial waste, slag as raw materials, shows excellent freeze–thaw resistance. Fu et al. (2011) investigated two freeze–thaw cycle damage models, the dynamic elasticity modulus attenuation model, and the accumulative freeze–thaw damage model for geopolymer concrete. The relative dynamic elasticity modulus reaches about 90% after 300 times freeze–thaw cycles (Fu et al., 2011). The geopolymer concrete also has little mass loss due to freeze–thaw damage. Moreover, the final hydration products of geopolymer concrete are mostly C–S–H(I) with low Ca/Si, alkaline aluminosilicate, and zeolite minerals (Fu et al., 2011). Consequently, water hardly penetrates through the symmetrical and compact structure. It is not easy for geopolymer concrete to become frozen and saturated.

Though geopolymer concrete is considered more sustainable than OPC concrete, there are worries about high costs that have hindered market adoption of the concrete (Assi, Carter, et al., 2018; Hasanbeigi et al., 2010; McLellan et al., 2011). The cost of geopolymer concrete may be up to 93%–139% of the cost of OPC concrete (McLellan et al., 2011). The activating solution of geopolymer concrete mixture is the main cost driver, roughly 80% (Abdollahnejad et al., 2015). Furthermore, Assi, Carter, et al. (2018) investigated the cost differences between silica-based activating geopolymer concrete and OPC concrete. The authors estimated the standard difference in mixture costs of geopolymer concrete and OPC concrete is around 17%. Moreover, the authors concluded that SH is responsible for up to 80% of the total cost of geopolymer concrete. Consequently, the cost of geopolymer concrete can be reduced by reducing the SH concentration at the same time as keeping an acceptable level of performance from an engineering standpoint (Abdollahnejad et al., 2015; Assi, Carter, et al., 2018; Rajini et al., 2020). Also, the cost of geopolymer concrete can be reduced by up to 55% by eliminating the external heat (Assi, Carter, et al., 2018).

4.4 Concluding remarks

As a summary of the conducted review and survey, several points about the mechanical and durability properties of geopolymer/sustainable concrete can be drawn as follows:

- Geopolymer/sustainable concrete showed superior mechanical properties. With a compressive strength range from 35 to 106 MPa. Other mechanical properties such as modulus of elasticity and fracture properties were similar to the CC.

- The external heat is essential for many geopolymers/sustainable concrete mixtures. Precast/prestressed concrete might be a useful application for that concrete. However, several studies showed that the need for external heat was eliminated by using Portland cement/blast furnace slag as a partial replacement.
- Fire resistance, sulfate attack resistance, and freeze–thaw performance were excellent for sustainable/geopolymer concrete in comparison with Portland cement concrete. Permeability and voids ratio was much lower than its counterpart, and it can be much reduced by using some fine materials such as ultrafine FA, silica fume, and fine blast furnace slag.
- Geopolymers are likely to have increased applications in the future. This is partly due to their improved mechanical and durability properties compared with traditional concrete. Cement manufacturing produces large amounts of CO_2 into the atmosphere and the use of geopolymer can fully replace the cement; thereby the pollution would tend to substantially decrease. Further innovation into the ingredients used in the production of geopolymer would reduce the pollution even further.

References

Abdollahnejad, Z., Pacheco-Torgal, F., Félix, T., Tahri, W., & Aguiar, J. B. (2015). Mix design, properties and cost analysis of fly ash-based geopolymer foam. *Construction and Building Materials*, *80*, 18–30. Available from https://doi.org/10.1016/j.conbuildmat.2015.010.063.

Abdulkareem, O. A., Al Bakri, A. M. M., Kamarudin, H., Nizar, I. K., & Saif, A. A. (2014). Effects of elevated temperatures on the thermal behavior and mechanical performance of fly ash geopolymer paste, mortar and lightweight concrete. *Construction and Building Materials*, *50*, 377–387. Available from https://doi.org/10.1016/j.conbuildmat.2013.090.047.

ACI Committee. (1991). *Standard practice for selecting proportions for normal, heavyweight, and mass concrete: (ACI 211.1-91)*.

ACI Committee 363. 2010. "363R-10: Report on High-Strength Concrete." Technical. American Concrete Institute. https://www.concrete.org/publications/internationalconcreteabstractsportal/m/details/id/51663590.

ACI Committee 318. (2019). *ACI 318-19: Building code requirements for structural concrete: Commentary on building code requirements for structural concrete (ACI 318R-19)*.

Adak, D., Sarkar, M., & Mandal, S. (2014). Effect of nano-silica on strength and durability of fly ash based geopolymer mortar. *Construction and Building Materials*, *70*, 453–459. Available from https://doi.org/10.1016/j.conbuildmat.2014.070.093.

Aggarwal, P., Singh, R. P., Aggarwal, Y., & Hussain, R. R. (2015). Use of nano-silica in cement based materials—A review. *Cogent Engineering*, *2*(1). Available from https://doi.org/10.1080/23311916.2015.1078018.

Ahmed, M. F., Nuruddin, M. F., & Shafiq, N. (2011). Compressive strength and workability characteristics of low-calcium fly ash-based self-compacting geopolymer concrete. *International Journal of Civil and Environmental Engineering*, *5*, 64–70.

Alanazi, H., Yang, M., Zhang, D., & Gao, J. (2016). Early strength and durability of metakaolin-based geopolymer concrete. *Magazine of Concrete Research*, *69*. Available from https://doi.org/10.1680/jmacr.16.00118.

Alsalman, A., Kareem, R., Dang, C. N., Martí-Vargas, J. R., & Micah Hale, W. (2022). Prediction of modulus of elasticity of UHPC using maximum likelihood estimation method. *Structures*, *35*, 1308–1320. Available from https://doi.org/10.1016/j.istruc.2021.110.002.

Ariffin, M. A. M., Bhutta, M. A. R., Hussin, M. W., Tahir, M. M., & Aziah, N. (2013). Sulfuric acid resistance of blended ash geopolymer concrete. *Construction and Building Materials*, *43*, 80–86. Available from https://doi.org/10.1016/j.conbuildmat.2013.010.018.

Assi, L., Carter, K., Deaver, E., Anay, R., & Ziehl, P. (2018). Sustainable concrete: Building a greener future. *Journal of Cleaner Production*, *198*, 1641–1651. Available from https://doi.org/10.1016/j.jclepro.2018.070.123.

Assi, L. N., Deaver, E., & Ziehl, P. (2018). Effect of source and particle size distribution on the mechanical and microstructural properties of fly ash-based geopolymer concrete. *Construction and Building Materials*, *167*, 372–380. Available from https://doi.org/10.1016/j.conbuildmat.2018.010.193.

Assi, L. N., Carter, K., Deaver, E., & Ziehl, P. (2020). Review of availability of source materials for geopolymer/sustainable concrete. *Journal of Cleaner Production*, *263*. Available from https://doi.org/10.1016/j.jclepro.2020.121477.

Australian Standard. (2009). *Concrete structures*.

Cao, Y., Tao, Z., Pan, Z., Murphy, T. D., & Wuhrer, R. (2017). *Fire resistance of fly ash-based geopolymer concrete blended with calcium aluminate cement. Proceedings of the 1st international conference on structural engineering research* (pp. 210–220). .

Chi, M., & Huang, R. (2013). Binding mechanism and properties of alkali-activated fly ash/slag mortars. *Construction and Building Materials*, *40*, 291–298. Available from https://doi.org/10.1016/j.conbuildmat.2012.110.003.

Chindaprasirt, P., Chareerat, T., Hatanaka, S., & Cao, T. (2011). High-strength geopolymer using fine high-calcium fly ash. *Journal of Materials in Civil Engineering*, *23*(3), 264–270. Available from https://doi.org/10.1061/(asce)mt.1943-5533.0000161.

Davidovits, J. (1994). Properties of geopolymer cements. *First International Conference on Alkaline Cements and Concretes*, *1*, 131–149.

Deb, P. (2012). *Durability of fly ash based geopolymer concrete*. Thesis, Curtin University.

Detphan, S., & Chindaprasirt, P. (2009). Preparation of fly ash and rice husk ash geopolymer. *International Journal of Minerals, Metallurgy and Materials.*, *16*(6), 720–726. Available from https://doi.org/10.1016/S1674-4799(10)60019-2.

Duan, P., Yan, C., & Zhou, W. (2016). Influence of partial replacement of fly ash by metakaolin on mechanical properties and microstructure of fly ash geopolymer paste exposed to sulfate attack. *Ceramics International*, *42*(2), 3504–3517. Available from https://doi.org/10.1016/j.ceramint.2015.100.154.

Duxson, P., Fernández-Jiménez, A., Provis, J. L., Lukey, G. C., Palomo, A., & van Deventer, J. S. J. (2007). Geopolymer technology: The current state of the art. *Journal of Materials Science*, *42*(9), 2917–2933. Available from https://doi.org/10.1007/s10853-006-0637-z.

Fang, G., Ho, W. K., Tu, W., & Zhang, M. (2018). Workability and mechanical properties of alkali-activated fly ash-slag concrete cured at ambient temperature. *Construction and Building Materials*, *172*, 476–487. Available from https://doi.org/10.1016/j.conbuildmat.2018.040.008.

Farhan, K. Z., Johari, M. A. M., & Demirboğa, R. (2020). Assessment of important parameters involved in the synthesis of geopolymer composites: A review. *Construction and Building Materials*, *264*. Available from https://doi.org/10.1016/j.conbuildmat.2020.120276.

Firdaus., Yunus, I., & Rosidawani. (2017). Contribution of fineness level of fly ash to the compressive strength of geopolymer mortar. *MATEC Web of Conferences*, *103*. Available from https://doi.org/10.1051/matecconf/201710301026.

Fu, Y., Cai, L., & Yonggen, W. (2011). Freeze-thaw cycle test and damage mechanics models of alkali-activated slag concrete. *Construction and Building Materials*, *25*(7), 3144–3148. Available from https://doi.org/10.1016/j.conbuildmat.2010.120.006.

Garg, A., Singhal, D., & Parveen. (2020). Review on the durability properties of sustainable alkali activated concrete. *Materials Today: Proceedings*, *33*, 1643–1649. Available from https://doi.org/10.1016/j.matpr.2020.060.370.

Görhan, G., Aslaner, R., & Şinik, O. (2016). The effect of curing on the properties of metakaolin and fly ash-based geopolymer paste. *Composites Part B: Engineering*, *97*, 329–335. Available from https://doi.org/10.1016/j.compositesb.2016.050.019.

Gunasekara, C., Setunge, S., Law, D. W., Willis, N., & Burt, T. (2018). Engineering properties of geopolymer aggregate concrete. *Journal of Materials in Civil Engineering*, *30*(11), 04018299.

Hadi, M. N. S., Farhan, N. A., & Sheikh, M. N. (2017). Design of geopolymer concrete with GGBFS at ambient curing condition using Taguchi method. *Construction and Building Materials*, *140*, 424–431. Available from https://doi.org/10.1016/j.conbuildmat.2017.020.131.

Hardjito, D., & Rangan, B. V. (2005). *Development and Properties of Low-Calcium Fly Ash-Based Geopolymer Concrete*. Curtin University of Technology. Available from https://espace.curtin.edu.au/handle/20.500.11937/5594.

Hardjito, D., Wallah, S. E., Sumajouw, D. M. J., & Rangan, B. V. (2004). The stress-strain behaviour of fly ash-based geopolymer concrete. *Development in Mechanics of Structures and Materials*, *35*, 831–834.

Hardjito, D., Wallah, S. E., Sumajouw, D. M. J., & Rangan, B. V. (2015). Fly ash-based geopolymer concrete. *Australian Journal of Structural Engineering*, *6*(1), 77–86. Available from https://doi.org/10.1080/13287982.2005.11464946.

Hasanbeigi, A., Menke, C., & Price, L. (2010). The CO_2 abatement cost curve for the Thailand cement industry. *Journal of Cleaner Production*, *18*(15), 1509–1518. Available from https://doi.org/10.1016/j.jclepro.2010.060.005.

He, J., Jie, Y., Zhang, J., Yu, Y., & Zhang, G. (2013). Synthesis and characterization of red mud and rice husk ash-based geopolymer composites. *Cement and Concrete Composites*, *37*(1), 108–118. Available from https://doi.org/10.1016/j.cemconcomp.2012.110.010.

Hwang, C. L., & Huynh, T. P. (2015). Effect of alkali-activator and rice husk ash content on strength development of fly ash and residual rice husk ash-based geopolymers. *Construction and Building Materials*, *101*, 1–9. Available from https://doi.org/10.1016/j.conbuildmat.2015.100.025.

Isa, M. N., & Awang, H. (2023). Characteristics of palm oil fuel ash geopolymer mortar activated with wood ash lye cured at ambient temperature. *Journal of Building Engineering*, *66*. Available from https://doi.org/10.1016/j.jobe.2023.105851.

Jamkar, S. S., Ghugal, Y. M., & Patankar, S. V. (2013). Effect of fly ash fineness on workability and compressive strength of geopolymer concrete. *Indian Concrete Journal*, *87*(4), 57–62.

Jegan, M., Annadurai, R., & Rajkumar, P. R. K. (2023). A state of the art on effect of alkali activator, precursor, and fibers on properties of geopolymer composites. *Case Studies in Construction Materials*, *18*. Available from https://doi.org/10.1016/j.cscm.2023.e01891.

Jindal, B. B., Jangra, P., & Garg, A. (2020). Effects of ultra fine slag as mineral admixture on the compressive strength, water absorption and permeability of rice husk ash based geopolymer concrete. *Materials Today: Proceedings*, *32*, 871–877. Available from https://doi.org/10.1016/j.matpr.2020.040.219.

Jindal, B. B., Singhal, D., Sharma, S. K., Ashish, D. K., & Parveen. (2017). Improving compressive strength of low calcium fly ash geopolymer concrete with alccofine. *Advances in Concrete Construction*, *5*(1), 17–29. Available from https://doi.org/10.12989/acc.2017.5.1.17.

Junaid, M. T., Kayali, O., Khennane, A., & Black, J. (2015). A mix design procedure for low calcium alkali activated fly ash-based concretes. *Construction and Building Materials*, *79*, 301–310. Available from https://doi.org/10.1016/j.conbuildmat.2015.010.048.

Kareem, R. S., Dang, C. N., & Hale, W. M. (2021). Flexural behavior of concrete beams cast with high-performance materials. *Journal of Building Engineering*, *34*. Available from https://doi.org/10.1016/j.jobe.2020.101912.

Kareem, R. S., Assi, L. N., Alsalman, A., & Al-Manea, A. (2021). Effect of supplementary cementitious materials on RC concrete piles. *AIP Conference Proceedings*, *2404*(1), 080036. Available from http://doi.org/10.1063/5.0070671, 15517616.

Khale, D., & Chaudhary, R. (2007). Mechanism of geopolymerization and factors influencing its development: A review. *Journal of Materials Science*, *42*(3), 729–746. Available from https://doi.org/10.1007/s10853-006-0401-4.

Kim, Y. Y., Lee, B. J., Saraswathy, V., & Kwon, S. J. (2014). Strength and durability performance of alkali-activated rice husk ash geopolymer mortar. *Scientific World Journal*, *2014*. Available from https://doi.org/10.1155/2014/209584.

Kong, D. L. Y., & Sanjayan, J. G. (2010). Effect of elevated temperatures on geopolymer paste, mortar and concrete. *Cement and Concrete Research*, *40*(2), 334–339. Available from https://doi.org/10.1016/j.cemconres.2009.100.017.

Kumar, P., Pankar, C., Manish, D., & Santhi, A. S. (2018). Study of mechanical and microstructural properties of geopolymer concrete with GGBS and Metakaolin. *Materials Today: Proceedings*, *14*, 28127–28135. Available from https://doi.org/10.1016/j.matpr.2018.10.054, 22147853.

Kumar, S., Kumar, R., & Mehrotra, S. P. (2010). Influence of granulated blast furnace slag on the reaction, structure and properties of fly ash based geopolymer. *Journal of Materials Science*, *45*(3), 607–615. Available from https://doi.org/10.1007/s10853-009-3934-5.

Kusbiantoro, A., Nuruddin, M. F., Shafiq, N., & Qazi, S. A. (2012). The effect of microwave incinerated rice husk ash on the compressive and bond strength of fly ash based geopolymer concrete. *Construction and Building Materials*, *36*, 695–703. Available from https://doi.org/10.1016/j.conbuildmat.2012.060.064.

Li, Z., & Liu, S. (2007). Influence of slag as additive on compressive strength of fly ash-based geopolymer. *Journal of Materials in Civil Engineering*, *19*(6), 470–474. Available from https://doi.org/10.1061/(ASCE)0899-1561(2007)19:6(470).

Mallikarjuna Rao, G., & Gunneswara Rao, T. D. (2018). A quantitative method of approach in designing the mix proportions of fly ash and GGBS-based geopolymer concrete. *Australian Journal of Civil Engineering*, *16*(1), 53–63. Available from https://doi.org/10.1080/14488353.2018.1450716.

McLellan, B. C., Williams, R. P., Lay, J., Van Riessen, A., & Corder, G. D. (2011). Costs and carbon emissions for geopolymer pastes in comparison to ordinary Portland cement. *Journal of Cleaner Production*, *19*(9–10), 1080–1090. Available from https://doi.org/10.1016/j.jclepro.2011.020.010.

Mehta, A., & Siddique, R. (2017). Properties of low-calcium fly ash based geopolymer concrete incorporating OPC as partial replacement of fly ash. *Construction and Building Materials*, *150*, 792–807. Available from https://doi.org/10.1016/j.conbuildmat.2017.060.067.

Mehta, P. K., & Monteiro, P. J. M. (2006). *Concrete: Microstructure, properties, and materials*. McGraw-Hill.

Munshi, S., & Sharma, R. P. (2016). Experimental investigation on strength and water permeability of mortar incorporate with rice straw ash. *Advances in Materials Science and Engineering, 2016*, 16878442. Available from https://doi.org/10.1155/2016/9696505.

Nagral, M. R., Ostwal, T., & Chitawadagi, M. V. (2014). Effect of curing temperature and curing hours on the properties of geo-polymer concrete. *International Journal of Computational Engineering Research, 4*, 1–11.

Nath, P., & Sarker, P. K. (2015). Use of OPC to improve setting and early strength properties of low calcium fly ash geopolymer concrete cured at room temperature. *Cement and Concrete Composites, 55*, 205–214. Available from https://doi.org/10.1016/j.cemconcomp.2014.080.008.

Nath, P., & Sarker, P. K. (2017). Flexural strength and elastic modulus of ambient-cured blended low-calcium fly ash geopolymer concrete. *Construction and Building Materials, 130*, 22–31. Available from https://doi.org/10.1016/j.conbuildmat.2016.110.034.

Nazari, A., & Sanjayan, J. G. (2015). Synthesis of geopolymer from industrial wastes. *Journal of Cleaner Production, 99*, 297–304. Available from https://doi.org/10.1016/j.jclepro.2015.030.003.

Nazari, A., Bagheri, A., & Riahi, S. (2011). Properties of geopolymer with seeded fly ash and rice husk bark ash. *Materials Science and Engineering A, 528*(24), 7395–7401. Available from https://doi.org/10.1016/j.msea.2011.060.027.

Nuruddin, M. F., Qazi, S. A., Kusbiantoro, A., & Shafiq, N. (2011). Utilisation of waste material in geopolymeric concrete. *Proceedings of Institution of Civil Engineers: Construction Materials, 164*(6), 315–327. Available from https://doi.org/10.1680/coma.2011.164.60.315.

Okoye, F. N., Durgaprasad, J., & Singh, N. B. (2015). Fly ash/Kaolin based geopolymer green concretes and their mechanical properties. *Data in Brief, 5*, 739–744. Available from https://doi.org/10.1016/j.dib.2015.100.029.

Okoye, F. N., Prakash, S., & Singh, N. B. (2017). Durability of fly ash based geopolymer concrete in the presence of silica fume. *Journal of Cleaner Production, 149*, 1062–1067. Available from https://doi.org/10.1016/j.jclepro.2017.020.176.

Olivia, M., & Nikraz, H. R. (2011). Strength and water penetrability of fly ash geopolymer concrete. *Journal of Engineering and Applied Sciences, 6*(7), 70–78.

Olivia, M., & Nikraz, H. S. (2008). *Water penetrability of low calcium fly ash geopolymer concrete. Proceedings of the international conference on construction and building technology* (pp. 517–530).

Olivia, M., & Nikraz, H. (2012). Properties of fly ash geopolymer concrete designed by Taguchi method. *Materials & Design (1980-2015), 36*, 191–198. Available from https://doi.org/10.1016/j.matdes.2011.10.036.

Pacheco-Torgal, F., Castro-Gomes, J., & Jalali, S. (2008). Alkali-activated binders: A review. Part 1. Historical background, terminology, reaction mechanisms and hydration products. *Construction and Building Materials, 22*(7), 1305–1314. Available from https://doi.org/10.1016/j.conbuildmat.2007.100.015.

Parthiban, K., & Mohan, K. S. R. (2017). Influence of recycled concrete aggregates on the engineering and durability properties of alkali activated slag concrete. *Construction and Building Materials, 133*, 65–72. Available from https://doi.org/10.1016/j.conbuildmat.2016.120.050.

Patankar, S. V., Jamkar, S. S., & Ghugal, Y. M. (2013). Effect of water-to-geopolymer binder ratio on the production of fly ash based geopolymer concrete. *International Journal of Advanced Technology in Civil Engineering, 2*, 79–83.

Pavithra, P., Srinivasula Reddy, M., Dinakar, P., Hanumantha Rao, B., Satpathy, B. K., & Mohanty, A. N. (2016). A mix design procedure for geopolymer concrete with fly ash. *Journal of Cleaner Production*, *133*, 117–125. Available from https://doi.org/10.1016/j.jclepro.2016.050.041.

Pouhet, R., & Cyr, M. (2016). Formulation and performance of flash metakaolin geopolymer concretes. *Construction and Building Materials*, *120*, 150–160. Available from https://doi.org/10.1016/j.conbuildmat.2016.050.061.

Rajini, B., Rao, A. V. N., & Sashidhar, C. (2020). Cost analysis, geopolymer, over, concrete. *engrXiv*. Available from https://doi.org/10.31224/osf.io/3mxgz.

Reddy, D. V., Edouard, J. B., & Sobhan, K. (2013). Durability of fly ash-based geopolymer structural concrete in the marine environment. *Journal of Materials in Civil Engineering*, *25*(6), 781–787. Available from https://doi.org/10.1061/(ASCE)MT.1943-5533.0000632.

Rendell, F., Jauberthie, R., & Grantham, M. (2002). *Deteriorated concrete*. Thomas Telford Publishing. Available from 10.1680/dc.31197.

Ruiz-Santaquiteria, C., Skibsted, J., Fernández-Jiménez, A., & Palomo, A. (2012). Alkaline solution/binder ratio as a determining factor in the alkaline activation of aluminosilicates. *Cement and Concrete Research*, *42*(9), 1242–1251. Available from https://doi.org/10.1016/j.cemconres.2012.050.019.

Sarker, P. K., Haque, R., & Ramgolam, K. V. (2013). Fracture behaviour of heat cured fly ash based geopolymer concrete. *Materials and Design*, *44*, 580–586. Available from https://doi.org/10.1016/j.matdes.2012.080.005.

Sata, V., Sathonsaowaphak, A., & Chindaprasirt, P. (2012). Resistance of lignite bottom ash geopolymer mortar to sulfate and sulfuric acid attack. *Cement and Concrete Composites*, *34*(5), 700–708. Available from https://doi.org/10.1016/j.cemconcomp.2012.010.010.

Shaikh, F. U. A. (2016). Mechanical and durability properties of fly ash geopolymer concrete containing recycled coarse aggregates. *International Journal of Sustainable Built Environment*, *5*(2), 277–287. Available from https://doi.org/10.1016/j.ijsbe.2016.050.009.

Shi, X. S., Collins, F. G., Zhao, X. L., & Wang, Q. Y. (2012). Mechanical properties and microstructure analysis of fly ash geopolymeric recycled concrete. *Journal of Hazardous Materials*, *237–238*, 20–29. Available from https://doi.org/10.1016/j.jhazmat.2012.070.070.

Shinde, P., Patankar, S., & Sayyad, A. (2018). Investigation on effects of fineness of fly ash and alkaline ratio on mechanical properties of geopolymer concrete. *Research on Engineering Structures & Materials*, *4*, 61–71.

Statista. 2023. U.S. Cement Production: Portland and Masonry 2023. Statista. Accessed March 21, 2024. https://www.statista.com/statistics/219329/us-production-of-portland-and-masonery-cement/.

Sujatha, T., Kannapiran, K., & Nagan, S. (2012). Strength assessment of heat cured geopolymer concrete slender column. *Asian Journal of Civil Engineering*, *13*(5), 635–646.

Sumajouw, M., & Rangan, B. V. (2006). *Low-calcium fly ash-based geopolymer concrete: Reinforced beams and columns. Research Report GC 3 Faculty of Engineering Curtin*. Australia: University of Technology Perth. Available from https://espace.curtin.edu.au/bitstream/handle/20.500.11937/23928/19466_downloaded_stream_558.pdf?sequence = 2&isAllowed = y.

Türkmen, İ., Karakoç, M. B., Kantarcı, F., Maraş, M. M., & Demirboğa, R. (2016). Fire resistance of geopolymer concrete produced from Elazığ ferrochrome slag. *Fire and Materials*, *40*(6), 836–847. Available from https://doi.org/10.1002/fam.2348.

Venkatesan, R. P., & Pazhani, K. C. (2016). Strength and durability properties of geopolymer concrete made with ground granulated blast furnace slag and black rice husk ash. *KSCE*

Journal of Civil Engineering, *20*(6), 2384–2391. Available from https://doi.org/10.1007/s12205-015-0564-0.

Vidyadhara, V., & Ranganath, R. V. (2023). Upcycling of pond ash in cement-based and geopolymer-based composite: A review. *Construction and Building Materials*, *379*. Available from https://doi.org/10.1016/j.conbuildmat.2023.130949.

Vora, P. R., & Dave, U. V. (2013). Parametric studies on compressive strength of geopolymer concrete. *Procedia Engineering*, *51*, 210–219. Available from https://doi.org/10.1016/j.proeng.2013.01.030.

Xie, T., & Ozbakkaloglu, T. (2015). Behavior of low-calcium fly and bottom ash-based geopolymer concrete cured at ambient temperature. *Ceramics International*, *41*(4), 5945–5958. Available from https://doi.org/10.1016/j.ceramint.2015.010.031.

Xie, T., Visintin, P., Zhao, X., & Gravina, R. (2020). Mix design and mechanical properties of geopolymer and alkali activated concrete: Review of the state-of-the-art and the development of a new unified approach. *Construction and Building Materials*, *256*. Available from https://doi.org/10.1016/j.conbuildmat.2020.119380.

Yang, K. H., & Song, J. K. (2009). Workability loss and compressive strength development of cementless mortars activated by combination of sodium silicate and sodium hydroxide. *Journal of Materials in Civil Engineering*, *21*(3), 119–127. Available from https://doi.org/10.1061/(ASCE)0899-1561(2009)21:3(119).

Yashwanth Reddy, M., & Harihanandh, M. (2023). Experimental studies on strength and durability of alkali activated slag and coal bottom ash based geopolymer concrete. *Materials Today: Proceedings*. Available from https://doi.org/10.1016/j.matpr.2023.030.644.

Zhang, B., Cheng, Y., & Zhu, H. (2023). Bond performance between BFRP bars and alkali-activated seawater coral aggregate concrete. *Engineering Structures*, *279*. Available from https://doi.org/10.1016/j.engstruct.2023.115596.

Zhang, B., Yan, B., & Li, Y. (2023). Study on mechanical properties, freeze–thaw and chlorides penetration resistance of alkali activated granulated blast furnace slag-coal gangue concrete and its mechanism. *Construction and Building Materials*, *366*. Available from https://doi.org/10.1016/j.conbuildmat.2022.130218.

Zhang, M. H., Islam, J., & Peethamparan, S. (2012). Use of nano-silica to increase early strength and reduce setting time of concretes with high volumes of slag. *Cement and Concrete Composites*, *34*(5), 650–662. Available from https://doi.org/10.1016/j.cemconcomp.2012.020.005.

Zhuang, H. J., Zhang, H. Y., & Xu, H. (2017). China Resistance of geopolymer mortar to acid and chloride attacks. *Procedia Engineering*, *210*, 126–131. Available from https://doi.org/10.1016/j.proeng.2017.11.057.

Zhuang, X. Y., Chen, L., Komarneni, S., Zhou, C. H., Tong, D. S., Yang, H. M., Yu, W. H., & Wang, H. (2016). Fly ash-based geopolymer: Clean production, properties and applications. *Journal of Cleaner Production*, *125*, 253–267. Available from https://doi.org/10.1016/j.jclepro.2016.030.019.

Sustainable alkali-activated construction materials from construction and demolition waste

Anil Kul[1,2], *Emircan Ozcelikci*[1,2], *Gurkan Yildirim*[2,3], *Musab Alhawat*[3] *and Ashraf Ashour*[3]
[1]Institute of Science, Hacettepe University, Beytepe, Ankara, Turkey, [2]Department of Civil Engineering, Hacettepe University, Ankara, Turkey, [3]Faculty of Engineering and Digital Technologies, University of Bradford, Bradford, United Kingdom

5.1 Introduction

The construction industry is one of the most significant contributors of economic growth in many countries worldwide. The sector's importance is evident in the vast number of infrastructure projects being developed, ranging from housing and commercial buildings to transportation and energy systems. The increasing population, migration, and urbanization have propelled the demand for new construction activities, leading to a significant expansion of the industry. However, this growth has come with its challenges, primarily related to the sector's environmental impact. The construction industry is known to consume a significant amount of energy and natural resources, emit greenhouse gases, and generate substantial waste, all of which having considerable detrimental effect on the environment. In straightforward terms, the buildings sector is a major consumer of energy, accounting for roughly 30% of total global energy consumption (International Energy Agency, 2022). Additionally, this sector is responsible for approximately 27% of global carbon dioxide (CO_2) emissions, resulting from the day-to-day operation of buildings.

One of the most critical factors contributing to the industry's environmental impact is the use of ordinary Portland cement (OPC), the primary binder used in the manufacture of concrete materials. The production of OPC is energy intensive and releases a considerable amount of CO_2 into the atmosphere, contributing to global warming. The use of cement in construction accounts for about 7%–8% of global greenhouse gas emissions, making it one of the most significant contributors of climate change (Amran et al., 2022). Moreover, the extraction of raw materials used in cement production can cause significant ecological damage, including deforestation, soil erosion, and biodiversity loss (Hossain et al., 2017). Given the extensive use of OPC in the construction industry, it is crucial to find ways to reduce its environmental impact. One promising approach is the development of

Sustainable Concrete Materials and Structures. DOI: https://doi.org/10.1016/B978-0-443-15672-4.00005-X

alternative binding materials that require less energy to produce and emit less greenhouse gases. Alkali activation is a potential environmentally friendly solution that can be adopted by the industry. It involves the activation of aluminosilicate materials using alkali solutions, such as sodium hydroxide or potassium hydroxide, resulting in the formation of alkali-activated materials (AAMs; Zhang et al., 2014). AAMs have similar properties to OPC but require less energy to produce, emit fewer greenhouse gases, and have higher durability and fire resistance (Bernal et al., 2013; Jiang et al., 2020; Neupane et al., 2018; Pouhet & Cyr, 2015). They can be produced by using industrial waste materials, such as fly ash, slag, and other mining and metallurgical wastes, which can potentially reduce the amount of waste being sent to landfills.

Another environmental challenge facing the construction industry is waste management. Construction and demolition activities generate significant amounts of construction and demolition waste (CDW), including concrete, bricks, asphalt, metals, wood, and plastics. In addition to the types of waste mentioned, CDW includes hazardous materials, such as asbestos and lead, which require careful handling to prevent health risks to the workers and public. In the European Union, CDW accounts for approximately 25%–30% of all waste generated, with an estimated 850 Mt of CDW generated each year (Eurostat, 2016). In the United States, the construction industry generates approximately 570 million tons of CDW each year, representing around 23% of all waste generated (US EPA, 2020). This waste, if not appropriately managed, can have severe environmental impacts, such as soil and water contamination, air pollution, and habitat destruction. The improper disposal of CDW in landfills also consumes valuable land resources leading to the release of harmful gases, including methane, which contributes to climate change. One solution is to adopt a circular economy approach by promoting the reuse and recycling of CDW materials (Zhang et al., 2022). The cost of recycling a ton of CDW, which includes concrete, brick, and masonry debris, is approximately $21 per ton, whereas landfilling costs about $136 per ton (Lennon, 2005). CDW has been identified as a promising candidate for alkali activation in recycling activities due to its favorable aluminosilicate content (Alhawat et al., 2022; Robayo-Salazar et al., 2020). Although the potential of CDW-based AAMs as an alternative to OPC is promising, current literature indicates that research in this area is limited. Furthermore, it is necessary to achieve the optimum recycling of each CDW component through proper formulation for effective waste management. To achieve this, a comprehensive understanding of the properties and characteristics of each CDW component is essential. In recent years, there has been increasing research in the development of sustainable and efficient recycling methods for different types of CDW, such as concrete, asphalt, wood, and plastics (Choudhary et al., 2020; Sevim, Alakara, et al., 2023; Sevim, Demir, et al., 2023; Tuğluca et al., 2023; Xiao et al., 2018; Yıldırım et al., 2021). These studies aim to enhance the utilization of CDW in the construction industry while minimizing its environmental impact. Overall, continued efforts in CDW management and recycling practices will be necessary to promote the circular economy approach in the construction industry and contribute to sustainable development.

This chapter aims to examine the latest research and applications in the alkali activation technology, covering a broad range of materials including binders, mortars, concretes, 3D printable composites, engineered geopolymer composites (EGC), and insulation materials based on CDW to facilitate the active participation of CDWs into circular economy in the construction industry. The work comprehensively assesses the fresh, mechanical, and durability properties of the materials with a focus on the analysis of all parameters that can potentially impact their ultimate performance. As CDW is an issue on global scale, this state-of-the-art chapter aims to provide insights into sustainable waste management solutions and the potential use of CDW as raw materials to produce construction materials in an effort to reducing the carbon footprint of the construction industry.

5.2 Beyond Portland cement: exploring alternatives for sustainable construction

5.2.1 Ordinary Portland cement

OPC is the primary binder used in the production of traditional concrete. It typically accounts for approximately 10%–15% of the total volume of the concrete, which is the second most used material in the world after water, and is used in a wide range of applications, including building construction, infrastructural development, and transportation systems (Vijayan et al., 2020). On the other hand, OPC manufacturing is an energy-intensive process that causes considerable harm to the environment. Its adverse environmental impacts include greenhouse gas emissions, air pollution, and water pollution. The main source of emissions is the calcination process that releases CO_2 from limestone. Significant energy inputs are also required during the handling, processing, transportation, and curing of the raw materials and cement further exacerbating the emissions. Emissions lead to air pollution through the release of particulate matter (PM),sulphur oxidess (SO_2), and nitrogen oxides (NO_x), which can cause respiratory and cardiovascular problems. Handling and processing of raw materials and cement can also cause dust and soil damage. Water pollution results from the discharge of process water and wastewater containing heavy metals and other pollutants.

Globally, around 4.2 billion metric tons of OPC were produced in 2019 (Garside, 2020), requiring energy equivalent to the energy output of over 175 nuclear power plants or 5.5 million wind turbines. According to the Environmental Protection Agency (EPA), the production of 1 ton of OPC generates approximately 1 ton of CO_2 emissions (US EPA, 2020). This means that the production of approximately 4.2 billion metric tons of OPC resulted in the emission of around 4.2 billion metric tons of CO_2. To put this into perspective, the total global CO_2 emissions from all sources in 2019 were around 36.4 billion metric tons (Friedlingstein et al., 2022). Thus the production of OPC accounted for approximately 11.5% of the total global CO_2 emissions in 2019, and lion's share of this contribution comes from

China with more than 800 million tons (Fig. 5.1). To mitigate the environmental impacts of cement production, alternative low-carbon and carbon-neutral cement technologies are being developed, along with improving energy efficiency and adopting best practices for waste management and pollution control. However, significant efforts are still required to address the adverse environmental impacts of cement production on a global scale.

5.2.2 Alternative materials

The necessity of valorization of alternative materials in the construction industry arises from the negative environmental impact of OPC production and the depletion of natural resources. By utilizing alternative materials with high pozzolanic activity, the amount of OPC required in construction materials can be reduced or even eliminated, resulting in a significant reduction in carbon footprint and waste generation associated with cement production. Pozzolanic reaction refers to the chemical reaction between the amorphous silica and alumina present in pozzolanic materials and $Ca(OH)_2$ produced during the OPC hydration (Shi & Day, 2000). This reaction results in the formation of additional calcium silicate hydrates (CSH) and calcium aluminate hydrates (CAH), which contribute to the strength and durability of

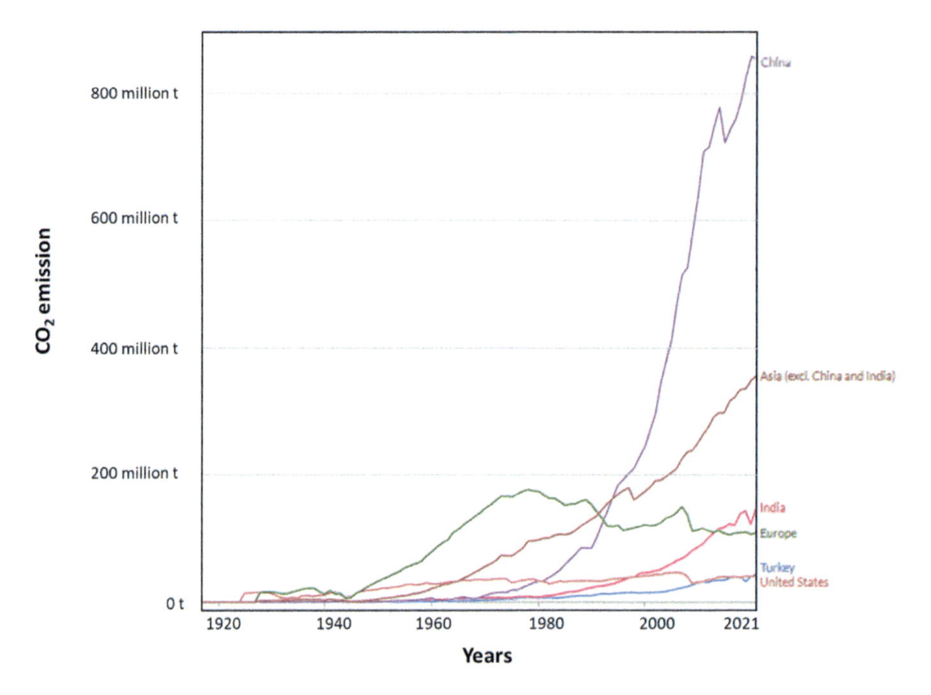

Figure 5.1 Annual CO_2 emissions from cement.
Source: From Friedlingstein, P., Jones, M. W., Andrew, R. M., Gregor, L., Hauck, J., & Zheng, B. (2022). Global carbon budget 2022. *Earth System Science Data, 14*(11), 4811–4900.

cementitious materials. The pozzolanic reaction occurs in the presence of moisture and alkalinity, which activates the pozzolanic material and facilitates its reaction with $Ca(OH)_2$. The reaction is exothermic and takes place slowly, typically over a period of weeks to months, depending on the type and amount of pozzolanic material used.

One of the most commonly used alternative materials in cementitious materials is fly ash, which is a by-product of coal-fired power plants. In 2018, the global production of fly ash was estimated to be around 780 million tons (Singh et al., 2019). The use of fly ash in cementitious materials has been shown to improve their compressive strength, durability, and reduce the permeability. Other alternative materials such as slag, rice husk ash, and calcined clay have also been studied for their potential use in cementitious materials (Malhotra & Mehta, 1996). Slag is a by-product of the steel industry, while rice husk ash and calcined clay are agricultural waste products. The use of these materials in cementitious composites has several advantages, including increased strength and durability, reduced permeability, and improved resistance to chemical attack and environmental degradation (Hossain et al., 2016; Juenger & Siddique, 2015; Juenger et al., 2019). Additionally, the use of these materials can help reduce the environmental impact of the construction industry by diverting waste materials from landfills and reducing the energy consumption and carbon footprint associated with OPC production.

While these mainstream wastes have shown promise as replacements for OPC, they also come with their own set of challenges. One issue is the variability of these materials from source to source, which can affect the final product's properties. Additionally, some sources of these materials may contain contaminants, such as heavy metals, that can pose environmental and health risks if not properly managed (Cinquepalmi et al., 2008). Furthermore, the availability and cost of these materials can be a challenge, as they are often by-products of industrial processes and may not be consistently produced in large quantities. This can lead to fluctuations in supply and pricing, which can make it difficult for construction industry to plan and execute projects that rely on these materials. Therefore researchers have been exploring the potential of alternative aluminosilicate materials, including those derived from CDW, mining waste, etc. (Veiga Simão et al., 2022). It is important to recognize that incorporating alternative materials in construction is not a perfect solution, and there may exist certain challenges and limitations to their widespread adoption. For example, there may be regulatory barriers or economic constraints that hinder the use of certain materials in some regions or applications (Habert et al., 2011). However, the increasing awareness of the environmental impact of the construction industry and the development of more sustainable and innovative building materials suggest that the use of alternative materials will continue to be an important area of research and development in the coming years.

5.2.3 Alkali activation mechanism

The usage of alternative materials as a replacement for OPC has gained significant attention in recent years, particularly in the context of geopolymerization and alkali activation processes. Alkali activation refers to the process of synthesizing cementitious materials through the activation of aluminosilicate precursor materials by strong alkali solutions. The mechanism of alkali activation involves a series of complex chemical reactions between the precursor material and the alkaline activator solution, leading to the formation of amorphous, three-dimensional silicate networks (Palomo et al., 1999). This is achieved without the high-temperature calcination typically required in the production of OPC, which results in significant energy savings and a reduced carbon footprint.

The five main phases of alkali activation include dissolution, speciation equilibrium, gelation, reorganization, and polymerization and hardening as illustrated in Fig. 5.2 (Duxson et al., 2007). During the dissolution phase, the alkaline activator solution interacts with the precursor material, leading to the release of reactive species. These reactive species can include dissolved silica, alumina, and calcium ions, among others. One of the primary reactions that occur during the speciation equilibrium phase is the hydrolysis of the alkaline activator, which results in the formation of hydroxide ions (OH^-) and/or silicate species. The hydroxide ions and silicate species then react with the cations released from the precursor material, leading to the formation of various soluble and insoluble species. For instance, calcium ions (Ca^{2+}) released from the precursor material can react with hydroxide ions to form calcium hydroxide ($Ca[OH]_2$), which is a soluble species. The gelation phase involves the reaction of the speciated species with each other to form an amorphous gel-like substance. This gel-like substance undergoes further structural reorganization in the reorganization phase, leading to the formation of a stable, three-dimensional network including CAH, CSH, and sodium aluminosilicate hydrates (NASH). The hardening phase can continue for an extended period, typically lasting several days to weeks, depending on the curing conditions and the composition of the precursor material and the alkaline activator used. During this time, the final product undergoes further structural reorganization and polymerization, resulting in

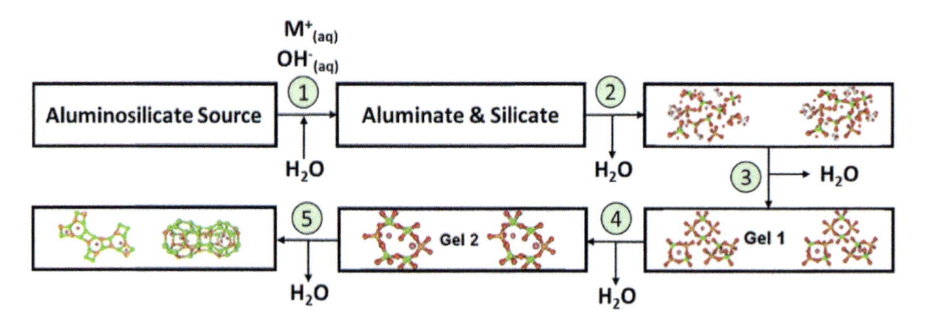

Figure 5.2 Alkali activation mechanism; (1) dissolution, (2) speciation equilibrium, (3) gelation, (4) reorganization, and (5) polymerization and hardening.
Source: From Duxson, P., Fernández-Jiménez, A., Provis, J. L., Lukey, G. C., Palomo, A., & van Deventer, J. S. J. (2007). Geopolymer technology: The current state of the art. *Journal of Materials Science*, *42*(9), 2917–2933. https://doi.org/10.1007/s10853-006-0637-z.

the development of compressive strength and durability characteristics that are similar to those of traditional cementitious materials (Bernal et al., 2012).

5.3 Construction and demolition waste: a promising alternative to ordinary Portland cement

CDW is a complex waste stream that includes various materials and substances, resulting from construction, renovation, and demolition activities. CDW is generated from different sources such as residential, commercial, and industrial buildings, infrastructure, and other structures. The composition of CDW varies depending on many factors, such as the type and characteristics of the structure, the building materials used, and the construction methods employed. For instance, CDW from a residential building may contain wood, plaster, tiles, bricks, plastics, and metals, while CDW from an industrial facility may contain hazardous materials such as asbestos, lead, and chemicals (Huang et al., 2018).

The activities leading to the generation of CDW, typically cause formation of large quantities of waste, although often data on CDW are not collected routinely or consistently, so most published information about its amount is based on estimates that need to be interpreted with caution. These estimates include approximately 8.2 billion tons of CDW generated across the EU in 2012 (Eurostat, 2015), 77 million tons in Japan, 33 million tons in China, 17 million tons in India in 2010, and nearly 7 million tons in both Dubai in 2011 and Abu Dhabi in 2013, respectively (UN Environment Programme, 2015). CDW is frequently the largest proportion of overall waste generated, accounting for 34% of urban waste produced in The Organization for Economic Cooperation and Development nations and the amount of CDW being generated is rapidly increasing due to the rate of infrastructure development around the world (UN Environment Programme, 2015). The construction industry is a significant contributor of CDW generation globally. According to the Global Waste Management Outlook report, the construction and demolition sector generates about 30% of the world's total waste and consumes up to 40% of the world's raw materials (UN Environment Programme, 2015). Moreover, CDW generation is expected to increase due to the rapid urbanization, population growth, and economic development in emerging markets. For example, in China, the construction industry is one of the fastest growing industries and a significant contributor to CDW generation, accounting for over 50% of the country's total waste (Huang et al., 2018).

The generation of waste is a common consequence of both human-made construction and demolition activities as well as natural disasters. The amount and composition of the waste produced depend on the type and severity of the disaster and the nature of the built environment. For instance, recent natural disasters such as the 2023 Turkey earthquake, 2015 Nepal earthquake, 2010 Haiti earthquake, 2005 Hurricane Katrina, and 2004 Indian Ocean tsunami led to the production of significant volumes of waste, which overwhelmed existing waste management systems. The debris generated from these disasters often creates obstacles that hinder the efforts of rescuers and emergency services to reach survivors. Among the materials commonly found in disaster debris are natural aggregates, construction, and demolition waste (such as concrete, bricks, timber, and metal), vehicles and boats, and electrical goods and appliances. However, these materials can all be recycled and have a significant market in developing countries (World Bank & United Nations, 2010). For instance, they can be utilized for landfill cover, as aggregate for concrete, fill for land reclamation, compost for fertilization, and slope stabilization (Channell et al., 2009). Some of these materials can even be employed to generate energy in a beneficial way (EPA, 2008; Yepsen, 2008).

The increase in the generation of CDW has led to growing concerns regarding its environmental impact, including the depletion of natural resources, land use, energy consumption, greenhouse gas emissions, and pollution. Landfills that receive CDW contribute to the depletion of land resources as they take up significant amounts of space. Additionally, the incineration of CDW results in the release of hazardous air pollutants and greenhouse gases such as CO_2 and methane, contributing to climate change (Liikanen et al., 2019). Moreover, transportation of CDW requires significant energy consumption and contributes to greenhouse gas emissions. To address the environmental concerns associated with CDW, many countries have implemented regulations and policies for its management. For instance, the European Union has adopted a circular economy approach to CDW management, with a target of 70% CDW recycling by 2020 (European Commission, 2020). Similarly, in the United States, the EPA has established guidelines for the beneficial use of CDW, such as using recycled CDW as a substitute for natural aggregates in construction (US EPA, 2020).

The use of CDW in concrete industry has been of subject to extensive research, but its present application is mostly limited to low-tech methods. Studies have shown that replacing natural aggregates with crushed concrete for structural purposes should be limited to 25%–30% (Etxeberria et al., 2007; Mohammed Ali et al., 2020), while replacing up to 50% of natural sand with CDW-based fine particles is unlikely to significantly alter concrete properties (Halicka et al., 2013). However, utilizing 100% recycled concrete aggregate (RCA) could reduce the mechanical properties of concrete and increase its absorption capacity compared to standard concrete. Several techniques have been suggested to enhance the quality of CDW-based aggregates, including mechanical grinding, acid soaking, polymer emulsion, and accelerated carbonation. Nonetheless, economic constraints hinder the broad adoption of CDW reuse as a viable option.

The crushing of CDW produces a significant amount of fine powder. However, few studies have investigated the use of this powder as a partial replacement for cement. The incorporation of concrete waste powders into geopolymer mortar significantly impairs fluidity and compressive strength. It is preferable not to exceed 15% as a replacement rate to OPC (Kim & Choi, 2012). The use of up to 30% CDW powder in conventional concrete instead of OPC can facilitate hydration and reduce permeability and porosity, but higher ratios may have significant negative effects, particularly on properties like frost resistance and chloride ion permeability

(Ma et al., 2019). Utilizing fine powder from recycled concrete as an alternative source for producing Portland cement clinker has been found to be partially viable (Schoon et al., 2015). In conclusion, CDW can be effectively used in concrete production by applying appropriate techniques to enhance its quality and considering its limitations to achieve optimal results.

Using recycled and reused CDW as a source for geopolymer materials can be a sustainable solution to reduce ecological impact and decrease the demand for ordinary OPC. Recent studies have shown that CDW can be an excellent source of aluminosilicates for geopolymer binders, especially since demolished structures generate vast quantities of fine silt particles rich in crystalline aluminosilicates (Demiral et al., 2022; Ilcan et al., 2023; Ozcelikci, Yildirim, et al., 2023). However, considering the variability of CDW, there is a need for much more in-depth research in developing CDW-based geopolymer materials to comprehensively reveal the effects of the materials' variability on the mechanical, fresh, and structural properties of ultimate materials.

5.4 Construction and demolition waste–based alkali-activated materials: current applications and future directions

5.4.1 Construction and demolition waste–based alkali-activated materials: binders

In recent years, there has been a notable upsurge in research efforts geared toward developing AAMs employing components derived from CDW, including clay-based masonry (such as red clay bricks, hollow bricks, and roof tiles), concrete and mortar, and waste glass. These materials have shown promise, particularly due to their suitable aluminosilicate nature. Literature suggests that a minimum $SiO_2 + Al_2O_3 + Fe_2O_3$ content of over 70% is essential to achieving successful alkali activation (Leroy et al., 2019) by adopting the ASTM standard related to pozzolanic activity of coal fly ash and raw or calcined natural pozzolan (ASTM C618, 2019). Notably, masonry-originated CDW components such as red clay bricks, hollow bricks, and roof tiles are considered ideal due to their chemical composition, whereas concrete and glass are known to exhibit relatively weak aluminosilicate content. In one study, CDW-based roof tiles, red clay bricks, and hollow bricks activated with 12.5 M NaOH and subjected to thermal curing achieved compressive strengths of approximately 40 MPa, while glass achieved a strength of about 30 MPa. This was attributed to the lower Al_2O_3 content of glass (1.27%) and its relatively coarse particle size distribution compared to other materials (Ulugöl et al., 2021). Another study demonstrated that concrete with a total $SiO_2 + Al_2O_3$ content of 36.36% activated with 19 M NaOH and cured at 125°C for 48 hours achieved a compressive strength of 34 MPa. When 25% of concrete was replaced with glass, the compressive strength increased to 36 MPa. However, the low aluminosilicate content of concrete was compensated by the high SiO_2 content of glass, and the contribution of the concrete to the compressive strength remained limited. Notably, the addition of bricks and tiles to the mixture along with the concrete and glass significantly increased the compressive strength of the resulting pastes up to 54 MPa (Özçelikci, 2020). In another study, glass activated solely with NaOH solutions designed to involve 12% Na achieved a strength of 43 MPa, while concrete remained at 9.8 MPa. However, by adding hollow bricks and roof tiles to the concrete, the compressive strength reached 59.9 MPa (Kul, 2019). According to the findings of these studies, a general conclusion can be drawn that the masonry-originated portion of the CDWs have significant reactivity in the highly alkaline medium due to the balanced SiO_2 and Al_2O_3 content; however, for the glass and concrete, the lack of adequate Al_2O_3 and SiO_2, respectively, caused lower reactivity. These results also indicate that mixed use of CDW-based components in alkali activation leads to better performance than single use, as reported in the literature and compiled in Fig. 5.3.

In addition to the aluminum silicate content, achieving a balance in the Si/Al, Na/Si, and Ca/Si ratios is of great importance for reaction kinetics in alkali activation. A recent review paper reported that SiO_2 content of bricks and tiles is generally in the range of 53%–70% and 57%–70%, respectively, while the Al_2O_3 content is reported to be in the range of 11%–20% and 10%–18%, respectively (Alhawat et al., 2022). These materials have a Si/Al ratio of approximately 5 when used alone, but typically result in the formation of NASH type gel structures through alkali activation. However, due to low Ca content, there is a lack of CSH/CAH type gel structures that can contribute to mechanical performance in the system. This can be achieved by adding concrete with a higher CaO content (20%–

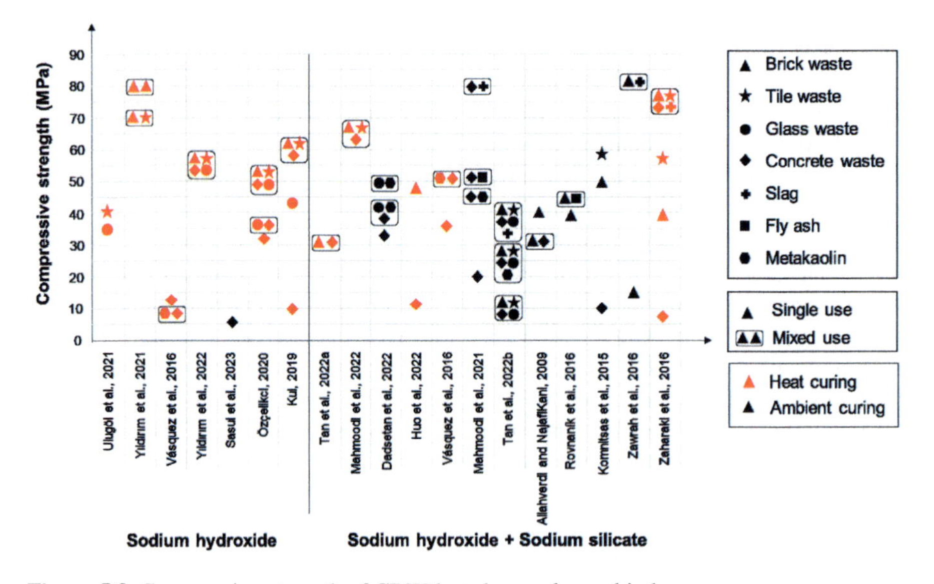

Figure 5.3 Compressive strength of CDW-based geopolymer binders.

31%) to the mixture, but this leads to a decrease in the Si/Al ratio, which is ultimately compensated for by adding glass with a high SiO_2 content (70%–75%) to the mixture to achieve ionic balance and higher compressive strengths. For instance, in a study where geopolymer was developed using brick, ceramic, and concrete waste, an increase in Na/Si along with Si/Al molar ratio was reported to result in shorter setting times (Mahmoodi et al., 2022). This can be attributed to the likely acceleration of the dissolution of aluminosilicates from CDW materials at increased Na_2O concentrations, leading to multiple condensation, compression, and hardening periods. The same study also indicated that the optimum Si/Al ratio may change to obtain maximum compressive strengths and optimum mechanical performance was obtained at well-balanced molar ratios of Si/Al = 10.2 and Na/Si = 0.18. Additionally, it has been observed that a higher Si/Al ratio enhances the compressive strength of CDW-based geopolymers. The high Si content strongly affects the formation of alkali aluminosilicate gels, while the presence of Al in the precursor determines the chemical structure and network formation. Since Si–O–Si bonds are stronger than Al–O–Al and Al–O–Si bonds, an increase in available silica in the alkali activation system can significantly alter the composition and structure of the produced gels, resulting in more polymerized and densely packed reaction products with better mechanical properties as depicted in Fig. 5.4 (Tan et al., 2022). It has been reported that the Ca/Si ratio is also an important factor in determining the mechanical performance, with a positive effect of increasing Ca/Si ratio at low Si/

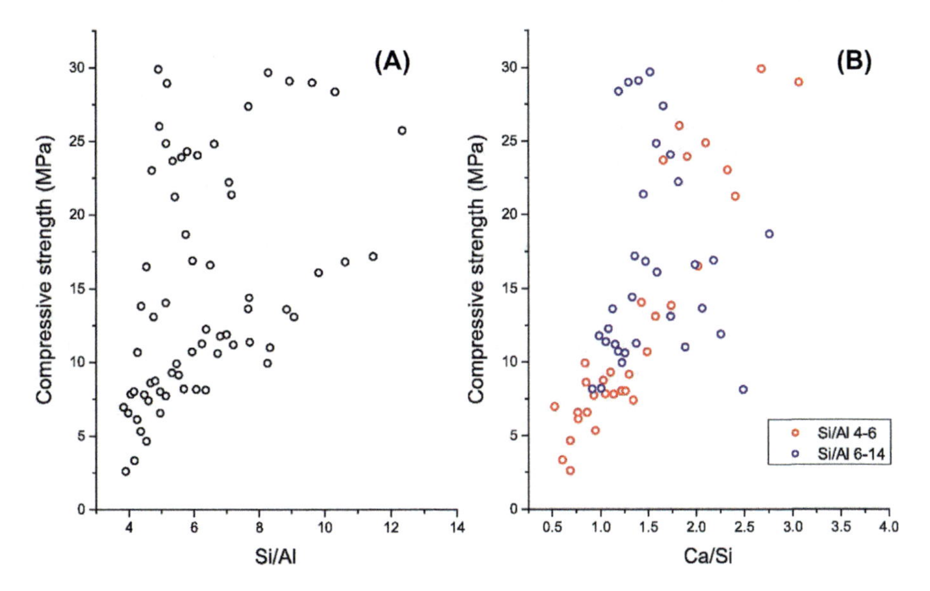

Figure 5.4 Compressive strength of CDW-based geopolymers with different (A) Si/Al and (B) Ca/Si ratios.
Source: From Tan, J., Cai, J., & Li, J. (2022). Recycling of unseparated construction and demolition waste (UCDW) through geopolymer technology. *Construction and Building Materials*, 341. https://doi.org/10.1016/j.conbuildmat.2022.127771.

Al ratios (i.e., 4–6), while at high Si/Al ratios (i.e., 6–14), the Ca/Si ratio has been found to have a threshold value of 1.0–1.5 in terms of contributing to compressive strength. Considering the related literature on the effects of the amount and ratios of Si, Al, Ca, and Na oxides, it should be stated that some thresholds can be determined according to the mechanical performances of geopolymers; however, variations in the source material, production technique of alkali activator, different calculation methodologies for the determination of ratios that considers precursor, and/or alkali activator individually or collectively can cause discrepancies in the decision of the optimum ingredient quantity.

Recently, numerous studies have been conducted on the development of CDW-based geopolymers with various combinations of materials with high pozzolanic activity such as fly ash, metakaolin, and slag to enhance the mechanical performance (Mahmoodi et al., 2021; Rovnaník et al., 2016; Zawrah et al., 2016). It was reported in such a study that a geopolymer based on waste-fired clay bricks reached 15 MPa after 90 days of ambient curing, while a 60% slag replacement led compressive strength to increase to the level of 83 MPa (Zawrah et al., 2016). The improvement was attributed to the formation of additional CSH gels in the matrix due to hydration reactions of calcium oxide from the slag, which enhanced the mechanical performance in addition to the aluminosilicate network in the geopolymer. In a different investigation, the compressive strength of geopolymer binders produced only with concrete waste reached 20 MPa within 28 days. However, the compressive strengths rose to about 48, 30, 43, and 82 MPa after the substitution of 45% of Class-C fly ash, Class-F fly ash, metakaolin, and slag, respectively (Mahmoodi et al., 2021). The authors attributed the excellent performance of slag to the presence of high CaO content, which contributed to the formation of CAH/CSH structures in the matrix. On the other hand, the other supplementary cementitious materials (SCMs) improved the mechanical performance by facilitating CAH/NASH gel formations at different degrees. According to Rovnaník et al. (2016), geopolymer binders created by activating brick powder with sodium silicate and sodium hydroxide attained a compressive strength of 41 MPa after 90 days. However, when 25% of the brick powder was replaced with fly ash, an increase of about 12.5% in compressive strength was observed. This enhancement is associated with the larger specific surface area, higher pozzolanic activity, and more densely packed microstructure of fly ash. Overall, the mechanical performance of CDW-based AAMs can be significantly improved by the addition of SCMs such as fly ash, slag, metakaolin, and others, which contribute to the formation of additional CAH/CSH type bonds and gel structures within the matrices, resulting in enhanced compressive strength and more compact microstructures.

5.4.2 Construction and demolition waste–based alkali-activated materials: mortars and concretes

The use of various alkali agents for the alkali activation of CDW-based materials has led to promising developments in the production of alkali-activated binders.

Furthermore, recent studies have focused on the development of structural mortars and concretes by incorporating recycled aggregates derived from CDW-based components. In order to evaluate the characteristics of these materials, various long-term durability tests have been conducted alongside with the mechanical tests such as compressive strength, flexural and tensile strengths, as well as tests for drying shrinkage, water absorption, efflorescence, sulfate resistance, and freeze–thaw resistance. This section provides a comprehensive overview of the tests and material–performance relationships that have been conducted in the literature for the development of CDW-based mortars and concretes.

In a recent study, the relationship between compressive strength, water absorption, drying shrinkage, and porosity of alkali-activated mortars containing CDW-based materials with 100% fine recycled aggregate and various alkali activator ratios and slag substitutions was investigated (Fig. 5.5) (Ozcelikci, Kul, et al., 2023). The study reported that geopolymer mortars developed entirely with CDW-based materials reached a compressive strength of about 30 MPa, while mixtures with 20% slag substitution reached a compressive strength of about 50 MPa. The findings showed that compressive strength was inversely proportional to water

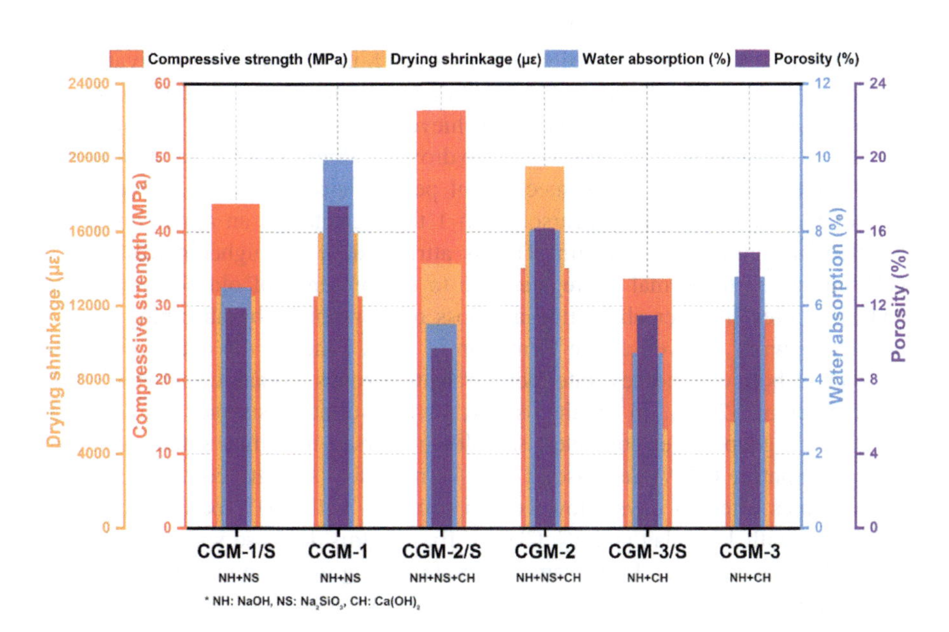

Figure 5.5 Mechanical and durability properties of CDW-based geopolymer mortars (CGM-1/2/3: 100% CDW based; CGM-1S/2S/3S: 20% slag substituted. Alkaline activator combinations of the mixtures are presented below the mixture names on the horizontal axis). *Source*: From Ozcelikci, E., Kul, A., Gunal, M. F., Ozel, B. F., Yildirim, G., Ashour, A., & Sahmaran, M. (2023). A comprehensive study on the compressive strength, durability-related parameters and microstructure of geopolymer mortars based on mixed construction and demolition waste. *Journal of Cleaner Production*, *396*. https://doi.org/10.1016/j.jclepro.2023.136522.

absorption, drying shrinkage, and porosity, and there was a linear relationship between drying shrinkage and porosity, with a significant increase in drying shrinkage observed with the use of sodium silicate. Additionally, slag substitution was reported to reduce the water absorption, porosity, and drying shrinkage while significantly improving compressive strength. Unlike OPC-based systems, AAMs have a high amount of free water in their systems since mixing water does not directly participate in the formation of the gel (Tchakouté et al., 2016), resulting in high drying shrinkage over a long period of time. This trend is also observed in CDW-based alkali-activated systems where the increased porosity allows both water and unreacted alkalis to continuously leach from the system (Firdous & Stephan, 2021). The so-called "efflorescence" occurs when the alkalis leach out of the system and react with CO_2 in the atmosphere, forming white crystalline structures such as calcite and natrite (Firdous & Stephan, 2021). While efflorescence is typically a visual problem in CDW-based AAMs, it is an indicator of both internal matrix defects and an imbalance in the alkali activator. This issue can be effectively addressed using some SCMs such as slag, which not only improves the pore structure refinement through its self-cementing properties at an early age but also has the ability to bind alkalis at a high level (Zhang et al., 2013).

In a study conducted by Yildirim et al. (2023), a total of nine geopolymer concrete mixtures were prepared using different combinations of alkali activators and additions of fly ash and slag. After a 28-day ambient curing period, the highest compressive strength of 40.1 MPa was achieved. The simultaneous use of sodium silicate, calcium hydroxide, and sodium hydroxide, along with the addition of slag, significantly contributed to the mechanical performance. However, increasing the ratio of recycled aggregate/precursor from 1 to 2 resulted in an approximate 29% decrease in compressive strength. This is attributed to the higher porosity of the RCA compared to normal aggregates due to the presence of old adhered mortar, leading to greater water absorption and loss of workability by absorbing some of the mixing water, as well as the formation of weak zones in the matrix due to the presence of both old and new interfacial transition zones (Akbarnezhad et al., 2011; Allujami et al., 2022). In another study, in the sulfate resistance test of 100% CDW-based geopolymer concretes, samples immersed in sulfate solutions showed an approximately 6.29% decrease in their compressive strength compared to reference samples after 28 days. However, in mixtures with CDW-based components substituted with 15% slag and 5% Class-F fly ash, this decrease was at the level of 9.13% (Özçelikci, 2020). In the same study, following the rapid chloride permeability test, 100% CDW-based geopolymer concrete specimens had an average chloride ion penetrability of 2670 C after 28 days of curing, whereas the mixtures with CDW-based components substituted with 15% slag and 5% Class-F fly ash, the same value decreased to 2250 C. Although CDW-based geopolymer concretes show promising durability performance, especially when combined with certain SCMs, further investigations are necessary for these mixtures to fully replace OPC concretes in structural applications.

Few studies have been recently conducted to analyze the structural behavior of CDW-based concretes (Akduman et al., 2021; Aldemir et al., 2022; Kocaer & Aldemir, 2022).

In one of these studies (Akduman et al., 2021), reinforced beams produced with CDW-based components were investigated to assess their structural performance. It was found that the load and displacement behaviors of geopolymer concrete beams were similar to those of OPC concrete beams. However, the use of recycled aggregates reduced the normalized energy dissipation capacities by approximately 30%. Moreover, CDW-based geopolymer concretes exhibited highly ductile performance in flexural-dominant behavior, and both types of specimens with normal or RCAs exhibited similar failure patterns. In another study that investigated the shear behavior of beams produced with CDW-based geopolymer concretes (Aldemir et al., 2022), it was reported that, regardless of the material used, all specimens with aspect ratios (a/d) of 0.50 and 1.00 failed in shear, as evidenced by brittle load-deflection and moment-curvature responses. Moreover, the inclusion of recycled aggregates had little influence on the failure mechanism for aspect ratios less than 1.65, which corresponded to shear-dominated regions (Fig. 5.6). This was attributed

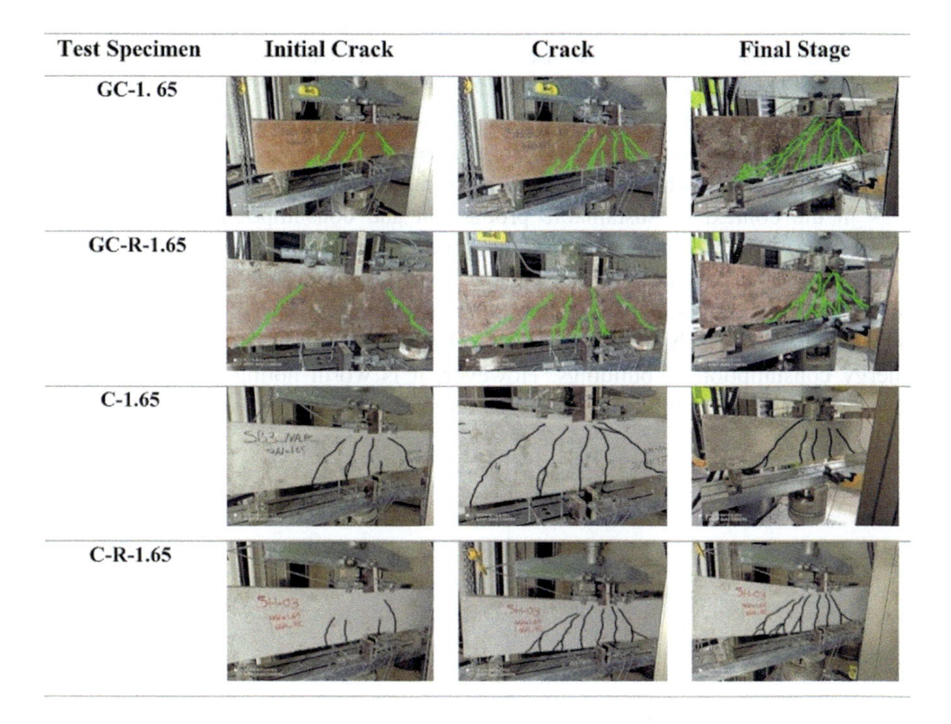

Figure 5.6 Pattern of crack propagation of CDW-based geopolymer and OPC concrete beams. *C*, Conventional concrete; *C-R*, conventional concrete with recycled concrete aggregates; *GC*, geopolymer concrete; *GC-R*, geopolymer concrete with recycled concrete aggregates.
Source: From Aldemir, A., Akduman, S., Kocaer, O., Aktepe, R., Sahmaran, M., Yildirim, G., Almahmood, H., & Ashour, A. (2022). Shear behaviour of reinforced construction and demolition waste-based geopolymer concrete beams. *Journal of Building Engineering*, *47*. https://doi.org/10.1016/j.jobe.2021.103861.

to the fact that the shear resisting mechanism in reinforced concrete specimens was not significantly dependent on the tensile strength of concrete.

In another study conducted by Kocaer and Aldemir, a novel stress–strain model has been developed to accurately assess the flexural capacity of geopolymer structural elements derived from CDW (Kocaer & Aldemir, 2022). The main objective of the study was to model the unique mechanical behavior of geopolymer concretes and to develop a mathematical formulation to expand the use of geopolymer structural elements, which are still being evaluated according to conventional concrete structural element codes. The study commenced by formulating a distinctive stress–strain model capable of capturing the compressive behavior of geopolymer concrete, drawing upon recent experimental findings regarding the flexural behavior of geopolymer concrete beam specimens. The ensuing evaluation yielded highly promising results, with an absolute mean percentage error of 5.13% irrespective of the mode of failures. Besides, the ratios of predicted and experimental ultimate moment capacity revealed the better capability of the developed model to predict the mechanical behavior of the geopolymer concrete compared to the ACI Committee (2011), which is used for conventional concrete-based systems (Fig. 5.7). This outcome underscores the efficacy and reliability of the proposed approach, substantiating its potential as a robust tool for accurately predicting the flexural capacity of geopolymer structural elements.

In today's world, where the obligation to increase the energy efficiency of buildings in the context of sustainable production and consumption has gained great importance, the importance of developing building materials with improved thermal insulation properties is increasing (EPBD, 2020). Integrating geopolymer technology into the production of building materials with improved thermal insulation properties can be a holistic solution for both recycling waste and reducing the energy consumption of buildings. However, it is worth noting that the corpus of research dedicated to this particular area remains notably scarce. The only example

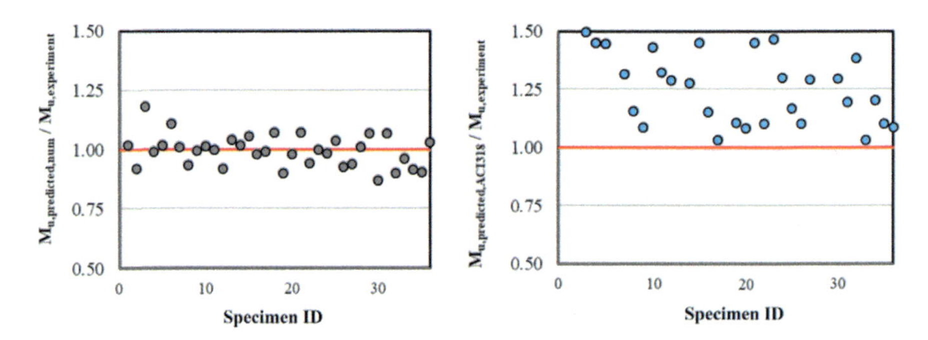

Figure 5.7 Comparative analysis of estimation performances for the test soft database (left: developed model, right: ACI318 code).
Source: From Kocaer, O., & Aldemir, A. (2022). Compressive stress–strain model for the estimation of the flexural capacity of reinforced geopolymer concrete members. *Structural Concrete*, *24*, 5102–5121.

found in the literature is a study conducted by Kvočka et al. (2020), that examined the thermal conductivity performance of high-density geopolymer mortars containing 50% CDW aggregates, by mass originated from fired clay, mortar, and concrete rubble, as well as the environmental impacts of facade cladding panels produced with these mixtures. Thermal performance of a geopolymer containing 40% CDW-based wood particles was also investigated as an alternative. The binder phase in these mixtures consisted of a combination of metakaolin, ground granulated blast furnace slag, and Class-F fly ash, while the alkali activator phase was potassium silicate for the high-density geopolymer and sodium silicate for the geopolymer containing wood waste particles. The findings from the study revealed that the high-density geopolymer mixture exhibited a 28-day thermal resistance of 2.037 (m^2 K/W), an average compressive strength of 37 MPa, and a tensile strength of 3 MPa. It was also noted that the apparent density was 1.89 g/cm^3, and the open porosity was approximately 30%. For the mixture containing waste wood particles, the apparent density was determined to be around 1.0 g/cm^3 under dry conditions and an average of 1.15 g/cm^3 under environmental conditions (indoor environment). The flexural strength and elastic modulus measured under three-point bending were approximately 5.6 and 2.02×10^3 MPa, respectively. Furthermore, based on a "cradle-to-gate" life cycle assessment comparison, it was concluded that prefabricated geopolymeric facade cladding panels could be considered an environmentally friendly construction product. Considering an alternative heat production scenario, the environmental impact of panels made from geopolymer mortars was found to be up to 100% lower in individual impact categories compared to panels made from natural materials (especially marble, glass, and aluminum) (Fig. 5.8). These compelling findings conclusively affirm that geopolymer-based materials not only embody inherent environmental friendliness but also outshine their technically competitive counterparts, particularly in the case of facade panels. Moreover, the design attributes of these panels, which facilitate effortless disassembly and recyclability, hold immense promise for the development of novel products incorporating a substantial proportion of recycled geopolymers, thereby further attenuating the environmental burden. However, there is a need for further research and investigation into the production of CDW-based geopolymer materials that encompass all these advantageous properties.

5.4.3 Multifunctional building material concepts for construction and demolition waste–based alkali-activated materials

5.4.3.1 Three-dimensional concrete printing

Three-dimensional (3D) printing is a new manufacturing technology that involves computer-controlled systems to produce solid objects layer-by-layer based on digital models (Wangler et al., 2019). Compared to traditional processes, 3D printing offers increased accuracy and convenience, improves production efficiency, and reduces labor consumption (Buswell et al., 2018). Furthermore, 3D printing can contribute to construction digitalization, which has gained relevance due to its

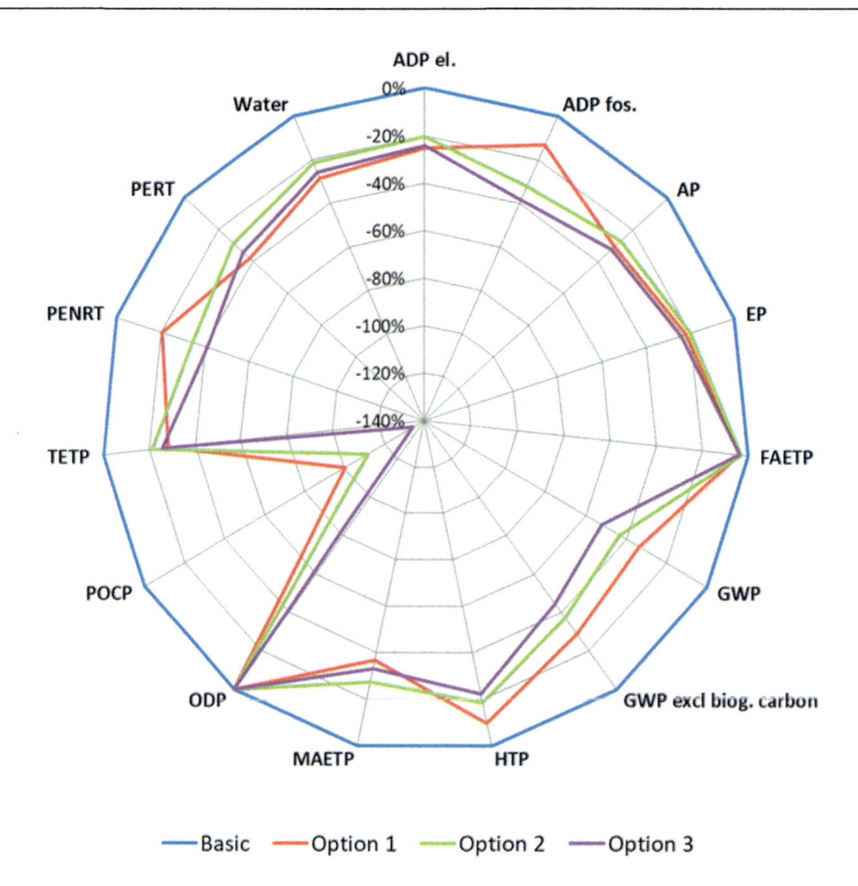

Figure 5.8 Reduced environmental impact for different heat generation scenarios compared to the basic panel production process.
Source: From Kvočka, D., Lešek, A., Knez, F., Ducman, V., Panizza, M., Tsoutis, C., & Bernardi, A. (2020). Life cycle assessment of prefabricated geopolymeric façade cladding panels made from large fractions of recycled construction and demolition waste. *Materials, 13* (18). https://doi.org/10.3390/ma13183931.

potential to deal with health and war crises. It is projected that 3D printing construction methods can significantly decrease waste and time in construction while creating high-end technology-based jobs and addressing labor workforce limitations in the industry.

The most prevalent 3D printing technology in the construction industry is 3D cementitious materials printing, usually involving mortars. 3D concrete printing (3DCP) enables the production of cement-based building elements or particular architectural elements designed on a computer and extruded from a 3D printer's nozzle to create a physical object. The use of 3DCP eliminates the need for formwork and temporary structures, which contribute to a considerable amount of waste generation, materials consumption, time, and labor costs (Şahin & Mardani-Aghabaglou, 2022).

Moreover, 3DCP enables greater architectural freedom, providing enormous potential to revolutionize the construction industry.

The adaptation of this innovative technology to the construction industry has recently received considerable attention, and many studies have been carried out to develop 3D printable geopolymer mixtures (Chen et al., 2022; Chougan et al., 2020; Ishwarya et al., 2019; Panda et al., 2019; Zhang et al., 2018). However, research on the compatibility of CDW-based geopolymer mixtures with 3DCP technique has remained limited (Demiral et al., 2022; Ilcan et al., 2023, 2022; Pasupathy et al., 2023; Şahin et al., 2021). In the study conducted by Şahin et al. (2021), rheological properties and extrudability performances of CDW-based geopolymers produced by the activation of a combination of CDW-based brick and glass with alkali activators, including sodium hydroxide (NaOH), sodium silicate (Na_2SiO_3) and calcium hydroxide ($Ca(OH)_2$) were investigated. In both individual and combined forms, NaOH was used from 5 to 30 M, $Ca(OH)_2$ from 2% to 10%, and Na_2SiO_3 in ratios of 0.5 and 1 with respect to NaOH. According to the experimental results, using NaOH as the sole alkaline activator had varying impacts on the rheological properties of CDW-based geopolymers depending on the molarity. The flowability increased and buildability and vane shear stress decreased as the molarity of NaOH increased from 6.25 to 11.25 M. However, flowability decreased after 11.25 M due to sticky gel formation. $Ca(OH)_2$ increased viscosity, buildability, and vane shear stress, while Na_2SiO_3 decreased viscosity and increased flowability, but shortened setting/open time. Single use of NaOH did not provide adequate compressive strength development, while $Ca(OH)_2$ increased compressive strength, and Na_2SiO_3 increased it in ternary combinations. Mixtures with 6.25 M NaOH and 6%–10% $Ca(OH)_2$ were suitable for 3D printing. Vane shear tests were not sensitive enough to show the effect of $Ca(OH)_2$, while the ram extruder was more effective (Fig. 5.9).

In the study conducted by Ilcan et al. (2022), the rheological properties of entirely CDW-based geopolymer mortar mixtures were investigated with empirical tests, such as flow table, buildability, and vane shear, to determine the suitability of the mixtures for 3D-Additive Manufacturing (3D-AM) using a laboratory-scale 3D printer. It was observed that NaOH had a significant impact on the rheological properties, with a distinct trend inversion at a specific molarity level. The inclusion of $Ca(OH)_2$ increased viscosity, reduced flowability index, and improved buildability and vane shear stress. The addition of Na_2SiO_3 resulted in a decrease in the initial yield stress, increased flowability, and accelerated hardening. Furthermore, the study found that incorporating RCAs did not affect the rheological properties and compressive strength of the mixtures, making them suitable for 3D-AM. The performance of the mixtures was evaluated based on various parameters including printability, flowability, viscosity, buildability, extrudability, and open time, with the mortar mixture exhibiting low viscosity showing precise dimensional conformity in the printed end product (Fig. 5.10).

Ilcan et al. (2023) also investigated the rheological properties of 3D printable geopolymer mortar mixtures adopted from the abovementioned study by using different test protocols, namely flow curve, constant shear rate, varied shear

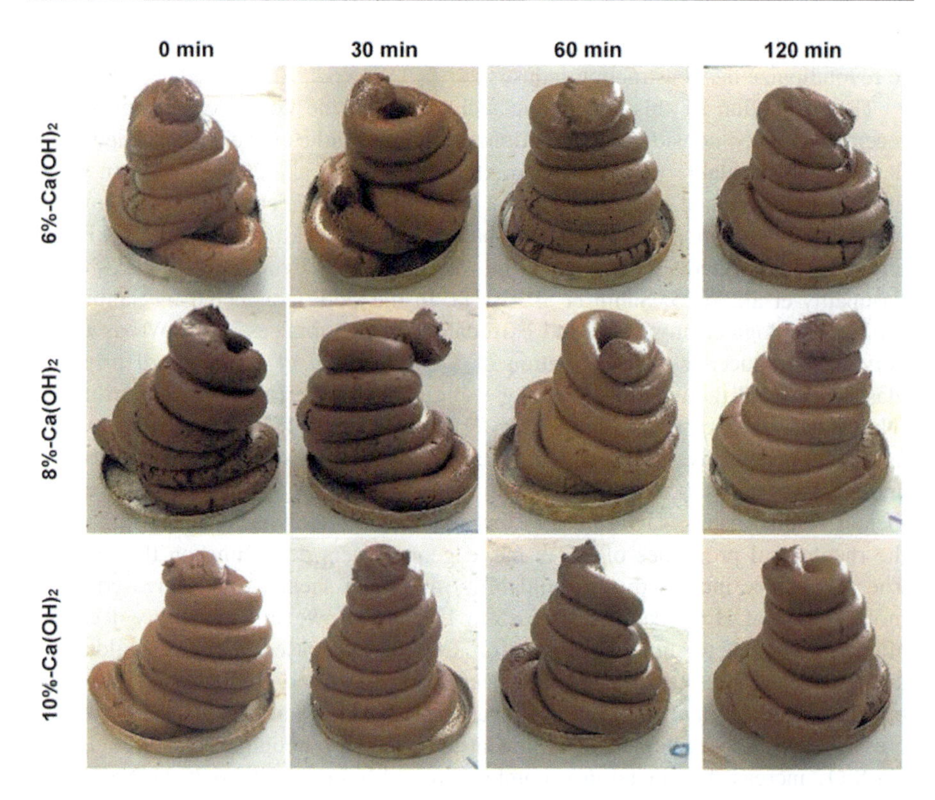

Figure 5.9 Ram extruded samples of different CDW-based geopolymer mixtures. *Source*: From Şahin, O., İlcan, H., Ateşli, A. T., Kul, A., Yıldırım, G., & Şahmaran, M. (2021). Construction and demolition waste-based geopolymers suited for use in 3-dimensional additive manufacturing. *Cement and Concrete Composites, 121*. https://doi.org/10.1016/j.cemconcomp.2021.104088.

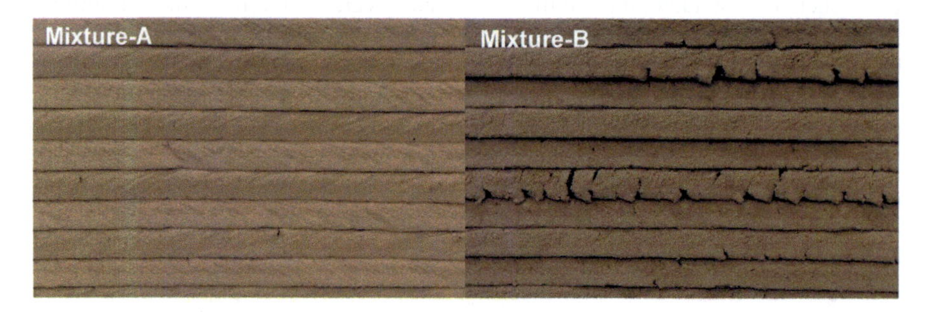

Figure 5.10 Visuals of multilayer specimens of 3D printed geopolymer mixtures. Mixture A is characterized by a low viscosity, whereas Mixture B exhibits a high viscosity profile. *Source*: From Ilcan, H., Sahin, O., Kul, A., Yildirim, G., & Sahmaran, M. (2022). Rheological properties and compressive strength of construction and demolition waste-based geopolymer mortars for 3D-printing. *Construction and Building Materials, 328*. https://doi.org/10.1016/j.conbuildmat.2022.127114.

rate, and three interval thixotropy testing, and evaluated the effects of different alkali activator types and concentrations on the rheological properties. It was observed that an increase in NaOH initially reduced the static and dynamic flow stresses and plastic viscosity, but beyond a certain level, these parameters increased. On the other hand, increasing $Ca(OH)_2$ content resulted in a continuous increment trend in these rheological parameters. The constant shear rate test showed that the static flow stress was higher than the dynamic flow stress, and both values decreased with increasing shear rate. The molarity of NaOH influenced these stress values, reversing the trend at a certain point, while the use rate of $Ca(OH)_2$ caused a continuous increase. Thixotropy performance was influenced by the molarity of NaOH in mixtures activated with NaOH. The three interval thixotropy testing revealed that an increase in $Ca(OH)_2$ up to a certain level had a positive effect on thixotropy. The ram extrusion test, which evaluated the extrusion capability, flow properties, and extruded filament quality, aligned with other rheological tests in terms of stress values (Fig. 5.11). Yield stress below 0.55 MPa resulted in defect-free and continuous flow, while values above 0.70 MPa led to discontinuities and defects. It was also observed that various failures occurred even with prolonged resting times during printing.

Demiral et al. investigated the anisotropy (direction-dependency) in mechanical performance and bonding properties of geopolymer mortars made entirely of CDW and manufactured by 3D-AM (Demiral et al., 2022). The results showed that the mechanical properties were significantly affected by the alkaline activator content. Elevating the NaOH molarity resulted in improved compressive and flexural strengths, yet it was associated by a decrease in splitting and direct tensile strengths of 3D printed specimens. While the incorporation of 4% $Ca(OH)_2$ positively influenced the mechanical properties, excessive $Ca(OH)_2$ adversely affected viscosity hindering its beneficial impact. Direction-dependent tests revealed anisotropic

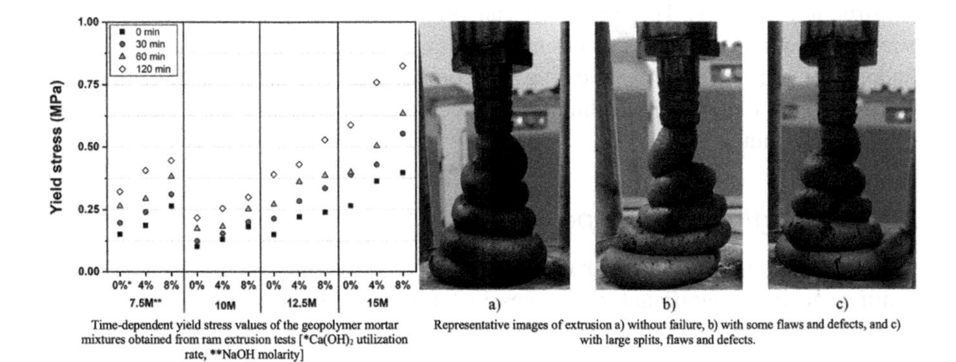

Figure 5.11 Quantitative and illustrative results of ram extruder experiments.
Source: From Ilcan, H., Sahin, O., Kul, A., Ozcelikci, E., & Sahmaran, M. (2023). Rheological property and extrudability performance assessment of construction and demolition waste-based geopolymer mortars with varied testing protocols. *Cement and Concrete Composites*, *136*. https://doi.org/10.1016/j.cemconcomp.2022.104891.

behavior; however, the perpendicular-loaded 3D printed specimens showed similar or slightly better performance than the mold-casted specimens indicating that the bond zone had little influence on the perpendicular loading direction. Anisotropic properties were observed in compressive and flexural strength tests, with variations of up to 20% and 30%, respectively. Related outputs were associated with the degree of geopolymerization, which is directly responsible of the quality of interfacial transition zone and porosity in the layer bond zone (Fig. 5.12). This study suggests that enhancing the bond adhesion between consecutive layers can reduce the anisotropic behavior of printed structures. The rheological properties and matrix performance of the mixtures can be optimized to improve the adhesion.

Pasupathy et al., who investigated the influence of brick waste content on the fresh and hardened properties of 3D printed slag-fly ash based geopolymer mortars, stated that the incorporation of brick waste led to enhancements in the fresh properties, including increased yield strength and viscosity (Pasupathy et al., 2023). However, higher concentrations of brick waste (above 10%) resulted in reduced hardened properties, specifically in terms of compressive strength and interlayer strength of 3D printed concrete. Besides, anisotropic behavior was observed in the compressive strength of 3D printed samples, with the highest strength reported in the printing direction and the lowest strength in the lateral direction.

Despite the progress made in developing 3D printable geopolymer mixtures, there remains a limited understanding of the compatibility of CDW-based geopolymer mixtures with the 3D-AM technique. Further research is necessary to investigate the rheological behavior of CDW-based geopolymer mixtures and their compatibility with the 3D-AM technique. Such research can contribute to the development of sustainable construction materials and innovative construction methods, while reducing the carbon footprint of the construction industry. The successful development and implementation of CDW-based geopolymer mixtures for 3D-AM has the potential to deliver significant environmental and economic benefits for the construction industry. This approach can offer a promising solution to waste disposal and natural resource depletion challenges. Therefore there is a need for continuous research efforts for revealing the compatibility of CDW-based geopolymer mixtures with the 3D-AM technique to facilitate their sustainable adoption in the construction industry.

5.4.3.2 Engineered geopolymer composites

Concrete structures, primarily due to their inherent susceptibility to tensile cracking, continue to pose significant challenges in terms of resilience, durability, and sustainability (Li, 1993). In response, the field of construction materials has witnessed the emergence of a revolutionary solution, called engineered cementitious composites (ECC). Known as strain-hardening cementitious composites, ECC represents a paradigm shift in the field of fiber-reinforced concrete possessing exceptional ductility and a remarkable tensile elongation capacity in excess of 2% (Li, 2019). Unlike its ultra-high-performance fiber-reinforced concrete counterparts, which rely primarily on optimizing particle packing density, ECC embodies a deliberate

Figure 5.12 Representative general and close-up views of (A) nonporous and (B) porous bond zone.
Source: From Demiral, N. C., Ozkan Ekinci, M., Sahin, O., Ilcan, H., Kul, A., Yildirim, G., & Sahmaran, M. (2022). Mechanical anisotropy evaluation and bonding properties of 3D-printable construction and demolition waste-based geopolymer mortars. *Cement and Concrete Composites*, *134*. https://doi.org/10.1016/j.cemconcomp.2022.104814.

engineering approach that manages the intricate interaction between fibers, matrices and fiber/matrix interfaces to enable multiple cracking mechanisms under tensile loading (Li, 2019).

The outstanding properties of ECC have enabled it to be used in a wide range of large-scale structures, from awe-inspiring bridges to towering skyscrapers effectively extending their service life and providing functional enhancements (Li et al., 2020). Compositionally, ECC has a distinctive blend of constituents, including OPC, SCMs such as fly ash, silica sand, and short fibers (Afroughsabet et al., 2016). However, despite the clear benefits of ECC, there is a notable drawback in the form of its high cement content, necessitated by the negligence of coarse aggregates. This increases the cement content which has cost implications and a significant carbon footprint necessitating a concerted search for sustainable alternatives.

In response to the aforementioned demand, the emergence of EGC represents a compelling prospect with transformative implications. EGC represents a remarkable fusion of the beneficial properties of ECC toward reducing the inherent brittleness usually associated with geopolymers and hold potential for reducing the environmental impact associated with conventional building materials (Provis, 2018). In addition, EGC not only outperforms ECC in terms of sustainability but also exhibit comparable tensile and flexural properties under static loading conditions, while outperforming ECC in terms of mechanical performance under dynamic loading scenarios (Cai et al., 2021; Nematollahi et al., 2017). These alternative materials have a formulation based on the fundamental principles of micromechanical design theory, which allows for a wide range of customization options, including different matrices, fiber reinforcements, and carefully tailored mixing and curing methods. Similar to other multifunctional advanced construction materials, CDWs possess remarkable potential to be used in the development of matrices for EGC. However, the dearth of dedicated studies addressing this specific topic is evident, underscoring a noteworthy research gap that necessitates further exploration and investigation. The study conducted by Ulugöl et al. represents the sole investigation to date that focuses on exploring the autogenous self-healing capability of EGCs with entirely CDW-based matrices that employed masonry units, concrete, and glass from CDWs in the mixed form as precursors, and recycled concrete as fine aggregates (Ulugöl et al., 2021). To assess the impact of incorporating ground granulated blast furnace slag into CDW-based mixtures, additional mixtures were prepared with the slag substitution while maintaining a constant Si/Al ratio. The degree of self-healing was examined by monitoring the changes in the transport properties (i.e., electrical impedance and water absorption rate) as well as microstructural characteristics. According to the findings, the self-healing capability of CDW-based EGCs was related to the chemical composition of the materials used in the mixtures, particularly the presence of calcium and sodium derived from the raw materials and alkaline activators. The addition of slag to CDW-based EGCs accelerated the early-stage reactions of geopolymeric mixtures, resulting in a slower but more stable self-healing process. Self-healing products such as $CaCO_3$ and Na_2CO_3 were detected in the healed microcracks, indicating ongoing carbonation and geopolymerization. Moreover, the self-healing behavior of EGCs exhibited similarities to

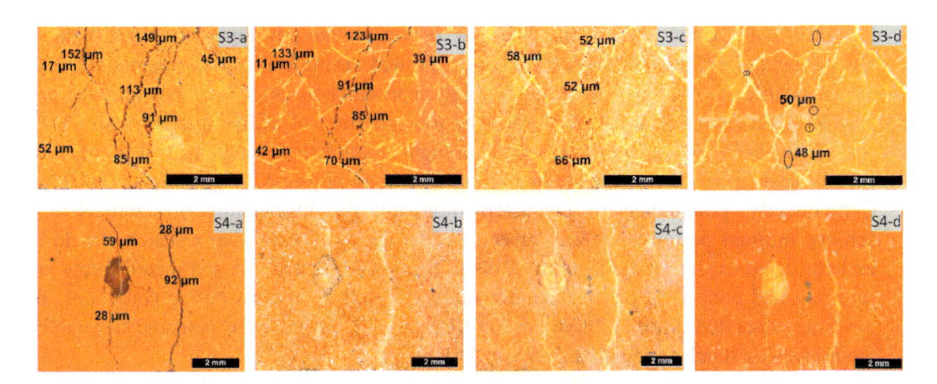

Figure 5.13 Optical microscopy images of different EGC specimens on the (A) 7th, (B) 28th, (C) 56th, and (D) 90th days showing the effect of self-healing. *S3*, CDW + SCM-based mixture; *S4*, fully CDW-based mixture.
Source: From Ulugöl, H., Günal, M. F., Yaman, İ. Ö., Yıldırım, G., & Şahmaran, M. (2021). Effects of self-healing on the microstructure, transport, and electrical properties of 100% construction- and demolition-waste-based geopolymer composites. *Cement and Concrete Composites*, *121*. https://doi.org/10.1016/j.cemconcomp.2021.104081.

OPC-based composites. Another important outcome of this study was the variation in the results of the self-healing characterization techniques, exhibiting incoherence when using water absorption and electrical impedance tests, primarily due to multiple factors affecting these tests, particularly electrical impedance. In contrast, microscopic investigations and microstructural analyses provided clearer evidence of self-healing (Fig. 5.13). Given the cumulative outcomes of the aforementioned investigations, it is strongly advised to augment the existing analyses with supplementary test methodologies to undertake a comprehensive and exhaustive assessment of the self-healing phenomenon in EGCs. By employing a diverse range of test methods, a more holistic understanding can be achieved regarding the effectiveness, mechanisms, and intricacies underlying the self-healing capabilities exhibited by these innovative construction materials.

5.5 Concluding remarks and recommendations for the future

This chapter offers a comprehensive examination of the most recent research works in the field of alkali activation technology applied specifically to CDW-based materials encompassing a diverse array of materials such as binders, mortars, concretes, 3D printable composites, and engineered geopolymer composites. The primary objective is to enable the seamless integration of CDW into the circular economy, specifically within the construction industry.

- It can be inferred that CDW offers a viable option for the development of AAMs as an alternative to OPC pastes. The utilization of thermal curing can enable the attainment of commendable compressive strengths reaching approximately 80 MPa, while ambient curing can yield strengths around 50 MPa. The incorporation of SCMs together with CDWs contributes to the further enhancement of strength characteristics. However, it is crucial to acknowledge that CDWs exhibit diverse chemical and physical compositions worldwide. Consequently, comprehensive characterization and the formulation of approaches to achieve optimal performance are imperative. To this end, extensive research should be undertaken to explore the effects of CDW utilization in both single and mixed forms, the selection and concentration of alkali activators, as well as optimal curing conditions. By devoting concerted efforts to these areas of inquiry, CDWs have the potential to establish themselves as standardized alkali-activated binder on a global scale.
- The utilization of CDW-based alkali-activated binders in mortar and concrete mixtures with the incorporation of recycled aggregates from concrete waste is still at an early stage in the literature. To fully replace conventional OPC-based concrete, extensive long-term durability analyses should be performed, focusing on crucial aspects such as drying shrinkage, water absorption, permeability, freeze–thaw resistance, and resistance to chemical agents in CDW-based mortars and concretes. Furthermore, it is imperative to perform comprehensive characterizations of structural elements like columns, beams, and walls manufactured using CDW-based materials, encompassing evaluations of shear, flexural behavior, creep, crack patterns, and seismic performance. These efforts should be complemented by diverse demonstration activities that attract attention from the construction industry, stakeholders, academia, and governments, thereby facilitating the integration of CDW into the construction sector. Additionally, further research and development endeavors are warranted to overcome remaining challenges, optimize the performance of CDW-based materials, and foster their widespread adoption, thereby contributing to sustainable construction practices.
- Further research is essential to explore the compatibility of CDW-based geopolymer mixtures with 3D-AM techniques and investigate their rheological behavior, mechanical properties, and bonding performances for suitability in construction applications. Successful development and implementation of CDW-based geopolymer mixtures for 3D-AM can deliver significant environmental and economic benefits, addressing waste disposal and resource depletion challenges and promoting circular economy principles in construction. Ongoing research on CDW-based geopolymer mixtures' compatibility with 3D-AM is crucial for their sustainable integration into the construction sector, fostering a smarter, more sustainable, and inclusive society.
- Although CDW-based EGCs have gained recent attention, the existing literature in this area is limited, leaving many aspects open to exploration. A comprehensive investigation of the mechanical, durability, and self-healing properties of CDW-based EGCs will enable the attainment of comparable performance to traditional ECCs and EGCs, but with reduced environmental impacts. This research will facilitate the widespread adoption of CDW-based EGCs and their potential to achieve lower environmental impacts while maintaining comparable performance to ECCs and traditional EGCs based on mainstream SCMs.
- The utilization of geopolymers based on CDW exhibits significant potential for enhancing the energy efficiency of buildings. This innovative strategy, characterized by its low embodied energy and capacity for material recycling and reduced resource utilization, has garnered acknowledgment within the area of sustainable construction practices. However, despite its promising potential, the full realization of this technology has yet to be achieved due to a significant research gap and a lack of extensive investigations in this field.

Acknowledgments

This project has received funding from the European Union's Horizon research and innovation program under the Marie Skłodowska-Curie grant agreement no. 894100.

References

ACI Committee (2011). *ACI 318 Building code requirements, structural concrete and commentary*, 318.

Afroughsabet, V., Biolzi, L., & Ozbakkaloglu, T. (2016). High-performance fiber-reinforced concrete: A review. *Journal of Materials Science*, *51*(14), 6517–6551. Available from https://doi.org/10.1007/s10853-016-9917-4.

Akbarnezhad, A., Ong, K. C. G., Zhang, M. H., Tam, C. T., & Foo, T. W. J. (2011). Microwave-assisted beneficiation of recycled concrete aggregates. *Construction and Building Materials*, *25*(8), 3469–3479. Available from https://doi.org/10.1016/j.conbuildmat.2011.03.038.

Akduman, Ş., Kocaer, O., Aldemir, A., Şahmaran, M., Yıldırım, G., Almahmood, H., & Ashour, A. (2021). Experimental investigations on the structural behaviour of reinforced geopolymer beams produced from recycled construction materials. *Journal of Building Engineering*, *41*. Available from https://doi.org/10.1016/j.jobe.2021.102776.

Aldemir, A., Akduman, S., Kocaer, O., Aktepe, R., Sahmaran, M., Yildirim, G., Almahmood, H., & Ashour, A. (2022). Shear behaviour of reinforced construction and demolition waste-based geopolymer concrete beams. *Journal of Building Engineering*, *47*. Available from https://doi.org/10.1016/j.jobe.2021.103861.

Alhawat, M., Ashour, A., Yildirim, G., Aldemir, A., & Sahmaran, M. (2022). Properties of geopolymers sourced from construction and demolition waste: A review. *Journal of Building Engineering*, *50*. Available from https://doi.org/10.1016/j.jobe.2022.104104.

Allujami, H. M., Abdulkareem, M., Jassam, T. M., Al-Mansob, R. A., Ng, J. L., & Ibrahim, A. (2022). Nanomaterials in recycled aggregates concrete applications: Mechanical properties and durability. A review. *Cogent Engineering*, *9*(1), 23311916. Available from https://doi.org/10.1080/23311916.2022.2122885.

Amran, M., Makul, N., Fediuk, R., Lee, Y. H., Vatin, N. I., Lee, Y. Y., & Mohammed, K. (2022). *Global carbon recoverability experiences from the cement industry*, . *Case Studies in Construction Materials* (17). Elsevier Ltd. Available from 10.1016/j.cscm.2022.e01439.

ASTM C618 (2019). *ASTM C618-19: Standard specification for coal fly ash and raw or calcined natural pozzolan for use in concrete*. ASTM International, Unpublished content.

Bernal, S., Gutiérrez, R., & Rodríguez, E. (2013). Alkali-activated materials: Cementing a sustainable future. *Ingeniería y Competitividad*, *15*, 211–223. Available from https://doi.org/10.25100/IYC.V15I2.2608.G3434.

Bernal, S. A., Mejía De Gutiérrez, R., & Provis, J. L. (2012). Engineering and durability properties of concretes based on alkali-activated granulated blast furnace slag/metakaolin blends. *Construction and Building Materials*, *33*, 99–108. Available from https://doi.org/10.1016/j.conbuildmat.2012.01.017.

Buswell, R. A., Leal de Silva, W. R., Jones, S. Z., & Dirrenberger, J. (2018). 3D printing using concrete extrusion: A roadmap for research. *Cement and Concrete Research*, *112*, 37–49. Available from https://doi.org/10.1016/j.cemconres.2018.05.006.

Cai, J., Pan, J., Han, J., Lin, Y., & Sheng, Z. (2021). Impact behaviours of engineered geopolymer composite exposed to elevated temperatures. *Construction and Building Materials, 312*. Available from https://doi.org/10.1016/j.conbuildmat.2021.125421.

Channell, M., Graves, M.R., Medina, V.F., Morrow, A.B., Brandon, D., & Nestler, C.C. (2009). *Enhanced tools and techniques to support debris management in disaster response missions*. Report number: ERDC/EL TR-09-12.

Chen, Y., Liu, C., Cao, R., Chen, C., Mechtcherine, V., & Zhang, Y. (2022). Systematical investigation of rheological performance regarding 3D printing process for alkali-activated materials: Effect of precursor nature. *Cement and Concrete Composites, 128*. Available from https://doi.org/10.1016/j.cemconcomp.2022.104450.

Choudhary, J., Kumar, B., & Gupta, A. (2020). Utilization of solid waste materials as alternative fillers in asphalt mixes: A review. *Construction and Building Materials, 234*. Available from https://doi.org/10.1016/j.conbuildmat.2019.117271.

Chougan, M., Hamidreza Ghaffar, S., Jahanzat, M., Albar, A., Mujaddedi, N., & Swash, R. (2020). The influence of nano-additives in strengthening mechanical performance of 3D printed multi-binder geopolymer composites. *Construction and Building Materials, 250*. Available from https://doi.org/10.1016/j.conbuildmat.2020.118928.

Cinquepalmi, M. A., Mangialardi, T., Panei, L., Paolini, A. E., & Piga, L. (2008). Reuse of cement-solidified municipal incinerator fly ash in cement mortars: Physico-mechanical and leaching characteristics. *Journal of Hazardous Materials, 151*(2–3), 585–593. Available from https://doi.org/10.1016/j.jhazmat.2007.06.026.

Demiral, N. C., Ozkan Ekinci, M., Sahin, O., Ilcan, H., Kul, A., Yildirim, G., & Sahmaran, M. (2022). Mechanical anisotropy evaluation and bonding properties of 3D-printable construction and demolition waste-based geopolymer mortars. *Cement and Concrete Composites, 134*. Available from https://doi.org/10.1016/j.cemconcomp.2022.104814.

Duxson, P., Fernández-Jiménez, A., Provis, J. L., Lukey, G. C., Palomo, A., & van Deventer, J. S. J. (2007). Geopolymer technology: The current state of the art. *Journal of Materials Science, 42*(9), 2917–2933. Available from https://doi.org/10.1007/s10853-006-0637-z.

EPA. (2008). *Planning for natural disaster debris*. Office of Solid Waste and Emergency Response, Office of Solid Waste.

EPBD. (2020). *Energy performance of buildings directive*, EPBD.

Etxeberria, M., Vázquez, E., Marí, A., & Barra, M. (2007). Influence of amount of recycled coarse aggregates and production process on properties of recycled aggregate concrete. *Cement and Concrete Research, 37*(5), 735–742. Available from https://doi.org/10.1016/j.cemconres.2007.02.002.

European Commission. (2020). *Circular economy action plan*. https://environment.ec.europa.eu/strategy/circular-economy-action-plan_en.

Eurostat. (2015). *Waste statistics*.

Eurostat. (2016). *Eurostat statistics for waste flow generation*.

Firdous, R., & Stephan, D. (2021). Impact of the mineralogical composition of natural pozzolan on properties of resultant geopolymers. *Journal of Sustainable Cement-Based Materials, 10*(3), 149–164. Available from https://doi.org/10.1080/21650373.2020.1809028.

Friedlingstein, P., Jones, M. W., Andrew, R. M., Gregor, L., Hauck, J., & Zheng, B. (2022). Global carbon budget 2022. *Earth System Science Data, 14*(11), 4811–4900.

Garside, M. (2020). *Major countries in worldwide cement production 2015–2019*. Statista International.

Habert, G., d'Espinose de Lacaillerie, J. B., & Roussel, N. (2011). An environmental evaluation of geopolymer based concrete production: Reviewing current research trends.

Journal of Cleaner Production, *19*(11), 1229−1238. Available from https://doi.org/10.1016/j.jclepro.2011.03.012.

Halicka, A., Ogrodnik, P., & Zegardlo, B. (2013). Using ceramic sanitary ware waste as concrete aggregate. *Construction and Building Materials*, *48*, 295−305. Available from https://doi.org/10.1016/j.conbuildmat.2013.06.063.

Hossain, M. M., Karim, M. R., Hasan, M., Hossain, M. K., & Zain, M. F. M. (2016). Durability of mortar and concrete made up of pozzolans as a partial replacement of cement: A review. *Construction and Building Materials*, *116*, 128−140. Available from https://doi.org/10.1016/j.conbuildmat.2016.04.147.

Hossain, M. U., Poon, C. S., Lo, I. M. C., & Cheng, J. C. P. (2017). Comparative LCA on using waste materials in the cement industry: A Hong Kong case study. *Resources, Conservation and Recycling*, *120*, 199−208. Available from https://doi.org/10.1016/j.resconrec.2016.12.012.

Huang, B., Wang, X., Kua, H., Geng, Y., Bleischwitz, R., & Ren, J. (2018). Construction and demolition waste management in China through the 3R principle. *Resources, Conservation and Recycling*, *129*, 36−44. Available from https://doi.org/10.1016/j.resconrec.2017.09.029.

Ilcan, H., Sahin, O., Kul, A., Ozcelikci, E., & Sahmaran, M. (2023). Rheological property and extrudability performance assessment of construction and demolition waste-based geopolymer mortars with varied testing protocols. *Cement and Concrete Composites*, *136*. Available from https://doi.org/10.1016/j.cemconcomp.2022.104891.

Ilcan, H., Sahin, O., Kul, A., Yildirim, G., & Sahmaran, M. (2022). Rheological properties and compressive strength of construction and demolition waste-based geopolymer mortars for 3D-Printing. *Construction and Building Materials*, *328*. Available from https://doi.org/10.1016/j.conbuildmat.2022.127114.

International Energy Agency, (IEA). (2022). Tracking Buildings 2022. Paris: International Energy Agency. Available at: https://www.iea.org/reports/tracking-buildings-2021.

Ishwarya, G., Singh, B., Deshwal, S., & Bhattacharyya, S. K. (2019). Effect of sodium carbonate/sodium silicate activator on the rheology, geopolymerization and strength of fly ash/slag geopolymer pastes. *Cement and Concrete Composites*, *97*, 226−238. Available from https://doi.org/10.1016/j.cemconcomp.2018.12.007.

Jiang, X., Xiao, R., Zhang, M., Hu, W., Bai, Y., & Huang, B. (2020). A laboratory investigation of steel to fly ash-based geopolymer paste bonding behavior after exposure to elevated temperatures. *Construction and Building Materials*, *254*. Available from https://doi.org/10.1016/j.conbuildmat.2020.119267.

Juenger, M. C. G., & Siddique, R. (2015). Recent advances in understanding the role of supplementary cementitious materials in concrete. *Cement and Concrete Research*, *78*, 71−80. Available from https://doi.org/10.1016/j.cemconres.2015.03.018.

Juenger, M. C. G., Snellings, R., & Bernal, S. A. (2019). Supplementary cementitious materials: New sources, characterization, and performance insights. *Cement and Concrete Research*, *122*, 257−273. Available from https://doi.org/10.1016/j.cemconres.2019.05.008.

Kim, Y. J., & Choi, Y. W. (2012). Utilization of waste concrete powder as a substitution material for cement. *Construction and Building Materials*, *30*, 500−504.

Kocaer, O., & Aldemir, A. (2022). Compressive stress−strain model for the estimation of the flexural capacity of reinforced geopolymer concrete members. *Structural Concrete*, *24*, 5102−5121.

Kul, A. (2019) *New generation geopolymer binders incorporating construction demolition wastes*. Hacettepe University.

Kvočka, D., Lešek, A., Knez, F., Ducman, V., Panizza, M., Tsoutis, C., & Bernardi, A. (2020). Life cycle assessment of prefabricated geopolymeric façade cladding panels made from large fractions of recycled construction and demolition waste. *Materials, 13* (18). Available from https://doi.org/10.3390/ma13183931.

Lennon, M. (2005). *Recycling construction and demolition wastes: A guide for architects and contractors*. US EPA.

Leroy, M. N. L., Dupont, F. M. C., & Elie, K. (2019). Valorization of wood ashes as partial replacement of Portland cement: Mechanical performance and durability. *European Journal of Scientific Research, 151*(4), 468–478.

Li, V. C. (1993). From micromechanics to structural engineering—The design of cementitious composites for civil engineering applications. *Doboku Gakkai Rombun-Hokokushu/Proceedings of the Japan Society of Civil Engineers* (471), 1–12. Available from https://doi.org/10.2208/jscej.1993.471_1.

Li, V. C. (2019). *Bendable concrete for sustainable and resilient infrastructure.. Engineered cementitious composites (ECC)* (pp. 1–419). Springer. Available from 10.1007/978-3-662-58438-5.

Li, V. C., Bos, F. P., Yu, K., McGee, W., Ng, T. Y., Figueiredo, S. C., Nefs, K., Mechtcherine, V., Nerella, V. N., Pan, J., van Zijl, G. P. A. G., & Kruger, P. J. (2020). On the emergence of 3D printable engineered, strain hardening cementitious composites (ECC/SHCC). *Cement and Concrete Research, 132*. Available from https://doi.org/10.1016/j.cemconres.2020.106038.

Liikanen, M., Grönman, K., Deviatkin, I., Havukainen, J., Hyvärinen, M., Kärki, T., Varis, J., Soukka, R., & Horttanainen, M. (2019). Construction and demolition waste as a raw material for wood polymer composites—Assessment of environmental impacts. *Journal of Cleaner Production, 225*, 716–727. Available from https://doi.org/10.1016/j.jclepro.2019.03.348.

Ma, Z., Li, W., Wu, H., & Cao, C. (2019). Chloride permeability of concrete mixed with activity recycled powder obtained from C&D waste. *Construction and Building Materials, 199*, 652–663. Available from https://doi.org/10.1016/j.conbuildmat.2018.12.065.

Mahmoodi, O., Siad, H., Lachemi, M., Dadsetan, S., & Sahmaran, M. (2021). Development of normal and very high strength geopolymer binders based on concrete waste at ambient environment. *Journal of Cleaner Production, 279*. Available from https://doi.org/10.1016/j.jclepro.2020.123436.

Mahmoodi, O., Siad, H., Lachemi, M., Dadsetan, S., & Şahmaran, M. (2022). Optimized application of ternary brick, ceramic and concrete wastes in sustainable high strength geopolymers. *Journal of Cleaner Production, 338*. Available from https://doi.org/10.1016/j.jclepro.2022.130650.

Malhotra, V. M., & Mehta, P. K. (1996). *Pozzolanic and cementitious materials* (1). Taylor & Francis.

Mohammed Ali, A. A., Zidan, R. S., & Ahmed, T. W. (2020). Evaluation of high-strength concrete made with recycled aggregate under effect of well water. *Case Studies in Construction Materials, 12*. Available from https://doi.org/10.1016/j.cscm.2020.e00338.

Nematollahi, B., Sanjayan, J., Qiu, J., & Yang, E. H. (2017). Micromechanics-based investigation of a sustainable ambient temperature cured one-part strain hardening geopolymer composite. *Construction and Building Materials, 131*, 552–563. Available from https://doi.org/10.1016/j.conbuildmat.2016.11.117.

Neupane, K., Chalmers, D., & Kidd, P. (2018). High-strength geopolymer concrete-properties, advantages and challenges. *Advances in Materials, 7*(2), 15–25.

Özçelikci, E. (2020). *Development of geopolymer concretes with construction demolition waste* [MSc., thesis]. Hacettepe University.

Ozcelikci, E., Kul, A., Gunal, M. F., Ozel, B. F., Yildirim, G., Ashour, A., & Sahmaran, M. (2023). A comprehensive study on the compressive strength, durability-related parameters and microstructure of geopolymer mortars based on mixed construction and demolition waste. *Journal of Cleaner Production, 396*. Available from https://doi.org/10.1016/j.jclepro.2023.136522.

Ozcelikci, E., Yildirim, G., Alhawat, M., Ashour, A., & Sahmaran, M. (2023). An investigation into durability aspects of geopolymer concretes based fully on construction and demolition waste. In A. Ilki, D. Cavunt, & Y. S. Cavunt (Eds.), *Building for the future: Durable, sustainable, resilient fib symposium 2023* (349). Springer Science and Business Media LLC. Available from 10.1007/978-3-031-32519-9_36.

Palomo, A., Grutzeck, M. W., & Blanco, M. T. (1999). Alkali-activated fly ashes: A cement for the future. *Cement and Concrete Research, 29*(8), 1323–1329. Available from https://doi.org/10.1016/S0008-8846(98)00243-9.

Panda, B., Unluer, C., & Tan, M. J. (2019). Extrusion and rheology characterization of geopolymer nanocomposites used in 3D printing. *Composites Part B: Engineering, 176*. Available from https://doi.org/10.1016/j.compositesb.2019.107290.

Pasupathy, K., Ramakrishnan, S., & Sanjayan, J. (2023). 3D concrete printing of eco-friendly geopolymer containing brick waste. *Cement and Concrete Composites, 138*. Available from https://doi.org/10.1016/j.cemconcomp.2023.104943.

Pouhet, R., & Cyr, M. (2015). Alkali–silica reaction in metakaolin-based geopolymer mortar. *Materials and Structures, 48*(3), 571–583. Available from https://doi.org/10.1617/s11527-014-0445-x.

Provis, J. L. (2018). Alkali-activated materials. *Cement and Concrete Research, 114*, 40–48. Available from https://doi.org/10.1016/j.cemconres.2017.02.009.

Robayo-Salazar, R. A., Valencia-Saavedra, W., & Mejía de Gutiérrez, R. (2020). Construction and demolition waste (CDW) recycling—As both binder and aggregates—In alkali-activated materials: a novel re-use concept. *Sustainability, 12*(14). Available from https://doi.org/10.3390/su12145775.

Rovnaník, P., Řezník, B., & Rovnaníková, P. (2016). Blended alkali-activated fly ash/brick powder materials. *Procedia Engineering, 151*, 108–113. Available from https://doi.org/10.1016/j.proeng.2016.07.397.

Şahin, H. G., & Mardani-Aghabaglou, A. (2022). Assessment of materials, design parameters and some properties of 3D printing concrete mixtures: A state-of-the-art review. *Construction and Building Materials, 316*. Available from https://doi.org/10.1016/j.conbuildmat.2021.125865.

Şahin, O., İlcan, H., Ateşli, A. T., Kul, A., Yıldırım, G., & Şahmaran, M. (2021). Construction and demolition waste-based geopolymers suited for use in 3-dimensional additive manufacturing. *Cement and Concrete Composites, 121*. Available from https://doi.org/10.1016/j.cemconcomp.2021.104088.

Schoon, J., De Buysser, K., Van Driessche, I., & De Belie, N. (2015). Fines extracted from recycled concrete as alternative raw material for Portland cement clinker production. *Cement and Concrete Composites, 58*, 70–80. Available from https://doi.org/10.1016/j.cemconcomp.2015.01.003.

Sevim, O., Alakara, E. H., Demir, I., & Bayer, I. R. (2023). Effect of magnetic water on properties of slag-based geopolymer composites incorporating ceramic tile waste from construction and demolition waste. *Archives of Civil and Mechanical Engineering, 23*. Available from https://doi.org/10.1007/s43452-023-00649-z.

Sevim, O., Demir, I., Alakara, E. H., & Bayer, I. R. (2023). Experimental evaluation of new geopolymer composite with inclusion of slag and construction waste firebrick at elevated temperatures. *Polymers, 15*, 20734360. Available from https://doi.org/10.3390/polym15092127.

Shi, C., & Day, R. L. (2000). Pozzolanic reaction in the presence of chemical activators: Part II. Reaction products and mechanism. *Cement and Concrete Research, 30*(4), 607–613. Available from https://doi.org/10.1016/S0008-8846(00)00214-3.

Singh, N., Kumar, P., & Goyal, P. (2019). Reviewing the behaviour of high volume fly ash based self compacting concrete. *Journal of Building Engineering, 26*. Available from https://doi.org/10.1016/j.jobe.2019.100882.

Tan, J., Cai, J., & Li, J. (2022). Recycling of unseparated construction and demolition waste (UCDW) through geopolymer technology. *Construction and Building Materials, 341*. Available from https://doi.org/10.1016/j.conbuildmat.2022.127771.

Tchakouté, H. K., Rüscher, C. H., Kong, S., & Ranjbar, N. (2016). Synthesis of sodium waterglass from white rice husk ash as an activator to produce metakaolin-based geopolymer cements. *Journal of Building Engineering, 6*, 252–261. Available from https://doi.org/10.1016/j.jobe.2016.04.007.

Tuğluca, M. S., Özdoğru, E., İlcan, H., Özçelikci, E., Ulugöl, H., & Şahmaran, M. (2023). Characterization of chemically treated waste wood fiber and its potential application in cementitious composites. *Cement and Concrete Composites, 137*. Available from https://doi.org/10.1016/j.cemconcomp.2023.104938.

Ulugöl, H., Günal, M. F., Yaman, İ. Ö., Yıldırım, G., & Şahmaran, M. (2021). Effects of self-healing on the microstructure, transport, and electrical properties of 100% construction- and demolition-waste-based geopolymer composites. *Cement and Concrete Composites, 121*. Available from https://doi.org/10.1016/j.cemconcomp.2021.104081.

UN Environment Programme. (2015). *Global waste management outlook*. https://www.unep.org/resources/report/global-waste-management-outlook.

US EPA. (2020). *Inventory of US greenhouse gas emissions and sinks: 1990–2019*. https://www.epa.gov/ghgemissions/inventory-us-greenhouse-gas-emissions-and-sinks-1990-2019.

Veiga Simão, F., Chambart, H., Vandemeulebroeke, L., Nielsen, P., Adrianto, L. R., Pfister, S., & Cappuyns, V. (2022). Mine waste as a sustainable resource for facing bricks. *Journal of Cleaner Production, 368*. Available from https://doi.org/10.1016/j.jclepro.2022.133118.

Vijayan, D. S., Dineshkumar Arvindan, S., & Janarthanan, T. S. (2020). Evaluation of ferrock: A greener substitute to cement. *Materials Today: Proceedings, 22*, 781–787. Available from https://doi.org/10.1016/j.matpr.2019.10.147.22147853.

Wangler, T., Roussel, N., Bos, F. P., Salet, T. A. M., & Flatt, R. J. (2019). Digital concrete: A review. *Cement and Concrete Research, 123*. Available from https://doi.org/10.1016/j.cemconres.2019.105780.

World Bank & United Nations. (2010). *Natural hazards, unnatural disasters: The economics of effective prevention*. The World Bank.

Xiao, J., Ma, Z., Sui, T., Akbarnezhad, A., & Duan, Z. (2018). Mechanical properties of concrete mixed with recycled powder produced from construction and demolition waste. *Journal of Cleaner Production, 188*, 720–731. Available from https://doi.org/10.1016/j.jclepro.2018.03.277.

Yepsen, R. (2008). Generating biomass fuel from disaster debris. *Biocycle, 49*(7), 51–55.

Yıldırım, G., Kul, A., Özçelikci, E., Şahmaran, M., Aldemir, A., Figueira, D., & Ashour, A. (2021). Development of alkali-activated binders from recycled mixed masonry-originated waste. *Journal of Building Engineering, 33*. Available from https://doi.org/10.1016/j.jobe.2020.101690.

Yildirim, G., Ozcelikci, E., Alhawat, M., & Ashour, A. (2023). *Development of concrete mixtures based entirely on construction and demolition waste and assessment of parameters influencing the compressive strength, . International RILEM conference on synergising expertise towards sustainability and robustness of cement-based materials and concrete structures* (43). Springer Science and Business Media LLC. Available from 10.1007/978-3-031-33211-1_45.

Zawrah, M. F., Gado, R. A., Feltin, N., Ducourtieux, S., & Devoille, L. (2016). Recycling and utilization assessment of waste fired clay bricks (Grog) with granulated blast-furnace slag for geopolymer production. *Process Safety and Environmental Protection, 103*, 237–251. Available from https://doi.org/10.1016/j.psep.2016.08.001.

Zhang, C., Hu, M., Di Maio, F., Sprecher, B., Yang, X., & Tukker, A. (2022). An overview of the waste hierarchy framework for analyzing the circularity in construction and demolition waste management in Europe. *Science of The Total Environment, 803*. Available from https://doi.org/10.1016/j.scitotenv.2021.149892.

Zhang, D. W., Wang, Dm, Lin, X. Q., & Zhang, T. (2018). The study of the structure rebuilding and yield stress of 3D printing geopolymer pastes. *Construction and Building Materials, 184*, 575–580. Available from https://doi.org/10.1016/j.conbuildmat.2018.06.233.

Zhang, Z., Provis, J. L., Reid, A., & Wang, H. (2014). Geopolymer foam concrete: An emerging material for sustainable construction. *Construction and Building Materials, 56*, 113–127. Available from https://doi.org/10.1016/j.conbuildmat.2014.01.081.

Zhang, Z., Wang, H., Provis, J.L., & Reid, A. (2013). *Efflorescence: A critical challenge for geopolymer applications*. Concrete Institute of Australia's biennial national conference. 1–10.

Recycled materials used for sustainable pervious concrete

Joseph Assaad[1] *and Jamal Khatib*[2,3]
[1]Department of Civil and Environmental Engineering, University of Balamand, Al Koura, Lebanon, [2]Faculty of Engineering, Beirut Arab University, Beirut, Lebanon, [3]Faculty of Science and Engineering, University of Wolverhampton, Wolverhampton, United Kingdom

6.1 Introduction

Pervious concrete (PC) is a specialty type of permeable concrete that contains a specific aggregate gradation leading to a large number of pores (Ghosh et al., 2015; Giustozzi, 2016). Nowadays, PC mixtures are recommended by many agencies such as the American Environmental Protection Agency to provide sustainable and efficient solutions related to poor management of stormwater runoff (ACT Committee 522, 2010). The PC is widely used in pavement applications, commonly known as permeable pavements, in emerging countries such as Italy, France, Germany, Netherlands, and the United States (El-Maaty, 2016; Ghosh et al., 2015).

The typical pore sizes found in PC generally vary between 2 and 8 mm, while the total pore volume fraction ranges between 10% and 35% (El-Maaty, 2016). Nevertheless, since the PC possesses a porous microstructure, it is characterized by relatively lower strengths when compared to ordinary concrete mixtures. According to current literature (El-Maaty, 2016; Giustozzi, 2016), the compressive strengths could vary between 2.5 and 25 MPa, while the permeability measurements ranging from 3×10^{-5} to 3×10^{-2} m/s. Yu et al. (2023) conducted a review on the pore structure of PC mixtures, and their effect on the mechanical and physical properties. A method to obtain a good balance between strength and permeability is proposed. Chen et al. (2023) investigated the frost resistance of PC mixtures subjected to freeze–thaw cycles, and concluded that the induced stressed can be reduced if the curing duration is extended and air void content is reduced. The stresses created by freeze–thaw cycles are increased with the increase in the saturation degree.

The PC material is composed of Portland cement, supplementary cementitious materials (SCMs), coarse aggregates, admixtures, fibers, and water (ACT Committee 522, 2010; Akand et al., 2016; Nasser Eddine et al., 2023; Tennis et al., 2004). It is generally produced to exhibit an almost zero slump, with discontinuous aggregate grading. The main objective of this review is to present the state-of-the-art on the various ingredients used for PC production.

Sustainable Concrete Materials and Structures. DOI: https://doi.org/10.1016/B978-0-443-15672-4.00006-1

6.2 Supplementary cementitious materials

The strength of PC mainly originates from the bond achieved between the cement paste and aggregates. However, the production of cement releases large amounts of CO_2 into the atmosphere, thus affecting global environments (Assaad, 2017; Mehta & Monteiro, 2014). In order to reduce the related environmental negative impacts, many researchers investigated the effect of incorporating SCMs to partially replace the cement binder. Many SCM materials are known by their pozzolanic properties that foster reactivity with the cement hydration products (i.e., calcium hydroxide [$Ca(OH)_2$]) to produce additional calcium silicate hydrate and calcium silicate aluminate hydrate, just like those obtained from the cement hydration (AlArab et al., 2020; Assaad & El Mir, 2020; Debnath & Sarkar, 2019; Mehta & Monteiro, 2014).

If adequate curing regime is adopted, the use of SCM can greatly contribute to increased PC strength and durability, while reducing the utilization of cement. The SCM materials include a wide range of alumino-silicates such as fly ash, silica fume, blast furnace slag, metakaolin, rice husk ash, desulfurized waste, gypsum, and photogypsum (Assaad & Saba, 2020; Dvorkin et al., 2017; Khatib et al., 2013, 2014, 2016). For example, the fly ash is a fine by-product resulting from burning coal in thermoelectric plants. The particles are amorphous and spherically shaped mainly silica and alumina based, albeit their chemical properties such as the amount of calcium oxide vary depending upon the type of coal used (Saboo et al., 2019). The silica fume is an amorphous material formed from the silicon and ferrosilicon alloys production. It is an ultrafine powder essentially composed of SiO_2; it contains spherical particles with an average diameter of 100 nm (Ghosh et al., 2015). Metakaolin is produced from the calcination of high-purity clay in a temperature range of 700°C–850°C (Saboo et al., 2019). It can be produced to achieve minimum variation in its composition, high purity, together with increased degree of pozzolanicity (Saboo et al., 2019; Wild & Khatib, 1997). The rice husk ash is a term used to characterize all types of ash produced by burning rice husks, in temperatures ranging between 350°C and 1100°C (Hesami et al., 2014). The effect of various SCM along with the cement replacement rate on typical properties and performance of PC mixtures are summarized in Table 6.1.

Yang and Jiang (2003) used 6% silica fume to produce PC mixtures; the coarse aggregate size varied between 3–5 and 5–10 mm, while the superplasticizer dosage was adjusted to reach the targeted workability. The compressive strengths ranged from 13.8 to 57.2 MPa, and the mixture with 3–5 mm aggregate size, 6% silica fume, and 0.8% superplasticizer reached the highest strength values. This later mixture showed a satisfactory water penetration with a permeability coefficient of 0.0017 m/s (Yang & Jiang, 2003).

Lian and Zhuge (2010) used coarse aggregates ranging between 4.75 and 9.5 mm and different water/cement ratio (w/c) of 0.28, 0.32, and 0.36. The silica fume ranged from 7% to 10% of the binder content. The authors concluded that PC mixtures containing 7% silica fume and 0.8% superplasticizer exhibited higher 28 days compressive strengths of 33.2 MPa, compared to those with only 10% of silica

Table 6.1 Typical references that used SCM in PC production.

SCM type	Cement replacement rate (%)	Remarks	References
Silica fume	Up to 20	– Decreases porosity and permeability due to better compacity – Increases strength, durability, and resistance to fatigue	Yang and Jiang (2003), Lian and Zhuge (2010), Costa (2019), Zhong and Wille (2015), Seeni et al. (2023), Tang et al. (2022)
Fly ash	Up to 75	– Microfiller effect reduces porosity and permeability – Improves strength and resistance to freeze–thaw – Can be detrimental at rates higher than 50%	Kim et al. (2016), Um Maguesvari and Sundararajan (2017), Sri Ravindrarajah and Kassis (2014), Nazeer et al. (2023), Mohammed et al. (2018)
Metakaolin and rice husk ash	Up to 12	– Better compacity, leading to reduced permeability	Saboo et al. (2019), Hesami et al. (2014), Arshad (2018)

fume (22 MPa) as well as those produced without silica fume or superplasticizer (17 MPa). The corresponding permeability coefficients were 0.00398, 0.00613, and 0.00851 m/s, respectively. The authors noted that the recommended permeability rate for PC mixtures should range between 0.002 and 0.0054 m/s, leading to minimal influence on compressive strength (Lian & Zhuge, 2010).

Costa (2019) used 8% silica fume with w/c of 0.26 and 0.30 to produce PC. Results showed that the use of silica fume in mixes with 0.3-w/c led to 1 MPa decrease in compressive strength, yet, mixes with 0.26-w/c achieved two times higher compressive strengths. This was attributed to a greater admixture content, which densified the paste and decreased porosity, leading to increased compressive strength. For permeability, values of 0.89 cm/s were found for PC prepared with 0.3-w/c, while this was 0.76 cm/s for mixtures with 0.26-w/c. Specimens with 0.26-w/c and silica fume achieved 33% increase in splitting tensile strength, when compared to mixtures prepared without silica fume.

Zhong and Wille (2015) produced PC mixtures with the help of microsilica and silica powder. They reported that the compressive strength could reach a value as high as 40 MPa, which was associated with a decrease in permeability. The concrete exhibiting a compressive strength greater than 50 MPa and permeability coefficient greater than 1 mm/s was found to meet the specified criteria for PC. Seeni et al. (2023) conducted an experimental study on the influence of silica fume on the properties of PC. Their results indicated that using 15% silica fume as cement replacement enhanced the compressive, split tensile, and flexural strengths by about 20%, 13%, and 14%, respectively. Also, using this percentage of silica fume (i.e., 15%) produced PC with adequate physical and hydrological properties.

Tang et al. (2022) used silica fume, ultrafine silica powder, polypropylene fiber, and lightweight coarse aggregate to design PC mixtures and assess their mechanical properties. The results showed that the compressive strength of PC is highly influenced by the size of the lightweight coarse aggregate. As for the flexural and splitting tensile strengths, the fine aggregate content was the most important factor. The authors showed that the compressive, flexural, and splitting tensile strengths of PC varied within 4.80−7.78, 1.19−1.86, and 0.78−1.11 MPa, respectively. Overall, the mechanical properties and durability of the prepared PC mixtures containing silica fume, ultrafine silica powder, and polypropylene fiber were better than those possessing a general composition (Tang et al., 2022).

Kim et al. (2016) used 20% fly ash as partial cement replacement to produce PC. While the compressive strength of the control mixture was around 9.4 MPa, the incorporation of fly ash was found to reduce the strength by about 9.6%. The permeability was obtained by subjecting the PC to various levels of load intensity varying from 0% to 75%. Kim et al. (2016) noticed that the greater the load application, the lower becomes the PC permeability coefficient.

Um Maguesvari and Sundararajan (2017) used 10%− 20% fly ash to partially replace the cement in PC mixtures. The binder content in the mixtures varied from 250 to 400 kg/m^3. It was found that the compressive strengths varied between 5.7 and 8.83 MPa, while the splitting tensile strengths was between 1.45 and 1.86 MPa. The permeability ranged from 0.00641 to 0.0119 m/s. The authors noticed that the permeability coefficients decrease as paste content increases, while PC containing 10% fly ash showed higher permeability compared to equivalent PC with 20% fly ash (Um Maguesvari & Sundararajan, 2017).

Sri Ravindrarajah and Kassis (2014) used both fly ash and silica fume to produce PC. Besides the control mixtures, specimens containing 24.75% fly ash, 7.4% silica fume, and others with 8.2% fly ash and 7.6% silica fume were tested. The w/c was between 0.3 and 0.32. The resulting compressive strengths varied between 12.8 and 22.5 MPa. It was found that replacement of cement by SCM could increase the compressive strength, since the lowest values were found in mixtures without SCM. The highest strength corresponded to the PC containing only silica fume. However, a significant reduction in permeability was observed when incorporating SCM in the mixtures. The highest permeability coefficient (0.00663 m/s) was obtained for the control mixture, whereas the lowest value (0.00387 m/s) was found for the combined mixture. Also, Nazeer et al. (2023) assessed the performance of PC containing fly ash and silica fume. When cement was replaced with 10% fly ash and 10% silica fume, the compressive strength increased by about 95% at 120 days of curing while maintaining adequate permeability. The shear resistance as well as the split tensile and flexural strengths were also enhanced.

Mohammed et al. (2018) attempted to incorporate a large volume of fly ash (0%–70%) into the binder, while the nanosilica varied from 0% to 3%. They observed that the nanosilica increased compressive strength, as this refined the cement paste microstructure. Such particles have the benefits of activating fly ash, which helps reducing the voids' internal surface and improving water permeability. However, the presence of fly ash led to a decrease in the PC compressive strength. In addition, an increase in cement percentage replacement by fly ash was found to lead to higher permeability coefficients.

Saboo et al. (2019) used fly ash (0%–20%) and metakaolin (2%) in PC production. It was observed that the increased fly ash and metakaolin contents led to reduced porosity values. Mixtures with 2% metakaolin addition, without fly ash, showed an increase in the concrete density by about 6.5%. Saboo et al. (2019) noticed that the permeability decreases with the increase in fly ash, which was attributed to the density increase together with reduced porosity.

Hesami et al. (2014) produced PC with different rice husk ash contents varying from 0% to 12%. The fine and coarse aggregates were set at predefined fixed values, while the amounts of superplasticizer were adjusted. The w/c for the PC mixtures varied from 0.27 to 0.4. It was demonstrated that the pozzolanic properties of rice husk ash improved the physical structure and denseness of the cement paste. This SCM reacts with the calcium hydroxide released from the cement hydration, thus densifying the paste connectivity with the aggregate which would improve the interfacial transition zone (ITZ) between the coarse aggregate and cement paste.

Similar results were obtained by Arshad (2018) who replaced the cement by 10% rice husk ash which was ground for various durations. The compressive strength was found to increase as the grinding period increases. The highest strength of 19 MPa was found in PC containing rice husk ground for 63 hours. The replacement levels between 10%–12% of this SCM were found ideal to obtain a PC with good mechanical properties and adequate water permeability coefficient.

From the foregoing, it is clear that SCMs can be efficiently used in PC mixtures, which helps reducing the CO_2 footprint due to Portland cement. SCMs include a wide range of alumino-silicates such as fly ash, silica fume, blast furnace slag, metakaolin, and rice husk ash. Such materials are known by their pozzolanic properties that foster reactivity with the cement hydrates and contribute to increased PC strength and durability.

6.3 Coarse aggregates

The coarse aggregate fraction constitutes the main structure of PC mixtures, thus directly affected compressive strength, porosity, and water permeability (Fig. 6.1).

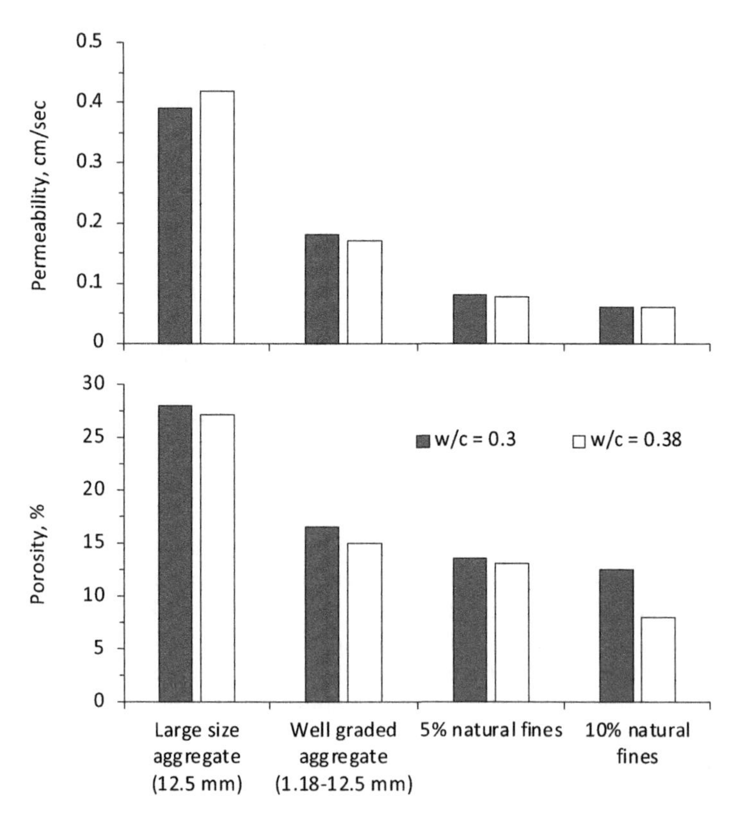

Figure 6.1 Effect of aggregate gradation on porosity and permeability of PC mixtures having 0.3 or 0.38 w/c (Vijayalakshmi, 2021).

Generally, the good performance of PC could be achieved when the coarse aggregate uses a single grading size having either 10–20 or 5–10 mm (ACT Committee 522, 2010; Ghosh et al., 2015; Mehta & Monteiro, 2014). The coarse aggregate types include conventional aggregate (such as gravel, limestone, dolomite, sandstone, etc.), particular aggregate types (such as lightweight or recycled concrete aggregate), and industrial by-product aggregates (such as steel slag or copper slag) (ACT Committee 522, 2010; Assaad et al., 2020; Kevern & Nowasell, 2018; Toghroli et al., 2020; Yeih & Chang, 2019).

Öz (2018) reported that the acidic pumice natural lightweight aggregate is expected to lead to lower PC density and strength, when compared to the crushed stone concrete. Therefore, the mechanical strengths decreased with increasing the replacement of crushed stones by the pumice, leading to better water permeability and resistance to surface abrasion (Öz, 2018). Debnath and Sarkar (2020) showed that the overburnt waste bricks can constitute a potential source of aggregate when producing PC mixtures, leading to reduced amount of virgin stone aggregates.

Yeih and Chang (2019) revealed that PC prepared using electric arc furnace slag aggregate and sulfo-aluminate cement releases significant amounts of alumina, ferric oxide, and trisulfate which severely degrade the PC mechanical strengths. Conversely, the copper slag that is free from reactive silicates does not pose any problem related to volumetric stability whether in expansion or in shrinkage. The copper slag also increases the density fresh and hardened PC. Lori et al. (2019) found that the shrinkage and deformation decreased when the natural aggregate is replaced by copper slag aggregate. Kevern and Nowasell (2018) concluded that the prewetted lightweight aggregate such as the expanded shales, slates, or clays could provide increased degree of cement hydration, which thereby leads to reduced shrinkage and improvement in the PC strengths.

Khankhaje et al. (2017) demonstrated that the use of lightweight aggregates could potentially improve the acoustic and thermal insulation of pavement applications made using PC mixtures. Aliabdo et al. (2018) showed that the recycled concrete aggregate adversely affects the PC strengths including the durability and abrasion resistance. This was mostly attributed to poorer aggregate–mortar interfaces resulting from the fine recycled aggregate fraction as well as inherent porosity and weaker adhesion between the recycled aggregates and cement paste (Aliabdo et al., 2018).

Chen and Jin (2023) reviewed some research studies on the production of PC using recycled aggregate. They found that it is possible to prepare PC with recycled aggregate that meet the standard requirements for road applications. Nasser Eddine et al. (2023) came to the same conclusions where the coarse aggregate was partially or fully replaced by municipal solid waste incineration bottom ash. Zhu and Jiang (2023) carried out some experiments on the performance of PC containing waste rubber. The presence of waste rubber caused a reduction in mechanical properties (e.g., compressive and splitting tensile strengths). Also, the waste rubber weakened the bond between the aggregate and matrix, although the ductility of PC mixtures improved with waste rubber additions.

Tang et al. (2023) explored the effect of carbonation on the properties recycled aggregate PC, and concluded that the produced concrete is characterized by adequate permeability and enhanced compressive strength. The added benefit is that carbonated PC will help reducing carbon dioxide emissions into the atmosphere by retaining it inside the concrete. Ibrahim et al. (2020) reported that sustainable PC can be produced by combining the natural aggregates with recycled ones, which was demonstrated through microscopic analysis. Compared to natural aggregate, Lang et al. (2019) showed that the steel slag has a higher impact and crushing strengths. They reported that the steel slag has excellent antiskid performance and higher bonding strength with cementitious materials. Yet, the PC produced with steel slag aggregates becomes more sensitive to the binder type and content (Lang et al., 2019).

The literature review unanimously converge that PC mixtures can be produced using different types of coarse aggregate, provided a discontinuous in their gradation to ensure porous microstructure and facilitate water permeation. The coarse aggregates include conventional aggregate (such as gravel, limestone, dolomite, sandstone, etc.), particular aggregate types (such as lightweight or recycled concrete aggregate), and industrial by-product aggregates (such as steel slag or copper slag). Typical PC compressive strengths vary between 2.5 and 25 MPa, while the permeation range from 3×10^{-5} to 3×10^{-2} m/s.

6.4 Chemical admixtures and polymers

The use of chemical admixtures is common in PC production since this ensures adequate workability together with adjusted setting times and strength development. Typical admixtures include superplasticizers (Chen et al., 2013; Kim et al., 2016; Lian & Zhuge, 2010), viscosity modifying agents (Brake et al., 2016; Zhong & Wille, 2015), rheology modifying admixtures (Deo & Neithalath, 2011), retarding admixtures (ACT Committee 522, 2010; Mehta & Monteiro, 2014), air entraining admixtures (Ghafoori & Dutta, 1995), and a combination of resins and polymeric admixtures. The mode of function of each of these admixtures including their influence on concrete workability and cement hydration processes are well documented in literature (ACT Committee 522, 2010; Mehta & Monteiro, 2014).

Jang et al. (2015) used a geopolymeric approach (i.e., alkali-activated fly ash and slag paste) together with bottom ash coarse aggregate to develop a porous PC mix design. The concrete mixture was free of cement and considered to be eco-friendly and sustainable for the construction pavement industry. Similarly, Sun and Vollpracht (2018) developed a PC made without cement; the geopolymer binder is composed of are alkali-activated slag and metakaolin material. Akand et al. (2018) showed that the chemical treatment using silane polymer emulsions can be quite suitable for PC production as it enhances the membrane-forming layer around the cement paste, thus improving the concrete water permeability without altering the strength development.

Tabatabaeian et al. (2019) invested the possibility of producing high-performance PC mixtures with the aid of polyester and epoxy resins as polymeric composites. Those resins completely substituted the ordinary cement, which led to significantly enhanced mechanical properties together with optimized water permeability and resistance against the freeze–thaw attack. Yang et al. (2020) explored the influence of dispersible powders such as the vinyl acetate and ethylene copolymer on the ITZs developed between the aggregates and cement paste. Such polymers found to improve workability and air entrainment; however, the plastic viscosity of fresh concrete tended to increase because of the higher solution stickiness. The cement hydration and strength of ITZ were slightly curtailed with such polymer additions, thus altered the pore characteristic and denseness of the matrix microstructure.

Dai (2020) and Borhan and Al Karawi (2020) attempted using some adhesion admixtures (such as polymeric latexes) to improve the gluing and bonding effects between the cement paste and aggregate particles (Assaad & Issa, 2017). Generally, it was shown that such adhesion admixtures consumed more fine cement, which reduced the segregation phenomenon of coarse aggregates due to increased adhesiveness of the fresh cement paste. Borhan and Al Karawi (2020) reported that the styrene butadiene rubber polymer is particularly useful for PC production as it exhibits several positive effects on strength and permeability, as compared to PC mixtures prepared without polymers.

6.5 Fibers

The use of fibers in PC mixtures is efficient to improve the mechanical strength and durability properties (Ahmed et al., 2019; Kevern et al., 2014; Khalil & Assaad, 2021; Rodin et al., 2018). In certain applications, an increase in compressive strength can be observed with fibers, whereas several scholars reported that the fibers could negatively alter the compressive strength of highly porous PC mixtures (El-Hassan & Kianmehr, 2017; Mussado et al., 2017). The durability of PC, especially the one related to abrasion and freeze–thaw resistance can be improved with the use of fibers (Mussado et al., 2017).

In general, the use of metallic fibers is restricted in PC production due their proneness to corrosion (Rodin et al., 2018). From the other hand, the polypropylene fibers are much better for PC; these are characterized by a high melting point, better chemical stability, hydrophobic surface, and less cost. During mixing, the polypropylene fibers will form clusters, thus requiring enough mixing energy for proper distribution. Polypropylene fibers exhibited limited or no physiochemical adhesion with the cement paste; however, they essentially work on bridging the cracks upon initiation to limit their opening during loading (Kevern et al., 2014). The incorporation of polypropylene fibers is reported to improve the flexural and splitting tensile properties of PC, just like what happens for the ordinary concrete (Mussado et al., 2017).

Ibrahim et al. (2019) showed that the low-density fibers such as glass fibers could lead to reduced PC unit weight, together with lower mechanical performance. Emiko et al. (2013) noticed that the steel fibers could improve the compressive and flexural strengths of PC mixtures. Nevertheless, the clustering of fibers should be avoided by proper mixing means other than the rotary drum mixers (Emiko et al., 2013; Kevern et al., 2014). Bonicelli et al. (2014) reported that the fibers could reduce the void content of PC by closing up voids. Yet, the balling effect and poor cohesive nature of fibers could compensate such phenomenon, thus leading to increased void ratio which increases water permeability.

Vijayalakshmi (2021) carried out a comprehensive review on the use of cured carbon fiber composite materials (CCFCM) in PC mixtures. In fact, large quantities of such waste materials are generated each year, which could be disposal in PC applications. Generally, it has been found that the CCFCM specimens exhibit increased tensile, flexural, and compressive strengths. The short fibers (13–19 mm) could be added at only 0.3 to 0.9 kg/m^3 dosage, which may enhance the compressive and splitting tensile strengths by 24%. However, when longer fibers (50 mm) are used, no changes were observed in the PC strengths. The author reported that the microfiber reinforcement could delay the crack generation but with limited capability to restrain large deformation especially after cracking.

From the foregoing, it seems that the incorporation of fibers potentially improves the concrete mechanical properties, yet without hampering the water permeation required for PC mixtures. Fibers of different natures (i.e., metallic, polypropylene, glass, carbon, etc.) are known by their action to bridge across cracks, thus reducing their propagation and enhancing the postcracking concrete resistance (Matar & Assaad, 2019). The fibers can also enhance the concrete toughness, which can be of particular interest to PC mixtures possessing high porous nature.

6.6 Pollution removal

PC has the potential to reduce the amount of urban runoff contaminants. Teymouri et al. (2023) examined the addition of iron slag to PC mixtures on the level of urban runoff pollution. The use of iron slag caused a reduction in the amount of urban runoff contaminants, namely the chemical oxygen demand, total suspended solid, and turbidity. In another study, Ahmadi et al. (2023) studied the effect of alkali-activated slag-based PC on the level of runoff contamination. Activated carbon and zeolite were added to the PC mixtures. The levels of Pb, Cu, and Cd were greatly reduced in the runoff. This was attributed to the adsorption characteristics of PC mixtures. Also, Anwar et al. (2023) conducted some research on the use alkali-activated fly ash and slag to produce PC and found an optimum mixture where the stormwater infiltration is maximized.

The use of PC containing photocatalysts in pavement applications have the potential to remove the pollutants from the air, especially in urban and congested areas (Ballari et al., 2010; Jabali et al., 2023; Shen et al., 2012). The photocatalytic compounds are known by their capacity to absorb organic and inorganic particles existing in the atmosphere, which helps removing the harmful pollutants such as nitrogen oxides or other polycyclic aromatic hydrocarbons. Titanium dioxide (TiO_2) is one of the photocatalytic semiconductors that can be incorporated in building materials to purify the quality of surrounding air (Cros et al., 2015; Osborn et al., 2014). The TiO_2 decomposes the gaseous pollutants especially the nitrogen oxides in presence of the sunlight, thus removing the emission pollutants. The rough surface and porous PC materials could retain more TiO_2 particles, contributing to highly efficient alternative to a photocatalytic effect during sunny days. Many scholars found from laboratory results an increment in the NOx removal efficiency of PC materials hovering about 3.6% per 25 mm of depth (Cros et al., 2015; Osborn et al., 2014). The best results were obtained when the PC contained about 31% air voids in its structure.

Different TiO_2 grades possessing various photocatalytic efficiencies are tested in PC pavements including other applications such as plasters, paints, and stuccos (Asadi et al., 2012; Cros et al., 2015; Hassan et al., 2012). Hence, for instance, the application of stucco yielded about 81% NOX removal, in comparison with clear paint and white paint products which exhibited 52% and 42% NOx removal, respectively. Nevertheless, it is worth noting that the PC pavements are much better retainers photocatalytic compounds because of their higher porosity and presence of voids. Nevertheless, the PC pavements can lose part of their photocatalytic efficiency over times because they are more likely subjected to wear traffic and rain water (Asadi et al., 2012; Hassan et al., 2012).

6.7 Concluding remarks

PC has a potential for contribution toward sustainable development including the development of pavements and urban drainage systems. Waste, recycled, industrial by-products, and SCMs can be incorporated in PC mixtures, thus reducing the amount of cement and pollution through the reduction in carbon dioxide emissions into the atmosphere. The use of PC in pavement applications and roadwork can reduce the contamination level in the runoff water, thus contributing toward reduced air and soil pollutions. Future research should examine the durability properties of PC mixtures containing waste and recycled materials including wetting, drying, freeze–thaw, and contamination levels, allowing wider utilization in the construction industry.

References

ACT Committee 522. (2010). *ACI report on pervious concrete*. American Concrete Institute.

Ahmadi, Z., Behfarnia, K., Faghihian, H., Soltaninia, S., Behravan, A., & Ahmadi, S. (2023). Application of pervious alkali-activated slag concrete to adsorb runoff contaminants. *Construction and Building Materials*, *375*, 130998. Available from https://doi.org/10.1016/j.conbuildmat.2023.130998.

Ahmed, A. M., Hussein, A. H., & Hammood, M. T. (2019). Recycling of disposal polypropylene blister tablets and strapping ties as fibre reinforcement for pervious concrete. *IOP Conference Series: Materials Science and Engineering*, *584*, 012031. Available from https:/doi.org/10.1088/1757-899X/584/1/012031.

Akand, L., Yang, M., & Wang, X. (2018). Effectiveness of chemical treatment on polypropylene fibers as reinforcement in pervious concrete. *Construction and Building Materials*, *163*, 32–39. Available from https://doi.org/10.1016/j.conbuildmat.2017.12.068.

Akand, L., Yang, M., & Gao, Z. (2016). Characterization of pervious concrete through image based micromechanical modelling. *Construction and Building Materials*, *114*, 547–555. Available from https:/doi.org/10.1016/j.conbuildmat.2016.04.005.

AlArab, A., Hamad, B., Chehab, G., & Assaad, J. J. (2020). Use of ceramic-waste powder as value-added pozzolanic material with improved thermal properties. *Journal of Materials in Civil Engineering*, *32*(9), 04020243. Available from https://doi.org/10.1061/(ASCE)MT.1943-5533.0003326.

Aliabdo, A. A., Abd Elmoaty, A. E. M., & Fawzy, A. M. (2018). Experimental investigation on permeability indices and strength of modified pervious concrete with recycled concrete aggregate. *Construction and Building Materials, 193*, 105–127. Available from https://doi.org/10.1016/j.conbuildmat.2018.10.182.

Anwar, F. H., El-Hassan, H., Hamouda, M., & El-Mir, A. (2023). Evaluation of pervious geopolymer concrete pavements performance for effective stormwater infiltration and water purification using Taguchi method. *Materials Today: Proceedings*. Available from https://doi.org/10.1016/j.matpr.2023.02.350.

Arshad, M. F., et al. (2018). Effect of nano black rice husk ash on the chemical and physical properties of porous concrete pavement. *Journal of Southwest Jiaotong University, 53*. Available from https://doi.org/10.3969/j.issn.0258-2724.2018.015.

Asadi, S., Hassan, M., Kevern, J., & Rupnow, T. (2012). Development of photocatalytic pervious concrete pavement for air and storm water improvements. *Transportation Research Record: Journal of the Transportation Research Board., 2290*, 161–167. Available from https://doi.org/10.3141/2290-21.

Assaad, J. J., & El Mir, A. (2020). Durability of polymer-modified lightweight flowable concrete made using expanded polystyrene. *Construction and Building Materials, 249*, 118764. Available from https://doi.org/10.1016/j.conbuildmat.2020.118764.

Assaad, J. J., & Issa, C. A. (2017). Effect of recycled acrylic-based polymers on bond stress-slip behavior in reinforced concrete structures. *Journal of Materials in Civil Engineering, 29*(1), 04016173. Available from https://doi.org/10.1061/(asce)mt.1943-5533.0001700.

Assaad, J. J., & Saba, M. (2020). Suitability of metakaolin-based geopolymers for masonry plastering. *ACI Materials Journal, 117*(6), 269–279. Available from https://doi.org/10.14359/51725991.

Assaad, J. J., Matar, P., & Gergess, A. (2020). Effect of quality of recycled aggregates on bond strength between concrete and embedded steel reinforcement. *Journal of Sustainable Cement-Based Materials, 9*(2), 94–111. Available from https://doi.org/10.1080/21650373.2019.1692315.

Assaad, J. J. (2017). Effect of energy and temperature on performance of alkanolamine processing additions. *Minerals Engineering, 102*, 30–41. Available from https://doi.org/10.1016/j.mineng.2016.12.007.

Ballari, M. M., Hunger, M., Hüsken, G., & Brouwers, H. J. H. (2010). Modelling and experimental study of the NOx photocatalytic degradation employing concrete pavement with titanium dioxide. *Catalysis Today., 151*, 71–76. Available from https://doi.org/10.1016/j.cattod.2010.03.042.

Bonicelli, A., Giustozzi, F., Crispino, M., & Borsa, M. (2014). Evaluating the effect of reinforcing fibres on pervious concrete volumetric and mechanical properties according to different compaction energies. *European Journal of Environmental and Civil Engineering, 19*(2), 184–198. Available from https:/doi.org/10.1080/19648189.2014.939308.

Borhan, T. M., & Al Karawi, R. J. (2020). Experimental investigations on polymer modified pervious concrete. *Case Studies in Construction Materials*. Available from https://doi.org/10.1016/j.cscm.2020.e00335.

Brake, N. A., Allahdadi, H., & Adam, F. (2016). Flexural strength and fracture size effects of pervious concrete. *Construction and Building Materials, 113*, 536–543. Available from https:/doi.org/10.1016/j.conbuildmat.2016.03.045.

Chen, J., Zhao, C., Liu, Q., Shi, X., & Sun, Z. (2023). Investigation on frost heaving stress (FHS) of porous cement concrete exposed to freeze-thaw cycles. *Cold Regions Science and Technology, 205*103694. Available from https://doi.org/10.1016/j.coldregions.2022.103694.

Chen, W., & Jin, R. (2023). Pervious concrete adopting recycled aggregate for environmental sustainability. In Y. Xu, & R. Jin (Eds.), *Multi-functional concrete with recycled aggregates* (pp. 289–306). Woodhead Publishing. Available from https://doi.org/10.1016/B978-0-323-89838-6.00008-6.

Chen, Y., Wang, K., Wang, X., & Zhou, W. (2013). Strength, fracture and fatigue of pervious concrete. *Construction and Building Materials*, *42*, 97–104. Available from https:/doi.org/10.1016/j.conbuildmat.2013.01.006.

Costa, F.B. P. (2019). *Análise e desenvolvimento de misturas de concreto permeável para aplicação em pavimentação. Tese* [Doutorado em Engenharia]. Programa de Pós-graduação em Engenharia Civil, Universidade Federal do Rio Grande do Sul, Porto Alegre.

Cros, C. J., Terpeluk, A. L., Burris, L. E., Crain, N. E., Corsi, R. L., & Juenger, M. C. G. (2015). Effect of weathering and traffic exposure on removal of nitrogen oxides by photocatalytic coatings on roadside concrete structures. *Materials and Structures*, *48*, 3159–3171. Available from https://doi.org/10.1617/s11527-014-0388-2.

Dai, Z., Li, H., Zhao, W., Wang, X., Wang,, H., Zhou, H., & Yang, B. (2020). Multi-modified effects of varying admixtures on the mechanical properties of pervious concrete based on optimum design of gradation and cement-aggregate ratio. *Construction and Building Materials*, *233*, 117178. Available from https://doi.org/10.1016/j.conbuildmat.2019.117178.

Debnath, B., & Sarkar, P. P. (2019). Permeability prediction and pore structure feature of pervious concrete using brick as aggregate. *Construction and Building Materials*, *213*, 643–651. Available from https://doi.org/10.1016/j.conbuildmat.2019.04.099.

Debnath, B., & Sarkar, P. P. (2020). Characterization of pervious concrete using over burnt brick as coarse aggregate. *Construction and Building Materials*, *242*, 118154. Available from https://doi.org/10.1016/j.conbuildmat.2020.118154.

Deo, O., & Neithalath, N. (2011). Compressive response of pervious concretes proportioned for desired porosities. *Construction and Building Materials*, *25*, 4181–4189. Available from https:/doi.org/10.1016/j.conbuildmat.2011.04.055.

Dvorkin, L., Lushnikova, N., Sonebi, M., & Khatib, J. (2017). Properties of modified phosphogypsum binder. *Academic Journal of Civil Engineering*, *3*(2), 96–102. Available from https://doi.org/10.26168/icbbm2017.13.

El-Hassan, H., & Kianmehr, P. (2017). Sustainability assessment and physical characterization of pervious concrete pavement made with GGBS. *MATEC Web of Conferences*, *120*, 1–11. Available from https:/doi.org/10.1051/matecconf/201712007001.

El-Maaty, A. E. A. (2016). Establishing a balance between mechanical and durability properties of pervious concrete pavement. *American Journal of Traffic and Transportation Engineering*, *1*(2), 13–25. Available from https:/doi.org/10.11648/j.ajtte.20160102.11.

Emiko, L., Hwee, T. K., & Fang, F. T. (2013). High-strength high-porosity pervious concrete pavement. *Advanced Materials Research*, *723*, 361–367. Available from https:/doi.org/10.4028/http://www.scientific.net/AMR.723.361.

Ghafoori, N., & Dutta, S. (1995). Laboratory investigation of compacted no-fines concrete for paving materials. *Journal of Materials in Civil Engineering*, *7*, 183–191.

Ghosh, S. K., Chaudhury, A., Datta, R., & Bera, D. K. (2015). A review on performance of pervious concrete using waste materials. *International Journal of Research in Engineering and Technology*, *4*.

Giustozzi, F. (2016). Polymer-modified pervious concrete for durable and sustainable transportation infrastructures. *Construction and Building Materials*, *111*, 502–512. Available from https://doi.org/10.1016/j.conbuildmat.2016.02.136.

Hassan, M., Asadi, S., Kevern, J.T., & Rupnow, T. (2012). Nitrogen oxide reduction and nitrate measurements on TiO2 photocatalytic pervious concrete pavement. In: *Construction challenges in a flat world, proceedings of the 2012 construction research congress*, 21–23 May, West Lafayette, IN, pp. 1920–1930.

Hesami, S., Ahmadi, S., & Nematzadeh, M. (2014). Effects of rice husk ash and fiber on mechanical properties of pervious concrete pavement. *Construction and Building Materials*, *53*, 680–691.

Ibrahim, H. A., Goh, Y., Ng, Z. A., Yap, S. P., Mo, K. H., Yuen, C. W., & Abutaha, F. (2020). Hydraulic and strength characteristics of pervious concrete containing a high volume of construction and demolition waste as aggregates. *Construction and Building Materials*, *253*, 119251. Available from https://doi.org/10.1016/j.conbuildmat.2020.119251.

Ibrahim, H. A., Mahdi, M. B., & Abbas, B. J. (2019). Performance evaluation of fibre and silica fume on pervious concrete pavements containing waste recycled concrete aggregate. *International Journal of Advancements in Technology*, *10*(2), 1–9. Available from https:/doi.org/10.24105/0976-4860.10.230.

Jabali, Y., Assaad, J. J., & Aouad, G. (2023). Photocatalytic activity and mechanical properties of cement slurries containing titanium dioxide. *Buildings*, *13*, 1046. Available from https://doi.org/10.3390/buildings13041046.

Jang, J. G., Ahn, Y. B., Souri, H., & Lee, H. K. (2015). A novel eco-friendly porous concrete fabricated with coal ash and geopolymeric binder: Heavy metal leaching characteristics and compressive strength. *Construction and Building Materials*, *79*, 173–181. Available from https://doi.org/10.1016/j.conbuildmat.2015.01.058.

Kevern, J. T., & Nowasell, Q. C. (2018). Internal curing of pervious concrete using lightweight aggregates. *Construction and Building Materials*, *161*, 229–235. Available from https://doi.org/10.1016/j.conbuildmat.2017.11.055.

Kevern, J. T., Biddle, D., & Cao, Q. (2014). Effects of macrosynthetic fibres on pervious concrete properties. *Journal of Materials in Civil Engineering*, *27*(9). Available from https:/doi.org/10.1061/(ASCE)MT.1943-5533.0001213.

Khalil, N., & Assaad, J. J. (2021). Bond properties between smooth carbon fibre-reinforced polymer bars and ultra-high performance concrete modified with polymeric latexes and fibres. *European Journal of Environmental and Civil Engineering*, *26*(13), 6211–6228. Available from https://doi.org/10.1080/19648189.2021.1934554.

Khankhaje, E., Rafieizonooz, M., Salim, M. R., Mirza, J., Salmiati., & Hussin, M. W. (2017). Comparing the effects of oil palm kernel shell and cockle shell on properties of pervious concrete pavement. *International Journal of Pavement Research and Technology*, *10*(5), 383–392. Available from https://doi.org/10.1016/j.ijprt.2017.05.003.

Khatib, J. M., Mangat, P. S., & Wright, L. (2014). Pore size distribution of cement pastes containing fly ash-gypsum blends cured for 7 days. *Korean Society of Civil Engineering (KSCE)*, *18*(4), 1091–1096. Available from https://doi.org/10.1007/s12205-014-0136-8.

Khatib, J. M., Wright, L., & Mangat, P. S. (2013). Effect of fly ash-gypsum blend on porosity and pore size distribution of cement pastes. *Journal of Advances in Applied Ceramics, Structural, Functional and Bioceramics*, *112*(4), 197–201. Available from https://doi.org/10.1179/1743676112Y.0000000032.

Khatib, J. M., Wright, L., & Mangat, P. S. (2016). Mechanical and physical properties of concrete containing FGD waste. *Magazine of Concrete Research*, *68*(11), 550–560. Available from 10.1680/macr.15.00092.

Kim, Y. J., Gaddafi, A., & Yoshitake, I. (2016). Permeable concrete mixed with various admixtures. *Materials & Design*, *100*, 110–119.

Lang, L., Duan, H., & Chen, B. (2019). Properties of pervious concrete made from steel slag and magnesium phosphate cement. *Construction and Building Materials*, *209*, 95–104. Available from https://doi.org/10.1016/j.conbuildmat.2019.03.123.

Lian, C., & Zhuge, Y. (2010). Optimum mix design of enhance permeable concrete – An experimental investigation. *Construction and Building Materials*, *24*, 2664–2671.

Lori, A. R., Hassani, A., & Sedghi, R. (2019). Investigating the mechanical and hydraulic characteristics of pervious concrete containing copper slag as coarse aggregate. *Construction and Building Materials*, *197*, 130–142. Available from https://doi.org/10.1016/j.conbuildmat.2018.11.230.

Matar, P., & Assaad, J. J. (2019). Concurrent effects of recycled aggregates and polypropylene fibers on workability and key strength properties of self-consolidating concrete. *Construction and Building Materials*, *199*, 492–500. Available from https://doi.org/10.1016/j.conbuildmat.2018.12.091.

Mehta, P. K., & Monteiro, P. J. M. (2014). *Concreto: microestrutura, propriedades e materiais* (2o ed., pp. 594–602). Ibracon.

Mohammed, B. S., Liew, M. S., Alaloul, W. S., Khed, V. C., Hoong, C. Y., & Adamu, M. (2018). Properties of nano-silica modified pervious concrete. *Case Studies in Construction Materials*, *8*, 409–422.

Mussado, J. L. A., Toralles, B. M., & Sandoval, G. F. B. (2017). Performance of pervious concrete reinforced with polypropylene fibres. *Mix Sustentável*, *3*(4), 195–197.

Nasser Eddine, Z., Khatib, J., El Kordi, A., & Assi, L. (2023). Performance of a pervious concrete pavement containing municipal solid waste incineration bottom ash: A Lebanese case study. *International Journal of Pavement Research and Technology*, 1–14. Available from https://doi.org/10.1007/s42947-023-00320-z.

Nazeer, M., Kapoor, K., & Singh, S. P. (2023). Strength, durability and microstructural investigations on pervious concrete made with fly ash and silica fume as supplementary cementitious materials. *Journal of Building Engineering*, *69*, 106275. Available from https://doi.org/10.1016/j.jobe.2023.106275.

Osborn, D., Hassan, M., Asadi, S., & White, J. R. (2014). Durability quantification of TiO_2 surface coating on concrete and asphalt pavements. *Journal of Materials in Civil Engineering*, *26*, 331–337. Available from https://doi.org/10.1061/(ASCE)MT.1943-5533.0000816.

Öz, H. Ö. (2018). Properties of pervious concretes partially incorporating acidic pumice as coarse aggregate. *Construction and Building Materials*, *166*, 601–609. Available from https://doi.org/10.1016/j.conbuildmat.2018.02.010.

Rodin, H., Rangelov, M., Nassiri, S., & Englund, K. (2018). Enhancing mechanical properties of pervious concrete using carbon fibre composite reinforcement. *Journal of Materials in Civil Engineering*, *30*, 3.

Saboo, N., Shivhare, S., Kori, K. K., & Chandrappa, A. K. (2019). Effect of fly ash and metakaolin on pervious concrete properties. *Construction and Building Materials*, *223*, 322–328.

Seeni, B. S., Madasamy, M., Chellapandian, M., & Arunachelam, N. (2023). Effect of silica fume on the physical, hydrological and mechanical properties of pervious concrete. *Materials Today: Proceedings*. Available from https://doi.org/10.1016/j.matpr.2023.03.473.

Shen, S., Burton, M., Jobson, B., & Haselbach, L. (2012). Pervious concrete with titanium dioxide as a photocatalyst compound for a greener urban road environment. *Construction and Building Materials*, *35*, 874–883. Available from https://doi.org/10.1016/j.conbuildmat.2012.04.097.

Sri Ravindrarajah, R., & Kassis, S.J., (2014). Effect of supplementary cementitious materials on the properties of pervious concrete with fixed porosity. *23rd Australasian conference on the mechanics of structures and materials*, Byron Bay, Australia, 10 p.

Sun, Z. L., & Vollpracht, A. (2018). Pervious concrete made of alkali activated slag and geopolymers. *Construction and Building Materials*, *189*, 797–803. Available from https://doi.org/10.1016/j.conbuildmat.2018.09.067.

Tabatabaeian, M., Khaloo, A., & Khaloo, H. (2019). An innovative high performance pervious concrete with polyester and epoxy resins. *Construction and Building Materials*, *228*, 116820. Available from https://doi.org/10.1016/j.conbuildmat.2019.116820.

Tang, B., Fan, M., Yang, Z., Sun, Y., & Yuan, L. (2023). A comparison study of aggregate carbonation and concrete carbonation for the enhancement of recycled aggregate pervious concrete. *Construction and Building Materials*, *371*, 130797. Available from https://doi.org/10.1016/j.conbuildmat.2023.130797.

Tang, C. W., Cheng, C. K., & Ean, L. W. (2022). Mix design and engineering properties of fiber-reinforced pervious concrete using lightweight aggregates. *Applied Sciences*, *12*(1), 524. Available from https://doi.org/10.3390/app12010524.

Tennis, P.D., Leming, M.L., & Akers, D.J. (2004). *Pervious concrete pavements*. EB302.02, Portland Cement Association, Skokie, Illinois and National Ready Mixed Concrete Association, SAD, 36 pp.

Teymouri, E., Wong, K. S., & Pauzi, N. N. (2023). Iron slag pervious concrete for reducing urban runoff contamination. *Journal of Building Engineering*, *70*, 106221. Available from https://doi.org/10.1016/j.jobe.2023.106221.

Toghroli, A., Mehrabi, P., Shariati, M., Trung, N. T., Jahandari, S., & Rasekh, H. (2020). Evaluating the use of recycled concrete aggregate and pozzolanic additives in fiber-reinforced pervious concrete with industrial and recycled fibers. *Construction and Building Materials*, *252*, 118997. Available from https://doi.org/10.1016/j.conbuildmat.2020.118997.

Um Maguesvari, M., & Sundararajan, T. (2017). Influěncia das cinzas volantes e agregados finos nas caracteŕisticas do concreto permeável. *International Journal of Applied Engineering Research*, *12*, 1598–1609.

Vijayalakshmi, R. (2021). Recent studies on the properties of pervious concrete; a sustainable solution for pavements and water treatment. *Civil and Environmental Engineering Reports*, *3*(31). Available from https://doi.org/10.2478/ceer-2021-0034.

Wild, S., & Khatib, J. M. (1997). Portlandite consumption in metakaolin cement pastes and mortars. *Cement and Concrete Research*, *27*(1), 137–146.

Yang, J., & Jiang, G. (2003). Experimental study on properties of pervious concrete pavement materials. *Cement and Concrete Research*, *33*, 381–386.

Yang, X., Liu, J., Li, H., & Ren, Q. (2020). Performance and ITZ of pervious concrete modified by vinyl acetate and ethylene copolymer dispersible powder. *Construction and Building Materials*, *235*, 117532. Available from https://doi.org/10.1016/j.conbuildmat.2019.117532.

Yeih, W., & Chang, J. J. (2019). The influences of cement type and curing condition on properties of pervious concrete made with electric arc furnace slag as aggregates. *Construction and Building Materials*, *197*, 813–820. Available from https://doi.org/10.1016/j.conbuildmat.2018.08.178.

Yu, F., Guo, J., Liu, J., Cai, H., & Huang, Y. (2023). A review of the pore structure of pervious concrete: Analyzing method, characterization parameters and the effect on performance. *Construction and Building Materials*, *365*, 129971. Available from https://doi.org/10.1016/j.conbuildmat.2022.129971.

Zhong, R., & Wille, K. (2015). Material design and characterization of high performance pervious concrete. *Construction and Building Materials*, *98*, 51–60.

Zhu, X., & Jiang, Z. (2023). Reuse of waste rubber in pervious concrete: Experiment and DEM simulation. *Journal of Building Engineering*, *71*, 106452. Available from https://doi.org/10.1016/j.jobe.2023.106452.

Sustainable recycled aggregate concrete materials and structures

7

Solomon Debbarma[1], Beng Wei Chong[2], Xijun Shi[2], Surender Singh[3] and Alexander S. Brand[4]
[1]Department of Civil Engineering, Indian Institute of Technology Bombay, Mumbai, Maharashtra, India, [2]Ingram School of Engineering, Texas State University, San Marcos, TX, United States, [3]Department of Civil Engineering, Indian Institute of Technology Madras, Chennai, Tamil Nadu, India, [4]Charles Edward Via, Jr. Department of Civil and Environmental Engineering, Virginia Polytechnic Institute and State University, Blacksburg, VA, United States

7.1 Introduction

Aggregate is an indispensable ingredient of concrete materials, typically accounting for around 70%–80% of the total volume of concrete. The enormous demand for concrete in new and existing construction requires an increasingly higher amount of aggregate materials. The depleting trend of many quality aggregate sources (e.g., Langer, 2011; Meininger & Stokowski, 2011) leads to a continuous increase in aggregate cost caused by higher energy consumption during aggregate production and transportation along with higher expenses to regulate environment-related issues. As such, alternative sources of aggregates that include recycled aggregates from various sources have the potential to be a supplement to current virgin aggregate reserves. The construction industry is now putting sustainability into their business strategies and recycled aggregate concrete (RAC) is one of their environmental management tools for achieving the sustainable development goals. Construction, automotive, plastic, steel manufacturing, and glass industries generate significant quantities of by-products that can be used as virgin aggregate replacements to produce Portland cement concrete (PCC). Table 7.1 presents the projected global waste generation and the amount of waste recycled and landfilled.

RAC is a viable solution to minimizing waste and reducing energy consumption as well as reducing the overall cost of a construction project. Significant research has been carried out to explore the potential of recycled aggregates for various civil engineering applications. However, their wide application in the field is limited because of socio-economic-enviro barriers and due to some technical difficulties. In this chapter, RAC performance at fresh and hardened concrete is reviewed.

Sustainable Concrete Materials and Structures. DOI: https://doi.org/10.1016/B978-0-443-15672-4.00007-3

Table 7.1 Waste statistics and overall recycling rate.

Type of waste	Total generated (in thousand tons)	Total recycled (in thousand tons)	Recycling rate	Total landfilled (in thousand tons)
Construction and demolition wastes	1424	1419	99%	5
Plastics	1001	57	6%	944
Used slag	169	166	99%	2
Glass	73	11	14	63
Scrap tires	26	25	95	1
Others (stones, ceramics, etc.)	249	30	N.A.	219

Source: From National Environment Agency Report for Singapore, 2022.

7.2 Types, sources, and need of recycled aggregates

7.2.1 Recycled concrete aggregates

Recycled concrete aggregates (RCAs) are produced from construction & demolition (C&D) waste, which is one of the most voluminous wastes generated worldwide. In the United States, the C&D waste produced from building demolition alone is estimated to be 123 million tons per year (Malešev et al., 2010). For countries such as China and India, the annual estimated production of C&D waste is around 1002 and 530 million tons per year, respectively (Akhtar & Sarmah, 2018). The increased amount of C&D waste not only demands more landfill space but also poses a threat to public and environmental safety by generating a considerable amount of dust and leachate (Cook et al., 2022). To overcome the challenges associated with natural aggregate shortage and the disposal of C&D waste, RCA has been actively promoted for use in PCC as virgin aggregate replacements (Danish & Mosaberpanah, 2022; McNeil & Kang, 2013; Wang et al., 2021). RCA is produced by a single-state or two-stage crushing of demolished concrete followed by screening and removal of contaminants such as reinforcing steel, wood, and gypsum (Malešev et al., 2010). This process can be carried out in stationary and or mobile plants. The current method for producing valuable recycled aggregates from C&D wastes is not only technically feasible but can also be adopted without a huge initial investment (Tam et al., 2018). A distinct feature of RCA is that it consists of a layer of residual mortar, which is the adhered mortar remaining on the original aggregates after crushing concrete (see Fig. 7.1). Typically, RCA contains 30%–60% of adhered mortar (Safiuddin et al., 2013). A greater amount of residual mortar is attached to the smaller size fractions of coarse aggregates. RCA is similar to crushed rock in particle shape, but the type of crushing equipment influences the gradation and

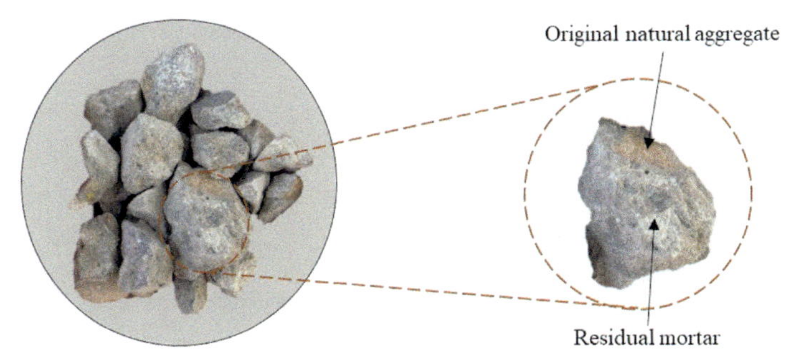

Figure 7.1 Recycled concrete aggregates.

other characteristics of crushed concrete (Prajapati et al., 2021). Though the physical and mechanical properties of RCAs are inferior to natural aggregates, they satisfy the requirements of American Society for Testing and Materials (ASTM), European Standards (EN), and Bureau of Indian Standards (BIS) specifications. RCA can be processed to a similar particle size distribution as the natural aggregates; however, they exhibit lower specific gravity, higher water absorption, and low impact and crushing strengths (Bairagi et al., 1990).

7.2.2 Reclaimed asphalt pavement

Reclaimed asphalt pavement (RAP) is the material produced from the milling or demolition of existing asphalt pavement for repair, maintenance, rehabilitation activities, etc. The increasing maintenance and rehabilitation of asphalt concrete pavement has generated excess RAP in some states. According to a survey, the possession of excess RAP has been reported by 91% of U.S. contractors (Hansen & Copeland, 2015). The expanding RAP stockpiling not only requires higher fees for space and regulation but also poses a threat to both environment and public safety. Although the use of RAP to make hot mix asphalt is a common practice, the replacement levels of RAP (i.e., 20%–25% (Hansen & Copeland, 2015)) in this application are not adequate to solve the RAP stockpiling issue. The use of RAP in PCC was found to be an effective way to promote greater consumption of RAP and mitigate RAP stockpiling issues (Shi, Mukhopadhyay, & Zollinger, 2018). Contractors have been motivated to use RAP as aggregate replacements to produce PCC wherever virgin aggregate shortage is a critical issue. A significant amount of laboratory research has been conducted worldwide on the utilization of RAP to make PCC along with a few field-level case studies (i.e., Berry et al., 2013; Singh & Ransinchung, 2020). Lab-based research on the use of RAP in PCC probably dates to the 1970s (Patankar & Williams, 1970), after which a considerable amount of effort has been continuously made around the world (Al-Oraimi et al., 2009; Debbarma, Singh, & Ransinchung, 2019; Delwar et al., 1997; Huang et al., 2006; Okafor, 2010; Singh et al., 2017b). The rapidly growing interest in exploring

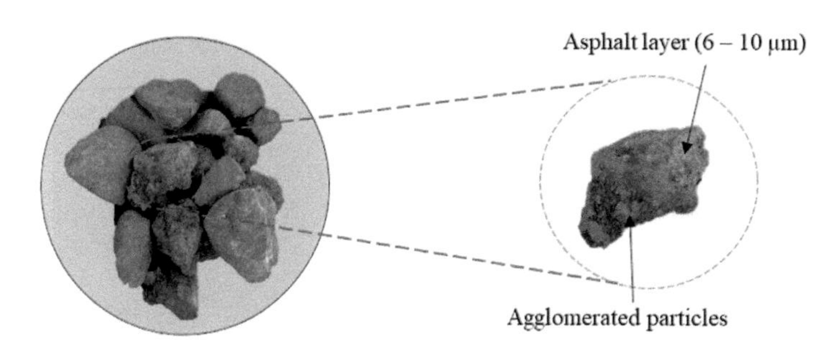

Figure 7.2 Reclaimed asphalt pavement aggregates. RAP aggregates consist of an oxidized asphalt layer of about 6–10 μm. In addition, finer particles are generally agglomerated into larger-sized particles. The effect of agglomerated particles is discussed in Section 7.4.

alternative ways to use RAP has motivated several Departments of Transportation and Toll Highway Authorities in the United States to support research projects focusing on using RAP in PCC (Berry et al., 2013; Brand, Roesler, et al., 2012; Mukhopadhyay & Shi, 2017; Tia et al., 2012). A visual representation of the RAP aggregates is shown in Fig. 7.2.

7.2.3 Waste tire rubber

Unrecycled tire waste is a global problem because scrap tires are nonbiodegradable, combustible, and can leach toxic substances into the ground. Approximately 1.3 billion automobile vehicles are currently used globally, and about 5.2 billion tires will be on the verge of disposal after being exhausted by these vehicles. By 2030, approximately 1.5 billion tires are estimated to be discarded as waste annually. According to the US scrap tire markets report, 4.46 million tons of scrap tires were generated in 2019 in the United States, and 0.6 million tons were disposed of in landfills (U.S. Tire Manufacturers Association, 2020). These pose serious concerns about environmental and public safety, in addition to the huge wastage of valuable materials that can be recycled and reused. Traditionally, these scrap tires are used for fuel production in many energy recovery plants and cement industries due to their high carbon content. A sustainable way to recycle tire waste is to utilize it in concrete production as a construction material. Ganjian et al. (2009) have classified the scrap tires into three categories viz., (1) shredded or chipped rubber that replaces coarse aggregates, (2) crumb rubber that replaces fine aggregates, and (3) ground rubber that can partially replace cement. The specific gravity of tire rubber aggregates typically ranges between 0.6 and 1.5 which makes it a potential lightweight aggregate for concrete.

7.2.4 Steel slag

Steel slag is a by-product of the steel-making industry produced during the extraction of impurities from steel with lime, dolomite, and other auxiliary materials. The U.S. and China combined generate about 16–100 million tons of steel slag every year (Dong et al., 2021). Depending on the steel-making process, steel can be broadly categorized into blast oxygen furnace (BOF) slag, electric arc furnace (EAF) slag, and ladle furnace (LF) slag. The different steel-making process and production of steel slags is shown in Fig. 7.3. Compared to BOF slag and EAF slag which consists of raw materials such as iron, steel scrap, lime, and dolomite, the LF slag is used for secondary refining. Worldwide, 71% of steel is produced by BOF, 29% is produced by EAF (World Steel Association, 2022), and an estimated 126 kg of BOF slag and 169 kg of EAF slag is produced for every ton of steel produced (World Steel Association, 2021). Therefore, knowing that 1.9 billion tons of steel are produced globally every year (World Steel Association, 2022), an estimated 263 million tons of steel slag are generated per year. Japan has the highest recycling rate of steel slag being recycled every year at approximately 98.4% (Dong et al., 2021), while countries such as India and China recycle only about 30% and 40% of the total steel slag generated. In 2018, the U.S. recycled over 60% of steel slag as a road base material. Likewise, Europe produced nearly 22 million tons of steel slag in 2012 and recycled 40% of them as road construction material (Dong et al., 2021).

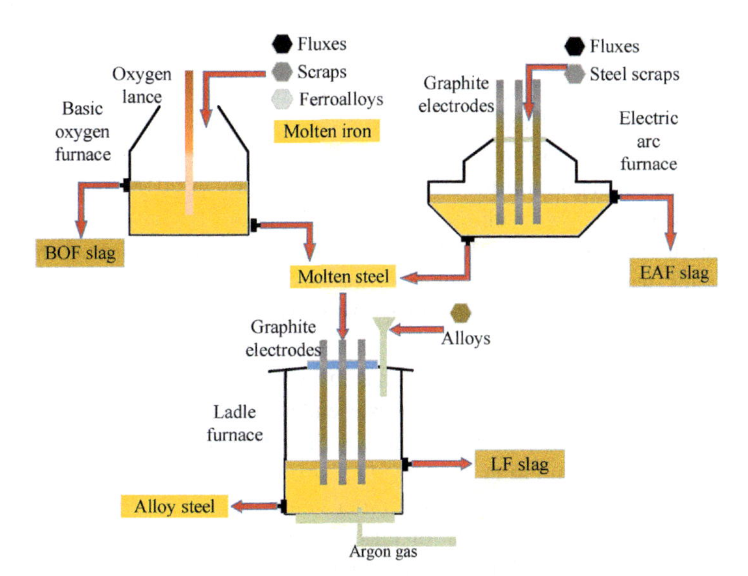

Figure 7.3 Different steel-making and production processes.
Source: Reprinted with Permission From Dong, Q., Wang, G., Chen, X., Tan, J., & Gu, X. (2021). Recycling of steel slag aggregate in Portland cement concrete: An overview. *Journal of Cleaner Production, 282*, 124447. https://doi.org/10.1016/j.jclepro.2020.124447.

Because of its similar characteristics to cement clinker, fine ground steel slag can be used as an inert filler in cement-based materials, while coarse aggregates can be directly used as a replacement for natural aggregates. Steel slag is a crushed product with a black color stone appearance; it has good angularity and a rough surface texture. Compared to natural aggregates, it has a higher specific gravity (typically ranges between 2.9 and 3.4), higher abrasion resistance, lower aggregate crushing value, and better resistance to fragmentation. Because of these excellent intrinsic properties, a considerable amount of effort has been made in using steel slag as a virgin aggregate replacement for concrete production. In spite of these benefits, the use of steel slag in concrete is limited as it contains free CaO and MgO that can lead to concrete volume instability, since these oxides will expand volumetrically by 92% and 120% when hydrated to form $Ca(OH)_2$ and $Mg(OH)_2$, respectively (Erlin & Jana, 2003). The total steel furnace slag (SFS) expansion can be upwards of 10% or more (Emery, 1982), since the free CaO contents in BOF and EAF slags can range from 1% to 10% and 0%–4%, respectively (Balcázar et al., 1999). The major oxides in steel slag are CaO (20%–60%), SiO_2 (10%–40%), and FeO (5%–40%) while the contents of other oxides including MgO, Al_2O_3, MnO, P_2O_5, and free CaO are relatively lower (Yildirim & Prezzi, 2011). The tricalcium silicate and dicalcium silicate contents are typically higher in BOF slag, hence showing better cementitious properties compared to EAF slag.

7.2.5 Waste plastics

Plastic is one of the most used materials in the world, and the production of plastic has grown at a pace that exceeds most materials such as aluminum and steel. In 1950, approximately 2 million metric tons of plastic were produced globally. This number soared exponentially to 322 million metric tons in 2015, and the trend was only expected to continue with the versatility of plastic as a material in all aspects of applications (Brooks et al., 2018). According to a study conducted in 2016, the United States was believed to generate the most plastic waste in both the total amount and per capita (Law et al., 2020). However, plastic recycling in the United States has lagged behind other nations. In 2018, the United States Environmental Protection Agency (USEPA) reported that the U.S. recycled only about 8% of its plastic waste, and about 80% of the waste ended up in landfill. Even worse, this number has seen a decline in recent years, with a 2022 report from an environmental advocacy group estimating that the U.S. recycled less than 5% of its plastic waste (Greenpeace, 2022). Meanwhile, other countries and regions such as Europe and China recycled 30% and 25% of plastic waste, respectively. The numbers far exceeded the United States but still were uncomforting in the face of massive plastic production across the globe (Geyer et al., 2017). The United States is also a leading exporter of plastic waste to China and other Southeast Asian developing countries, where the waste remained mostly landfilled due to the inadequacy of technology and infrastructure for effective waste management. In 2017, China issued a ban on plastic import, while all prominent plastic importers such as

Indonesia, Malaysia, and Vietnam have tightened their control on this plastic in the form of greater inspection, limitation, and taxation.

As the world scrambles to find a solution to the ever-growing problem of plastic waste, the accumulated impacts from the rudimental management of plastic waste begin to surface. Today, microplastics (MPs) fills the air, soil, and water of every corner of the earth. From the landfill, plastic particles contaminated the air of the surrounding environment. At the same time, macroplastics will degrade to form MPs over a long span of time, causing infertility in the soil as well as a decline in agricultural yield (Sajjad et al., 2022). Moreover, MP had been known as an effective carrier of lethal chemical pollutants and bacteria, furthering the contamination of the environment. In the water, studies had detected the presence of MP in the seawater and marine creatures of the Tropical Eastern Pacific, the Galápagos archipelago, and the Gulf of Mexico (Alfaro-Núñez et al., 2021; Wojnowska-Baryła et al., 2022). An abundance of MP had also been discovered in the deep sea sediments of the ocean seafloor (Barrett et al., 2020).

MP accumulation had been linked to the health of organs, reproductive capability, and the feeding behavior of fishes and mammals (Zolotova et al., 2022). Humans had certainly not been spared by the prevalence of MP contamination in our environment. The human body easily picked up MP from breathing air, drinking water, and even the intake of crops and seafood due to the MP in our soil and water. The health effect of MP ingestion is an ongoing and complicated topic that is yet to be fully answered. However, exposure to MP had been linked to oxidative stress, DNA damage, and inflammation (Jassim, 2023). Apart from the potential health effect due to the direct ingestion of plastic itself, MP may also carry other hazardous chemicals and heavy metals that are detrimental to human health (Campanale et al., 2020). In 2022, scientists managed to find traces of plastic in human blood (Leslie et al., 2022), renewing concerns about the invasion of plastic waste as a major public health risk.

One approach to tackling the prevalence of plastic waste is to utilize waste plastic as a construction material. Transforming construction material usage into an outlet of waste management is an elegant solution, as it also reduces the consumption of other highly sought, nonrenewable materials such as stone and sand, on top of producing unique composite materials that may fit the requirement of the ever-evolving construction practices. Plastic categorization followed the convention established by the US Society of Plastic (SPI) as shown in Fig. 7.4, as different types of plastic possessed different properties. To date, a significant number of studies on waste plastic in concrete applications had been conducted to varying degrees of success.

7.2.6 Waste glass

Glass is the most generated nonbiodegradable waste after end-of-life tires and plastic around the globe (Ferdous et al., 2021). About 130 million tons of glass waste is generated every year globally (Ferdous et al., 2021). In the United States, the glass generation in all products was nearly 12 million tons in 2018, which was 4.2% of

Recycling No.	Symbol	Abbreviation	Polymer Name	Uses Once Recycled
1	1 PETE	PETE or PET	Polyethylene terephthalate	Polyester fibers, thermoformed sheet, strapping and soft drink bottles.
2	2 HDPE	HDPE	High-density polyethylene	Bottles, grocery bags, recycling bins, agricultural pipe, base cups, car stops, playground equipment and plastic lumber.
3	3 V	PVC or V	Polyvinyl chloride	Pipe, fencing and nonfood bottles.
4	4 LDPE	LDPE	Low-density polyethylene	Plastic bags, six-pack rings, various containers, dispensing bottles, wash bottles, tubing and various molded laboratory equipment.
5	5 PP	PP	Polypropylene	Auto parts, industrial fibers, food containers and dishware.
6	6 PS	PS	Polystyrene	Desk accessories, cafeteria trays, toys, videocassettes and cases, and insulation board and other expanded polystyrene products (e.g., Styrofoam).
7	7 OTHER	Other	Other plastics, including acrylic, acrylonitrile butadiene styrene, fiberglass, nylon, polycarbonate and polylactic acid.	

Figure 7.4 SPI classification of plastics.

all municipal solid wastes (MSWs) generation (USPA, 2022). The amount of glass recycled was 3.1 million tons in that year, for a recycling rate of 31.3% while approximately 7.6 million tons of MSW glass was landfilled, which was 5.2% of all MSW landfilled that year (USPA, 2022). The ecology and ecosystem are negatively impacted by disposing of it in landfills since it is nonbiodegradable (Bisht & Ramana, 2018). Consequently, there is much to investigate the feasibility of waste glass in various civil engineering applications. According to literature, recycled waste glass could be used for backfilling, tiles, masonry blocks, paving blocks, cement concrete, asphalt concrete, cement concrete (Mostofinejad et al., 2020), subbase (Saberian et al., 2020), and other ornamental uses (Kou & Poon, 2009). Crushed waste glass has the potential to function as a pozzolanic material due to its chemical composition, which includes significant amounts of silicon and calcium with an amorphous structure (Jani & Hogland, 2014). However, depending on the composition, coarse particles of glass can cause expansion due to the alkali-silica reaction (ASR) in concrete (Saccani & Bignozzi, 2010). Waste glass has also been researched to be utilized as a partial replacement for natural fine aggregates to produce PCC mixes (Ismail & AL-Hashmi, 2009; Soliman & Tagnit-Hamou, 2017). The specific gravity of waste glass aggregates typically ranges between 2.19 and 2.89 (Ismail & AL-Hashmi, 2009; Park et al., 2004; Ramakrishnan et al., 2017; Soliman & Tagnit-Hamou, 2017). The unit weight of waste glass used in replacement for natural fine aggregates (4–16 mm) was reported as 1493 kg/m^3 (Topçu & Canbaz, 2004).

Glass particles are angular in shape and have a rough surface (Nodehi et al., 2023; Shao et al., 2000). The combined effect of pozzolanic activity and the good particle-to-particle interlocking of the glass particles can significantly enhance the strength and durability of the concrete. Consequently, concrete structures involving glass materials are reportedly durable and can withstand extreme environmental conditions. For instance, the use of waste glass aggregates not only decreases the thermal conductivity (because of the low specific heat of the glass compared to natural sand and because of the pozzolanic activity of glass (Poutos et al., 2008)) of the concrete but makes it more energy-efficient and environmentally friendly. In spite of its numerous benefits, the recycling rate of waste glass was reportedly 11% only (see Table 7.1), according to the National Environment Agency Report.

7.3 Effect of recycled concrete aggregate on concrete performance

Incorporation of RCA of different origins and qualities can produce concretes with comparable workability and strength to normal concrete (Brand et al., 2015; Butler et al., 2013). RCA concrete can be mixed and finished as easily as natural aggregate concrete. However, saturated RCA is generally desired for RCA concrete to obtain the same workability as natural aggregate concrete. Contrary, it has been reported that greater angularity and surface roughness of RCA particles can reduce the workability of concrete to make it more difficult to finish properly (Safiuddin et al., 2013). This decrease in workability becomes more prominent at higher RCA incorporation levels. However, the use of high-range water-reducing admixtures can improve the workability of the RCA concrete mixes. In addition, the two-stage mixing approach for RCA has been shown to improve workability and mechanical properties (Tam et al., 2005; Tam & Tam, 2007, 2008). A number of studies have demonstrated that the moisture condition of the RCA will affect concrete performance, with many studies suggesting that a presaturated condition is beneficial for macroscopic properties (Brand et al., 2015; De Oliveira & Vazquez, 1996; Pickel et al., 2017) and microstructural development (Brand & Roesler, 2018a; Le et al., 2017; Leite & Monteiro, 2016). Because some of the residual mortar rubs off during mixing to produce additional fines in the concrete mix, these fines absorb some of the mixing water and thus reduce the bleeding in concrete (Safiuddin et al., 2013). In addition, the combined effect of higher fines in the concrete mix and the increased angularity and surface roughness at higher RCA content greatly contributes to reduced bleeding and increased cohesiveness of the RCA concrete (Safiuddin et al., 2011).

Several methods viz. American Concrete Institute (ACI) method, Indian Standards (IS) method, relative ratio method (RRL) method, Surface, and Angularity Index method can be used for designing RCA concretes. The ACI method which can produce concrete grades up to 30 MPa has been proven as one of the most acceptable methods for the mix design of RCA concrete. Later, Fathifazl et al. (2009) proposed the equivalent mortar volume (EMV) method of mix design for RCA concrete. In this method, RCA is considered a two-phase material consisting of old natural aggregates and residual mortar. This method considers the residual mortar as a part of the total mortar in the RCA concrete. The RCA concrete produced using the ERM method can exhibit similar or better fresh and hardened concrete properties compared to the concretes designed using traditional mix design methods. In fact, a significant amount of cement and fine aggregates can be reduced using the ERM mix design method. Another method that considers the particle packing approach was recently proposed by Pradhan et al. (2017). The particle packing method is a two-stage mix design approach that aims to achieve the maximum possible packing density. Identifying the coarse-to-fine aggregate ratio to obtain the maximum packing density followed by the determination of paste content are the key steps in this method. In addition to the reduction in cement requirement, it is also possible to produce concretes utilizing 100% of RCAs using this method.

RAC produced using RCA can have a detrimental or comparable performance to normal concrete depending on its source, size, incorporation level, gradation, and physical properties. A study by Pani et al. (2020) shows that the properties of the parent concrete have little or no impact on the mechanical properties of RAC. Instead, the type of cement utilized and the mix design of the RAC play a significant role in its mechanical performance (Lotfi et al., 2015). Butler et al. (2013) studied the effect of RCA produced from different sources, on the mechanical strength of concrete. The authors concluded that it is possible to produce an RAC using different quality RCA with comparable workability and compressive strength compared to the concretes made without RCA. In most studies, the incorporation of coarse RCA generally leads to compressive strength reductions in the range of 3%–35% when compared to the concretes made without any RCA (Huda & Shahria Alam, 2015; Moallemi Pour & Alam, 2016; Padmini et al., 2009; Topçu & Şengel, 2004). The reduction in the compressive strength increases as the RCA incorporation level increases (Ahmad et al., 2021). In addition, the use of 100% RCA (coarse + fine) has been reported to show a much higher compressive strength reduction of about 40% compared to normal concrete. The effect of RCAs on the flexural strength and split tensile strength has also been reported with a percentage reduction in the range of 10%–25% (Chen et al., 2003; Katz, 2003; Topçu & Şengel, 2004) and 10%–20% (Katz, 2003; Rakshvir & Barai, 2006; Silva et al., 2015), respectively. Likewise, several researchers also reported a decrease in the modulus of elasticity (MOE) of RAC, and the percentage decrease becomes much higher as the RCA replacement level increases (Chen et al., 2003; Rahal, 2007; Xiao et al., 2005).

The reduction in the mechanical performance of the RAC is primarily due to the higher porosity in the residual/adhered mortar of the RCA. As a result, the adhered mortar can be considered the weakest point in RCA concrete which greatly influences its overall mechanical strength. In addition, the poor interfacial transition zone (ITZ) bonding between the RCA and the cement/mortar paste can also lead to drastic mechanical strength reduction compared to normal concrete. RAC made with RCAs typically has three different ITZs: (1) old ITZ between original aggregate and original cement mortar paste, (2) new ITZ between the adhered old mortar

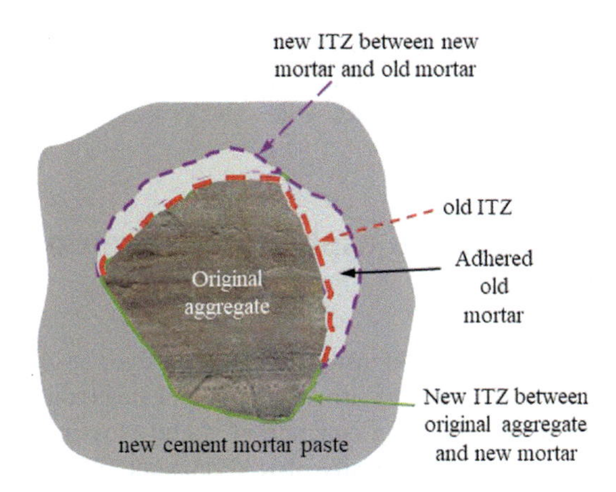

Figure 7.5 Schematic diagram of different ITZs in RCA concrete.

and the new cement mortar paste, and (3) new ITZ between the original aggregate and the new cement mortar paste. A schematic representation is shown in Fig. 7.5.

RCA concrete has been reported to exhibit lower or similar durability performance compared to normal concrete. Nandanam et al. (2021) demonstrated that ultra-high performance self-compacting concrete can be produced by utilizing 100% of RCAs and mineral admixtures such as fly ash, metakaolin, and ground granulated blast furnace slag. However, most studies reported that permeability index, chloride conductivity, and sorptivity increase as the RCA replacement increases (Olorunsogo & Padayachee, 2002). Domingo-Cabo et al. (2009) demonstrated that drying shrinkage of RCA concrete is generally higher than normal concrete and increases as the RCA replacement level increases. This is due to the increased adhered mortar content introduced in the concrete mixture. Because of the higher paste content and fewer natural aggregates in the RCA concrete, there is less restraint, and hence, shows higher levels of shrinkage. Contrary, it has been reported that RCA concrete can show similar levels of drying shrinkage compared to normal concrete (Adams et al., 2016). Available literature shows that RCA concrete can have poor freeze−thaw resistance, corrosion resistance, sulfate resistance, carbonation resistance, and acid resistance, respectively. The magnitude of the effect of RCA on the fresh and hardened properties is greatly influenced by several factors such as water-to-cement ratio, RCA content, the type and size of RCA, the physical properties of RCA, the quality of parent concrete of RCA, moisture condition of RCA, curing condition, cement content, and air entrainment. Generally, up to 30% of coarse RCA can be incorporated without significantly affecting the mechanical properties and durability of the concrete. Higher RCA replacement levels could also produce concretes of comparable mechanical strength and durability to normal concrete. However, it is recommended to evaluate the performance of the RCA concrete depending on its application rather than simple mix proportioning to achieve its target compressive strength. Several methods to improve the performance of RCA concrete include improving the RCA quality, adjusting the w/c ratio, incorporating pozzolanic materials, using new mixing techniques, extending the curing approach, and presoaking RCA in water or a cementitious slurry.

7.4 Effect of reclaimed asphalt pavement on concrete performance

RAP aggregates consist of a low-density thin asphalt layer ($\sim$6–10 μm) engulfing its surface. The low-density coating around the aggregates reduces the overall density of the concrete, thus, producing concrete of slightly lower density compared to natural aggregate concretes. The density of RAP concrete typically ranges between 2150 and 2400 kg/m^3. Brand and Roesler (2015) demonstrated the use of a coarser fraction of RAP (up to 50% by vol. replacement) produced concrete with a density of 5%–6% lower than the concrete made with natural aggregates only. Delwar et al. (1997) also reported that the use of any fractions of RAP (coarse, fine, or a combination of coarse and fine) showed less than a 3% variation in the unit weight of the concrete. Similarly, Hossiney et al. (2010) observed less than 2% variation in the concrete's density when 40% of natural aggregate was replaced by coarse RAP. When the complete replacement of natural aggregate by RAP was studied, not more than a 6% variation in the unit weight was noticed (Singh et al., 2017c).

Due to their hydrophobic nature, RAP aggregates generally have a lower water absorption compared to natural aggregates (Boussetta et al., 2020; Debbarma, Singh, & Ransinchung, 2019; Shi et al., 2017). However, it has been reported that RAP aggregates can have a higher water absorption possibly due to the agglomeration of finer particles into coarser-sized particles (Shi et al., 2020). Consequently, several researchers demonstrated the incorporation of RAP aggregates (coarse, fine, or combined fractions) can significantly reduce the workability (measured in terms of initial slump) of the fresh concrete mix (Mukhopadhyay & Shi, 2017; Shi et al., 2017; Singh et al., 2018, 2017a). This is because of the combined effect of viscous asphalt coating and high amounts of agglomerated particles on the coarser RAP particles that act as tiny water reservoirs (see Fig. 7.6), thus negatively affecting the workability of the concrete mixture. However, other authors have found that RAP can increase workability (Brand & Roesler, 2015), possibly due to the hydrophobic nature of the asphalt coating. In lean concrete mixes, like roller-compacted concrete (RCC), the agglomeration effect of RAP particles contributes to higher water demand, thus, resulting in higher optimum moisture content values (Debbarma, Ransinchung, & Singh, 2019; Debbarma et al., 2019; Debbarma & Ransinchung, 2021a). In addition, RAP procured using full-depth reclamation can produce aggregates with an additional layer of dust contaminants which can reduce the concrete's slump significantly (Singh et al., 2018).

No clear trends have been reported in the air content of RAP concrete mixes. The incorporation of a coarser fraction of RAP has been reportedly found to increase the air content by about 17%–50% (Brand & Roesler, 2015, 2016; Huang

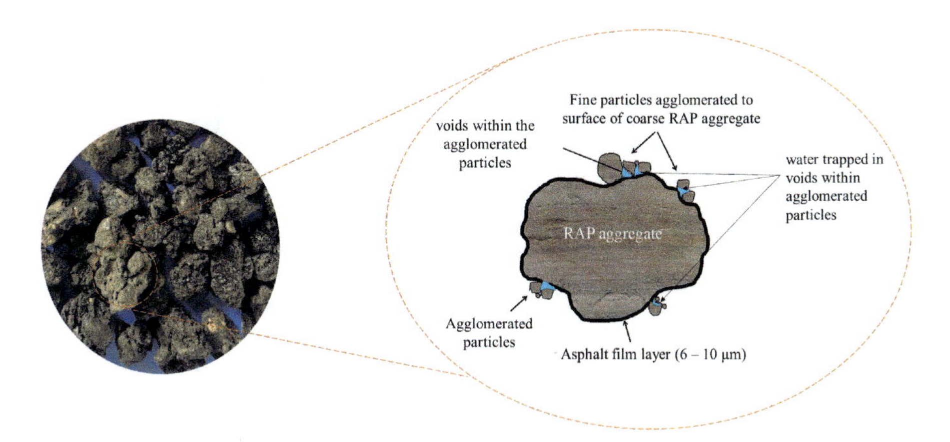

Figure 7.6 Schematic representation of agglomeration effect of RAP particles.

et al., 2005; Singh et al., 2017a). Contrary, Shi et al. (2017) concluded the addition of 40% of coarse RAP (gap- and dense-graded) could reduce the air content considerably. Brand, Smith, et al. (2012) did not discern a clear trend when comparing the fresh air content of unprocessed RAP to processed (sieved or washed) RAP, even though the air entrainment admixture dosage was equivalent for all mixtures. More research considering the aggregate shape properties, asphalt concentration, and absorption is needed to understand the effect of RAP on the air content of concrete mixes.

The incorporation of RAP causes considerable changes in concrete mechanical properties, which will lead to significant impacts on pavement behavior—concrete pavement is a major application for RAP concrete. One immediate concern regarding the use of RAP in concrete pavement is the reduction in strength. More specifically, concrete flexural strength is one of the most important inputs for pavement slab thickness design, and the use of RAP leads to a reduction in the flexural strength of RAP–PCC. Fortunately, it has been reported that the flexural strength reduction is not significant as the compressive strength reduction, and the flexural strength reduction can be controlled within 25% if not more than 40% coarse RAP is used to replace the same volume of virgin coarse aggregate in concrete. To compensate for the reduced strength, a thicker pavement slab may need to be considered in the design for RAP–PCC pavement. However, Brand et al. (2014) demonstrated that a thicker slab may not be required; despite the reduced flexural strength, the concrete with RAP was found to have a higher slab capacity than the virgin aggregate concrete, which was argued to be attributable to the improved fracture properties of the concrete with RAP. One pavement design case study found that a two-lift concrete slab design with RAP in the bottom lift would not require a different total thickness design relative to conventional concrete (Gillen et al., 2012).

It was observed that the asphalt film around RAP particles is the primary weak zone in the RAP–PCC as crack easily propagates through the asphalt film under force during strength testing (Mukhopadhyay & Shi, 2017). As seen in Fig. 7.7, the crack propagates around the aggregate in natural aggregate concrete while it

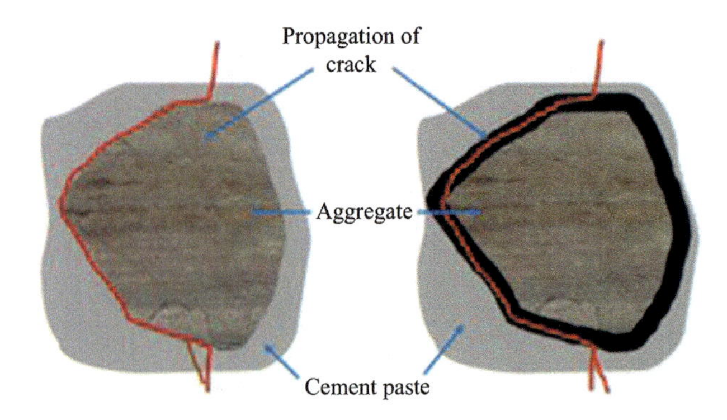

Figure 7.7 Crack propagation in concrete made with RAP aggregates.

propagates through the asphalt film in RAP concrete mixes. Brand and Roesler (2017b) reported the asphalt-cohesion failure as the dominant mode of failure in RAP concrete mixes. This means after the propagation of cracks, some portion of asphalt remains on the aggregate while some remain on the cement paste/mortar, thus reducing the overall strength of the concrete (see Fig. 7.8). It is also shown that the asphalt volumetric content in the concrete mixture is highly relevant to RAP–PCC properties. Accordingly, a term called total asphalt volumetric fraction (TAVF) was used to represent asphalt content contributed by the RAP used to make concrete (Shi et al., 2017). TAVF is defined as the volume fraction of the asphalt in the total aggregate mixture (virgin aggregates plus RAP aggregates).

The asphalt in RAP (even though aged) is less stiff than cement matrix and virgin aggregates, but Brand and Roesler (2015) found that this effect was minimal by measuring the temperature-dependent dynamic modulus of concrete with RAP. Concrete with RAP does have a lower MOE relative to virgin aggregate concrete, but it is not caused by the reduced stiffness of the asphalt, especially considering that the total volume of asphalt in a concrete mix with RAP would be a fraction of one percent. Rather, the reduced MOE demonstrated by Brand and Roesler (2017a) was caused by the larger, more porous ITZ that is formed around the RAP aggregates. The larger, more porous ITZ might be attributable to the organic compounds that leach from the asphalt exposed to the concrete pore solution (Behravan et al., 2023).

Concrete pavement slab with reduced MOE is anticipated to have better-cracking resistance due to the lower stress level (Shi, Mukhopadhyay, et al., 2019). In addition, the higher viscoelasticity of RAP-based concrete could potentially lead to higher creep in the concrete structures, which further relaxes stress (Shi et al., 2020). On the other hand, the low MOE could cause higher slab deflections, though. The higher differential energy caused by the deflection differential between the unloaded and loaded slabs results in a higher amount of base erosion, which eventually leads to higher slab faulting (Shi, Mukhopadhyay, et al., 2019). This finding has been validated in a field study in Oklahoma. The field evaluation findings based on an RCA pavement section in

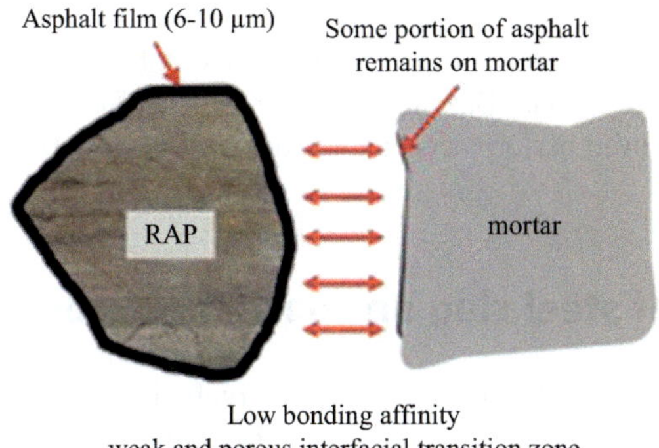

Figure 7.8 Schematic diagram of asphalt-cohesion failure in RAP concrete.

Oklahoma can be applicable to the RAP–PCC pavement case because RAP–PCC and RCA-PCC behave similarly in terms of having reduced MOE and modulus of rupture (MOR) and increased coefficient of thermal expansion (CoTE). Falling weight deflectometer results from the RCA-pavement field study showed that slabs built with the RCA exhibited higher slab deflection differential compared to control slabs. Distress survey data confirmed that there existed higher joint faulting in the RCA concrete pavement section relative to the control section (Shi, Mukhopadhyay, & Zollinger, 2019). To account for the higher base erosion caused by the softer slab, the use of stronger base materials is highly recommended for pavements built with recycled aggregate-based concrete slabs.

The increased CoTE and drying shrinkage of the RAP–PCC could lead to lower concrete pavement performance. According to the simulation studies by the authors (Shi, Zollinger, & Mukhopadhyay, 2018; Shi, Mukhopadhyay, et al., 2019), the increased CoTE causes higher flexural stress levels in concrete labs, leading to a higher chance of fatigue cracking. Reducing joint spacing turned out to be effective in reducing slab stress. Blending recycled aggregates with aggregate having a low CoTE (such as limestone, granite, or basalt) and using shrinkage-reducing admixture can also be useful to mitigate the problem.

The change in the thermal properties will not cause a significant change in the temperature profile in the pavement for the RAP–PCC mixtures (Shi, Mukhopadhyay, et al., 2019). Therefore the effect of RAP on pavement performance due to the pavement temperature change is minimal. However, it should be noted that pavement slabs built with a low thermal conductivity material will develop an increased temperature gradient, causing higher thermal stress and deflection in the slab. It is also shown that pavement with lower thermal conductivity could have a hotter pavement surface during summer and a colder surface during winter, and this may lead to a higher urban heat island effect and extra snow and ice formation, respectively (Shi, Rew, et al., 2019).

In addition, the RAP–PCC exhibits improved fracture properties (Brand et al., 2014; Brand & Roesler, 2015, 2016; Moaveni et al., 2016; Shi, Mirsayar, et al., 2019).

The size effect theory suggests that, compared to flexural strength, the concrete fracture properties are more relevant to the performance of intermediate and large-sized structures such as pavements (Bažant & Oh, 1983). Therefore RAP−PCC pavement is likely to have improved performance, as was demonstrated by Brand et al. (2014).

7.5 Effect of steel slag on concrete performance

The performance of steel slag concrete is largely influenced by several factors including the mix design, cement type, and the properties of the steel slag aggregates. A review by Brand and Fanijo (2020) demonstrated the significant variability of the results in the literature. Because of the rough surface texture of EAF slag, the workability is decreased which makes it difficult to place and compact the fresh concrete. To obtain the desired workability, a high-range water-reducing admixture is generally required (Dong et al., 2021). EAF slag can lead to increased bleeding and segregation due to its high angularity and rough surface texture. Steel slag concrete is heavy weight and can have densities typically 10%−30% higher than natural aggregate concretes (Dong et al., 2021). Lai et al. (2021) demonstrated that BOF slag concrete has a higher density (2413−2543 kg/m^3) compared to normal concrete (2388 kg/m^3). Because it is heavier than normal concrete, BOF slag concrete could be beneficial to earth retaining structures where the heavy weight of concrete can increase the stability and basements or off-shore structure where the weight can improve the floating resistance.

Concretes made with properly treated steel slag aggregates generally have higher concrete strength if they are well-designed. The use of BOF slag as a replacement for natural coarse/fine aggregate was reported to improve the compressive strength of the concrete (Lai et al., 2021). This improvement was more prominent at later curing ages primarily due to the cementitious behavior of the coarser BOF slag aggregate since it contained tricalcium silicate and dicalcium silicates. In RACs, the introduction of steel slag aggregates as a 67% replacement of RCA has also been observed to significantly enhance the compressive strength, flexural strength, and MOE of RCA concretes (Qasrawi, 2014). In another study by Anastasiou et al. (2014), the use of mixed C&D wastes as fine aggregate and EAF slag as coarse aggregate was reported to produce concrete with a compressive strength of 30 MPa and shows adequate durability for low-grade applications. In addition, the incorporation of higher steel slag has been noticed to produce concretes with higher concrete strength. The improvement in the performance of concrete due to the addition of steel slag is primarily due to the improved bond between the steel slag and cement/mortar paste, higher angularity, rough surface texture of steel slag, and denser concrete matrix. Moreover, the cementitious materials on the surface of steel slag can enhance the ITZ between the aggregate and bulk cement paste, thereby, improving the concrete's strength, although the properties of the ITZ are strongly dependent on the type and composition of the steel slag (Brand & Roesler, 2018b). In addition, some steel slags will have a surface layer of a different composition

(Brand & Roesler, 2018b; Coomarasamy & Walzak, 1995; Kawamura et al., 1983; Pang et al., 2015), which has been argued to detrimentally affect the concrete performance.

Contrarily, some studies demonstrated a decrease in the concrete strength which is mainly attributed to the volume expansion caused by the reaction of excess free CaO in the steel slag aggregates (Lam et al., 2017, 2018). Limited research on the durability of steel slag concrete shows contradictory findings. Steel slag aggregates can show detrimental effects on the durability of concrete due to their poor quality and improper mix design while some studies reported comparable or slightly better chloride resistance and carbonation resistance than normal concrete (Ortega-López et al., 2018; Ting et al., 2020; Wang et al., 2020). On the other hand, steel slag concrete can have two to three times better freeze–thaw resistance than normal concrete but can have lower resistance to acid and salt corrosion (Pang et al., 2016). This is mainly due to the reaction between cementitious materials and free CaO with acid and salt slowly resulting in the formation of expansive hydration products. Nevertheless, this detrimental effect can be mitigated using certain mineral additives and fibers to improve the permeability, and thus, strength. Another way to improve the performance of steel slag concrete is the pretreatment (preliminary treatment followed by posttreatment) of steel slag aggregates before their use in concrete. The preliminary treatment is generally done to reduce the free CaO and MgO contents and can be achieved by several methods including natural air cooling, water spray, instantaneous slag chill, water quenching, air quenching and modifying process, natural weathering, and accelerated carbonation (Dong et al., 2021). The posttreatment is usually done for volume stability and the most common methods include natural aging, SiO_2 modification, autoclaved process, carbonation, acidizing, and sealing the concrete surface (Dong et al., 2021).

7.6 Effect of waste tire rubber on concrete performance

Waste tire rubber concrete is different from normal concrete because it has lower unit weight, higher ductility, better freeze–thaw resistance, and improved thermal and sound conductivities. A major drawback of waste tire rubber concrete is its poor workability and low mechanical strength. However, a number of researchers have proposed processing techniques to mitigate the reduced properties (Tran et al., 2022). It has been reported that the workability decreases as the grade and incorporation levels of the crumb rubber increase. This is likely due to the reduced flowability of the larger particles which can be explained by reduced interparticle friction between the rubber particles and cement grains, thus, lowering the unit weight of the mix (Thomas & Gupta, 2016). Contrarily, some researchers reported that crumb rubber exhibits hyper-fluid behavior while normal concrete exhibits fluid behavior, thus, enhancing the workability of the concrete (Aiello & Leuzzi, 2010). The use of high-range water-reducing admixtures can limit the reduction in workability due to the addition of rubber particles (Youssf et al., 2014). With the increase in waste tire rubber content, higher compressive strength reduction is prominent. This is because of the inferior properties of the waste tire rubber, higher voids, and poor interfacial bonding between the aggregate and the cement paste/mortar (Ganjian et al., 2009; Gesoğlu et al., 2014a, 2014b). The addition of tire rubber ash up to 10% reported an increase in compressive strength (Al-Akhras & Smadi, 2004). Similar to compressive strength, a reduction in flexural strength was also reported due to poor interfacial bonding (Ganjian et al., 2009; Su et al., 2015). The presence of rubber prevents concrete from a brittle failure rather it leads to deformation without completely losing cohesion or strength (Yilmaz & Degirmenci, 2009). The addition of waste tire rubber decreases the material's overall stiffness and rigidity, lowering the elasticity modulus (Ganjian et al., 2009; Gesoğlu et al., 2014a, 2014b). Due to internal tensions and shrinkage, the rubber particles have a lower CoTE than the surrounding cement paste/mortar. It was observed rubber particles may hinder calcium silicate hydrates forming, which may be caused by organic compounds leaching from the rubber (Tran et al., 2023), and the reduced calcium silicate hydrate may cause increased shrinkage (Hunag et al., 2016; Uygunoğlu & Topçu, 2010). On the contrary, some researchers observed that waste tire rubber holds concrete constituent particles together, hence, preventing cracking (Raghavan et al., 1998). In addition, concrete-containing waste tire rubber was observed to have good freeze-and-thaw resistance (Gesoğlu et al., 2014a, 2014b; Zhu et al., 2012). Concrete-containing waste rubber tire has several other advantages such as low heat conductivity and increased sound absorption. Also, concrete-containing waste tire rubber is highly resistant to chloride ion penetration (Gupta et al., 2014; Thomas et al., 2015).

It was reported that a maximum replacement level of 20% could be used to prevent severe strength reduction (Holmes et al., 2014). Poor interfacial bonding between the aggregate and cement paste/mortar was observed to be the main cause of loss in strength. Chemical modifiers, surface treatments, or specialized adhesives could be used to strengthen the bonding interface between aggregate and cement paste/mortar, resulting in improved load transfer and higher strength (Dong et al., 2021). In addition, cementitious elements, such as fly ash or silica fume, can be used to improve the pozzolanic reactions and strength development (Ganjian et al., 2009; Gesoğlu et al., 2014a, 2014b; Onuaguluchi & Panesar, 2014).

7.7 Effect of waste plastics on concrete performance

Polyethylene terephthalate (PET) plastic is the most common form of plastic waste used in bottled water and food packaging (Aslani et al., 2021). It is also the easiest plastic to recycle, hence the rate of PET recycling is often the highest among plastic categories. In the construction industry, PET plastic was often explored as an alternative to coarse or fine aggregate, depending on the size that it had been shredded into. Concrete with PET plastic aggregate had demonstrated a multitude of superior qualities compared to conventional concrete. Tensile strength, ductility, and durability to volume change are among the improvements that had been reported. These advantages are common for concrete with waste plastic as they are closely associated with the properties of plastic material. It is also worth noting that those qualities perfectly compensated for the weaknesses of concrete, namely its low tensile-to-compression ratio, and brittleness. Despite that, it also incurred significant loss to the compressive strength of concrete, chiefly due to the incompatibility of plastic particle morphology to form strong bonds with other concrete constituents. To optimize the application of PET plastic as an aggregate and mitigate the downside, the detailed analysis indicated that a moderate proportion of fine aggregate replacement was the recommended approach in such an application (Chong & Shi, 2023).

The improvement of flexural and splitting tensile strength was common observations for concrete utilizing other forms of plastic (Abeysinghe et al., 2021). Plastic of higher quality, such as high-density polyethylene (HDPE) and polyvinyl chloride (PVC) had also been largely attempted. HDPE is the third largest group of produced and discarded plastic after the lower-graded categories, while PVC is a more rigid plastic often used in piping applications. However, despite their better load-bearing capacity compared to low-graded plastic such as PET, the incorporation of shredded HDPE and PVC into concrete material had failed to circumvent the decline of compressive strength. In both cases, it was reported that the compressive strength of concrete decreased with the proportion of aggregate replacement. However, the reduction induced by PVC was noted to be minor, where up to 30% fine aggregate replacement caused only approximately an 8% compressive strength loss (Mohammed et al., 2019), as compared to PET fine aggregate which had been reported to compromise more than half of concrete's compressive strength at the same replacement level (Bamigboye et al., 2022).

Another widely explored option was expanded polystyrene (EPS), a low-density, porous polymer. Making the best use of its lightweight nature, EPS had been incorporated as aggregate, fine aggregate, and fiber to reduce the density of concrete and masonry (Ferrándiz-Mas et al., 2016). The serviceability of concrete was enhanced by the addition of EPS. Due to the ductility of EPS, concrete-resisted dynamic and cyclic load to a greater extent, and suffered less damage from sulfate attack because EPS provided more room for ettringite-induced expansion (You et al., 2019). Demirboga and Kan (2012) exploited the porous nature of EPS and produced concrete with 70% less thermal conductivity than conventional concrete. This opened up other applications beyond concrete, with ESP-incorporated sandwich walls, panels, and other nonload-bearing constructs that may provide enhanced energy-saving properties within buildings (Kim & You, 2015). Moreover, EPS concrete also had superior acoustic insulation, adding to its repertoire of utilities in construction applications. As with other plastic materials in concrete, the reduction of load-bearing strength remained a concern, but it was a reasonable trade-off for a material of lower density value.

Despite all its appeals, concrete with plastic aggregate was faced with a few challenges. The most obvious downside is the loss of compressive strength, which was shared by concrete incorporated with all the above categories of plastic. The strength lost was attributed to a few reasons. Compared to conventional aggregates such as rock and sand, plastic trades its load-bearing capacity for flexibility, resulting in concrete with lower strength. Plastic material also had a nonexistent water absorption capacity, and studies had claimed that such an attribute caused the accumulation of water around the ITZ, weakening the bond between plastic particles and cement paste (Chong & Shi, 2023). In addition, the smooth texture and morphology of plastic particles were claimed to exacerbate such phenomenon, even though the same properties may allow concrete to achieve better workability in the fresh state. Scanning electron microscope (SEM) of concrete with plastic aggregate largely corroborated those claims, with a clear gap between plastic particles and cement paste constituting a zone of weakness for concrete failure as shown in Fig. 7.9.

Further studies proposed two methods to mitigate the strength lost by minimizing the proportion of replacement and the particle size of plastic aggregate. By reducing the size of the plastic particles, the weakness zone in the ITZ was subsequently reduced (Osubor et al., 2019; Saikia & De Brito, 2014). Meanwhile, a smaller particle size aggregate induced filling effect, reducing concrete porosity, contributing to better microstructure packing, and hence enhancing strength development. An example of optimized plastic aggregate utilization was performed by Thorneycroft et al. (2018), who observed a minor strength gain using 5% PET plastic aggregate the size of 0.5 mm. At 10% replacement, concrete with comparable strength to control was reported. Meanwhile, higher-grade plastic such as PVC caused only about 8% strength loss at up to an impressive 30% coarse aggregate replacement proportion (Mohammed et al., 2019). Still, studies generally recommend using plastic as fine aggregate for the best result.

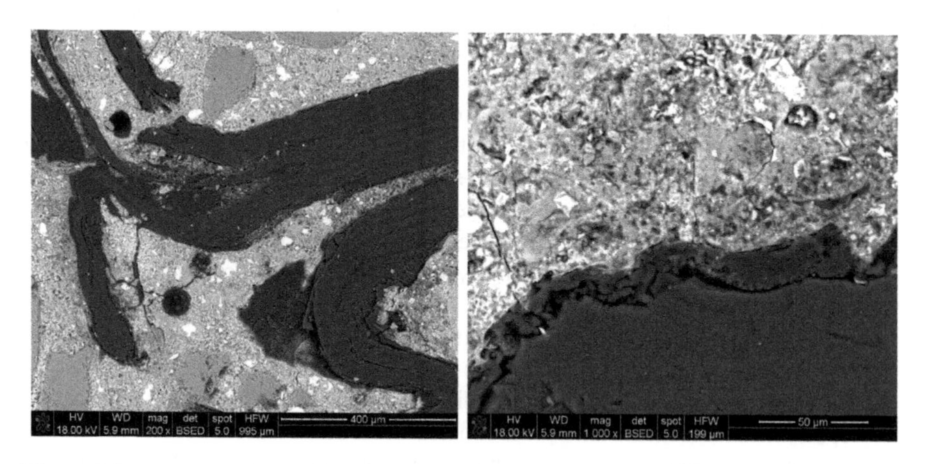

Figure 7.9 SEM images of PET mortar sample (left) at 200 × magnification (right) at 1000 × magnification.
Source: Images of Sugarcane Crop and Sugarcane Bagasse taken From Payá, J., Monzó, J., Borrachero, M. V., Tashima, M. M., & Soriano, L. (2018). Bagasse ash. In *Waste and supplementary cementitious materials in concrete: Characterisation, properties and applications* (pp. 559–598). Elsevier. https://doi.org/10.1016/B978-0-08-102156-9.00017-1.

Another concern fell on the fire rating of plastic aggregate concrete. Low-graded plastic such as PET has a melting point between 280°C and 320°C (Islam et al., 2016), while HDPE and EPS ignite at around 490°C (Prasittisopin et al., 2022). This raised concern not only about the failure of concrete during fire but also the toxic styrene that may be emitted. This is compounded by the fact that smoke inhalation was the major cause of injury and death in a fire incident. Polymer with fire retarding properties or coating was proposed to get around the weakness, but more study would be required before mass application of plastic aggregate concrete became widely possible. In addition, it was also feared that the incorporation of plastic material in concrete would complicate end-of-life management of concrete, as the plastic particles would eventually return to pollute the environment in the form of waste concrete.

7.8 Effect of waste glass on concrete performance

The incorporation of waste glass as a replacement for natural fine aggregates has been reported to negatively affect the workability of fresh concrete (Chen et al., 2006; Ismail & AL-Hashmi, 2009; Limbachiya, 2009; Tuaum et al., 2018). This is because the smaller-sized glass particles increase the surface-to-volume ratio to promote higher water absorption. Contrary, Ali and Al-Tersawy (2012) and Mardani-Aghabaglou et al. (2015) demonstrated the initial slump increased by 35% as the incorporation level of waste glass increased from 0% to 30%. This is due to the dense microstructure and smooth surface of the glass particles which facilitates reduced absorption and friction. The increase was also due to the dense microstructure and smooth surface of glass reducing water absorption and friction. The replacement of waste glass for coarse aggregate was observed to decrease the slump (Topçu & Canbaz, 2004). This is because as the size of the glass particle is increased, the surface-to-volume ratio decreases; thus, there is less paste absorbed on the surface of glass particles for lubricating adjacent glass particles. In general, employing glass particles with a content of up to 20%–30% improves flowability. The flowability of waste glass with a diameter of up to 100 mm increases with particle size. However, using coarse glass particles can make it less flowable. Based on the available literature, it was observed that natural fine aggregates can be replaced by waste glass by up to 20% to improve the workability of the fresh concrete.

The effect of waste glass on the compressive strength of concrete varies depending on the glass type, particle size, aggregate/cement replacement, and dosage rates. Contradictory findings were reported in the literature. Replacing 10% of Portland cement with waste glass was noticed to improve the compressive strength by about 22% (Kim et al., 2018). It has been observed that waste glass powders with a size smaller or equal to that of cement grains are more advantageous because they have a greater surface area that can facilitate cement hydration due to the nucleation effect and pozzolanic action, thus densifying the concrete microstructure to improve the concrete compressive strength. Contrary, Bijeljic et al. (2018), Gorospe et al. (2019), Kashani et al. (2019), Liu et al. (2020), Naaamandadin et al. (2020), Schwarz et al. (2008), Wang and Huang (2010), and Yang et al. (2019) reported the incorporations of 25%–30% of waste glass as a fine aggregate replacement result in compressive strength reductions by about 7%–29%. However, the use of waste glass as a coarse aggregate replacement has a negative effect with a compressive strength reduction of about 8%–55% being reported by several researchers (Arabi et al., 2019; Bostanci, 2020; Gerges et al., 2018; Hunag et al., 2015). Replacing natural fines with waste glass not only improves the compressive strength but also improves the flexural strength of the concrete (Ismail & AL-Hashmi, 2009; Turgut & Yahlizade, 2009). But this improvement in flexural strength is limited to a replacement level of not more than 20%, beyond which a decrease in flexural strength was reported (Ali & Al-Tersawy, 2012).

Replacing Portland cement with waste glass has been reported to lower the chloride ion permeability (Chen et al., 2006; Kamali & Ghahremaninezhad, 2016). This is because the smaller size glass particles refine the concrete microstructure to make it denser (due to pozzolanic activity as well as act as inert filler), thus reducing the permeability. However, as the replacement level increased, a mass loss of about 50% after freeze–thaw cycles was observed mainly due to the dilution effect (Abendeh et al., 2015).

ASR is a chemical interaction between the silica in the aggregate and the hydroxyl ions in the concrete pore water (West, 1996). Due to high silica concentration, waste glass powder becomes highly susceptible to alkali attack. Concrete could be damaged by the swelling brought on by this reaction and the stresses develop microcracks (Meyer & Xi, 1999; Mirzahosseini & Riding, 2015). However, replacing 30% of Portland cement with waste glass powder (45–75 μm) has been reported to decrease the ASR of concrete (Pereira-De-Oliveira et al., 2012; Shao et al., 2000). Some of the measures used to mitigate ASR are changing the chemical composition of glass, grinding glass into particles smaller than 0.3 mm, sealing concrete to keep moisture out, and utilizing mineral admixtures (Meyer & Xi, 1999).

7.9 New recycled materials

7.9.1 Electronic wastes

In the past few decades, the electrical and electronic industries have seen a booming market at a rocket pace. This has led to a significant increase in the generation of electrical/electronic solid waste discarded after the end of its life. These wastes often termed as "e-wastes" are toxic and not biodegradable. Each year, several tons of e-wastes end up in landfills or in open lands which becomes a major environmental issue. When dumped in landfills, these e-wastes may contaminate the groundwater and soil too. Also, e-wastes containing harmful chemicals can percolate into the ground and pollute the surface water or groundwater aquifers. Incinerating these e-wastes also disposes of harmful chemicals known as "dioxins" and is often considered to contribute to environmental pollution (Luhar & Luhar, 2019). One of the best solutions to recycle these e-wastes is by converting it to a nonhazardous material mainly associated with the judicious utilization in concrete applications.

As per statistics, it is estimated that approximately 50 million metric tons of e-waste is generated on an annual basis by the end of 2018 (UN Report, 2017). The USEPA estimated that in the year 2009, e-wastes such as discarded televisions, computers, cell phones, printers, scanners, and faxes accounted for about 2.37 million tons (USEPA, 2022). Approximately 25% of these e-wastes was collected for recycling, while the remaining ended up in landfills (USEPA, 2022). By the end of 2010, the e-wastes generated by the United States alone accounted for about 3 million tons of e-waste. China also contributed to about 2.3 million tons of e-waste generation as per the 2010 estimate. Developing countries like India have also emerged as one of the leading countries booming e-waste generation by about 500% during the year 2007–2020 (UN Report, 2017). It has become quite essential to implement the concept of the 4Rs, that is, Reduce, Reuse, Recover, and Recycle, to minimize the issue of e-wastes hazards. A possible solution to recycle and reuse these e-wastes is to use it as an aggregate replacement in various concreting applications where the contamination of plastic waste may not make any significant difference to the concrete properties. Although e-wastes have poor cohesive properties that may lead to reduced compressive strength and MOE, there are some benefits that can be enfolded by their use in concrete (Makri et al., 2019). For instance, e-wastes have good flexibility and adjustability, lightweight nature, high resistance and durability in chemicals, better thermal and insulation properties, ease of design, and low cost (Makri et al., 2019; UN Report, 2017). These properties make e-wastes a potential material for aggregate replacement in various concrete applications.

7.9.2 Recycled steel fiber

Unrecycled tire waste is a global problem because scrap tires are nonbiodegradable. Scrap tires are also combustible and can leach toxic substances into the ground. According to the US scrap tire markets report, 4.2 million tons of scrap tires were generated in 2017 in the United States and 0.6 million tons were disposed of in landfills (U.S. Tire Manufacturers Association, 2020). These pose serious concerns about environmental and public safety, in addition to the huge wastage of valuable materials that can be recycled and reused. A sustainable way to recycle tire waste is to utilize it in construction materials. The use of ground rubber for civil engineering applications is one of the top markets for scrap tires. For instance, quieter and more durable roads have been successfully built using ground rubber-modified asphalt and cement concrete (Lo Presti, 2013; Siddique & Naik, 2004). During the production of ground rubber, steel fibers with irregular shapes and varying dimensions are extracted. With the rapid development of the tire recycling industry, waste tire processors have used better processing technology to extract recycled steel fibers (RSFs) with good quality (e.g., little rubber residue adhered to fiber surfaces). About 15% of the total weight of the tire waste accounts for the amount of the RSF that can be extracted from scrap tires. Unfortunately, RSF has conventionally been treated as ready-to-melt steel, which is a less economical and environmentally friendly application in comparison to direct reuse. The extraction of RSFs from scrap tires not only solves the disposal issue of the used scrap tires but also serves as an alternative, sustainable and economical solution to replace some amount of the industry-made manufactured steel fiber (MSF) to produce a high-quality fiber-reinforced concrete (FRC). However, the RSFs can be easily damaged during the recycling process, and loose particles attached to the surface do not provide enough anchorage (Mastali, Dalvand, Sattarifard & Abdollahnejad, & Illikainen, 2018). This results in a decrease in the postpeak response in terms of energy absorption which is mainly attributed to the irregular geometry of the RSFs compared to the smooth-surfaced MSFs. But the hybrid inclusion of hooked-end MSFs and RSFs (dosage of MSFs typically in the range of 15%−30%) can exhibit better flexural response (Mastali, Dalvand, Sattarifard, & Illikainen, 2018). In addition, a higher aspect ratio (ratio of length to diameter) of RSFs has been found to significantly improve the ductility and flexural response of the concrete (Liew & Akbar, 2020).

The fiber geometry of RSFs does not greatly influence the concrete compressive strength with inconsistent trends being observed in the results presented by different researchers but has a significant effect on the tensile strength of recycled steel FRC. For example, the inclusions of 0.6−1.2 mm diameter and 50 mm long RSFs were found to improve the tensile strength by about 40%−68% (Aghaee et al., 2015; Sengul, 2016). More importantly, the fiber volume and fiber length have a great influence on the tensile strength of recycled steel FRC. Previous research reveals that the inclusions of 25 and 26.17 mm long fibers cause the tensile strength to increase by about 23% and 45% (Skarżyński & Suchorzewski, 2018), whereas the use of short fibers of 14 and 16.5 mm resulted in the tensile strength reduction by 10% and 21%, respectively. Based on the literature, it can be concluded that the tensile strength of steel FRC is greatly influenced by the fiber geometry precisely the average length of the RSFs. However, RSFs can vary from source to source due to variability in the recycling techniques making it difficult to achieve uniform fiber geometrics. Therefore more research on the hybridization of MSFs and RSFs as a promising and sustainable approach to producing a high-quality FRC is needed.

7.9.3 Copper slag

With the global increase in the production of copper, recycling copper slag as a construction material is a vital step toward achieving sustainability in pavement construction. It is reported that the production of 1 ton of copper generates, approximately 2−3 tons of copper slag (Shi et al., 2008). In the United States, the amount of copper slag produced is about 4 million tons, and in Japan, it is about 2 million tons per year (Ayano & Sakata, 2000; Collins & Ciesielski, 1994). The current options for proper management of copper slag include recycling, recovering metal, production of value-added products, and disposal in slag dumps or stockpiles, which is not an environmentally friendly option (Shi et al., 2008). Given the huge supply of copper slag, it is important that the copper slag is used in various civil engineering applications to promote sustainability and lower the carbon footprint. Copper slag is a by-product obtained during the matte smelting and refining of copper. The major oxides of copper slag are Fe_2O_3 and SiO_2, typically in the range of 35%–60% and 25%–40% (Shi et al., 2008). It is reported that using copper slag as a pozzolanic material for 5%–15% of the replacement of Portland cement could produce concretes with similar or slightly better mechanical strength as compared to normal concrete (Arino & Mobasher, 1999; Moura et al., 1999; Zain et al., 2004). Contrary, several researchers reported that the mechanical strength of concretes made with copper slag as a coarse and/or fine aggregate replacement is almost similar to or even higher than that of conventional concretes. In fact, the use of copper slag as a fine aggregate replacement material has been noticed to greatly improve the abrasion resistance of cement mortar mixtures. One major challenge is the higher chances of bleeding of the fresh concrete due to the heavy specific weight and the glass-like smooth surface properties of the irregular grain shape of the copper slag aggregates (Wang et al., 2021). In addition, the smaller the particle size of copper slag, the longer the setting time of the fresh concrete will be. It is evident that more research needs to be carried out to understand the various effects of copper slag on the performance of concrete mixtures.

7.9.4 Zinc waste

Zinc waste or more commonly known as jarosite, from the zinc industry, has been recognized as a hazardous waste material under the Basel Convention, European Commission, and Schedule II of MoEF (2008) owing to its heavy metal concentration: copper, lead, zinc, sulfur, cadmium, and chromium (Gupta & Sachdeva, 2019; Mehra et al., 2016a, 2016b; MoEF, 2008; Pappu et al., 2006). Jarosites are generally yellowish in color (Dent, 1986), and morphologically pseudocubic-shaped euhedral crystals (Hochella et al., 1999; Lin, 1997) with a few exceptions of being hexagonally prismatic- or plate-shaped crystals (Gillott, 1980). They are a basic hydrous sulfate of potassium and iron with a chemical formula $KFe^{3+}{}_3(SO_4)_2(OH)_6$ and a molecular weight of 500.81 g (Mees & Stoops, 2010). In addition, jarosites are hydrophobic in nature and insoluble, and therefore, regarded as a hazardous material (Pappu et al., 2006). India alone produces about 1.02 million kilograms of zinc waste annually (Mehra et al., 2016b) while the European countries produce nearly 6 million kilograms of zinc waste annually alone (Asokan et al., 2010; Pappu et al., 2006; Romero & Rincón, 1997). Because of the presence of toxic metals in the zinc waste, their disposal in landfills can contaminate the air, water, and nearby soils leading to the formation of acidic mine drainages or infillings in acidic sulfate soil (Mees & Stoops, 2010; Mehra et al., 2016b; Slager et al., 1970). With the huge generation of zinc waste every year, it becomes inevitable to develop sustainable strategies for its utilization in an effective manner.

Only a handful of researchers attempted to investigate the several effects of jarosite on the fresh and hardened properties of cement concrete mixes. Most researchers reported that replacing Portland cement with jarosite at a 20% replacement level did not significantly affect the cement concrete mixture properties. Debbarma and

Ransinchung (2021b) reported mechanical strength enhancement of RCC mixtures up to 15% jarosite replacement level. In a different study by Mehra et al. (2016b), the substitution of natural sand with jarosite at 5%–15% replacement levels exhibited 13%–50% and 8%–12% improvement in compressive strength and flexural strength days. It has been reported that SiO_2 which is an essential constituent of a material to show pozzolanic is almost negligible in jarosite particles (Debbarma & Ransinchung, 2021b; Ray et al., 2020). Therefore the improvement in the mechanical strength is primarily due to the "filler effect" provided by the jarosite particles, that is, a good degree of pore refinement at the microlevel, thus densifying the concrete matrix to improve its mechanical strength (Debbarma et al., 2020; Debbarma & Ransinchung, 2021b). At higher replacement levels, the reduction in the mechanical strengths becomes prominent because the deficiency in the calcium silicate hydrate phases is too large to be compensated by the pozzolanic reaction and the filling ability of the jarosite particles (Ray et al., 2020). In addition, the presence of zinc oxide in the jarosite particles could retard the setting time, thereby leading to a loss in the compressive strength at higher replacement levels.

A major concern of utilizing jarosite particles as a fine aggregate/cement replacement in cement concrete mixes is the leaching of heavy toxic metals. However, concrete mixes made with jarosite particles have been reported to show satisfactory leaching potential that is well within the USEPA (1992) limits. According to USEPA, the permissible limits for heavy elements such as zinc, lead, copper, cadmium, and iron are 500, 5, 5, 1, and 30 ppm, respectively. Based on a study by Mehra et al. (2016a), the leaching values of jarosite-incorporated concrete mixes were found to be 4, 3, 0.1, 0.2, and 27 ppm, which are well within the permissible USEPA limits. Therefore zinc waste has the potential to be utilized as a construction material in PCC mixes to yield strength benefits as well as economic and environmental benefits. Other applications of jarosite include the manufacturing of bricks in combination with fly ash (Patrick Mubiayi et al., 2018; Wang & Vipulanandan, 1996). Fly ash has been reported as a suitable material to immobilize the hazardous toxic heavy elements present in the jarosite particles. The presence of lead (Pb) retards the initial and final setting times of Portland cement to a great extent by interfering with the normal hydration reactions (Wang & Vipulanandan, 1996). Due to the physiochemical characteristics of fly ash, it immobilizes the lead and other heavy toxic elements from the PCC mixes made with jarosite particles. It has been demonstrated that the combination of jarosite and fly ash particles could produce bricks with compressive strength within the stipulated strength requirement for standard building bricks (Patrick Mubiayi et al., 2018).

7.9.5 Sugarcane bagasse ash

Nonconventional supplementary cementitious materials (SCMs) such as the use of agricultural wastes (e.g., rice husk ash, sugarcane bagasse ash [SCBA], corn stover ash [CSA], sawdust) have been widely studied for use in PCC mixes. One promising agro-SCM is SCBA, which is obtained after the controlled burning of bagasse (fibrous residue typically 40%–45% of sugarcane left after crushing and extraction of its juice) for the production of fuel for heat generation (Loh et al., 2013). The burning of bagasse in the process typically generates about 8%–10% of SCBA (Modani & Vyawahare, 2013), which is rich in amorphous silica (~55%–78%) that lends its good pozzolanic properties (Thomas, Yang, Bahurudeen, et al., 2021). The use of SCBA as a partial replacement for Portland cement in PCC mixes can lead to enhanced properties in terms of low heat of hydration, strength gain, low permeability, and similar drying shrinkage, respectively (Bahurudeen & Santhanam, 2014).

Of the 121 countries that produce sugarcane crop, 15 countries including Brazil, India, Thailand, Pakistan, China, Bangladesh, South Africa, Cuba, Columbia, Philippines, Australia, Mexico, the United States, Myanmar, and Argentina produce about 87.4% of the total worldwide sugarcane crop (Jahanzaib Khalil et al., 2021). A total area of 20.42 million hectare of land is cultivated sugarcane in the world, and the worldwide production of sugarcane is estimated to be more than 1500 million tons per year (Jahanzaib Khalil et al., 2021). In the United States, sugarcane is produced in Florida, Louisiana, and Texas (USDA, 2021). The acreage of sugarcane for sugar in the United States rose from an average of 704,000 acres in the first half of the 1980s to 903,400 acres in the fiscal year 2020–21 (USDA, 2021). For each ton of sugarcane produced, nearly 300 kg of bagasse is generated; and for each ton of bagasse burnt, 0.62% of SCBA is produced (Thomas, Yang, Bahurudeen, et al., 2021), with an approximate yearly production of 11.42 metric tons (Cordeiro et al., 2004). With an enormous production every year, it is a challenge to handle SCBA both environmentally and economically. The current practice for the disposal of SCBA is landfilling as refuse, leading to heavy metal concentration (e.g., Ag, As, Ba, Cd, Cr, Hg, and Pb) but beyond the permissible limits. Because SCBA is rich in amorphous silica, alumina, and ferric oxide, it can be used as a good pozzolana in cement-based materials.

The performance of SCBA-based concrete mostly depends on the processing methods (burning, grinding, chemical activation, sieving, and a combination of the above methods) adopted to produce the SCBA. In general, SCBA is burned at varying temperatures ranging between 600°C and 900°C (Thomas, Yang, Bahurudeen, et al., 2021). Several researchers recommended that calcination at 700°C produces SCBA with desirable pozzolanic properties and can be used as cement replacement material (Bahurudeen et al., 2015; Bahurudeen & Santhanam, 2015; Thomas, Yang, Bahurudeen, et al., 2021). Concrete produced with SCBA (up to 25% replacement level) has shown similar or marginally better strength performance compared to normal concrete at all curing ages (Bahurudeen & Santhanam, 2014). In addition, the incorporation of SCBA in concrete can reduce its heat of hydration, reduce its thermal conductivity, improve its resistance to chloride and gas penetration, improve its surface resistivity, and show similar drying shrinkage behavior compared to normal concrete (Bahurudeen et al., 2015; Bahurudeen & Santhanam, 2014; 2015; Thomas, Yang, Bahurudeen, et al., 2021). However, SCBA is hygroscopic in nature, and the requirement for water-reducing admixture increases as the SCBA replacement level increases (Thomas, Yang, Bahurudeen, et al., 2021). It has been observed that water absorption of SCBA concrete increases as the replacement level of SCBA increases, thus, affecting the water permeability of the concrete

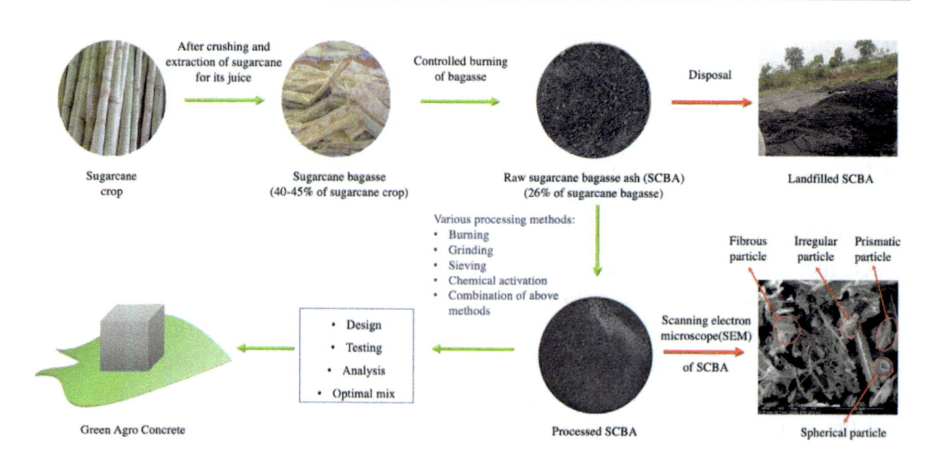

Figure 7.10 Concept of agro-based concrete.
Source: Images of Sugarcane Crop and Sugarcane Bagasse taken From Payá, J., Monzó, J., Borrachero, M. V., Tashima, M. M., & Soriano, L. (2018). Bagasse ash. In *Waste and supplementary cementitious materials in concrete: Characterisation, properties and applications* (pp. 559–598). Elsevier. https://doi.org/10.1016/B978-0-08-102156-9.00017-1.

considerably. Nevertheless, the use of SCBA in concrete enfolds several advantages, and therefore, its application can be extended to different sectors in the construction industry. In addition, it is important to understand the long-term durability (e.g., shrinkage, creep, carbonation) of SCBA-based concrete, and therefore, more detailed studies are required. Because SCBA is rich in silica and alumina, its feasibility as a precursor material in one-part geopolymer concrete can also be explored to find an optimal mix and coordination mechanism. A detailed concept of agro-based concrete made with SCBA is shown in Fig. 7.10.

7.9.6 Corn stover ash

Several agricultural-based ashes (e.g., rice husk ash, sugarcane bagasse ash) rich in silica and alumina that display pozzolanic activity have been used as SCMs in concrete mixtures (Jittin et al., 2020; Sandhu & Siddique, 2017; Shakouri et al., 2020). By 2030, the total primary agricultural residue in the United States is estimated at 320 million dry tons, with 85% of this quantity composed of corn stover (Perlack, 2011). Maize corn, with an annual production of over 1.09 billion metric is the most-produced grain worldwide. The United States stands as the largest maize producer, accounting for 43% of the world's maize production, and approximately 15% of the corn grain production is the crop residue or the agricultural waste product named corn cob (Thomas, Yang, Mo, et al., 2021). There are several limitations to the application of corn cob ash in concrete which make it a less favorable alternative to SCM since corn cob is nutritious and has a higher food value as a cattle feed. On the other hand, limited work has been done on using CSA as an alternative SCM in concrete. The ash content of corn stover ranges from 3% to 8% which is nearly comparable with the ash content of bituminous coal, the most common type of fuel used in generating power and heat in the cement production process, ranging from 6% to 12% (Shakouri et al., 2020). It is estimated that the annual production of CSA in the United States would be approximately 6.3 million metric tons per year and this amount can roughly offset 7% of the currently produced cement in the United States (Shakouri et al., 2020). This in turn can reduce CO_2 emissions by approximately 5 million metric tons of CO_2. Hence, CSA can be used as a promising SCM source in the concrete industry if proven to improve or meet the desirable properties of concrete. Previous research shows that untreated CSA should not be used as an SCM due to its extremely high K_2O, P_2O_5, Cl contents, high loss of ignition (LOI), and concrete strength reduction (Shakouri et al., 2020). However, acid- and water-washing of CSA have been shown to improve the quality of the ash, thus, comparable concrete properties. In terms of sustainability, water-washing of CSA is preferred but the high K_2O content of the ash can exhibit a negative impact on the hydration process and create durability issues such as ASR in the concrete. Due to the limited studies available on the use of CSA as an alternative SCM source in concrete, detailed research to understand the effect of CSA on the cement hydration kinetics, compressive strength, flexural strength, and durability of the concrete is needed.

7.9.7 Seashell wastes

From 1990 to 2018, global aquaculture production has increased significantly by 527% and it is projected to reach 109 million tons in 2030, an increase of 32% over 2018 (The State of World Fisheries and Aquaculture, 2020). China, Indonesia, Peru, India, the Russian Federation, the United States, and Vietnam are the largest producers and account for almost 50% of the total global aquaculture production. With such an enormous production, the generation of waste from aquaculture is inevitable with most of the wastes being disposed of in landfills. One such potential waste material that is abundantly available is waste seashells. Seashell wastes primarily consist of oyster shells, clam shells, scallop shells, mussel shells, and cockle shells (Mo et al., 2018). Only a small fraction of these seashells are utilized as fertilizers and handicrafts but their restriction is limited due to the amount that can be used, soil solidification, and economic problems (Yang et al., 2005). Most seashell wastes are discarded as wastes and disposed of in landfills with China alone disposing about 10 million tons of seashell wastes in landfills annually (Mo et al., 2018). One sustainable solution is to use these wastes as an alternative cementitious/aggregate material in concrete mixtures.

Previous research showed that the main component of seashell waste is calcium carbonate ($CaCO_3$), which is similar to limestone, primarily consisting of calcium oxide (CaO), so, it can be a potential material that can be used as an inert material in concrete (Mo et al., 2018). However, proper treatment of the seashell waste is required to ensure its quality and the amount that can be used as a fine aggregate replacement in concrete mixtures. An unclear trend is reported in the literature when seashell waste was used as cement or aggregate replacement material. Some studies reported that seashell powder does not react chemically with cement and only acts as a filler material (Mo et al., 2018) while some reported that $CaCO_3$ can be converted into CaO at high-temperature pretreatments (Felipe-Sesé et al., 2011; Martínez-García et al., 2017; Mohamed et al., 2012). Likewise, some studies reported that replacing cement with seashell powder improves the workability (Lertwattanaruk et al., 2012), shrinkage (Lertwattanaruk et al., 2012), and insulation properties (Khankhaje et al., 2017; Lertwattanaruk et al., 2012) of cement-based materials. Whereas, some studies reported that seashells as a fine aggregate replacement can affect the concrete strength (Nguyen et al., 2017), porosity, and water absorption (Cuadrado-Rica et al., 2016). Contrary, an unclear and contrasting trend was observed in the concrete strength when seashell powder was used as a cement replacement material.

From the literature, it is still unclear what amount of seashell wastes can be used in concrete as a cement or aggregate replacement material, the effect of pretreatment techniques, and its effect on the workability, strength, durability, and insulation properties of concrete. More research in this direction is needed to demonstrate seashell wastes as a potential construction material in concrete. At the same time, this will help conserve the environment through the reduced usage of virgin materials in the concrete industry.

7.10 Conclusion

Depletion of quality virgin aggregates, an increasing amount of waste material stockpiles, and the high disposal cost of waste materials necessities the use of recycled aggregates in PCC mixes. It is possible to produce RACs utilizing recycled aggregates with comparable or marginally better mechanical strength and durability performance than natural aggregate concrete.

- RCA is one of the potential recycled aggregates that can be substituted as a partial replacement to natural aggregates in the production of strong and durable PCC mixes for various civil engineering applications. Depending on their application, the performance of RCA concrete can be controlled by adjusting the w/c ratio, RCA content, the type and size of RCA, the physical properties of RCA, the quality of parent concrete, the moisture condition of RCA, curing condition, cement content, air entrainment, using mineral admixtures, removing the adhered mortar, and strengthening the adhered mortar in the RCA, respectively.
- While the use of RAP aggregates is anticipated to have better-cracking resistance and improved fracture properties, the presence of asphalt coating around the RAP aggregates could reduce the overall strength of the concrete. However, natural coarse aggregates can be replaced with coarse RAP aggregates up to 40% replacement level without much affecting the mechanical strength and durability of PCC mixes.
- Likewise, waste tire rubber has been shown to improve ductility, and enhance thermal conductivity and sound insulation properties; however, shown to have a detrimental effect on mechanical strength. Regardless, up to 20% of waste tire rubber can be utilized in PCC mixes without causing a significant strength loss.

- Waste plastic aggregates such as PET, HDPE, PVC, and EPS can be used to produce lightweight concrete with superior acoustic properties relative to normal concrete. The use of waste plastics in concrete is restricted to not more than 30% replacement level because of the severe strength reduction at higher replacement levels.
- Contrary, steel slag is proven to improve the mechanical strength of PCC mixes; however, care should be given to the volume expansion caused by the reaction of the excess free CaO present in the steel slag particles.
- Waste glass is another potential recycled material for use as aggregate/cement replacement in PCC mixes to improve their mechanical strength and durability performance. As a natural fine aggregate replacement, glass waste of up to 20% can be utilized to attain fresh and hardened properties improvement while glass waste powder of up to 30% cement replacement can provide the same benefits.

This chapter strongly highlights the importance and benefits of different recycled aggregates as a potential aggregate/cement replacement without compromising the mechanical strength and durability of concrete. In addition, other waste materials such as e-waste, RSFs, copper slag, zinc waste, SCBA, CSA, and seashell waste also have the potential to be utilized as aggregate/cement replacement or fiber reinforcement to produce sustainable concrete mixes. The present review will help the concrete industry to embark on sustainable construction technologies for a sustainable infrastructure utilizing recycled materials.

References

Abendeh, R., Baker, M. B., Salem, Z. A., & Ahmad, H. (2015). The feasibility of using milled glass wastes in concrete to resist freezing-thawing action. *International Journal of Civil and Environmental Engineering*, *9*(8).

Abeysinghe, S., Gunasekara, C., Bandara, C., Nguyen, K., Dissanayake, R., & Mendis, P. (2021). Engineering performance of concrete incorporated with recycled high-density polyethylene (HDPE)—A systematic review. *Polymers*, *13*(11). Available from https://doi.org/10.3390/polym13111885.

Adams, M. P., Fu, T., Cabrera, A. G., Morales, M., Ideker, J. H., & Isgor, O. B. (2016). Cracking susceptibility of concrete made with coarse recycled concrete aggregates. *Construction and Building Materials*, *102*, 802–810. Available from https://doi.org/10.1016/j.conbuildmat.2015.110.022.

Aghaee, K., Yazdi, M. A., & Tsavdaridis, K. D. (2015). Investigation into the mechanical properties of structural lightweight concrete reinforced with waste steel wires. *Magazine of Concrete Research.*, *67*(4), 197–205. Available from https://doi.org/10.1680/macr.14.00232.

Ahmad, Ayaz, Chaiyasarn, Krisada, Farooq, Furqan, Ahmad, Waqas, Suparp, Suniti, & Aslam, Fahid (2021). Compressive strength prediction via gene expression programming (GEP) and artificial neural network (ANN) for concrete containing RCA. *Buildings*, *11* (8), 324. Available from https://doi.org/10.3390/buildings11080324.

Aiello, M. A., & Leuzzi, F. (2010). Waste tyre rubberized concrete: Properties at fresh and hardened state. *Waste Management.*, *30*(8–9), 1696–1704. Available from https://doi.org/10.1016/j.wasman.2010.02.005.

Akhtar, A., & Sarmah, A. K. (2018). Construction and demolition waste generation and properties of recycled aggregate concrete: A global perspective. *Journal of Cleaner Production*, *186*, 262–281. Available from https://doi.org/10.1016/j.jclepro.2018.030.085.

Al-Akhras, N. M., & Smadi, M. M. (2004). Properties of tire rubber ash mortar. *Cement and Concrete Composites*, *26*(7), 821–826. Available from https://doi.org/10.1016/j.cemconcomp.2004.01.004.

Alfaro-Núñez, A., Astorga, D., Cáceres-Farías, L., Bastidas, L., Soto Villegas, C., Macay, K., & Christensen, J. H. (2021). Microplastic pollution in seawater and marine organisms across the Tropical Eastern Pacific and Galápagos. *Nature Research, Denmark Scientific Reports*, *11*(1). Available from https://doi.org/10.1038/s41598-021-85939-3.

Ali, E. E., & Al-Tersawy, S. H. (2012). Recycled glass as a partial replacement for fine aggregate in self compacting concrete. *Construction and Building Materials*, *35*, 785–791. Available from https://doi.org/10.1016/j.conbuildmat.2012.040.117.

Al-Oraimi, S., Hassan, H. F., & Hago, A. (2009). Recycling of reclaimed asphalt pavement in Portland cement concrete. *Journal of Engineering Research*, *6*(1), 37–45. Available from https://doi.org/10.24200/tjer.vol6iss1pp37-45.

Anastasiou, E., Georgiadis Filikas, K., & Stefanidou, M. (2014). Utilization of fine recycled aggregates in concrete with fly ash and steel slag. *Construction and Building Materials*, *50*, 154–161. Available from https://doi.org/10.1016/j.conbuildmat.2013.090.037.

Arabi, N., Meftah, H., Amara, H., Kebaïli, O., & Berredjem, L. (2019). Valorization of recycled materials in development of self-compacting concrete: Mixing recycled concrete aggregates – Windshield waste glass aggregates. *Construction and Building Materials*, *209*, 364–376. Available from https://doi.org/10.1016/j.conbuildmat.2019.030.024.

Arino, A. M., & Mobasher, B. (1999). Effect of copper slag on the strength, and toughness of cementitious mixtures. *ACI Materials Journal*, *96*(1), 68–75.

Aslani, H., Pashmtab, P., Shaghaghi, A., Mohammadpoorasl, A., Taghipour, H., & Zarei, M. (2021). Tendencies towards bottled drinking water consumption: Challenges ahead of polyethylene terephthalate (PET) waste management. *Health Promotion Perspectives*, *11*(1), 60–68. Available from https://doi.org/10.34172/hpp.2021.09.

Asokan, P., Saxena, M., & Asolekar, S. R. (2010). Recycling hazardous jarosite waste using coal combustion residues. *Materials Characterization*, *61*(12), 1342–1355. Available from https://doi.org/10.1016/j.matchar.2010.090.005.

Ayano, T., & Sakata, K. (2000). *Durability of concrete with copper slag fine aggregate*. American Concrete Institute, ACI Special Publication. https://www.concrete.org/topicsinconcrete/topicdetail/special%20publication.

Bahurudeen, A., Kanraj, D., Gokul Dev, V., & Santhanam, M. (2015). *Performance evaluation of sugarcane bagasse ash blended cement in concrete, Cement and Concrete Composites* (59, pp. 77–88). Available from 10.1016/j.cemconcomp.2015.030.004.

Bahurudeen, A., & Santhanam, M. (2014). Performance evaluation of sugarcane bagasse ash-based cement for durable concrete. *Proceedings of the 4th International Conference on the Durability of Concrete Structures* (ICDCS), Purdue University, India, pp. 275–281. Available from https://doi.org/10.5703/1288284315412.

Bahurudeen, A., & Santhanam, M. (2015). Influence of different processing methods on the pozzolanic performance of sugarcane bagasse ash. *Cement and Concrete Composites*, *56*, 32–45. Available from https://doi.org/10.1016/j.cemconcomp.2014.11.002.

Bairagi, N. K., Vidyadhara, H. S., & Ravande, K. (1990). Mix design procedure for recycled aggregate concrete. *Construction and Building Materials*, *4*(4), 188–193. Available from https://doi.org/10.1016/0950-0618(90)90039-4.

Balcázar, N., Kühn, M., Baena, J.M., Formoso, A., & Piret, J. (1999). *Summary report on RTD in iron and steel slags: Development and perspectives*. Proceedings No. EUR 19066 EN.

Bamigboye, G. O., Tarverdi, K., Wali, E. S., Bassey, D. E., & Jolayemi, K. J. (2022). Effects of dissimilar curing systems on the strength and durability of recycled PET-modified concrete. *Silicon*, *14*(3), 1039–1051. Available from https://doi.org/10.1007/s12633-020-00898-0.

Barrett, J., Chase, Z., Zhang, J., Holl, M. M. B., Willis, K., Williams, A., Hardesty, B. D., & Wilcox, C. (2020). *Microplastic pollution in deep-sea sediments from the Great Australian Bight*, *Frontiers in Marine Science* (7). Available from 10.3389/fmars.2020.576170.

Bažant, Z. P., & Oh, B. H. (1983). Crack band theory for fracture of concrete. *Matériaux et Constructions*, *16*(3), 155–177. Available from https://doi.org/10.1007/BF02486267.

Behravan, A., Lowry, M., Ashraf-Khorasani, M., Tran, T. Q., Feng, X., & Brand, A. S. (2023). Effect of pretreatment on reclaimed asphalt pavement aggregates for minimizing the impact of leachate on cement hydration. *Cement and Concrete Research*. Available from https://doi.org/10.1016/j.cemconres.2023.107305.

Berry, M., Stephens, J., Bermel, B., Hagel, A., Schroeder, D. (2013). *Feasibility of reclaimed asphalt pavement as aggregate in Portland cement concrete pavement, phase II: Field demonstration*. US. Department of Transportation.

Bijeljic, J., Ristic, N., Grdic, Z., Toplicic-Curcic, G., & Stojkovic, N. (2018). Influence of used waste cathode ray tube glass on alkali silicate reaction and mechanical properties of mortar mixtures. *Architecture and Civil Engineering*, *16*(3), 437–448. Available from https://doi.org/10.2298/fuace180704020b.

Bisht, K., & Ramana, P. V. (2018). Sustainable production of concrete containing discarded beverage glass as fine aggregate. *Construction and Building Materials*, *177*, 116–124. Available from https://doi.org/10.1016/j.conbuildmat.2018.05.119.

Boussetta, I., El Euch Khay, S., & Neji, J. (2020). Experimental testing and modelling of roller compacted concrete incorporating RAP waste as aggregates. *European Journal of Environmental and Civil Engineering*, *24*(11), 1729–1743. Available from https://doi.org/10.1080/19648189.2018.1482792.

Bostanci, S. C. (2020). Use of waste marble dust and recycled glass for sustainable concrete production. *Journal of Cleaner Production*, *251*, 119785. Available from https://doi.org/10.1016/j.jclepro.2019.119785.

Brand, A. S., Amirkhanian, A. N., & Roesler, J. R. (2014). *Flexural capacity of full-depth and two-lift concrete slabs with recycled aggregates*, *Transportation Research Record* (2456, pp. 64–72). Available from 10.3141/2456-07.

Brand, A. S., & Fanijo, E. O. (2020). A review of the influence of steel furnace slag type on the properties of cementitious composites. *Applied Sciences (Switzerland)*, *10*(22), 1–27. Available from https://doi.org/10.3390/app10228210.

Brand, A. S., & Roesler, J. (2018a). Interfacial transition zone of cement composites with recycled concrete aggregate of different moisture states. *Advances in Civil Engineering Materials*, *7*(1), 87–102. Available from https://doi.org/10.1520/ACEM20170090.

Brand, A. S., & Roesler, J. R. (2015). Ternary concrete with fractionated reclaimed asphalt pavement. *ACI Materials Journal*, *112*(1), 155–163. Available from https://doi.org/10.14359/51687176.

Brand, A. S., & Roesler, J. R. (2016). Expansive and concrete properties of SFS-FRAP aggregates. *Journal of Materials in Civil Engineering*, *28*(2), 08991561. Available from https://doi.org/10.1061/(ASCE)MT.1943-5533.0001403.

Brand, A. S., & Roesler, J. R. (2017a). Bonding in cementitious materials with asphalt-coated particles: Part I – The interfacial transition zone. *Construction and Building Materials*, *130*, 171–181. Available from https://doi.org/10.1016/j.conbuildmat.2016.100.019.

Brand, A. S., & Roesler, J. R. (2017b). Bonding in cementitious materials with asphalt-coated particles: Part II – Cement-asphalt chemical interactions. *Construction and Building Materials*, *130*, 182–192. Available from https://doi.org/10.1016/j.conbuildmat.2016.10.013.

Brand, A. S., & Roesler, J. R. (2018b). *Interfacial transition zone of cement composites with steel furnace slag aggregates, Cement and Concrete Composites* (86, pp. 117–129). Available from 10.1016/j.cemconcomp.2017.11.012.

Brand, A.S., Roesler, J.R., Al-Qadi, I.L., & Shangguan, P.. (2012). *Fractionated reclaimed asphalt pavement (FRAP) as a coarse aggregate replacement in a ternary blended concrete pavement* [Research Report ICT-12-008]. Illinois Center for Transportation..

Brand, A. S., Roesler, J. R., & Salas, A. (2015). Initial moisture and mixing effects on higher quality recycled coarse aggregate concrete. *Construction and Building Materials*, *79*, 83–89. Available from https://doi.org/10.1016/j.conbuildmat.2015.01.047.

Brand, A.S., Smith, R., Roesler, J.R., Al-Qadi, I.L., & Gillen, S.L. (2012). Fresh and hardened properties of concrete with fractionated reclaimed asphalt pavement. In *10th international conference on concrete pavements*, Québec City, Canada, 8–12 July.

Brooks, A. L., Wang, S., & Jambeck, J. R. (2018). The Chinese import ban and its impact on global plastic waste trade. *Science Advances*, *4*(6). Available from https://doi.org/10.1126/sciadv.aat0131.

Butler, L., West, J. S., & Tighe, S. L. (2013). Effect of recycled concrete coarse aggregate from multiple sources on the hardened properties of concrete with equivalent compressive strength. *Construction and Building Materials*, *47*, 1292–1301. Available from https://doi.org/10.1016/j.conbuildmat.2013.05.074.

Campanale, C., Massarelli, C., Savino, I., Locaputo, V., & Uricchio, V. F. (2020). A detailed review study on potential effects of microplastics and additives of concern on human health. *International Journal of Environmental Research and Public Health*, *17*(4). Available from https://doi.org/10.3390/ijerph17041212.

Chen, C. H., Huang, R., Wu, J. K., & Yang, C. C. (2006). Waste E-glass particles used in cementitious mixtures. *Cement and Concrete Research*, *36*(3), 449–456. Available from https://doi.org/10.1016/j.cemconres.2005.120.010.

Chen, H. J., Yen, T., & Chen, K. H. (2003). Use of building rubbles as recycled aggregates. *Cement and Concrete Research*, *33*(1), 125–132. Available from https://doi.org/10.1016/S0008-8846(02)00938-9.

Chong, B. W., & Shi, X. (2023). Meta-analysis on PET plastic as concrete aggregate using response surface methodology and regression analysis. *Journal of Infrastructure Preservation and Resilience*, *4*(1). Available from https://doi.org/10.1186/s43065-022-00069-y.

Collins, R.J., Ciesielski, S.K., & Mason, L.S. (1994). *Recycling and use of waste materials and by-products in highway construction: A synthesis of highway practice* [Final report]. Transportation Research Board.

Cook, E., Velis, C. A., & Black, L. (2022). *Construction and demolition waste management: A systematic scoping review of risks to occupational and public health, Frontiers in Sustainability* (3). Available from 10.3389/frsus.2022.924926.

Coomarasamy, A., & Walzak, T. L. (1995). Effects of moisture on surface chemistry of steel slags and steel slag-asphalt paving mixes. *Transportation Research Record*, *1492*, 85–95.

Cordeiro, G. C., Filho, R. D. T., Fairbairn, E. M. R., Tavares, L. M. M., & Oliveira, C. H. (2004). Influence of mechanical grinding on the pozzolanic activity of residual sugarcane bagasse ash. *International RILEM Conference on the Use of Recycled Materials in Building and Structures*, 731–740.

Cuadrado-Rica, H., Sebaibi, N., Boutouil, M., & Boudart, B. (2016). Properties of ordinary concretes incorporating crushed queen scallop shells. *Materials and Structures/ Materiaux et Constructions*, *49*(5), 1805–1816. Available from https://doi.org/10.1617/ s11527-015-0613-7.

Danish, A., & Mosaberpanah, M. A. (2022). A review on recycled concrete aggregates (RCA) characteristics to promote RCA utilization in developing sustainable recycled aggregate concrete (RAC). *European Journal of Environmental and Civil Engineering*, *26*(13), 6505–6539. Available from https://doi.org/10.1080/19648189.2021.1946721.

Debbarma, S., Ransinchung, G., & Singh, S. (2020). Zinc waste as a substitute for Portland cement in roller-compacted concrete pavement mixes containing RAP aggregates. *Journal of Materials in Civil Engineering*, *32*(8). Available from https://doi.org/ 10.1061/(ASCE)MT.1943-5533.0003278.

Debbarma, S., & Ransinchung, G. D. (2021a). *Achieving sustainability in roller compacted concrete pavement mixes using reclaimed asphalt pavement aggregates – state of the art review, Journal of Cleaner Production* (287). Available from 10.1016/j.jclepro.2020.125078.

Debbarma, S., Ransinchung, G. D., & Singh, S. (2019). Feasibility of roller compacted concrete pavement containing different fractions of reclaimed asphalt pavement. *Construction and Building Materials*, *199*, 508–525. Available from https://doi.org/ 10.1016/j.conbuildmat.2018.12.047.

Debbarma, S., & Ransinchung, G. D. R. N. (2021b). Effect of natrojarosite as replacement to portland cement in roller-compacted concrete pavement mixtures. *ACI Materials Journal*, *118*(1), 91–100. Available from https://doi.org/10.14359/51728281.

Debbarma, S., Singh, S., & Ransinchung, G. D. (2019). *Laboratory investigation on the fresh, mechanical, and durability properties of roller compacted concrete pavement containing reclaimed asphalt pavement aggregates, Transportation Research Record Journal of the Transportation Research Board* (2673, pp. 652–662). Available from 10.1177/0361198119849585.

De Oliveira, M. B., & Vazquez, E. (1996). The influence of retained moisture in aggregates from recycling on the properties of new hardened concrete. *Waste Management*, *16(*(1-3), 113–117. Available from 10.1016/S0956-053X(96)00033-5.

Delwar, M., Fahmy, M., & Taha, R. (1997). Use of reclaimed asphalt pavement as an aggregate in portland cement concrete. *ACI Materials Journal*, *94*(3), 251–256.

Demirboga, R., & Kan, A. (2012). Thermal conductivity and shrinkage properties of modified waste polystyrene aggregate concretes. *Construction and Building Materials*, *35*, 730–734. Available from https://doi.org/10.1016/j.conbuildmat.2012.040.105.

Dent, D. (1986). *Acid sulphate soils: A baseline for research and development.* International Institute for Land Reclamation and Improvement.

Domingo-Cabo, A., Lázaro, C., López-Gayarre, F., Serrano-López, M. A., Serna, P., & Castaño-Tabares, J. O. (2009). Creep and shrinkage of recycled aggregate concrete. *Construction and Building Materials*, *23*(7), 2545–2553. Available from https://doi.org/ 10.1016/j.conbuildmat.2009.02.018.

Dong, Q., Wang, G., Chen, X., Tan, J., & Gu, X. (2021). Recycling of steel slag aggregate in Portland cement concrete: An overview. *Journal of Cleaner Production*, *282*, 124447. Available from https://doi.org/10.1016/j.jclepro.2020.124447.

Emery, J. J. (1982). Extending Aggreg. Resour. *Slag utilization in pavement construction* (774). ASTM Special Technical Publication.

Erlin, B., & Jana, D. (2003). Forces of hydration that can cause havoc in concrete. *Concrete International.*, *25*, 51–57.

Fathifazl, G., Abbas, A., Razaqpur, A. G., Isgor, O. B., Fournier, B., & Foo, S. (2009). New mixture proportioning method for concrete made with coarse recycled concrete aggregate. *Journal of Materials in Civil Engineering*, *21*(10), 601–611. Available from https://doi.org/10.1061/(ASCE)0899-1561(2009)21:10(601).

Felipe-Sesé, M., Eliche-Quesada, D., & Corpas-Iglesias, F. A. (2011). The use of solid residues derived from different industrial activities to obtain calcium silicates for use as insulating construction materials. *Ceramics International*, *37*(8), 3019–3028. Available from https://doi.org/10.1016/j.ceramint.2011.050.003.

Ferdous, W., Manalo, A., Siddique, R., Mendis, P., Zhuge, Y., Wong, H. S., Lokuge, W., Aravinthan, T., & Schubel, P. (2021). *Recycling of landfill wastes (tyres, plastics and glass) in construction – A review on global waste generation, performance, application and future opportunities, Resources, Conservation and Recycling* (173). Available from 10.1016/j.resconrec.2021.105745.

Ferrándiz-Mas, V., Sarabia, L. A., Ortiz, M. C., Cheeseman, C. R., & García-Alcocel, E. (2016). Design of bespoke lightweight cement mortars containing waste expanded polystyrene by experimental statistical methods. *Materials and Design*, *89*, 901–912. Available from https://doi.org/10.1016/j.matdes.2015.100.044.

Ganjian, E., Khorami, M., & Maghsoudi, A. A. (2009). Scrap-tyre-rubber replacement for aggregate and filler in concrete. *Construction and Building Materials*, *23*(5), 1828–1836. Available from https://doi.org/10.1016/j.conbuildmat.2008.09.020.

Gerges, N. N., Issa, C. A., Fawaz, S. A., Jabbour, J., Jreige, J., & Yacoub, A. (2018). Recycled glass concrete: Coarse and fine aggregates. *European Journal of Engineering and Technology Research*, *3*(1), 1–9. Available from https://doi.org/10.24018/ejeng.2018.3.10.533.

Gesoğlu, M., Güneyisi, E., Khoshnaw, G., & İpek, S. (2014a). Abrasion and freezing–thawing resistance of pervious concretes containing waste rubbers. *Construction and Building Materials*, *73*, 19–24. Available from https://doi.org/10.1016/j.conbuildmat.2014.09.047.

Gesoğlu, M., Güneyisi, E., Khoshnaw, G., & İpek, S. (2014b). Investigating properties of pervious concretes containing waste tire rubbers. *Construction and Building Materials*, *63*, 206–213. Available from https://doi.org/10.1016/j.conbuildmat.2014.04.046.

Geyer, R., Jambeck, J. R., & Law, K. L. (2017). Production, use, and fate of all plastics ever made. *Science Advances*, *3*(7). Available from https://doi.org/10.1126/sciadv.1700782.

Gillen, S.L., Brand, A.S., Roesler, J.R., & Vavrik, W.R. (2012). Sustainable long-life composite concrete pavement for Illinois Tollway. In *International conference on long-life concrete pavements.*

Gillott, J. E. (1980). Use of the scanning electron microscope and Fourier methods in characterization of microfabric and texture of sediments. *Journal of Microscopy*, *120*(3), 261–277. Available from https://doi.org/10.1111/j.1365-2818.1980.tb04147.x.

Gorospe, K., Booya, E., Ghaednia, H., & Das, S. (2019). Effect of various glass aggregates on the shrinkage and expansion of cement mortar. *Construction and Building Materials*, *210*, 301–311. Available from https://doi.org/10.1016/j.conbuildmat.2019.030.192.

Gupta, T., Chaudhary, S., & Sharma, R. K. (2014). Assessment of mechanical and durability properties of concrete containing waste rubber tire as fine aggregate. *Construction and Building Materials*, *73*, 562–574. Available from https://doi.org/10.1016/j.conbuildmat.2014.090.102.

Gupta, T., & Sachdeva, S. N. (2019). Investigations on jarosite mixed cement concrete pavements. *Arabian Journal for Science and Engineering*, *44*(10), 8787–8797. Available from https://doi.org/10.1007/s13369-019-03801-1.

Greenpeace. (2022). *Circular claims fail flat again* [Full report].).

Hansen, K.R., & Copeland, A. (2015). *Annual asphalt pavement industry survey on recycled materials and warm-mix asphalt usage*.

Hochella, M. F., Moore, J. N., Golla, U., & Putnis, A. (1999). A TEM study of samples from acid mine drainage systems: Metal-mineral association with implications for transport. *Geochimica et Cosmochimica Acta*, *63*(19-20), 3395–3406. Available from https://doi.org/10.1016/S0016-7037(99)00260-4.

Holmes, N., Browne, A., & Montague, C. (2014). Acoustic properties of concrete panels with crumb rubber as a fine aggregate replacement. *Construction and Building Materials*, *73*, 195–204. Available from https://doi.org/10.1016/j.conbuildmat.2014.09.107.

Hossiney, N., Tia, M., & Bergin, M. J. (2010). Concrete containing RAP for use in concrete pavement. *International Journal of Pavement Research and Technology*, *3*(5), 251–258.

Huang, B., Shu, X., & Burdette, E. G. (2006). Mechanical properties of concrete containing recycled asphalt pavements. *Magazine of Concrete Research*, *58*(5), 313–320. Available from https://doi.org/10.1680/macr.2006.58.5.313UnitedStates.

Huang, B., Shu, X., & Li, G. (2005). Laboratory investigation of Portland cement concrete containing recycled asphalt pavements. *Cement and Concrete Research*, *35*(10), 2008–2013. Available from https://doi.org/10.1016/j.cemconres.2005.05.002.

Huda, S. B., & Shahria Alam, M. (2015). Mechanical and freeze-thaw durability properties of recycled aggregate concrete made with recycled coarse aggregate. *Journal of Materials in Civil Engineering*, *27*(10). Available from https://doi.org/10.1061/(ASCE)MT.1943-5533.0001237.

Hunag, L. J., Wang, H. Y., & Wang, S. Y. (2015). A study of the durability of recycled green building materials in lightweight aggregate concrete. *Construction and Building Materials*, *96*, 353–359. Available from https://doi.org/10.1016/j.conbuildmat.2015.080.018.

Hunag, L. J., Wang, H. Y., & Wu, Y. W. (2016). Properties of the mechanical in controlled low-strength rubber lightweight aggregate concrete (CLSRLC. *Construction and Building Materials*, *112*, 1054–1058. Available from https://doi.org/10.1016/j.conbuildmat.2016.03.016.

Islam, M. J., Meherier, M. S., & Islam, A. K. M. R. (2016). Effects of waste PET as coarse aggregate on the fresh and harden properties of concrete. *Construction and Building Materials*, *125*, 946–951. Available from https://doi.org/10.1016/j.conbuildmat.2016.08.128.

Ismail, Z. Z., & AL-Hashmi, E. A. (2009). Recycling of waste glass as a partial replacement for fine aggregate in concrete. *Waste Management*, *29*(2), 655–659. Available from https://doi.org/10.1016/j.wasman.2008.08.012.

Jahanzaib Khalil, M., Aslam, M., & Ahmad, S. (2021). Utilization of sugarcane bagasse ash as cement replacement for the production of sustainable concrete – A review. *Construction and Building Materials*, *270*, 121371. Available from https://doi.org/10.1016/j.conbuildmat.2020.121371.

Jani, Y., & Hogland, W. (2014). Waste glass in the production of cement and concrete – A review. *Journal of Environmental Chemical Engineering*, *2*(3), 1767–1775. Available from https://doi.org/10.1016/j.jece.2014.030.016.

Jassim, A. K. (2023). Effect of microplastic on the human health. In E.-S. Salama (Ed.), *Advances and challenges in microplastics*. IntechOpen. Available from http://doi.org/10.5772/intechopen.107149.

Jittin, V., Bahurudeen, A., & Ajinkya, S. D. (2020). Utilisation of rice husk ash for cleaner production of different construction products. *Journal of Cleaner Production, 263*, 121578. Available from https://doi.org/10.1016/j.jclepro.2020.121578.

Kamali, M., & Ghahremaninezhad, A. (2016). An investigation into the hydration and microstructure of cement pastes modified with glass powders. *Construction and Building Materials, 112*, 915–924. Available from https://doi.org/10.1016/j.conbuildmat.2016.020.085.

Kashani, A., Ngo, T. D., & Hajimohammadi, A. (2019). *Effect of recycled glass fines on mechanical and durability properties of concrete foam in comparison with traditional cementitious fines, . Cement and Concrete Composites* (99, pp. 120–129). . Available from 10.1016/j.cemconcomp.2019.03.004.

Katz, A. (2003). Properties of concrete made with recycled aggregate from partially hydrated old concrete. *Cement and Concrete Research, 33*(5), 703–711. Available from https://doi.org/10.1016/S0008-8846(02)01033-5.

Kawamura, M., Torii, K., Hasaba, S., Nicho, N., & Oda, K. (1983). *Applicability of basic oxygen furnace slag as a concrete aggregate* (pp. 1123–1141). SP - American Concrete Institute.

Khankhaje, E., Salim, M. R., Mirza, J., Salmiati, M. W., Hussin, R., Khan, M., & Rafieizonooz. (2017). *Properties of quiet pervious concrete containing oil palm kernel shell and cockleshell, Applied Acoustics* (122, pp. 113–120). Available from 10.1016/j.apacoust.2017.02.014.

Kim, J. H., & You, Y. C. (2015). Composite behavior of a novel insulated concrete sandwich wall panel reinforced with GFRP shear grids: Effects of insulation types. *Materials, 8* (3), 899–913. Available from https://doi.org/10.3390/ma8030899.

Kim, S. K., Hanif, A., & Jang, I. Y. (2018). Incorporating liquid crystal display (LCD) glass waste as supplementary cementing material (SCM) in cement mortars-Rationale based on hydration, durability, and pore characteristics. *Materials, 11*(12). Available from https://doi.org/10.3390/ma11122538.

Kou, S. C., & Poon, C. S. (2009). Properties of self-compacting concrete prepared with recycled glass aggregate. *Cement and Concrete Composites, 31*(2), 107–113. Available from https://doi.org/10.1016/j.cemconcomp.2008.12.002.

Lai, M. H., Zou, Jiajun, Yao, Boyu, Ho, J. C. M., Zhuang, Xin, & Wang, Qing (2021). Improving mechanical behavior and microstructure of concrete by using BOF steel slag aggregate. *Construction and Building Materials, 277*, 122269. Available from https://doi.org/10.1016/j.conbuildmat.2021.122269.

Lam, M. N. T., Jaritngam, S., & Le, D. H. (2017). Roller-compacted concrete pavement made of Electric Arc Furnace slag aggregate: Mix design and mechanical properties. *Construction and Building Materials, 154*, 482–495. Available from https://doi.org/10.1016/j.conbuildmat.2017.07.240.

Lam, M. N. T., Le, D. H., & Jaritngam, S. (2018). Compressive strength and durability properties of roller-compacted concrete pavement containing electric arc furnace slag aggregate and fly ash. *Construction and Building Materials, 191*, 912–922. Available from https://doi.org/10.1016/j.conbuildmat.2018.10.080.

Langer, W.H. (2011). *Aggregate resource availability in the conterminous united states, including suggestions for addressing shortages, quality, and environmental concerns* [Report 2011-1119]. U.S. Geological Survey.

Law, K. L., Starr, N., Siegler, T. R., Jambeck, J. R., Mallos, N. J., & Leonard, G. H. (2020). The United States' contribution of plastic waste to land and ocean. *Science Advances, 6* (44). Available from https://doi.org/10.1126/sciadv.abd0288.

Le, T., Le Saout, G., Garcia-Diaz, E., Betrancourt, D., & Rémond, S. (2017). Hardened behavior of mortar based on recycled aggregate: Influence of saturation state at macro-

and microscopic scales. *Construction and Building Materials*, *141*, 479–490. Available from https://doi.org/10.1016/j.conbuildmat.2017.020.035.

Leite, M. B., & Monteiro, P. J. M. (2016). *Microstructural analysis of recycled concrete using X-ray microtomography*, *Cement and Concrete Research* (81, pp. 38–48). Available from 10.1016/j.cemconres.2015.11.010.

Lertwattanaruk, P., Makul, N., & Siripattarapravat, C. (2012). Utilization of ground waste seashells in cement mortars for masonry and plastering. *Journal of Environmental Management*, *111*, 133–141. Available from https://doi.org/10.1016/j.jenvman.2012.060.032.

Leslie, H. A., van Velzen, M. J. M., Brandsma, S. H., Vethaak, A. D., Garcia-Vallejo, J. J., & Lamoree, M. H. (2022). *Discovery and quantification of plastic particle pollution in human blood*, *Environment International* (163). Available from 10.1016/j.envint.2022.107199.

Liew, K. M., & Akbar, Arslan (2020). The recent progress of recycled steel fiber reinforced concrete. *Construction and Building Materials*, *232*, 117232. Available from https://doi.org/10.1016/j.conbuildmat.2019.117232.

Limbachiya, M. C. (2009). Bulk engineering and durability properties of washed glass sand concrete. *Construction and Building Materials*, *23*(2), 1078–1083. Available from https://doi.org/10.1016/j.conbuildmat.2008.050.022.

Lin, Z. (1997). Mobilization and retention of heavy metals in mill-tailings from Garpenberg sulfide mines, Sweden. *Science of the Total Environment*, *198*(1), 13–31. Available from https://doi.org/10.1016/S0048-9697(97)05433-8.

Liu, Tiejun, Wei, Huinan, Zou, Dujian, Zhou, Ao, & Jian, Hongshu (2020). Utilization of waste cathode ray tube funnel glass for ultra-high performance concrete. *Journal of Cleaner Production*, *249*, 119333. Available from https://doi.org/10.1016/j.jclepro.2019.119333.

Lo Presti, D. (2013). Recycled tyre rubber modified bitumens for road asphalt mixtures: A literature review. *Construction and Building Materials*, *49*, 863–881. Available from https://doi.org/10.1016/j.conbuildmat.2013.09.007.

Loh, Y. R., Sujan, D., Rahman, M. E., & Das, C. A. (2013). *Review sugarcane bagasse – the future composite material: A literature review*, *Resources, Conservation and Recycling* (75, pp. 14–22). Available from 10.1016/j.resconrec.2013.03.002.

Lotfi, S., Eggimann, M., Wagner, E., Mróz, R., & Deja, J. (2015). Performance of recycled aggregate concrete based on a new concrete recycling technology. *Construction and Building Materials*, *95*, 243–256. Available from https://doi.org/10.1016/j.conbuildmat.2015.07.021.

Luhar, S., & Luhar, I. (2019). Potential application of E-wastes in construction industry: A review. *Construction and Building Materials*, *203*, 222–240. Available from https://doi.org/10.1016/j.conbuildmat.2019.01.080.

Makri, C., Hahladakis, J. N., & Gidarakos, E. (2019). Use and assessment of "e-plastics" as recycled aggregates in cement mortar. *Journal of Hazardous Materials*, *379*, 120776. Available from https://doi.org/10.1016/j.jhazmat.2019.120776.

Malešev, M., Radonjanin, V., & Marinković, S. (2010). Recycled concrete as aggregate for structural concrete production. *Sustainability.*, *2*(5), 1204–1225. Available from https://doi.org/10.3390/su2051204.

Mardani-Aghabaglou, A., Tuyan, M., & Ramyar, K. (2015). Mechanical and durability performance of concrete incorporating fine recycled concrete and glass aggregates. *Materials and Structures/Materiaux et Constructions*, *48*(8), 2629–2640. Available from https://doi.org/10.1617/s11527-014-0342-3.

Martínez-García, C., González-Fonteboa, B., Martínez-Abella, F., & Carro- López, D. (2017). Performance of mussel shell as aggregate in plain concrete. *Construction and*

Building Materials, *139*, 570–583. Available from https://doi.org/10.1016/j.conbuildmat.2016.090.091.

Mastali, M., Dalvand, A., Sattarifard, A. R., Abdollahnejad, Z., & Illikainen, M. (2018). Characterization and optimization of hardened properties of self-consolidating concrete incorporating recycled steel, industrial steel, polypropylene and hybrid fibers. *Composites Part B: Engineering*, *151*, 186–200. Available from https://doi.org/10.1016/j.compositesb.2018.060.021.

Mastali, M., Dalvand, A., Sattarifard, A. R., & Illikainen, M. (2018). Development of eco-efficient and cost-effective reinforced self-consolidation concretes with hybrid industrial/recycled steel fibers. *Construction and Building Materials*, *166*, 214–226. Available from https://doi.org/10.1016/j.conbuildmat.2018.010.147.

McNeil, K., & Kang, T. H. K. (2013). Recycled concrete aggregates: A review. *International Journal of Concrete Structures and Materials*, *7*(1), 61–69. Available from https://doi.org/10.1007/s40069-013-0032-5.

Mees, F., & Stoops, G. (2010). *Sulphidic and sulphuric materials. Interpretation of micromorphological features of soils and regoliths* (pp. 543–568). Elsevier. Available from 10.1016/B978-0-444-53156-8.00024-6.

Mehra, P., Gupta, R. C., & Thomas, B. S. (2016a). Properties of concrete containing jarosite as a partial substitute for fine aggregate. *Journal of Cleaner Production*, *120*, 241–248. Available from https://doi.org/10.1016/j.jclepro.2016.01.015.

Mehra, P., Gupta, R. C., & Thomas, B. S. (2016b). Assessment of durability characteristics of cement concrete containing jarosite. *Journal of Cleaner Production*, *119*, 59–65. Available from https://doi.org/10.1016/j.jclepro.2016.010.055.

Meininger, R. C., & Stokowski, S. J. (2011). Wherefore art thou aggregate resources for highways? *Public Roads*, *75*(2).

Meyer, C., & Xi, Y. (1999). Use of recycled glass and fly ash for precast concrete. *Journal of Materials in Civil Engineering*, *11*(2), 89–90. Available from https://doi.org/10.1061/(ASCE)0899-1561(1999)11:2(89).

Mirzahosseini, M., & Riding, K. A. (2015). *Influence of different particle sizes on reactivity of finely ground glass as supplementary cementitious material (SCM), Cement and Concrete Composites* (56, pp. 95–105). Available from 10.1016/j.cemconcomp.2014.10.004.

Mo, K. H., Alengaram, U. J., Jumaat, M. Z., Lee, S. C., Goh, W. I., & Yuen, C. W. (2018). Recycling of seashell waste in concrete: A review. *Construction and Building Materials*, *162*, 751–764. Available from https://doi.org/10.1016/j.conbuildmat.2017.12.009.

Moallemi Pour, S., & Alam, M. S. (2016). *Investigation of compressive bond behavior of steel rebar embedded in concrete with partial recycled aggregate replacement, Structures* (7, pp. 153–164). Available from 10.1016/j.istruc.2016.06.010.

Moaveni, M., Cetin, S., Brand, A. S., Dahal, S., Roesler, J. R., & Tutumluer, E. (2016). *Machine vision based characterization of particle shape and asphalt coating in Reclaimed Asphalt Pavement, Transportation Geotechnics* (6, pp. 26–37). Available from 10.1016/j.trgeo.2016.01.001.

Modani, P. O., & Vyawahare, M. R. (2013). Utilization of bagasse ash as a partial replacement of fine aggregate in concrete. *Procedia Engineering*, *51*, 25–29. Available from https://doi.org/10.1016/j.proeng.2013.01.00718777058.

MoEF. (2008). *Hazardous waste (management, handling and transboundary movement rules)*. Ministry of Environment and Forests.

Mohamed, M., Yousuf, S., & Maitra, S. (2012). Decomposition study of calcium carbonate in cockle shell. *Journal of Engineering Science and Technology*, *7*(1), 1–10.

Mohammed, A. A., Mohammed, I. I., & Mohammed, S. A. (2019). Some properties of concrete with plastic aggregate derived from shredded PVC sheets. *Construction and Building Materials, 201*, 232–245. Available from https://doi.org/10.1016/j.conbuildmat.2018.120.145.

Mostofinejad, D., Hosseini, S. M., Nosouhian, F., Ozbakkaloglu, T., & Nader Tehrani, B. (2020). *Durability of concrete containing recycled concrete coarse and fine aggregates and milled waste glass in magnesium sulfate environment, Journal of Building Engineering* (29). Available from 10.1016/j.jobe.2020.101182.

Moura, W., Masuero, A., Dal Molin, D., & Vilela, A. (1999). *Concrete performance with admixtures of electrical steel slag and copper slag concerning mechanical properties* (pp. 81–100). American Concrete Institute, ACI Special Publication 01932527.

Mukhopadhyay, A., & Shi, X. (2017). *Validation of RAP and/or RAS in hydraulic cement concrete* [Technical report No. 0-6855-1]. Texas A&M Transportation Institute.

Naaamandadin, N.A., Abdul Aziz, I.S., Mustafa, W.A., & Santiagoo, R. (2020). Mechanical properties of the utilisation glass powder as partial replacement of cement in concrete. *Lecture Notes in Mechanical Engineering* (pp. 221-229).

Nandanam, K., Biswal, U. S., & Dinakar, P. (2021). Effect of fly ash, GGBS, and metakaolin on mechanical and durability properties of self-compacting concrete made with 100% coarse recycled aggregate. *Journal of Hazardous, Toxic, and Radioactive Waste, 25*(2). Available from https://doi.org/10.1061/(ASCE)HZ.2153-5515.0000595.

Nguyen, D. H., Boutouil, M., Sebaibi, N., Baraud, F., & Leleyter, L. (2017). Durability of pervious concrete using crushed seashells. *Construction and Building Materials, 135*, 137–150. Available from https://doi.org/10.1016/j.conbuildmat.2016.12.219.

Nodehi, M., Ren, J., Shi, X., Debbarma, S., & Ozbakkaloglu, T. (2023). Experimental evaluation of alkali-activated and Portland cement-based mortars prepared using waste glass powder in replacement of fly ash. *Construction and Building Materials, 394*, 132124. Available from https://doi.org/10.1016/j.conbuildmat.2023.132124.

Okafor, F. O. (2010). Performance of recycled asphalt pavement as coarse aggregate in concrete. *Leonardo Electronic Journal of Practices and Technologies, 9*(17), 47–58.

Olorunsogo, F. T., & Padayachee, N. (2002). Performance of recycled aggregate concrete monitored by durability indexes. *Cement and Concrete Research, 32*(2), 179–185. Available from https://doi.org/10.1016/S0008-8846(01)00653-6.

Onuaguluchi, O., & Panesar, D. K. (2014). Hardened properties of concrete mixtures containing pre-coated crumb rubber and silica fume. *Journal of Cleaner Production, 82*, 125–131. Available from https://doi.org/10.1016/j.jclepro.2014.06.068.

Ortega-López, V., Fuente-Alonso, J. A., Santamaría, A., San-José, J. T., & Aragón, Á. (2018). Durability studies on fiber-reinforced EAF slag concrete for pavements. *Construction and Building Materials, 163*, 471–481. Available from https://doi.org/10.1016/j.conbuildmat.2017.12.121.

Osubor, Stanley O., Salam, Kamoru A., & Audu, Taiwo M. (2019). Effect of flaky plastic particle size and volume used as partial replacement of gravel on compressive strength and density of concrete mix. *Journal of Environmental Protection, 10*(06), 711–721. Available from https://doi.org/10.4236/jep.2019.106042.

Padmini, A. K., Ramamurthy, K., & Mathews, M. S. (2009). Influence of parent concrete on the properties of recycled aggregate concrete. *Construction and Building Materials, 23*(2), 829–836. Available from https://doi.org/10.1016/j.conbuildmat.2008.030.006.

Pang, B., Zhou, Z., Cheng, X., Du, P., & Xu, H. (2016). ITZ properties of concrete with carbonated steel slag aggregate in salty freeze-thaw environment. *Construction and*

Building Materials, *114*, 162–171. Available from https://doi.org/10.1016/j.conbuildmat.2016.030.168.

Pang, B., Zhou, Z., & Xu, H. (2015). Utilization of carbonated and granulated steel slag aggregate in concrete. *Construction and Building Materials*, *84*, 454–467. Available from https://doi.org/10.1016/j.conbuildmat.2015.03.008.

Pani, L., Francesconi, L., Rombi, J., Mistretta, F., Sassu, M., & Stochino, F. (2020). Effect of parent concrete on the performance of recycled aggregate concrete. *Sustainability (Switzerland)*, *12*(22), 1–17. Available from https://doi.org/10.3390/su12229399.

Pappu, A., Saxena, M., & Asolekar, S. R. (2006). Jarosite characteristics and its utilisation potentials. *Science of the Total Environment*, *359*(1-3), 232–243. Available from https://doi.org/10.1016/j.scitotenv.2005.040.024.

Park, S. B., Lee, B. C., & Kim, J. H. (2004). Studies on mechanical properties of concrete containing waste glass aggregate. *Cement and Concrete Research*, *34*(12), 2181–2189. Available from https://doi.org/10.1016/j.cemconres.2004.02.006.

Patankar, V. D., & Williams, R. I. (1970). Bitumen in dry lean concrete. *Highways and Traffic Engineering*, *38*.

Patrick, M. M., Elizabeth, M. M., & Titilayo, E. A. (2018). Characterization, leachate characteristics and compressive strength of Jarosite/clay/fly ash bricks. *Materials Today: Proceedings*, *5*(9), 17802–17811. Available from https://doi.org/10.1016/j.matpr.2018.06.105.

Pereira-De-Oliveira, L. A., Castro-Gomes, J. P., & Santos, P. M. S. (2012). The potential pozzolanic activity of glass and red-clay ceramic waste as cement mortars components. *Construction and Building Materials*, *31*, 197–203. Available from https://doi.org/10.1016/j.conbuildmat.2011.120.110.

Perlack, R. (2011). *Crop residues and agricultural wastes*. U.S. Department of Energy.

Pickel, D., Tighe, S., & West, J. S. (2017). Assessing benefits of pre-soaked recycled concrete aggregate on variably cured concrete. *Construction and Building Materials*, *141*, 245–252. Available from https://doi.org/10.1016/j.conbuildmat.2017.020.140.

Poutos, K. H., Alani, A. M., Walden, P. J., & Sangha, C. M. (2008). Relative temperature changes within concrete made with recycled glass aggregate. *Construction and Building Materials*, *22* (4), 557–565. Available from https://doi.org/10.1016/j.conbuildmat.2006.110.018.

Pradhan, S., Kumar, S., & Barai, S. V. (2017). Recycled aggregate concrete: Particle Packing Method (PPM) of mix design approach. *Construction and Building Materials*, *152*, 269–284. Available from https://doi.org/10.1016/j.conbuildmat.2017.06.171.

Prajapati, R., Gettu, R., & Singh, S. (2021). Thermomechanical beneficiation of recycled concrete aggregates (RCA). *Construction and Building Materials*, *310*, 125200. Available from https://doi.org/10.1016/j.conbuildmat.2021.125200.

Prasittisopin, L., Termkhajornkit, P., & Kim, Y. H. (2022). Review of concrete with expanded polystyrene (EPS): Performance and environmental aspects. *Journal of Cleaner Production*, *366*, 132919. Available from https://doi.org/10.1016/j.jclepro.2022.132919.

Qasrawi, H. (2014). The use of steel slag aggregate to enhance the mechanical properties of recycled aggregate concrete and retain the environment. *Construction and Building Materials*, *54*, 298–304. Available from https://doi.org/10.1016/j.conbuildmat.2013.12.063.

Raghavan, D., Huynh, H., & Ferraris, C. F. (1998). Workability, mechanical properties, and chemical stability of a recycled tyre rubber-filled cementitious composite. *Journal of Materials Science*, *33*(7), 1745–1752. Available from https://doi.org/10.1023/A:1004372414475.

Rahal, K. (2007). Mechanical properties of concrete with recycled coarse aggregate. *Building and Environment*, *42*(1), 407–415. Available from https://doi.org/10.1016/j.buildenv.2005.07.033.

Rakshvir, M., & Barai, S. V. (2006). Studies on recycled aggregates-based concrete. *Waste Management and Research*, *24*(3), 225–233. Available from https://doi.org/10.1177/0734242X06064820.

Ramakrishnan, K., Pugazhmani, G., Sripragadeesh, R., Muthu, D., & Venkatasubramanian, C. (2017). Experimental study on the mechanical and durability properties of concrete with waste glass powder and ground granulated blast furnace slag as supplementary cementitious materials. *Construction and Building Materials*, *156*, 739–749. Available from https://doi.org/10.1016/j.conbuildmat.2017.080.183.

Ray, S., Daudi, L., Yadav, H., & Ransinchung, G. D. (2020). Utilization of Jarosite waste for the development of sustainable concrete by reducing the cement content. *Journal of Cleaner Production*, *272*, 122546. Available from https://doi.org/10.1016/j.jclepro.2020.122546.

Romero, M., & Rincón, J. M. (1997). Microstractural characterization of a goethite waste from zinc hydrometallurgical process. *Materials Letters*, *31*(1-2), 67–73. Available from https://doi.org/10.1016/S0167-577X(96)00235-2.

Saberian, M., Li, J., Boroujeni, M., Law, D., & Li, C. Q. (2020). *Application of demolition wastes mixed with crushed glass and crumb rubber in pavement base/subbase, Resources, Conservation and Recycling* (156). Available from 10.1016/j.resconrec.2020.104722.

Saccani, A., & Bignozzi, M. C. (2010). ASR expansion behavior of recycled glass fine aggregates in concrete. *Cement and Concrete Research*, *40*(4), 531–536. Available from https://doi.org/10.1016/j.cemconres.2009.090.003.

Safiuddin, M., Alengaram, U. J., Rahman, M. M., Salam, M. A., & Jumaat, M. Z. (2013). Use of recycled concrete aggregate in concrete: A review. *Journal of Civil Engineering and Management*, *19*(6), 796–810. Available from https://doi.org/10.3846/13923730.2013.799093.

Safiuddin, M. D., Salam, M. A., & Jumaat, M. Z. (2011). Effects of recycled concrete aggregate on the fresh properties of self-consolidating concrete. *Archives of Civil and Mechanical Engineering*, *11*(4), 1023–1041. Available from https://doi.org/10.1016/s1644-9665(12)60093-4.

Saikia, N., & De Brito, J. (2014). *Mechanical properties and abrasion behaviour of concrete containing shredded PET bottle waste as a partial substitution of natural aggregate, Construction and Building Materials* (52, pp. 236–244). Available from 10.1016/j.conbuildmat.2013.11.049.

Sajjad, M., Huang, Q., Khan, S., Khan, M. A., Liu, Y., Wang, J., Lian, F., Wang, Q., & Guo, G. (2022). Microplastics in the soil environment: A critical review. *Environmental Technology & Innovation*, *27*, 102408. Available from https://doi.org/10.1016/j.eti.2022.102408.

Sandhu, R. K., & Siddique, R. (2017). Influence of rice husk ash (RHA) on the properties of self-compacting concrete: A review. *Construction and Building Materials*, *153*, 751–764.

Schwarz, N., Cam, H., & Neithalath, N. (2008). Influence of a fine glass powder on the durability characteristics of concrete and its comparison to fly ash. *Cement and Concrete Composites*, *30*(6), 486–496. Available from https://doi.org/10.1016/j.cemconcomp.2008.02.001.

Sengul, O. (2016). Mechanical behavior of concretes containing waste steel fibers recovered from scrap tires. *Construction and Building Materials*, *122*, 649–658. Available from https://doi.org/10.1016/j.conbuildmat.2016.060.113.

Shakouri, M., Exstrom, C. L., Ramanathan, S., Suraneni, P., & Vaux, J. S. (2020). *Pretreatment of corn stover ash to improve its effectiveness as a supplementary cementitious material in concrete, Cement and Concrete Composites* (112). Available from 10.1016/j.cemconcomp.2020.103658.

Shao, Y., Lefort, T., Moras, S., & Rodriguez, D. (2000). Studies on concrete containing ground waste glass. *Cement and Concrete Research, 30*(1), 91–100. Available from https://doi.org/10.1016/S0008-8846(99)00213-6.

Shi, C., Meyer, C., & Behnood, A. (2008). Utilization of copper slag in cement and concrete. *Resources, Conservation and Recycling, 52*(10), 1115–1120. Available from https://doi.org/10.1016/j.resconrec.2008.06.008.

Shi, X., Grasley, Z., Hogancamp, J., Brescia-Norambuena, L., Mukhopadhyay, A., & Zollinger, D. (2020). Microstructural, mechanical, and shrinkage characteristics of cement mortar containing fine reclaimed asphalt pavement. *Journal of Materials in Civil Engineering, 32*(4), 04020050. Available from https://doi.org/10.1061/(ASCE)MT.1943-5533.0003110.

Shi, X., Mirsayar, M., Mukhopadhyay, A., & Zollinger, D. (2019). *Characterization of two-parameter fracture properties of Portland cement concrete containing reclaimed asphalt pavement aggregates by semicircular bending specimens, Cement and Concrete Composites.* (95, pp. 56–69). Available from 10.1016/j.cemconcomp.2018.10.013.

Shi, X., Mukhopadhyay, A., & Liu, K. W. (2017). Mix design formulation and evaluation of Portland cement concrete paving mixtures containing reclaimed asphalt pavement. *Construction and Building Materials, 152*, 756–768. Available from https://doi.org/10.1016/j.conbuildmat.2017.060.174.

Shi, X., Mukhopadhyay, A., & Zollinger, D. (2018). *Sustainability assessment for Portland cement concrete pavement containing reclaimed asphalt pavement aggregates, Journal of Cleaner Production* (192, pp. 569–581). Available from 10.1016/j.jclepro.2018.05.004.

Shi, X., Mukhopadhyay, A., & Zollinger, D. (2019). Long-term performance evaluation of concrete pavements containing recycled concrete aggregate in Oklahoma. *Transportation Research Record: Journal of the Transportation Research Board, 2673*(5), 429–442. Available from https://doi.org/10.1177/0361198119839977.

Shi, X., Mukhopadhyay, A., Zollinger, D., & Huang, K. (2019). Performance evaluation of jointed plain concrete pavement made with Portland cement concrete containing reclaimed asphalt pavement. *Road Materials and Pavement Design, 22*(1), 59–81.

Shi, X., Rew, Y., Ivers, E., Seon, S.-C., Stenger, E. M., & Park, P. (2019). Effects of thermally modified asphalt concrete on pavement temperature. *International Journal of Pavement Engineering, 20*(6), 669–681. Available from https://doi.org/10.1080/10298436.2017.1326234.

Shi, X., Zollinger, D., & Mukhopadhyay, A. K. (2018). Punchout study for continuously reinforced concrete pavement containing reclaimed asphalt pavement using pavement ME models. *International Journal of Pavement Engineering*, 1–14.

Siddique, R., & Naik, T. R. (2004). Properties of concrete containing scrap-tire rubber – an overview. *Waste Management, 24*(6), 563–569. Available from https://doi.org/10.1016/j.wasman.2004.010.006.

Silva, R. V., De Brito, J., & Dhir, R. K. (2015). Tensile strength behaviour of recycled aggregate concrete. *Construction and Building Materials, 83*, 108–118. Available from https://doi.org/10.1016/j.conbuildmat.2015.030.034.

Singh, S., & Ransinchung, G. D. R. N. (2020). Laboratory and field evaluation of RAP for cement concrete pavements. *Journal of Transportation Engineering, Part B: Pavements, 146*(2), 04020011. Available from https://doi.org/10.1061/jpeodx.0000162.

Singh, S., Ransinchung, G. D. R. N., & Kumar, P. (2017a). Laboratory investigation of concrete pavements containing fine RAP aggregates. *Journal of Materials in Civil Engineering, 30.*

Singh, S., Ransinchung, G. D. R. N., & Kumar, P. (2017b). An economical processing technique to improve RAP inclusive concrete properties. *Construction and Building Materials, 148*, 734–747. Available from https://doi.org/10.1016/j.conbuildmat.2017.05.030.

Singh, S., Ransinchung, G. D. R. N., & Kumar, P. (2017c). Feasibility study of RAP aggregates in cement concrete pavements. *Road Materials and Pavement Design, 20*(1), 151–170. Available from https://doi.org/10.1080/14680629.2017.1380071.

Singh, S., Ransinchung, G. D. R. N., & Kumar, P. (2018). Laboratory investigation of concrete pavements containing fine RAP aggregates. *Journal of Materials in Civil Engineering, 30*(2). Available from https://doi.org/10.1061/(asce)mt.1943-5533.0002124.

Skarżyński, Ł., & Suchorzewski, J. (2018). Mechanical and fracture properties of concrete reinforced with recycled and industrial steel fibers using Digital Image Correlation technique and X-ray micro computed tomography. *Construction and Building Materials, 183*, 283–299. Available from https://doi.org/10.1016/j.conbuildmat.2018.060.182.

Slager, S., Jongmans, A. G., & Pons, L. J. (1970). Morphology of some tropical alluvial clay soils. *Journal of Soil Science, 21*(2), 233–241. Available from https://doi.org/10.1111/j.1365-2389.1970.tb01172.x.

Soliman, N. A., & Tagnit-Hamou, A. (2017). Using glass sand as an alternative for quartz sand in UHPC. *Construction and Building Materials, 145*, 243–252. Available from https://doi.org/10.1016/j.conbuildmat.2017.03.187.

Su, H., Yang, J., Ling, T. C., Ghataora, G. S., & Dirar, S. (2015). *Properties of concrete prepared with waste tyre rubber particles of uniform and varying sizes, . Journal of Cleaner Production* (91, pp. 288–296). . Available from 10.1016/j.jclepro.2014.12.022.

Tam, V. W. Y., Gao, X. F., & Tam, C. M. (2005). Microstructural analysis of recycled aggregate concrete produced from two-stage mixing approach. *Cement and Concrete Research, 35*(6), 1195–1203. Available from https://doi.org/10.1016/j.cemconres.2004.100.025.

Tam, V. W. Y., Soomro, M., & Evangelista, A. C. J. (2018). A review of recycled aggregate in concrete applications (2000–2017. *Construction and Building Materials, 172*, 272–292. Available from https://doi.org/10.1016/j.conbuildmat.2018.03.240.

Tam, V. W. Y., & Tam, C. M. (2007). Assessment of durability of recycled aggregate concrete produced by two-stage mixing approach. *Journal of Materials Science, 42*(10), 3592–3602.

Tam, V. W. Y., & Tam, C. M. (2008). Diversifying two-stage mixing approach (TSMA) for recycled aggregate concrete: TSMAs and TSMAsc. *Construction and Building Materials, 22* (10), 2068–2077. Available from https://doi.org/10.1016/j.conbuildmat.2007.070.024.

Food and Agriculture Organization of the United States. (2020). *The state of world fisheries and aquaculture*. https://www.fao.org/state-of-fisheries-aquaculture.

Thomas, B. S., & Gupta, R. C. (2016). *A comprehensive review on the applications of waste tire rubber in cement concrete, Renewable and Sustainable Energy Reviews* (54, pp. 1323–1333). Available from 10.1016/j.rser.2015.10.092.

Thomas, B. S., Gupta, R. C., Mehra, P., & Kumar, S. (2015). Performance of high strength rubberized concrete in aggressive environment. *Construction and Building Materials, 83*, 320–326. Available from https://doi.org/10.1016/j.conbuildmat.2015.030.012.

Thomas, B. S., Yang, J., Bahurudeen, A., Abdalla, J. A., Hawileh, R. A., Hamada, H. M., Nazar, S., Jittin, V., & Ashish, D. K. (2021). Sugarcane bagasse ash as supplementary cementitious material in concrete – a review. *Materials Today Sustainability, 15*, 100086. Available from https://doi.org/10.1016/j.mtsust.2021.100086.

Thomas, B. S., Yang, J., Mo, K. H., Abdalla, J. A., Hawileh, R. A., & Ariyachandra, E. (2021). *Biomass ashes from agricultural wastes as supplementary cementitious materials or aggregate replacement in cement/geopolymer concrete: A comprehensive review, Journal of Building Engineering* (40). Available from 10.1016/j.jobe.2021.102332.

Thorneycroft, J., Orr, J., Savoikar, P., & Ball, R. J. (2018). Performance of structural concrete with recycled plastic waste as a partial replacement for sand. *Construction and Building Material, 161*, 63–69. Available from https://doi.org/10.1016/j.conbuildmat.2017.11.127.

Tia, M., Hossiney, N., Su, Y.-M., Chen, Y., & Do, T.A. (2012). Properties of concrete containing recycled asphalt pavement. In International Concrete Sustainability Conference. pp. 1-13.

Ting, M. Z. Y., Wong, K. S., Rahman, M. E., & Selowara Joo, M. (2020). *Mechanical and durability performance of marine sand and seawater concrete incorporating silicomanganese slag as coarse aggregate, Construction and Building Materials* (254). Available from 10.1016/j.conbuildmat.2020.119195.

Topçu, I. B., & Canbaz, M. (2004). Properties of concrete containing waste glass. *Cement and Concrete Research, 34*(2), 267–274. Available from https://doi.org/10.1016/j.cemconres.2003.070.003.

Topçu, I. B., & Şengel, S. (2004). Properties of concretes produced with waste concrete aggregate. *Cement and Concrete Research, 34*(8), 1307–1312. Available from https://doi.org/10.1016/j.cemconres.2003.12.019.

Tran, T.Q., Li, S., Ji, B., Zhao, X., Rahat, M.H. H., Nguyen, T.-N., Le, B.-C., & Zhang, A.S.. (2023). *Mitigation of zinc and organic carbon leached from end-of-life tire rubber in cementitious composites*. In review.

Tran, T. Q., Skariah Thomas, B., Zhang, W., Ji, B., Li, S., & Brand, A. S. (2022). *A comprehensive review on treatment methods for end-of-life tire rubber used for rubberized cementitious materials, . Construction and Building Materials* (359). . Available from 10.1016/j.conbuildmat.2022.129365.

Tuaum, A., Shitote, S., & Oyawa, W. (2018). Experimental study of self-compacting mortar incorporating recycled glass aggregate. *Buildings, 8*(2), 15. Available from https://doi.org/10.3390/buildings8020015.

Turgut, P., & Yahlizade, E. S. (2009). Research into concrete blocks with waste glass. *International Journal of Environmental Science and Engineering., 3*, 186–192.

UN Report. (2017). *United Nations system-wide response to tackling E-waste report.*

USDA. (2021). *U.S. sugar production.* https://www.ers.usda.gov/topics/crops/sugar-sweeteners/background/#:~:text=In the United States%2C sugarcane,acres in FY 2020%2F21.

USEPA. (1992). *Toxicity characteristics leaching procedure*. Method 1311. United States Environmental Protection Agency.

USEPA. (2022). *Cleaning up electronic waste (e-waste).* https://www.epa.gov/international-cooperation/cleaning-electronic-waste-e-waste.

USPA. (2022). *Glass: Material-specific data.* https://www.epa.gov/facts-and-figures-about-materials-waste-and-recycling/glass-material-specific-data.

U.S. Tire Manufacturers Association. (2020). *U.S. scrap tire management summary.*

Uygunoğlu, T., & Topçu, I. B. (2010). The role of scrap rubber particles on the drying shrinkage and mechanical properties of self-consolidating mortars. *Construction and*

Building Materials, *24*(7), 1141–1150. Available from https://doi.org/10.1016/j.conbuildmat.2009.120.027.

Wang, B., Yan, L., Fu, Q., & Kasal, B. (2021). A comprehensive review on recycled aggregate and recycled aggregate concrete. *Resources, Conservation and Recycling*, *171*, 105565. Available from https://doi.org/10.1016/j.resconrec.2021.105565.

Wang, H. Y., & Huang, W. L. (2010). A study on the properties of fresh self-consolidating glass concrete (SCGC. *Construction and Building Materials*, *24*(4), 619–624. Available from https://doi.org/10.1016/j.conbuildmat.2009.08.047.

Wang, R., Shi, Q., Li, Y., Cao, Z., & Si, Z. (2021). A critical review on the use of copper slag (CS) as a substitute constituent in concrete. *Construction and Building Materials*, *292*, 123371. Available from https://doi.org/10.1016/j.conbuildmat.2021.123371.

Wang, S., Zhang, G., Wang, B., & Wu, M. (2020). Mechanical strengths and durability properties of pervious concretes with blended steel slag and natural aggregate. *Journal of Cleaner Production*, *271*, 122590. Available from https://doi.org/10.1016/j.jclepro.2020.122590.

Wang, S. Y., & Vipulanandan, C. (1996). Leachability of lead from solidified cement-fly ash binders. *Cement and Concrete Research*, *26*(6), 895–905. Available from https://doi.org/10.1016/0008-8846(96)00070-1.

West, G. (1996). *Alkali-aggregate reaction in concrete road and bridges*. ICE.

Wojnowska-Baryła, I., Bernat, K., & Zaborowska, M. (2022). Plastic waste degradation in landfill conditions: The problem with microplastics, and their direct and indirect environmental effects. *International Journal of Environmental Research and Public Health*, *19*(20). Available from https://doi.org/10.3390/ijerph192013223.

World Steel Association. (2021). *Steel industry co-products*.

World Steel Association. (2022). *World steel in figures*.

Xiao, J., Li, J., & Zhang, C. (2005). Mechanical properties of recycled aggregate concrete under uniaxial loading. *Cement and Concrete Research*, *35*(6), 1187–1194. Available from https://doi.org/10.1016/j.cemconres.2004.090.020.

Yang, E. I., Yi, S. T., & Leem, Y. M. (2005). Effect of oyster shell substituted for fine aggregate on concrete characteristics: Part I. Fundamental properties. *Cement and Concrete Research*, *35*(11), 2175–2182. Available from https://doi.org/10.1016/j.cemconres.2005.030.016.

Yang, S., Ling, T. C., Cui, H., & Poon, C. S. (2019). Influence of particle size of glass aggregates on the high temperature properties of dry-mix concrete blocks. *Construction and Building Materials*, *209*, 522–531. Available from https://doi.org/10.1016/j.conbuildmat.2019.03.131.

Yildirim, I. Z., & Prezzi, M. (2011). Chemical, mineralogical, and morphological properties of steel slag. *Advances in Civil Engineering*, *2011*. Available from https://doi.org/10.1155/2011/463638.

Yilmaz, A., & Degirmenci, N. (2009). Possibility of using waste tire rubber and fly ash with Portland cement as construction materials. *Waste Management*, *29*(5), 1541–1546. Available from https://doi.org/10.1016/j.wasman.2008.110.002.

You, Q., Miao, L., Li, C., Fang, H., & Liang, X. (2019). Study on the fatigue and durability behavior of structural expanded polystyrene concretes. *Materials*, *12*(18). Available from https://doi.org/10.3390/ma12182882.

Youssf, O., Elgawady, M. A., Mills, J. E., & Ma, X. (2014). An experimental investigation of crumb rubber concrete confined by fibre reinforced polymer tubes. *Construction and Building Materials*, *53*, 522–532. Available from https://doi.org/10.1016/j.conbuildmat.2013.120.007.

Zain, M. F. M., Islam, M. N., Radin, S. S., & Yap, S. G. (2004). Cement-based solidification for the safe disposal of blasted copper slag. *Cement and Concrete Composites*, *26*(7), 845–851. Available from https://doi.org/10.1016/j.cemconcomp.2003.08.002.

Zhu, X., Miao, C., Liu, J., & Hong, J. (2012). China Influence of crumb rubber on frost resistance of concrete and effect mechanism. *Procedia Engineering*, *27*, 206–213. Available from https://doi.org/10.1016/j.proeng.2011.12.445.

Zolotova, N., Kosyreva, A., Dzhalilova, D., Fokichev, N., & Makarova, O. (2022). Harmful effects of the microplastic pollution on animal health: a literature review. *PeerJ*, *10*, e13503. Available from https://doi.org/10.7717/peerj.13503.

Crumb rubber in sustainable self-compacting concrete

Rafat Siddique and Amandeep Singh Sidhu
Civil Engineering Department, Thapar Institute of Engineering and Technology, Patiala, India

8.1 Introduction

The end-of-life of tires (ELTs) quantity around the world has been steadily increasing in the past decades and the rapid modernization among developing countries will further push these numbers higher due to the increased demand of personal and public vehicles. According to European Tire and Rubber Manufacturers Association (ETRMA, 2021) report, the estimated vehicle sales was about 79 million worldwide in the year 2020 out of which 16.5 million were commercial vehicles. The waste tire generation from these vehicles is around 1 billion (Goldstein Market Research, 2020) each year, which would require some novel technologies for the disposal of these ELTs in a cleaner and environmentally friendly method. Presently, vast area of land is needed for their storage which could become a potential fire hazard or a breeding ground for mosquitoes, rodents, etc. in the long run (Amari et al., 1999; Blackman & Palma, 2002; Jang et al., 1998; Wójtowicz & Serio, 1996).

The ELTs have been successfully managed in the developed countries such as European nations and United States where a high percentage i.e., greater than 90% are recovered and reused. Around 50% of the scrap tires generated in United States are used as fuel in various industries (ETRMA, 2019; EPA, 2010). Pyrolysis may be a potential technique which is relatively a cleaner way for the energy recovery from waste tire rubber, but it has other challenges such as high energy needs from the inefficient heat transfer for driving the process which raises the costs and the process becomes uneconomical (Hita et al., 2016; Martínez et al., 2013; Roy et al., 1990). The pyrolysis-derived fuel may also have high emissions of carbon monoxide, unburnt hydrocarbon, and smoke emissions (İlkılıç & Aydın, 2011). The recycling rate for tires around the globe have been declining from the 1960s due to cheap raw oil imports which reduced recycling rate from 50% to 20% from 1900s to 1960s, which further reduced to around 2% worldwide by the year 1995 (Frank & Musacchio, 2008; Reschner, 2008).

The discarded tires may be used in civil engineering–related applications ranging from retention walls to drainage basins. The tire rubber in finer form can be used as a partial substitute for sand in concrete or rubber ash may be used in cement mortar (Angelin et al., 2017; del Río Merino et al., 2007; Djadouni et al., 2019; Reschner, 2008).

Sustainable Concrete Materials and Structures. DOI: https://doi.org/10.1016/B978-0-443-15672-4.00008-5

The concrete can be classified into normal strength concrete, high-strength concrete, ultra-high-performance concrete based on the strength attributes, self-compacting concrete (SCC) based on the flowability characteristics, geopolymer concrete based the binding material incorporated into the concrete and fiber-reinforced concrete when fibers are used as reinforcements. The SCC has a highly flowable nature, which is a prerequisite for such concrete type as it helps to fill the formwork thus excluding the need for any type of external vibration (Gettu et al., 2004; Siddique, 2019). The waste tire rubber can be incorporated in SCC according to the size variations (Aslani, Ma, Yim Wan, & Tran Le, 2018; Hilal, 2017; Mishra & Panda, 2015a) and can provide benefit in terms of lower density and reduced brittleness and stiffness of concrete (X. Li et al., 2021; Raj et al., 2011). The usage of crumb tire rubber in SCC mixes may incur higher initial cost (Kelechi, Adamu, Mohammed, Obianyo, et al., 2022), but the environmental impact is also significant in terms of waste disposal and resource reusage which offsets the monetary losses eventually. The typical chemical composition of crumb rubber (CR) is shown in Table 8.1. The thermogravimetry analysis (TGA) curve of the tire rubber is shown in Fig. 8.1. The tire rubber can be used as fine and coarse aggregate based on its size and can be categorized mainly into four classes (Najim & Hall, 2010). The classification of waste tire rubber is presented in Fig. 8.2 which is based on its various sizes. The potential of SCC

Table 8.1 Typical chemical composition of tire rubber.

Chemical composition	**Weight (%)**					
	Sukontasukkul (2009)	**Gupta et al. (2016)**	**Fraile-Garcia et al. (2018)**	**López-Zaldívar et al. (2017)**	**Cao (2007)**	**Dong et al. (2018)**
Carbon black	28	87.51	31.3	30–38	29.5	80.41
Zinc	–	1.76	–	–	–	–
Sulfur	–	1.08	3.23	0–5	–	2.39
Oxygen	–	9.23	–	–	–	10.77
Silicon	–	0.20	–	–	–	–
Magnesium	–	0.14	–	–	–	–
Aluminum	–	0.08	–	–	–	–
Hydrogen		–	–	–	–	5.75
Steel	14.5	–	–	–	–	–
Ash content	5.1	–	5.43	3–7	6	–
Acetone extract		–	7.3	–	15.5	–
Rubber (synthetic and natural)	41	–	38.3	40–55	49	49
Filler, accelerators, etc.	16.5	–	14.21	–	–	–

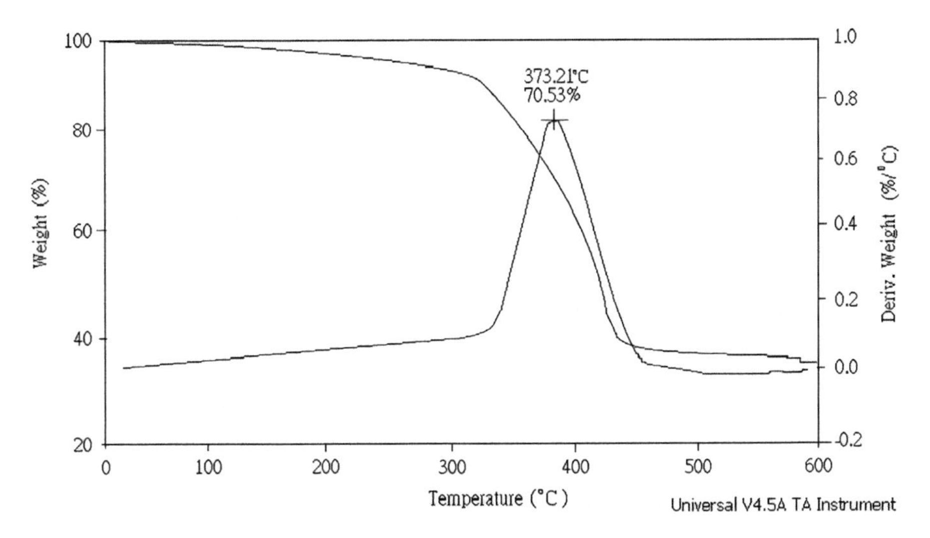

Figure 8.1 TGA of tire rubber.
Source: From Yung, W. H., Yung, L. C., & Hua, L. H. (2013). A study of the durability properties of waste tire rubber applied to self-compacting concrete. *Construction and Building Materials, 41*, 665–672. https://doi.org/10.1016/j.conbuildmat.2012.11.019.

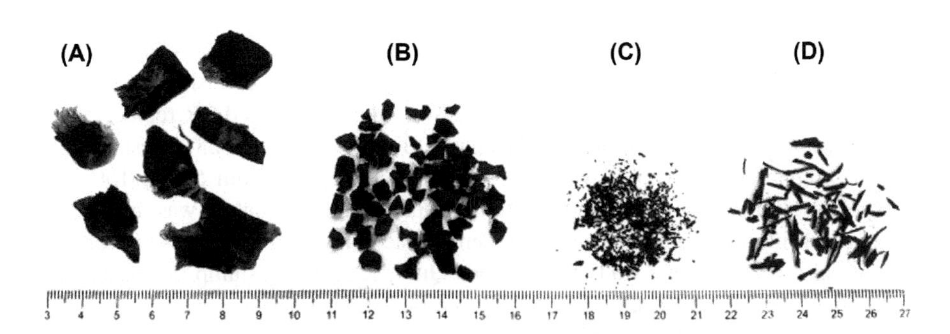

Figure 8.2 Tire rubber classification: (A) chipped, (B) crumb, (C) granular, and (D) fiber.
Source: From Najim, K. B., & Hall, M. R. (2010). A review of the fresh/hardened properties and applications for plain-(PRC) and self-compacting rubberised concrete (SCRC). *Construction and Building Materials, 24*(11), 2043–2051. https://doi.org/10.1016/j.conbuildmat.2010.04.056.

rubberized concrete is massive due to the recycling and monetary benefits it offers. The properties of rubberized SCC are summarized below, which include compressive strength, split tensile strength, flexural strength, stress–strain performance and modulus of elasticity along with durability aspect consisting of shrinkage, water permeability, resistance to chemical attack (sulfate, chloride and acidic solutions), and reinforcement bond strength. Elevated temperature studies related to the usage of CR in SCC are also discussed in this chapter.

8.2 Fresh properties

8.2.1 Slump flow and T_{500} time period

The rubberized SCC has improved workability as per Takada (2000), the CR replacement of 33.3% provided a slump of 700 mm which is well above the standard requirement of 650 mm for SCC. Fakhri et al. (2021), Raj et al. (2011), and Valizadeh et al. (2020) also reiterated these findings suggesting that CR may reduce the slump flow value in concrete, but these values generally lie above the recommended 650 mm value even when supplementary cementitious materials (SCMs) like silica fume, fly ash (FA) are added. The study by Topçu and Bilir (2009) also presented the advantage of CR as fine aggregate replacement in improving the slump and reducing T_{500} value. Such results were also linked to the type of admixture used and therefore, its effect should also be considered along with CR addition during the mix design of SCC. The slump flow values for rubberized and nonrubberized mixes were similar in a study by Angelin et al. (2020) which demonstrated that CR does not affect workability by any considerable amount.

A study by Turatsinze and Garros (2008), however, concluded that CR reduced workability and a significant increase of superplasticizer amount was needed for maintaining a constant slump. The 25% replacement had a slump flow value of above 600 mm while the superplasticizer content increased by 2.6-fold to achieve such workability. The concrete mixes had a segregation ratio ranging from 3.1% to 8.7% which was well below the SCC limit of 15% and therefore, can be categorized as SCC. Güneyisi (2010) in his study, also had to increase the superplasticizer content by roughly 2.9 times at 25% rubber replacement to achieve a slump of around 500 mm. The slump loss may be relatively lower when finer sized CR replacement is compared to larger sized rubber replacement (Thakare et al., 2023). Hesami et al. (2016) compared SCC mixes with CR replacement ranging from 5% to 15%. The mixes lost slump flow values and additional superplasticizer content was necessitated with increasing CR content to maintain comparable slump flow. The loss of workability was also recorded in a study by Tian and Qiu (2022) where slump flow reduced to 620 mm (30% CR replacement) from 740 mm of reference mix.

Findings reported by Aslani, Ma, Yim Wan, and Tran Le (2018), Khalil et al. (2015), N. Li, Long, et al. (2019), Liu et al. (2021), Rahman et al. (2012), Wanasinghe et al. (2022), and Yang et al. (2019) showed that SCC mixes experienced loss of slump flow diameter with increased percentage of rubber and decrease in size of rubber. Coarse aggregate replacement tends to possess lower workability whereas finer aggregate replacement of different sizes have similar but lower slump loss compared to the coarse aggregate replacement. The slump flow values of different sized CR-SCC mixes are shown in Fig. 8.3. Aslani and Khan (2019) in the study of ultra-high-performance SCC using CR as fine aggregate and coarse aggregate replacement showed that results achieved were similar to previous studies where loss of workability was relatively higher for coarse aggregate replacement. Majority of the SCC mixes tested fell into the SF2 category and therefore, can be successfully used to prepare SCC.

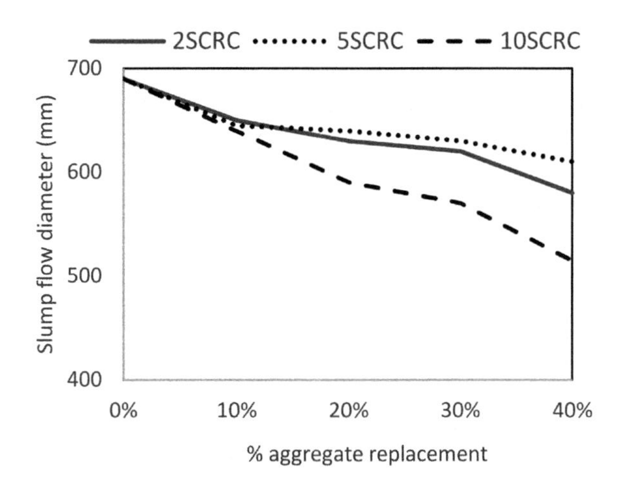

Figure 8.3 Slump flow values of different sized rubber-SCC.
Source: From Aslani, F., Ma, G., Yim Wan, D. L., & Tran Le, V. X. (2018). Experimental investigation into rubber granules and their effects on the fresh and hardened properties of self-compacting concrete. *Journal of Cleaner Production, 172*, 1835–1847. https://doi.org/10.1016/j.jclepro.2017.12.003.

CR addition particularly have negative effect on the viscosity of SCC mix by increasing the friction with the walls of V-funnel and additionally raising the ceiling of the critical energy point (Rahman et al., 2012). Güneyisi (2010) directly tried to measure the viscosity as a function of the at variable mixture rotational speed. The viscosity metric of the SCC mix was highly dependent on the rotation speed of the concrete mixer and it was suggested to implement high shear rate while SCC mix preparation, to achieve under 5000 centipoise (cP) value for mixes. The SCC mortar mixes with CR are known to have high apparent viscosity which may hinder its movement reducing overall workability (Anil Thakare et al., 2020). The study by Bušić et al. (2020) and Tian and Qiu (2022) used 0%–30% substitution of CR, and the results seemed to corroborate previously mentioned findings as the T_{500} test results in the experiments showed steady rise in the time period from 1.9 to 6.5 seconds and 3.62 to 6.42 seconds, respectively. The filling ability results for 15%–30% CR replacement could lie in the VS2 range as opposed to VS1 category of SCC up to 10% replacement. Ayub et al. (2022) in a study achieved higher flow time (T_{500}) as the CR replacement increased which suggest low workability. The study by Hesami et al. (2016), however, resulted in the SCC mixes which all belonged to the VS2 category where replacement of CR was performed up to 15%. The experiment by Yang et al. (2019) on SCC containing rubber aggregate sized 1–5 mm had increased flow time as percentage rubber increased from 5% to 30%, but the mixes still fell under VS1 category hence having good filling ability.

The effect of size difference of CR on slump flow values is an important factor as finer rubber particles leads to higher decline of workability. The surface roughness and high surface area of rubber particles tends to reduces slump flow of SCC.

The slump flow of mixes stays in the range of 610–590 mm at 30% CR replacement (N. Li, Long, et al., 2019). The study by Lv et al. (2019) which implemented lightweight aggregates along with CR also demonstrated lower slump flow due to increased percentage of CR. The usage of finer particles (<4 mm) as coarse aggregate may be a feasible method to achieve high workability even at higher replacement level of CR (Jiang & Zhang, 2022).

8.2.2 Segregation resistance and V-funnel test

A study by Yang et al. (2019) tested segregation of CR (0%–30%) SCC mixes. The results indicated a higher segregation among the mixes with greater content of CR. The percentage of segregation for all the mixes lied in the range 2.5%–6.2%. However, experimental results from Tian and Qiu (2022) depicted that CR possesses antisegregation character which reduced segregation from 18.6% (reference) to 5.6% (replacement of 30%). The CR fibers (0–4.5 mm sized) also have similar characteristics relating to the segregation of SCC mix, which lie in segregation resisting class, that is, SR2 (Thakare et al., 2023). The segregation ratio in an experiment by Lv et al. (2019) consisting of CR were also well within range of standard SCC mixes ranging from 10.8% to 3.2%. It can be concluded from the above studies that CR may be a viable method to reduce segregation of SCC mix. The V-funnel testing for different grades of rubberized SCC was suggested to have low flowability in a study by Raj et al. (2011), as CR can worsen flow time by more than 30%. The V-funnel results from a study by Lv et al. (2019), where rubber particles were used as fine aggregates is shown in Fig. 8.4. It increased time of flow can be clearly observed from the graph in case of CR incorporated concrete mixes.

8.2.3 Passing ability and J-ring/ L-box

The passing ability of SCC is a significant factor for satisfactory placement during concrete-reinforced construction works (Siddique, 2019). Therefore, the testing of such a parameter becomes important before SCC usage at construction site. The passing ability of SCC is measured using two test methods, that is, J-ring test ad L-box test. These tests try to simulate the passing behavior of the SCC through reinforcement and provide an estimate of its passing ability.

Aslani, Ma, Yim Wan, and Tran Le (2018) experimented with tire rubber in SCC with varied sizes ranging from 1 to 3 mm, 2 to 5 mm, and 5 to 10 mm replacing fine and coarse aggregates. The J-ring diameter value reduced for all the mixes consisting of both fine and coarse tire rubber aggregates, but the percentage decrease for coarser aggregate was minimal at 9% for replacement of 40%. The passing ability of SCC is severely hindered by CR and higher addition of superplasticizer is required to counter such effect. The slump loss is directly related to an increase in rubber content and decrease in the size of rubber aggregates (N. Li, Long, et al., 2019). The findings by Bušić et al. (2020) also showed that the passing ability of CR-SCC as per J-ring test mostly lies in range of 10–15 mm, whereas the EFNARC guidelines sets the higher limit at 10 mm, therefore such concrete mixes

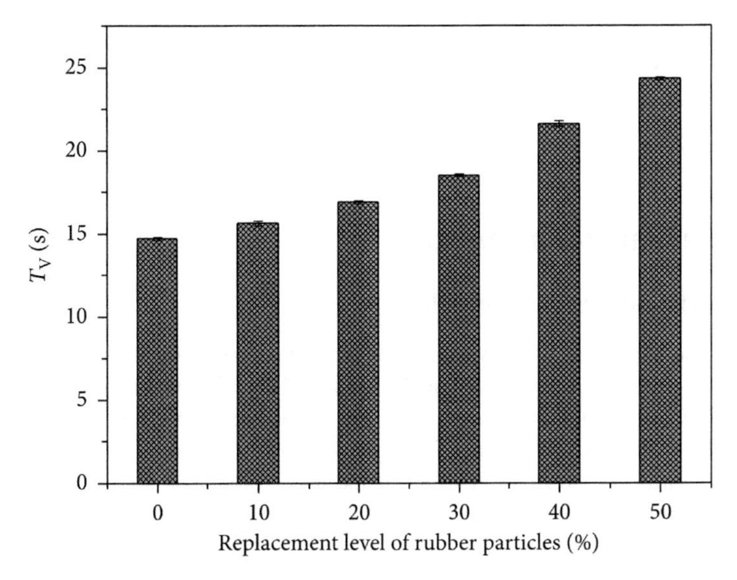

Figure 8.4 V-funnel flow time comparison.
Source: From Lv, J., Du, Q., Zhou, T., He, Z., & Li, K. (2019). Fresh and mechanical properties of self-compacting rubber lightweight aggregate concrete and corresponding mortar. *Advances in Materials Science and Engineering, 2019*. https://doi.org/10.1155/2019/8372547.

failed to satisfy the criteria due to CR addition. The extra dosage of superplasticizer was recommended in such a case to improve the concrete's passing ability. The fresh SCC mix containing CR of about 10%–30% can have J-ring flow values in the range 670–650 mm as per studies by Liu et al. (2021) and Yang et al. (2019).

The experiment by Rahman et al. (2012) tested the passing ability of SCC with CR and compared results the with reference SCC mix using L-box test. The CR seemed to hinder the passing ability as the values for L-box test in the such SCC mixes was found in the range of 0.28–0.40 as opposed to the high passing ability of reference mix which exhibited passing ability ratio of 0.85. The additional usage of superplasticizer also seemed to be ineffective in enhancing the passing ability rather it affected SCC mix negatively. The passing ability of SCC with CR substitution of sand varying from 0% to 15% was found to be adequate even when 10% of cement was replaced SF as per findings by Bušić et al. (2020). Similar loss of passing ability was also seen in study by Angelin et al. (2020), where L-box value reduced from 0.85 (reference mix) to 0.83 for 15% of rubber content in SCC. Lv et al. (2019) also reported that L-box value was compromised due to CR, the values were found to be 0.98 and 0.82 at 0% and 50% replacement of fine aggregates. Hamza et al. (2018), Khalil et al. (2015), and Zaouai et al. (2020) also found the passing ability of SCC mix to be hampered due CR introduction in the mix.

However, study by Bignozzi and Sandrolini (2006) opposed the above findings where J-ring testing depicts better passing ability for CR incorporated SCC as compared to reference SCC mix. Similarly, the finding by Wanasinghe et al. (2022)

also suggested that J-ring flow diameter values can be improved with rubber incorporation as both fine and coarse aggregate replacement. The experimental study by Mallek et al. (2021) was inconclusive regarding passing ability (L-box testing) as values did not follow any particular pattern.

8.3 Mechanical properties

8.3.1 Compressive strength

The high performance SCC with rubber aggregates of size 5–10 mm was developed by Aslani, Ma, Yim Wan, and Muselin (2018). The CR usage was accompanied with varying percentage of recycled aggregates. The compressive strength of the rubberized mixes at 7 and 28 days was lowered due to the weak interfacial transition zone (ITZ) formation which acted as a voids. The studies by Ayub et al. (2022), Gesoğlu and Güneyisi (2011), Lv et al. (2020), Rahman et al. (2012), Shaji et al. (2021), and Tian and Qiu (2022) also depicted the inferiority of the rubber-SCC mixes in terms of compressive strength. The rubber aggregates are responsible for a nonhomogeneous structure due to its lightweight nature which reduces its ability to resist compressive stresses. Thakare et al. (2023) in their study presented a strong relationship between density and compressive strength and concluded that density can be a reliable indicator of compressive strength in rubberized SCC. The study by Mishra and Panda (2015b) and Yung et al. (2013), however, suggested that 5% of waste tire rubber was an optimum amount for fine as well as coarse aggregate replacement which may improve the compressive strength slightly. Topçu and Bilir (2009) suggested 180 kg/m^3 of rubber particles sized 0–4 mm as ideal quantity to achieve a balance of strength and fresh properties in SCC. The finer CR particles (sized $<$4 mm) as coarse aggregate are unable to improve the compressive strength even when providing better packing, as compressive strength is highly dependent on coarse aggregate properties and its volume decreases in the SCC mix due to finer sized CR replacement (Jiang & Zhang, 2022). The study from Bušić and Miličević (2020) suggested the ideal CR level to be 10% to satisfy both strength and fresh properties criteria for SCC mix.

The study by Bušić et al. (2020) and Naji Hilal et al. (2018) compared different variation of CR and SF in SCC mixes. The losses in compressive strength were evident due to CR addition but SF addition seemed to mitigate these losses. Bušić et al. (2020) suggested a binary SCC mix of 5% CR and 10% SF as optimum when considering overall monetary gains. The study also developed compressive strength prediction models which could reproduce reliable results. The design of experiment (DoE) statistical analysis can also provide the relational significance among factors such as rubber content, curing time, and rubber size (granulometry) on the compressive strength of rubberized SCC (Werdine et al., 2021). The other DOE method, that is, Taguchi method can also have reasonable accuracy in prediction of compressive strength and fresh properties of SCC incorporating rubber aggregates (Emara et al., 2018). The other method to predict compressive strength is using nonlinear regression model which takes into account the rubber content and the size of aggregates replaced, which was developed by Jiang and Zhang (2022). The relation developed by the authors can reliably predict compressive strength with around 20% error rate. The compressive strength prediction equation is as follows (Eq. (8.1)):

$$f_{cr} = 1.04 f_{c0} e^{-k_f k_a k_S \rho_{ru}} \tag{8.1}$$

where f_{c0} and f_{cr} represent compressive strength for reference and rubberized SCC concrete respectively. ρ_{ru} is the rubber content (by volume), k_f, k_a, k_S are parameters which take into account the compressive strength of SCC, aggregate type and SCM type used respectively in the concrete mix. These parameters are defined as follows:

$$k_f = 0.07 \times f_{c0} + 3 \; k_a = \left\{ \begin{array}{c} 1, d_{r,\max} < d_{f,\max} \\ 1.23, d_{r,\max} > d_{f,\max} \end{array} \right\} \quad k_s = \left\{ \begin{array}{c} 1, NoSCM's \\ 1.25, SCM'sexceptFlyash \end{array} \right\} \tag{8.2}$$

where

- $d_{r,max}$ is maximum aggregate size
- $d_{f,max}$ is maximum size of sand.

The study by Angelin et al. (2020) found the combined effect of expanded clay (constant amount in the mix) and varying CR on the compressive strength of SCC. The compressive strength loss was recorded throughout the increasing CR content in SCC mixes, where compressive strength at 28 days reduced from 60.5 MPa to 22 MPa. The effect of CR and FA on SCC mix in a study by Güneyisi (2010) showed that at 28 days testing, the CR surely has a negative effect on compressive strength. The addition of FA in the SCC mix did not improve the strength but rather showed even lower values along with delayed setting time as the FA quantity increased in mix from 0% to 60%. However, SF may be implemented in the rubberized SCC mix to improve its strength characteristics based on a study by Khed et al. (2022). The precoating of CR with crushed dune sand was also found be beneficial for the strength improvement in study by Zaouai et al. (2020). The surface modification successfully enhanced the compressive strength by 14% with 20% substitution of dune sand-treated CR.

The study by Yang et al. (2019) experimented with dynamic load testing based on American Society for Testing and Materials (ASTM) C39/C39M on rubberized SCC to attain quasistatic compressive strength under vibrations. The results of the testing were similar to previous studies where compressive strength loss has been observed as an outcome of CR addition in SCC mixes. The loss of quasistatic compressive strength was about 54% when the crumb replacement level was kept at 30%. The author also compared the dynamic compressive strength at different air pressures to quasistatic compressive behavior where increased air pressure was followed by improvement in dynamic compressive strength. It was also observed that the crack propagation in samples deviated from central propagation axis providing jagged surface at failure, it was due to the reason that CR can act as a "hole" which eliminates stress concentration.

A similar study from Zhuang et al. (2022) also determined the dynamic compressive strength with different stain rates. Results showed that dynamic compression decreased proportional to each increase in CR content in the SCC mix. Higher strain rate on the tested samples showed lower loss in dynamic compressive strength. The higher strain rate limits the energy accumulation time under impact loading which gave rise to hysteresis phenomenon hence improvement was noted in dynamic behavior of SCC. The improved dynamic compressive strength from CR substitution was also mentioned in a study by Chen et al. (2022).

The quasistatic testing was also performed in a study by Zhang et al. (2021) and dynamic increase factor (DIF) was derived from the ratio of dynamic compression strength to static compression strength. Eq. (8.3) as presented below was obtained from the experimental data:

$$\mathrm{DIF} = 0.40397(\varepsilon)^{0.41775}\gamma \tag{8.3}$$

where $\log\gamma = 5.156\alpha - 0.492$; $\alpha = \left(2+3f_{cu}/4\right)^{-1}$ and f_{cu} is the quasistatic compressive strength. The R^2 for the above presented relation was 0.8288. Eq. 8.3 provide more accurate results than the formula recommended by (CEB-FIP, 1994).

The compressive strength from different studies is presented in Table 8.2.

8.3.2 Splitting tensile strength

The split tensile strength is directly affected by the CR content introduced in SCC mix. The split tensile strength loss can vary from 20% to 50% at 5% to 15% CR substitution for sand, respectively (Angelin et al., 2020).

The importance of size variation of CR in SCC on split tensile strength is clearly reflected in a study by Jiang and Zhang (2022). The loss in split tensile was relative to the compressive strength loss, the results were favored the usage of CR (<4 mm) as a fine aggregate rather than coarse aggregate. Jiang and Zhang (2022) found the relation of split tensile strength and compressive strength to be almost linear in nature as shown in Fig. 8.5. The usage of steel fibers can help alleviate the problem of loss of split tensile strength to a small extent (Aslani & Gedeon, 2019).

The study by Najim and Hall (2013) compared the efficacy of different pretreatments of CR in retaining lost tensile strength of rubberized SCC. The mortar coated method helped regain lost split tensile strength by 19% when compared to similar single replacement level of combined 38% of both fine (19%) and coarse aggregate (19%). The dynamic split tensile strength of SCC mixes with CR decreases with higher replacement level as per study by Yang et al. (2019). The dynamic split tensile strength however increased throughout the replacement in case of a special condition, that is increased air pressure.

The empirical relationship which may be used to predict split tensile strengths in rubberised SCC as suggested by Najim and Hall (2012) is based on linear relationship and is presented as below as Eq. (8.4):

$$F(t) = 0.0535(f_{cu}) + 1.886 \tag{8.4}$$

where $F(t)$ is the predicted tensile strength and f_{cu} is the compressive strength.

Table 8.2 Compressive strength comparison from different studies.

Author	SCM/additives	w/b ratio	Tire rubber size	Replacement material	Replacement (%) by volume	Compressive strength approx. (MPa) (28 days)
Hilal (2017)	FA[a]	0.35	–	–	0	72.3
		0.35	<1 mm	Sand	5	67.7
		0.35	(Mesh 18)	Sand	10	63.9
			1–4 mm		15	60.5
			(Mesh 5)		20	57.3
					25	49.7
					5	63.5
					10	59.3
					15	53.9
					20	45.6
					25	39.3
Yung et al. (2013)	Slag and FA[a]	0.35	50 Mesh	Sand	0	32.0
		0.35	30 Mesh	Sand	5	30.9
					10	23.5
					15	23.4
					20	21.9
					5	28.9
					10	24.8
					15	26.9
					20	22.8
Bignozzi and Sandrolini (2006)	Calcium carbonate	0.53	0.05–2.0 mm	Sand	0	33.0
					22.2	24.7
					33.3	20.2
Najim and Hall (2012)	PFA[b]	0.36	2–6 mm	Sand	0	55.8
		0.36	2–6 mm	Coarse aggregate	5	37.1

(*Continued*)

Table 8.2 (Continued)

Author	SCM/additives	w/b ratio	Tire rubber size	Replacement material	Replacement (%) by volume	Compressive strength approx. (MPa) (28 days)
					10	32.1
					15	25.1
					5	93.9
					10	33.0
					15	23.2
Turatsinze and Garros (2008)	Calcareous filler	0.40	4–10 mm	Coarse aggregate	0	44.3
					10	30.0
					15	20.0
					20	15.3
					25	11.2
Aslani, Ma, Yim Wan, and Tran Le (2018)	FA[a],	0.45	1–3 mm	Sand	0	50.3
	GGBS[c],	0.45	2–5 mm	Sand	10	35.7
	SF[d],	0.45	5–10 mm	Coarse aggregate	20	29.8
	Tire rubber in SSD[e] condition	0.45			30	25.5
					40	19.6
					10	40.6
					20	33.2
					30	29.4
					40	26.3
					10	31.4
					20	22.2
					30	19.3
					40	16.4
N. Li, Long, et al. (2019)	FA[a] and slag	0.35	2–4 mm	Sand	0	43.0
		0.35	1–2 mm	Sand	30	27.9
		0.35	0–0.3 mm	Sand	10	36.5

					20	32.7
					30	28.6
					30	23.2
Bušić et al. (2020)	–	0.4	0.5–3.5 mm	Sand	0	43.9
	SF[d] (10%)	0.4	0.5–3.5 mm	Sand	5	39.0
					10	31.0
					15	28.0
					20	23.1
					25	13.4
					30	12.9
					0	66.3
					5	60.7
					10	42.9
					15	33.1
					20	25.1
					25	24.4
					30	19
Lv et al. (2020)	FA[a] and shale ceramsite	0.35	0–10 mm	Sand	0	37.3
					10	35.2
					20	31.2
					30	29.8
					40	24.1
					50	20.5
Najim and Hall (2013)	Cement paste precoating	0.36	2–6 mm	Sand (50%) + Aggregate (50%)	38% (12%, w/w)	373345
	NaOH treatment (20 min)	0.36	2–6 mm	Sand (50%) + Aggregate (50%)	38% (12%, w/w)	
	Mortar precoat	0.36	2–6 mm	Sand (50%) + Aggregate (50%)	38% (12%, w/w)	

(Continued)

Table 8.2 (Continued)

Author	SCM/additives	w/b ratio	Tire rubber size	Replacement material	Replacement (%) by volume	Compressive strength approx. (MPa) (28 days)
Wanasinghe et al. (2022)	SF[d], GGBS[c], and FA[a]	0.45	2–5 mm	Sand	0	39.4
		0.45	5–10 mm	Aggregate	10	37.5
					20	27.0
					30	26.8
					40	18.2
					10	24.8
					20	26.4
					30	11.1
					40	11.6
Thakare et al. (2023)	FA[a]	0.35	0.60–1.18 mm	Sand	0	72.3
		0.35	1.18–2.36 mm	Sand	10	51.2
		0.35	2.36–4.75 mm	Sand	20	37.4
		0.35			30	24.3
					10	43.8
					20	36.4
					30	22.2
					10	45.2
					20	35.0
					30	19.4
Raj et al. (2011)	FA[a]	0.37	Max 4.75 mm	Sand	0	62.1
					5	57.8
					10	53.6
					15	50.7
					20	37.8

[a]Fly ash.
[b]Pulverized fly ash.
[c]Granulated ground blast furnace slag.
[d]Silica fume.
[e]Saturated surface dry.

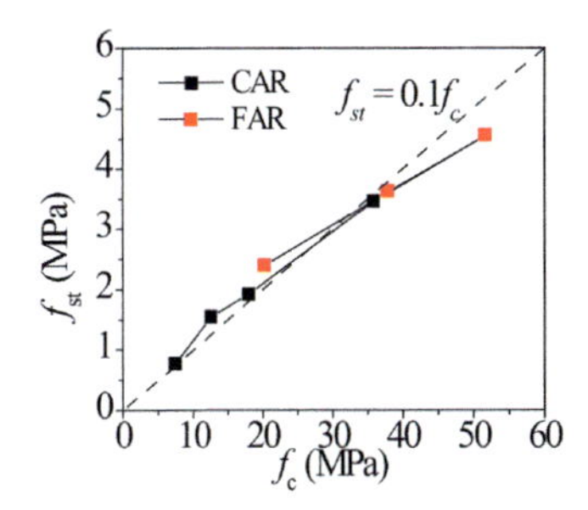

Figure 8.5 Split tensile strength versus compressive strength of rubberized SCC for coarse aggregate rubber (CAR) and fine aggregate rubber (FAR).
Source: From Jiang, Y., & Zhang, S. (2022). Experimental and analytical study on the mechanical properties of rubberized self-compacting concrete. *Construction and Building Materials, 329*. https://doi.org/10.1016/j.conbuildmat.2022.127177.

Valizadeh et al. (2020) evaluated the tensile strength of the SCC mixes with CR as aggregate replacement. The split tensile was found to possess indirect relationship to CR quantity and size used in the experiment. The value of tensile strength reduced by 7% and 36.5% when substitution of fine aggregate was 10 and 20%, respectively, in SCC mix. The loss values were even higher, that is, 8.65% and 38.8% with coarse aggregate substitution at similar replacement level of 10% and 20%, respectively. The split tensile strength is highly influenced by the size of CR added and testing age of the SCC (Wanasinghe et al., 2022). The split tensile strength achieved by previous studies at different CR replacement level is presented in Table 8.3.

8.3.3 Flexural strength

N. Li, Long, et al. (2019) in a study calculated various fracture parameters from the three-point flexural testing. The energy absorption in terms of G_f (fracture energy parameter) showed slight improvement with CR while peak load reduced drastically which was due to high energy absorption of rubberized SCC. The rubber particle of size 1−2 mm at 30% sand replacement yielded D_u (Deformability parameter) of 82 mm which was 61% of reference SCC mix. The load deflection curve for beams tested under three-point loading is shown in Fig. 8.6. The study by Fakhri et al. (2021) examined the relationship among fracture toughness (K_{IC}) and flexure strength for different sized CR incorporated in SCC. The relation established among the parameters, that is, K_{IC} and flexure strength was found to be linear in nature.

Role of CR on the flexural strength with varying CR and SF in SCC mix was tested in a study by Bušić et al. (2020). The optimum SF content was 10% with 30% CR in SCC mix depicting least loss of flexural strength of 29% compared to other tested mixes. The flexural strength percentage loss was relatively lower compared to compressive strength percentage loss throughout the varying CR in SCC. In an experimental setup by Najim and Hall (2013) the modulus of rupture (R_t) of

Table 8.3 Splitting tensile strength of rubberized SCC mixes.

Author	SCM/ additives/ pretreatment	Tire rubber size	w/b ratio	Replacement material	Replacement (%) by volume	Splitting tensile strength approx. (MPa) (28 days)
Najim and Hall (2012)	PFA[a]	2–6 mm	0.36	Sand	0	4.4
			0.36	Aggregate	5	3.1
			0.36		10	3.4
					15	2.5
					5	3.7
					10	3.1
					15	2.3
Aslani, Ma, Yim Wan, and Tran Le (2018)	Fly ash, GGBS[b], silica fume,	1–3 mm	0.45	Sand	0	3.7
	Tire rubber in SSD[c] condition	2–5 mm	0.45	Sand	10	3.4
		5–10 mm	0.45	Coarse aAggregate	20	2.8
			0.45		30	2.7
					40	2.2
					10	4.3
					20	3.1
					30	3.1
					40	2.5
					10	2.9
					20	2.7
					30	2.5
					40	1.8
Hesami et al. (2016)	Limestone powder	Max size 4.75 mm	0.39	Sand	0	4.9
					5	4.82
					10	4.63
					15	4.2

Khalilpasha et al. (2012)	Limestone powder	3–5 mm	0.39	Sand (w/w)[f]	0	4
		3–5 mm	0.39	Coarse aggregate (w/w)[f]	4	3.2
					8	3
					12	2.7
					4	3.2
					8	2.8
					12	2.6
Khalil et al. (2015)	Cement kiln dust	Max 2 mm	0.45	Sand	0	3.4
					10	3.2
					20	3.0
					30	3.0
					40	2.4
Tian and Qiu (2022)	FA[d], nanosilica	0.15–4.75 mm	0.29	Sand	0	4.1
					10	3.1
					20	2.6
					30	2.3
Yang et al. (2022)	FA[d], SF[e]	~ 0.75–4 mm	0.32	Sand	0	4.1
					10	3.0
					20	2.4
					30	2.2

[a]Pulverized fuel ash.
[b]Granulated ground blast furnace slag.
[c]Saturated surface dry.
[d]Fly ash.
[e]Silica fume.
[f]Weight to weight replacement.

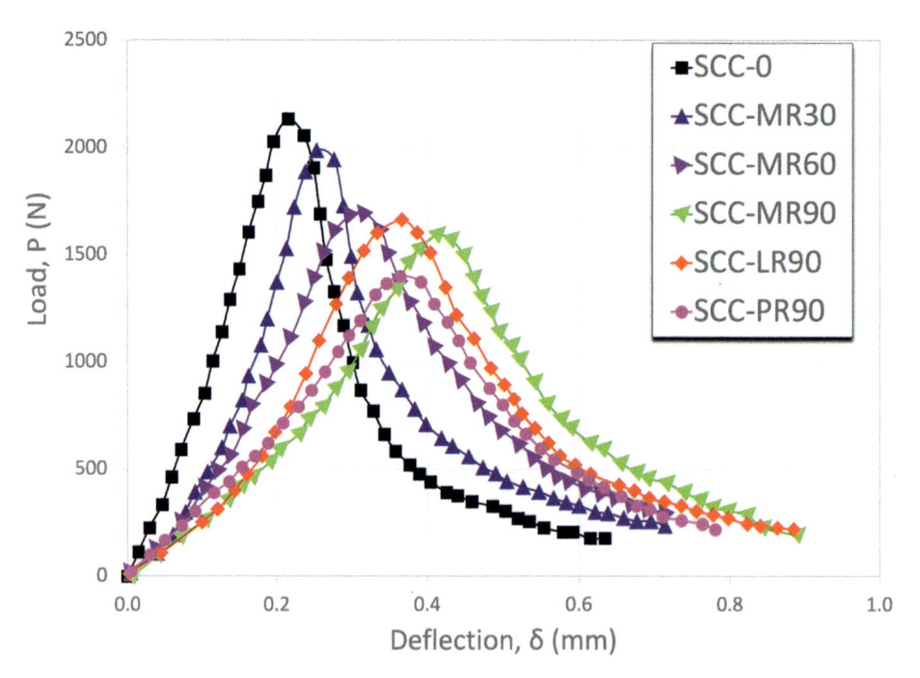

Figure 8.6 P–δ curve of three-point beam loading.
Source: Data from Li, N., Long, G., Ma, C., Fu, Q., Zeng, X., Ma, K., Xie, Y., & Luo, B. (2019). Properties of self-compacting concrete (SCC) with recycled tire rubber aggregate: A comprehensive study. *Journal of Cleaner Production, 236*. https://doi.org/10.1016/j.jclepro.2019.117707.

SCC where CR was precoated with mortar was 10.5% improved compared to the reference SCC mix. The prewashed CR samples had insignificant effect on R_t value. The load deflection curve from flexural testing was used to determine toughness indices at certain preset deflections. Among the various CR treatments, mortar treatment proved superior to the NaOH and cement coating pretreatments.

Fakhri et al. (2021) and Raj et al. (2011) concluded from their study that flexural strength in terms of modulus of rupture decreased with CR rubber addition and Fakhri et al. (2021) also advised that usage of SF in the SCC mix to regain lost flexural strength. The study by Wanasinghe et al. (2022) found an increased toughness index in flexural testing for almost every CR mix consisting of 10 mm sized rubber particles. The best performance of toughness index was found at 91 days, belonging to 10% CR-SCC mix. The flexural strength of rubberized SCC typically decreases as the CR percentage increase whereas it increases with the testing age of the SCC specimen. The dynamic flexural strength is a direct function of the pressure and inversely related to the CR content in SCC as depicted by a study from Yang et al. (2019). The poor ITZ is the chief reason for such loss in the dynamic flexural strength.

However, certain studies suggests that flexural strength of SCC can be improved by implementing CR. The flexural strength of lightweight SCC was tested at 7, 28,

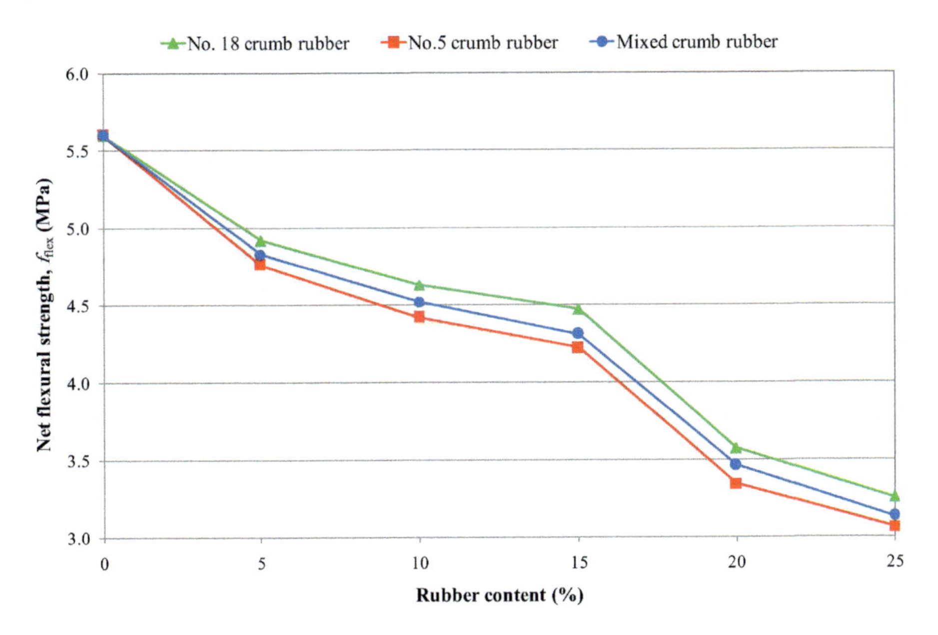

Figure 8.7 Flexural strength of rubberized concrete at different rubber content. No. 18 (size <1 mm), No. 5 (size 1–4 mm).
Source: From Hilal, N. N. (2017). Hardened properties of self-compacting concrete with different crumb rubber size and content. *International Journal of Sustainable Built Environment, 6*(1), 191–206. https://doi.org/10.1016/j.ijsbe.2017.03.001.

and 90 days by Lv et al. (2019) and the rough surface of rubber particles was attributed for the improvement in flexural strength. The typical flexural strength trend of different sized rubber particles in a study by Hilal (2017) is shown in Fig. 8.7. The study by Ganesan et al. (2013) also had findings which showed an increase of flexural strength of SCC as the CR (0–4.75 mm sized) was added to mix, with optimum 15% replacement. The author explained such behavior from the perspective that rubber particles have relatively better tensile load resisting ability. Therefore it is suggested that flexural strength of rubberized SCC is highly dependent on the size and content of rubber as well as the testing method followed as per Naji Hilal et al. (2018).

Valizadeh et al. (2020) assessed the relation of the compressive strength with flexural strength. It was concluded that most of the SCC codes and studies provide excess value of flexural strength from the equations provided, when these prediction models were applied to rubberized SCC mixes. The authors suggested new equations to accommodate the effect of CR on the SCC. The proposed Eqs. (8.5) and (8.6) by the authors are presented as follows:

$$\text{Cubic specimen: } f'_{cr} = 0.36 \times f_c^{\prime 0.445} \tag{8.5}$$

$$\text{Cylindrical specimen: } f'_{cr} = 0.67 \times f_c^{\prime 0.28} \tag{8.6}$$

where f'_{cr} represent the flexural strength (MPa) and f'_c represents the compressive strength (MPa).

8.3.4 Modulus of elasticity/stress–strain characteristics/toughness

The modulus of elasticity (*E*) value is typically proportional to the compressive strength of concrete (Neville, 2011) barring the changes due to aggregate type and proportion of the mix. The relation of compressive strength and E in rubberized SCC was study by Raj et al. (2011). The load-displacement curve of rubberized SCC depicted higher deformability with maximum stress concentration at 0.5% strain. The rubberized SCC is capable of sustaining cyclic loading due to inherent resistance to complete fracture at maximum stress (Bignozzi & Sandrolini, 2006). The load-displacement curve for rubberized SCC depicts how CR provides delayed microcrack formation resulting in a better premicrocrack strain capacity. The CR load deflection curves at different replacement level of CR are presented in Fig. 8.8. The stress–strain response in a study by Khalil et al. (2015) portrayed a loss of strain energy due to decreased area under the curve, this is due to the reason that

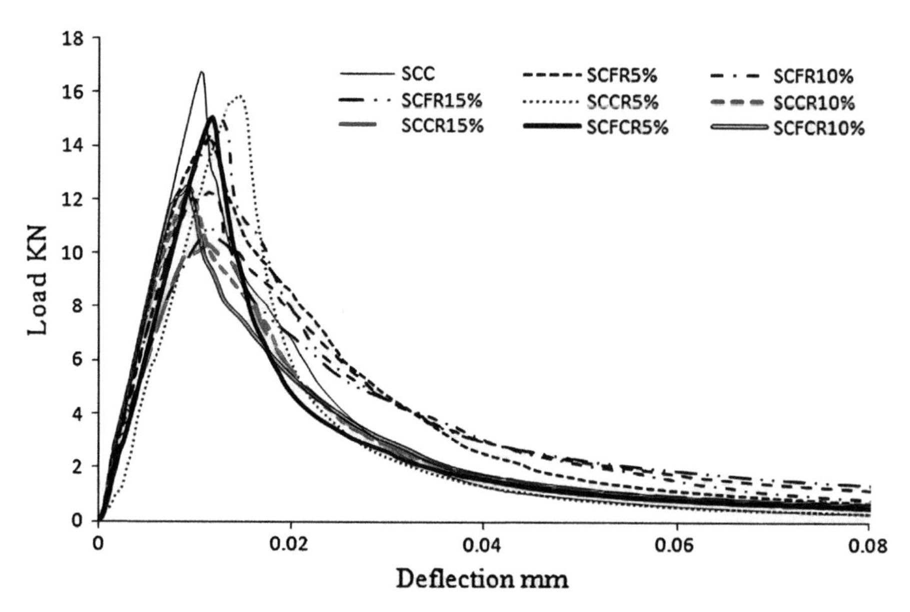

Figure 8.8 Load deflection curve of SCC incorporating crumb rubber.
SCFR5%, SCC with fine aggregate 5% replacement of CR; *SCFR10%*, SCC with fine aggregate 10% replacement of CR; *SCFR15%*, SCC with fine aggregate 15% replacement of CR; *SCCR5%*, SCC with coarse aggregate 5% replacement of CR; *SCCR10%*, SCC with coarse aggregate 10% replacement of CR; *SCCR15%*, SCC with coarse aggregate 15% replacement of CR; *SCFCR5%*, SCC with fine and coarse aggregate combined 5% replacement of CR; *SCFCR10%*, SCC with fine and coarse aggregate combined 10% replacement of CR; *SCFCR15%*, SCC with coarse aggregate 15% replacement of CR.
Source: From Najim, K. B., & Hall, M. R. (2012). Mechanical and dynamic properties of self-compacting crumb rubber modified concrete. *Construction and Building Materials, 27* (1), 521–530. https://doi.org/10.1016/j.conbuildmat.2011.07.013.

rubberized concrete inherently shows high deformation at low-stress levels. The drastic loss in E value for the rubber particles of size 4–10 mm incorporated in SCC as aggregate replacement was also measured in a study by Turatsinze and Garros (2008). The results depicted the highly dependent nature of E on aggregate type as 25% usage of the rubber aggregate reduced E value by over 70% measuring around 10 GPa. A similar study by Aslani, Ma, Yim Wan, and Tran Le (2018) also confirmed the high strain of rubberised SCC at relatively low-stress application. The peak strain obtained were in the range of 1000–1500 and 1500–2000 $\mu\varepsilon$ for CR of size 2 and 5 mm, respectively, in SCC as fine aggregate replacement. The peak strain in a study by Jiang and Zhang (2022) also confirms achievement of similar values varying from approximately 1700–1300 $\mu\varepsilon$ for 20%–80% CR as coarse aggregate replacement:

The modulus of elasticity of SCC dropped with percentage increase of CR as per study by Bušić et al. (2020). The E value recorded was 56% lower for 30% CR-SCC when compared to reference mix exclusive of CR. The CR replacement up to 20% provided E above 20 GPa even without the usage of SF in the mix. Lv et al. (2019) also suggested that E value is lower if fine aggregate was substituted by CR in lightweight SCC.

The CR has tendency to reduce the elastic modulus and brittleness index of SCC due to high deformable nature as depicted in a study by Bignozzi and Sandrolini (2006) and Raj et al. (2011). The dynamic modulus of SCC is reduced by 27.5% when 33.3% replacement of CR was performed by equal volume of sand. Thakare et al. (2023) also found out that the loss of dynamic modulus of elasticity for rubberized SCC is around 45% at 30% rubber incorporation in the mix. The study by Jiang and Zhang (2022), Najim and Hall (2012), Raj et al. (2011), and Turatsinze and Garros (2008) also corroborated these findings where increased CR replacement resulted in a lower modulus of elasticity (E) showing a linear relationship with replacement level of CR for fine and coarse aggregate replacement in SCC. The fine aggregate replacement shows better E values as compared to coarse aggregate replacement for every replacement percentage. The E value may be successfully further enhanced by using treatment of CR such as mortar coating, water soaking, NaOH treatment, etc. as proposed by Miličević et al. (2021) and Najim and Hall (2013). The elastic modulus decrease was also seen in a study by Yang et al. (2019), where E value loss in SCC was 40% for replacement level of 30% by CR (sized 1–5 mm).

The study by Rahman et al. (2012) focused on predicting the dynamic modulus of elasticity using mathematical relations. The different equations had varied efficacy in prediction of the value; however, it was concluded that SCC mixes had 10%–20% lower dynamic elasticity modulus as consequence of rubber particles addition. The dynamic modulus as per N. Li, Long, et al. (2019) study decreased linearly with increase of CR (1–2 mm) and value loss was around 32.6% at 30% replacement of sand in SCC.

The study by Wang et al. (2022) used CR in the size range 0.56–2 mm and steel fibers of 13 mm length with aspect ratio (l/d) of 75 in self-compacting mix to evaluate fracture behavior. Fracture energy improvement of 43.4% was recorded at 10%

rubber replacement along with significant improved postcrack energy absorption prior to failure. The results were validated by comparing outcomes of smaller beam size with larger beam size. In both cases the rubber incorporated samples had relatively comparable initial fracture energy and the stress intensity was also increased with rubber introduction in SCC. The 15% rubber replacement by volume of sand seemed to provide best postcracking characteristics along with 0.2% steel fibers demonstrating the superiority of the coupling (rubber and steel fiber) in arresting the cracks.

N. Li, Long, et al. (2019) used SCC mix with water/binder (w/b) ratio of 0.35 with varying plasticizer content to study the stress–strain characteristics due to inclusion of CR in the mix. The inherent highly deformable nature of CR tends to provide low compressive stress resistance while also portraying a large critical strain and ultimate strain which seems to provide better deformation, therefore, delaying the onset of cracks (Najim & Hall, 2013). The study also found that the constant increment in strain rate leads to multiple cracks that causes the samples to deteriorate in multiple fragments under failure. The toughness index derived from stress–strain response shows improved values over higher rubber replacement levels.

The toughness testing performed by Zhuang et al. (2022) on SCC containing 10% and 20% CR showed improvement of toughness index with CR at all the strain rates. Fakhri et al. (2021) and Khalilpasha et al. (2012) also achieved better toughness of SCC with both fine and coarse aggregate replacement of CR (sized 3–5 mm and 4.75–12.5 mm). Fakhri et al. (2021) also concluded from the study that 20% CR is the optimum substitution level in a SCC mix. The toughness improvement was attributed to the deformable nature of CR which results in higher strain in concrete. The study by Zhang et al. (2021) showed that the toughness index increase is insignificant up to 10% CR replacement but the increase is abrupt at 15% CR replacement in SCC. Jiang and Zhang (2022) in a study showed similar toughness pattern with 20% replacement of CR regarded as the optimum content to achieve highest toughness index of 1.978 in SCC. The toughness index relation to the compressive strength in the aforementioned study was shown to be nonmonotonic in nature.

8.3.5 Reinforcement bond strength

The bond strength evaluation was performed on 16 mm diameter reinforcement bars under tensile loading by Hilal (2017) for crumb rubber SCC. The bond strength reduced as the CR replacement increased for all the tested sizes of the CR. The highest bond loss was tested for Mesh No.5 concrete samples where bond strength reduced 87%. The result of reinforcement bond strength at various replacement levels is shown in Fig. 8.9.

8.3.6 Damping/fatigue behavior/fracture behavior/impact resistance

The CR having different sizes is capable of improving the damping response in SCC mixes where damping ratio and damping coefficient both are improved around 2.3 times at 15% replacement level of fine aggregates (Najim & Hall, 2012).

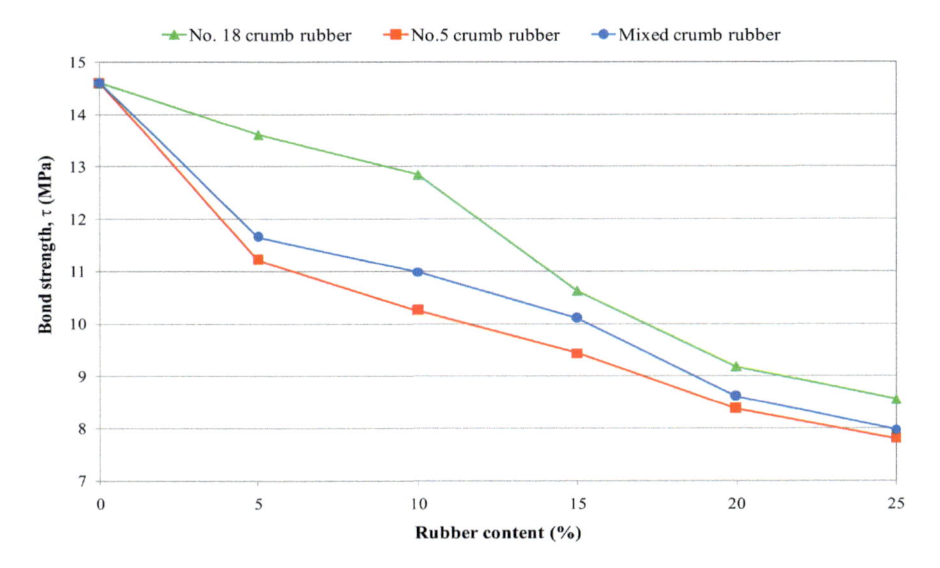

Figure 8.9 Bond strength change in SCC mixes with respect to crumb rubber sizes and content. No.18—Mesh size 18 Crumb rubber; No. 5–Mesh size 5 Crumb rubber; Mixed CR—Mesh size 18 and mesh size 5 combined.
Source: From Hilal, N. N. (2017). Hardened properties of self-compacting concrete with different crumb rubber size and content. *International Journal of Sustainable Built Environment, 6*(1), 191–206. https://doi.org/10.1016/j.ijsbe.2017.03.001.

The rubber particles can have subpar performance in fatigue limit strength in SCC as per study by Lv et al. (2020). However, the stress level showed dual nature by depicting increasing trend initially but decreasing in later response. The rubber substitution in SCC contributes to better fatigue-related performance of concrete with optimum value achieved at 30% replacement. Fatigue life of 2×10^6 corresponded to a peak stress level of 0.68 at 30% substitution of sand in the aforemntioned study.

Eqs. (8.7)–(8.12) (Lv et al., 2020) dictated the fatigue behavior with failure probability of 0.5 for different replacement levels:

$$\text{SCLC:} \quad l_g S = 0.117 - 0.047 l_g N_f \tag{8.7}$$

$$\text{SCRLC10:} \quad l_g S = 0.163 - 0.057 l_g N_f \tag{8.8}$$

$$\text{SCRLC20:} \quad l_g S = 0.139 - 0.051 l_g N_f \tag{8.9}$$

$$\text{SCRLC:} \quad l_g S = 0.111 - 0.044 l_g N_f \tag{8.10}$$

$$\text{SCRLC40:} \quad l_g S = 0.174 - 0.055 l_g N_f \tag{8.11}$$

$$\text{SCRLC50:} \quad l_g S = 0.227 - 0.066 l_g N_f \tag{8.12}$$

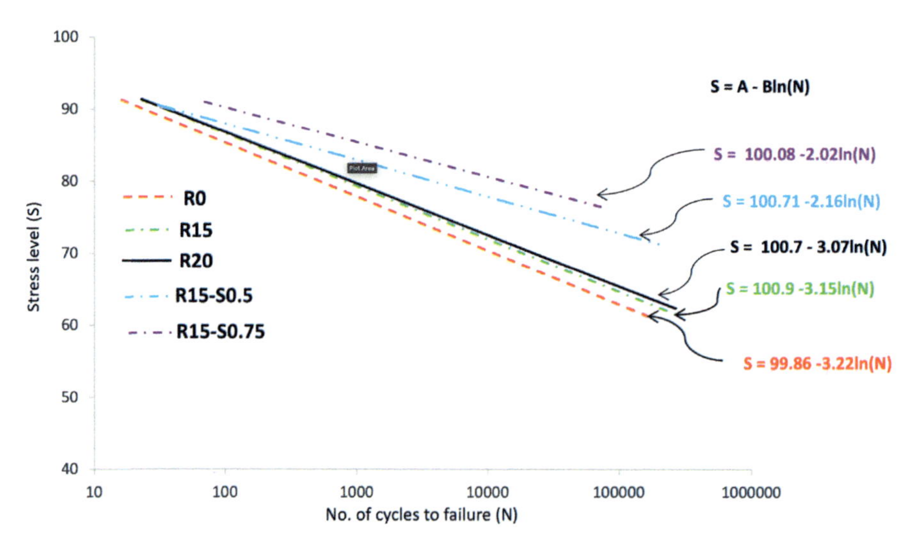

Figure 8.10 Stress level versus cycles before failure of rubberized SCC.
Source: Data from Ganesan, N., Bharati Raj, J., & Shashikala, A. P. (2013). Flexural fatigue behavior of self compacting rubberized concrete. *Construction and Building Materials, 44*, 7–14. https://doi.org/10.1016/j.conbuildmat.2013.02.077.

where

S is stress level = P_{min}/P_{max}
N_f is fatigue lives
l_gS and l_gN_f were evaluated from correlation curves.

The study by Ganesan et al. (2013) focused on the fatigue behavior of rubberized SCC originating under flexural loading. The fatigue behavior was tested at different stress levels varying in the range 90%–60% of the ultimate flexural load of the samples. The fatigue was related to the number of loading cycles sustained before failure. The results from the study favored the higher CR proportions in SCC mix which sustained above 263,000 cycles as compared to 163,276 cycles sustained at 60% stress level by reference non rubberised mix. The addition of steel fibers (0.5% and 0.75% of the mix volume) in the mix can offer further improvement in fatigue life of SCC. The relationship among stress level and loading cycles (N) to failure point in the study is presented in Fig. 8.10. The rubber particle incorporation provides better fatigue as per Chen et al. (2019) but its application in structural and pavement SCC is limited due to nonunderstanding of true mechanisms behind fatigue.

The rubber incorporated SCC may have high variability in terms of fatigue behavior. The rubberized SCC concrete due to high variability in rubber dispersion among the sample does not provide a regular pattern regarding the fatigue-related performance. The fatigue failure-strain in rubberized SCC is similar to the monotonic postpeak strain under similar stress level according to a study by Chen et al. (2021).

Table 8.4 Average number of blows (N) for each mix to assess impact resistance.

Mix	Khalil et al. (2015)		Thakare et al. (2023)	
	First crack (N)	Ultimate impact failure (N)	First crack (N)	Ultimate impact failure (N)
Reference	25	29	86	102
10% rubberized	45	52	122	143
20% rubberized	63	70	194	226
30% rubberized	75	81	291	338
40% rubberized	63	71	–	–

The failure mode in compressive strength highlights the CR response in providing evenly distributed cracks, which are finer in nature but are numerically higher, as CR opposes crack propagation (Jiang & Zhang, 2022).

Impact resistance of concrete is prerequisite characteristics to bear any instantaneous loading. The study by Khalil et al. (2015) and Shaji et al. (2021) attempted to understand the results from various CR incorporating SCC mixes. First crack and ultimate impact resistance were found to be related to number of blows. The CR did assist in improving the overall impact resistance as was seen from the increased number of blows required to produce initial crack and ultimate failure. Table 8.4 represents the impact values from different studies depicting the impact resistance. The dynamic impact factor of stress (DIF-σ) and strain (DIF-ε) are strain rate dependent and will be higher for high strain rates (Chen et al., 2022).

The study by Thakare et al. (2023) implemented a setup to define and compare the impact rebound energy for rubberized SCC. The setup consisted of dropping a 250-g weighted steel ball on cylindrical disk of diameter 150 mm and thickness 65 mm from height of 1 m. The difference in the initial ball drop and the rebound height was used to calculate rebound impact energy. The mixes consisted of various sizes CR rubber ranging from 0.60 to 4.75 mm which were designated as fine, medium, and coarse sized. The replacement performed was 10%, 20%, and 30% of sand with CR. The CR-SCC mixes successfully improved upon the rebound energy criteria which relates to the improved flexibility of SCC mix.

8.4 Durability properties

8.4.1 Shrinkage

Shrinkage of concrete can lead to the formation of cracks which affects the overall durability and service life of a structure. Therefore it is recommended to perform the shrinkage testing of concrete before its application.

The shrinkage values of rubberized SCC can vary from 35% to 95% higher, for 5% to 20% CR (Mesh 30 and Mesh 50) replacement of sand (Yung et al., 2013). The autogenous shrinkage also gets enlarged as a direct consequence of CR introduction in SCC (Yang et al., 2022). This effect is a direct consequence of low modulus of elasticity of CR providing high deformation characteristics and high pore formation in CR-SCC resulting in higher evaporation and hence increased shrinkage (Elchalakani, 2015; Nagataki & Yonekura, 1984; Wittman, 1982).

The aggregate replacement by crumb tire rubber in SCC under restrained shrinkage condition is capable of resisting crack formation by delaying the crack initiation. The rubber aggregate addition in SCC reduces its potential for cracking from high category to moderate–low category range (ASTM C1581, 2004; Turatsinze & Garros, 2008).

Study by Lv et al. (2019) measured the shrinkage of SCC mixes by varying the tire rubber percentage up to 50%. The shrinkage in SCC was found to be comparable up to 10% CR replacement, but increased at higher replacement level. The volume instability was introduced in the mix from the inherent high deformation under load application. The results of shrinkage using predictive modeling was estimated by Lv et al. (2019) and are shown in Fig. 8.11. Based on the experimental data a prediction model was developed based on the previous model, that is, China

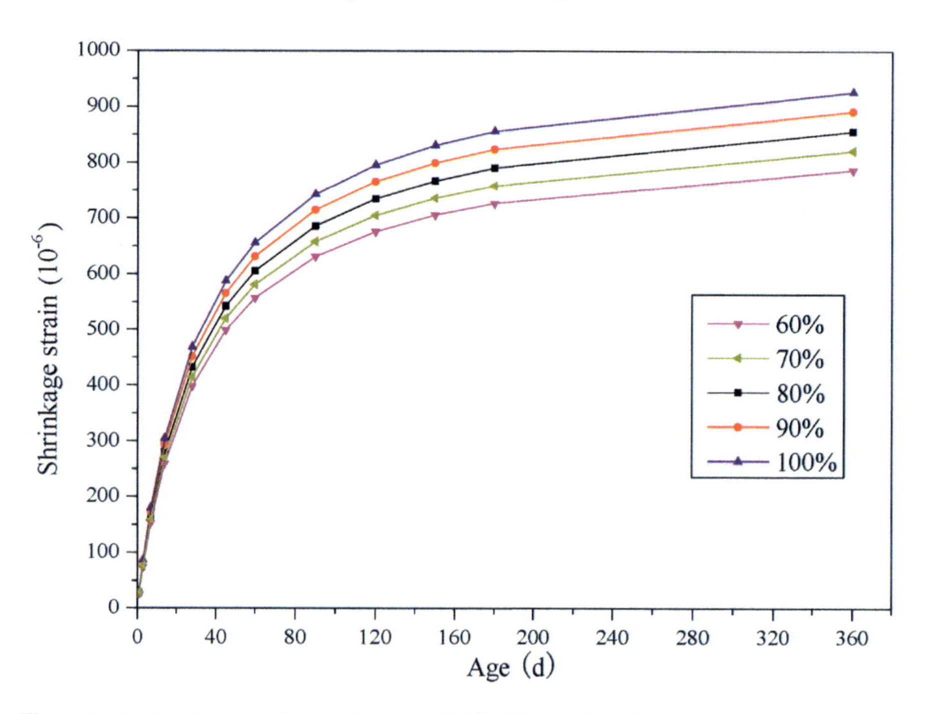

Figure 8.11 Shrinkage model of rubberized SCC with varying rubber content.
Source: From Lv, J., Zhou, T., Li, K., & Sun, K. (2019). Shrinkage properties of self-compacting rubber lightweight aggregate concrete: Experimental and analytical studies. *Materials, 12*(24), 4059. https://doi.org/10.3390/ma12244059.

Academy of Building Research model. The developed model was highly accurate with $R^2 = 0.99$. Eq. (8.13) shows the relation of shrinkage with time and rubber replacement level in SCC.

$$\varepsilon(t) = \frac{t}{47.8 + 1.47t} \times 10^{-3} \times \beta_6 \tag{8.13}$$

where $\varepsilon(t)$ is shrinkage value, t is time period in days, and β_6 is coefficient based on percentage of rubber incorporated (values vary from 1 to 1.2).

In a study by Younis et al. (2017) tests were performed on 100 mm × 100 mm × 400 mm samples with 200 mm as effective length in regard to finding the shrinkage of rubberized SCC. The shrinkage for the mixes tends to gain high rate after initial testing of 7 days. The shrinkage of rubberized samples was found to be on the higher side as the percentage of replacement increased. The 28 days testing showed a substantial difference in the reference sample and the CR (30%) sample, as the percentage increase in shrinkage was around 187% in the latter. The author suggested that the shrinkage tendency of rubberized SCC was due to the fact that CR induce high porosity in the mix which enable water loss through micro- and macropores to the surrounding environment. Low stiffness of CR also contributed toward shrinkage effect. The CR, however, proves to be better in terms of restrained shrinkage as the crack width developed under stress in initial and final stages seemed to be reduced with higher replacement quantity.

8.4.2 Water permeation characteristics (water absorption/porosity/sorptivity/water permeability)

The water absorption constitutes a direct relationship with the porosity of the concrete (Hall & Raymond Yau, 1987). The CR-SCC has porous structure due to the air entrapment inside CR and poor ITZ formation which increases the water absorption (Bignozzi & Sandrolini, 2006; Taha et al., 2003). The air content existing in rubberized SCC mortar is also high as it can increase up to 139.15% with rubber particles 25% content (Anil Thakare et al., 2020). The studies by Gesoğlu and Güneyisi (2011), Mallek et al. (2021), and Turatsinze and Garros (2008) also suggested an increased water porosity as the replacement of CR in SCC varied from 0% to 25% replacement. The porosity increase presented in these studies was approximately 30% higher to reference mix at 25% CR replacement of sand. According to experiments performed by Uygunoğlu and Topçu (2010) the self-compacting mortar mix also tends to have increased apparent porosity as the sand is replaced with fine rubber particles (1–4 mm), the increase in value for SCC mortar is around 35% at 25% CR replacement. The increase in terms of percentage change can be as high as 41% at high replacement level of 50% of the sand. The air content for SCC incorporating CR can be as high as 11.4% at 30% sand replacement which directly related to the increased porosity of the mix (N. Li, Long, et al., 2019). Hesami et al. (2016) suggested in a study that void increase in the rubberized SCC causes elevated water absorption. Water absorption reported in the study were 1.36%, 1.49%, 1.60%, and 1.72% at 0%, 5%, 10% and 15% CR replacement level, respectively. Kelechi, Adamu, Mohammed, Ibrahim, and Obianyo (2022) also mentioned the role of CR in elevating the water absorption in a study where water absorption was 62.6% higher due to void increase when rubber content was 20% in SCC. Angelin et al. (2020) also corroborated these findings from the study where void index was particularly high after incorporating CR in SCC mix.

The research by N. Li, Long, et al. (2019) evaluated the CR impact on the sorptivity of SCC. The results however, indicated an improved response to water ingress as the sorptivity height in SCC samples decreased from 1.71 to 1.04 mm for 0% to 30% replacement, respectively. The hydrophobic nature of CR along with an increased length of capillary channel deter upward water movement inside SCC samples. Similar findings were also found in this study related to water absorption which was valued lower for all rubberized SCC mixes with varied sizes compared to reference SCC mix, implying an overall better permeability performance.

The sorptivity values were found to be consistently lower as the percentage of CR increased in the SCC mix in a study by Tian and Qiu (2022). The values pertaining to 48-hour testing period were 0.83%, 0.77%, 0.63%, and 0.60% for CR substitution of 0%, 10%, 20%, and 30% in the tested samples, respectively. Fig. 8.12 shows the sorptivity (capillary water absorption) results with testing age for all the mixes which include CR and hydrophobic agent (HA) from a study by Tian and Qiu (2022).

Gesoğlu and Güneyisi (2011) tested the effect of CR addition on sorptivity performance of SCC mix. The sorptivity values corresponding to reference mix and 25% CR mix were 0.078 and 0.106 mm/$\sqrt{}$mm, respectively, displaying around 36% worse performance. The sorptivity performance for all CR mixes did, however, improved marginally with age. Inclusion of FA helped improve sorptivity of

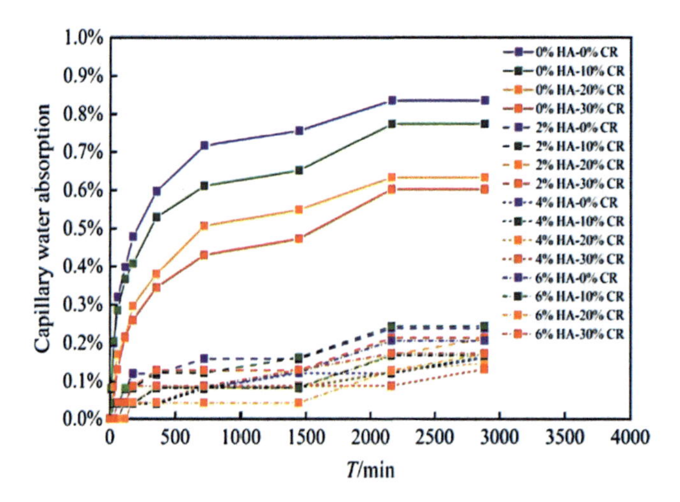

Figure 8.12 CR and HA effect on sorptivity rate of SCC.
Source: From Tian, L., & Qiu, L.c. (2022). Preparation and properties of integrally hydrophobic self-compacting rubberized concrete. *Construction and Building Materials, 338*, 127641. https://doi.org/10.1016/j.conbuildmat.2022.127641.

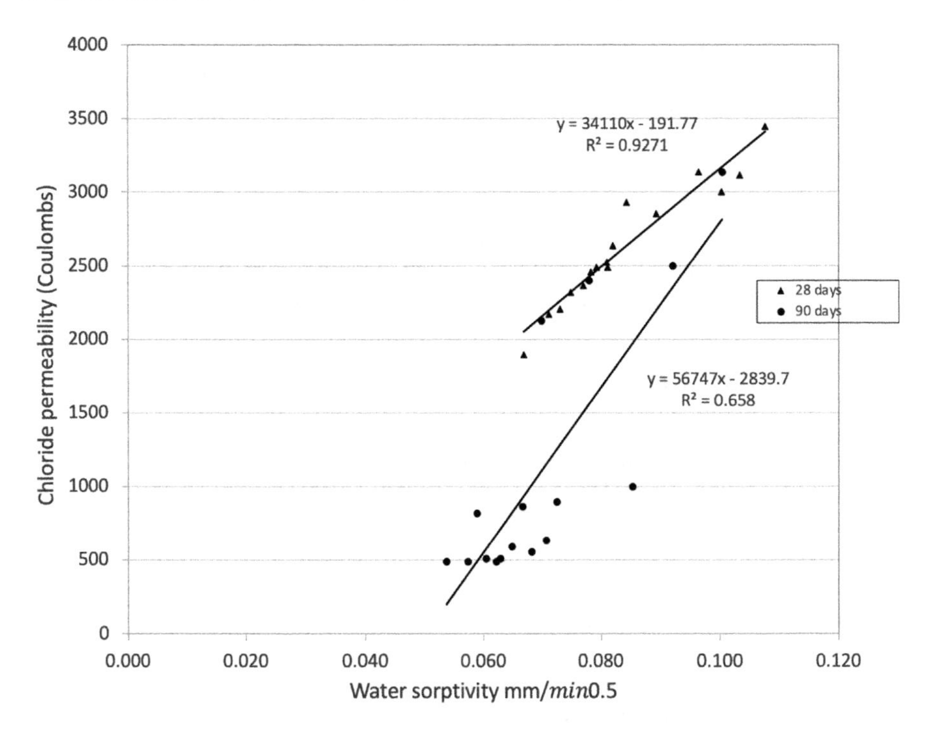

Figure 8.13 Sorptivity versus chloride ion permeability in rubberized SCC.
Source: Data from Gesoğlu, M., & Güneyisi, E. (2011). Permeability properties of self-compacting rubberized concretes. *Construction and Building Materials, 25*(8), 3319–3326. https://doi.org/10.1016/j.conbuildmat.2011.03.021.

rubberized SCC mix. The relationship of water sorptivity with chloride ingress was developed, where 28 days testing values had a strong relationship (Fig. 8.13).

The study by Thakare et al. (2023) focused on comparing various CR categories designated as fine (0.60–1.18 mm), medium (1.18–2.36 mm), and coarse (2.36–4.75 mm) based on size and which replaced sand by 10%, 20%, and 30%. The water permeability of the rubberized samples had higher values out of which the highest penetration depth of 57.80 mm was recorded for coarse-sized rubberized SCC, that is, an increase of 93.9% over the reference mix. The pores around the rubber particles are responsible for easiness of water penetration under pressure even when CR itself is hydrophobic in nature (Tian & Qiu, 2022).

8.4.3 Abrasion resistance

The concrete abrasion resistance performance is dependent on aggregates types, surface finish, curing method, internal pore structure (Hesami et al., 2016; Naik et al., 1995; Sadegzadeh et al., 1987). The abrasion is also a function of the compressive strength as higher compressive strength tends to improve abrasion resistance (H. Li et al., 2006; Singh & Siddique, 2012; Yazici & Sezer, 2007).

The abrasion resistance of the rubberized (0–4.75 mm) SCC up to 15% replacement was studied by Hesami et al. (2016). The impact of rubber addition was directly observed with the loss of abrasion index of rubberized concrete samples. The lower stiffness of CR was the contributing factor for the loss in abrasion resistance. Abrasion resistance index lied in the range 1.28–1.03 up to 15% replacement of sand by CR in SCC. The study by Ridgley et al. (2018) found that rubberized SCC has higher tendency of abrasion and the abrasion wear depth was found to be highest in case of the 40% (highest substitution level tested in the study) rubberized samples which was approximately 160% higher than the reference SCC mix.

8.4.4 Sulfate attack

The sulfate resistance of CR-SCC concrete at 5% concentration of $MgSO_4$ was performed by Kelechi, Adamu, Mohammed, Ibrahim, and Obianyo (2022). The extended exposure to sulfate solution led to growth of sulfate-related crystals (ettringite) which exert excessive stresses leading to fine cracks formation. The condition was further aggravated from CR addition due to high porosity leading to more ingress of sulfate solution. In a similar study by Uche et al. (2022) it was found that CR has tendency to increase the damage incurred when exposed to sulfate environment ($MgSO_4$). The CR testing was performed up to 20% sand substitution and the results showed that at this particular replacement weight loss was 2.04%, representing an increase of about 53.9% in comparision to reference mix.

The fine particles up to a size of mesh 30 (0.6 mm) may enhance resistance to sulfate attack at low replacement level up to 5% as per a study by Yung et al. (2013).

8.4.5 Acid attack

The acid resistance of CR-based SCC was tested in a study by Kelechi, Adamu, Mohammed, Obianyo, et al. (2022). The rubberized concrete had poor performance in terms of weight reduction when exposed to acidic solution over 3, 7, and 28 days. The percentage reduction in weight was around 2.2% higher for 20% CR mix at 28 days compared to the nonrubberized reference mix. The graphene oxide addition to cementitious material composites helped offset some of harmful effect of acid attack and can be used as a remedial material to control damage in rubberised SCC (Sabapathy et al., 2020). The study by Saleh et al. (2022), however, stated that the CR addition can provide resistance to ingress of ions from the acidic solution thereby improving durability of rubberized SCC.

8.4.6 Chloride permeability

The chloride permeability may slighly decrease in the rubberized SCC as a direct consequence of high resistivity of CR as suggested in rapid chloride permeability test (RCPT) by Li et al. (2019). The SCC with finer particles of CR tends to perform worse in terms of chloride permeability at equivalent replacement level (N. Li, Long, et al., 2019). The charge passed in all the mixes in the study up to 30% falls

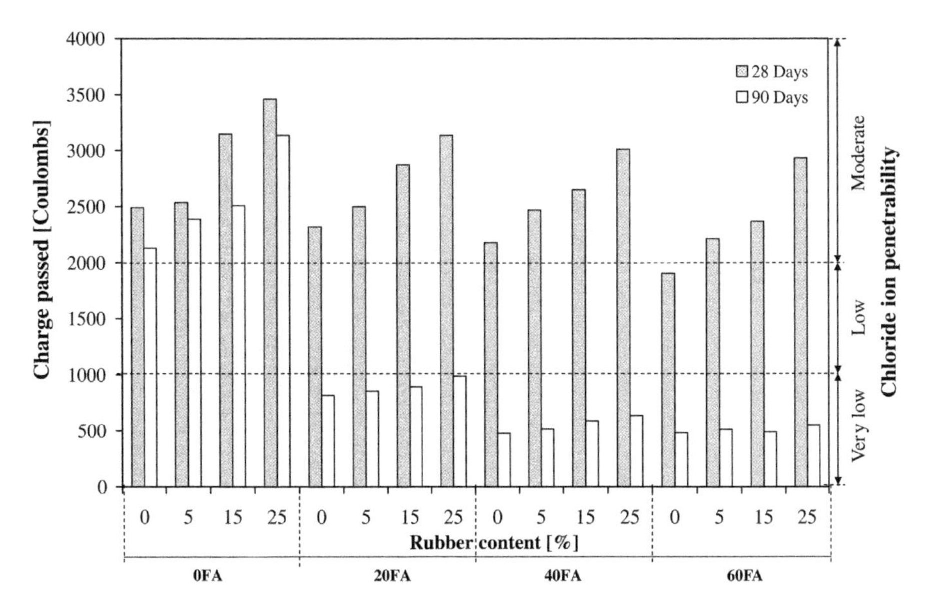

Figure 8.14 RCPT results showing variation of crumb rubber and fly ash.
Source: From Gesoğlu, M., & Güneyisi, E. (2011). Permeability properties of self-compacting rubberized concretes. *Construction and Building Materials, 25*(8), 3319–3326. https://doi.org/10.1016/j.conbuildmat.2011.03.021.

below 1000 which makes such concrete highly immune to chloride attack as specified by ASTM C1202 (2012). However, Gesoğlu and Güneyisi (2011) also studied the chloride permeability resistance in SCC and found it to be a negatively affected by the amount of CR replacement. The permeability may improved over time due to the increased hydration products formation. The performance of all the SCC mixes improved vastly over 90 days testing period when FA was also incorporated into the mix, shifting the mixes from moderate to very low category, that is, charged passed was below 1000 Coulombs in RCPT. The results regarding RCPT of the study are shown in Fig. 8.14. Chloride diffusion was found to rise with increasing CR percentage in SCC in a study by Mallek et al. (2021) leading to loss of chloride resistance. Chloride penetration depth increased by roughly 35% at the replacement level of 30% of sand mainly due to increased porosity and lack of internal packing.

8.5 Nondestructive testing

8.5.1 Ultrasonic pulse velocity

Ultrasonic pulse velocity (UPV) values of concrete are highly dependent on the concrete density, testing age, aggregate type, size, etc. and can provide insight into the porosity, discontinuities or cracking in concrete (Lorenzi et al., 2007; Saint-Pierre et al., 2016). The UPV values in concrete above 4500 m/s are considered

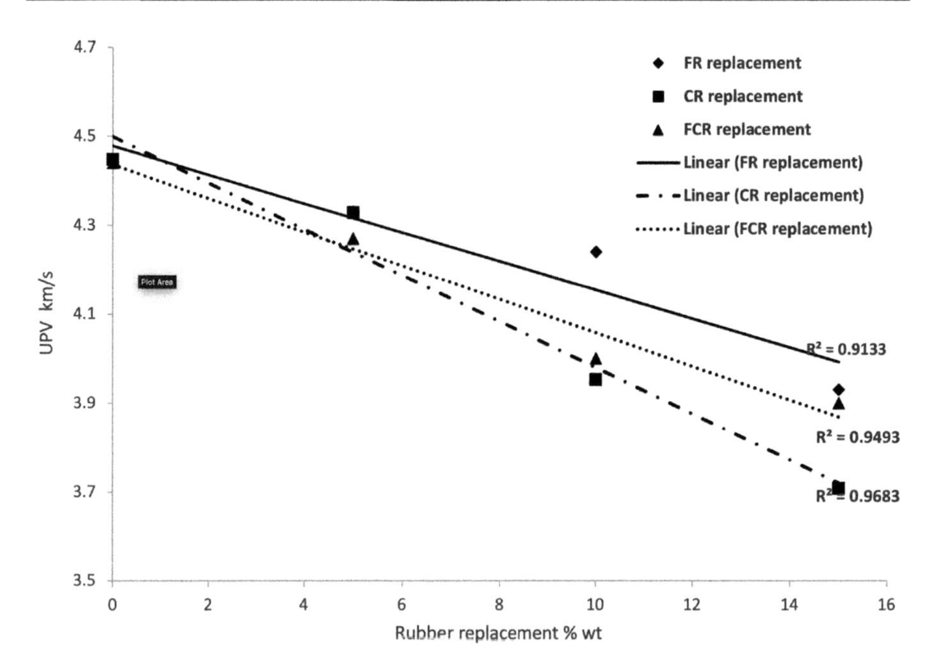

Figure 8.15 UPV variation in different size crumb rubber-SCC.
Source: From Najim, K. B., & Hall, M. R. (2012). Mechanical and dynamic properties of self-compacting crumb rubber modified concrete. *Construction and Building Materials, 27* (1), 521–530. https://doi.org/10.1016/j.conbuildmat.2011.07.013.

excellent and in range of 3600–4500 m/s are considered good (Mendes et al., 2020; Saint-Pierre et al., 2016).

The UPV values of rubberized SCC seems to be follow an inverse relation to the CR percentage increase. The UPV above 4000 m/s is achievable at 5% waste tire addition in rubber form in SCC (Yung et al., 2013). The usage of CR in SCC reduces UPV but such mixes are still able to achieve UPV > 3.6 km/s up to 15% CR replacement (Najim & Hall, 2012) which may be classified as good category according to studies by Mohammed et al. (2011) and Solís-Carcaño and Moreno (2008). The graph comparing response of different sized CR aggregates is shown in Fig. 8.15.

The loss in UPV was also observed by Rahman et al. (2012) in a study where CR was added in SCC and testing was performed at 7 and 28 days period. The UPV improved over the increasing curing period as an effect of hydration reaction. The UPV values improve upon cement slurry and mortar pretreatment of CR as observed in a study by Najim and Hall (2013). All the design mixes in the study with CR provided UPV in excess of 4.0 km/s which can be classified as good grade of concrete (Feldman, 1977; Leslie & Cheesman, 1949).

8.5.2 Electrical resistance

The electrical surface resistance of CR-SCC is highly dependent on the CR content added. The electrical resistance lies in the range of 34.5–44.88 kΩ-cm for 5%–20%

rubber replacement as compared to normal concrete resistance of 20 kΩ-cm (Yung et al., 2013).

The study from Mallek et al. (2021) tested the porosity of the rubberized SCC mix at 10%, 20%, and 30% replacement and related their performance to electrical resistivity. The crumb was successful in reducing the SCC conductivity, and electrical resistivity possessed an inverse relation to concrete porosity.

8.6 Elevated temperature studies

CR possess the noteworthy ability to enhance the concrete thermal conductivity, which may be used high-temperature–related applications (Hamza et al., 2018).

Kelechi, Adamu, Mohammed, Ibrahim, and Obianyo (2022) studied the combined effect of CR, FA, and calcium carbide aggregates on properties of SCC at elevated temperatures. The w/b ratio was variable for different mixes varying from 0.37 to 0.43 with total cement content fixed at 520 kg/m^3 for the reference mix. The CR that was used as fine aggregate replacement at 10% and 20%, provided poor performance in strength characteristics. The calcium carbide addition in the mix helped recover some of the compressive strength. The viable temperature for the application of rubberized SCC was found to be 400°C, above which weight losses were dramatic. The weight loss percentage of various mixes from the study is shown in Fig. 8.16.

The study related to the high-temperature performance of ultra-high-performance concrete was undertaken by Aslani and Khan (2019) where SCC mixes incorporating tire rubber (2–5 and 5–10 mm) were subjected to temperature up to 600°C. The compressive strength and split tensile strength both suffered heavy loss at 600°C as result of disintegration and burning of rubber particles present in the SCC mix. The samples at 100°C and 300°C, however, depicted some improvements in compressive strength from hydration rate enhancement and as of melting of CR consecutively. The finer CR (2–5 mm)-based SCC mixes were more prone to spalling as compared to its counterpart larger sized 5–10 mm rubberized SCC.

Mohammed et al. (2019) used high-strength self-compacting rubberized concrete with w/b ratio of 0.45 and binder content of 450 kg/m^3 consisting of cement, granulated ground blast furnace slag and silica fume. It was concluded from the results that residual strength properties of rubberized concrete can provide satisfactory results up to 25% replacement level.

8.7 Microstructure analysis

8.7.1 Scanning electron microscopy analysis

Scanning electron microscopy (SEM) image analysis is an imaging technique which can be implemented for studying of the microstructure of concrete by using high magnification.

The morphology of the failure surface in rubberized SCC was studied by X. Li, Chen, et al. (2019). The results concentrated on the failure pattern and its origin.

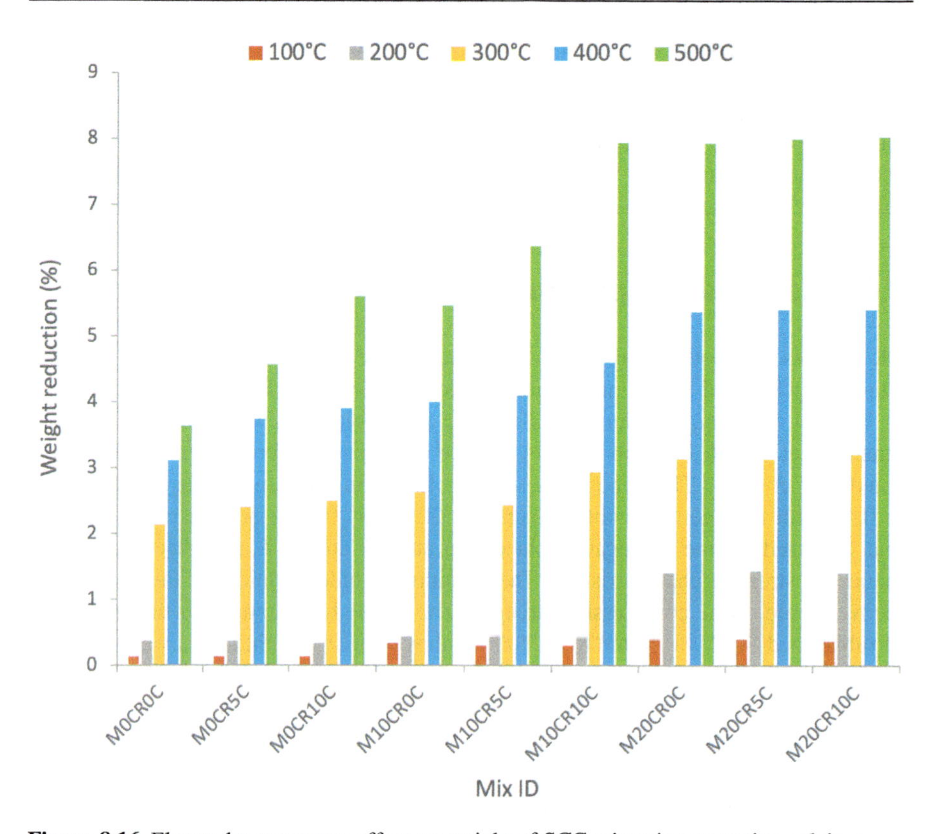

Figure 8.16 Elevated temperature effect on weight of SCC mixes incorporating calcium carbide waste (C) and crumb rubber (CR). *M0CR0C*, control mix; *M0CR5C*, 0% CR with 5% calcium carbide; *M0CR10C*, 0% CR with 10% calcium carbide; *M10CR0C*, 10% CR with 0% calcium carbide; *M10CR5C*, 10% CR with 5% calcium carbide; *M10CR5C*, 10% CR with 10% calcium carbide; *M20CR0C*, 20% CR with 0% calcium carbide; *M20CR5C*, 20% CR with 5% calcium carbide; *M20CR10C*, 20% CR with 10% calcium carbide.
Source: From Kelechi, S. E., Adamu, M., Mohammed, A., Ibrahim, Y. E. & Obianyo, I. I. (2022). Durability performance of self-compacting concrete containing crumb rubber, fly ash and calcium carbide waste. *Materials, 15*(2), 488. https://doi.org/10.3390/ma15020488.

The failure surface for reference mix (without rubber) displayed crack formation at the ITZ existing between aggregate and mortar circumventing the aggregates. Rubber aggregates addition modified the crack pattern as voids in the crack surface became smaller but higher in number. The cracks in the rubberized concrete are likely to develop at rubber-mortar ITZ matrix rather than aggregate mortar ITZ as the energy required is lower in the earlier case.

The study by Thakare et al. (2023) also explored the SEM image technique for studying the bonding performance of CR with cement matrix. The poor ITZ was evident from the SEM images showing wide gap between CR and cement matrix. The presence of microcrack and microvoids was also seen in the neighborhood of

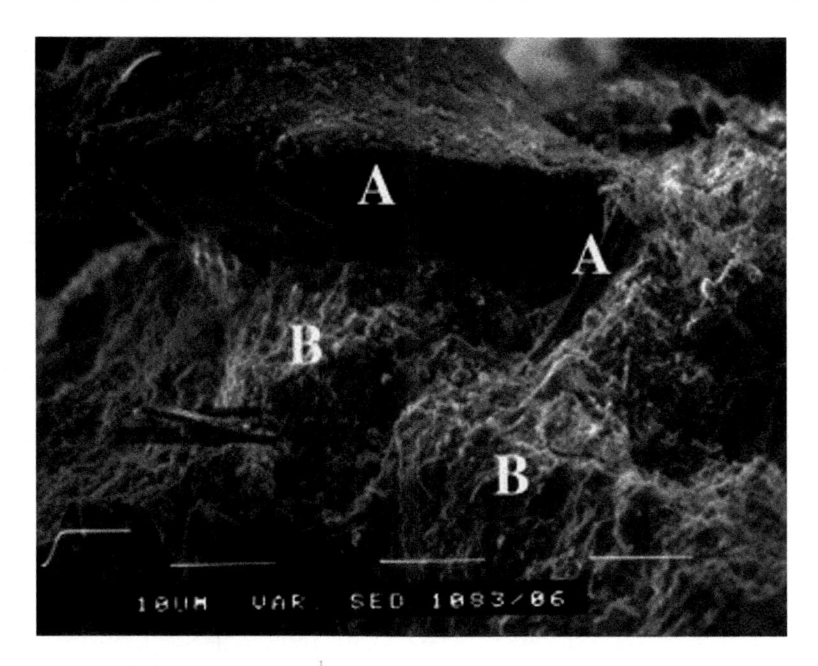

Figure 8.17 SEM image showing (A) Rubber particle and (B) Cement matrix.
Source: From Bignozzi, M. C., & Sandrolini, F. (2006). Tyre rubber waste recycling in self-compacting concrete. *Cement and Concrete Research, 36*(4), 735–739. https://doi.org/10.1016/j.cemconres.2005.12.011.

CR particles. The increased porosity as a direct result of cracks and cavities formed proves to be detrimental when considering durability aspect of concrete. The SEM image of the rubber particles with cement matrix is shown in Fig. 8.17. Zaouai et al. (2020) in a study showed that the ITZ of the CR rubber can be successfully improved with pretreatment of crushed dune sand prior to its usage in the SCC mix, as a fine aggregate replacement.

8.7.2 X-ray computed tomography

The X-ray computed tomography (XCT) is used to create two-dimensional (2D) image of an object at several orientations. The data from these images is then further processed to create a 3D volumetric image of the object using computer software. This technique may be implemented in the testing of concrete to identify the cracks, voids, and even dimensioning of the samples (Sun et al., 2012). The X-ray tomography not only can provide the information about the internal flaws but also can help identify any geometrical deviation of the sample from the supposed geometry to be analyzed (Obaton et al., 2020). Due to its nondestructive nature, XCT is a very great technique to extract internal structural information of the given material (Gao et al., 2015).

The study by Liu et al. (2021) investigated the partial replacement of CR in SCC at 10%–30% replacement level using the XCT to calculate pore amount in the microstructure and rubber particles quantitatively. The study was able to demonstrate the successful applicability of the XCT technique to construct the SCC concrete microstructure. The 3D rendered images from the study are shown in Fig. 8.18. The values of the mean porosity reduced by 64.8%, 34.9%, and 61.1% for 10%, 20% and 30% replacement of CR, respectively, as measured by XCT. The pore number was also estimated for the samples and its values were 112666, 30706, 56859, and 23533 for 0%, 10%, 20%, and 30% CR replacement, respectively,

Figure 8.18 XCT-3D rendered images: (A) NSCC, (B) RSCC10, (C) RSCC20, and (D) RSCC30. *NSCC*, Normal self-compacting concrete; *RSCC10%*, 10% rubberized self-compacting concrete; *RSCC20%*, 20% rubberized self-compacting concrete; *RSCC30%*, 30% rubberized self-compacting concrete.
Source: From Liu, Z., Chen, X., Wu, P., & Cheng, X. (2021). Investigation on microstructure of self-compacting concrete modified by recycled grinded tire rubber based on X-ray computed tomography technology. *Journal of Cleaner Production, 290*, 125838. https://doi.org/10.1016/j.jclepro.2021.125838.

indicating a lowered pore formation from CR introduction in SCC mix contradicting previous study from Bušić et al. (2018), Ismail and Hassan (2016), and Uygunoğlu and Topçu (2010). The reason for such discrepancy lied in the fact that in many cases the porosity testing was done on fresh state whereas SCC mixes settles over time removing air bubbles. Another reason may also be the microscale pores occurring at ITZ typically have very small sizes and are ignored in XCT to achieve a good resolution, also previous studies measured porosity indirectly from the water absorption testing which may not coincide with the actual porosity.

8.8 Conclusions

The chapter focused on the varied outcomes of the addition of CR into SCC providing the state-of-the-art review. The results of SCC mixes containing different sizes and quantity of CR can be summarized as follows:

- The CR in SCC can be cheaply implemented to achieve a lightweight concrete.
- The CR will typically lead to loss of workability but the other factors such as size of CR, mix design, and w/b ratio play important role in determining the workability of SCC mix. Higher dosage of superplasticizer may alleviate workability loss but lab testing should be performed to determine the optimum quantity of superplasticizer before usage.
- The CR introduction in SCC mix reduces its strength properties, that is, compressive strength, split tensile strength, flexural strength etc. However, by application of different pretreatment techniques on rubber particles, usage of fibers, and adding SCMs in the SCC mix can counter such effect to varied degree. The optimum pretreatment technique may be based on the evaluation of the monetary benefits offered. The amount of SCMs and fibers can also be used based on the availability and the required characteristics of the SCC mix.
- The deformable nature of the CR is responsible for higher strain under compressive loading which improves the toughness of SCC.
- The shrinkage in SCC is highly sensitive to CR content where, increase in CR can lead to high shrinkage in SCC mixes, but the crack formation under restrained shrinkage particularly can be controlled by CR addition. The abrasion resistance offered by the rubberised SCC is inferior to regular SCC mix.
- The CR particle in SCC leads to poor ITZ having porous nature surrounding the rubber particle.
- The optimum CR percentage varies based on the various strength and durability parameters typically lying in the range of 5%–15%.
- The CR can be used in SCC under elevated temperature conditions up to a temperature range of 400°C.

References

ASTM C1202. (2012). *Standard test method for electrical indication of concrete's ability to resist chloride ion penetration*. American Society for Testing and Materials, pp. 1–8.

ASTM C1581. (2004). *Standard test method for determining age at cracking and induced tensile stress characteristics of mortar and concrete under restrained shrinkage*. American Society for Testing and Materials.

Amari, T., Themelis, N. J., & Wernick, I. K. (1999). Resource recovery from used rubber tires. *Resources Policy*, *25*(3), 179–188. Available from https://doi.org/10.1016/S0301-4207(99)00025-2.

Angelin, A. F., Lintz, R. C. C., Gachet-Barbosa, L. A., & Osório, W. R. (2017). The effects of porosity on mechanical behavior and water absorption of an environmentally friendly cement mortar with recycled rubber. *Construction and Building Materials*, *151*, 534–545. Available from https://doi.org/10.1016/j.conbuildmat.2017.060.061.

Angelin, A. F., Cecche Lintz, R. C., Osório, W. R., & Gachet, L. A. (2020). Evaluation of efficiency factor of a self-compacting lightweight concrete with rubber and expanded clay contents. *Construction and Building Materials*, *257*. Available from https://doi.org/10.1016/j.conbuildmat.2020.119573.

Anil Thakare, A., Siddique, S., Sarode, S. N., Deewan, R., Gupta, V., Gupta, S., & Chaudhary, S. (2020). A study on rheological properties of rubber fiber dosed self-compacting mortar. *Construction and Building Materials*, *262*, 120745. Available from https://doi.org/10.1016/j.conbuildmat.2020.120745.

Aslani, F., & Gedeon, R. (2019). Experimental investigation into the properties of self-compacting rubberised concrete incorporating polypropylene and steel fibers. *Structural Concrete*, *20*(1), 267–281. Available from https://doi.org/10.1002/suco.201800182.

Aslani, F., & Khan, M. (2019). Properties of high-performance self-compacting rubberized concrete exposed to high temperatures. *Journal of Materials in Civil Engineering*, *31*(5). Available from https://doi.org/10.1061/(ASCE)MT.1943-5533.0002672.

Aslani, F., Ma, G., Yim Wan, D. L., & Tran Le, V. X. (2018). Experimental investigation into rubber granules and their effects on the fresh and hardened properties of self-compacting concrete. *Journal of Cleaner Production*, *172*, 1835–1847. Available from https://doi.org/10.1016/j.jclepro.2017.120.003.

Aslani, F., Ma, G., Yim Wan, D. L., & Muselin, G. (2018). Development of high-performance self-compacting concrete using waste recycled concrete aggregates and rubber granules. *Journal of Cleaner Production*, *182*, 553–566. Available from https://doi.org/10.1016/j.jclepro.2018.020.074.

Ayub, T., Khan, S. U., & Mahmood, W. (2022). Mechanical properties of self-compacting rubberised concrete (SCRC) containing polyethylene terephthalate (PET) fibres. *Journal of Science and Technology - Transactions of Civil Engineering*, *46*(2), 1073–1085. Available from https://doi.org/10.1007/s40996-020-00568-6.

Bignozzi, M. C., & Sandrolini, F. (2006). Tyre rubber waste recycling in self-compacting concrete. *Cement and Concrete Research*, *36*(4), 735–739. Available from https://doi.org/10.1016/j.cemconres.2005.120.011.

Blackman, A., & Palma, A. (2002). Scrap tires in Ciudad Juárez and El Paso: Ranking the risks. *Journal of Environment and Development*, *11*(3), 247–266. Available from https://doi.org/10.1177/107049602237157.

Bušić, R., & Miličević, I., Influence of waste tire rubber on fresh and hardened properties of self-compacting rubberized concrete (SCRC). In: Mechtcherine, V., Khayat, K., Secrieru, E. (Eds.), *International Conference on Rheology and Processing of Construction Materials International Symposium on Self-Compacting Concrete*. RheoCon SCC 2019. RILEM Bookseries. Vol. 23 (2020). Available from https://doi.org/10.1007/978-3-030-22566-7_1.

Bušić, R., Miličević, I., Šipoš, T., & Strukar, K. (2018). Recycled rubber as an aggregate replacement in self-compacting concrete—Literature overview. *Materials*, *11*(9), 1729. Available from https://doi.org/10.3390/ma11091729.

Bušić, R., Benšić, M., Miličević, I., & Strukar, K. (2020). Prediction models for the mechanical properties of self-compacting concrete with recycled rubber and silica fume. *Materials*, *13* (8), 1821. Available from https://doi.org/10.3390/ma13081821.

Cao, W. (2007). Study on properties of recycled tire rubber modified asphalt mixtures using dry process. *Construction and Building Materials*, *21*(5), 1011–1015. Available from https://doi.org/10.1016/j.conbuildmat.2006.020.004.

CEB-FIP. (1994). Committee Euro-International for Concrete-International Federation for Prestressing. In *CEB-FIP model code 1990: Design code 1994*, London: Thomas Telford.

Chen, C., Chen, X., & Zhang, J. (2021). Experimental study on flexural fatigue behavior of self-compacting concrete with waste tire rubber. *Mechanics of Advanced Materials and Structures*, *28*(16), 1691–1702. Available from https://doi.org/10.1080/15376494.2019.1701152.

Chen, J., Zhuang, J., Shen, S., & Dong, S. (2022). Experimental investigation on the impact resistance of rubber self-compacting concrete. *Structures*, *39*, 691–704. Available from https://doi.org/10.1016/j.istruc.2022.030.057.

Chen, X., Liu, Z., Guo, S., Huang, Y., & Xu, W. (2019). Experimental study on fatigue properties of normal and rubberized self-compacting concrete under bending. *Construction and Building Materials*, *205*, 10–20. Available from https://doi.org/10.1016/j.conbuildmat.2019.010.207.

del Río Merino, M., Santa Cruz Astorqui, J., & Cortina, M. G. (2007). Viability analysis and constructive applications of lightened mortar (rubber cement mortar). *Construction and Building Materials*, *21*(8), 1785–1791. Available from https://doi.org/10.1016/j.conbuildmat.2006.050.014.

Djadouni, H., Trouzine, H., Gomes Correia, A., & Miranda, T. F. d. S. (2019). Life cycle assessment of retaining wall backfilled with shredded tires. *International Journal of Life Cycle Assessment*, *24*(3), 581–589. Available from https://doi.org/10.1007/s11367-018-1475-3.

Dong, R., Zhao, M., Xia, W., Yi, X., Dai, P., & Tang, N. (2018). Chemical and microscopic investigation of co-pyrolysis of crumb tire rubber with waste cooking oil at mild temperature. *Waste Management*, *79*, 516–525. Available from https://doi.org/10.1016/j.wasman.2018.080.024.

ETRMA. (2019). *Brussels European Tyre and Rubber Manufacturer's Association*. Unpublished content. https://www.etrma.org/wp-content/uploads/2021/05/20210520_ETRMA_PRESS-RELEASE_ELT-2019.pdf.

ETRMA. (2021). *2021 European Tyre and Rubber Industry Statistics*. https://www.etrma.org/wp-content/uploads/2021/11/20211030-Statistics-booklet-2021VF.pdf.

Elchalakani, M. (2015). High strength rubberized concrete containing silica fume for the construction of sustainable road side barriers. *Structures*, *1*, 20–38. Available from https://doi.org/10.1016/j.istruc.2014.060.001.

Emara, M., Eid, F. M., Nasser, A., & Safaan, M. (2018). Prediction of self-compacting rubberized concrete mechanical and fresh properties using taguchi method. *Journal of Civil & Environmental Engineering*, *8*. Available from https://doi.org/10.4172/2165-784X.1000301.

EPA. (2010). *Scrap tires: Handbook on recycling applications and management for the U.S. and Mexico*. EPA-United States Environmental Protection Agency, Washington, DC.

Fakhri, M., Yousefian, F., Amoosoltani, E., Aliha, M. R. M., & Berto, F. (2021). Combined effects of recycled crumb rubber and silica fume on mechanical properties and mode I fracture toughness of self-compacting concrete. *Fatigue and Fracture of Engineering Materials and Structures*, *44*(10), 2659–2673. Available from https://doi.org/10.1111/ffe.13521.

Feldman, R. F. (1977). *Non-destructive testing of concrete CBD-187*. NRC-IRC.

Fraile-Garcia, E., Ferreiro-Cabello, J., Mendivil-Giro, M., & Vicente-Navarro, A. S. (2018). Thermal behaviour of hollow blocks and bricks made of concrete doped with waste tyre rubber. *Construction and Building Materials*, *176*, 193–200. Available from https://doi.org/10.1016/j.conbuildmat.2018.050.015.

Frank, Z., & Musacchio, A. (2008). *The international natural rubber market, 1870–1930*. EH.Net Encyclopedia. Economic History Association. http://eh.net/encyclopedia/the-international-natural-rubber-market-1870-1930/.

Ganesan, N., Bharati Raj, J., & Shashikala, A. P. (2013). Flexural fatigue behavior of self compacting rubberized concrete. *Construction and Building Materials*, *44*, 7–14. Available from https://doi.org/10.1016/j.conbuildmat.2013.020.077.

Gao, L., Ni, F., Luo, H., & Charmot, S. (2015). Characterization of air voids in cold in-place recycling mixtures using X-ray computed tomography. *Construction and Building Materials*, *84*, 429–436. Available from https://doi.org/10.1016/j.conbuildmat.2015.030.081.

Gesoğlu, M., & Güneyisi, E. (2011). Permeability properties of self-compacting rubberized concretes. *Construction and Building Materials*, *25*(8), 3319–3326. Available from https://doi.org/10.1016/j.conbuildmat.2011.030.021.

Gettu, R. Gomes, P. C. C., Aguila, L., & Josa, A. (2004). *High-strength self-compacting concrete with fly ash: Development and utilization* (pp. 507–522). American Concrete Institute. https://www.concrete.org/topicsinconcrete/topicdetail/special%20publication SP-221.

Goldstein Market Research. (2020). *Goldstein Market Research 2020* global tire recycling market analysis *2025: Opportunity*, demand, growth and forecast *2017-2025*. Goldstein Market Research. Unpublished content. https://www.goldsteinresearch.com/report/global-tire-market-outlook-2024-global-opportunity-and-demand-analysis-market-forecast-2016-2024.

Gupta, T., Chaudhary, S., & Sharma, R. K. (2016). Mechanical and durability properties of waste rubber fiber concrete with and without silica fume. *Journal of Cleaner Production*, *112*, 702–711. Available from https://doi.org/10.1016/j.jclepro.2015.070.081.

Güneyisi, E. (2010). Fresh properties of self-compacting rubberized concrete incorporated with fly ash. *Materials and Structures/Materiaux et Constructions*, *43*(8), 1037–1048. Available from https://doi.org/10.1617/s11527-009-9564-1.

Hall, C., & Raymond Yau, M. H. (1987). Water movement in porous building materials-IX. The water absorption and sorptivity of concretes. *Building and Environment*, *22*(1), 77–82. Available from https://doi.org/10.1016/0360-1323(87)90044-8.

Hamza, B., Belkacem, M., Said, K., & Walid, Y. (2018). Performance of self-compacting rubberized concrete. *MATEC Web of Conferences.*, *149*, 01070. Available from https://doi.org/10.1051/matecconf/201814901070.

Hesami, S., Salehi Hikouei, I., & Emadi, S. A. A. (2016). Mechanical behavior of self-compacting concrete pavements incorporating recycled tire rubber crumb and reinforced with polypropylene fiber. *Journal of Cleaner Production*, *133*, 228–234. Available from https://doi.org/10.1016/j.jclepro.2016.040.079.

Hilal, N. N. (2017). Hardened properties of self-compacting concrete with different crumb rubber size and content. *International Journal of Sustainable Built Environment*, *6*(1), 191–206. Available from https://doi.org/10.1016/j.ijsbe.2017.030.001.

Hita, I., Arabiourrutia, M., Olazar, M., Bilbao, J., Arandes, J. M., & Castaño, P. (2016). Opportunities and barriers for producing high quality fuels from the pyrolysis of scrap tires. *Renewable and Sustainable Energy Reviews*, *56*, 745–759. Available from https://doi.org/10.1016/j.rser.2015.11.081.

İlkılıç, C., & Aydın, H. (2011). Fuel production from waste vehicle tires by catalytic pyrolysis and its application in a diesel engine. *Fuel Processing Technology*, *92*(5), 1129–1135. Available from https://doi.org/10.1016/j.fuproc.2011.01.009.

Ismail, M. K., & Hassan, A. A. A. (2016). Use of metakaolin on enhancing the mechanical properties of self-consolidating concrete containing high percentages of crumb rubber. *Journal of Cleaner Production*, *125*, 282–295. Available from https://doi.org/10.1016/j.jclepro.2016.030.044.

Jang, J. W., Yoo, T. S., Oh, J. H., & Iwasaki, I. (1998). Discarded tire recycling practices in the United States, Japan and Korea. *Resources, Conservation and Recycling*, *22*(1-2), 1–14. Available from https://doi.org/10.1016/S0921-3449(97)00041-4.

Jiang, Y., & Zhang, S. (2022). Experimental and analytical study on the mechanical properties of rubberized self-compacting concrete. *Construction and Building Materials*, *329*. Available from https://doi.org/10.1016/j.conbuildmat.2022.127177.

Kelechi, S. E., Adamu, M., Mohammed, A., Obianyo, I. I., Ibrahim, Y. E., & Alanazi, H. (2022). Equivalent CO_2 emission and cost analysis of green self-compacting rubberized concrete. *Sustainability (Switzerland)*, *14*(1), 20711050. Available from https://doi.org/10.3390/su14010137.

Kelechi, S. E., Adamu, M., Mohammed, A., Ibrahim, Y. E., & Obianyo, I. I. (2022). Durability performance of self-compacting concrete containing crumb rubber, fly ash and calcium carbide waste. *Materials*, *15*(2). Available from https://doi.org/10.3390/ma15020488.

Khalil, E., Abd-Elmohsen, M., & Anwar, A. M. (2015). Impact resistance of rubberized self-compacting concrete. *Water Science*, *29*(1), 45–53. Available from https://doi.org/10.1016/j.wsj.2014.120.002.

Khalilpasha, M. H., Sadeghi-Nik, A., Lotfi-Omran, O., Omran, O. L., Kimiaeifard, K., & Molla, M. A. (2012). Su2stainable development using recyclable rubber in self-compacting concrete. In *2012 third international conference on construction in developing countries (ICCIDC–III)*, Bangkok, Thailand, pp. 580–585.

Khed, V. C., Mahalakshmi, S. H. V., & Achara, B. E. (2022). Optimization of manufactured-sand (M-sand) and silica fume built-in self-compacting rubber create. *Iranian Journal of Science and Technology - Transactions of Civil Engineering*, *46*(3), 2217–2233. Available from https://doi.org/10.1007/s40996-022-00821-0.

Leslie, J. R., & Cheesman, W. J. (1949). An ultrasonic method of studying deterioration and cracking in concrete structures. *Journal of the American Concrete Institute*, *21*, 17–36.

Li, H., Zhang, M., & Ou, J. (2006). Abrasion resistance of concrete containing nano-particles for pavement. *Wear*, *260*(11–12), 1262–1266. Available from https://doi.org/10.1016/j.wear.2005.080.006, 00431648.

Li, N., Long, G., Ma, C., Fu, Q., Zeng, X., Ma, K., Xie, Y., & Luo, B. (2019). Properties of self-compacting concrete (SCC) with recycled tire rubber aggregate: A comprehensive study. *Journal of Cleaner Production*, *236*. Available from https://doi.org/10.1016/j.jclepro.2019.117707.

Li, X., Chen, X., Jivkov, A. P., & Zhang, J. (2019). 3D mesoscale modeling and fracture property study of rubberized self-compacting concrete based on uniaxial tension test. *Theoretical and Applied Fracture Mechanics, 104*. Available from https://doi.org/10.1016/j.tafmec.2019.102363.

Li, X., Chen, X., Jivkov, A. P., & Hu, J. (2021). Investigation of tensile fracture of rubberized self-compacting concrete by acoustic emission and digital image correlation. *Structural Control and Health Monitoring*, *28*(8). Available from https://doi.org/10.1002/stc.2744.

Liu, Z., Chen, X., Wu, P., & Cheng, X. (2021). Investigation on micro-structure of self-compacting concrete modified by recycled grinded tire rubber based on X-ray computed tomography technology. *Journal of Cleaner Production, 290*. Available from https://doi.org/10.1016/j.jclepro.2021.125838.

Lorenzi, A., Tisbierek, F.T., Carlos, L., & Filho, S. (2007). Ultrasonic pulse velocity analysis in concrete specimens. In *4th Pan American conference for NDT*, Buenos Aires, Argentina. https://www.ndt.net/?id = 4685.

Lv, J., Du, Q., Zhou, T., He, Z., & Li, K. (2019). Fresh and mechanical properties of self-compacting rubber lightweight aggregate concrete and corresponding mortar. *Advances in Materials Science and Engineering, 2019*. Available from https://doi.org/10.1155/2019/8372547.

Lv, J., Zhou, T., Du, Q., & Li, K. (2020). Experimental and analytical study on uniaxial compressive fatigue behavior of self-compacting rubber lightweight aggregate concrete. *Construction and Building Materials*, 237. Available from https://doi.org/10.1016/j.conbuildmat.2019.117623.

Lv, Jing, Zhou, Tianhua, Li, Kunlun, & Sun, Kai (2019). Shrinkage properties of self-compacting rubber lightweight aggregate concrete: Experimental and analytical studies. *Materials*, *12*(24), 4059. Available from https://doi.org/10.3390/ma12244059.

López-Zaldívar, O., Lozano-Díez, R., Herrero del Cura, S., Mayor-Lobo, P., & Hernández-Olivares, F. (2017). Effects of water absorption on the microstructure of plaster with end-of-life tire rubber mortars. *Construction and Building Materials*, *150*, 558–567. Available from https://doi.org/10.1016/j.conbuildmat.2017.060.014.

Mallek, J., Daoud, A., Omikrine-Metalssi, O., & Loulizi, A. (2021). Performance of self-compacting rubberized concrete against carbonation and chloride penetration. *Structural Concrete*, *22*(5), 2720–2735. Available from https://doi.org/10.1002/suco.202000687.

Martínez, J. D., Puy, N., Murillo, R., García, T., Navarro, M. V., & Mastral, A. M. (2013). Waste tyre pyrolysis – a review. *Renewable and Sustainable Energy Reviews*, *23*, 179–213. Available from https://doi.org/10.1016/j.rser.2013.020.038.

Mendes, S. E. S., Oliveira, R. L. N., Cremonez, C., Pereira, E., Pereira, E., & Medeiros-Junior, R. A. (2020). Mixture design of concrete using ultrasonic pulse velocity. *International Journal of Civil Engineering*, *18*(1), 113–122. Available from https://doi.org/10.1007/s40999-019-00464-9.

Miličević, I., Nyarko, M. H., Bušić, R., Radosavljević, J. S., Prokopijević, M., & Vojisavljević, K. (2021). Effect of rubber treatment on compressive strength and modulus of elasticity of self-compacting rubberized concrete. *International Journal of Structural and Construction Engineering*, *15*, 131–134.

Mishra, M., & Panda, K. C. (2015a). An experimental study on fresh and hardened properties of self compacting rubberized concrete. *Indian Journal of Science and Technology*, *8* (29). Available from https://doi.org/10.17485/ijst/2015/v8i29/86799.

Mishra, M., & Panda, K. C. (2015b). Influence of rubber on mechanical properties of conventional and self compacting concrete. In V. Matsagar (Ed.), *Advances in structural engineering* (pp. 1785–1794). Springer India. Available from https://doi.org/10.1007/978-81-322-2187-6_136.

Mohammed, B. S., Azmi, N. J., & Abdullahi, M. (2011). Evaluation of rubbercrete based on ultrasonic pulse velocity and rebound hammer tests. *Construction and Building Materials*, *25*(3), 1388–1397. Available from https://doi.org/10.1016/j.conbuildmat.2010.090.004.

Mohammed, M., Kadir, M. A. A., & Shukor, N. H. B. A. (2019). High strength self compacting concrete incorporating crumb rubber fibre exposed to elevated temperatures. *International Journal of Recent Technology and Engineering*, *8*(1), 500–508. Available from https://www.ijrte.org/wp-content/uploads/papers/v8i1C2/A10830581C219.pdf.

Nagataki, S., & Yonekura, A. (1984). The mechanisms of drying shrinkage and creep of concrete. *Concrete Library of JSCE*, *3*, 177–191.

Naik, T. R., Singh, S. S., & Hossain, M. M. (1995). Abrasion resistance of high-strength concrete made with Class C fly ash. *ACI Materials Journal*, *92*(6), 649–659.

Naji Hilal, N., & Mahmoud Hama, S. (2018). The effect of particle size distribution (PSD) of rubbrized self-compacting concrete (RSCC). *Journal of Engineering and Sustainable Development*, *22*(2), 13–22. Available from https://doi.org/10.31272/jeasd.2018.2.46.

Najim, K. B., & Hall, M. R. (2010). A review of the fresh/hardened properties and applications for plain- (PRC) and self-compacting rubberised concrete (SCRC). *Construction and Building Materials*, *24*(11), 2043–2051. Available from https://doi.org/10.1016/j.conbuildmat.2010.040.056.

Najim, K. B., & Hall, M. R. (2012). Mechanical and dynamic properties of self-compacting crumb rubber modified concrete. *Construction and Building Materials*, *27*(1), 521–530. Available from https://doi.org/10.1016/j.conbuildmat.2011.07.013.

Najim, K. B., & Hall, M. R. (2013). Crumb rubber aggregate coatings/pre-treatments and their effects on interfacial bonding, air entrapment and fracture toughness in self-compacting rubberised concrete (SCRC). *Materials and Structures/Materiaux et Constructions*, *46*(12), 2029–2043. Available from https://doi.org/10.1617/s11527-013-0034-4.

Neville, A. M. (2011). *Properties of concrete*. Pearson.

Obaton, A.-F., Klingaa, C., Rivet., Mohaghegh, K., Baier, s, Andreasen., Carli, L., & Chiffre. (2020). Reference standards for XCT measurements of additively manufactured parts. *e-Journal of Nondestructive Testing*, *25*(2). Available from https://doi.org/10.58286/25111.

Rahman, M. M., Usman, M., & Al-Ghalib, A. A. (2012). Fundamental properties of rubber modified self-compacting concrete (RMSCC. *Construction and Building Materials*, *36*, 630–637. Available from https://doi.org/10.1016/j.conbuildmat.2012.04.116, 09500618.

Raj, B., Ganesan, N., & Shashikala, A. P. (2011). Engineering properties of self-compacting rubberized concrete. *Journal of Reinforced Plastics and Composites*, *30*(23), 1923–1930. Available from https://doi.org/10.1177/0731684411431356.

Reschner, K. (2008). *Scrap tire recycling: A summary of prevalent disposal and recycling methods*. EnTire Engineering.

Ridgley, K. E., Abouhussien, A. A., Hassan, A. A. A., & Colbourne, B. (2018). Evaluation of abrasion resistance of self-consolidating rubberized concrete by acoustic emission analysis. *Journal of Materials in Civil Engineering*, *30*(8). Available from https://doi.org/10.1061/(ASCE)MT.1943-5533.0002402.

Roy, C., Labrecque, B., & de Caumia, B. (1990). Recycling of scrap tires to oil and carbon black by vacuum pyrolysis. *Resources, Conservation and Recycling*, *4*(3), 203–213. Available from https://doi.org/10.1016/0921-3449(90)90002-L.

Sabapathy, L., Mohammed, B. S., Al-Fakih, A., Wahab, M. M. A., Liew, M. S., & Amran, Y. H. M. (2020). Acid and sulphate attacks on a rubberized engineered cementitious composite containing graphene oxide. *Materials, 13*(14). Available from https://doi.org/10.3390/ma13143125.

Sadegzadeh, M., Page, C. L., & Kettle, R. J. (1987). Surface microstructure and abrasion resistance of concrete. *Cement and Concrete Research, 17*(4), 581–590. Available from https://doi.org/10.1016/0008-8846(87)90131-1.

Saint-Pierre, F., Philibert, A., Giroux, B., & Rivard, P. (2016). Concrete quality designation based on ultrasonic pulse velocity. *Construction and Building Materials, 125*, 1022–1027. Available from https://doi.org/10.1016/j.conbuildmat.2016.080.158.

Saleh, F. A. H., Kaid, N., Ayed, K., Kerdal, D.-E. , Chioukh, N., & Leklou, N. (2022). Influence of waste tyre rubber of different aggregate forms and sizes on the sustainable behaviour of self-compacting sand concrete in aggressive environment. *Journal of Rubber Research, 25*(2), 89–104. Available from https://doi.org/10.1007/s42464-022-00160-9.

Shaji, R., Ramkrishnan, R., & Sathyan, D. (2021). Strength characteristics of crumb rubber incorporated self-compacting concrete. *Materials Today: Proceedings, 46*, 4741–4745. Available from https://doi.org/10.1016/j.matpr.2020.100.306.

Siddique, R. (2019). *Self-compacting concrete: Materials, properties and applications* (pp. 1–411). India Elsevier Inc. Available from http://www.sciencedirect.com/science/book/9780128173695.

Singh, G., & Siddique, R. (2012). Abrasion resistance and strength properties of concrete containing waste foundry sand (WFS). *Construction and Building Materials, 28*(1), 421–426. Available from https://doi.org/10.1016/j.conbuildmat.2011.080.087.

Solís-Carcaño, R., & Moreno, E. I. (2008). Evaluation of concrete made with crushed limestone aggregate based on ultrasonic pulse velocity. *Construction and Building Materials, 22*(6), 1225–1231. Available from https://doi.org/10.1016/j.conbuildmat.2007.010.014.

Sukontasukkul, P. (2009). Use of crumb rubber to improve thermal and sound properties of pre-cast concrete panel. *Construction and Building Materials, 23*(2), 1084–1092. Available from https://doi.org/10.1016/j.conbuildmat.2008.050.021.

Sun, W., Brown, S. B., & Leach R. (2012). An overview of industrial X-ray computed tomography 32. National Physical Laboratory. http://eprintspublications.npl.co.uk/id/eprint/5385.

Taha, M.R., El-Dieb, A., & Abd El-Wahab. Fracture toughness of concrete incorporating rubber tire particles. International conference on performance of construction materials: A new era of building. 18–20 February, Cairo, Egypt (2003).

Takada, K. (2000). Part XII-Test method description. In Å. Skarendahl, & Ö. Petersson (Eds.), *Self-compacting concrete - State-of-the-art report of RILEM TC 174-SCC* (pp. 117–141). RILEM Publications SARL.

Thakare, A. A., Singh, A., Gupta, T., & Chaudhary, S. (2023). Effect of size variation of fibre-shaped waste tyre rubber as fine aggregate on the ductility of self-compacting concrete. *Environmental Science and Pollution Research, 30*(8), 20031–20051. Available from https://doi.org/10.1007/s11356-022-23488-6.

Tian, L., & Qiu, L. (2022). Preparation and properties of integrally hydrophobic self-compacting rubberized concrete. *Construction and Building Materials, 338*. Available from https://doi.org/10.1016/j.conbuildmat.2022.127641.

Topçu, I. B., & Bilir, T. (2009). Experimental investigation of some fresh and hardened properties of rubberized self-compacting concrete. *Materials and Design, 30*(8), 3056–3065. Available from https://doi.org/10.1016/j.matdes.2008.120.011.

Turatsinze, A., & Garros, M. (2008). On the modulus of elasticity and strain capacity of self-compacting concrete incorporating rubber aggregates. *Resources, Conservation and Recycling*, *52*(10), 1209–1215. Available from https://doi.org/10.1016/j.resconrec.2008.060.012.

Uche, O. A., Kelechi, S. E., Adamu, M., Ibrahim, Y. E., Alanazi, H., & Okokpujie, I. P. (2022). Modelling and optimizing the durability performance of self consolidating concrete incorporating crumb rubber and calcium carbide residue using response surface methodology. *Buildings*, *12*(4). Available from https://doi.org/10.3390/buildings12040398.

Uygunoğlu, T., & Topçu, İ. B. (2010). The role of scrap rubber particles on the drying shrinkage and mechanical properties of self-consolidating mortars. *Construction and Building Materials*, *24*(7), 1141–1150. Available from https://doi.org/10.1016/j.conbuildmat.2009.12.027.

Valizadeh, A., Hamidi, F., Aslani, F., & Shaikh, F. U. A. (2020). The effect of specimen geometry on the compressive and tensile strengths of self-compacting rubberised concrete containing waste rubber granules. *Structures*, *27*, 1646–1659. Available from https://doi.org/10.1016/j.istruc.2020.070.069.

Wanasinghe, D., Aslani, F., & Dai, K. (2022). Effect of age and waste crumb rubber aggregate proportions on flexural characteristics of self-compacting rubberized concrete. *Structural Concrete*, *23*(4), 2041–2060. Available from https://doi.org/10.1002/suco.202000597.

Wang, J., Dai, Q., & Si, R. (2022). Experimental and numerical investigation of fracture behaviors of steel fiber-reinforced rubber self-compacting concrete. *Journal of Materials in Civil Engineering*, *34*(1). Available from https://doi.org/10.1061/(ASCE)MT.1943-5533.0004010.

Werdine, D., Oliver, G. A., de Almeida, F. A., de Lourdes Noronha, Mirian, & Gomes, Guilherme Ferreira (2021). Analysis of the properties of the self-compacting concrete mixed with tire rubber waste based on design of experiments. *Structures*, *33*, 3461–3474. Available from https://doi.org/10.1016/j.istruc.2021.06.076.

Wittman, F. H. (1982). *Fundamental research on creep and shrinkage of concrete*. Springer. Available from https://doi.org/10.1007/978-94-010-3716-7.

Wójtowicz, M. A., & Serio, M. A. (1996). Pyrolysis of scrap tires: Can it be profitable? *Chemtech*, *26*(10), 48–53.

Yang, G., Chen, X., Guo, S., & Xuan, W. (2019). Dynamic mechanical performance of self-compacting concrete containing crumb rubber under high strain rates. *KSCE Journal of Civil Engineering*, *23*(8), 3669–3681. Available from https://doi.org/10.1007/s12205-019-0024-3.

Yang, G., Wang, J., Li, H., Yao, T., Wang, Y., Hu, Z., Jin, M., & Liu, J. (2022). Creep behavior of self-compacting rubberized concrete at early age. *Journal of Materials in Civil Engineering*, *34*(3). Available from https://doi.org/10.1061/(ASCE)MT.1943-5533.0004091.

Yazici, Ş., & Sezer, G. İ. (2007). Abrasion resistance estimation of high strength concrete. *Journal of Engineering Sciences*, *13*(1), 1–6.

Younis, K. H., Naji, H. S., & Najim, K. B. (2017). Cracking tendency of self-compacting concrete containing crumb rubber as fine aggregate. *Key Engineering Materials*, *744*, 55–60. Available from https://doi.org/10.4028/www.scientific.net/KEM.744.55.

Yung, W. H., Yung, L. C., & Hua, L. H. (2013). A study of the durability properties of waste tire rubber applied to self-compacting concrete. *Construction and Building Materials*, *41*, 665–672. Available from https://doi.org/10.1016/j.conbuildmat.2012.11.019.

Zaouai, S., Tafraoui, A., Makani, A., & Benmerioul, F. (2020). Hardened and transfer properties of self-compacting concretes containing pre-coated rubber aggregates with crushed dune sand. *Journal of Rubber Research*, *23*(1), 5–12. Available from https://doi.org/10.1007/s42464-019-00030-x.

Zhang, J., Chen, C., Li, X., Chen, X., & Zhang, Y. (2021). Dynamic mechanical properties of self-compacting rubberized concrete under high strain rates. *Journal of Materials in Civil Engineering*, *33*(2). Available from https://doi.org/10.1061/(ASCE)MT.1943-5533.0003560.

Zhuang, J., Xu, R., Pan, C., & Li, H. (2022). Dynamic stress–strain relationship of steel fiber-reinforced rubber self-compacting concrete. *Construction and Building Materials*, *344*. Available from https://doi.org/10.1016/j.conbuildmat.2022.128197.

Sustainable cementitious composites with recycled aggregates and fibers

Hocine Siad[1], *Mohamed Lachemi*[1], *Mustafa Sahmaran*[2], *Maziar Zareechian*[1] and *Waqas Latif Baloch*[1]
[1]Department of Civil Engineering, Toronto Metropolitan University, Toronto, ON, Canada,
[2]Department of Civil Engineering, Hacettepe University, Ankara, Çankaya, Turkey

9.1 Introduction

It is crucial to address the urgency of the current environmental crisis, particularly with regard to global warming and resource depletion facing the planet (Pierrehumbert, 2019). Among the contributors to this environmental degradation, concrete production stands out prominently. Cementitious materials, as the fundamental building ingredients, have retained their dominance in the construction industry since the early 1900s and continue to play a pivotal role in shaping modern society. Indeed, the present annual consumption of concrete exceeds 10 billion tones (Mahmoodi et al., 2021), surpassing the combined utilization of various construction materials, including steel, plastic, wood, and aluminum, by more than double. This is also shown to be responsible for the release of around 8% of the overall carbon dioxide (CO_2) emissions (Dadsetan et al., 2022), exerting a substantial negative impact on the global greenhouse effect.

In addition to Portland cement (PC) the extraction of raw materials necessary for concrete production also has a significant detrimental impact on the undesirable environmental burden of concrete. Sand and gravel, accounting for approximately 60%–75% of the overall concrete volume, serve as key raw materials in the majority of concrete compositions (Marinković et al., 2023). They are usually derived from natural sources through various methods and mining stages. These typically involve site preparation, followed by excavation of the aggregate and subsequent crushing and processing stages. Although aggregate production typically has milder environmental consequences compared with PC, it still presents significant risks such as the potential conversion of agricultural lands into temporary excavation sites, the generation of excessive noise during preparation activities, and the emission of dust leading to a decline in air quality. It should be emphasized that the magnitude of these

Sustainable Concrete Materials and Structures. DOI: https://doi.org/10.1016/B978-0-443-15672-4.00009-7

impacts can worsen when mining operations are carried out in areas where the stability of the surrounding rocks is already compromised, especially in regions prone to landslides and/or unstable slopes. Furthermore, it is estimated that during the production of 1 ton of natural aggregate (NA), approximately 20 kg of CO_2 is released as a result of various processes such as quarrying, crushing, and transportation of the aggregates (Marinković et al., 2023). Consequently, as for PC, significant efforts have been dedicated to addressing the environmental concerns associated with traditional aggregates used in concrete production. One of the encouraging ecological methods involves reutilizing accumulated construction and demolition wastes (CDWs), where concrete, attached mortar, bricks, and ceramics constitute over 70% of the total weight of these wastes (Mahmoodi et al., 2023). The resulted recycled aggregates (RAs) from crushing and sieving of CDWs have been widely utilized to partially or completely substitute NA. This practice has been more successful in the case of concrete waste, which was incorporated into numerous construction projects worldwide. In addition to the use of RAs several industrial and agricultural waste materials, such as plastic wastes, rubber tires, leather by-products, paper mill wastes, industrial sludge, mine tailings, slag, fly ash, as well as coconut shells (CSs) and oil palm shells (OPSs), were served as alternatives to NA in concrete production (Kishore & Gupta, 2020).

While there has been significant research focused on making cement more environmental friendly and utilizing RA, insufficient emphasis has been placed on investigating the possibilities of utilizing recycled industrial fibers in the concrete compositions (Anandamurthy et al., 2017). However, recognizing the possibility of recycled fibers in converting wastes into useful materials, an increasing number of scholars have recently started investigating the use of recycled waste materials as substitutes for conventional fibers in fiber-reinforced concretes (FRCs). Conventional synthetic fibers improve the performance of concrete but are derived from nonrenewable and costly sources. Furthermore, these fibers lack biodegradability and contribute to waste generation and adverse environmental impacts when disposed of Li et al. (2023). The available research thus far suggests that recycled fibers obtained from different waste sources can be utilized in the manufacturing of FRCs and provide cost benefits in comparison to traditional fibers (Merli et al., 2020; Siad et al., 2019).

This chapter aims to provide a critical, comprehensive, and up-to-date evaluation of the cleaner production of cementitious composites incorporating RAs and recycled fibers, with a particular emphasis on the research gaps concerning the classification and optimization of these valuable components. The first part focuses on the evaluation of various types of recycled waste materials as aggregates and their impact on both the fresh and hardened properties of concretes. The parameters that influence the recycling efficiency of wastes as aggregates, as well as the approaches for enhancing their performance, are also examined. In the second phase this chapter delves into the type and properties of recycled fibers utilized in cementitious composites, examining their reported impacts on cementitious composites. Lastly, a throughout discussion of the advantages and disadvantages of incorporating these materials as well as future opportunities for research is provided.

9.2 Waste recycled aggregates

9.2.1 Current literature background on recycled aggregates in cementitious composites

Fig. 9.1 shows the use of RA in the literature. Among the keywords related to concrete properties, important ones include mechanical properties, durability, water absorption, sustainable development, and microstructure. This suggests that a considerable portion of studies on RA has focused on recycled concrete aggregate (RCA), while the utilization of recycled plastics or agricultural waste has received relatively less attention compared with CDW materials.

9.3 Influential properties of waste recycled aggregates in concrete

It is to highlight that both the origin of waste materials and the process of crushing waste recycled aggregates (WRAs) significantly influence their physical (including apparent density, particle size distribution, pore structure, and surface morphology), chemical, and mechanical characteristics (Kim, 2022; Kumar & Singh, 2023). These factors play a crucial role in determining the impact on the fresh and

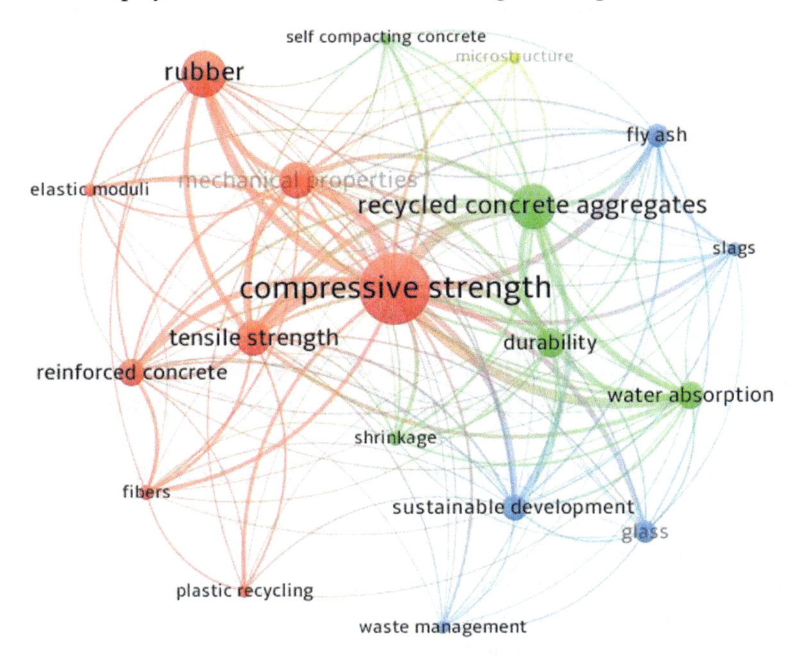

Figure 9.1 The networks' diagrams of the most existed literature about recycled aggregates. This figure illustrates the use of recycled aggregate in the literature.

hardened properties of WRAs in concretes. CDW-based aggregates typically exhibited lower densities and higher water absorption compared with NAs due to the presence of porous particles and adhesive mortars/pastes in their composition. However, the type of CDW was also reported to highly influence these data, especially when waste brick is present with or without waste concrete and ceramics. In addition to the possible attached mortar the production of brick involves high-temperature calcination of clay, resulting in the formation of rough and porous surfaces of bricks (Li, Pei, et al., 2019). In the case of RA from waste concrete the attached mortar layer can occupy different portions, sometimes over 35% of the RA's total volume (Zhan et al., 2022). The presence of voids is a characteristic feature of the layer of old mortar found in RA, primarily composed of cement paste attached to loose sand particles. The amplified occurrence of this layer has been demonstrated to have a notable impact on augmenting the water absorption capacity of the RA (Zhan et al., 2022). Thus coarse RA tends to have a higher water absorption capacity compared to that of fine RA, since the proportion of adhered mortar decreases as the nominal size of the aggregate increases. However, research studies have indicated that, in general, both coarse and fine RAs can reach more than 12 times higher water absorption ratio than NA (Guo et al., 2020). Glass waste, which can be from CDW, residential or industrial sources, has typically less specific gravity than normal sand (NS), with an average of 2.2 for waste glass as compared to 2.6–2.7 for NA (Jani & Hogland, 2014). In addition, it was shown with water absorption of around 14% lower than NA (Jani & Hogland, 2014). Rubber and plastic waste aggregates share common characteristics such as being lightweight with lower specific gravity and water absorption than NA (Li et al., 2020). Similarly, both the agriculture waste materials of CS and OPS aggregates have higher water absorption than all other WRAs because of their high surface porosity, however, with lower specific gravity, in the range of 1.12–1.56 for CS and 1.17–1.4 for OPS (Chinnu et al., 2021). It is mentioned that the water absorption and specific gravity of NA were reported in the range of 2.6–2.7 and 1.3%–1.8%, respectively (Li et al., 2020).

9.4 Assessment of the effect of recycled aggregates on the fresh properties of cementitious composites

9.4.1 Recycled construction and demolition waste aggregate

RCAs present inherent challenges due to their high porosity and heightened absorption capacity, and therefore the rheological properties and workability of the incumbent mixes are adversely affected by these attributes. There is also a concern over the optimal volumetric quantity of RCA in concrete. If used in excess, potential issues may include rapid slump loss, fluidity, stability maintenance, and complexity of rheological behavior, which all contribute to early crack initiation in the incumbent mixes (Duan & Hou, 2023). For instance, self-compacting concrete (SCC) containing fine RCA was evaluated for its rheological attributes, revealing its suitability for optimal passing and filling capacities up to 20% RCA (Carro-López et al., 2015). Nevertheless, exceeding half the NA replacement ratio adversely impacted SCC flowability and workability.

Cementitious matrices are also strongly affected by aggregate morphology and composition. In particular, aggregates derived from crushed recycled concrete have been found to enhance surface area at the expense of workability (de Andrade Salgado et al., 2022). On the other hand, smooth and rounded aggregates produce superior flowability and workability in concrete. Presently, in concrete mixes superplasticizers (SPs) are used to achieve certain slump or flowability. However, the presence of fine RCA in concrete formulations complicates the performance of SP, preventing the achievement of desired rheological characteristics (Silva et al., 2018). As a result, a blend of SP, inhibitor, and retarder has been suggested, which was more compatible with fine RCA than the sole use of conventional polycarboxylate-based SP, as shown in Fig. 9.2 (Cartuxo et al., 2015; Li et al., 2021).

It is important to note that the higher water absorption capability of RCA can lead to water absorption from the cement paste, thus affecting workability and altering the precise control of the water-to-cement ratio (de Andrade Salgado & de Andrade Silva, 2022). In experimental investigations comparing various RCA types under different environmental conditions concrete mixtures exposed to drier surroundings displayed a reduction in slump and dynamic yield stress, while exhibiting an accelerated increase in plastic viscosity and static yield stress over 45 minutes (Duan & Hou, 2023). Although presaturation in water was found to improve the RCA-concrete workability, particularly for coarse RCAs, the response of recycled concrete is further affected by RCA's presaturation time and conditions. While the optimum presaturation time was 7 days or longer, adding a quantity of water equal to the water absorption +5% of RA and using it after 24 hours was found to cause better workability (Nedeljković et al., 2021). For an accurate assessment of RCA-concrete workability,

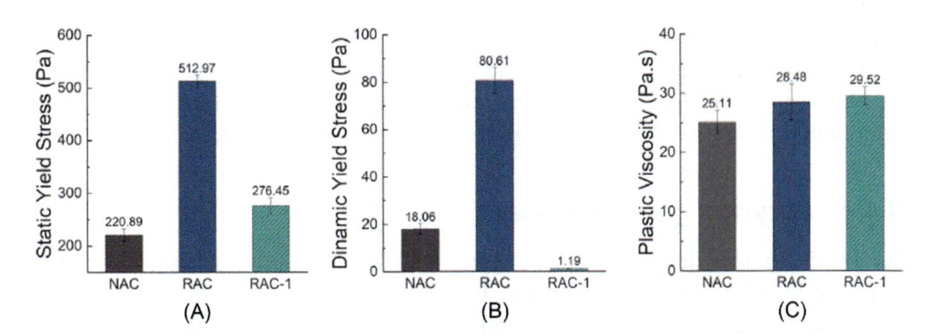

Figure 9.2 The results of rheological parameters of concrete with ordinary recycled aggregates concrete (RAC) and developed (RAC-1) polycarboxylate superplasticizer.
Source: From Li, B., Hou, S., Duan, Z., Li, L., & Guo, W. (2021). Rheological behavior and compressive strength of concrete made with recycled fine aggregate of different size range. *Construction and Building Materials, 268*, 121172.

no matter the SP type used, it is inadequate to rely solely on slump measurements and slump flow (Joseph et al., 2023; Ray et al., 2021). In spite of the fact that slump tests are a common empirical method for evaluating the workability of RCA concrete detailed studies of their rheological properties remain limited.

Recycled waste ceramic aggregates (RWCAs) were also shown by most authors to pose significant challenges due to their inherent characteristics, including elevated porosity, moisture absorption, angular shape, and rough surface texture. RWCA-incorporated concrete exhibits a notable reduction in workability due to these factors. Compromised workability is further worsened by adhered mortars, impurities, and cement and aggregate substitutions. Water absorption, bulk density, and overall density of concrete are also significantly correlated with RWCA content. This is due to the lower density of RWCAs compared with conventional aggregates. To compensate for the diminished workability, it is imperative to incorporate plasticizing admixtures (Juan-Valdés et al., 2021). The higher the amount of RWCAs in the concrete mixture, the greater the water demand, and the greater the risk of water loss through bleeding when SPs are added in higher dosages (Zareei, Ameri, Shoaei, et al., 2019). It should be noted, however, that the workability of concrete mixes incorporating both fine and coarse RWCAs varies from moderate to high, whereas it substantially declines when a complete substitution of coarse RWCAs is used (Kumar & Singh, 2023). As a consequence of their glazed surface, angular shape, and increasing porosity, coarse RWCAs require more water than fine aggregates (Ray et al., 2021). To maintain the desired workability, the water-to-cement ratio must be adjusted taking RWCA's inherent water absorption capacity into consideration (Pitarch et al., 2019). By the use of supplementary cementitious materials (SCMs) such as fly ash, flowability has been shown to increase in the RWCA concrete (Meena et al., 2022). Since, fly ash particles may improve frictional bonding between RWCA and concrete, due to their smooth and spherical surfaces.

Recycled waste brick aggregates (RWBAs) are also known to absorb more water and are more porous than conventional aggregates, which result in a noticeable reduction in fresh density and workability in RWBA concrete (Wong et al., 2018). An increase in dynamic yield stress and plastic viscosity is observed over time with an increase in RWBA content in the concrete mixture. Prewetting of RWBAs is often used to improve the fresh-mix properties (Duan & Hou, 2023).

9.4.2 Recycled glass aggregate

Incorporating recycled waste glass aggregate (RWGA) into concrete has proven effective in enhancing its workability. Potentially, this positive effect is caused by the weaker cohesion between the smooth exterior surface of RWGA and the cement matrix (Omoding et al., 2021; Qaidi et al., 2022). By incorporating RWGA, high workability can be achieved with little or no extra water or SP, an attractive opportunity for applications that require highly workable mixes. Nevertheless, the particle size distribution of RWGAs can highly govern their positive effect on the workability of cementitious composites. The coarser the RWGA, the higher the workability

(Chandra Paul et al., 2018). In addition, some other investigations indicated reduced workability at increased RWGA content. According to Qaidi et al. (2022), this behavior is found principally when the glass particles have an angular shape with a high aspect ratio. The fresh density of RWGA concrete was stated to decrease as the RWGA proportion increased in concretes (Mohajerani et al., 2017). NAs have a higher specific gravity and density than RWGA, which explains the density decrease, even though literature also demonstrates contrasting results based on the type of RWGA (Liu et al., 2020). Thus for RWGA to be used to its full potential in concrete applications, further research and experimentation are required, particularly in workability-critical projects.

9.4.3 Recycled rubber aggregate

Most of the previous research has observed a decrease in concrete density, with reductions of around 10%–30% based on the incorporation level of recycled rubber aggregates (RRAs) between 10% and 40% (Alyousef et al., 2021; Asutkar et al., 2017; Mei et al., 2022). Although this decrease in density was shown to be mainly related to the RRA bulk density, reported at least two times lower than NA (Alyousef et al., 2021), its particle size and substitution method are highly influential in achieving lightweight RRA-based concretes. Also, most experiments about RRA concretes have shown a decreased workability at increased RRA amounts, regardless of RRA size (Ram Kumar et al., 2019). The substitution of normal aggregates with RRA at 20%–100% resulted in slump decrements between 19% and 93% (Siddika et al., 2019), with a higher effect when coarse RRA is included (Fig. 9.3; Mahmood & Kockal, 2020). The main reason for the total reduction is because of the higher water absorption of RRA than NA, which can be overcome by improving the particle size distribution and its surface absorbent characteristics (Mei et al., 2022). Increasing the amount of SP can be another solution to improve the workability of RRA-cementitious composites, though an optimized concentration is necessary to avoid bleeding.

9.4.4 Recycled plastic aggregate

A number of plastics have been investigated as aggregates in concrete, including polystyrene (PS), polyethylene (PE), polypropylene (PP), polyolefin (PO), polyethylene terephthalate (PET), high-density polyethylene (HDPE), polyvinyl chloride (PVC), expanded polystyrene (EPS), and low-density polyethylene (LDPE). The general trend was presented to cause a discernible decrease in density, leading also to a substantial reduction in workability (Alyousef et al., 2021; Colangelo et al., 2016). However, plastic aggregate shapes were confirmed to influence this result, since the angularity of aggregates decreased slump values, whereas the sphericity and regular shape particles showed improved flowability (Mahmood & Kockal, 2020). The type of recycled plastic aggregates (RPA) can also affect the general tendency of results. When the feasibility of using three types of RPA of PET, HDPE, and PP were investigated, although the higher replacement of all of the

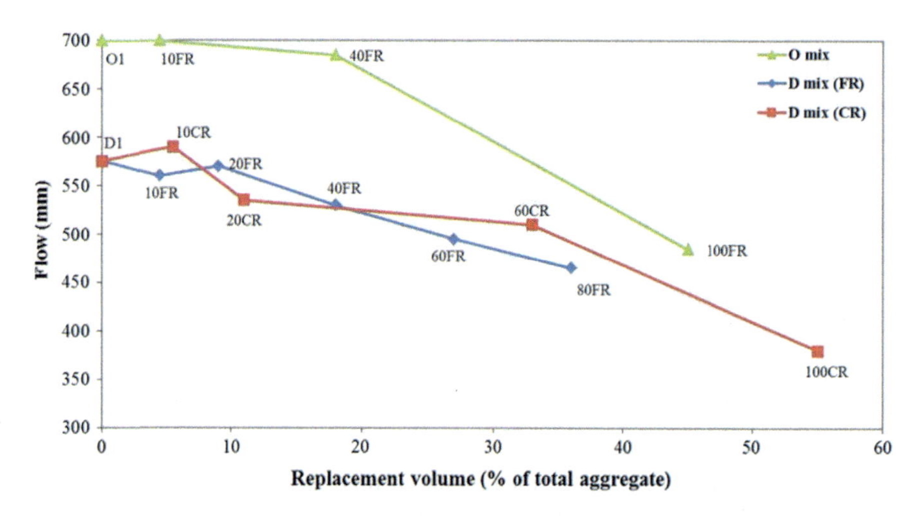

Figure 9.3 Workability results of the mixtures containing fine rubber and coarse rubber as function of the total aggregate volume replacement. This figure shows the effect of using fine and coarse rubber aggregates on the workability results of concrete compositions.
Source: Modified from Zareei, S. A., Ameri, F., Bahrami, N., Shoaei, P., Musaeei, H. R., & Nurian, F. (2019). Green high strength concrete containing recycled waste ceramic aggregates and waste carpet fibers: Mechanical, durability, and microstructural properties. *Journal of Building Engineering, 26*. https://doi.org/10.1016/j.jobe.2019.100914.

types of RPA led to lower workability, the enhanced difference was noticed for PET than HDPE and PP. Due to the flaky morphology of PET, researchers found that an increased concentration of PET waste may also adversely affect concrete's workability (Abu-Saleem et al., 2021). It should be noted that RPA naturally has a smoother surface texture than NAs, and develops a weaker bond with the host matrix in fresh state. It is possible, however, to enhance the RPA/matrix bond through surface modification techniques such as granulation or foaming, giving rise to promising possibilities for improving the overall strength of the bond which might also improve both fresh and hardened properties (Alyousef et al., 2021).

9.4.5 Agricultural waste aggregate

Agricultural waste materials, such as OPSs and CSs, have been investigated extensively in the development of sustainable concretes, with corn cob, rice husk, and date seed considered in a limited number of publications. Agricultural by-products appear to have a significant impact on fresh-mix properties such as workability and density. Concrete density is significantly reduced when OPS and CS are substituted for NAs, underscoring their potential for the development of lightweight concrete (Haddadian et al., 2023). In general, OPS enhances concrete flowability, especially at lower replacement percentages. OPS content above the optimum level (between 20% and 30% according to Chinnu et al., 2021), however, adversely affects concrete workability.

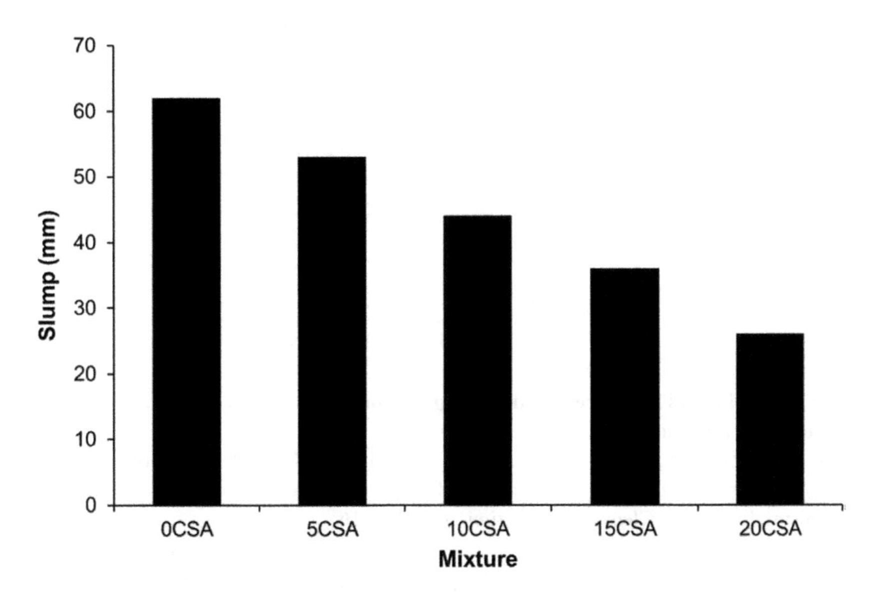

Figure 9.4 Slump of concrete compositions with higher coconut shell aggregates.

Source: From Ong, S. W., Lau, T. L., Yeong, T. W., Anwar, M. P., & Elleithy, W. (2018). A feasibility study on partially substituted coarse aggregate with oil palm shell in coconut fiber reinforced concrete. *International Journal of Engineering &Technology, 7*, 62.

In opposite findings it was reported that the increased percentages of OPS result in stiffer concrete and lower slump values. This is primarily due to the finer particle size and larger surface area of palm oil fuel ash compared with sand, leading to increased water requirement and decreased workability (Khankhaje et al., 2016). Thus the effect of OPS aggregates on the workability of concretes is dependent on their physical characteristics, such as particle shape, porosity after crushing, and water absorption (Chinnu et al., 2021). Similarly, contradicting results were shown for CS aggregates. While the CS aggregates were found to enhance the concrete workability because of their smooth-textured surfaces (Fanijo et al., 2020), they also have been stated to highly decrease the slump due to their higher water absorption, as shown in Fig. 9.4 (Ong et al., 2018). Consequently, maintaining satisfactory workability for OPS and CS composites requires careful consideration of optimum replacement levels, particle size, and surface quality of these agricultural aggregates.

9.5 Assessment of the effect of recycled aggregates on the hardened properties of cementitious composites

9.5.1 Recycled construction and demolition waste aggregate

The composition, texture, and structure of the aggregates profoundly influence the strength properties of concrete. Fig. 9.5 shows the formation of a triple interfacial

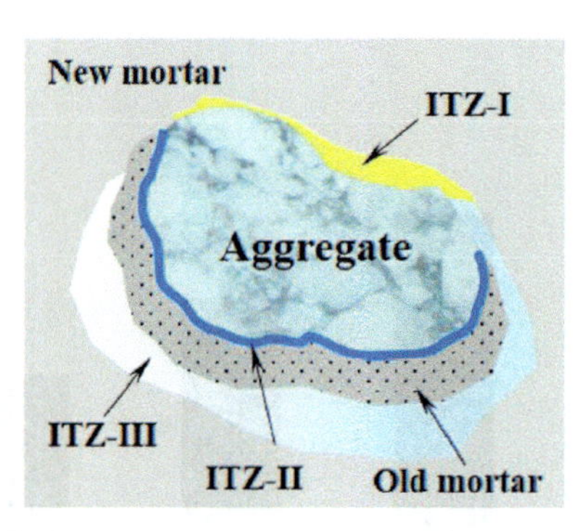

Figure 9.5 Triple ITZs in recycled aggregate concrete: ITZ-I, ITZ-II, and ITZ-III.
ITZ, interfacial transition zone
Source: From Zhan, P., Xu, J., Wang, J., Zuo, J., & He, Z. (2022). A review of recycled aggregate concrete modified by nanosilica and graphene oxide: Materials, performances and mechanism. *Journal of Cleaner Production, 375*, 134116.

transition zone (ITZ) in concrete following RA inclusion. The first ITZ develops between aggregates and freshly laid mortar. A similar transition zone forms at the interface between aggregates and old mortar. Also, an additional interface is developed between the old and new mortars. Under applied stresses, load-bearing patterns may vary between ITZs, a fusion that is principally responsible for the extended weak bonding mechanism, and consequently reduced CDW concrete strengths (Zhan et al., 2022).

It is, however, imperative to note that the efficiency of RCA varies significantly based on the type of aggregates being repurposed. Concrete waste exhibits high productivity due to its uniformity and predictable mechanical properties. In contrast, RWBA and RWCA may vary in size and shape, affecting their recycling efficiency. Even so, angular and rough-shaped CDW aggregates may provide enhanced concrete mechanical properties due to their ability to refine the microstructural bonds with the cement paste. This was also found to be beneficial for concrete durability since it reduces freeze–thaw damage, alkali-silica reactions (ASRs), and other durability properties (Mohanta & Murmu, 2022). Other factors that determine the strength of the host concrete include RCA content and size, water absorption capacity of RCAs, sourced concrete strength and water-to-cement ratio, and eventual surface quality/cleanness (Chen et al., 2023; Tran, Dang, et al., 2022). The increased RCAs and related higher effective water-to-cement ratio can lead to a poor connection between old and new mortar (ITZ interaction), which can negatively affect concrete's compressive strength (Bidabadi et al., 2020). Compressive strength can be worsened by using coarser RCAs (Nedeljković et al., 2021), though it was possible to achieve greater performance at coarser RCA particles by implementing an intensive crushing process to remove the previously attached paste/mortars (Rashid et al., 2020). The presence of impurities and irregularities within RAs may result in a reduction in strength for composites, particularly if a higher amount of RAs is used. CDW concrete can show increased degradation with enhanced age CDW aggregates due to the possible presence of contaminants (Mohanta & Murmu, 2022). The higher water content necessary for flowability in concrete mixes with RCAs was commonly presented to adversely affect the RCA/paste bond and associated compressive strength (Behera et al., 2019). Recycled concrete strength is, however, solely determined not only by aggregate properties but also by the properties of paste surrounding the CDW aggregates, such as the type of cement, the volume, and the presence of SCMs (Da Silva & Andrade, 2022).

As for durability, concrete using RCAs has shown higher shrinkage during drying and high chloride penetration, while freeze–thaw and sulfate attack resistance was similar/or greater than concrete using NAs (Guo et al., 2018; Nedeljković et al., 2021). Consequently, shrinkage cracks can form during the curing process when high paste/mortar is attached to RCAs (Talamona & Hai Tan, 2012). Similarly, the filler effect and high porosity of RCAs were described to increase the chloride penetration compared with NA (Yildirim et al., 2015). Freezing–thawing performances were according to the specific porous microstructure, water absorption capacity, and the amount of old mortar attached to the RCAs. Thus based on the RCA content, some studies reported comparable freeze–thaw performance than conventional concrete and others reported improved performance to natural concrete (Evangelista & De Brito, 2019; Guo et al., 2020). The high-water absorption of RCA was beneficial for the concrete resistance under sulfate, though up to an optimum content that has to be determined for concretes with different RCA characteristics (Nedeljković et al., 2021).

The feasibility of using ceramic waste as a substitute for both fine and coarse aggregates without negatively affecting the mechanical strengths of concrete has been confirmed in many studies (Magbool, 2022). Specifically, fine RWCAs can replace natural sand completely in concrete by seamlessly integrating it into the mix. Also, the mechanical properties of concrete produced by pulverizing RWCAs were further enhanced (Ray et al., 2021). The improved strengths were explained by the greater RWCA/paste interconnection due to the pozzolanic activity, angular shape, and rough surfaces of RWCA particles. The optimal replacement level for coarse RWCAs is generally found to be between 20% and 30% (Ray et al., 2021). There is a gradual decline in compressive strength beyond this range due to irregularities in the ceramic waste aggregates, which reduced the effective interfacial bonding with the concrete matrix.

It has been demonstrated that concrete mixtures containing higher RWCA contents are more susceptible to capillary absorption (Senhadji et al., 2019). Increasing water absorption is attributed to the RWCA effect on enhancing the capillary porosity, particularly at coarse RWCAs/paste. Furthermore, as shown in Fig. 9.6 (Medina et al., 2016), even at optimum content, RWCA's addition increased the chloride penetrability of recycled concrete, though lower chloride penetration was found when using finer RWCAs, reflecting also minimal corrosion risk than NA (Ray et al., 2021). Similar to chloride penetrability, the porous bond with the surrounding paste resulted in lower freeze/thaw resistance at higher RWCA mount in concrete (Magbool, 2022).

As a consequence of its pronounced porosity, RWBA was found by most authors to negatively influence mechanical strengths and increase the water absorption and

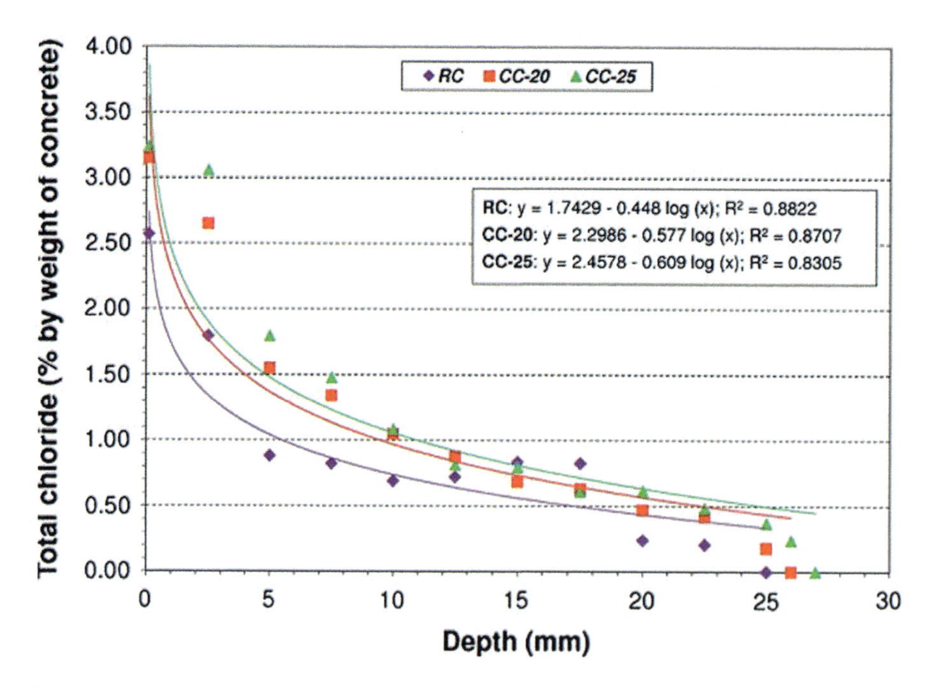

Figure 9.6 Chloride penetrability results for the mixture with 20% (CC-20) and 25% (CC-25) compared with the reference mixture (RC).

Source: Modified from Zareei, S. A., Ameri, F., Bahrami, N., Shoaei, P., Musaeei, H. R., & Nurian, F. (2019). Green high strength concrete containing recycled waste ceramic aggregates and waste carpet fibers: Mechanical, durability, and microstructural properties. *Journal of Building Engineering, 26*. https://doi.org/10.1016/j.jobe.2019.100914.

permeability. This was registered even when fibers, such as steel fiber in Fig. 9.7, were included (Ji et al., 2022). However, some studies revealed that the effect was negligible up to an optimum RWBA content between 10% and 20%, as the potential pozzolanic activity of fine RWBAs may have helped offsetting the negative influence of its surface defects (Wong et al., 2018). The increased RWBA replacement has also been reported to increase chloride penetration and carbonation of recycled concretes (Mohanta & Murmu, 2022). In contrast to conventional concrete RWBA concrete demonstrated reduced drying shrinkage and enhanced resistance to freeze–thaw cycles due to its enhanced porosity (Zhu & Zhu, 2020).

9.5.2 Recycled glass aggregate

Research that examined the effects of using recycled glass waste aggregates (RGWAs) in concrete agreed about the reduced compressive strength at increased size and amount of RGWAs (Qaidi et al., 2022). As glass particles have smooth surfaces and sharp edges, their ITZ with mortar is weakened. As well, with RGWA

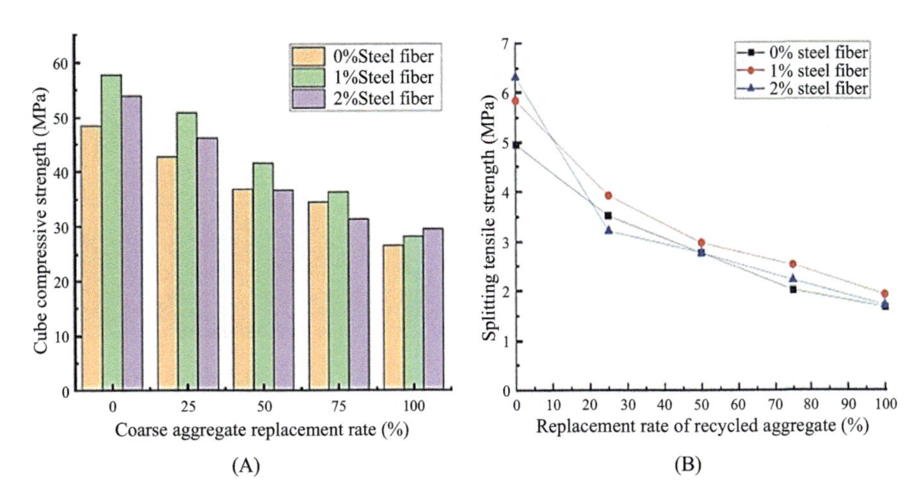

Figure 9.7 The outputs of (A) compressive strength and (B) tensile strength of the recycled waste brick aggregate-containing mixtures. This figure shows the effect of coarse recycled aggregates on the compressive and tensile strengths of steel fibers reinforced concrete.
Source: Modified from Zareei, S. A., Ameri, F., Bahrami, N., Shoaei, P., Musaeei, H. R., & Nurian, F. (2019). Green high strength concrete containing recycled waste ceramic aggregates and waste carpet fibers: Mechanical, durability, and microstructural properties. *Journal of Building Engineering, 26*. https://doi.org/10.1016/j.jobe.2019.100914.

content and particle size increasing, the negative effect of RGWA becomes more pronounced in relation to the expanded risk of ASR. An increase in microstructural cracks and voids has also been reported at increased RGWA contents, specifically when poor gradation and/or larger particle size of glass particles is included (Fig. 9.8; Liu et al., 2022; Omoding et al., 2021). Although some studies have presented contrasting results, most investigations have found that RGWA can improve the mechanical strength of concrete when fine particles are used, particularly at lower than 1.18 mm size (Khan et al., 2020). The pozzolanic reaction of fine RGWA has been identified as an essential contributor factor in addition to its angular shape that helped enhancing the ITZ (Khan et al., 2020). An optimum range of 15%−30% was mentioned, though the replacement of fine RGWA was possible up to 100% (Mohajerani et al., 2017; Siad et al., 2017). This effect was further improved when treated roughened RGWA surfaces were utilized, whereas the effect of color was negligible in most research (Khan et al., 2020). While most of the available studies agreed about the ASR risk related to the higher particle size of RGWA, there are discrepancies about the maximum size that can be used safely in cementitious composites. RGWA coarser than 600 μm can generate ASR problems (Siad et al., 2017), as a result of the chemical reaction between their amorphous silica content and alkalis (Na^{+}, K^{+}, and Ca^{2+}) in cementitious networks. ASR can first cause expansion due to the formation of a hygroscopic gel with the property to absorb water and enlarge, then cracks might develop and deteriorate the in-service

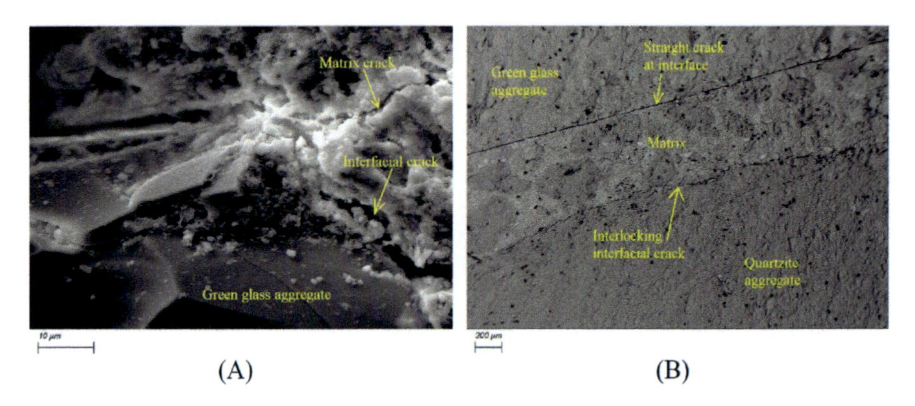

(A) (B)

Figure 9.8 Glass aggregate-cement paste interface: (A) 10 μm and (B) 200 μm. This figure presents the glass aggregate-cement paste interface properties.
Source: From Omoding, N., Cunningham, L. S., & Lane-Serff, G. F. (2021). Effect of using recycled waste glass coarse aggregates on the hydrodynamic abrasion resistance of concrete. *Construction and Building Materials, 268*, 121177.

performance of cementitious composites. The addition of RGWA to concrete has shown promise in enhancing its resistance to capillary water absorption, chloride penetration, and sulfate attack, especially at lower than 600 μm particle size and 30% content (Danish et al., 2023; Liu et al., 2018).

Glass aggregate-cement paste interface properties: (1) 10 μm and (2) 200 μm (Omoding et al., 2021).

9.5.3 Recycled rubber aggregate

As a result of their lower strength and stiffness than traditional aggregates, RRAs can adversely affect the mechanical properties of the incumbent matrix. Numerous factors, including the amount, surface roughness, size, and shape of RRA, can influence this reduction. Particularly, larger particle sizes, smoother surfaces, higher quantities, and irregular shape RRA resulted in a greater decrease in compressive strength (Angelin et al., 2020). Also, while, the inclusion of rubber was anticipated to act as an obstacle in the growth of cracks, RRA had limited effectiveness in bridging the tensile fractures when used with sizes ranging from 1.18 to 2.36 mm (Guo et al., 2017). Naturally, the brittleness of RRA/cement interface increases its susceptibility to fracture paths under flexural loads. Consequently, it is generally recommended to incorporate RRA in higher proportions only in nonstructural applications, and where low strength requirements prevail. A significant impact was also found on the overall properties of concrete with regard to the size of the RRA particles. It was possible to improve the mechanical properties by refining the microstructure on the use of finer RRA, due mostly to its filler effect. The finer RRA particles also caused a more effective reduction in water absorption capacity, chloride ion penetration, and freeze–thaw, whereas it was important to limit the content to around 5% when a maximum size of 5 mm was incorporated (Li, Zhang, et al., 2019). Similarly, an enhanced RRA particle size was linked with a higher drying shrinkage in concretes (Siddika et al., 2019).

9.5.4 Recycled plastic aggregate

A range of factors controlled the effect of recycled plastic waste aggregates (RPWAs) on the mechanical strengths of cementitious composites, including RPWA's amount, type, size, shape, and surface properties. In general, the increased quantity and particle size of RPWA reduced the compressive and flexural strengths significantly. This was connected to the repelled water and related lower adhesion between aggregate and mortar, as well as the smooth surface layer as a result of the reduced formation of reaction products on the plastic aggregates (Li, Zhang, et al., 2019). Thus to mitigate the significant strength loss, it was recommended to limit the amount and size of plastic aggregates to less than 20% and 2 mm, respectively (Alyousef et al., 2021). The spherical angular shape of RPWA was also found to reduce its negative effect on the strength decrement as compared to that flat angular form (Li et al., 2020). No studies compared the effect of different RPWA types; however, in a recapitulating review by Alyousef et al. (2021), PP delivered lower strength reduction after the heat-treated shaped PET (Fig. 9.9). In another review paper by Sharma and Bansal (2016) they revealed that the strengths of PP aggregates were greater than those of equivalent contents of shredded PET particles. However, Abu-Saleem et al. (2021) found that the use of PET from 10% to 30% caused larger strengths than PP and HDPE aggregates, in direct relation to the compressive yield strength of these materials.

Extensive research has also revealed that NA substitution by RPWA is associated with increased entrapped air within the matrix, leading to enhanced drying shrinkage, porosity, and water absorption properties of concretes, regardless of their particle shape (Almeshal et al., 2020). Further, this also compromises the chloride resistance, although the results remained in general within a moderate range compared with NA (Medina et al., 2016). Intriguingly, the introduction of plastic waste has proven to increase the electrical resistance of RPWA composites, mainly due to its insulation properties (Faraj et al., 2020). However, the optimum needs to be confirmed for each RPWA since conflicting results are reported about the slope of increment/reduction.

9.5.5 Agricultural waste aggregate

Although they have been widely incorporated to produce lightweight concretes, agricultural by-products, such as OPS, CS, corn cob, rice husk, and date seed waste into concrete matrices negatively affected their mechanical properties (Chinnu et al., 2021; Shafigh et al., 2014). These phases weaken the concrete microstructure, resulting in poor performance as a result of the insufficient interfacial bonding with the host matrix, especially at higher replacement levels. Yet, CS-based concretes demonstrated significantly superior performance than those containing OPS

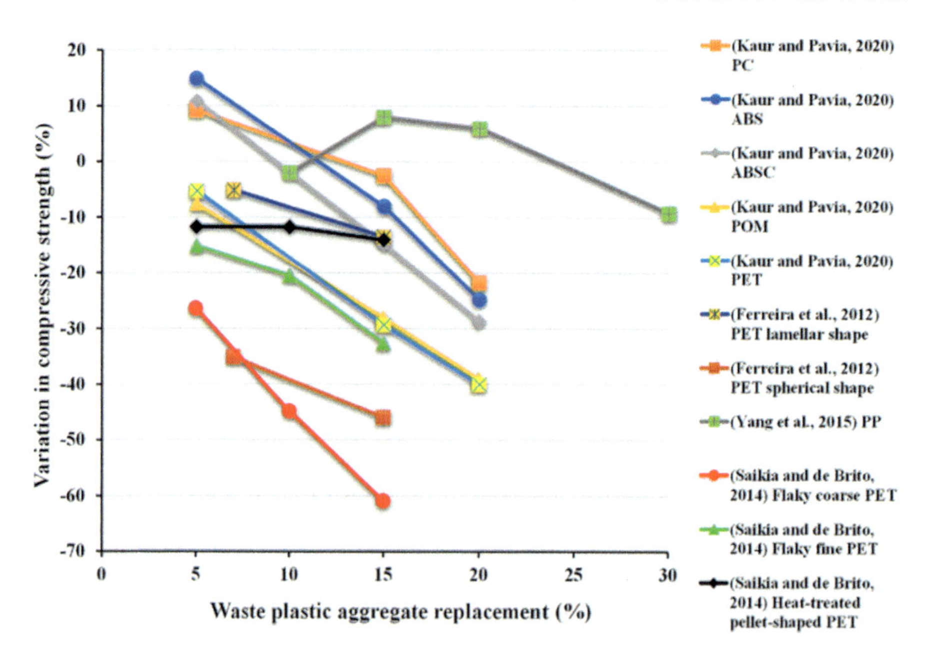

Figure 9.9 Influence of recycled plastic waste aggregate replacement on the 28-day compressive strengths as compared to natural aggregates.

Source: Modified from Zareei, S. A., Ameri, F., Bahrami, N., Shoaei, P., Musaeei, H. R., & Nurian, F. (2019). Green high strength concrete containing recycled waste ceramic aggregates and waste carpet fibers: Mechanical, durability, and microstructural properties. Journal of Building Engineering, 26. Available from https://doi.org/10.1016/j.jobe.2019.100914.

aggregates, since the average reported strengths at 100% replacement of OPS and CS aggregates were in the range of 20–25 and 28–37 MPa, respectively (Prusty & Patro, 2015). An inconsistency that can be related to the different morphological characteristics of OPS and CS. Indeed, OPS surface is smoother and more porous than OPS, consequently providing different effective interfacial properties with the host matrix. However, it was possible to improve the bond between OPS aggregates and cement matrix by using reduced size particles lower than 8–9.5 mm, old than new wastes or crushed OPS (Shafigh et al., 2014).

Comprehensive investigations also underscored the elevated drying shrinkage and water absorption of concrete composites incorporating higher OPS and CS aggregate contents (Chinnu et al., 2021; Mo et al., 2020); a consequence of increased surficial porosity and water absorption of their grains. Nevertheless, even with OPS replacement levels reaching up to 40%, the resulting concrete structures still maintain water absorption values within the acceptable threshold for high-quality concrete formulations. Similarly, chloride penetration was stated to increase with the enhanced inclusion of OPS and CS, with lower increments in CS than OPS concretes, though both matrices were in the moderate range of 3000–4000 coulombs (Mo et al., 2020).

9.6 Improving the performance of cementitious composites with recycled aggregates

A variety of approaches have been instituted to improve the physical and mechanical characteristics of WRA and to narrow performance gaps between concrete mixtures containing NAs and those containing RAs. In addition to the mechanical (crushing, shredding, compaction, surface abrasion, etc.) and energy (thermal treatment, approaches applied on recycled plastic/rubber) some of the other known treatment methods applied for WRA included preexposure to chemicals, using two-phase casting, incorporating pozzolanic nano- and micropowders, introducing polymer emulsions, using heat treatment, cleaning with water, and applying carbonation (Bahraq et al., 2022; Fernando et al., 2022; Hamada et al., 2020; Ouyang et al., 2023). Chemical conditioning, pozzolanic and nanomodifications, and polymer coating methods were involved in most pretreated WRAs, whereas the two-phase casting, presoaking in water, and carbonation techniques were more useful for recycled CDW aggregates. Indeed, a cement slurry layer was formed, leading the recycled CDW aggregate properties to improve when the mixing water was introduced at two identical quantities at different times (Bahraq et al., 2022). Almost the same principle is valid for polymer emulsions, where the new adhesive polymer layer reduces the water absorption of WRA, thereby improving their effects at fresh and hardened states (Yan et al., 2023). Premixing in fresh cement paste with or without Na_2SiO_3 was also used to cover WRA with a solid-paste layer, though this was more for RPA and RRA (Alyousef et al., 2021; Assaggaf et al., 2021). While the preimmersion in acids, such as sulfuric, hydrochloric, phosphoric, and acetic acids, can dissolve the previously hardened paste on CDW aggregates and increase its ITZ characteristics in cementitious composites, NaOH, and calcium hypochlorite as well as silane coupling agent, acetone and ethanol were applied successfully to increase the surface adhesion performance of RRA (Alyousef et al., 2021). Calcium hypochlorite and a mix of bleach plus NaOH were also used on RPA and resulted in promising physical alterations for their surfaces (Khern et al., 2020). The pozzolanic and/or nanomodification are known practices to pretreat RA. They used common SCMs like slag, fly ash, silica fume and/or nanosilica and graphene oxide and carbon nanotubes resulted in physically and mechanically improve the properties of RA (Faysal et al., 2020; Fernando et al., 2022; Mosallam et al., 2022; Reches, 2018). Heat treatment, usually with a temperature range of 60°C–130°C, was found beneficial in reducing the porosity and improving the surface texture of OPS and CS agricultural waste aggregates (Mo et al., 2020). In addition, it was usefully tested on plastic aggregates, like EPS which was exposed to heat treatment at 130°C for 15 minutes to improve its fresh and hardened properties in concretes (Chinnu et al., 2021).

It is worth noting that, in addition to the mentioned methods, researchers have recently shown interest in exploring alternative approaches such as ultraviolet treatment, gamma radiation, and microwave techniques. Another innovative

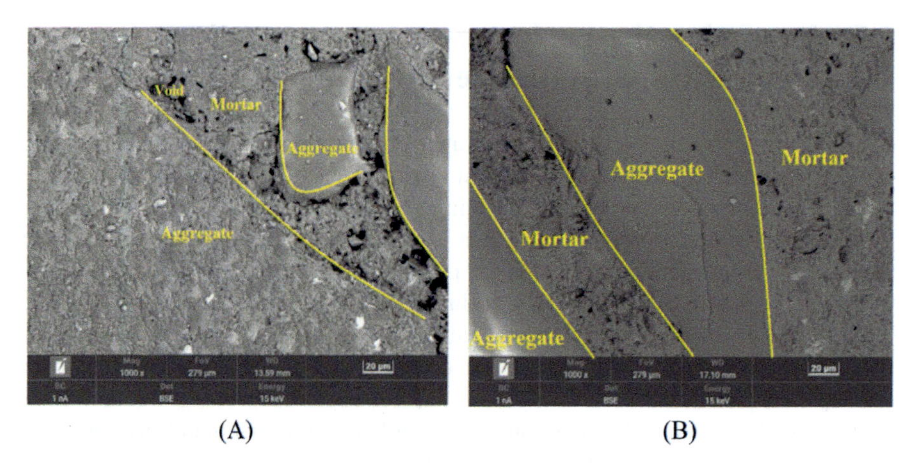

Figure 9.10 Interfacial transition zone of RCA and carbonated RCA at 70°C. *RCA*, recycled concrete aggregate.

Source: From Lu, Z., Tan, Q., Lin, J. & Wang, D. (2022). Properties investigation of recycled aggregates and concrete modified by accelerated carbonation through increased temperature. *Construction and Building Materials, 341*, 127813.

approach that is increasingly being investigated particularly for recycled CDWs referred to as carbonation, involving the exposure of WRA to high CO_2 to form porosity-filling $CaCO_3$ and Si gel products (Bahraq et al., 2022). However, this was found with an advanced positive effect on the recycled concrete microstructure when combined with high-temperature carbonation, as shown in Fig. 9.10 (Lu et al., 2022).

9.7 Use of recycled fibers in cementitious composites

Cementitious composites have brittle properties, making them vulnerable to cracking even under minimal stress. To mitigate this issue, the approach of integrating reinforcing fibers into concrete has gained extensive application. Various types of recycled fibers were integrated into cementitious composites to improve their mechanical properties as well as their sustainability. The list includes the following:

- Recycled steel fibers (RSFs): The fibers are made of recycled materials such as postconsumer steel and discarded metal, with waste tires being a major source (Fig. 9.11A and B).
- Recycled plastic fibers (RPFs): Acquired from recycled sources like postconsumer plastics or industrial waste sites (Fig. 9.11C and D).
- Recycled carbon fibers: These fibers are obtained from recycled carbon-based materials, including carbon fiber composites or carbonized waste products (Fig. 9.11E).
- Recycled glass fibers: Manufactured from recycled glass materials, these fibers serve as reinforcement in cementitious composites (Fig. 9.11F).
- Recycled cellulosic fibers: Derived from recycled paper, cardboard, or agricultural waste.

Figure 9.11 (A, B) Recycled steel fibers (Frazão et al., 2022; Mastali & Dalvand, 2016), (C) recycled polyethylene fibers obtained from the shredded plastic bottles (Anandan & Alsubih, 2021), (D) white recycled polypropylene fibers (Małek et al., 2020), (E) recycled carbon fibers (Sun et al., 2015), (F) recycled glass fiber (Ali & Ali Qureshi, 2019), (G) textile waste fiber (Sadrolodabaee et al., 2021), (H) recycled carpet fibers (Zareei, Ameri, Bahrami, et al., 2019), and (I) recycled fabric fibers (Qin et al., 2019). This figure shows the state of different prepared recycled fibers.
Source: From Mastali, M., & Dalvand, A. (2016). Use of silica fume and recycled steel fibers in self-compacting concrete (SCC). *Construction and Building Materials, 125*, 196–209. https://doi.org/10.1016/j.conbuildmat.2016.08.046.

- Recycled textile fibers (RTFs): Refers to fibers produced from recycled or reclaimed materials, typically sourced from used textiles or textile waste such as discarded carpet or fabric (Fig. 9.11G–I).

According to the literature, the most commonly used recycled fibers are steel fibers and plastic fibers with several important parameters associated with the concrete properties, including mechanical properties, durability, fresh properties, and microstructure (Fig. 9.12).

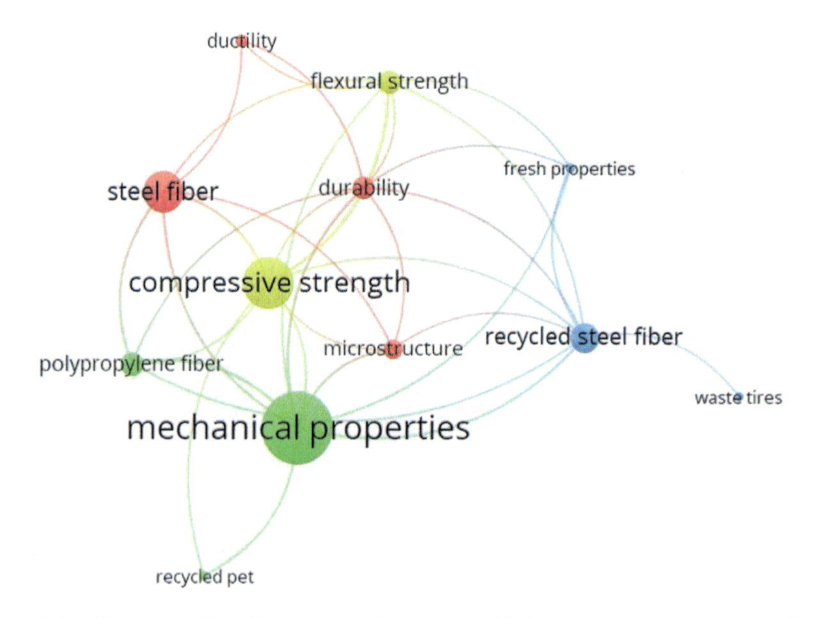

Figure 9.12 The networks' diagrams of the most studied parameters for recycled fibers.

9.7.1 Overview of the type and properties of fibers used in cementitious composites

9.7.1.1 Recycled steel fibers

In recent years several researchers have focused their attention on recycling the existed waste steel materials to be reused in civil engineering applications. RSFs were the most recycled fibers incorporated in concretes, revealing the greatest technical outcomes as compared to the other recycled fibers (Kumaresan et al., 2022). Typically, most RSFs incorporated in concretes were sourced from postconsumer vehicle tires, which produce nearly half-billion waste tires with around 9%–17% of steel content (Ali et al., 2023; Zhang et al., 2022). A substantial amount of recycled tire steel fibers (TSF) can be added to structural and nonstructural concretes, making it not only a viable alternative to normal steel fibers (NSFs) but also a cost-effective material for sustainable construction practices. In the preparation process of TSF they are separated from tires using rupture, low-temperature treatment, and thermal technology. In rupture waste tires are fragmented into small segments, and then steel wires are removed from the tire particles by magnetic substance. A low-temperature treatment procedure requires the waste tire segments to be broken after being frozen in temperatures as low as 80°C (Yao et al., 2023), a magnetic force is subsequently used to separate the steel fibers from the rubber. In thermal technique waste tires undergo pyrolysis or microwave degradation, and steel fibers are effectively separated from carbon under high temperature and pressure. An example of

Figure 9.13 Recycled tire steel fibers obtained using different techniques. (A) Rupture method, (B) low-temperature method, and (C) pyrolysis method. This figure shows various recycled steel fibers.
Source: From Zhang, P., Wang, C., Wu, C., Guo, Y., Li, Y., & Guo, J. (2022). A review on the properties of concrete reinforced with recycled steel fiber from waste tires. *Reviews on Advanced Materials Science, 61*(1), 276–291. https://doi.org/10.1515/rams-2022-0029.

TSFs extracted based on the three previous preparation methods is shown in Fig. 9.13.

RSF from the cable industry has also been reused effectively in concretes and presented equivalent properties to TSFs (Nath et al., 2021; Yao et al., 2023). Nevertheless, it is worth highlighting that researchers have paid minimal attention to the reutilization of RSFs from CDWs, though the importance of this subject for the optimum recycling of FRC after the demolition operations.

9.7.1.2 Recycled plastic fibers

Since plastic is not dissolvable when landfilled or dumped, the recycling of this material as a fiber in sustainable cementitious composites can result in numerous economic and ecological advantages for the construction industry. There are different types of plastic waste that can be recycled as fibers, including PET, LDPE/linear LDPE, polylactic acid, HDPE, PP, PVC, EPS, LDPE, PS, and polyamide (PL) (Gu & Ozbakkaloglu, 2016; Kumaresan et al., 2022). PET thread-based bottles are one of the most reused plastic fibers, which have been incorporated into concretes at varying concentrations without compromising their fresh and hardened properties as compared with the newly manufactured plastic fibers (Bahij et al., 2020; Ojeda & Mercante, 2023). Furthermore, carpet waste generated from discarded and recycled consumer carpets is the source of a high amount of RPFs such as PP and nylon, though with higher challenges in their production than those processed from beverage bottles by cutting (Kumaresan et al., 2022). Although the automotive, aerospace, and wind energy sectors are primarily generating carbon fiber waste, they also serve as sources for synthetic fiber recycling. This is projected to grow in the coming years since more than 6000 aircrafts are estimated to be disposed by 2030 and over 700,000 tons of wind turbine blade waste is projected to generated by 2035 (Akbar & Liew, 2020). Recycling of plastic-based fiber-reinforced polymer (FRP) waste materials into concrete has also been investigated in very recent years, considering the many structures containing this element around the world, which can be usually processed into polyester and PET-type plastic fibers (Baturkin et al., 2021). Syringe plastic waste and personal protective equipment (PPE), such as face masks, protective suits, and safety foot shoes, were also processed into fibers and investigated with different amounts in concrete compositions. This was particularly throughout and following the COVID-19 pandemic, where there has been a significant increase in waste from the healthcare sector, consisting of millions of tons of used PPE and syringes (Fig. 9.14; El Aal et al., 2022). Fishing net wastes are significant components of regenerated recycled synthetic fibers since PET, HDPE, or PL materials are commonly used in fishing nets, which can generally be found floating on water bodies for varying distances and depths (Kumaresan et al., 2022).

9.7.1.3 Recycled textile fibers

Textile waste can include materials originated from municipal, industrial, or commercial sources, with carpets, postconsumer clothing, packaging, mattresses, upholstery, mattresses, and packaging, contributing heavily to textile waste generation (Echeverria et al., 2019). While plastic-based fibers such as acrylic, PE, PP, and nylon can be derived from these waste materials, they are also serving as significant sources of purely textile materials, including cotton, wool, silk, linen, hemp, jute remnants, and synthetic elastane polymer fibers. Furthermore, the automotive industry such as car interiors and waste tire textiles are increasingly being utilized for the production of RTF. Although recycling postconsumer textile waste can require a complex process, including sorting, cleaning, separation, and treatment (Pakravan et al., 2018), simple cutting techniques with knife or rotary mill with blades are enough in the preparation of most RTFs (Fig. 9.15; Tang et al., 2023). In general, the resulted purely RTFs were classified with high/appropriate ductility and reduced tensile strength and elastic modulus, thus they are commonly downcycled in nonstructural cementitious composites due to their lightweight and intrinsically porous nature (Juanga-Labayen et al., 2022). However, Kevlar-type RTFs, as a pure textile fabric, are considered as a high-strength synthetic fiber, consequently categorized with high-end fibers, which were usefully incorporated in structural concretes. Moreover, they can display a rough surface and crimped geometry when processed mechanically, offering a surface texture that differs significantly from virgin fibers (Tran, Gunasekara, et al., 2022).

9.8 Fresh properties of cementitious composites with recycled fibers

9.8.1 Recycled steel fibers

The incorporation of NSFs into concrete mixtures adversely affects their air content and flowability (Onaizi et al., 2021), which highlights the importance of considering the fresh-state properties when including RSF. Although the general use of RSF

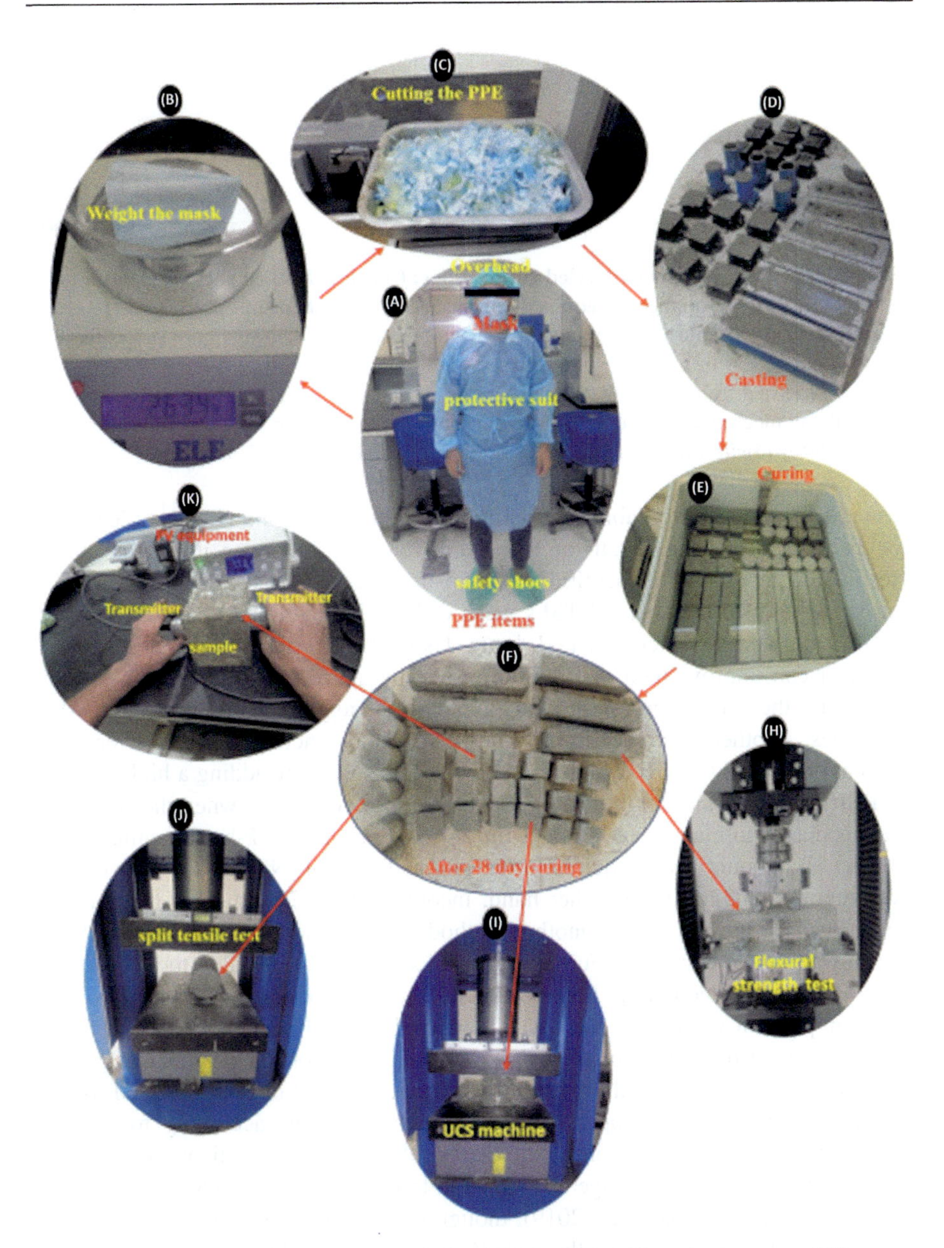

Figure 9.14 Experimental process of the use of waste PPE fibers in concrete. This figure presents the use of waste PPE fibers from the protective suits used during COVID-19. *PPE*, personal protective equipment.
Source: From El Aal, A. A., Abdullah, G. M. S., Qadri, S. M. T., Abotalib, A. Z., & Othman, A. (2022). Advances on concrete strength properties after adding polypropylene fibers from health personal protective equipment (PPE) of COVID-19: Implication on waste management and sustainable environment. Physics and Chemistry of the Earth, Parts A/B/C, 128. Available from https://doi.org/10.1016/j.pce.2022.103260.

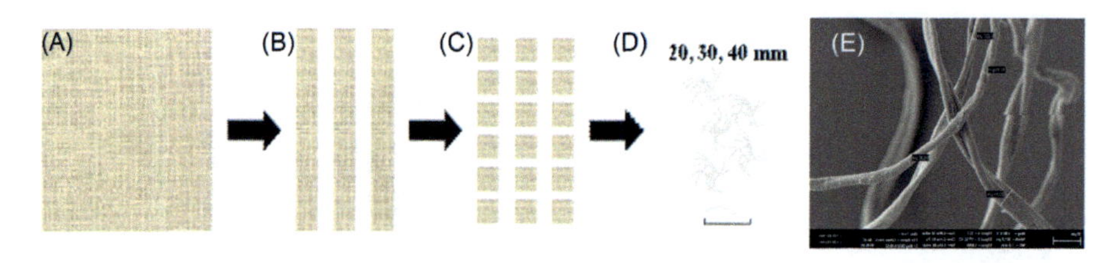

Figure 9.15 Preparation of recycled cotton fiber: (A) recycled bed sheet, (B) cut into strips, (C) cut into squares of required lengths, (D) recycled cotton fibers, and (E) Scanning electron microscope image of recycled cotton fibers. This figure shows the preparation process of recycled cotton fiber.
Source: Modified from Tang, W., Monaghan, R., & Sajjad, U. (2023). Investigation of physical and mechanical properties of cement mortar incorporating waste cotton fibres. *Sustainability (Switzerland), 15*(11). https://doi.org/10.3390/su15118779.

resulted in reduced workability and fluidity, similar to NSFs (Reches, 2018), the inconsistent shape and length of RSF were found to influence its effects on the fresh characteristics of cementitious composites. According to Zhang et al. (2022), the studied waste tire-sourced TSFs were with high length and diameter ranges of 0.8–60 mm and diameter of 0.1–1.4 mm. Using TSFs with a consistent length and diameter can improve their distribution inside the matrix and reduce their cohesive force with the surrounding paste, thus avoiding the agglomeration effect of TSF in concretes. Another difference in integrating RSF into concretes was its lower aspect ratio as compared to NSF, which was shown to be solved by adding a higher dosage of RSF and SP in the mix (Ouyang et al., 2023). Nevertheless, when the SP dosage was increased to maintain consistent fluidity in TSF- and NSF-based composites, a higher SP amount was required after the addition of NSF than TSF (Qin & Kaewunruen, 2022). On the other hand, incorporating SCMs such as the spherical shape fly ash particles can be another method to help reducing the effect of RSF on the flowability of concretes. Furthermore, silica fume was revealed to result in less flow diameter reduction in RSF-based SCC compositions as compared to their effect in plain SCC (Mastali & Dalvand, 2016). Like NSF, the increased addition of RSF in cementitious composites caused enhanced air content percentage, particularly when used with irregular morphology. Hence, a limit of 60 kg/m^3 RSF is generally recommended for appropriate air content amounts, according to Qin and Kaewunruen (2022). Interestingly, the use of RSF in some investigations generated lower increase in air percentage as compared to the newly manufactured steel fibers (Table 9.1; Samarakoon et al., 2019), though the findings from the synergic utilization of RSF and NSF suggest otherwise (Guerini et al., 2018).

9.8.2 Recycled plastic fibers

Numerous investigations have examined the impact of utilizing RPFs on the density, workability, and air content of cementitious composites. There is consistent evidence that increasing RPFs results in a significant reduction in the flowability of

Table 9.1 Properties of fresh concrete prepared with tire steel and normal steel fibers.

Mix ID	0% RC (reinforced concrete)	0.5% RFRC (recycled fibre reinforced concrete)	1% RFRC	0.5% SFRC (steel fibre reinforced concrete)	1% SFRC
Slump	Not applicable	215	183	165	50
Flow (mm)	535	430	385	358	205
Density	2360	2400	2320	2360	2370
Air content (%)	0.3	1	2.5	2.1	2.6

Source: From Samarakoon, S. M., Samindi, M. K., Ruben, P., Wie Pedersen, J., & Evangelista, L. (2019). Mechanical performance of concrete made of steel fibers from tire waste. *Case Studies in Construction Materials, 11*. https://doi.org/10.1016/j.cscm.2019.e00259.

concrete (Ahmed et al., 2021; Bhogayata & Arora, 2017; Thomas & Moosvi, 2020). Although, not well investigated in the literature, the decrement level in slump/flow look influenced by the type of RPFs, in addition to their dosage, length, aspect ratio, and rugosity. For example, in the case of recycled PET fibers, the slump reduction was not significant at a concentration lower than 0.5% (Duan et al., 2023), while PP- and HDPE-based RPFs presented a clear decrement effect of more than 35% and 40% at lower amounts of 0.25% and 0.40%, respectively (Ahmed et al., 2021; Bahij et al., 2020). Notably, the workability of cementitious mixes decreased as the content of RPFs increased (Fig. 9.16), though it conversely enhanced their resistance to segregation because of increased concrete viscosity (Bhogayata & Arora, 2017). Likewise, for the effect of fiber length, there is an agreement about the significant slump reduction at increased RPFs' length. Nevertheless, this was with lower effect at reduced length-to-diameter ratio (Duan et al., 2023), since the inclusion of RPFs with higher aspect ratio and surface area led to a more obstruction of the flow in the cementitious mixtures (Khatab et al., 2019). Furthermore, the RPF irregular shapes were explained with slight influence on the workability, though it enhanced the cohesiveness and rigidity of the concrete mix (Al-Hadithi & Abbas, 2018; Marthong & Marthong, 2016). The inclusion of greater SP amounts was not highly effective in reducing the negative influence of RPFs on the flow/slump of mixtures. On the contrary, the use of SCMs such as fly ash or a combination of fly ash/slag resulted in an improvement of slump flow and filing ability of the mixtures in comparison to the reference mixtures, even at lower SP contents (Anandan & Alsubih, 2021; Jain et al., 2021). Although no clear agreement was reported on the impact of RPFs on the density of cementitious composites, the general trend supported a reduced density and increased air content, particularly when RPF was higher than 1.5% (Bahij et al., 2020).

9.8.3 Recycled textile fibers

According to previous studies, the inclusion of pure RTFs has been found to negatively affect the workability and air content of the mixtures, specifically at higher amounts (Candamano et al., 2021; Tran, Gunasekara, et al., 2022). For example,

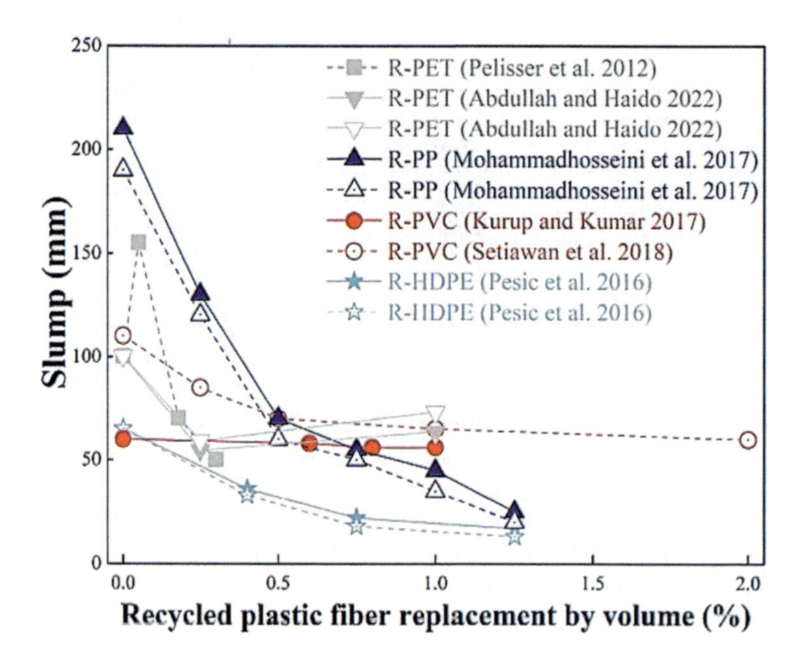

Figure 9.16 Slump/flow of cementitious composites at increased recycled plastic fiber amounts. This figure presents the slump flow at increased recycled plastic fibers. *Source*: Modified from Zareei, S. A., Ameri, F., Bahrami, N., Shoaei, P., Musaeei, H. R., & Nurian, F. (2019). Green high strength concrete containing recycled waste ceramic aggregates and waste carpet fibers: Mechanical, durability, and microstructural properties. *Journal of Building Engineering, 26*. https://doi.org/10.1016/j.jobe.2019.100914.

when three different recycled textile fabrics were examined at concentrations between 0.1% and 0.5% (Kumaresan et al., 2022), the increased use of Kevlar and Nomex-based pure RTFs caused a flowability reduction of up to 94% and 68%, respectively, as compared to Nylon fibers with lower reductions of up to around 33%. In addition, the increased RTF length exhibited lower flowability at a constant fiber content, and two-dimensional woven fibers showed higher flowability compared with bundled fibers due to their smaller contact areas with the cement matrix (Kumaresan et al., 2022). By contrast, Tang et al. (2023) found that the enhanced lengths of cotton-type RTF resulted in no significant effect on the flow table diameter of cementitious composites. However, the high-water absorption of recycled cotton, wool, and hydraulic hackled hemp fibers was observed to cause a significant decrement in workability (Fig. 9.17; Alyousef et al., 2019; Candamano et al., 2021; Tang et al., 2023), even when SP amount was increased. Furthermore, the increased incorporation of lightweight RTFs reduced the fresh density of the cementitious mixtures, with the weight reduction was not significant at lower inclusions (Awal & Mohammadhosseini, 2016; Candamano et al., 2021). It is to be mentioned that the rheology properties of pure RTFs in concrete can be considered as

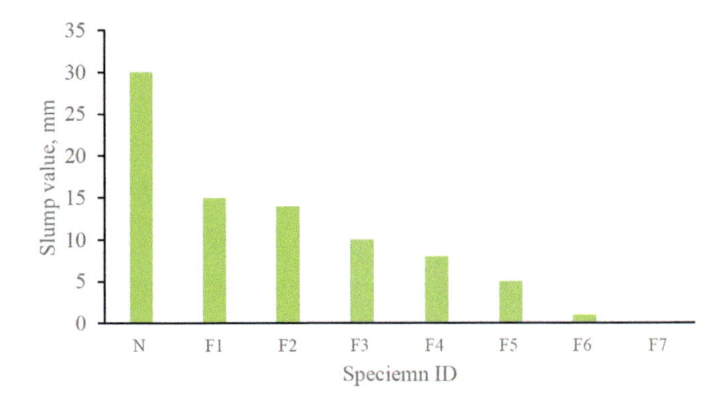

Figure 9.17 Slump values of concrete incorporating 0.5%–6% of recycled wool fibers. This figure shows the slump flow at increased recycled wool fibers.
Source: From Alyousef, R., Aldossari, K., Ibrahim, O., Mustafa, H., & Jabr, A. (2019). Effect of sheep wool fiber on fresh and hardened properties of fiber reinforced concrete. *International Journal of Civil Engineering and Technology, 10*, 190–199.

an interesting subject for future research because of their eventual utilization in three-dimensional printing technology (Rajeev et al., 2023).

9.8.4 Hardened properties of cementitious composites with recycled fibers

9.8.4.1 Recycled steel fibers

RSF effect on the compressive strength of cementitious composites was shown in most studies with reductions or increments of less than 5% up to a concentration of 4% (Fig. 9.18; Qin & Kaewunruen, 2022). This was equivalent to NSF, indicating an insignificant impact of RSF on the compressive strength of cementitious composites. However, the clear discrepancy in the results can be seen in the literature, especially at RSF contents ranging from 0.5% to 2%. While some studies reported an increase in compressive strengths (Ming et al., 2021; Samarakoon et al., 2019), other researchers pointed out that RSF amounts can negatively affect the strength (Kumaresan et al., 2022; Zhang et al., 2022). In addition to the possible different properties and level of cleaning of RSFs the method of mixing was also confirmed to influence the dispersion and consequently the compressive strengths of RSF concretes. According to Ming et al. (2021), the utilization of a planetary shear mixer can help improve the dispersion of RSF and associated compressive strengths. For instance, as compared to the regular concrete mixer compositions, a strength enhancement of around 8% was found at a lower RSF content of 0.03% when a shear mixer was used (Leone et al., 2018). Moreover, the incorporation of RSF in SCC with an appropriate compaction aspect can avoid the strength reduction, even at higher RSF contents of more than 2% (Qin & Kaewunruen, 2022).

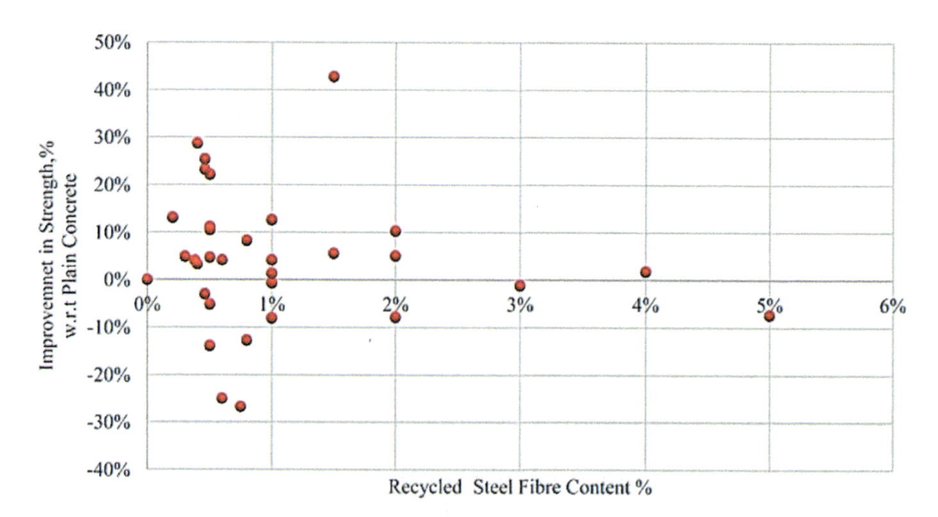

Figure 9.18 Rate of change in compressive strength of RSFs in cementitious composites. This figure shows the change in compressive strengths when RSFs are included into cementitious composites. *RSFs*, recycled steel fibers.
Source: Modified from Zareei, S. A., Ameri, F., Bahrami, N., Shoaei, P., Musaeei, H. R., & Nurian, F. (2019). Green high strength concrete containing recycled waste ceramic aggregates and waste carpet fibers: Mechanical, durability, and microstructural properties. *Journal of Building Engineering, 26*. https://doi.org/10.1016/j.jobe.2019.100914.

Contrary to compressive strengths, there was agreement about the positive effect of RSFs on improving the tensile and flexural strengths of cementitious composites as compared to plain mixtures. In addition, when compared to NSF, there were similar or greater flexural properties at a concentration of 1.5% or lower, with limited differences at higher fiber contents (Qin & Kaewunruen, 2022). The hybrid inclusion of RSF and NSF was presented to further increase the flexural characteristics, including first-cracking, optimal, and postcracking strengths. For example, the tensile strength of concrete was increased by 18% on synergic incorporation of RSFs and industrial steel fibers, according to Samarakoon et al. (2019). The length of RSF was also shown to be influential on the level of flexural/tensile strength improvements of cementitious composites. However, while Sengul (2016) found that smaller fiber diameters increased tensile strengths, a review table displayed by Ming et al. (2021) showed generally that both flexural and tensile strengths increased at higher RSF lengths. Equivalent pull-out performances were reported between RSF- and NSF-based concretes, though the presence of impurities was predicted to increase the roughness of RSF and its bonding to the mortar matrix (Samarakoon et al., 2019).

Limited literature was noticed about the durability of RSF cementitious composites as compared to those of NSF, especially those related to the concrete resistance under different chemical exposures. Most research about RSF focused on chloride and freeze–thaw effects (Alsaif et al., 2018; Liew & Akbar, 2020; Simalti

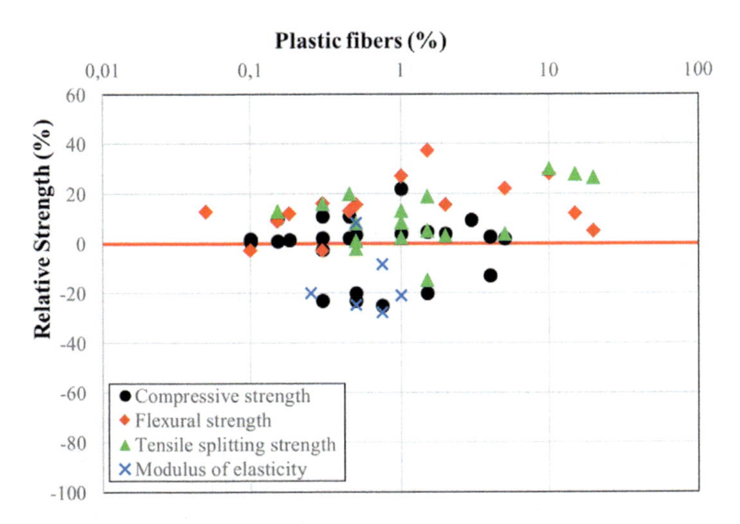

Figure 9.19 Development of the compressive strength for the mixtures containing various percentages of RPF. This figure reveals the different compressive strengths for cementitious composites containing various RPF amounts. *RPF*, recycled plastic fiber. *Source*: From Bahij, S., Omary, S., Feugeas, F., & Faqiri, A. (2020). Fresh and hardened properties of concrete containing different forms of plastic waste – A review. *Waste Management, 113*, 157–175. https://doi.org/10.1016/j.wasman.2020.05.048.

& Singh, 2021; Zhang et al., 2022). A clear gap in research was registered about the influence of RSF on the fire/high-temperature resistance of cementitious composites, considering the importance of this parameter for FRC elements. The inclusion of RSFs in concrete, however, was shown with a positive outcome on the studied durability properties, though the impurities on RSFs were stated to generate a higher water absorption effect, leading to increased freeze–thaw degradation as compared to NSF (Zhang et al., 2022).

9.8.4.2 Recycled plastic fibers

The compressive strengths of plain cementitious composites have been generally found to reduce at higher RPF amounts (Fig. 9.19; Bhogayata & Arora, 2017; Mehvish et al., 2020). The significant reductions were noticed more at RPF concentrations of 10 kg/m^3 and higher (Ojeda & Mercante, 2023), as well as enhanced binder volume and water/binder ratio. In addition to the increased porous structure of the matrix the weaker fiber-to-matrix ITZ was associated with the RPF hydrophobic surfaces (Bahij et al., 2020). The type of RPF, however, seems highly affecting its impact on the compressive strength reductions of concrete, with important decrements observed in studies investigating cementitious composites with recycled PET, PP, and poly(vinyl alcohol) (PVA) fibers of varying lengths and dosages (Ahmed et al., 2021). Conversely, recycled HDPE, LDPE, and fishing net waste plastic fibers have demonstrated no or various increasing effects on concrete's compressive strength (Duan et al., 2023; Kumaresan et al., 2022; Pešić et al., 2016). It is also important to note that a uniform shape was important to reach a lower negative effect of RPFs were used at different dosages (Borg et al., 2016). The simultaneous inclusion of RPFs and SCMs (fly ash, slag, and silica fume) exhibited a remarkable synergistic effect, significantly enhancing the mechanical properties of the material that surpasses the individual advantages gained from SCMs or RPFs alone (Dong et al., 2022). The simultaneous addition of 1.0% RPFs and 50% SCMs, including slag, fly ash, and silica fume, resulted in 35.3% and 16.7%, in 7- and 28-day compressive strength, respectively, when compared to conventional SCC mixes. Thomas and Moosvi (2020) reported that by the synergic incorporation of RPFs and Metakaolin, the compressive strength increased by 10.7% compared with the plain mixture.

Different studies have found varying effects of RPF on the flexural and tensile strengths of the host matrices; however, the most reported trends were in line with the increased results up to a certain optimum, highlighting the significance of considering the RPF type, shape, and length when comparing their influence on the flexural properties of RPF concretes. For example, while PET-RPF was found to enhance the flexural strengths when used up to 1.5% at a length of 20 mm, it was also shown with an optimum of 1% at higher lengths of 30 and 50 mm (Bahij et al., 2020; Borg et al., 2016). Also, PP-RPF was indicated to increase the flexural strengths when its dosage enhanced up to 0.45% and 1% at 12 and 27.1−32.6 mm lengths, respectively, whereas the fishing net waste fiber was with enhanced effect at 1.5% and 1% at 77 and 83.33 aspect ratios, respectively (Kumaresan et al., 2022). According to Duan et al. (2023), curly shaped RPF proved a greater influence on flexural strengths than straight RPF. In addition, PET revealed a better effect than PE, PVA was better than nylon fiber, PP fiber was greater than PVA, and PVC/HDPE caused higher strengths compared with PET/ LDPE-based RPF. When compared to their newly manufactured counterparts, HDPE-RPF generated lower flexural properties, however, PP-RPF exhibited similar postcracking reinforcement capabilities as virgin PP fiber (Pešić et al., 2016). Finally, improving the flexural and tensile properties of RPF concrete was possible through different surface treatment methods, like the chemical exposure to alkaline or saline solutions, physical crimping for advanced roughening, and application of Gamma radiation (Usman et al., 2018).

Limited research has been conducted on the durability of RPFs. Regarding the water absorption, the optimal volume fraction and aspect ratio of RPF were necessary for lower matrix porosity and related reduction in absorption and permeability. An opposite effect was stated when resulting in higher porous system by exceeding the optimum or reducing the aspect ratio of RPF. For example, the single incorporation of RPFs in the cementitious composites with an optimum volume content of 1%, reduced the water sorptivity by 22% in comparison to the reference mixture (Huynh et al., 2023). Also, it was generally informed that RPFs reduced the chloride penetration of the mixtures, especially when incorporated with shorter size or/and with the addition of styrene butadiene rubber or SCMs (Bhogayata & Arora, 2018).

The incorporation of RPFs was also highly effective in preventing the spalling of concrete specimens when exposed to acid attacks (Thomas & Moosvi, 2020).

9.8.4.3 Recycled textile fibers

Pure RTFs have varying effects on the mechanical properties of cementitious composites. This is mainly in accordance with the type and concentration of RTF, though most low-end RTFs caused reduced compressive strengths at higher inclusions. The Nomex-based high-end RTFs showed negligible effect on the compressive strength, unlike Kevlar which improved the strength value at 0.3% dosage and reduced it at higher amount of 0.5%. Irrespective of the type of RTF, the shorter-length Nomex and Kevlar fibers exhibited lower compressive strengths, while the woven fibers exhibited an increase in compressive strengths compared with the bundled fibers (Tran et al., 2023). However, the recycled cotton fibers caused a significant decrement in compressive, with increased differences than the plain matrix at longer curing times. It was explained by the high consumption of the available water for continuous hydration, the low fiber adhesion with the matrix, and balling effect beyond the optimum cotton RTFs (Tang et al., 2023). Opposite to Nomex and Kevlar RTFs, the compressive strengths further dropped at increased fiber size of cotton, especially when used at the range of 20–40 mm lengths (Alyousef et al., 2019). Similarly, the inclusion of carpet-sources RTFs with higher lengths has not positively affected the compressive strength. However, most RTFs contributed to improved tensile and flexural strengths, with greater outcomes at optimized fiber percentage. For instance, this was shown as 0.5% for carpet RTFs with 20 mm length and 2% and 0.8% for cotton RTFs with lengths of 70 and 30 mm, respectively. In addition, it was highlighted that the inclusion of RTFs' fabrics enhanced the resistance of mortar to brittle fractures, even when subjected to different types of loads (Tran, Gunasekara, et al., 2022). The combination of RTFs with SCMs, such as silica fume, fly ash, and palm oil fuel ash, yielded a synergistic effect, leading to increased tensile and flexural strength, along with improved ductility (Awal & Mohammadhosseini, 2016; Jóźwiak-Niedźwiedzka & Fantilli, 2020). Conversely, treating the surface of hemp with Gamma radiation resulted in insignificant changes in the tensile strength of cementitious composites (Jóźwiak-Niedźwiedzka & Fantilli, 2020).

When carpet-based RTFs with a length of 30 mm and varying volume fractions of 0.5%–2% were used, mixtures containing lower fiber dosage and shorter lengths exhibited higher drying shrinkage (Talamona & Hai Tan, 2012). Different than Kevlar RTFs, the inclusion of Nomex fibers negatively impacted the overall shrinkage of the mortar (Tran et al., 2023).

At an appropriate amount of carpet RTF, the durability of cementitious composites was expected to increase, since water absorption was significantly reduced (Tran, Gunasekara, et al., 2022). However, increasing fiber volume resulted in higher water permeability and porosity due to the increased void detection with enhanced use of carpet RTFs (Fig. 9.20; Zareei, Ameri, Bahrami, et al., 2019). An important reported limitation for sheep wool RTFs was their lower resistance in

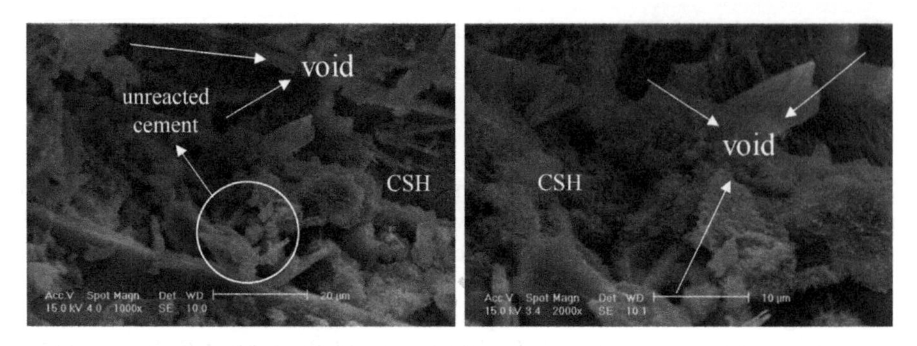

Figure 9.20 SEM image exhibiting the microstructure of the recycled-carpet-reinforced mixture . This figure shows SEM images of the microstructure of recycled-carpet-reinforced matrix. *SEM*, scanning electron microscope.
Source: Modified from Zareei, S. A., Ameri, F., Bahrami, N., Shoaei, P., Musaeei, H. R., & Nurian, F. (2019). Green high strength concrete containing recycled waste ceramic aggregates and waste carpet fibers: Mechanical, durability, and microstructural properties. *Journal of Building Engineering, 26*. https://doi.org/10.1016/j.jobe.2019.100914.

high-alkali environments, necessitating to be better incorporated in lower alkalinity pastes (Jóźwiak-Niedźwiedzka & Fantilli, 2020).

9.9 Challenges (benefits against disadvantages) and future work

In general, based on the previous literatures, it can be concluded that utilizing waste materials as aggregates or fibers resulted in the reduction of fresh and hardened properties that give more importance to the quality control of recycled materials and also further investigation to enhance waste-containing mixtures. Moreover, a significant impediment to the widespread adoption of recycled materials lies in the challenge of ensuring compliance with established standards and regulations. In addition, there exists a skeptical perception among stakeholders when it comes to replacing conventional materials with recycled waste materials.

However, the remarkable effect of incorporating waste materials on the sustainability of construction projects must not be overlooked. The utilization of waste and recycled materials as replacements for NAs or synthetic fibers helps mitigate the adverse effects of waste material accumulation, which can lead to landfill congestion and related environmental concerns such as water contamination and landslides. Furthermore, the addition of waste materials can offer a cost-effective solution due to their lower material costs and the resulting savings in transportation expenses, which incentivizes stakeholders to choose these materials. It is also worth noting that the production of such materials consumes less energy when compared to the extraction and production of natural materials from their sources.

Regarding the existing research gaps and recommendations for future work, it is important to highlight that the majority of experimental papers have primarily concentrated on assessing the workability and mechanical properties of mixtures incorporating RAs and recycled fibers. However, there is a noticeable dearth of investigations pertaining to the rheological properties and durability of these materials, particularly when it comes to recycled fibers. This indicates a significant opportunity for further exploration and in-depth analysis in these crucial areas of study. It is also noteworthy that there is a scarcity of experimental research concerning the utilization of RA and recycled fibers within the context of geopolymer concrete, which is known for its sustainable mix design and acceptable fresh and hardened properties. This underlines an additional critical area where further investigation and experimentation are needed to expand the understanding of sustainable construction practices. For instance, the synergic use of CDWs both as aggregates and precursors in geopolymers has the potential to lead to a significant enhancement in the sustainability of concrete structures, especially when considering its application in a one-part approach (just-add-water geopolymer).

9.10 Conclusions

This chapter aims to provide a comprehensive evaluation of recent publications on understanding and optimizing the use of RAs and recycled fibers in sustainable cementitious composites. It also addresses some of the actual research gaps according to existing studies. The following main conclusions can be summarized:

- The recycling efficiency of RAs is influenced by parameters such as waste material type, size, shape, density, and water absorption.
- Utilizing RCA affects negatively the rheological properties and workability of concrete due to higher porosity, water absorption, and the presence of old mortar. RCA can decrease compressive strength in cementitious composites, influenced by factors such as water absorption, original concrete strength, and quality of the aggregates. However, the finer RCAs' size can enhance strength through densification, internal curing, and improved interlocking.
- Recycled ceramic and brick aggregates also affect concrete workability due to porosity, water absorption, and surface texture. Ceramic waste aggregates improve strength and flexural properties, while recycled brick aggregates (RWBAs) generally decrease compressive strength.
- Recycled glass aggregate (RWGA) reduces density and improves workability, while RRA decreases density and negatively impacts workability. RGWA decreases compressive strength due to weak bonding and the presence of silica, but some studies report enhanced strength due to texture and pozzolanic reaction. RRAs negatively affect compressive and tensile strength due to lower strength and stiffness compared with NAs.
- Recycled plastic waste aggregate (RPWA) decreases concrete density and workability, with different types and shapes. RPWA generally decreases compressive strength due to disrupted interfaces, while larger plastic aggregates weaken the bond.

- Agricultural waste aggregates (e.g., OPSs, CSs) generally decrease concrete density and have varying effects on workability. These materials lead to lower mechanical properties.
- The workability of concrete with RAs can be improved through the use of SPs, adjustments in the water-to-cement ratio, and the addition of supplementary materials like fly ash.
- Two-stage mixing approach can improve the mechanical performance of cementitious composites with RAs. Other techniques include acid presoaking to remove adhered mortar, incorporation of pozzolans and nanomaterials to modify the microstructure, as well as techniques like polymer emulsions and carbonate biodeposition to reduce porosity. By combining these approaches, RA concrete has been shown to have improved mechanical properties and durability.
- Due to micropores and increased void content, recycled fibers (i.e., steel, plastic, and textile) reduce flowability and slump values. RSFs can enhance strength, but the effect is limited at lower fiber contents and excessive addition can have a negative impact. However, RPFs generally reduce compressive strength and have varying effects on tensile strength and Young's modulus.
- When used at higher dosages, RTFs may decrease compressive strength; however, optimized dosages and longer fibers can improve the flexural properties.

References

Abu-Saleem, M., Zhuge, Y., Hassanli, R., Ellis, M., & Levett, R. M. (2021). Evaluation of concrete performance with different types of recycled plastic waste for kerb application. *Construction and Building Materials*, *293*, 123477.

Ahmed, H. U., Faraj, R. H., Hilal, N., Mohammed, A. A., & Sherwani, A. F. H. (2021). Use of recycled fibers in concrete composites: A systematic comprehensive review. *Composites Part B: Engineering*, *215*. Available from https://doi.org/10.1016/j.compositesb.2021.108769, https://www.journals.elsevier.com/composites-part-b-engineering.

Akbar, A., & Liew, K. M. (2020). Assessing recycling potential of carbon fiber reinforced plastic waste in production of eco-efficient cement-based materials. *Journal of Cleaner Production*, *274*. Available from https://doi.org/10.1016/j.jclepro.2020.123001.

Al-Hadithi, A. I., & Abbas, M. A. (2018). The effects of adding waste plastic fibers on the flexural toughness of normal concrete. *Medwell Journals, Iraq Journal of Engineering and Applied Sciences*, *13*(24), 10282–10290. Available from https://doi.org/10.3923/jeasci.2018.10282.10290, http://docsdrive.com/pdfs/medwelljournals/jeasci/2018/10282-10290.pdf.

Ali, B., & Ali Qureshi, L. (2019). Influence of glass fibers on mechanical and durability performance of concrete with recycled aggregates. *Construction and Building Materials*, *228*. Available from https://doi.org/10.1016/j.conbuildmat.2019.116783.

Ali, B., Kurda, R., Ahmed, H., & Alyousef, R. (2023). Effect of recycled tyre steel fiber on flexural toughness, residual strength, and chloride permeability of high-performance concrete (HPC). *Journal of Sustainable Cement-Based Materials*, *12*(2), 141–157. Available from https://doi.org/10.1080/21650373.2021.2025165, http://www.tandfonline.com/toc/tscm20/current.

Almeshal, I., Tayeh, B. A., Alyousef, R., Alabduljabbar, H., Mustafa Mohamed, A., & Alaskar, A. (2020). Use of recycled plastic as fine aggregate in cementitious composites: A review. *Construction and Building Materials*, *253*. Available from https://doi.org/

10.1016/j.conbuildmat.2020.119146, https://www.journals.elsevier.com/construction-and-building-materials.

Alsaif, A., Bernal, S. A., Guadagnini, M., & Pilakoutas, K. (2018). Durability of steel fibre reinforced rubberised concrete exposed to chlorides. *Construction and Building Materials*, *188*, 130–142. Available from https://doi.org/10.1016/j.conbuildmat.2018.08.122.

Alyousef, R., Ahmad, W., Ahmad, A., Aslam, F., Joyklad, P., & Alabduljabbar, H. (2021). Potential use of recycled plastic and rubber aggregate in cementitious materials for sustainable construction: A review. *Journal of Cleaner Production*, *329*. Available from https://doi.org/10.1016/j.jclepro.2021.129736.

Alyousef, R., Aldossari, K., Ibrahim, O., Mustafa, H., & Jabr, A. (2019). Effect of sheep wool fiber on fresh and hardened properties of fiber reinforced concrete. *International Journal of Civil Engineering and Technology*, *10*, 190–199.

Anandamurthy, A., Guna, V., Ilangovan, M., & Reddy, N. (2017). A review of fibrous reinforcements of concrete. *Journal of Reinforced Plastics and Composites*, *36*(7), 519–552. Available from https://doi.org/10.1177/0731684416685168, http://jrp.sagepub.com/content/by/year.

Anandan, S., & Alsubih, M. (2021). Mechanical strength characterization of plastic fiber reinforced cement concrete composites. *Applied Sciences*, *11*(2). Available from https://doi.org/10.3390/app11020852.

Angelin, A. F., Lintz, R. C. C., Osorio, W. R., & Gachet, L. A. (2020). Evaluation of efficiency factor of a self-compacting lightweight concrete with rubber and expanded clay contents. *Construction and Building Materials*, *257*, 119573.

Assaggaf, R. A., Ali, M. R., Al-Dulaijan, S. U., & Maslehuddin, M. (2021). Properties of concrete with untreated and treated crumb rubber – A review. *Journal of Materials Research and Technology*, *11*, 1753–1798. Available from https://doi.org/10.1016/j.jmrt.2021.02.019, http://www.elsevier.com/journals/journal-of-materials-research-and-technology/2238-7854.

Asutkar, P., Shinde, S. B., & Patel, R. (2017). Study on the behaviour of rubber aggregates concrete beams using analytical approach. *Engineering Science and Technology, an International Journal*, *20*(1), 151–159. Available from https://doi.org/10.1016/j.jestch.2016.07.007, http://www.journals.elsevier.com/engineering-science-and-technology-an-international-journal/.

Awal, A. S. M. A., & Mohammadhosseini, H. (2016). Green concrete production incorporating waste carpet fiber and palm oil fuel ash. *Journal of Cleaner Production*, *137*, 157–166. Available from https://doi.org/10.1016/j.jclepro.2016.06.162.

Bahij, S., Omary, S., Feugeas, F., & Faqiri, A. (2020). Fresh and hardened properties of concrete containing different forms of plastic waste – A review. *Waste Management*, *113*, 157–175. Available from https://doi.org/10.1016/j.wasman.2020.05.048, http://www.elsevier.com/locate/wasman.

Bahraq, A. A., Jose, J., Shameem, M., & Maslehuddin, M. (2022). A review on treatment techniques to improve the durability of recycled aggregate concrete: Enhancement mechanisms, performance and cost analysis. *Journal of Building Engineering*, *55*. Available from https://doi.org/10.1016/j.jobe.2022.104713.

Baturkin, D., Hisseine, O. A., Masmoudi, R., Tagnit-Hamou, A., & Massicotte, L. (2021). Valorization of recycled FRP materials from wind turbine blades in concrete. *Resources, Conservation and Recycling*, *174*. Available from https://doi.org/10.1016/j.resconrec.2021.105807.

Behera, M., Minocha, A. K., & Bhattacharyya, S. K. (2019). Flow behavior, microstructure, strength and shrinkage properties of self-compacting concrete incorporating recycled fine aggregate. *Construction and Building Materials*, *228*. Available from https://doi.org/10.1016/j.conbuildmat.2019.116819.

Bhogayata, A. C., & Arora, N. K. (2017). Fresh and strength properties of concrete reinforced with metalized plastic waste fibers. *Construction and Building Materials*, *146*, 455–463. Available from https://doi.org/10.1016/j.conbuildmat.2017.04.095.

Bhogayata, A. C., & Arora, N. K. (2018). Workability, strength, and durability of concrete containing recycled plastic fibers and styrene-butadiene rubber latex. *Construction and Building Materials*, *180*, 382–395. Available from https://doi.org/10.1016/j.conbuildmat.2018.05.175.

Bidabadi, M. S., Akbari, M., & Panahi, O. (2020). Optimum mix design of recycled concrete based on the fresh and hardened properties of concrete. *Journal of Building Engineering*, *32*. Available from https://doi.org/10.1016/j.jobe.2020.101483, http://www.journals.elsevier.com/journal-of-building-engineering/.

Borg, R. P., Baldacchino, O., & Ferrara, L. (2016). Early age performance and mechanical characteristics of recycled PET fibre reinforced concrete. *Construction and Building Materials*, *108*, 29–47. Available from https://doi.org/10.1016/j.conbuildmat.2016.01.029.

Candamano, S., Crea, F., Coppola, L., De Luca, P., & Coffetti, D. (2021). Influence of acrylic latex and pre-treated hemp fibers on cement based mortar properties. *Construction and Building Materials*, *273*. Available from https://doi.org/10.1016/j.conbuildmat.2020.121720.

Carro-López, D., González-Fonteboa, B., De Brito, J., Martínez-Abella, F., González-Taboada, I., & Silva, P. (2015). Study of the rheology of self-compacting concrete with fine recycled concrete aggregates. *Construction and Building Materials*, *96*, 491–501. Available from https://doi.org/10.1016/j.conbuildmat.2015.08.091.

Cartuxo, F., de Brito, J., Evangelista, L., Jiménez, J. R., & Ledesma, E. F. (2015). Rheological behaviour of concrete made with fine recycled concrete aggregates – Influence of the superplasticizer. *Construction and Building Materials*, *89*, 36–47. Available from https://doi.org/10.1016/j.conbuildmat.2015.03.119.

Chandra Paul, S., Šavija, B., & Babafemi, A. J. (2018). A comprehensive review on mechanical and durability properties of cement-based materials containing waste recycled glass. *Journal of Cleaner Production*, *198*, 891–906. Available from https://doi.org/10.1016/j.jclepro.2018.07.095, https://www.journals.elsevier.com/journal-of-cleaner-production.

Chen, X., Vanoutrive, H., Gruyaert, E., & Li, J. (2023). *Properties of high-performance concrete with coarse recycled concrete aggregate for precast industry* (Vol. 40). Belgium: Springer Science and Business Media B.V. Available from https://www.springer.com/series/8781.

Chinnu, S., Minnu, S., Bahurudeen, A., & Senthilkumar, R. (2021). Recycling of industrial and agricultural wastes as alternative coarse aggregates: A step towards cleaner production of concrete. *Construction and Building Materials*, *287*, 123056.

Colangelo, F., Cioffi, R., Liguori, B., & Iucolano, F. (2016). Recycled polyolefins waste as aggregates for lightweight concrete. *Composites Part B: Engineering*, *106*, 234–241. Available from https://doi.org/10.1016/j.compositesb.2016.09.041.

Da Silva, S. R., & de Oliveira Andrade, J. J. (2022). A review on the effect of mechanical properties and durability of concrete with construction and demolition waste (CDW) and fly ash in the production of new cement concrete. *Sustainability*, *14*(11), 6740. Available from https://doi.org/10.3390/su14116740.

Dadsetan, S., Siad, H., Lachemi, M., Mahmoodi, O., & Sahmaran, M. (2022). Development of ambient cured geopolymer binders based on brick waste and processed glass waste.

Environmental Science and Pollution Research, *29*(53), 80755–80774. Available from https://doi.org/10.1007/s11356-022-21469-3, https://www.springer.com/journal/11356.

Danish, A., Mosaberpanah, M. A., Ozbakkaloglu, T., Salim, M. U., Khurshid, K., Bayram, M., Amran, M., Fediuk, R., & Qader, D. N. (2023). A compendious review on the influence of e-waste aggregates on the properties of concrete. *Case Studies in Construction Materials*, *18*(4), e01740.

de Andrade Salgado, F., & de Andrade Silva, F. (2022). Recycled aggregates from construction and demolition waste towards an application on structural concrete: A review. *Journal of Building Engineering*, *52*, 104452.

Dong, C., Zhang, Q., Chen, C., Jiang, T., Guo, Z., Liu, Y., & Lin, S. (2022). Fresh and hardened properties of recycled plastic fiber reinforced self-compacting concrete made with recycled concrete aggregate and fly ash, slag, silica fume. *Journal of Building Engineering*, *62*. Available from https://doi.org/10.1016/j.jobe.2022.105384.

Duan, Z., Deng, Q., Liang, C., Ma, Z., & Wu, H. (2023). Upcycling of recycled plastic fiber for sustainable cementitious composites: A critical review and new perspective. *Cement and Concrete Composites*, *142*. Available from https://doi.org/10.1016/j.cemconcomp.2023.105192.

Duan, Z., & Hou, S. (2023). *Rheological properties of fresh recycled concrete. Multifunctional concrete with recycled aggregates* (pp. 59–83). China: Elsevier. Available from https://www.sciencedirect.com/book/9780323898386.

Echeverria, C. A., Handoko, W., Pahlevani, F., & Sahajwalla, V. (2019). Cascading use of textile waste for the advancement of fibre reinforced composites for building applications. *Journal of Cleaner Production*, *208*, 1524–1536. Available from https://doi.org/10.1016/j.jclepro.2018.10.227, https://www.journals.elsevier.com/journal-of-cleaner-production.

El Aal, A. A., Abdullah, G. M. S., Qadri, S. M. T., Abotalib, A. Z., & Othman, A. (2022). Advances on concrete strength properties after adding polypropylene fibers from health personal protective equipment (PPE) of COVID-19: Implication on waste management and sustainable environment. *Physics and Chemistry of the Earth, Parts A/B/C*, *128*. Available from https://doi.org/10.1016/j.pce.2022.103260.

Evangelista, L., & De Brito, J. (2019). Durability of crushed fine recycled aggregate concrete assessed by permeability-related properties. *Magazine of Concrete Research*, *71*(21), 1142–1150. Available from https://doi.org/10.1680/jmacr.18.00093, http://www.icevirtuallibrary.com/content/serial/macr.

Fanijo, E., Babafemi, A. J., & Arowojolu, O. (2020). Performance of laterized concrete made with palm kernel shell as replacement for coarse aggregate. *Construction and Building Materials*, *250*, 118829.

Faraj, R. H., Hama Ali, H. F., Sherwani, A. F. H., Hassan, B. R., & Karim, H. (2020). Use of recycled plastic in self-compacting concrete: A comprehensive review on fresh and mechanical properties. *Journal of Building Engineering*, *30*. Available from https://doi.org/10.1016/j.jobe.2020.101283, http://www.journals.elsevier.com/journal-of-building-engineering/.

Faysal, R. M., Maslehuddin, M., Shameem, M., Ahmad, S., & Adekunle, S. K. (2020). Effect of mineral additives and two-stage mixing on the performance of recycled aggregate concrete. *Journal of Material Cycles and Waste Management*, *22*(5), 1587–1601. Available from https://doi.org/10.1007/s10163-020-01048-9, http://link.springer.de/link/service/journals/10163/index.htm.

Fernando, A., Selvaranjan, K., Srikanth, G., & Gamage, J. (2022). Development of high strength recycled aggregate concrete-composite effects of fly ash, silica fume and rice husk ash as pozzolans. *Materials and Structures*, *55*, 185.

Frazão, C., Barros, J., Bogas, J. A., García-Cortés, V., & Valente, T. (2022). Technical and environmental potentialities of recycled steel fiber reinforced concrete for structural applications. *Journal of Building Engineering, 45*. Available from https://doi.org/10.1016/j.jobe.2021.103579.

Gu, L., & Ozbakkaloglu, T. (2016). Use of recycled plastics in concrete: A critical review. *Waste Management, 51*, 19–42. Available from https://doi.org/10.1016/j.wasman.2016.03.005, http://www.elsevier.com/locate/wasman.

Guerini, V., Conforti, A., Plizzari, G., & Kawashima, S. (2018). Influence of steel and macro-synthetic fibers on concrete properties. *Fibers*, *6*(3), 20796439. Available from https://doi.org/10.3390/fib6030047, https://res.mdpi.com/fibers/fibers-06-00047/article_-deploy/fibers-06-00047-v2.pdf?filename = &attachment = 1.

Guo, H., Shi, C., Guan, X., Zhu, J., Ding, Y., Ling, T. C., Zhang, H., & Wang, Y. (2018). Durability of recycled aggregate concrete – A review. *Cement and Concrete Composites, 89*, 251–259. Available from https://doi.org/10.1016/j.cemconcomp.2018.03.008, http://www.sciencedirect.com/science/journal/09589465.

Guo, S., Dai, Q., Si, R., Sun, X., & Lu, C. (2017). Evaluation of properties and performance of rubber-modified concrete for recycling of waste scrap tire. *Journal of Cleaner Production, 148*, 681–689. Available from https://doi.org/10.1016/j.jclepro.2017.02.046.

Guo, Z., Chen, C., Lehman, D. E., Xiao, W., Zheng, S., & Fan, B. (2020). Mechanical and durability behaviours of concrete made with recycled coarse and fine aggregates. *European Journal of Environmental and Civil Engineering*, *24*(2), 171–189. Available from https://doi.org/10.1080/19648189.2017.1371083, http://www.tandfonline.com/loi/tece20.

Haddadian, A., Alengaram, U. J., Ayough, P., Mo, K. H., & Mahmoud Alnahhal, A. (2023). Inherent characteristics of agro and industrial by-products based lightweight concrete – A comprehensive review. *Construction and Building Materials, 397*. Available from https://doi.org/10.1016/j.conbuildmat.2023.132298.

Hamada, H. M., Skariah Thomas, B., Tayeh, B., Yahaya, F. M., Muthusamy, K., & Yang, J. (2020). Use of oil palm shell as an aggregate in cement concrete: A review. *Construction and Building Materials, 265*. Available from https://doi.org/10.1016/j.conbuildmat.2020.120357, https://www.journals.elsevier.com/construction-and-building-materials.

Huynh, T.-P., Ho Minh Le, T., & Vo Chau Ngan, N. (2023). An experimental evaluation of the performance of concrete reinforced with recycled fibers made from waste plastic bottles. *Results in Engineering, 18*. Available from https://doi.org/10.1016/j.rineng.2023.101205.

Jain, A., Sharma, N., Choudhary, R., Gupta, R., & Chaudhary, S. (2021). Utilization of non-metalized plastic bag fibers along with fly ash in concrete. *Construction and Building Materials, 291*, 123329.

Jani, Y., & Hogland, W. (2014). Waste glass in the production of cement and concrete - A review. *Journal of Environmental Chemical Engineering*, *2*(3), 1767–1775. Available from https://doi.org/10.1016/j.jece.2014.03.016.

Ji, Y., Zhang, H., & Li, W. (2022). Investigation on steel fiber strengthening of waste brick aggregate cementitious composites. *Case Studies in Construction Materials, 17*, e01240.

Joseph, H. S., Pachiappan, T., Avudaiappan, S., Maureira-Carsalade, N., Roco-Videla, Á., Guindos, P., & Parra, P. F. (2023). A comprehensive review on recycling of construction demolition waste in concrete. *Sustainability*, *15*(6). Available from https://doi.org/10.3390/su15064932.

Jóźwiak-Niedźwiedzka, D., & Fantilli, A. P. (2020). Wool-reinforced cement based composites. *Materials*, *13*(16). Available from https://doi.org/10.3390/ma13163590.

Juanga-Labayen, J. P., Labayen, I. V., & Yuan, Q. (2022). A review on textile recycling practices and challenges. *Textiles*, *2*(1), 174–188.

Juan-Valdés, A., Rodriguez-Robles, D., Garcia-Gonzalez, J., de Rojas Gómez, M. I. S., Guerra-Romero, M. I., & De BelieMorán-del Pozo, N. (2021). Mechanical and microstructural properties of recycled concretes mixed with ceramic recycled cement and secondary recycled aggregates. A viable option for future concrete. *Construction and Building Materials*, *270*, 121455.

Khan, M. N. N., Saha, A. K., & Sarker, P. K. (2020). Reuse of waste glass as a supplementary binder and aggregate for sustainable cement-based construction materials: A review. *Journal of Building Engineering*, *28*. Available from https://doi.org/10.1016/j.jobe.2019.101052, http://www.journals.elsevier.com/journal-of-building-engineering/.

Khankhaje, E., Salim, M. R., Mirza, J., Hussin, M. W., & Rafieizonooz, M. (2016). Properties of sustainable lightweight pervious concrete containing oil palm kernel shell as coarse aggregate. *Construction and Building Materials*, *126*, 1054–1065. Available from https://doi.org/10.1016/j.conbuildmat.2016.09.010.

Khatab, H. R., Mohammed, S. J., & Hameed, L. A. (2019). Mechanical properties of concrete contain waste fibers of plastic straps. *IOP Conference Series: Materials Science and Engineering*, *557*(1). Available from https://doi.org/10.1088/1757-899x/557/1/012059.

Khern, Y. C., Paul, S. C., Kong, S. Y., Babafemi, A. J., Anggraini, V., Miah, M. J., & Šavija, B. (2020). Impact of chemically treated waste rubber tire aggregates on mechanical, durability and thermal properties of concrete. *Frontiers in Materials*, *7*, 22968016. Available from https://doi.org/10.3389/fmats.2020.00090, journal.frontiersin.org/journal/materials.

Kim, J. (2022). Influence of quality of recycled aggregates on the mechanical properties of recycled aggregate concretes: An overview. *Construction and Building Materials*, *328*. Available from https://doi.org/10.1016/j.conbuildmat.2022.127071.

Kishore, K., & Gupta, N. (2020). Application of domestic & industrial waste materials in concrete: A review. *Materials Today: Proceedings*, *26*, 2926–2931. Available from https://doi.org/10.1016/j.matpr.2020.02.604.

Kumar, A., & Singh, G. J. (2023). Improving the physical and mechanical properties of recycled concrete aggregate: A state-of-the-art review. *Engineering Research Express*, *5* (1). Available from https://doi.org/10.1088/2631-8695/acc3df, https://iopscience.iop.org/journal/2631-8695.

Kumaresan, M., Sindhu Nachiar, S., & Anandh, S. (2022). Implementation of waste recycled fibers in concrete: A review. *Materials Today: Proceedings*, *68*, 1988–1994. Available from https://doi.org/10.1016/j.matpr.2022.08.228.

Leone, M., Centonze, G., Colonna, D., Micelli, F., & Aiello, M. A. (2018). Fiber-reinforced concrete with low content of recycled steel fiber: Shear behaviour. *Construction and Building Materials*, *161*, 141–155. Available from https://doi.org/10.1016/j.conbuildmat.2017.11.101.

Li, B., Hou, S., Duan, Z., Li, L., & Guo, W. (2021). Rheological behavior and compressive strength of concrete made with recycled fine aggregate of different size range. *Construction and Building Materials*, *268*, 121172.

Li, J., Pei, Y., Xu, P., & Chang, H. (2019). Determination of gradation of recycled mixed coarse aggregates for pavement base or subbase by crushing fractals. *Advances in Materials Science and Engineering*, *2019*. Available from https://doi.org/10.1155/2019/7963261, http://www.hindawi.com/journals/amse/.

Li, M., Chai, J., Zhang, X., Qin, Y., Ma, W., Duan, M., & Zhou, H. (2023). Quantifying the recycled nylon fibers influence on geometry of crack and seepage behavior of cracked concrete. *Construction and Building Materials*, *373*, 130853.

Li, X., Ling, T.-C., & Hung Mo, K. (2020). Functions and impacts of plastic/rubber wastes as eco-friendly aggregate in concrete – A review. *Construction and Building Materials*, *240*. Available from https://doi.org/10.1016/j.conbuildmat.2019.117869.

Li, Y., Zhang, S., Wang, R., & Dang, F. (2019). Potential use of waste tire rubber as aggregate in cement concrete – A comprehensive review. *Construction and Building Materials*, *225*, 1183–1201. Available from https://doi.org/10.1016/j.conbuildmat.2019.07.198, https://www.journals.elsevier.com/construction-and-building-materials.

Liew, K. M., & Akbar, A. (2020). The recent progress of recycled steel fiber reinforced concrete. *Construction and Building Materials*, *232*. Available from https://doi.org/10.1016/j.conbuildmat.2019.117232.

Liu, T., Qin, S., Zou, D., & Song, W. (2018). Experimental investigation on the durability performances of concrete using cathode ray tube glass as fine aggregate under chloride ion penetration or sulfate attack. *Construction and Building Materials*, *163*, 634–642. Available from https://doi.org/10.1016/j.conbuildmat.2017.12.135.

Liu, T., Wei, H., Zou, D., Zhou, A., & Jian, H. (2020). Utilization of waste cathode ray tube funnel glass for ultra-high performance concrete. *Journal of Cleaner Production*, *249*. Available from https://doi.org/10.1016/j.jclepro.2019.119333.

Liu, Z., Shi, C., Shi, Q., Tan, X., & Meng, W. (2022). Recycling waste glass aggregate in concrete: Mitigation of alkali-silica reaction (ASR) by carbonation curing. *Journal of Cleaner Production*, *370*. Available from https://doi.org/10.1016/j.jclepro.2022.133545.

Lu, Z., Tan, Q., Lin, J., & Wang, D. (2022). Properties investigation of recycled aggregates and concrete modified by accelerated carbonation through increased temperature. *Construction and Building Materials*, *341*, 127813.

Magbool, H. M. (2022). Utilisation of ceramic waste aggregate and its effect on eco-friendly concrete: A review. *Journal of Building Engineering*, *47*. Available from https://doi.org/10.1016/j.jobe.2021.103815.

Mahmood, R. A., & Kockal, N. U. (2020). Cementitious materials incorporating waste plastics: A review. *SN Applied Sciences*, *2*(12). Available from https://doi.org/10.1007/s42452-020-03905-6.

Mahmoodi, O., Siad, H., Lachemi, M., Dadsetan, S., & Sahmaran, M. (2021). Development of optimized binary ceramic tile and concrete wastes geopolymer binders for in-situ applications. *Journal of Building Engineering*, *43*. Available from https://doi.org/10.1016/j.jobe.2021.102906.

Mahmoodi, O., Siad, H., Lachemi, M., & Şahmaran, M. (2023). Comparative life cycle assessment analysis of mono, binary and ternary construction and demolition wastes-based geopolymer binders. *Materials Today: Proceedings*. Available from https://doi.org/10.1016/j.matpr.2023.05.658.

Małek, M., Jackowski, M., Łasica, W., & Kadela, M. (2020). Characteristics of recycled polypropylene fibers as an addition to concrete fabrication based on Portland cement. *Materials*, *13*(8), 1827.

Marinković, S., Josa, I., Braymand, S., & Tošić, N. (2023). Sustainability assessment of recycled aggregate concrete structures: A critical view on the current state-of-knowledge and practice. *Structural Concrete*, *24*(2), 1956–1979. Available from https://doi.org/10.1002/suco.202201245, http://onlinelibrary.wiley.com/journal/10.1002/(ISSN)1751-7648.

Marthong, C., & Marthong, S. (2016). An experimental study on the effect of PET fibers on the behavior of exterior RC beam-column connection subjected to reversed cyclic loading. *Structures*, *5*, 175–185. Available from https://doi.org/10.1016/j.istruc.2015.11.003, https://www.journals.elsevier.com/structures.

Mastali, M., & Dalvand, A. (2016). Use of silica fume and recycled steel fibers in self-compacting concrete (SCC). *Construction and Building Materials*, *125*, 196–209. Available from https://doi.org/10.1016/j.conbuildmat.2016.08.046.

Medina, C., Sánchez de Rojas, M. I., Thomas, C., Polanco, J. A., & Frías, M. (2016). Durability of recycled concrete made with recycled ceramic sanitary ware aggregate. Inter-indicator relationships. *Construction and Building Materials*, *105*, 480–486. Available from https://doi.org/10.1016/j.conbuildmat.2015.12.176.

Meena, R. V., Jain, J. K., Chouhan, H. S., & Beniwal, A. S. (2022). Use of waste ceramics to produce sustainable concrete: A review. *Cleaner Materials*, *4*. Available from https://doi.org/10.1016/j.clema.2022.100085, http://www.elsevier.com/locate/issn/2772-3976.

Mehvish, F., Ahmed, A., Saleem, M. M., & Saleem, M. A. (2020). Characterization of concrete incorporating waste polythene bags fibers. *Pakistan Journal of Engineering and Applied Sciences*, *26*, 93–101.

Mei, J., Xu, G., Ahmad, W., Khan, K., Amin, M. N., Aslam, F., & Alaskar, A. (2022). Promoting sustainable materials using recycled rubber in concrete: A review. *Journal of Cleaner Production*, *373*, 133927.

Merli, R., Preziosi, M., Acampora, A., Lucchetti, M. C., & Petrucci, E. (2020). Recycled fibers in reinforced concrete: A systematic literature review. *Journal of Cleaner Production*, *248*. Available from https://doi.org/10.1016/j.jclepro.2019.119207.

Ming, Y., Chen, P., Li, L., Gan, G., & Pan, G. (2021). A comprehensive review on the utilization of recycled waste fibers in cement-based composites. *Materials*, *14*(13). Available from https://doi.org/10.3390/ma14133643.

Mo, K. H., Thomas, B. S., Yap, S. P., Abutaha, F., & Tan, C. G. (2020). Viability of agricultural wastes as substitute of natural aggregate in concrete: A review on the durability-related properties. *Journal of Cleaner Production*, *275*. Available from https://doi.org/10.1016/j.jclepro.2020.123062, https://www.journals.elsevier.com/journal-of-cleaner-production.

Mohajerani, A., Vajna, J., Cheung, T. H. H., Kurmus, H., Arulrajah, A., & Horpibulsuk, S. (2017). Practical recycling applications of crushed waste glass in construction materials: A review. *Construction and Building Materials*, *156*, 443–467. Available from https://doi.org/10.1016/j.conbuildmat.2017.09.005.

Mohanta, N. R., & Murmu, M. (2022). Alternative coarse aggregate for sustainable and eco-friendly concrete—A review. *Journal of Building Engineering*, *59*, 105079.

Mosallam, S. J., Behbahani, H. P., Shahpari, M., & Abaeian, R. (2022). The effect of carbon nanotubes on mechanical properties of structural lightweight concrete using LECA aggregates. *Structures*, *35*, 1204–1218.

Nath, A. D., Datta, S. D., Hoque, M. I., & Shahriar, F. (2021). Various recycled steel fiber effect on mechanical properties of recycled aggregate concrete. *Journal of Building Pathology and Adaptation*. Available from https://doi.org/10.1108/IJBPA-07-2021-0102, http://www.emeraldinsight.com/journal/ijbpa.

Nedeljković, M., Visser, J., Šavija, B., Valcke, S., & Schlangen, E. (2021). Use of fine recycled concrete aggregates in concrete: A critical review. *Journal of Building Engineering*, *38*. Available from https://doi.org/10.1016/j.jobe.2021.102196.

Ojeda, J. P., & Mercante, I. T. (2023). Sustainability of recycling plastic waste as fibers for concrete: A review. *Journal of Material Cycles and Waste Management*. Available from https://doi.org/10.1007/s10163-023-01729-1.

Omoding, N., Cunningham, L. S., & Lane-Serff, G. F. (2021). Effect of using recycled waste glass coarse aggregates on the hydrodynamic abrasion resistance of concrete. *Construction and Building Materials*, *268*, 121177.

Onaizi, A. M., Huseien, G. F., Lim, N. H. A. S., Amran, M., & Samadi, M. (2021). Effect of nanomaterials inclusion on sustainability of cement-based concretes: A comprehensive review. *Construction and Building Materials*, *306*, 124850.

Ong, S. W., Lau, T. L., Yeong, T. W., Anwar, M. P., & Elleithy, W. (2018). A feasibility study on partially substituted coarse aggregate with oil palm shell in coconut fiber reinforced concrete. *International Journal of Engineering &Technology*, *7*, 62.

Ouyang, K., Liu, J., Liu, S., Song, B., Guo, H., Li, G., & Shi, C. (2023). Influence of pre-treatment methods for recycled concrete aggregate on the performance of recycled concrete: A review. *Resources, Conservation and Recycling*, *188*. Available from https://doi.org/10.1016/j.resconrec.2022.106717.

Pakravan, H. R., Asgharian Jeddi, A. A., Jamshidi, M., Memarian, F., & Saghafi, A. M. (2018). Properties of recycled carpet fiber reinforced concrete. *Use of Recycled Plastics in Eco-efficient Concrete*, 411–425. Available from https://doi.org/10.1016/B978-0-08-102676-2.00019-0, https://www.sciencedirect.com/book/9780081026762.

Pešić, N., Živanović, S., Garcia, R., & Papastergiou, P. (2016). Mechanical properties of concrete reinforced with recycled HDPE plastic fibres. *Construction and Building Materials*, *115*, 362–370. Available from https://doi.org/10.1016/j.conbuildmat.2016.04.050.

Pierrehumbert, R. (2019). There is no plan B for dealing with the climate crisis. *Bulletin of the Atomic Scientists*, *75*(5), 215–221. Available from https://doi.org/10.1080/00963402.2019.1654255.

Pitarch, A. M., Reig, L., Tomás, A. E., & López, F. J. (2019). Effect of tiles, bricks and ceramic sanitary-ware recycled aggregates on structural concrete properties. *Waste and Biomass Valorization*, *10*(6), 1779–1793. Available from https://doi.org/10.1007/s12649-017-0154-0, http://www.springer.com/engineering/journal/12649.

Prusty, J. K., & Patro, S. K. (2015). Properties of fresh and hardened concrete using agro-waste as partial replacement of coarse aggregate - A review. *Construction and Building Materials*, *82*, 101–113. Available from https://doi.org/10.1016/j.conbuildmat.2015.02.063.

Qaidi, S., Najm, H. M., Abed, S. M., Özkılıç, Y. O., Al Dughaishi, H., Alosta, M., Sabri, M. M. S., Alkhatib, F., & Milad, A. (2022). Concrete containing waste glass as an environmentally friendly aggregate: A review on fresh and mechanical characteristics. *Materials*, *15*(18). Available from https://doi.org/10.3390/ma15186222, http://www.mdpi.com/journal/materials.

Qin, X., & Kaewunruen, S. (2022). Environment-friendly recycled steel fibre reinforced concrete. *Construction and Building Materials*, *327*, 126967.

Qin, Y., Zhang, X., & Chai, J. (2019). Damage performance and compressive behavior of early-age green concrete with recycled nylon fiber fabric under an axial load. *Construction and Building Materials*, *209*, 105–114. Available from https://doi.org/10.1016/j.conbuildmat.2019.03.094.

Rajeev, P., Ramesh, A., Navaratnam, S., & Sanjayan, J. (2023). Using fibre recovered from face mask waste to improve printability in 3D concrete printing. *Cement and Concrete Composites*, *139*. Available from https://doi.org/10.1016/j.cemconcomp.2023.105047.

Ram Kumar, P., Anjan, B. K., & Arjun, V. (2019). Assessment of lightweight concrete using expanded polystyrene beads. *India International Journal of Innovative Technology and Exploring Engineering*, *8*(8), 3234–3237. Available from https://www.ijitee.org/wp-content/uploads/papers/v8i8/H7220068819.pdf.

Rashid, K., Rehman, M. U., de Brito, J., & Ghafoor, H. (2020). Multi-criteria optimization of recycled aggregate concrete mixes. *Journal of Cleaner Production, 276*. Available from https://doi.org/10.1016/j.jclepro.2020.124316, https://www.journals.elsevier.com/journal-of-cleaner-production.

Ray, S., Haque, M., Sakib, M. N., Mita, A. F., Rahman, M. D. M., & Tanmoy, B. B. (2021). Use of ceramic wastes as aggregates in concrete production: A review. *Journal of Building Engineering, 43*. Available from https://doi.org/10.1016/j.jobe.2021.102567.

Reches, Y. (2018). Nanoparticles as concrete additives: Review and perspectives. *Construction and Building Materials, 175*, 483–495. Available from https://doi.org/10.1016/j.conbuildmat.2018.04.214.

Sadrolodabaee, P., Claramunt, J., Ardanuy, M., & de la Fuente, A. (2021). Mechanical and durability characterization of a new textile waste micro-fiber reinforced cement composite for building applications. *Case Studies in Construction Materials, 14*. Available from https://doi.org/10.1016/j.cscm.2021.e00492.

Samarakoon, S. M. S. M. K., Ruben, P., Wie Pedersen, J., & Evangelista, L. (2019). Mechanical performance of concrete made of steel fibers from tire waste. *Case Studies in Construction Materials, 11*. Available from https://doi.org/10.1016/j.cscm.2019.e00259.

Sengul, O. (2016). Mechanical behavior of concretes containing waste steel fibers recovered from scrap tires. *Construction and Building Materials, 122*, 649–658. Available from https://doi.org/10.1016/j.conbuildmat.2016.06.113.

Senhadji, Y., Siad, H., Escadeillas, G., Benosman, S., Chihaoui, R., Mouli, M., & Lachemi, M. (2019). Physical, mechanical and thermal properties of lightweight composite mortars containing recycled polyvinyl chloride. *Construction and Building Materials, 195*, 198–207. Available from https://doi.org/10.1016/j.conbuildmat.2018.11.070.

Shafigh, P., Mahmud, H. B., Jumaat, M. Z., & Zargar, M. (2014). Agricultural wastes as aggregate in concrete mixtures - A review. *Construction and Building Materials, 53*, 110–117. Available from https://doi.org/10.1016/j.conbuildmat.2013.11.074, https://www.journals.elsevier.com/construction-and-building-materials.

Sharma, R., & Bansal, P. P. (2016). Use of different forms of waste plastic in concrete - A review. *Journal of Cleaner Production, 112*, 473–482. Available from https://doi.org/10.1016/j.jclepro.2015.08.042.

Siad, H., Lachemi, M., Ismail, M. K., Sherir, M. A. A., Sahmaran, M., & Hassan, A. A. A. (2019). Effect of rubber aggregate and binary mineral admixtures on long-term properties of structural engineered cementitious composites. *Journal of Materials in Civil Engineering, 31*(11), 08991561. Available from https://doi.org/10.1061/(ASCE)MT.1943-5533.0002894, http://ascelibrary.org/mto/resource/1/jmcee7/.

Siad, H., Lachemi, M., Sahmaran, M., Mesbah, H. A., Hossain, K. M. A., & Ozsunar, A. (2017). Potential for using recycled glass sand in engineered cementitious composites. *Magazine of Concrete Research, 69*(17), 905–918. Available from https://doi.org/10.1680/jmacr.16.00447, http://www.icevirtuallibrary.com/content/serial/macr.

Siddika, A., Mamun, M. A. A., Alyousef, R., Amran, Y. H. M., Aslani, F., & Alabduljabbar, H. (2019). Properties and utilizations of waste tire rubber in concrete: A review. *Construction and Building Materials, 224*, 711–731. Available from https://doi.org/10.1016/j.conbuildmat.2019.07.108, https://www.journals.elsevier.com/construction-and-building-materials.

Silva, R. V., de Brito, J., & Dhir, R. K. (2018). Fresh-state performance of recycled aggregate concrete: A review. *Construction and Building Materials, 178*, 19–31. Available from https://doi.org/10.1016/j.conbuildmat.2018.05.149.

Simalti, A., & Singh, A. P. (2021). Comparative study on performance of manufactured steel fiber and shredded tire recycled steel fiber reinforced self-consolidating concrete. *Construction and Building Materials, 266*. Available from https://doi.org/10.1016/j.conbuildmat.2020.121102.

Sun, H., Guo, G., Memon, S. A., Xu, W., Zhang, Q., Zhu, J. H., & Xing, F. (2015). Recycling of carbon fibers from carbon fiber reinforced polymer using electrochemical method. *Composites Part A: Applied Science and Manufacturing, 78*, 10–17. Available from https://doi.org/10.1016/j.compositesa.2015.07.015.

Talamona, D., & Hai Tan, K. (2012). Properties of recycled aggregate concrete for sustainable urban built environment. *Journal of Sustainable Cement-Based Materials, 1*(4), 202–210. Available from https://doi.org/10.1080/21650373.2012.754571.

Tang, W., Monaghan, R., & Sajjad, U. (2023). Investigation of physical and mechanical properties of cement mortar incorporating waste cotton fibres. *Sustainability (Switzerland), 15*(11). Available from https://doi.org/10.3390/su15118779, http://www.mdpi.com/journal/sustainability/.

Thomas, L. M., & Moosvi, S. A. (2020). Hardened properties of binary cement concrete with recycled PET bottle fiber: An experimental study. *Materials Today: Proceedings, 32*, 632–637. Available from https://doi.org/10.1016/j.matpr.2020.03.025.

Tran, N. P., Gunasekara, C., Law, D. W., Houshyar, S., Setunge, S., & Cwirzen, A. (2022). Comprehensive review on sustainable fiber reinforced concrete incorporating recycled textile waste. *Journal of Sustainable Cement-Based Materials, 11*(1), 28–42. Available from https://doi.org/10.1080/21650373.2021.1875273.

Tran, N. P., Gunasekara, C., Law, D. W., Houshyar, S., & Setunge, S. (2023). Utilization of recycled fabric-waste fibers in cementitious composite. *Journal of Materials in Civil Engineering, 35*(1). Available from https://doi.org/10.1061/(ASCE)MT.1943-5533.0004538, http://ascelibrary.org/mto/resource/1/jmcee7/.

Tran, V. Q., Dang, V. Q., & Ho, L. S. (2022). Evaluating compressive strength of concrete made with recycled concrete aggregates using machine learning approach. *Construction and Building Materials, 323*, 126578.

Usman, A., Sutanto, M. H., Napiah, M., Huang, Y. F., Tan, K. W., Ling, L., & Leong, K. H. (2018). Effect of recycled plastic in mortar and concrete and the application of gamma irradiation - A review. *E3S Web of Conferences, 65*. Available from https://doi.org/10.1051/e3sconf/20186505027.

Wong, C. L., Mo, K. H., Yap, S. P., Alengaram, U. J., & Ling, T. C. (2018). Potential use of brick waste as alternate concrete-making materials: A review. *Journal of Cleaner Production, 195*, 226–239. Available from https://doi.org/10.1016/j.jclepro.2018.05.193, https://www.journals.elsevier.com/journal-of-cleaner-production.

Yan, B., Zhang, X., Wang, Z., & Shi, Y. (2023). Study on the effect of pre-treatment of recycled aggregate on the durability of concrete. *Structural seismic and civil engineering research, Informa UK Limited*, 178–185. Available from https://doi.org/10.1201/9781003384342-23.

Yao, Y., Wu, B., Zhang, W., Fu, Y., & Kong, X. (2023). Experimental investigation on the impact properties and microstructure of recycled steel fiber and silica fume reinforced recycled aggregate concrete. *Case Studies in Construction Materials, 18*. Available from https://doi.org/10.1016/j.cscm.2023.e02213.

Yildirim, S. T., Meyer, C., & Herfellner, S. (2015). Effects of internal curing on the strength, drying shrinkage and freeze-thaw resistance of concrete containing recycled concrete aggregates. *Construction and Building Materials, 91*, 288–296. Available from https://doi.org/10.1016/j.conbuildmat.2015.05.045.

Zareei, S. A., Ameri, F., Bahrami, N., Shoaei, P., Musaeei, H. R., & Nurian, F. (2019). Green high strength concrete containing recycled waste ceramic aggregates and waste carpet fibers: Mechanical, durability, and microstructural properties. *Journal of Building Engineering*, *26*. Available from https://doi.org/10.1016/j.jobe.2019.100914, https://www.journals.elsevier.com/journal-of-building-engineering.

Zareei, S. A., Ameri, F., Shoaei, P., & Bahrami, N. (2019). Recycled ceramic waste high strength concrete containing wollastonite particles and micro-silica: A comprehensive experimental study. *Construction and Building Materials*, *201*, 11–32. Available from https://doi.org/10.1016/j.conbuildmat.2018.12.161, https://www.journals.elsevier.com/construction-and-building-materials.

Zhan, P., Xu, J., Wang, J., Zuo, J., & He, Z. (2022). A review of recycled aggregate concrete modified by nanosilica and graphene oxide: Materials, performances and mechanism. *Journal of Cleaner Production*, *375*, 134116.

Zhang, P., Wang, C., Wu, C., Guo, Y., Li, Y., & Guo, J. (2022). A review on the properties of concrete reinforced with recycled steel fiber from waste tires. *Reviews on Advanced Materials Science*, *61*(1), 276–291. Available from https://doi.org/10.1515/rams-2022-0029, https://www.degruyter.com/view/j/rams.

Zhu, L., & Zhu, Z. (2020). Reuse of clay brick waste in mortar and concrete. *Advances in Materials Science and Engineering*, *2020*. Available from https://doi.org/10.1155/2020/6326178, http://www.hindawi.com/journals/amse/.

Sustainable fiber-reinforced geopolymer composites

Hui Zhong and Mingzhong Zhang
Department of Civil, Environmental and Geomatic Engineering, University College London, London, United Kingdom

10.1 Introduction

Geopolymers synthesized by mixing aluminosilicate materials such as fly ash (FA) and granulated blast-furnace slag (GGBS) with alkaline activators such as sodium hydroxide (NaOH) and sodium silicate (Na_2SiO_3) are considered as a promising alternative to ordinary Portland cement with less environmental impact. However, similar to conventional cementitious materials, geopolymers are inherently brittle and prone to cracking under various loadings, which can be effectively mitigated by adding randomly distributed fibers into geopolymers to produce fiber-reinforced geopolymer composites (FRGC) (Ranjbar & Zhang, 2020).

A variety of constituent materials especially different binders and fibers have been used to produce FRGC. Among them, industrial by-products including FA and GGBS as well as synthetic fibers such as steel, carbon, and polypropylene (PP) fibers have been commonly adopted. Apart from conventional FRGC for basic engineering applications, some high-performance FRGC including engineered geopolymer composites with extraordinary tensile ductility (Zhong & Zhang, 2023) and ultra-high-performance geopolymer concrete with superior compressive strength (Liu et al., 2020) were developed for some special applications, e.g., blast-resistant structures. In the past decade, several review papers have been published to discuss the state-of-the-art of FRGC, mainly focusing on the effect of fibers in terms of fiber type, fiber shape, fiber aspect ratio, and fiber content on the fresh and mechanical properties as well as durability of FRGC (Abbas et al., 2022; Li, Shumuye, et al., 2022; Ranjbar & Zhang, 2020). However, these papers mostly focused on FRGC made from the commonly used aluminosilicate materials (especially FA), natural river sand or silica sand and synthetic fibers such as steel and polymeric fibers. In recent years, an increasing number of studies have attempted to further improve the sustainability of FRGC using some wastes or natural materials as binders (e.g., steel slag [SS] from the steel-making process and mine tailings), aggregates (e.g., crumb rubber from end-of-life tires) and fibers (e.g., recycled and plant fibers). Table 10.1 summarizes the constituent materials, mix design, and curing conditions of these sustainable FRGC. This chapter presents a systematic summary of recent advances in sustainable FRGC, with special focus on the effects of sustainable binders, aggregates, and fibers on the engineering properties and durability

Sustainable Concrete Materials and Structures. DOI: https://doi.org/10.1016/B978-0-443-15672-4.00010-3

Table 10.1 Summary of constituent materials, mix design, and curing condition collected from studies on sustainable fiber-reinforced geopolymer composites.

References	Binder (weight ratio)	Activator	L/b (or W/s or W/b)	Additive (Add/b)	Aggregate		Fiber		Curing regime
					Type	Agg/b	Type (%, by volume)	Type (L_f/d_f, mm/μm) [strength, MPa]	
Sustainable binder									
Salami et al. (2016, 2017)	POFA (1.0)	NaOH (8 M) + Na_2SiO_3	L/b (0.50–0.65)	SP (0, 0.05, 0.1)	Dune sand	1.8	PVA (2.0)	PVA (8/40) [1600]	65°C (1 day) + air curing at 25°C (1–26 days)
Kan et al. (2020)	GGBS (0.32–0.81) + SF (0.08–0.19) + IFA (0–0.6)	NaOH + Na_2SiO_3	L/b (0.53)	–	Sand	0.26	PE (–)	PE (12/24) [3000]	80°C (2 h) + air curing (28 days)
Zhao et al. (2021)	FA (0.23–0.30) + GGBS (0.52–0.70) + SS (0–0.25)	NaOH + Na_2SiO_3	L/b (0.52)	–	Sand	0.3	PE (–)	PE (12/24) [3000]	100°C (2 h) + air curing (28 days)
Yoo et al. (2022)	GGBS (0.5) + LCDG (0.5)	NaOH (4 M) + Na_2SiO_3	W/b (0.35)	SP (0.02)	Silica sand	0–1.0	PE (2.0)	PE (12/20) [–]	Air curing (7 days) + water curing at 85°C (1 day) + air curing (2 days)

Kan and Wang (2022)	GGBS (0.73, 0.81) + SF (0.17, 0.19) + RM (0, 0.1)	NaOH + Na_2SiO_3	L/b (0.53)	–	River sand	0.26	PE (1.9)	PE (12/24) [3000]	20°C and 70% RH (27 days), 80°C (2 h) + 20°C and 70% RH (27 days)
Li, Yin, et al. (2022)	C-CSMR (0.85, 1.0) + GGBS (0, 0.15)	NaOH + Na_2SiO_4	W/b (0.32, 0.36, 0.4)	–	Silica sand	0.2	PVA (2.0)	PVA (12/40) [1500]	20°C and 98% RH (27 days)
Adesina and Das (2023)	GCC (0–0.75) + GGBS (0.25–1.0)	Lime	W/b (0.27)	–	Silica sand	0.8	PVA (2.0)	PVA (–) [1600]	Moist curing (27 and 89 days)
Lee et al. (2023)	LCDG (0.5) + GGBS (0.5)	NaOH + Na_2SiO_3	W/b (0.35)	SP (0.02)	Silica sand	0.2	PE (2.0)	PE (6–18/20, 18/30) [3000]	Air curing (7 days) + water curing at 85°C (1 day) + air curing (2 days)
Sustainable aggregate									
Nematollahi et al. (2017)	FA (1.0)	NaOH (8 M) + Na_2SiO_3	L/b (0.45)	–	Silica sand, Expanded glass, Ceramic, Expanded perlite	0–0.6	PVA (2.0)	PVA (8/40) [1600]	60°C (1 day) + air curing (2 days)
Zhong et al. (2019)	FA (0.8) + GGBS (0.2)	NaOH (10 M) + Na_2SiO_3	L/b (0.4)	SP (0.01)	River sand + crumb rubber	2.0	RTS (0.5, 1.0)	RTS (23/220) [2570]	20°C and 50% RH (6 and 27 days)

(Continued)

Table 10.1 (Continued)

References	Binder (weight ratio)	Activator	L/b (or W/s or W/b)	Additive (Add/b)	Aggregate		Fiber		Curing regime
Lương et al. (2021)	GGBS (0.94, 1.0) + SF (0, 0.06)	$Ca(OH)_2$	W/s (0.18)	SP (0.03–0.06) + defoamer (0.001)	Crumb rubber	0, 0.06, 0.11	PE (1.75)	PE (18/12) [2700]	Water curing at 23°C (26 days)
Yaswanth et al. (2022a, 2022b)	GGBS (0.97) + RHA (0.03)	NaOH (8 M) + Na_2SiO_3	W/s (0.3)	SP (0.07)	Silica sand, Silica sand + copper slag	2.25–2.70	PVA (2.0)	PVA (6/39) [1604]	Air curing at 25°C (3–90 days)
Gholampour et al. (2022)	FA (0.5, 0.8) + GGBS (0.2, 0.5)	NaOH (12 M) + Na_2SiO_3	L/b (0.55, 0.75)	–	River sand, Lead smelter slag, Waste glass sand	2.4, 2.5, 3.1	Coir (1.0), Ramie (1.0, 2.0), Sisal (1.0), Hemp (1.0), Jute (1.0), Bamboo (1.0)	L_f < 2.5 mm	Air curing (6, 13, and 27 days)
Tahwia et al. (2022)	GGBS (0.76) + SF (0.24)	KOH + Na_2SiO_3	W/b (0.35)	–	River sand + waste glass sand	1.33, 1.34	Steel (3.0)	Steel (13/120) [–]	80°C (2 days) + air curing (4–117 days)

Lao et al. (2023)	FA (0.19, 0.76) + GGBS (0.19, 0.76) + SF (0.05)	Na_2SiO_3 (anhydrous) + waterglass	W/b (0.25)	Borax (0.038)	Sea sand	0.3	PE (2.0)	PE (18/24) [3000]	Air curing (27 days)
Nuaklong et al. (2023)	FA (1.0)	NaOH (10 M) + Na_2SiO_3	L/b (0.6)	–	Limestone + river sand, Limestone + river sand + granite industry waste	3.7	PP (0.5, 1.0)	PP (48/–) [640]	Air curing (6, 27, and 89 days)
Sustainable fiber									
Alomayri et al. (2013)	FA (1.0)	NaOH (8 M) + Na_2SiO_3	L/b (0.35)	–	–	–	Cotton (1.4%–4.1% by weight)	Cotton (–/200) [400]	105°C (3 h) + air curing (29 days)
Chen et al. (2014)	FA (1.0)	NaOH (10 M)	L/b (0.36)	–	–	–	Sweet sorghum (1.0%, 2.0%, 3.0% by weight)	–	60°C (7 days)
Malenab et al. (2017)	FA (1.0)	NaOH (12 M) + Na_2SiO_3	L/b (0.4)	–	–	–	Abaca (1.0% by weight)	Abaca (250/162) [210–450]	75°C (1 day) + air curing (28 days)
Nkwaju et al. (2019)	Lateritic soil (1.0)	NaOH (9 M) + Na_2SiO_3	L/b (0.6)	–	–	–	Sugarcane bagasse (1.5%–7.5% by weight)	Sugarcane bagasse (20–100/–) [–]	Bagged curing (27 days)

(Continued)

Table 10.1 (Continued)

References	Binder (weight ratio)	Activator	L/b (or W/s or W/b)	Additive (Add/b)	Aggregate		Fiber		Curing regime
Wongsa et al. (2020)	FA (1.0)	NaOH (10 M) + Na_2SiO_3	L/b (0.75)	–	River sand	2.75	Coir (0.5, 0.75, 1.0)	Coir (35–40/117) [–]	60°C (2 days) + 25°C and 50% RH (5 days)
Silva et al. (2020)	Milled fire clay brick (1.0)	NaOH + Na_2SiO_3	W/b (0.27)	–	–	–	Jute (0.5%–2.0% by weight), Sisal (0.5%–3.0% by weight)	Jute (10/53) [276], Sisal (10/137) [508]	65°C (3 days) + air curing (4 days)
Poletanovic et al. (2020)	FA (0.5, 1.0) + GGBS (0, 0.5)	Na_2SiO_3	L/b (0.79–0.90)	–	River sand	3.0	Hemp (0.5, 1.0)	Hemp (10/8–60) [270–900]	80°C (1 day) + air curing (27 days)
Nguyen and Mangat (2020)	GGBS (1.0)	NaOH + Na_2SiO_3	L/b (0.66)	Retarder (0.0076)	Sand	1.94	Rice straw (1.0%, 2.0%, 3.0% by weight)	–	20°C and 65% RH (1 day), 20°C and 65% RH (1 day) + water curing (13 and 27 days)
Wang et al. (2020)	FA (0.8) + GGBS (0.2)	NaOH (10 M) + Na_2SiO_3	L/b (0.4)	SP (0.01)	Silica sand	0.2	PVA (1.5, 1.75) + RTS (0.25, 0.5)	PVA (12/40) [1600], RTS (20/150) [2850]	20°C and 60% RH (6 and 27 days)

Lazorenko et al. (2020)	FA (1.0)	NaOH (8 M) + Na_2SiO_3	L/b (0.4)	–	–	–	Flaw tow (0.25%–1.0% by weight)	Flax tow (15/–) [2200]	75°C (1 day) + 20°C and 50% RH (28 days)
Zhong and Zhang (2021); and Zhong and Zhang (2022a, 2022b)	FA (0.8) + GGBS (0.2)	NaOH (10 M) + Na_2SiO_3	L/b (0.45)	SP (0.01)	Silica sand	0.2	PVA (1.0, 1.5, 1.75, 2.0) + RTP (0–1.0)	PVA (12/40) [1600], RTP (5.2/21.4) [761]	20°C and 95% RH (27 days)
Yanou et al. (2021)	Laterite soil (1.0)	NaOH (10 M) + Na_2SiO_3	L/b (0.6)	–	–	–	Sugarcane bagasse (1.5%–7.5% by weight)	–	Air curing (28 days)
Bayraktar et al. (2023)	GGBS (1.0)	Sodium metasilicate pentahydrate	W/b (0.35)	SP (0.02)	Expanded perlite	0.16	Hemp (0.5–3.0)	Hemp (10, 20/–) [–]	75°C (1 day) + 22°C and 65% RH (6 and 27 days)
Zhong and Zhang (2023)	FA (0.8) + GGBS (0.2)	NaOH (10 M) + Na_2SiO_3	L/b (0.45)	SP (0.01)	Silica sand	0.2	PVA (1.5, 1.75) + RTS (0.25, 0.5)	PVA (12/40) [1600], RTS (12.8/220) [2165]	20°C and 95% RH (27 days)

Add/b, Additive-to-binder ratio; *C-CSMR*, calcined cutter soil mixing residue; d_f, diameter of fiber; *FA*, fly ash; *GCC*, granulated calcium carbonate; *GGBS*, ground granulated blast-furnace slag; *IFA*, incineration fly ash; *LCDG*, liquid crystal display glass; L_f, length of fiber; *L/b*, liquid-to-binder ratio (the mass ratio of total liquid over the mass of solid from the binder); *MK*, metakaolin; *PE*, polyethylene; *PP*, polypropylene; *POFA*, palm oil fuel ash; *PVA*, polyvinyl alcohol; *RHA*, rice husk ash; *RM*, red mud; *RTS*, recycled tire steel; *RTP*, recycled tire polymer; *SF*, silica fume; *SS*, steel slag; *SP*, superplasticizer; *W/b*, water-to-binder ratio (the mass ratio of total water over the mass of solid from the binder); *W/s*, water-to-solid ratio (the mass ratio of total water over the mass of solids from the binder and alkaline activator).

of FRGC. The mix design and production process of these sustainable FRGC as well as their sustainability are also discussed.

10.2 Mix design and production of sustainable fiber-reinforced geopolymer composites

10.2.1 Constituent materials and mix design

Palm oil fuel ash (POFA) from palm oil industry (Salami et al., 2016, 2017) was used alone as the precursor to produce FRGC. SS (Zhao et al., 2021), incineration fly ash (IFA) from municipal solid waste incineration (Kan et al., 2020), liquid crystal display glass (LCDG) from waste electronics (Lee et al., 2023; Yoo et al., 2022), red mud (RM) from extraction of alumina (Kan & Wang, 2022), cutter soil mixing residue generated from the cutter soil mixing technique for ground improvement (Li, Yin, et al., 2022), and ground calcium carbonate (GCC) (Adesina & Das, 2023) were mostly combined with GGBS to synthesize FRGC. It should be noted that the cutter soil mixing residue was activated by alkaline solution to produce calcined cutter soil mixing residue (C-CSMR) before mixing with GGBS and fine silica sand (Li, Yin, et al., 2022). The adopted weight ratios of these materials had a wide range, from 0.05 to 1.0.

A blend of NaOH and Na_2SiO_3 was the most frequently used activator for FRGC, as shown in Table 10.1. A few studies have used solid activators (anhydrous Na_2SiO_3) to address the potential problems of utilizing liquid activators for large-scale applications (Bayraktar et al., 2023; Lao et al., 2023; Lương et al., 2021). A small amount of superplasticizer was generally used to adjust the flowability of the matrix (typically less than 2.0% by the mass of binder) as the rheology of the matrix can significantly affect the fiber distribution and orientation in FRGC (Zhong & Zhang, 2023).

Natural river sand and limestone were normally used as fine and coarse aggregates for ordinary FRGC. For engineered geopolymer composites, fine silica sand was commonly utilized as fine aggregate and coarser aggregates are prohibited since they would impair the tensile strain-hardening behavior (Khan et al., 2016). The adopted aggregate-to-binder ratio (Agg/b) for FRGC mostly ranged from 0.2 to 0.3. To improve the sustainability of FRGC and meanwhile reduce the amount of solid wastes (e.g., end-of-life tires), expanded glass, ceramic microsphere, expanded perlite, crumb rubber, copper slag, granite industry waste, sea sand, waste glass sand, and lead smelter slag were introduced to fully or partly substitute the commonly used fine aggregate in FRGC (Gholampour et al., 2022; Lao et al., 2023; Lương et al., 2021; Nematollahi et al., 2017; Nuaklong et al., 2023; Tahwia et al., 2022; Yaswanth et al., 2022a; Zhong et al., 2019).

As mentioned previously, natural fibers especially plant fibers have been used as reinforcement for geopolymers to reduce the environmental impact associated with the manufacture of synthetic fibers. Natural fibers have several advantages against synthetic fibers, such as renewable, cheap, and low carbon footprint (Zhang, Cai, et al., 2020), which mostly consist of cellulose, hemicellulose, and lignin (Santana et al., 2021). As seen in Table 10.1, many types of plant fibers were used in FRGC including coir, hemp, sisal, jute, cotton, abaca, flax, sweet sorghum, rice straw, and sugarcane bagasse fibers. It is worth mentioning that the chemical composition and properties of these plant fibers vary with the plant species, region, climate, and harvesting period (Santana et al., 2021). Some studies applied alkaline treatment to pretreat the plant fibers to remove the impurities and enhance the fiber–matrix interaction (Chen et al., 2014; Malenab et al., 2017; Nguyen & Mangat, 2020). Apart from plant fibers, recycled tire fibers including recycled tire steel (RTS) and recycled tire polymer (RTP) fibers were utilized to partly replace polyvinyl alcohol (PVA) fibers in FRGC to reduce the material cost and environmental impact (Wang et al., 2020; Zhong & Zhang, 2021). These tire fibers can be obtained from waste tires through mechanical shredding and cryogenic and thermal degradation of tires (Pilakoutas et al., 2004). The highest fiber dosage for FRGC was 3.0% by the mass of binder.

10.2.2 Production process

The mixing process of FRGC specimens depends on the form of used alkaline activator (liquid or solid) and the type of fiber. The liquid activator should be separately prepared before the mixing of FRGC. Most studies began with the dry mixing of binders and aggregates (if any), followed by adding liquid activator and additive (if any). When a flowable and consistent mixture was obtained, fibers were slowly and gradually added to avoid fiber clumping or balling. If the solid activator was used, the only difference was that it should be firstly dry mixed with the binder and aggregate. A few studies dry mixed the binders and fibers (jute, sisal, and sweet sorghum fibers) before the addition of liquid activators (Chen et al., 2014; Silva et al., 2020). To ensure acceptable fiber distribution and orientation, RTP fibers should be first mixed with part of the liquid activator and superplasticizer (Zhong & Zhang, 2021). Upon the completion of mixing, all fresh samples were poured into molds with different sizes. Most studies adopted ambient temperature curing or a combination of short-period elevated temperature curing and long-period ambient temperature air curing as the curing regime.

10.3 Engineering properties of sustainable fiber-reinforced geopolymer composites

10.3.1 Workability

The flow table test was the common approach to assessing the workability of FRGC mixes. The workability of FRGC containing expanded glass, ceramic microsphere, or expanded perlite was found to be 24%, 16%, and 19%, respectively, lower than that of FRGC with silica sand, which can be ascribed to the high water absorption of expanded glass (Nematollahi et al., 2017). In the presence of 0.5% RTS fiber, the spread diameter of FRGC was increased from 164 to 167 and 181 mm, respectively, when the river sand replacement level (by volume) by crumb rubber changed from 5% to 10% and 15% (Zhong et al., 2019), which can be attributed to the relatively smooth surface texture of crumb rubber as compared with sand and the hydrophobicity of crumb rubber. It is worth noting that the surface texture condition of crumb rubber is highly dependent on its recycling process. Owing to similar reasons, replacing sand with waste glass sand up to 22.5% (by volume) slightly enhanced the workability of FRGC containing 3.0% steel fiber (Tahwia et al., 2022). Nevertheless, an inconsistent finding was reported that fully replacing natural river sand with waste glass sand resulted in a 45% drop in flow value for FRGC containing 1.0% coir fiber, due to to the higher angularity and smaller particle size of waste glass sand relative to natural river sand (Gholampour et al., 2022). Replacing silica sand with 20% copper slag and natural river sand with 25%–50% granite industry waste was beneficial to the workability of FRGC (Nuaklong et al., 2023; Yaswanth et al., 2022a).

Fig. 10.1 presents the effect of fiber content on the workability of FRGC, indicating that similar to other cementitious and geopolymer composites, increasing the

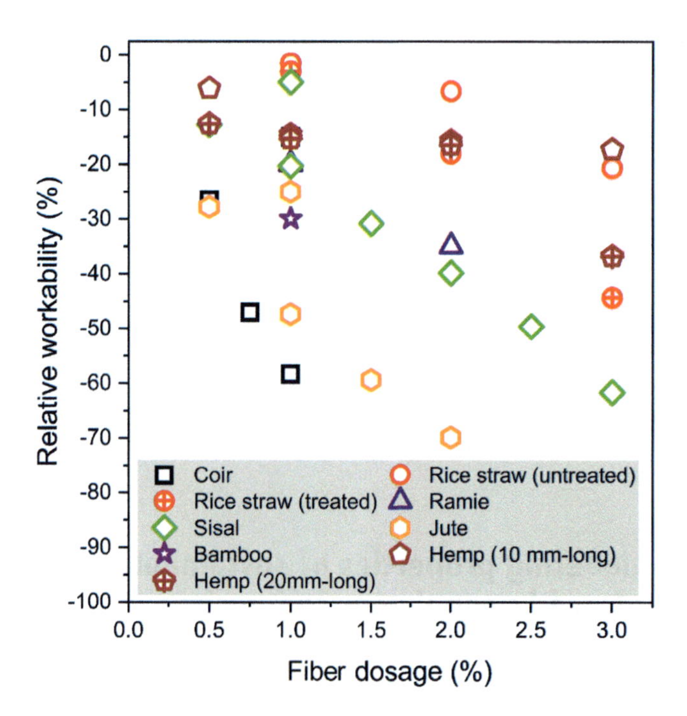

Figure 10.1 Effect of fiber dosage on workability of fiber-reinforced geopolymer composites.
Source: Data from Bayraktar et al. (2023), Gholampour et al. (2022), Nguyen and Mangat (2020), Silva et al. (2020), and Wongsa et al. (2020).

fiber dosage considerably weakened the workability of FRGC. The main reasons behind this are associated with the contact network between constituent materials, especially between paste/mortar and fibers, liquid absorption content by fibers, and critical fiber dosage (Khayat et al., 2019; Ranjbar & Zhang, 2020). By comparing the effect of different natural fiber types on the workability, FRGC with sisal fibers exhibited the highest flow value, while bamboo-FRGC had the poorest workability (Gholampour et al., 2022). Regardless of fiber content, the workability of FRGC with treated rice straw fibers was 3%–40% lower than that with untreated rice straw fibers, due to the higher surface area of treated rice straw fibers (Nguyen & Mangat, 2020). Replacing PVA fibers with either RTS or RTP fibers mostly did not benefit the workability of the resultant composites primarily due to the ununiform dimension of these recycled fibers (Wang et al., 2020; Zhong & Zhang, 2021, 2023). Using sieved RTS fibers to replace PVA fibers can substantially mitigate the workability loss of FRGC caused by as-received RTS fibers (Zhong et al., 2023).

10.3.2 Density

Owing to the lower density of crumb rubber as compared with binding materials, the density of FRGC with 5% and 10% crumb rubber was 5% and 10% lower than that of FRGC without any crumb rubber (Lương et al., 2021). FA can be combined with the expanded glass or ceramic microsphere to develop lightweight FRGC mixtures (Nematollahi et al., 2017). The effect of fiber dosage on the density of FRGC is shown in Fig. 10.2, indicating that the presence of plant fibers can increase, decrease or not change the density of the resultant composites. For instance, the density of FRGC was increased by 20% when 3.0% 10-mm-long hemp fiber reinforced the matrix (Bayraktar et al., 2023), while a 6.8% drop in density was captured for FRGC in the presence of 3.0% sweet sorghum fiber (Chen et al., 2014). It should be stated that the density of the reinforcing fiber is not the dominant factor affecting the density of the resultant composite, while the entrapped air content caused by the addition of fibers can significantly contribute to the reduced density (Ranjbar & Zhang, 2020), as confirmed by some studies that increasing the fiber content increased the porosity of the whole composite (Wongsa et al., 2020; Yanou et al., 2021).

10.3.3 Drying shrinkage

When the concrete specimen is exposed to a drying condition, drying shrinkage can occur which would promote the appearance of cracks and consequently affect the mechanical properties and durability of concrete. The drying shrinkage of geopolymers was found to be relatively higher than that of cement-based materials, which can be associated with the higher volume fraction of mesopores inside the geopolymers that would produce a higher compressive stress on their solid skeleton (Lee et al., 2014; Ma & Ye, 2015). As mentioned in Section 10.2.1, the coarse aggregates are inhibited in some FRGC mixes to ensure high tensile ductility and the

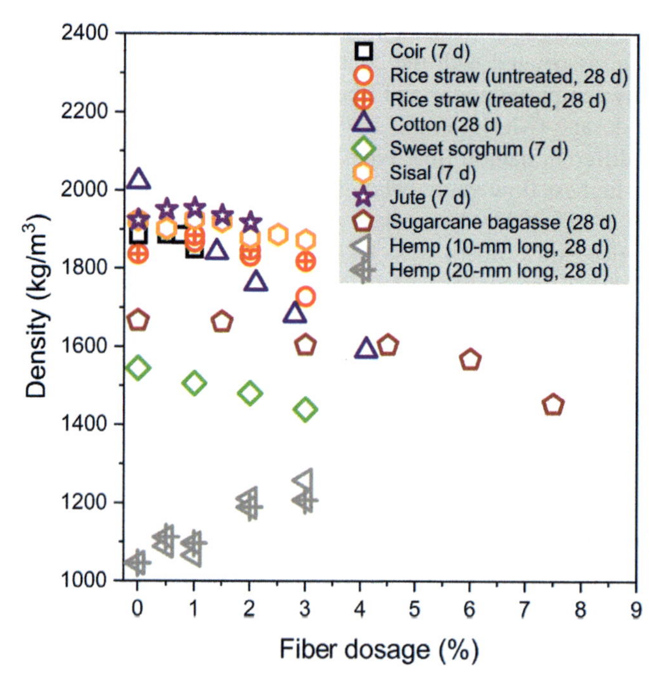

Figure 10.2 Effect of fiber dosage on density of fiber-reinforced geopolymer composites. *Source*: Data from Bayraktar et al. (2023), Nguyen and Mangat (2020), Wongsa et al. (2020), Alomayri et al. (2013), Chen et al. (2014), Silva et al. (2020), and Yanou et al. (2021).

adopted Agg/b was also low, which may pose challenges to the drying shrinkage resistance of these FRGC samples.

Fig. 10.3 displays the effect of sustainable fiber on the drying shrinkage of FRGC at various testing ages. A pronounced trend can be observed that irrespective of fiber type, the drying shrinkage of FRGC tended to be lower at a higher fiber dosage. The fiber–matrix interaction of FRGC can be enhanced with increasing curing age, which can improve the dimensional stability of FRGC as fibers and aggregates can be combined to form a strong skeleton (Fang et al., 2020). Besides, the fiber bridging behavior can prevent the formation of cracks caused by drying shrinkage and can also avoid or slow down moisture loss (Afroughsabet & Teng, 2020; Gao et al., 2018). Replacing PVA fibers with RTS or RTP fibers enhanced the drying shrinkage resistance of FRGC, which can be strongly related to the intrinsic properties of recycled fibers (i.e., high elastic modulus of RTS fibers and hydrophobic surface nature of RTP fibers) and the reduced porosity (Wang et al., 2020; Zhong & Zhang, 2021; Zhong et al., 2023). The presence of 1.0% treated rice straw fiber resulted in lower drying shrinkage for FRGC than that of 1.0% untreated rice straw fiber, while composites containing untreated rice straw fibers outperformed those incorporating treated rice straw fibers when the added fiber content was 2.0% or 3.0% (Nguyen & Mangat, 2020). For instance, the 10-day

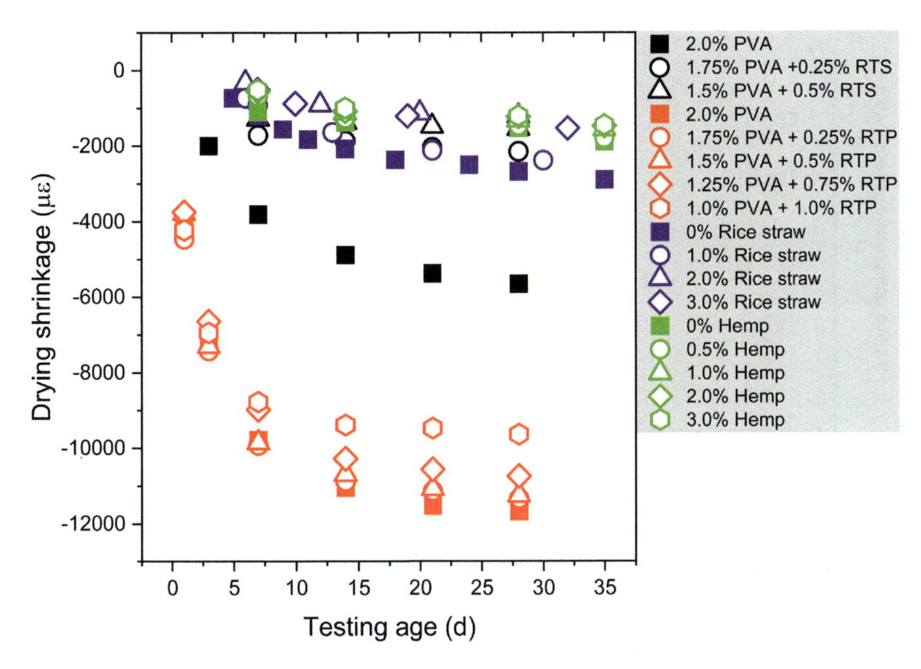

Figure 10.3 Drying shrinkage of fiber-reinforced geopolymer composites with different fiber type and dosage.
Source: Data from Zhong and Zhang (2021), Bayraktar et al. (2023), Nguyen and Mangat (2020), and Wang et al. (2020).

drying shrinkage of FRGC with 3.0% untreated rice straw fiber was 32% lower than that with the same fiber dosage of treated rice straw fiber.

10.3.4 Compressive behavior

10.3.4.1 Effect of binder

Fig. 10.4A presents the effect of binder proportion on the compressive strength of FRGC. The presence of sustainable binder materials mostly impaired the compressive strength of composites. For instance, utilizing IFA to replace 10% GGBS and silica fume (SF) increased the compressive strength of FRGC with polyethylene (PE) fibers by 29%, while the further replacement of IFA reduced the compressive strength by 28%−76% (Kan et al., 2020). It can be attributed to the competition between two different mechanisms, where the first mechanism was conducive to the strength of specimens due to the high alkaline metal and CaO contents of IFA, while the second mechanism weakened the strength of specimens due to the lower reactivity of IFA as compared with GGBS and the poisoning effects of heavy metal inside IFA. The first mechanism was dominant when the IFA dosage was lower whereas the second mechanism took over at a higher IFA dosage. Similarly, due to the simultaneous working mechanisms of positive and negative effects, only the use

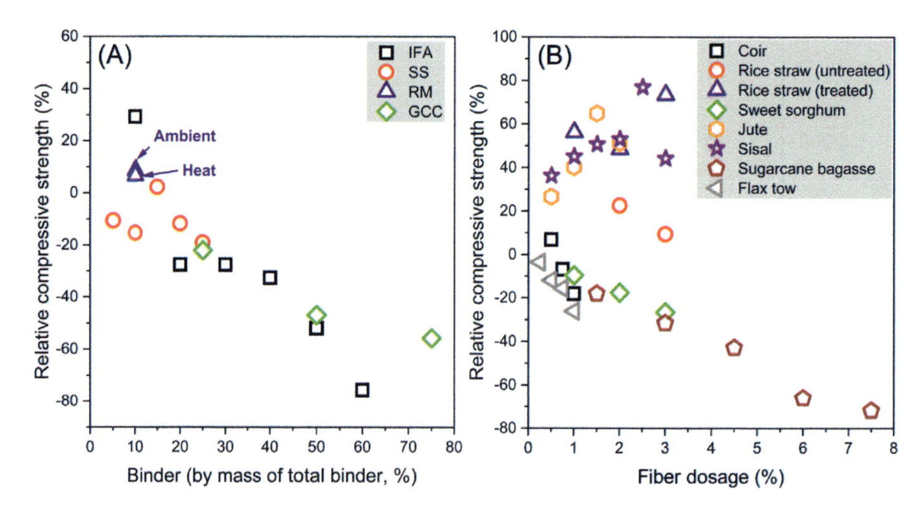

Figure 10.4 Effects of (A) binder (Kan et al., 2020; Kan & Wang, 2022; Zhao et al., 2021; Li, Yin, et al., 2022) and (B) fiber (Nguyen & Mangat, 2020; Wongsa et al., 2020; Chen et al., 2014; Lazorenko et al., 2020; Silva et al., 2020; Yanou et al., 2021) on compressive strength of fiber-reinforced geopolymer composites.
Source: Data from Kan et al. (2020), Kan and Wang (2022), Zhao et al. (2021), Li, Yin, et al. (2022), Nguyen and Mangat (2020), Wongsa et al. (2020), Chen et al. (2014), Lazorenko et al. (2020), Silva et al. (2020), and Yanou et al. (2021).

of a certain SS replacement level (15%) would not considerably weaken the compressive strength of FRGC (Zhao et al., 2021). Under both curing regimes (ambient and heat curing), the incorporation of RM had no significant influence on the compressive strength of FRGC (Kan & Wang, 2022).

10.3.4.2 Effect of aggregate

The presence of 5%–15% crumb rubber reduced the compressive strength of FRGC containing RTS or PE fibers (Lương et al., 2021; Zhong et al., 2019), which can be mainly assigned to the lower stiffness of crumb rubber compared to other matrix components and the weak bonding between the crumb rubber and the matrix (Ganjian et al., 2009; Wang et al., 2019). Replacing silica sand with 20%–40% copper slag led to a 25%–32% rise in compressive strength for FRGC made from GGBS and rice husk ash (RHA) primarily owing to the formation of ferrosialate (Yaswanth et al., 2022a). Due to the increased porosity caused by the high angularity of waste glass sand and its poor interaction with the matrix, the compressive strengths of FRGC at curing ages of 7–90 days were weakened when more sand was replaced with waste glass sand (Tahwia et al., 2022). Differently, it was reported that replacing 100% natural river sand with waste glass sand or lead smelter slag enhanced the compressive strength of FRGC (Gholampour et al., 2022). Using seawater to prepare sea sand FRGC led to a slightly lower compressive strength at 28 days than that prepared by using fresh water (Lao et al., 2023).

This has been explained as that the reaction between seawater and alkaline activator (sodium silicate solution) forms the magnesium silicate hydrate and amorphous SiO_2, reducing the alkali and soluble silica contents in the activator and thereby reducing the ultimate compressive strength of FRGC.

10.3.4.3 Effect of fiber

Fig. 10.4B illustrates the effect of fiber content on the compressive strength of FRGC. Increasing the plant fiber dosage can either increase or decrease the compressive strength of FRGC. The presence of coir (Wongsa et al., 2020), sweet sorghum (Chen et al., 2014), and sugarcane bagasse fibers (Yanou et al., 2021), as well as flaw tows (Lazorenko et al., 2020) reduced the compressive strength of FRGC possibly due to the increased porosity, poor fiber–matrix interaction, and insufficient fiber bridging capacity. By contrast, regardless of fiber pretreatment, the incorporation of 1.0%–3.0% rice straw fiber increased the compressive strength of geopolymers by 9%–73%, which can be attributed to the hydrophilic nature of rice straw fibers and excellent bond behavior between fibers and the matrix (Nguyen & Mangat, 2020). Treated rice straw fibers were more effective than untreated fibers in enhancing the compressive strength of FRGC mainly due to the high surface area of treated fibers that can absorb more liquid and can better improve the bonding behavior with the matrix. It was also found that the presence of jute and sisal fibers with various contents can considerably improve the compressive strength of geopolymers made from milled fire clay brick (Silva et al., 2020). Similar findings were observed for geopolymers reinforced with pretreated abaca fibers (Malenab et al., 2017) and hemp fibers with lengths of 10 and 20 mm (Bayraktar et al., 2023). A comparison of the effects of different natural fibers (coir, ramie, sisal hemp, jute and bamboo fibers) on the compressive strength of FRGC indicated that FRGC containing ramie fibers had the highest compressive strength, while FRGC with jute fibers exhibited the lowest compressive strength.

Similar to drying shrinkage, substituting PVA fibers with either RTS or RTP fibers resulted in a higher compressive strength for FA-GGBS–based FRGC (Wang et al., 2020; Zhong & Zhang, 2021; Zhong et al., 2023). Some typical compressive failure patterns of unreinforced geopolymers and FRGC mixes are illustrated in Fig. 10.5. Unreinforced geopolymers showed a brittle and catastrophic failure, while FRGC mixes can maintain the structural integrity of specimens.

10.3.5 Tensile behavior

10.3.5.1 Effect of binder

As aforementioned, a special class of FRGC called engineered geopolymer composites with a tensile strain capacity of up to 13.68% can be developed based on micromechanics theory (Li, 2019; Zhong & Zhang, 2023). Recently, sustainable binders including IFA, SS, C-CSMR, RM, and LCDG were applied to prepare such high-ductile geopolymer composites. As seen in Fig. 10.6A, the mixture with 100%

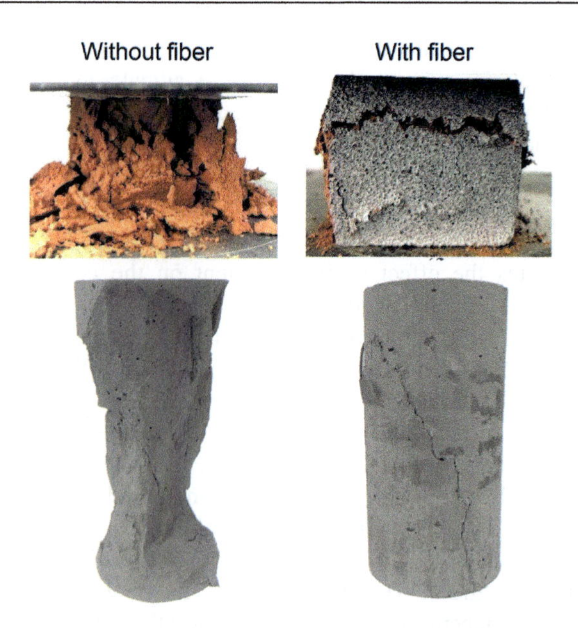

Figure 10.5 Compressive failure patterns of specimens with or without fibers.
Source: Data from Silva et al. (2020) and Zhong and Zhang (2022b).

C-CSMR possessed a tensile strain capacity of 5.13% which was 3.1 times greater than that with 85% C-CSMR and 15% GGBS (Li, Yin, et al., 2022). This can be strongly related to the higher matrix fracture toughness and superior fiber–matrix interaction of composites when 15% GGBS was added. Owing to the densified microstructure, the tensile strength of FRGC made from C-CSMR and GGBS was 41% higher than that made from only C-CSMR. Similar to compressive strength, the optimum SS and IFA replacement levels were 15% and 10%, respectively, considering the tensile strength and tensile strain capacity (Kan et al., 2020; Zhao et al., 2021). Consistent with compressive strength, the presence of RM did not substantially change the tensile strength of FRGC, while the tensile strain capacity of heat-cured FRGC was improved from 4.52% to 5.63% when 10% RM was used to replace GGBS and SF (Kan & Wang, 2022).

10.3.5.2 Effect of aggregate

As seen in Fig. 10.6B, adding 5% crumb rubber to FRGC led to better tensile strain capacity while retaining acceptable tensile strength (Lương et al., 2021). The improved tensile strain capacity was mainly due to the reduction of matrix fracture toughness (Li, 2019). Additionally, more active flaws can be introduced by the presence of crumb rubber, which would enhance the multiple cracking behavior of FRGC. Herein, active flaw stands for the flaw that can generate a new crack when the induced tensile stress is lower than the fiber bridging capacity. This can be also

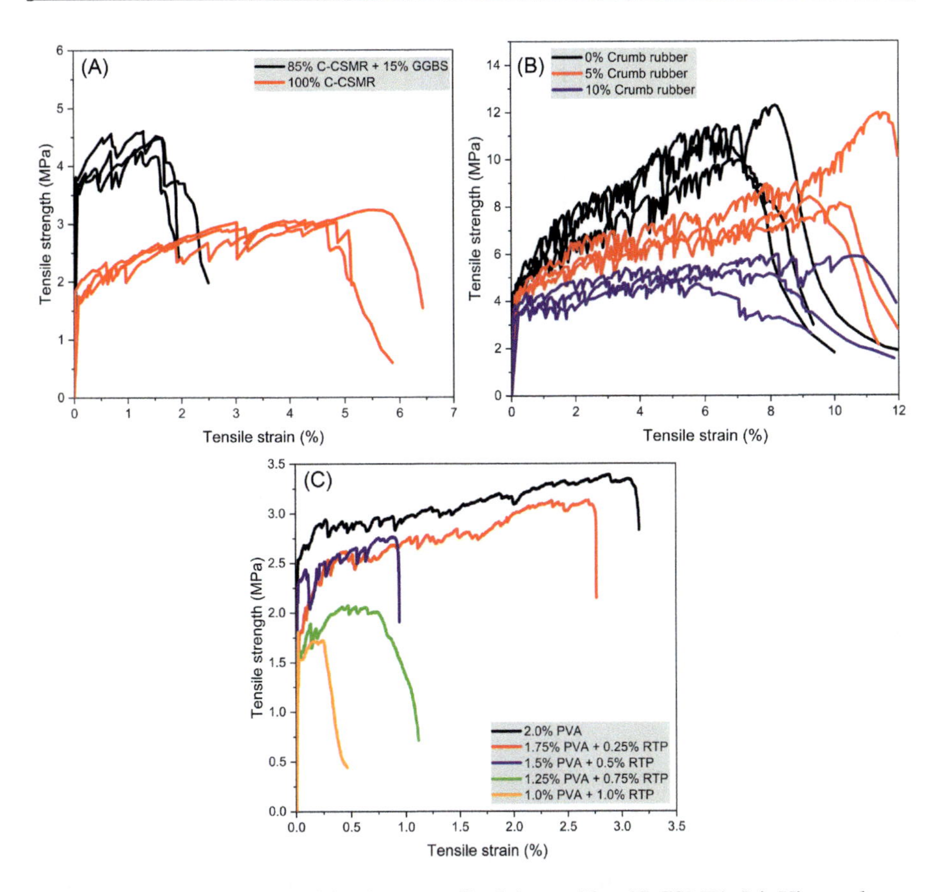

Figure 10.6 Effects of (A) calcined cutter soil mixing residue (C-CSMR) (Li, Yin, et al., 2022), (B) crumb rubber (Lương et al., 2021), and (C) recycled tire polymer (RTP) fiber (Zhong & Zhang, 2021) on tensile stress–strain response of fiber-reinforced geopolymer composites.
Source: Data from Li, Yin, et al. (2022), Lương et al. (2021), and Zhong and Zhang (2021).

evidenced in Fig. 10.7 that increasing Agg/b of LCDG-based FRGC consistently reduced its tensile strain capacity and crack number (Yoo et al., 2022). The matrix fracture toughness of FRGC went up with rising Agg/b which would increase the size of the active flaw (Li, 2019). As compared with FRGC made from silica sand, the tensile strength of FRGC made from expanded glass, ceramic microsphere, or expanded perlite was not enhanced (Nematollahi et al., 2017). The effect of copper slag on the tensile strength of GGBS-RHA–based FRGC was consistent with that on the compressive strength, where replacing 40% silica sand with copper slag led to the highest tensile strength for the composite (Yaswanth et al., 2022a). Converse to compressive strength, the tensile strength of FRGC was higher when seawater and sea sand replaced fresh water and washed sand (Lao et al., 2023). The highest improvement ratio was 11% when the FA:GGBS ratio of the composite was 80:20.

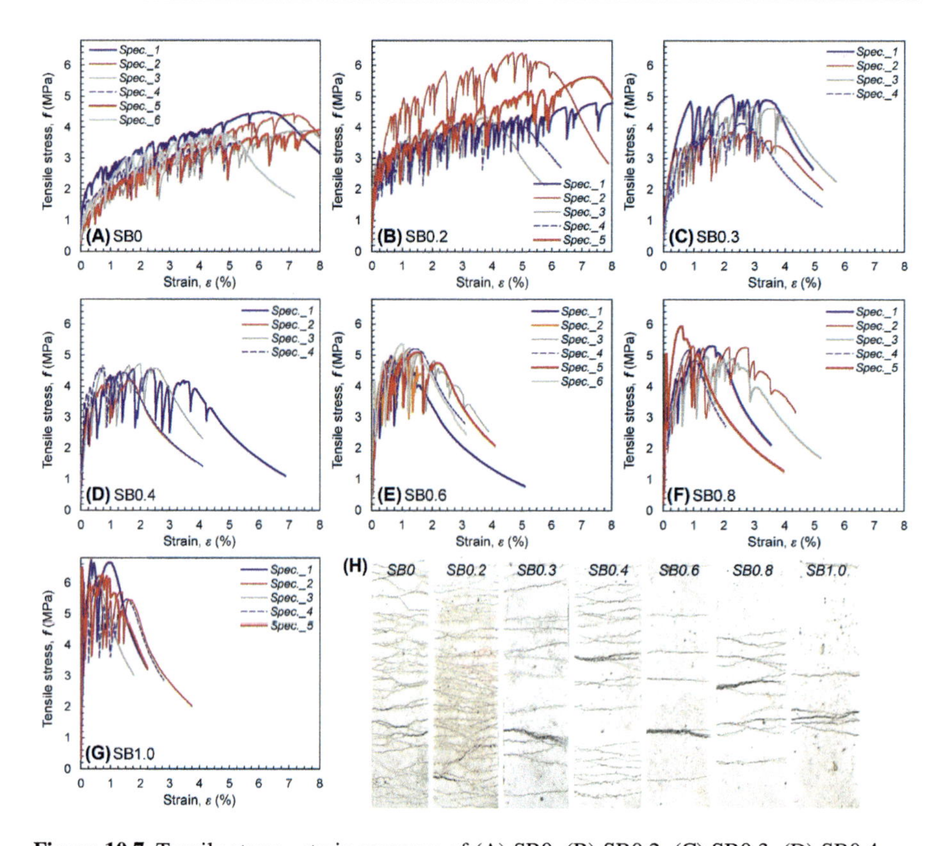

Figure 10.7 Tensile stress–strain response of (A) SB0, (B) SB0.2, (C) SB0.3, (D) SB0.4, (E) SB0.6, (F) SB0.8, (G) SB1.0, and (H) crack patterns (note: "SB0.2" means that the Agg/b of fiber-reinforced geopolymer composite is 0.2) (Yoo et al., 2022).
Source: Modified from Yoo, D.-Y., Lee, S. K., You, I., Oh, T., Lee, Y., & Zi, G. (2022). Development of strain-hardening geopolymer mortar based on liquid-crystal display (LCD) glass and blast furnace slag. *Construction and Building Materials, 331*. https://doi.org/10.1016/j.conbuildmat.2022.127334.

10.3.5.3 Effect of fiber

Different from compressive strength, most studies revealed that the presence of sustainable fibers including coir, sweet sorghum, jute, and sisal fibers can increase the tensile strength of geopolymers (Chen et al., 2014; Silva et al., 2020; Wongsa et al., 2020). Nevertheless, an opposite trend was captured that the presence of different dosages of flax tows led to an 8%–35% decline in splitting tensile strength for FRGC likely due to the random fiber distribution and orientation (Lazorenko et al., 2020). Replacing 0.25% PVA fiber with RTS fiber resulted in a 5.2% enhancement in tensile strength and a comparable tensile strain capacity for the composite compared to the mixture with 2.0% PVA fiber (Zhong et al., 2023). Apart from the high mechanical properties of RTS fibers, the improved fiber–matrix interaction due to the deformed shape of

RTS fibers can significantly contribute to the enhanced tensile strength of FRGC. However, both tensile strength and tensile strain capacity of FRGC declined when the RTS fiber replacement level was 0.5% due to the lower number of RTS fibers across a certain area caused by the larger diameter of RTS fibers compared to PVA fibers. Similarly, the uniaxial tensile behavior of FRGC was considerably impaired when the RTP fiber replacement level was higher than 0.25% (see Fig. 10.6C), primarily due to the insufficient fiber bridging capacity for FRGC when more RTP fibers were present (Zhong & Zhang, 2021). As seen in Fig. 10.8, PVA fiber had a better bonding with the matrix (more adhered matrix fragments) than RTP fiber (smooth surface), which can effectively bridge the crack and resist its growth.

10.3.6 Flexural behavior

Consistent with compressive strength (see Fig. 10.4A), replacing GGBS with GCC reduced the flexural load capacity of FRGC at both 28 and 90 days due to the low reactivity of GCC (Adesina & Das, 2023). When the RTS fiber dosage was 0.5%,

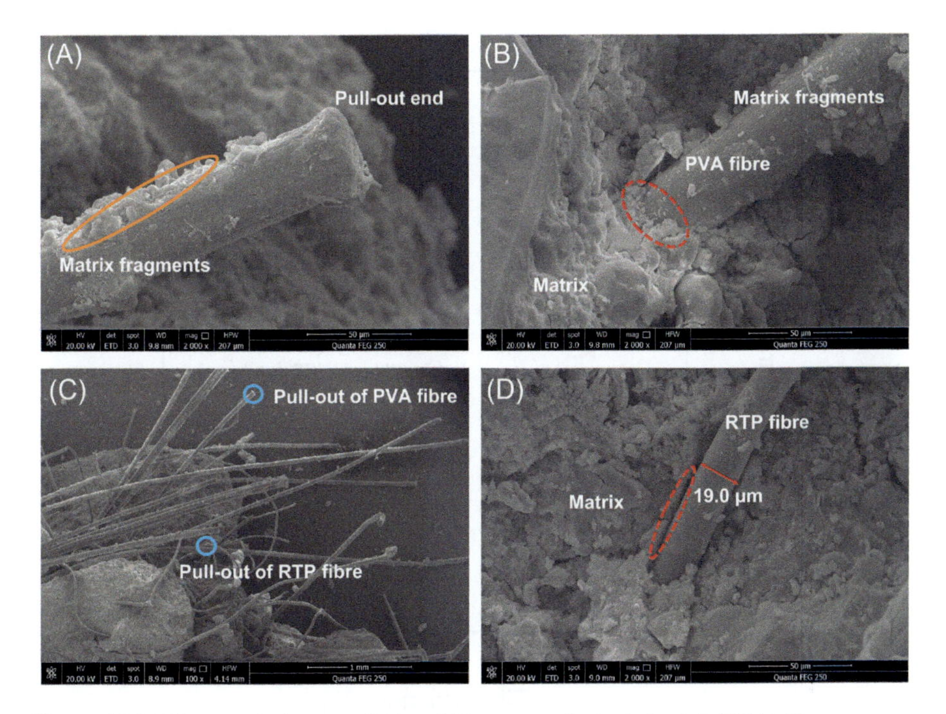

Figure 10.8 Microscopic images of (A and B) mono-polyvinyl alcohol (PVA) fiber-reinforced geopolymer composites and (C and D) hybrid fiber-reinforced geopolymer composites (Zhong & Zhang, 2021).
Source: Modified from Zhong, H., & Zhang, M. (2021). Effect of recycled tyre polymer fibre on engineering properties of sustainable strain hardening geopolymer composites. *Cement and Concrete Composites*, *122*. https://doi.org/10.1016/j.cemconcomp.2021.104167.

increasing crumb rubber volume fraction from 5% to 15% did not significantly alter the flexural strength of FRGC, while the flexural strength of FRGC with 1.0% RTS fiber dropped from 6.3 to 5.1 MPa when the crumb rubber content went up to 15% (Zhong et al., 2019). Because of the improved bond strength between aggregates and paste, replacing 25% natural river sand with granite industry waste considerably enhanced the flexural strength of FRGC (Bhargav & Syed Abdul Khadar, 2020; Nuaklong et al., 2023).

Fig. 10.9 shows the effect of fiber dosage on the flexural strength of FRGC, indicating that all plant fibers can effectively increase the flexural strength of FRGC. For instance, incorporating 0.5%–3.0% sisal fiber enhanced the flexural strength of geopolymers by 74%–368% (Silva et al., 2020). Under flexural loading, a crack can be generated at the weakest zone of a specimen, which can be bridged by the fibers at the crack interface. The stress can be sustained and transferred through the fiber bridging action, leading to the formation of additional microcracks. Partial replacement of PVA fibers with RTS fibers did not weaken the first cracking strength of FRGC but impaired the flexural strength at both 7 and 28 days (Wang et al., 2020). As mentioned above, the number of effective RTS fibers across a unit

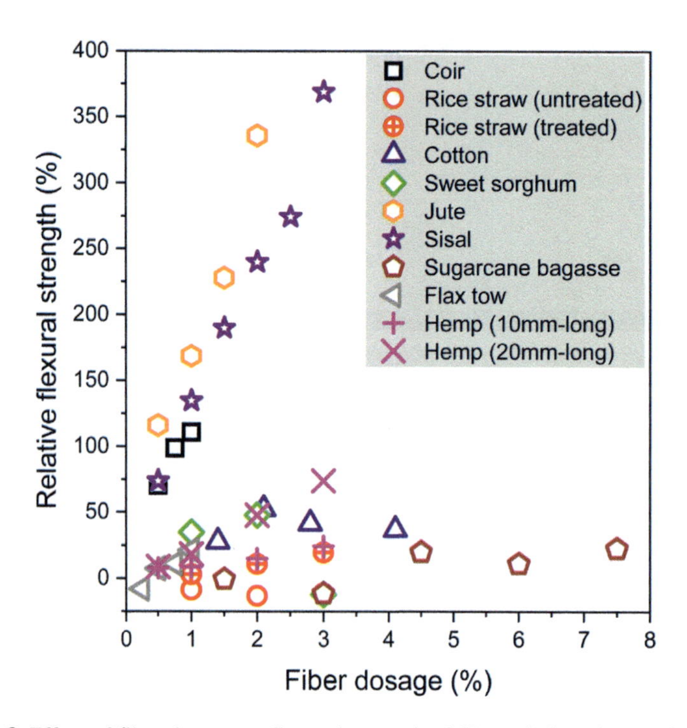

Figure 10.9 Effect of fiber dosage on flexural strength of fiber-reinforced geopolymer composites.

Source: Data from Alomayri et al. (2013), Bayraktar et al. (2023), Nguyen and Mangat (2020), Wongsa et al. (2020), Chen et al. (2014), Lazorenko et al. (2020), Nkwaju et al. (2019), and Silva et al. (2020).

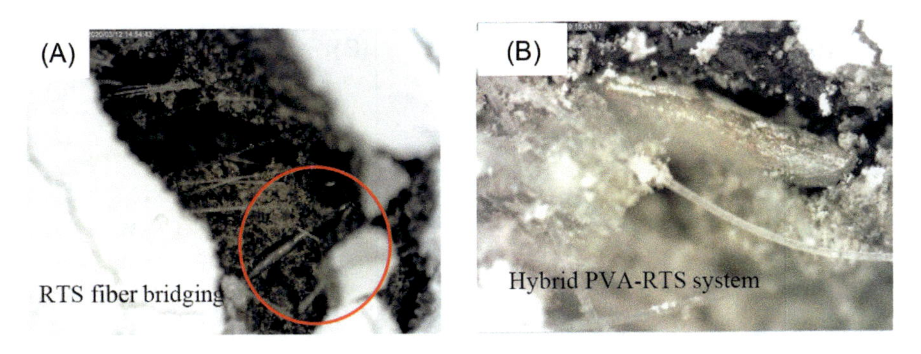

Figure 10.10 Microscopic images: (A) recycled tire steel (RTS) fiber bridging and (B) hybrid PVA and RTS fibers. *PVA*, Polyvinyl alcohol.
Source: Modified from Wang, Y., Chan, C. L., Leong, S. H., & Zhang, M. (2020). Engineering properties of strain hardening geopolymer composites with hybrid polyvinyl alcohol and recycled steel fibres. *Construction and Building Materials, 261*. https://doi.org/10.1016/j.conbuildmat.2020.120585.

area was smaller than that of PVA fibers, which reduced the fiber bridging capacity once a visible crack had been formed, as evidenced in Fig. 10.10. Besides, the ununiform geometry and dimension of RTS fibers would also affect the performance stability of FRGC under flexural loading.

10.3.7 Dynamic mechanical behavior

During the service life, concrete infrastructures would be subjected to a combination of static and dynamic loadings. Hence, understanding the dynamic mechanical properties of FRGC under various strain rates is important to guide future structural design. Different types of test methods have been developed to characterize the dynamic mechanical properties of concrete such as drop-weight impact, Charpy impact, and split Hopkinson pressure bar (SHPB) tests (Yoo & Banthia, 2019). The SHPB apparatus based on stress wave propagation has been commonly adopted to measure the compressive and splitting tensile behavior of concrete within a wide range of strain rate, from 10^{-1} to 10^3/s (Chen & Song, 2011; Lok & Zhao, 2004). The effects of replacing PVA fibers with RTS or RTP fibers on the dynamic compressive and splitting tensile properties of FRGC were presented and discussed in Zhong and Zhang (2022a, 2022b) and Zhong et al. (2023).

Part of the results is depicted in Fig. 10.11. Replacing 0.25%–0.5% PVA fiber with RTP fiber can lead to slightly better or comparable dynamic splitting tensile strength and energy absorption capacity for FRGC as compared to the mix with 2.0% PVA fiber (Zhong & Zhang, 2022a). For instance, the energy absorption capacity of FRGC with 0.25% RTP fiber was 17% higher than that with PVA fibers only when the strain rate was within 2–3.2/s. This can be caused by the combined effect of synergistic hybrid fiber effect in controlling the cracks, Stefan effect and pull-out behavior of RTP fibers under various strain rates. Similar findings were

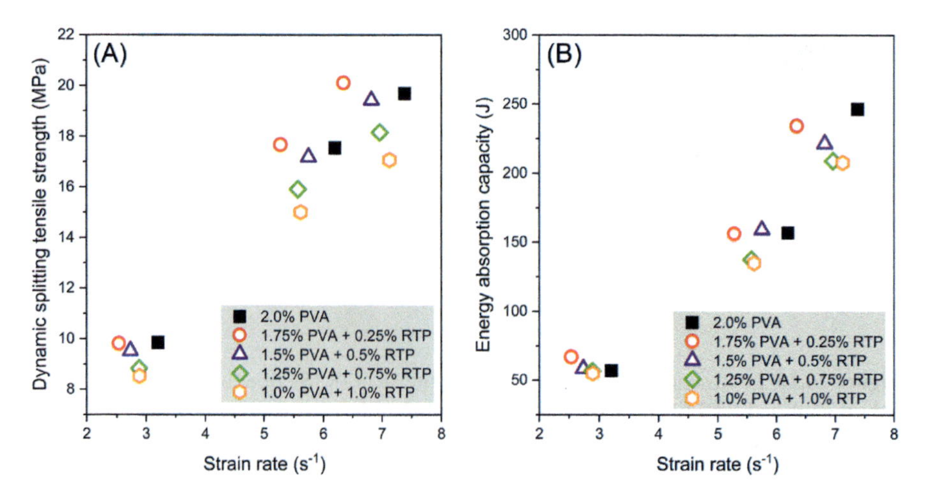

Figure 10.11 Effect of RTP fiber volume fraction on (A) dynamic splitting tensile strength and (B) energy absorption capacity of fiber-reinforced geopolymer composites. *RTP*, Recycled tire polymer.
Source: Data from Zhong, H., & Zhang, M. (2022a). Dynamic splitting tensile behaviour of engineered geopolymer composites with hybrid polyvinyl alcohol and recycled tyre polymer fibres. *Journal of Cleaner Production*, *379*. https://doi.org/10.1016/j.jclepro.2022.134779.

observed for dynamic compressive behavior (Zhong & Zhang, 2022b). Up to 20% enhancement in dynamic splitting tensile strength was captured for FRGC when RTS fibers were used to replace PVA fibers (Zhong et al., 2023).

10.4 Durability of sustainable fiber-reinforced geopolymer composites

10.4.1 Resistance to fire/elevated temperature

When concrete is exposed to fire or subjected to elevated temperatures, it will be damaged by the development of cracks (Lahoti et al., 2019; Wu et al., 2020). Incorporating some fibers with low melting points (e.g., PP fibers) can improve the concrete resistance to spalling due to the increased pore connectivity and permeability (Wu et al., 2020). A study on the effect of elevated temperature (250°C–750°C) on the compressive and flexural strengths of FRGC containing hemp fibers with different lengths indicated that the compressive and flexural strengths of all FRGC mixes were improved when the temperature changed from 22°C to 250°C, mainly due to the reaction between the carbon dioxide from burning cellulose and the alkalis of sodium metasilicate (Bayraktar et al., 2023). Based on the microscopic images, the number of reaction gels went up with the increasing hemp fiber dosage and length. After exceeding 250°C, both compressive and flexural strengths of all mixtures dropped.

The effect of fire exposure time (30–90 minutes) on the compressive strength of FRGC with various PP fiber dosages was investigated in Nuaklong et al. (2023). The visual appearances of the specimen with 50% granite industry waste and 1.0% PP fiber after different fire exposure time are illustrated in Fig. 10.12. After 30-minute fire exposure, the color of the tested specimen changed from dark brown to red mainly owing to the oxidation of the iron compounds in FA (Fig. 10.12B). Some channels left behind by melted PP fibers were also identified. More visible cracks can be seen on the surface of the specimen after 60-min fire exposure (Fig. 10.12C), followed by a complete deterioration of the specimen when the fiber exposure time was 90 minutes (Fig. 10.12D). In addition, the color of some remaining concrete components turned black, which can be ascribed to the burning carbon during the sintering process of concrete components (Lee et al., 2021). The inclusion of granite industry waste did not benefit the compressive strength of FRGC with or without a fire attack. For instance, the compressive strength of FRGC with

Figure 10.12 Visual observation of fiber-reinforced geopolymer composites containing granite waste (A) before and after fire exposure of (B) 30 min, (C) 60 min, and (D) 90 min. *Source*: Modified from Nuaklong, P., Hamcumpai, K., Keawsawasvong, S., Pethrung, S., Jongvivatsakul, P., Tangaramvong, S., Pothisiri, T., & Likitlersuang, S. (2023). Strength and post-fire performance of fiber-reinforced alkali-activated fly ash concrete containing granite industry waste. *Construction and Building Materials*, *392*. https://doi.org/10.1016/j.conbuildmat.2023.131984.

0.5% PP fiber after 30-min fire exposure dropped from 51 to 39 and 31 MPa, respectively, when the granite industry waste replacement level changed from 0% to 25% and 50%.

10.4.2 Resistance to chemical attack

Fig. 10.13 presents the effects of various sulfate solutions and exposure periods on the compressive strength of POFA-based FRGC made from 10 M NaOH solution. Regardless of the sulfate solution type, a pronounced strength loss can be observed for FRGC under different exposure levels (Salami et al., 2017). Mixes after exposure to 5% $MgSO_4$ experienced the least loss of compressive strength. Irrespective of copper slag replacement level, all FRGC mixes showed excellent resistance to seawater (Yaswanth et al., 2022a). When the specimens were subjected to sulfate exposure, using 40% copper slag to replace silica sand in FRGC led to a lower strength loss compared to the use of 60% copper slag to substitute silica sand.

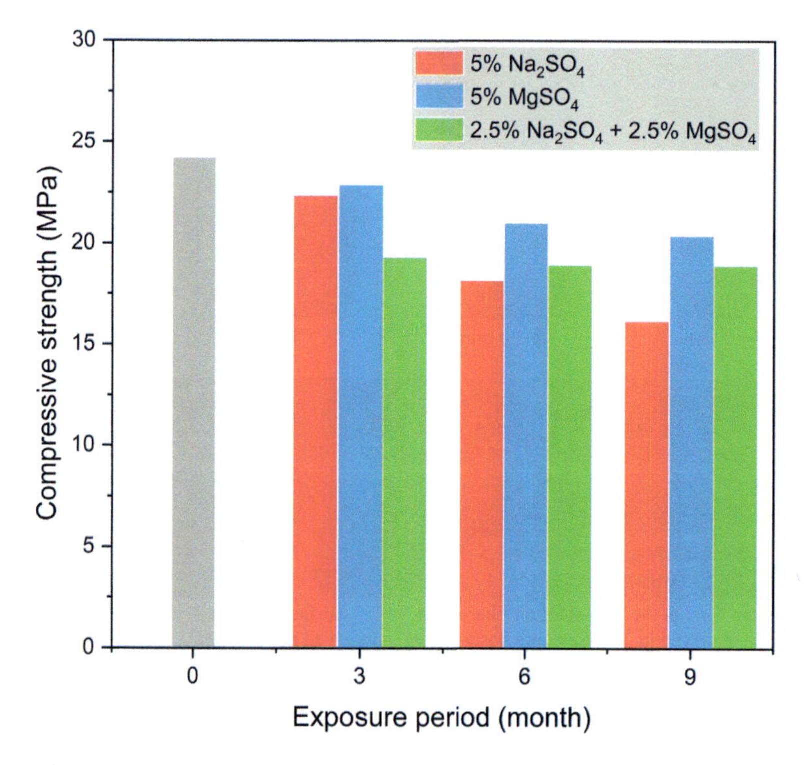

Figure 10.13 Effects of sulfate solution and exposure period on compressive strength of fiber-reinforced geopolymer composites containing palm oil fuel ash.
Source: Data from Salami, B. A., Megat Johari, M. A., Ahmad, Z. A., & Maslehuddin, M. (2017). Durability performance of palm oil fuel ash-based engineered alkaline-activated cementitious composite (POFA-EACC) mortar in sulfate environment. *Construction and Building Materials*, *131*, 229–244. https://doi.org/10.1016/j.conbuildmat.2016.11.048.

Under rapid chloride penetration, replacing silica sand with copper slag can considerably improve the resistance of FRGC (Yaswanth et al., 2022b). For instance, the chloride ion penetration of FRGC went down by 55% when 20% copper slag replaced silica sand.

10.4.3 Resistance to environmental loading

A few studies explored the mechanical properties of FRGC after exposure to wet and dry cycles (Nguyen & Mangat, 2020; Yanou et al., 2021). Considering the loss of mass and compressive strength, it was observed that utilizing 1.5% and 3.0% sugarcane bagasse fiber can effectively improve the resistance of geopolymers to 5, 10, and 20 wet-dry cycles due to the fiber bridging effect (Yanou et al., 2021). When the incorporated sugarcane bagasse fiber dosage reached 4.5%, 6.0%, and 7.5%, the resistance of resultant composites to wet-dry cycles was poorer than those of unreinforced geopolymers, which can be attributed to the increased porosity.

The effect of rice straw fiber content on the compressive and flexural strengths of FRGC with or without wet-dry cycles is shown in Fig. 10.14. Similar to compressive strength without wet-dry cycles, mixes with treated rice straw fibers outperformed those with untreated rice straw fibers (Fig. 10.14A) (Nguyen & Mangat, 2020). When the fiber dosage was 1.0%, the specimens with treated rice straw fibers had a compressive strength of around 44 MPa after 20 wet-dry cycles, which was even higher than that without any wet-dry cycles (about 42 MPa). Similar trends can be observed for flexural strength (Fig. 10.14B). Such superior

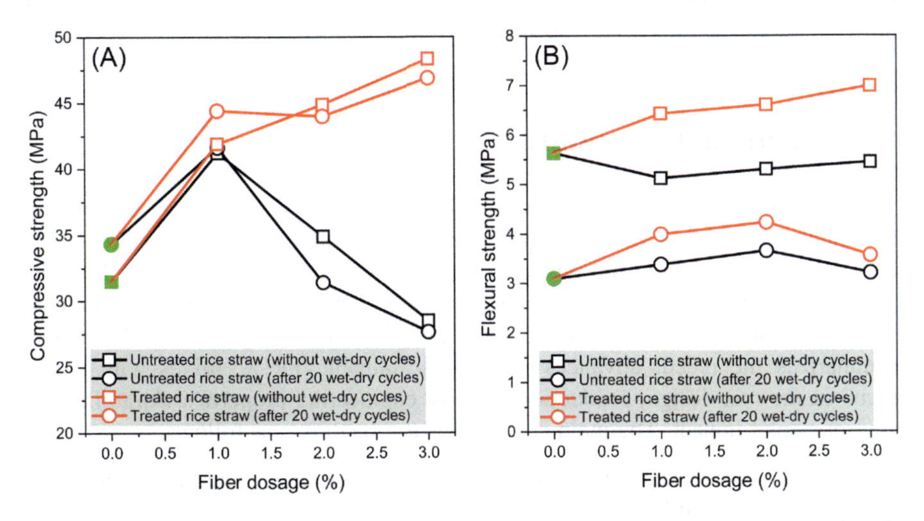

Figure 10.14 Effects of rice straw fiber and wet-dry cycle on (A) compressive strength and (B) flexural strength of fiber-reinforced geopolymer composites.
Source: Data from Nguyen, C. V., & Mangat, P. S. (2020). Properties of rice straw reinforced alkali activated cementitious composites. *Construction and Building Materials, 261*. https://doi.org/10.1016/j.conbuildmat.2020.120536.

performance can be strongly related to the fiber−matrix interaction. As compared with untreated rice straw fibers, some microscopic images indicated a better bonding behavior for the treated rice straw fibers with the matrix after 20 wet-dry cycles.

10.5 Sustainability assessment

Most of the existing studies adopted the cradle-to-gate boundary system to assess the environmental impacts of proposed materials. It was found that about 65% of the total embodied energy of FRGC mixes containing C-CSMR or blended C-CSMR and GGBS came from PVA fibers and precursors, while more than 50% of their total embodied carbon was derived from precursors (Li, Yin, et al., 2022). Replacing 15% C-CSMR with GGBS did not significantly change the embodied energy of FRGC but reduced the embodied carbon by 9.1%. As compared to fiber-reinforced cementitious composites (Zhang, Jaworska, et al., 2020), the embodied energy and embodied carbon of C-CSMR-based FRGC were 26% and 32% lower. It was reported that the mix containing 5% crumb rubber exhibited a lower embodied carbon, embodied energy, and material cost than the cement-based composites (Lương et al., 2021). The embodied carbon and embodied energy of FRGC made from seawater and sea sand were calculated as $526.2\ kg \cdot CO_2/m^3$ and $3899.8\ MJ/m^3$, respectively, which were 56% and 56% lower than those of cementitious composites containing sea sand (Huang et al., 2021; Lao et al., 2023). The presence of RTP fibers can decrease the material cost and embodied energy of FRGC by 8.6%−35% and 4.1%−16%, respectively (Zhong & Zhang, 2021).

10.6 Concluding remarks

A number of studies have been performed to understand the engineering properties and durability of FRGC made from commonly used aluminosilicate materials (e.g., FA), natural river sand and synthetic fibers (e.g., steel and PP fibers). In recent years, an increasing number of studies focused on further improving the sustainability of traditional FRGC using some waste materials and natural fibers. This chapter summarizes the recent advances in these sustainable FRGC mixes with special focus on the mix design, production process, engineering properties, durability, and environmental impact. The main conclusions can be drawn as follows:

1. Alternative binders (e.g., POFA and LCDG), aggregates (e.g., crumb rubber and granite industry waste), and fibers (e.g., RTS and RTP fibers) are utilized to produce sustainable FRGC. The mixing process and curing regime of these composites are similar to those of traditional FRGC. Replacing a certain content of natural river sand or silica sand with crumb rubber, copper slag, and granite industry waste is beneficial for the workability of FRGC. The effect of various plant fibers on the density of FRGC is inconsistent and the amount of entrapped air bubbles can significantly affect the density of FRGC. Replacing

PVA fibers with RTS or RTP fibers can enhance the drying shrinkage resistance of FRGC.

2. The presence of 10% IFA can increase the compressive strength of FRGC by 29% and about 25%−32% improvement in compressive strength is observed for FRGC when 20%−40% silica sand is replaced with copper slag. Some plant fibers such as rice straw and abaca fibers are effective in rising the compressive strength of plain geopolymers. The trends of tensile and flexural strengths are mostly consistent with that of compressive strength. Partial replacement of PVA fibers with RTP or RTS fibers can improve the dynamic compressive and splitting tensile properties of FRGC. Increasing the hemp fiber content and length can lead to the formation of more reaction gels, which can improve the compressive and flexural behavior of FRGC under various elevated temperatures.
3. Using 40% copper slag to replace silica sand in FRGC can enhance its resistance to sulfate and chloride attacks. Incorporating 1.5% and 3.0% sugarcane bagasse fiber and 1.0% treated rice straw fiber can effectively improve the resistance of FRGC to wet-dry cycles.
4. Compared to fiber-reinforced cementitious composites, the embodied carbon and embodied energy of some FRGC mixes discussed in this chapter are considerably lower.

References

Abbas, A.-G. N., Aziz., Abdan, K., Nasir., & Huseien, G. F. (2022). A state-of-the-art review on fibre-reinforced geopolymer composites. *Construction and Building Materials*, *330*. Available from https://doi.org/10.1016/j.conbuildmat.2022.127187.

Adesina, A., & Das, S. (2023). Feasibility study on the utilization of granulated calcium carbonate as precursor in alkali-activated fibre-reinforced composites. *Innovative Infrastructure Solutions*, *8*(99). Available from https://doi.org/10.1007/s41062-023-01064-2.

Afroughsabet, V., & Teng, S. (2020). Experiments on drying shrinkage and creep of high performance hybrid-fiber-reinforced concrete. *Cement and Concrete Composites*, *106*, 103481. Available from https://doi.org/10.1016/j.cemconcomp.2019.103481.

Alomayri, T., Shaikh, F. U. A., & Low, I. M. (2013). Thermal and mechanical properties of cotton fabric-reinforced geopolymer composites. *Journal of Materials Science*, *48*(19), 6746−6752. Available from https://doi.org/10.1007/s10853-013-7479-2.

Bayraktar, O. Y., Tobbala, D. E., Turkoglu, M., Kaplan, G., & Tayeh, B. A. (2023). *Hemp fiber reinforced one-part alkali-activated composites with expanded perlite: Mechanical properties, microstructure analysis and high-temperature resistance*, . *Construction and Building Materials* (363). . Available from 10.1016/j.conbuildmat.2022.129716.

Bhargav, M., & Syed Abdul Khadar, S. A. R. (2020). *Characterization of fibre R.C. beam made with partial replacement of sand with iron ore*, . *Materials Today: Proceedings* (45, pp. 6590−6595). . Available from 10.1016/j.matpr.2020.11.769.

Chen, R., Ahmari, S., & Zhang, L. (2014). Utilization of sweet sorghum fiber to reinforce fly ash-based geopolymer. *Journal of Materials Science*, *49*(6), 2548−2558. Available from https://doi.org/10.1007/s10853-013-7950-0.

Chen, W., & Song, B. (2011). *Split Hopkinson (Kolsky) bar: Design, testing and applications. Mechanical Engineering Series*. Springer US. Available from 10.1007/978-1-4419-7982-7.

Fang, C., Ali, M., Xie, T., Visintin, P., & Sheikh, A. H. (2020). The influence of steel fibre properties on the shrinkage of ultra-high performance fibre reinforced concrete.

Construction and Building Materials, *242*. Available from https://doi.org/10.1016/j.conbuildmat.2019.117993.

Ganjian, E., Khorami, M., & Maghsoudi, A. A. (2009). Scrap-tyre-rubber replacement for aggregate and filler in concrete. *Construction and Building Materials*, *23*(5), 1828–1836. Available from https://doi.org/10.1016/j.conbuildmat.2008.09.020.

Gao, S., Wang, Z., Wang, W., & Qiu, H. (2018). Effect of shrinkage-reducing admixture and expansive agent on mechanical properties and drying shrinkage of engineered cementitious composite (ECC). *Construction and Building Materials*, *179*, 172–185. Available from https://doi.org/10.1016/j.conbuildmat.2018.05.203.

Gholampour, A., Danish, A., Ozbakkaloglu, T., Yeon, J. H., & Gencel, O. (2022). Mechanical and durability properties of natural fiber-reinforced geopolymers containing lead smelter slag and waste glass sand. *Construction and Building Materials*, *352*. Available from https://doi.org/10.1016/j.conbuildmat.2022.129043.

Huang, B. T., Wu, J. Q., Yu, J., Dai, J. G., Leung, C. K. Y., & Li, V. C. (2021). *Seawater sea-sand engineered/strain-hardening cementitious composites (ECC/SHCC): Assessment and modeling of crack characteristics*, . *Cement and Concrete Research* (140). . Available from 10.1016/j.cemconres.2020.106292.

Kan, L., Shi, R., Zhao, Y., Duan, X., & Wu, M. (2020). Feasibility study on using incineration fly ash from municipal solid waste to develop high ductile alkali-activated composites. *Journal of Cleaner Production*, *254*. Available from https://doi.org/10.1016/j.jclepro.2020.120168.

Kan, L., & Wang, F. (2022). Mechanical properties of high ductile alkali-activated fiber reinforced composites incorporating red mud under different curing conditions. *Ceramics International*, *48*(2), 1999–2011. Available from https://doi.org/10.1016/j.ceramint.2021.09.285.

Khan, M. I., Fares, G., Mourad, S., & Abbass, W. (2016). Optimized fresh and hardened properties of strain-hardening cementitious composites: Effect of sand size and workability. *Journal of Materials in Civil Engineering*, *28*(12). Available from https://doi.org/10.1061/(ASCE)MT.1943-5533.0001665.

Khayat, K. H., Meng, W., Vallurupalli, K., & Teng, L. (2019). Rheological properties of ultra-high-performance concrete—An overview. *Cement and Concrete Research*, *124*. Available from https://doi.org/10.1016/j.cemconres.2019.105828.

Lahoti, M., Tan, K. H., & Yang, E. H. (2019). A critical review of geopolymer properties for structural fire-resistance applications. *Construction and Building Materials*, *221*, 514–526. Available from https://doi.org/10.1016/j.conbuildmat.2019.06.076.

Lao, J. C., Huang, B. T., Xu, L. Y., Khan, M., Fang, Y., & Dai, J. G. (2023). *Seawater sea-sand engineered geopolymer composites (EGC) with high strength and high ductility*, . *Cement and Concrete Composites* (138). . Available from 10.1016/j.cemconcomp.2023.104998.

Lazorenko, G., Kasprzhitskii, A., Kruglikov, A., Mischinenko, V., & Yavna, V. (2020). Sustainable geopolymer composites reinforced with flax tows. *Ceramics International*, *46*(8), 12870–12875. Available from https://doi.org/10.1016/j.ceramint.2020.01.184.

Lee, N. K., Jang, J. G., & Lee, H. K. (2014). Shrinkage characteristics of alkali-activated fly ash/slag paste and mortar at early ages. *Cement and Concrete Composites*, *53*, 239–248. Available from https://doi.org/10.1016/j.cemconcomp.2014.07.007.

Lee, K. H., Lee, K. G., Lee, Y. S., & Wie, Y. M. (2021). *Manufacturing and application of artificial lightweight aggregate from water treatment sludge*, . *Journal of Cleaner Production* (307). . Available from 10.1016/j.jclepro.2021.127260.

Lee, S. K., Oh, T., Banthia, N., & Yoo, D. Y. (2023). Optimization of fiber aspect ratio for 90 MPa strain-hardening geopolymer composites (SHGC) with a tensile strain capacity over 7.5%. *Cement and Concrete Composites*, *139*. Available from https://doi.org/10.1016/j.cemconcomp.2023.105055.

Liu, Y., Zhang, Z., Shi, C., Zhu, D., Li, N., & Deng, Y. (2020). Development of ultra-high performance geopolymer concrete (UHPGC): Influence of steel fiber on mechanical properties. *Cement and Concrete Composites*, *112*. Available from https://doi.org/10.1016/j.cemconcomp.2020.103670.

Li, W., Shumuye, E. D., Shiying, T., Wang, Z., & Zerfu, K. (2022). Eco-friendly fibre reinforced geopolymer concrete: A critical review on the microstructure and long-term durability properties. *Case Studies in Construction Materials*, *16*, e00894.

Li, Y., Yin, J., Yuan, Q., Huang, L., & Li, J. (2022). Greener strain-hardening cementitious composites (SHCC) with a novel alkali-activated cement. *Cement and Concrete Composites*, *134*, 104735.

Li, V. C. (2019). *Engineered cementitious composites (ECC): Bendable concrete for sustainable and resilient infrastructure* (pp. 1–419). Springer. Available from 10.1007/978-3-662-58438-5.

Lok, T. S., & Zhao, P. J. (2004). Impact response of steel fiber-reinforced concrete using a split Hopkinson pressure bar. *Journal of Materials in Civil Engineering*, *16*(1), 54–59. Available from https://doi.org/10.1061/(asce)0899-1561(2004)16:1(54).

Lương, Q. H., Nguyễn, H. H., Choi, J. I., Kim, H. K., & Lee, B. Y. (2021). Effects of crumb rubber particles on mechanical properties and sustainability of ultra-high-ductile slag-based composites. *Construction and Building Materials*, *272*. Available from https://doi.org/10.1016/j.conbuildmat.2020.121959.

Malenab, R. A. J., Ngo, J. P. S., & Promentilla, M. A. B. (2017). Chemical treatment of waste abaca for natural fiber-reinforced geopolymer composite. *Materials*, *10*(6). Available from https://doi.org/10.3390/ma10060579.

Ma, Y., & Ye, G. (2015). The shrinkage of alkali activated fly ash. *Cement and Concrete Research*, *68*, 75–82. Available from https://doi.org/10.1016/j.cemconres.2014.10.024.

Nematollahi, B., Ranade, R., Sanjayan, J., & Ramakrishnan, S. (2017). Thermal and mechanical properties of sustainable lightweight strain hardening geopolymer composites. *Archives of Civil and Mechanical Engineering*, *17*(1), 55–64. Available from https://doi.org/10.1016/j.acme.2016.08.002.

Nguyen, C. V., & Mangat, P. S. (2020). Properties of rice straw reinforced alkali activated cementitious composites. *Construction and Building Materials*, *261*. Available from https://doi.org/10.1016/j.conbuildmat.2020.120536.

Nkwaju, R. Y., Djobo, J. N. Y., Nouping, J. N. F., Huisken, P. W. M., Deutou, J. G. N., & Courard, L. (2019). Iron-rich laterite-bagasse fibers based geopolymer composite: Mechanical, durability and insulating properties. *Applied Clay Science*, *183*. Available from https://doi.org/10.1016/j.clay.2019.105333.

Nuaklong, P., Hamcumpai, K., Keawsawasvong, S., Pethrung, S., Jongvivatsakul, P., Tangaramvong, S., Pothisiri, T., & Likitlersuang, S. (2023). Strength and post-fire performance of fiber-reinforced alkali-activated fly ash concrete containing granite industry waste. *Construction and Building Materials*, *392*. Available from https://doi.org/10.1016/j.conbuildmat.2023.131984.

Pilakoutas, K., Neocleous, K., & Tlemat, H. (2004). Reuse of tyre steel fibres as concrete reinforcement. *Proceedings of the Institution of Civil Engineers: Engineering Sustainability*, *157*(3), 131–138. Available from https://doi.org/10.1680/ensu.2004.157.3.131.

Poletanovic, B., Dragas, J., Ignjatovic, I., Komljenovic, M., & Merta, I. (2020). Physical and mechanical properties of hemp fibre reinforced alkali-activated fly ash and fly ash/slag mortars. *Construction and Building Materials, 259*. Available from https://doi.org/10.1016/j.conbuildmat.2020.119677.

Ranjbar, N., & Zhang, M. (2020). Fiber-reinforced geopolymer composites: A review. *Cement and Concrete Composites, 107*. Available from https://doi.org/10.1016/j.cemconcomp.2019.103498.

Salami, B. A., Megat Johari, M. A., Ahmad, Z. A., & Maslehuddin, M. (2016). Impact of added water and superplasticizer on early compressive strength of selected mixtures of palm oil fuel ash-based engineered geopolymer composites. *Construction and Building Materials, 109*, 198–206. Available from https://doi.org/10.1016/j.conbuildmat.2016.01.033.

Salami, B. A., Megat Johari, M. A., Ahmad, Z. A., & Maslehuddin, M. (2017). Durability performance of palm oil fuel ash-based engineered alkaline-activated cementitious composite (POFA-EACC) mortar in sulfate environment. *Construction and Building Materials, 131*, 229–244. Available from https://doi.org/10.1016/j.conbuildmat.2016.11.048.

Santana, H. A., Amorim Júnior, N. S., Ribeiro, D. V., Cilla, M. S., & Dias, C. M. R. (2021). Vegetable fibers behavior in geopolymers and alkali-activated cement based matrices: A review. *Journal of Building Engineering, 44*. Available from https://doi.org/10.1016/j.jobe.2021.103291.

Silva, G., Kim, S., Bertolotti, B., Nakamatsu, J., & Aguilar, R. (2020). Optimization of a reinforced geopolymer composite using natural fibers and construction wastes. *Construction and Building Materials, 258*. Available from https://doi.org/10.1016/j.conbuildmat.2020.119697.

Tahwia, A. M., Heniegal, A. M., Abdellatief, M., Tayeh, B. A., & Elrahman, M. A. (2022). *Properties of ultra-high performance geopolymer concrete incorporating recycled waste glass, . Case Studies in Construction Materials* (17). . Available from 10.1016/j.cscm.2022.e01393.

Wang, Y., Chan, C. L., Leong, S. H., & Zhang, M. (2020). *Engineering properties of strain hardening geopolymer composites with hybrid polyvinyl alcohol and recycled steel fibres, . Construction and Building Materials* (261). . Available from 10.1016/j.conbuildmat.2020.120585.

Wang, J., Dai, Q., Si, R., & Guo, S. (2019). Mechanical, durability, and microstructural properties of macro synthetic polypropylene (PP) fiber-reinforced rubber concrete. *Journal of Cleaner Production, 234*, 1351–1364. Available from https://doi.org/10.1016/j.jclepro.2019.06.272.

Wongsa, A., Kunthawatwong, R., Naenudon, S., Sata, V., & Chindaprasirt, P. (2020). Natural fiber reinforced high calcium fly ash geopolymer mortar. *Construction and Building Materials, 241*. Available from https://doi.org/10.1016/j.conbuildmat.2020.118143.

Wu, H., Lin, X., & Zhou, A. (2020). A review of mechanical properties of fibre reinforced concrete at elevated temperatures. *Cement and Concrete Research, 135*. Available from https://doi.org/10.1016/j.cemconres.2020.106117.

Yanou, R. N., Kaze, R. C., Adesina, A., Nemaleu, J. G. D., Jiofack, S. B. K., & Djobo, J. N. Y. (2021). Performance of laterite-based geopolymers reinforced with sugarcane bagasse fibers. *Cameroon Case Studies in Construction Materials, 15*. Available from https://doi.org/10.1016/j.cscm.2021.e00762.

Yaswanth, K. K., Revathy, J., & Gajalakshmi, P. (2022a). Influence of copper slag on Mechanical, durability and microstructural properties of GGBS and RHA blended strain hardening geopolymer composites. *Construction and Building Materials*, *342*, 128042.

Yaswanth, K. K., Revathy, J., & Gajalakshmi, P. (2022b). Strength, durability and microstructural assessment of slag-agro blended based alkali activated engineered geopolymer composites. *Case Studies in Construction Materials*, *16*, e00920.

Yoo, D.-Y., & Banthia, N. (2019). Impact resistance of fiber-reinforced concrete – A review. *Cement and Concrete Composites*, *104*. Available from https://doi.org/10.1016/j.cemconcomp.2019.103389.

Yoo, D.-Y., Lee, S. K., You, I., Oh, T., Lee, Y., & Zi, G. (2022). Development of strain-hardening geopolymer mortar based on liquid-crystal display (LCD) glass and blast furnace slag. *Construction and Building Materials*, *331*. Available from https://doi.org/10.1016/j.conbuildmat.2022.127334.

Zhang, D., Jaworska, B., Zhu, H., Dahlquist, K., & Li, V. C. (2020). Engineered cementitious composites (ECC) with limestone calcined clay cement (LC3). *Cement and Concrete Composites*, *114*, 103766.

Zhang, Z., Cai, S., Li, Y., Wang, Z., Long, Y., Yu, T., & Shen, Y. (2020). High performances of plant fiber reinforced composites—A new insight from hierarchical microstructures. *Composites Science and Technology*, *194*, 108151.

Zhao, Y., Shi, T., Cao, L., Kan, L., & Wu, M. (2021). Influence of steel slag on the properties of alkali-activated fly ash and blast-furnace slag based fiber reinforced composites. *Cement and Concrete Composites*, *116*. Available from https://doi.org/10.1016/j.cemconcomp.2020.103875.

Zhong, H., Poon, E. W., Chen, K., & Zhang, M. (2019). Engineering properties of crumb rubber alkali-activated mortar reinforced with recycled steel fibres. *Journal of Cleaner Production*, *238*. Available from https://doi.org/10.1016/j.jclepro.2019.117950.

Zhong, H., Wang, Y., & Zhang, M. (2023). Quasi-static and dynamic mechanical properties of engineered geopolymer composites with hybrid PVA and recycled steel fibres. *Journal of Advanced Concrete Technology*, *21*(5), 405–420. Available from https://doi.org/10.3151/jact.21.405.

Zhong, H., & Zhang, M. (2022a). Dynamic splitting tensile behaviour of engineered geopolymer composites with hybrid polyvinyl alcohol and recycled tyre polymer fibres. *Journal of Cleaner Production*, *379*. Available from https://doi.org/10.1016/j.jclepro.2022.134779.

Zhong, H., & Zhang, M. (2022b). Effect of recycled polymer fibre on dynamic compressive behaviour of engineered geopolymer composites. *Ceramics International*, *48*(16), 23713–23730. Available from https://doi.org/10.1016/j.ceramint.2022.05.023.

Zhong, H., & Zhang, M. (2021). Effect of recycled tyre polymer fibre on engineering properties of sustainable strain hardening geopolymer composites. *Cement and Concrete Composites*, *122*. Available from https://doi.org/10.1016/j.cemconcomp.2021.104167.

Zhong, H., & Zhang, M. (2023). Engineered geopolymer composites: A state-of-the-art review. *Cement and Concrete Composites*, *135*. Available from https://doi.org/10.1016/j.cemconcomp.2022.104850.

Sustainable additive manufacturing of concrete with low-carbon materials

Shin Hau Bong and Hongjian Du
Department of Civil and Environmental Engineering, National University of Singapore, Singapore

11.1 Introduction

The construction industry is one of the largest industries in the global economy, contributing significantly to economic growth and infrastructure development around the world. According to a report released by McKinsey Global Institute (Barbosa et al., 2017), the construction-related spending was reported to be about 13% of the global gross domestic product in 2016. However, the construction industry's productivity growth has lagged behind that of the overall economy, with an annual average increase of only 1% over the past two decades (compared to 3.6% for manufacturing industry). This low productivity can be attributed to the labor-intensive nature of the industry, which relies heavily on manual labor. Despite the fact that modern construction equipment (e.g., hydraulic excavators, tower cranes, etc.) is extensively adopted in the construction industry nowadays, some construction activities such as placing concrete, building formwork, and installing steel reinforcement are still required a considerable amount of human labor. It is estimated that the construction industry employs approximately 7% of the global workforce (Barbosa et al., 2017). This situation can be further exacerbated by the aging of the world's population. Besides, increasing human labor also adding extra cost and compromising safety in the construction site.

Apart from the poor productivity and high labor demands, reducing the industry's carbon emissions is another particularly pressing issues that require immediate attention. Concrete is a common building material that has been widely used in countless large infrastructure projects around the world. The broad availability of its raw materials and low production cost make it as the second largest volume material used by human after water. As a result of the huge concrete demand, the manufacture of ordinary Portland cement (OPC)—the main binder for concrete production—contributes approximately 8% of the total global carbon dioxide emissions, making it one of the most significant contributors to climate change (Olivier et al., 2016). In addition, the production of OPC is also an energy-intensive activity, which required 3–5.4 GJ of energy for manufacturing 1 ton of OPC (Cement

Sustainable Concrete Materials and Structures. DOI: https://doi.org/10.1016/B978-0-443-15672-4.00011-5

Sustainability Initiative, 2016). Therefore the construction industry must find ways to reduce its carbon footprint to address its environmental impact.

Additive manufacturing (AM) of concrete, or three-dimensional concrete printing (3DCP), can be a potential solution to the aforementioned challenges. By utilizing digital design and automated construction processes, 3DCP can significantly improve the productivity and reduce the labor intensity, cost, wastage, and energy consumption associated with the conventional construction approach of casting concrete into the formwork (De Schutter et al., 2018; García de Soto et al., 2018; Weng et al., 2020). Among the various 3DCP technologies available, extrusion-based 3DCP technology is currently the most popular technology for fabricating concrete elements with complex geometry. It fabricates concrete elements by extruding the cementitious materials from a nozzle following a predefined printing path in a layer-by-layer buildup process. Moreover, this technology is easily scalable and has demonstrated its suitability for use in the construction industry through numerous large-scale projects (Xiao et al., 2021; Zhang, Wang, et al., 2019). Overall, this emerging technology has the potential to revolutionize the construction industry and create a more sustainable and efficiently built environment.

11.2 Additive manufacturing in construction industry

The construction industry is particularly well-suited to take advantage of the benefits offered by 3DCP. One of the major advantages of 3DCP in construction industry is that it requires significantly fewer labor due to its highly automated construction process as compared to the conventional construction approaches. With 3DCP, the construction process can run constantly with minimal human intervention, leading to increase productivity and reduce labor costs.

In addition, 3DCP can also fabricate concrete elements with less material and wastage due to its automated nature, which the material is only deposited where it is needed. This feature of depositing material only where it is needed can be further utilized to decrease the overall material cost through topology optimization (Vantyghem et al., 2020). Another outcome of utilizing 3DCP in the construction industry is that it can fabricate concrete elements with the absence of temporary formwork. The conventional construction approach of casting concrete requires the use of temporary formwork, which must be set up and disassembled for each project. This process is labor-intensive, time-consuming, and costly. The cost of temporary formwork, which is often made of timber or metal, is estimated to be about 35%–60% of the overall cost of concrete construction (Johnston, 2008). The temporary formwork is also a major source of wastage in construction as it would be eventually discarded after several times of use. Moreover, the geometric freedom of concrete elements is considerably limited by the formwork shape, unless a high cost is paid for the manufacture of bespoke formwork with complex geometry. By eliminating the need for formwork, construction projects can be completed more quickly and with less labor, resulting in significant cost savings. Recent case studies have demonstrated that 3DCP has the potential to significantly enhance productivity, reduce the overall construction cost, and minimize environmental impacts when compared to the conventional construction approaches (García de Soto et al., 2018; Weng et al., 2020). The study also showed that as the complexity of concrete elements' geometry increases, 3DCP can provide higher productivity without incurring additional cost (García de Soto et al., 2018).

Despite the many aforementioned benefits of 3DCP to the construction industry, one of the major challenges of the current 3DCP technology is that higher OPC content is typically used in the printable cementitious mixtures, as compared to the conventional mold-cast concrete (Mohan et al., 2021a, 2021b; Ur Rehman & Kim, 2021). Fig. 11.1 presents the binder content including OPC and supplementary cementitious materials (SCMs) (i.e., fly ash, ground granulated blast-furnace slag (GGBS), silica fume, etc.) of printable cementitious mixtures reported in the literature. As can be seen in the figure, most of the printable cementitious mixtures consist of higher OPC content in their mix design as compared to the 250–500 kg/m^3 of OPC content in the conventional mold-cast concrete (Damineli et al., 2010; Van Damme, 2018). The high OPC content in printable mixtures is often due to the low aggregate-to-binder ratio, which is necessary to ensure smooth pumping and extrusion processes without any blockages, surface defects, sand segregation issues, etc. (Le, Austin, Lim, Buswell, Gibb, & Thorpe, 2012; Mohan, Rahul, Van Tittelboom, & De Schutter, 2021). However, high OPC content can result in negative impacts such as high shrinkage, high heat of hydration, and high cost. Furthermore, higher OPC content in the mix design also increase the embodied carbon of the printable materials, which offset the environmental benefits offered by the 3DCP technology.

To further evaluate the OPC efficiency of the printable mixtures presented in Fig. 11.1, an index called binder intensity index (*bi*), proposed by (Damineli et al., 2010), was adopted to measure the total amount of binder (i.e., OPC) necessary to deliver one unit of a given performance indicator, which is 1 MPa of 28-day compressive strength for the printed specimens. Fig. 11.2 shows the *bi* of printable cementitious mixtures and conventional mold-cast concretes adopted from (Damineli et al., 2010). As shown in the figure, the *bi* of printable cementitious mixtures are mostly in the range of 10–20 $kg/(m^3\ MPa)$ and consist of more than 500 kg/m^3 of OPC content, which is higher than those of the conventional mold-cast concretes (5–15 $kg/(m^3\ MPa)$ achieved by 250–500 kg/m^3 of OPC content). This observation indicates that most of the printable mixtures reported in literature have less OPC efficiency than the conventional mold-cast concretes as additional of OPC is needed to add in the printable mixtures for achieving smooth printing operation. As OPC is the most carbon-intensive constituent in the mix design, from an environmental point of view, 3D printable cementitious materials are less environmentally friendly than conventional mold-cast concretes. Therefore, in this regard, researchers around the world are focusing on tackling this obstacle which could potentially hinder the development of 3DCP technology. There are two approaches to address this issue, namely complete replacement of OPC with an OPC-free binder and partial replacement of OPC with high volume of SCMs.

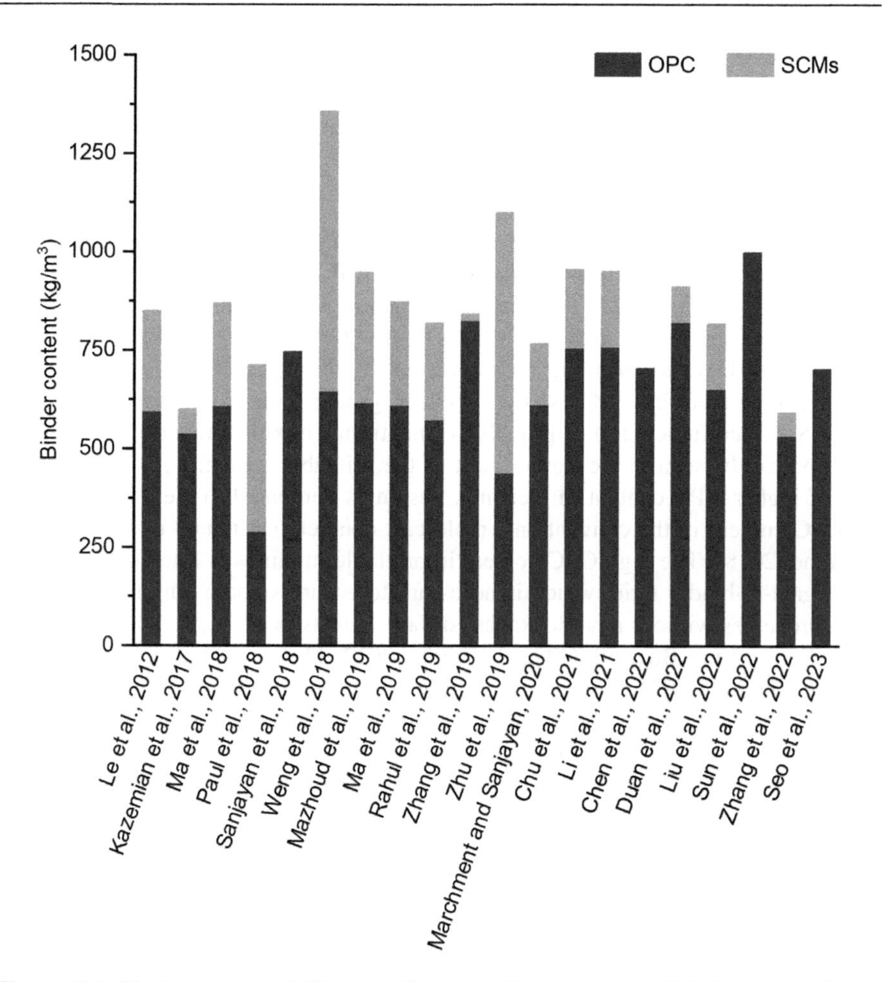

Figure 11.1 Binder content of 3D printable cementitious mixtures. Binder content (i.e., ordinary Portland cement (OPC) and supplementary cementitious materials (SCMs)) of 3D printable cementitious mixtures reported in the literature (Chen et al., 2022; Chu et al., 2021; Duan et al., 2022; Kazemian et al., 2017; Le, Austin, Lim, Buswell, Law, et al., 2012; L. G. Li et al., 2021; Liu et al., 2022; Ma et al., 2019; Ma, Li, et al., 2018; Marchment & Sanjayan, 2020; Mazhoud et al., 2019; Paul et al., 2018; Rahul et al., 2019; Sanjayan et al., 2018; Seo et al., 2023; Sun et al., 2022; Weng et al., 2018; Zhang, Jia, et al., 2022; Zhang, Zhang, et al., 2019; Zhu et al., 2019).

11.3 Ordinary Portland cement–free binder systems

Some of the most promising OPC-free binder systems that have been developed include alkali-activated materials (AAMs) (also commonly referred to as “geopolymer”) (Duxson et al., 2007; Provis, 2014), calcium sulfoaluminate (CSA) cement (Sharp et al., 1999; Winnefeld & Lothenbach, 2010), and magnesia-based cements

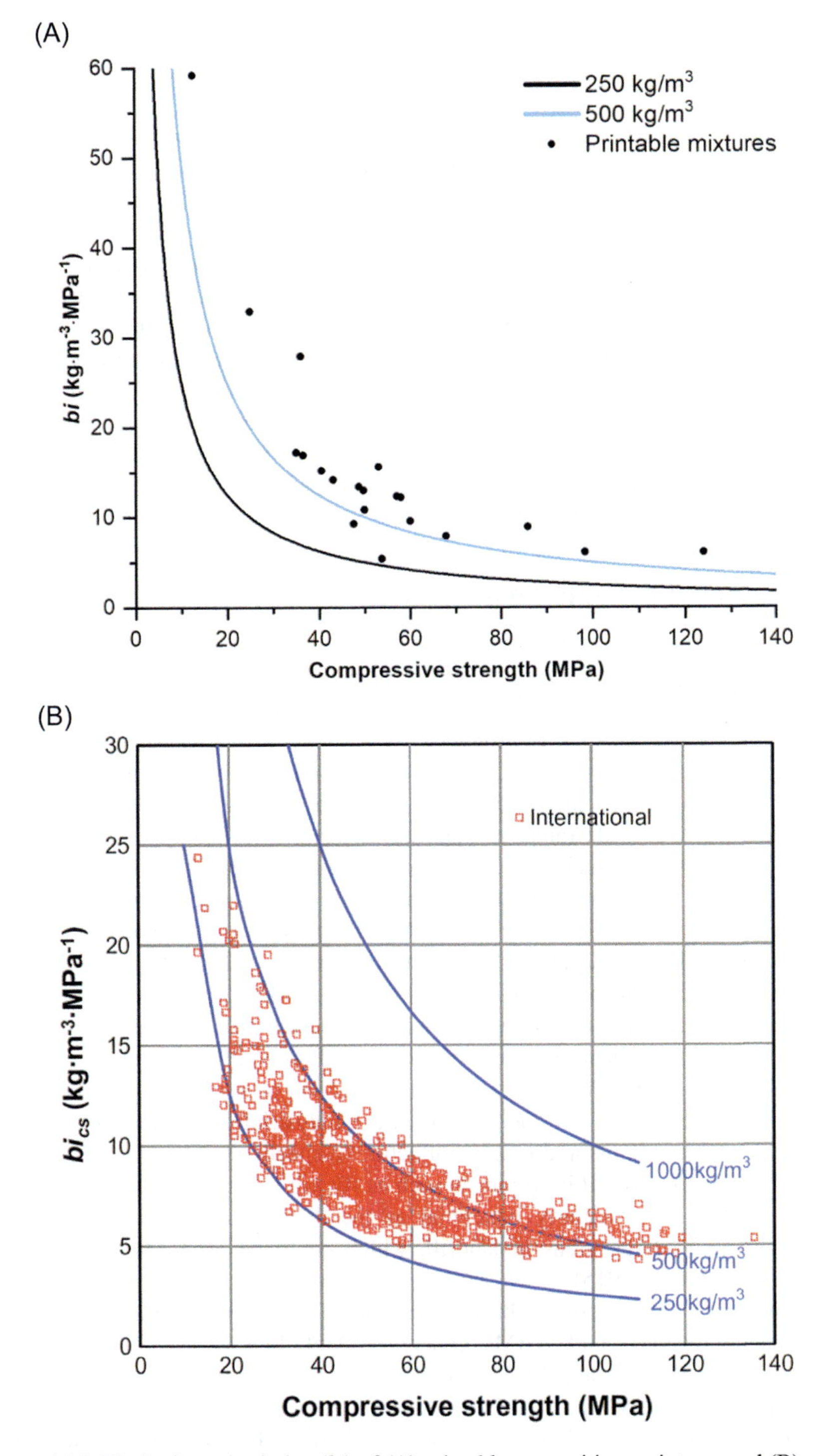

Figure 11.2 Binder intensity index (*bi*) of (A) printable cementitious mixtures and (B) conventional mold-cast concretes.
(B) *Source*: Adopted from Damineli, B. L., Kemeid, F. M., Aguiar, P. S., & John, V. M. (2010). Measuring the eco-efficiency of cement use. *Cement and Concrete Composites*, *32* (8), 555–562. https://doi.org/10.1016/j.cemconcomp.2010.07.009.

(Al-Tabbaa, 2013; Ruan & Unluer, 2016). However, regardless of which binder system is used, the material needs to fulfill several specific rheological requirements before it can be used as the printing materials in the 3DCP process (Bos et al., 2016; Le, Austin, Lim, Buswell, Gibb, & Thorpe, 2012; Roussel, 2018). Le, Austin, Lim, Buswell, Gibb, and Thorpe (2012) defined these rheological requirements as extrudability, buildability, pumpability, and open time. Specifically, the fresh material needs to have sufficient workability to be pumped to the extruder via the pumping hose and extruded through a nozzle attached to the extruder to form a small and continuous filament. In addition, the material needs to gain strength in short period of time to retain its shape after being extruded and sustain the weight of subsequent layers with minimum acceptable deformation. The following subsections provide an overview of each OPC-free binder system, highlighting their main advantages and challenges for 3DCP applications.

11.3.1 Alkali-activated materials

AAMs, which has been commonly referred to as "geopolymer," has attracted significant interest from the industry and academia as concern for environmental impacts and sustainability issues have become the main considerations in the construction industry nowadays. AAMs is produced by alkaline activation of aluminosilicate source materials (precursors) which consist of a considerable amount of aluminum (Al) and silicon (Si) in amorphous form (Duxson & Provis, 2008; Provis & Van Deventer, 2009). A wide range of aluminosilicate source materials can be used to manufacturing AAMs, such as feldspar, metakaolin, and industrial by-products like fly ash, GGBS, mining wastes, and rice husk ash (Detphan & Chindaprasirt, 2009; Duxson et al., 2007; C. Li et al., 2010). Alkaline activators are generally consisted of alkali metal hydroxide and/or alkali metal silicates. It should be noted that the alkaline activators can be in solution form or solid form. Examples of commonly used alkaline activators are sodium hydroxide, sodium silicate, potassium hydroxide, and potassium silicate (Palomo et al., 1999; Provis, 2009; Singh et al., 2015). Many academic studies have reported that AAMs can be produced by industrial by-products such as fly ash and GGBS, which are industrial by-products of coal-fired electric generating plants and iron manufacture, respectively (Duxson, 2009; Keyte, 2009). Utilizing industrial by-products such as fly ash as the raw materials to manufacture AAMs is particularly beneficial as the disposition of this industrial by-product occupies large areas in landfill, which could be used for other purposes (Sarker & McBeath, 2015).

Furthermore, several studies have been conducted to compare the environmental performance of OPC and AAMs concretes (McLellan et al., 2011; Turner & Collins, 2013; Yang et al., 2013). For instance, Yang et al. (2013) compared the CO_2 emissions of OPC and AAMs synthesized from different types of aluminosilicate source materials, including fly ash and GGBS. The authors considered the transportation of raw materials to concrete plan and the transportation of mixed concrete to construction site. They reported that the CO_2 emissions for OPC, OPC and supplemental cementitious materials blended, GGBS-based AAM and fly ash–based AAM concretes with a compressive strength of 40 MPa were 509, 400, 122, and 181 kg/m^3, respectively. Another study by Turner and Collins (2013) from Monash University, Australia reported that the carbon dioxide equivalent (CO_2-e) emissions of OPC and fly ash–based AAM concretes to be 354 and 320 kg CO_2-e/m^3, respectively, showing 9% difference. Similar to the typical castable concrete, literature also reported up to 79% lower CO_2 emissions of 3D printable mixture using AAMs as the binder as compared to that of the OPC-based mixture (Bong et al., 2021; Panda, Singh, Unluer, & Tan, 2019).

Apart from the environmental benefits mentioned above, AAMs can exhibit a variety of properties, including adjustable rheological properties and ability to developing higher strengths in a short period of time, which are favored in the layer-by-layer buildup 3DCP process (Nath & Sarker, 2014; Provis, 2014). Literature reveals that these properties are greatly influenced by the raw materials selection and mixture proportion used in formulating 3D printable AAMs mixture. For instance, increasing the GGBS content can result in an increase of static yield stress, plastic viscosity, and thixotropy, which are particularly beneficial to the extrudability and buildability of the mixture. This is mainly due to the angular shape (physical interlocking effect) and high calcium content (higher reactivity) of the GGBS particles. However, the dosage of GGBS should be carefully adjusted as the addition of GGBS can lead to a shorter setting time and thus the open time of the 3D printable mixture (Guo et al., 2020; Panda et al., 2018; Panda, Unluer, & Tan, 2019; Panda & Tan, 2018). Silica fume with an appropriate dosage in fly ash–based AAM mixture can also result in a significant improvement in static yield stress and thixotropy, which is mainly due to its fine particle size is beneficial for better packing of particles (Guo et al., 2020; Panda et al., 2018).

Besides the influences from aluminosilicate source materials, type, dosage, and blending proportion of alkaline activators can also greatly contribute to the rheological properties of the 3D printable AAMs mixtures. For example, among the commonly use sodium-based and potassium-based activators, literature showed that the AAMs mixtures activated by sodium-based activators exhibited higher flowability, lower yield stress, and longer open time as compared to that by potassium-based activators (Bong et al., 2019b, 2019c). This can be attributed to the higher viscosity and slower reactivity of the sodium-based activators used. A study also demonstrated that increasing molar ratio of the alkaline activator enhances the static yield stress and apparent viscosity of the AAMs mixtures which is mostly due to the rise in the alkaline activator solution's viscosity (Panda, Unluer, & Tan, 2019). Furthermore, higher dosage of alkaline activators can result in faster reactions which increases the static yield stress and shortens the open time of the AAMs mixtures (Bong et al., 2021; Muthukrishnan et al., 2021; Panda, Singh, Unluer, & Tan, 2019). A comprehensive overview of AAMs mixture design for 3DCP can be found in Lazorenko and Kasprzhitskii (2022) and Raza et al. (2022).

Despite the environmental advantages and its modifiable rheological properties, there are several hindrances could limit its development in commercial large-scale 3DCP. One of these obstacles is the limited and declining supply of aluminosilicate source materials for continued long-term usage particularly fly ash and GGBS, which are the most frequently used aluminosilicate source materials in AAMs binder production (Diaz-Loya et al., 2019; Elzeadani et al., 2022; Zunino et al., 2022). Furthermore, the high variation in the chemical composition, mineralogy, morphology, and particle size distribution of aluminosilicate source materials (e.g., fly ash) obtained from any single source/process from batch-to-batch or even from day-to-day could greatly affect the rheological and hardened properties of the resultant printable AAMs mixtures (Fernández-Jiménez & Palomo, 2003; Komljenović et al., 2010; Provis, 2014). In addition to the availability and quality issues of aluminosilicate source materials, the production of alkaline activators is considered to be highly energy- and carbon-intensive, which renders it as the most carbon- and energy-intensive constituent in the AAMs mixtures (Bong et al., 2021; Luukkonen et al., 2018). Furthermore, alkaline activators with high pH value are often used to facilitate faster yield stress development and higher machinal strengths, which rises safety concerns when used in large-scale construction (Nematollahi et al., 2015). The high pH of alkaline activators can also compromise the effectiveness of existing concrete admixtures (e.g., superplasticizer) in AAMs mixtures and may negatively impact their hardened properties (Alrefaei et al., 2020; Bong et al., 2019a; Zhang, Sun, et al., 2022). This creates additional challenge in adjusting the rheological properties of the AAMs mixtures for 3DCP process.

11.3.2 Calcium sulfoaluminate cement

CSA cement is another environmentally friendly alternative binder to OPC. The main constituent of CSA cement is calcium sulfoaluminate or ye'elimite ($3CaO \cdot 3Al_2O_3 \cdot CaSO_4$) (Sharp et al., 1999; Zhou et al., 2006). CSA cement is produced by calcining limestone, aluminous clay or bauxite, and calcium sulfate at temperatures ranging between 1250°C and 1350°C, which is lower than that required for typical OPC production (i.e., about 1450°C) (Glasser & Zhang, 2001; Zhou et al., 2006). The lower calcination temperature results in reduced energy consumption and lower CO_2 emissions during the production of CSA cement. A comprehensive analysis conducted by Hanein et al. (2018) found that the CSA clinker production emitted up to 35% less greenhouse gas compared to OPC clinker production. Additionally, the resultant CSA clinker is generally porous and fragile compared to the OPC clinker, making it easier to grind and further contributing to energy savings (Glasser & Zhang, 2001). Apart from the typical raw materials mentioned above, researchers have attempted to utilized various industrial by-products or waste materials, such as GGBS, low-calcium fly ash, phosphogypsum, bauxite fines, scrubber sludge or bag house dust for producing CSA clinker (Arjunan & Silsbee, 1999; Beretka et al., 1993; Sahu & Majling, 1994).

CSA cement is known for its rapid setting and high early-age strength, attributed to the fast hydration of CSA clinker in the presence of gypsum, leading to the formation of ettringite (Glasser & Zhang, 2001). The feature of gaining strength rapidly at a very early age is beneficial in enhancing the buildability of 3D printable mixtures (N. Khalil et al., 2017; Tao et al., 2023). For instance, a study conducted by N. Khalil et al. (2017) has demonstrated that the buildability of 3D printable mixture can be significantly enhanced by replacing a small portion (7%) of OPC in the mixture with CSA cement. The authors found that the rapid hydration of CSA cement in the mixture resulted in faster yield stress evolution and setting, thereby greatly enhancing the buildability of the fresh mixture. However, the exceptionally fast setting of CSA cement poses a challenge for its complete replacement of OPC in 3D printable mixtures. Subsequently, several research groups utilized various retarding agents, such as tartaric acid, borax, and, to mitigate the flash setting of CSA cement and successfully formulated 3D printable mixtures containing 100% CSA cement (M. Chen et al., 2018; M. Chen, Yang, et al., 2020; Mohan et al., 2021a, 2021b). It is worth noting that among the retarding agents reported in literature, sodium gluconate was found to effectively prolong the open time of the 3D printable mixture but resulted in a significant reduction in the 1-day compressive strength (Mohan et al., 2021a, 2021b).

There have been very limited studies on the rheological properties of CSA cement, primarily due to its rapid setting behavior. However, a few research groups have successfully studied the rheological properties of retarded CSA cement mixtures (M. Chen, Li, et al., 2020; M. Chen, Liu, et al., 2020; M. Chen, Yang, et al., 2020; Ke et al., 2020; Mohan et al., 2021a, 2021b). For example, Mohan et al. (2021a, 2021b) investigated the rheological behavior of the 3D printable CSA cement-based mixture with addition of borax as a retarder and compared it to the 3D printable OPC-based mixture. The authors found that the CSA cement-based mixture exhibited significantly higher plastic viscosity than the OPC-based mixture, resulting in slightly higher pumping pressure loss for CSA cement-based mixture. To mitigate this, the plastic viscosity was reduced by partially replacing CSA cement with limestone in the mixture, leading to a reduction in pumping pressure while slightly increasing the yield stress. Another study conducted by M. Chen, Yang, et al. (2020) investigated the rheological properties of 3D printable CSA cement mixture with metakaolin as a rheology modifier. The results indicated that the addition of metakaolin improved the rheological properties, including static and dynamic yielding behavior, as well as the thixotropy behavior of the 3D printable CSA cement mixture. These improvements are beneficial for enhancing the shape retention ability and buildability of the 3D printable CSA cement mixture, which was also demonstrated by the authors.

In addition to its setting characteristic, CSA cement generally exhibits lower shrinkage than OPC. The shrinkage behavior of CSA cement varies depending on the gypsum content added to the system, resulting in a range of shrinkage behaviors from lower volumetric shrinkage to expansive behavior. This is attributed to the expansive nature of ettringite, which forms within the matrix (Glasser & Zhang, 2001). The shrinkage compensating property of CSA cement is particularly important to 3D printed concrete elements as they are highly susceptible to shrinkage due to the formwork-free fabrication process and the high binder content found in typical 3D printable mixtures (Markin & Mechtcherine, 2023; Moelich et al., 2020; Shahmirzadi et al., 2021).

Although CSA cement appears to be a promising sustainable alternative binder to OPC, there are several obstacles that still needed to be overcome before its successful implementation in commercial large-scale 3DCP. One notable obstacle is the higher cost of CSA cement compared to OPC, primarily attributed to the high price of alumina-bearing raw materials such as bauxite (Pimraksa & Chindaprasirt, 2018). As a result, using CSA cement as the sole binder in 3D printable mixtures becomes less economically feasible. Unlike AAMs discussed in the previous section, CSA cement lacks international consensus on standards and nomenclature (Phair, 2006). This can create uncertainty and hesitancy among contractors, making them reluctant to adopt it in their projects including 3DCP project. Besides, ettringite as the dominant constitute in hydrated CSA cement, which is more prone to deterioration issues associated with carbonation (e.g., steel reinforcement corrosion) (Hargis et al., 2017; Quillin, 2001; Zhou & Glasser, 2000). Furthermore, the different hydration kinetics of CSA cement compared to OPC result in a lower calcium hydroxide content in the pore solution, leading to reduced alkalinity when compared to OPC. This lower alkalinity rises concerns regarding the ability of CSA cement to provide adequate corrosion protection for steel reinforcement (J. Ma et al., 2023; Zhou et al., 2006). Recently, some researchers have found that the lower alkalinity in CSA cement was sufficient to protect steel reinforcement from corrosion (Carsana et al., 2018; Koga et al., 2020). However, more studies are still needed to investigate the ability of CSA cement to passivate the steel reinforcement.

11.3.3 Magnesia-based cements

Another interesting OPC-free binder system is magnesia-based cements, where magnesium oxide (MgO) is used as the key reactive ingredient (Al-Tabbaa, 2013; Walling & Provis, 2016). MgO is mainly produced by the calcination of magnesite ($MgCO_3$)-based minerals, such as dolomite (Ruan & Unluer, 2016). The calcination temperature for $MgCO_3$ to form reactive MgO occurs in a range of 600–800°C depending on the presence of impurities and the physical properties of mined minerals (Flatt et al., 2012; Ruan & Unluer, 2016; Walling & Provis, 2016). Similar to the calcination of limestone in OPC production, the calcination of $MgCO_3$ emits CO_2 into the atmosphere. A study reported that the CO_2 emission of the decomposition of $MgCO_3$ into MgO is higher than that of limestone in OPC production (i.e., ~1.1 t/t vs ~0.8 t/t). However, the ability of CO_2 sequestration makes their net CO_2 emission to be about 70% lower than that of OPC (Ruan & Unluer, 2016). Additionally, the lower calcination temperature required in magnesia-based cements production has allowed the potential use of renewable energy sources which would further reduce the CO_2 emission above (Al-Tabbaa, 2013; Walling & Provis, 2016). Apart from calcination, MgO can also be obtained from saline lakes or seawater through precipitation. It is worth noting that magnesium is ranked as the third most abundant element present in seawater in a dissolved state (Shand, 2006). However, producing MgO from seawater is considered energy-intensive unless a highly concentrated by-product brine from the desalination process is used (Unluer, 2018).

Among the various magnesia-based cements developed (i.e., magnesium carbonate cement, magnesium phosphate cement, magnesium oxychloride cement, and magnesium oxysulfate cement), magnesium carbonate cement appears to be promising due to its ability to absorb CO_2 from the atmosphere to form a wide range of carbonates and hydroxycarbonates, which are the primary contributors to the strength development in magnesium carbonate cement-based mixtures (Dung et al., 2021; Hay & Celik, 2020; Walling & Provis, 2016). In addition, the resultant magnesium carbonates can be theoretically re-calcined to obtain the original starting material (i.e., MgO) (Phair, 2006). To date, there is only a study conducted by A. Khalil et al. (2020) on formulating magnesium carbonate cement-based mortar for 3D printing applications. The developed 3D printable mixture consisted of 97% magnesium carbonate cement and 3% caustic MgO as the main binder. A small amount of caustic MgO was added to enhance the early hydration of the mixture, and thereby better buildability. The authors also reported that the open time of the developed mixture was up to 60 minutes even with the addition of the caustic MgO. The compression test result showed that the 3D printed samples exhibited significantly higher strength than that of the mold-cast samples, which was associated with the increased carbonation rate as a result of the higher initial porosity of the 3D printed samples.

Apart from magnesium carbonate cement, magnesium phosphate cement, formed through an acid–base reaction between MgO and an acidic phosphate-based solution (i.e., ammonium or potassium phosphate), had been explored by several researchers to formulate 3D printable mixtures (Weng et al., 2019; Z. Zhao et al. 2021, 2022). Similar to CSA cement, magnesium phosphate cement is also known for its rapid setting characteristic after contacting with water, which is particular useful for achieving better buildability of 3D printable mixtures. However, high volume of SCMs and retarding agents are generally used to mitigate fresh setting issue and improve rheological properties for 3DCP applications. For instance, Weng et al. (2019) investigated the effects of fly ash, silica fume, and borax contents on the rheological properties and hardened properties of magnesium potassium phosphate cement-based mortar for 3D printing. Based on the results obtained, the authors concluded that the optimum mixture formulation consisted of 60 wt.% magnesia replaced by fly ash and an additional 10 wt.% silica fume with a borax-to-magnesia ratio of 1:4. Another group of researchers reported that the mixture with a magnesium-to-phosphate mass ratio of 3.0, 25 wt.% fly ash, and 40 wt.% borax was determined as the optimum mix design for 3D printing, which exhibited the minimum deformation (0.28%) and 3-day compressive strength of 32.6 MPa (Zhao et al., 2021).

Although magnesia-based cements have the potential to make concrete with a negative carbon footprint, the relatively higher price has hindered their large-scale 3DCP applications. One of the reasons is it is currently difficult to develop a cost-efficient industrial manufacturing process to obtain quality reactive MgO from raw minerals containing impurities. In addition, the limited global access to the raw minerals has also led to the higher price of magnesia-based cements (Unluer, 2018; Walling & Provis, 2016). Furthermore, the relatively lower pore solution pH value of magnesia-based cements indicates their poor ability to passivate the steel reinforcement which limits their application in 3D printed structural members (Hay & Celik, 2020). However, a contradictory result was reported by other researchers (Wang et al., 2022). Further detailed investigations are needed to understand the corrosion mechanism of steel reinforcement in magnesia-based cements.

11.4 Binders with high supplementary cementitious materials content

SCMs have been commonly used in most of the 3D printable mixtures reported in the literature, as well as commercially available 3D printable premixed mortars. They are used either as OPC replacements or rheology-modifying materials in 3D printable mixtures. Popular SCMs used in 3D printable mixtures include fly ash, GGBS, silica fume, etc. (Arunothayan et al., 2022; Le, Austin, Lim, Buswell, Gibb, & Thorpe, 2012; Nerella et al., 2019; Paul et al., 2018; Qian et al., 2021). Replacing OPC in 3D printable mixtures with these materials not only results in a significant reduction in embodied carbon of the mixtures but also offers a value-added means to utilize these solid wastes. Apart from reducing OPC usage, SCMs are beneficial in reducing the heat of hydration, which helps minimize the risk of thermal shrinkage, and improve the durability of printed concrete due to the denser microstructure (De Belie et al., 2018).

One significant advantage of incorporating SCMs in 3D printable mixtures is their ability to adjust and enhance the rheological properties of the mixture for better printing performance. For example, the inclusion of fly ash is beneficial in improving the workability of the mixture for better pumpability and extrudability, which is attributed to the spherical shape and glassy surface of fly ash particles, namely the ball-bearing effect (Sideris et al., 2018). Additionally, the addition of a small amount of silica fume can improve the grain size composition of the mixture, resulting in enhanced buildability (Panda & Tan, 2019). Furthermore, the incorporation of fine particles, such as fly ash and silica fume, into the mixture can greatly reduce the risk of segregation or phase separation issue during high-pressure pumping and extrusion (Lewis, 2018; Sideris et al., 2018). The mixtures containing high volumes of SCMs generally exhibit a relatively longer open time due to the slower pozzolanic reaction at early ages. A study also showed that fly ash, GGBS, and silica fume can effectively enhance the thixotropy property of 3D printable mixtures for a smoother printing process (Yuan et al., 2018).

Merely substituting a high volume of OPC with a single type of SCM can have detrimental effects on the rheological properties of the mixture, potentially leading to poor printing performance. For instance, Panda and Tan (2019) observed a significant drop in the static yield stress and apparent viscosity of the mixture with increasing fly ash content, which is unfavorable for shape retention ability and buildability. However, the authors found that incorporating a small amount of silica fume (up to 5%) into the mixture not only mitigated this negative effect but also

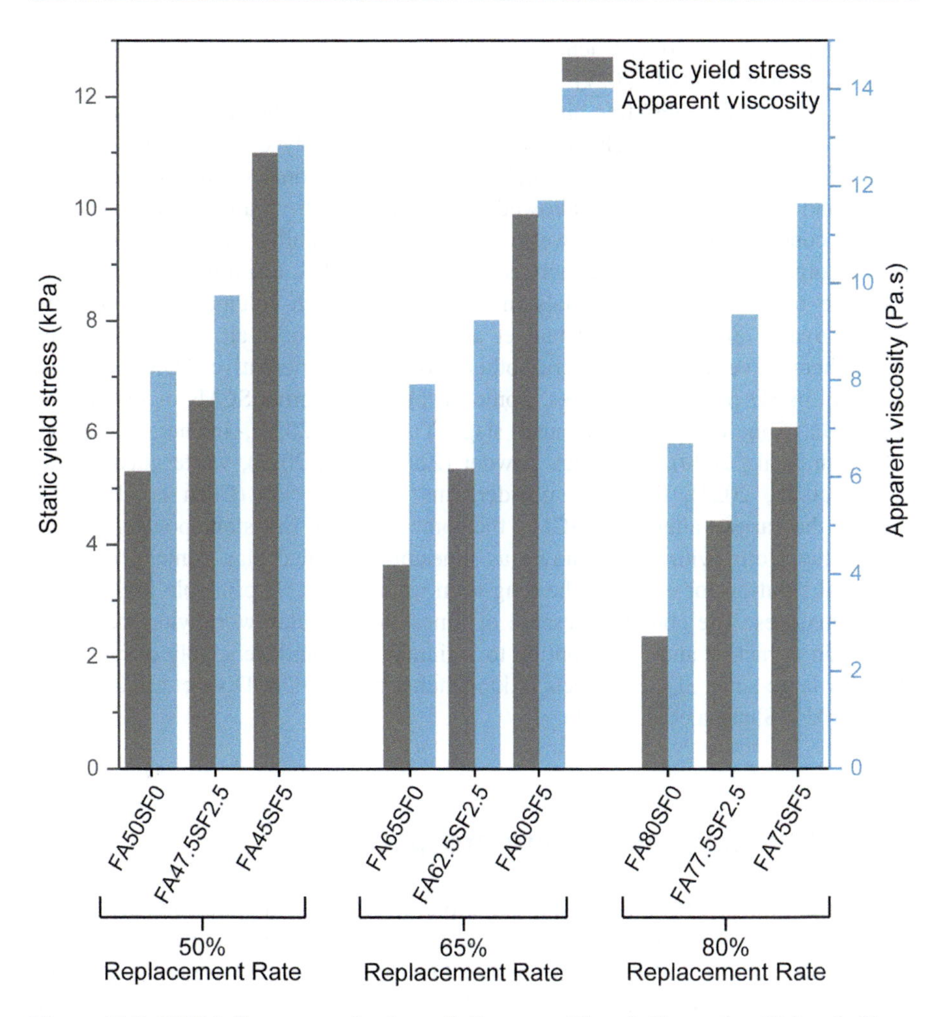

Figure 11.3 SCM influence on rheology. Influences of fly ash (denoted as FA) and silica fume (denoted as SF) contents on the rheology behaviors of high-volume SCM mixtures. *Source*: Adopted from Panda, B., & Tan, M. J. (2019). Rheological behavior of high volume fly ash mixtures containing micro silica for digital construction application. *Materials Letters*, *237*, 348–351. https://doi.org/10.1016/j.matlet.2018.11.131.

further enhanced the static yield stress and apparent viscosity (see Fig. 11.3). In another study, researchers also reported that the yield stress and buildability of a 3D printable high-volume fly ash mixture were significantly improved by incorporating a small amount of limestone powder (i.e., 5%) (Dey et al., 2022). These findings highlight the importance of carefully considering the combination of SCMs to optimize the rheological properties of 3D printable mixtures for achieving optimum printing performance. Due to the above reasons, a ternary binder system is more commonly adopted in formulating 3D printable mixtures with high SCMs content

(Arunothayan et al., 2022; Chen et al., 2021; Figueiredo et al., 2019; Nerella et al., 2019; Zahabizadeh et al., 2021).

Compared to the OPC-free binder systems discussed in the previous sections, 3D printable mixtures with high SCMs content are relatively easy to adopt in large-scale 3DCP projects, as various widely adopted international standards, such as EN 197-1 and ASTM C595, have allowed the usage of high-volume SCMs binders in current concrete production. However, the limited availability and declining supply of the commonly used SCMs, such as fly ash and GGBS, can hinder the long-term continue usage of these materials in 3D printable high-volume SCMs mixtures (Diaz-Loya et al., 2019; Elzeadani et al., 2022; Zunino et al., 2022). Therefore researchers have explored several other sustainable alternative SCMs for their usability in 3D printable mixtures. Some of these alternative SCMs include copper slag (Ma, Sun, et al., 2018), steel slag (Yu et al., 2023), incinerated fly ash (Rehman et al., 2020), clay brick powder (Zhao et al., 2023), waste glass powder (Deng et al., 2023), limestone powder, and calcined clay (Chen et al., 2021). Among the various alternative SCMs, the combination of limestone powder and calcined clay along with OPC, known as limestone calcined clay cement, can be a potential solution for ensuring the long-term usability of 3D printable high-volume SCM mixtures. The abundant reserve of limestone and clay worldwide makes this binder an attractive and viable option to maintain a sustainable supply of raw materials for large-scale 3DCP projects (Akindahunsi et al., 2020; Dixit et al., 2021; Lin et al., 2021; Sharma et al., 2021).

11.5 Summary and future prospects

3DCP technology offers promising solutions to the construction industry's challenges of low productivity, high labor demands, and high carbon emissions. Despite its potential benefits, the higher OPC content in 3D printable mixtures can result in negative impacts on the 3D printed concrete structures. This can undermine the economic and environmental benefits that 3DCP aims to provide. In this chapter, several promising OPC-free binder systems, including AAMs, CSA cement, and magnesia-based cements, as well as high-volume SCMs mixtures were reviewed within the context of 3DCP. The discussion centered on their main advantages, rheological properties, and limitations from the perspective of 3DCP applications.

While OPC-free binder systems offer a substantial decrease in carbon footprint, it is imperative to conduct further research to acquire a more comprehensive understanding of their rheological behaviors and long-term durability in the context of large-scale 3DCP applications. Moreover, the advancement of economically feasible industrial fabrication methodologies and the investigation into alternative raw materials, particularly those endowed with abundant global reserves, are essential to establish cost-competitive advantages and enhance the long-term viability of OPC-free binders. As the technology continues to evolve, it is crucial for researchers, practitioners, and policymakers to collaborate in developing the relevant standards or guidelines to further reinforce the acceptance and confidence in using these OPC-free binders in large-scale 3DCP projects. In the short term, addressing the challenges related to SCMs availability and exploring alternative options can provide a more stable supply chain and promote sustainable practices in 3DCP with low-carbon mixtures. In the ultimate trajectory, low-carbon binders should be seamlessly integrated into the current OPC manufacturing process to replace OPC clinker and should offer cost-competitive solution for large-scale and long-term use.

References

Akindahunsi, A. A., Avet, F., & Scrivener, K. (2020). The influence of some calcined clays from Nigeria as clinker substitute in cementitious systems. *Case Studies in Construction Materials, 13*, e00443. Available from https://doi.org/10.1016/j.cscm.2020.e00443.

Alrefaei, Y., Wang, Y. S., Dai, J. G., & Xu, Q. F. (2020). Effect of superplasticizers on properties of one-part $Ca(OH)_2/Na_2SO_4$ activated geopolymer pastes. *Construction and Building Materials, 241*. Available from https://doi.org/10.1016/j.conbuildmat.2019.117990.

Al-Tabbaa, A. (2013). Reactive magnesia cement. In F. Pacheco-Torgal, S. Jalali, J. Labrincha, & V. M. John (Eds.), *Eco-efficient concrete* (pp. 523−543). Woodhead Publishing. Available from http://www.sciencedirect.com/science/book/9780857094247, https://doi.org/10.1533/9780857098993.4.523.

Arjunan, P., Silsbee, M. R., & Roy, D. M. (1999). Sulfoaluminate-belite cement from low-calcium fly ash and sulfur-rich and other industrial by-products. *Cement and Concrete Research, 29*(8), 1305−1311. Available from https://doi.org/10.1016/S0008-8846(99)00072-1.

Arunothayan, A. R., Nematollahi, B., Ranade, R., Khayat, K. H., & Sanjayan, J. G. (2022). Digital fabrication of eco-friendly ultra-high performance fiber-reinforced concrete. *Cement and Concrete Composites, 125*. Available from https://doi.org/10.1016/j.cemconcomp.2021.104281.

Barbosa, F., Woetzel, J., Mischke, J., Ribeirinho, M., Sridhar, M., Parsons, M., Bertram, N., & Brown, S. (2017). *Reinventing construction: A route to higher productivity*. McKinsey Global Institute.

Beretka, J., de Vito, B., Santoro, L., Sherman, N., & Valenti, G. L. (1993). Hydraulic behaviour of calcium sulfoaluminate-based cements derived from industrial process wastes. *Cement and Concrete Research., 23*(5), 1205−1214. Available from https://doi.org/10.1016/0008-8846(93)90181-8.

Bong, S. H., Nematollahi, B., Nazari, A., Xia, M., & Sanjayan, J. (2019a). Efficiency of different superplasticizers and retarders on properties of 'one-part' fly ash-slag blended geopolymers with different activators. *Materials, 12*(20), 3410. Available from https://doi.org/10.3390/ma12203410.

Bong, S. H., Nematollahi, B., Nazari, A., Xia, M., & Sanjayan, J. G. (2019b). Fresh and hardened properties of 3D printable geopolymer cured in ambient temperature. In T. Wangler, & R. Flatt (Eds.), *First RILEM international conference on concrete and digital fabrication − digital concrete 2018. DC 2018. RILEM Bookseries* (19). Springer. Available from 10.1007/978-3-319-99519-9_1.

Bong, S. H., Nematollahi, B., Nazari, A., Xia, M., & Sanjayan, J. (2019c). Method of optimisation for ambient temperature cured sustainable geopolymers for 3D printing construction applications. *Materials*, *12*(6), 902. Available from https://doi.org/10.3390/ma12060902.

Bong, S. H., Xia, M., Nematollahi, B., & Shi, C. (2021). Ambient temperature cured 'just-add-water' geopolymer for 3D concrete printing applications. *Cement and Concrete Composites*, *121*, 104060. Available from https://doi.org/10.1016/j.cemconcomp.2021.104060.

Bos, F., Wolfs, R., Ahmed, Z., & Salet, T. (2016). Additive manufacturing of concrete in construction: Potentials and challenges of 3D concrete printing. *Virtual and Physical Prototyping*, *11*(3), 209–225. Available from https://doi.org/10.1080/17452759.2016.1209867.

Carsana, M., Canonico, F., & Bertolini, L. (2018). Corrosion resistance of steel embedded in sulfoaluminate-based binders. *Cement and Concrete Composites*, *88*, 211–219. Available from https://doi.org/10.1016/j.cemconcomp.2018.01.014.

Cement Sustainability Initiative (CSI). (2016). Cement Sustainability Initiative (CSI) Cement industry energy and CO_2 performance: Getting the numbers right (GNR), World Business Council for Sustainable Development (WBCSD).

Chen, M., Guo, X., Zheng, Y., Li, L., Yan, Z., Zhao, P., Lu, L., & Cheng, X. (2018). Effect of tartaric acid on the printable, rheological and mechanical properties of 3D printing sulphoaluminate cement paste. *Materials*, *11*(12), 2417. Available from https://doi.org/10.3390/ma11122417.

Chen, Y., He, S., Zhang, Y., Wan, Z., Çopuroğlu, O., & Schlangen, E. (2021). 3D printing of calcined clay-limestone-based cementitious materials. *Cement and Concrete Research*, *149*, 106553. Available from https://doi.org/10.1016/j.cemconres.2021.106553.

Chen, M., Li, L., Wang, J., Huang, Y., Wang, S., Zhao, P., Lu, L., & Cheng, X. (2020). Rheological parameters and building time of 3D printing sulphoaluminate cement paste modified by retarder and diatomite. *Construction and Building Materials*, *234*, 117391. Available from https://doi.org/10.1016/j.conbuildmat.2019.117391.

Chen, M., Liu, B., Li, L., Cao, L., Huang, Y., Wang, S., Zhao, P., Lu, L., & Cheng, X. (2020). Rheological parameters, thixotropy and creep of 3D-printed calcium sulfoaluminate cement composites modified by bentonite. *Composites Part B: Engineering*, *186*, 107821. Available from https://doi.org/10.1016/j.compositesb.2020.107821.

Chen, M., Yang, L., Zheng, Y., Huang, Y., Li, L., Zhao, P., Wang, S., Lu, L., & Cheng, X. (2020). Yield stress and thixotropy control of 3D-printed calcium sulfoaluminate cement composites with metakaolin related to structural build-up. *Construction and Building Materials*, *252*, 119090. Available from https://doi.org/10.1016/j.conbuildmat.2020.119090.

Chen, Y., Zhang, Y., Pang, B., Wang, D., Liu, Z., & Liu, G. (2022). Steel fiber orientational distribution and effects on 3D printed concrete with coarse aggregate. *Materials and Structures/Materiaux et Constructions.*, *55*(3). Available from https://doi.org/10.1617/s11527-022-01943-7.

Chu, S. H., Li, L. G., & Kwan, A. K. H. (2021). Development of extrudable high strength fiber reinforced concrete incorporating nano calcium carbonate. *Additive Manufacturing*, *37*, 101617. Available from https://doi.org/10.1016/j.addma.2020.101617.

Damineli, B. L., Kemeid, F. M., Aguiar, P. S., & John, V. M. (2010). Measuring the eco-efficiency of cement use. *Cement and Concrete Composites.*, *32*(8), 555–562. Available from https://doi.org/10.1016/j.cemconcomp.2010.07.009.

De Belie, N., Soutsos, M., & Gruyaert, E. (2018). *Properties of fresh and hardened concrete containing supplementary cementitious materials. State-of-the-art report of the RILEM Technical Committee 238-SCM, Working Group 4*. Springer. Available from http://doi.org/10.1007/978-3-319-70606-1.

Deng, Qi, Zou, Shuai, Xi, Yonghui, & Singh, Amardeep (2023). Development and characteristic of 3D-printable mortar with waste glass powder. *Buildings*, *13*(6), 1476. Available from https://doi.org/10.3390/buildings13061476.

De Schutter, G., Lesage, K., Mechtcherine, V., Nerella, V. N., Habert, G., & Agusti-Juan, I. (2018). Vision of 3D printing with concrete — Technical, economic and environmental potentials. *Cement and Concrete Research*, *112*, 25–36. Available from https://doi.org/10.1016/j.cemconres.2018.06.001.

Detphan, S., & Chindaprasirt, P. (2009). Preparation of fly ash and rice husk ash geopolymer. *International Journal of Minerals, Metallurgy and Materials.*, *16*(6), 720–726. Available from https://doi.org/10.1016/S1674-4799(10)60019-2.

Dey, D., Srinivas, D., Boddepalli, U., Panda, B., Gandhi, I. S. R., & Sitharam, T. G. (2022). 3D printability of ternary Portland cement mixes containing fly ash and limestone. *Materials Today: Proceedings*, *70*, 195–200. Available from https://doi.org/10.1016/j.matpr.2022.09.020.

Diaz-Loya, I., Juenger, M., Seraj, S., & Minkara, R. (2019). Extending supplementary cementitious material resources: Reclaimed and remediated fly ash and natural pozzolans. *Cement and Concrete Composites*, *101*, 44–51. Available from https://doi.org/10.1016/j.cemconcomp.2017.06.011.

Dixit, A., Du, H., & Pang, S. D. (2021). Performance of mortar incorporating calcined marine clays with varying kaolinite content. *Journal of Cleaner Production*, *282*, 124513. Available from https://doi.org/10.1016/j.jclepro.2020.124513.

Duan, Z., Li, L., Yao, Q., Zou, S., Singh, A., & Yang, H. (2022). Effect of metakaolin on the fresh and hardened properties of 3D printed cementitious composite. *Construction and Building Materials*, *350*, 128808. Available from https://doi.org/10.1016/j.conbuildmat.2022.128808.

Dung, N. T., Hay, R., Lesimple, A., Celik, K., & Unluer, C. (2021). Influence of CO_2 concentration on the performance of MgO cement mixes. *Cement and Concrete Composites*, *115*, 103826. Available from https://doi.org/10.1016/j.cemconcomp.2020.103826.

Duxson, P. (2009). *Geopolymer precursor design geopolymers: Structures, processing, properties and industrial applications* (pp. 37–49). Elsevier Ltd. Available from http://doi.org/10.1533/9781845696382.1.37.

Duxson, P., Fernández-Jiménez, A., Provis, J. L., Lukey, G. C., Palomo, A., & Van Deventer, J. S. J. (2007). Geopolymer technology: The current state of the art. *Journal of Materials Science*, *42*(9), 2917–2933. Available from https://doi.org/10.1007/s10853-006-0637-z.

Duxson, P., & Provis, J. L. (2008). Designing precursors for geopolymer cements. *Journal of the American Ceramic Society*, *91*(12), 3864–3869. Available from https://doi.org/10.1111/j.1551-2916.2008.02787.x.

Elzeadani, M., Bompa, D. V., & Elghazouli, A. Y. (2022). One part alkali activated materials: A state-of-the-art review. *Journal of Bu ilding Engineering.*, *57*, 104871. Available from https://doi.org/10.1016/j.jobe.2022.104871.

Fernández-Jiménez, A., & Palomo, A. (2003). Characterisation of fly ashes. Potential reactivity as alkaline cements. *Fuel*, *82*(18), 2259–2265. Available from https://doi.org/10.1016/s0016-2361(03)00194-7.

Figueiredo, S. C., Rodríguez, C. R., Ahmed, Z. Y., Bos, D. H., Xu, Y., Salet, T. M., Çopuroğlu, O., Schlangen, E., & Bos, F. P. (2019). An approach to develop printable strain hardening cementitious composites. *Materials & Design*, *169*.

Flatt, R. J., Roussel, N., & Cheeseman, C. R. (2012). Concrete: An eco material that needs to be improved. *Journal of the European Ceramic Society*, *32*(11), 2787–2798. Available from https://doi.org/10.1016/j.jeurceramsoc.2011.11.012.

García de Soto, B., Agustí-Juan, I., Hunhevicz, J., Joss, S., Graser, K., Habert, G., & Adey, B. T. (2018). Productivity of digital fabrication in construction: Cost and time analysis of a robotically built wall. *Automation in Construction*, *92*, 297–311. Available from https://doi.org/10.1016/j.autcon.2018.04.004.

Glasser, F. P., & Zhang, L. (2001). High-performance cement matrices based on calcium sulfoaluminate-belite compositions. *Cement and Concrete Research*, *31*(12), 1881–1886. Available from https://doi.org/10.1016/S0008-8846(01)00649-4.

Guo, X., Yang, J., & Xiong, G. (2020). Influence of supplementary cementitious materials on rheological properties of 3D printed fly ash based geopolymer. *Cement and Concrete Composites*, *114*, 103820. Available from https://doi.org/10.1016/j.cemconcomp.2020.103820.

Hanein, T., Galvez-Martos, J. L., & Bannerman, M. N. (2018). Carbon footprint of calcium sulfoaluminate clinker production. *Journal of Cleaner Production*, *172*, 2278–2287. Available from https://doi.org/10.1016/j.jclepro.2017.11.183.

Hargis, C. W., Lothenbach, B., Müller, C. J., & Winnefeld, F. (2017). Carbonation of calcium sulfoaluminate mortars. *Cement and Concrete Composites*, *80*, 123–134. Available from https://doi.org/10.1016/j.cemconcomp.2017.03.003.

Hay, R., & Celik, K. (2020). Hydration, carbonation, strength development and corrosion resistance of reactive MgO cement-based composites. *Cement and Concrete Research*, *128*, 105941. Available from https://doi.org/10.1016/j.cemconres.2019.105941.

Johnston, D. W. (2008). *Design and construction of concrete formwork. Concrete construction engineering handbook* (2nd ed, pp. 233–282). CRC Press. Available from https://www.taylorfrancis.com/books/e/9781420007657.

Kazemian, A., Yuan, X., Cochran, E., & Khoshnevis, B. (2017). Cementitious materials for construction-scale 3D printing: Laboratory testing of fresh printing mixture. *Construction and Building Materials*, *145*, 639–647. Available from https://doi.org/10.1016/j.conbuildmat.2017.04.015.

Ke, G., Zhang, J., Xie, S., & Pei, T. (2020). Rheological behavior of calcium sulfoaluminate cement paste with supplementary cementitious materials. *Construction and Building Materials*, *243*, 118234. Available from https://doi.org/10.1016/j.conbuildmat.2020.118234.

Keyte, L. M. (2009). *Fly ash glass chemistry and inorganic polymer cements. Geopolymers: Structures, processing, properties and industrial applications* (pp. 15–36). Elsevier Ltd. Available from 10.1533/9781845696382.1.15.

Khalil, A., Wang, X., & Celik, K. (2020). 3D printable magnesium oxide concrete: Towards sustainable modern architecture. *Additive Manufacturing*, *33*, 101145. Available from https://doi.org/10.1016/j.addma.2020.101145.

Khalil, N., Aouad, G., El Cheikh, K., & Rémond, S. (2017). Use of calcium sulfoaluminate cements for setting control of 3D-printing mortars. *Construction and Building Materials*, *157*, 382–391. Available from https://doi.org/10.1016/j.conbuildmat.2017.09.109.

Koga, G. Y., Comperat, P., Albert, B., Roche, V., & Nogueira, R. P. (2020). Electrochemical responses and chloride ingress in reinforced Belite-Ye'elimite-Ferrite (BYF) cement matrix exposed to exogenous salt sources. *Corrosion Science*, *166*. Available from https://doi.org/10.1016/j.corsci.2020.108469.

Komljenović, M., Baščarević, Z., & Bradić, V. (2010). Mechanical and microstructural properties of alkali-activated fly ash geopolymers. *Journal of Hazardous Materials*, *181* (1-3), 35–42. Available from https://doi.org/10.1016/j.jhazmat.2010.04.064.

Lazorenko, G., & Kasprzhitskii, A. (2022). Geopolymer additive manufacturing: A review. *Additive Manufacturing*, *55*, 102782. Available from https://doi.org/10.1016/j.addma.2022.102782.

Le, T. T., Austin, S. A., Lim, S., Buswell, R. A., Gibb, A. G. F., & Thorpe, T. (2012). Mix design and fresh properties for high-performance printing concrete. *Materials and Structures.*, *45*(8), 1221–1232. Available from https://doi.org/10.1617/s11527-012-9828-z.

Le, T. T., Austin, S. A., Lim, S., Buswell, R. A., Law, R., Gibb, A. G. F., & Thorpe, T. (2012). Hardened properties of high-performance printing concrete. *Cement and Concrete Research*, *42*(3), 558–566. Available from https://doi.org/10.1016/j.cemconres.2011.12.003.

Lewis, R. C. (2018). *Properties of fresh and hardened concrete containing supplementary cementitious materials silica fume: State-of-the-art report of the RILEM Technical Committee 238-SCM, Working Group 4*. Springer. Available from http://doi.org/10.1007/978-3-319-70606-1.

Li, C., Sun, H., & Li, L. (2010). A review: The comparison between alkali-activated slag (Si + Ca) and metakaolin (Si + Al) cements. *Cement and Concrete Research*, *40*(9), 1341–1349. Available from https://doi.org/10.1016/j.cemconres.2010.03.020.

Li, L. G., Xiao, B. F., Fang, Z. Q., Xiong, Z., Chu, S. H., & Kwan, A. K. H. (2021). Feasibility of glass/basalt fiber reinforced seawater coral sand mortar for 3D printing. *Additive Manufacturing*, *37*, 101684. Available from https://doi.org/10.1016/j.addma.2020.101684.

Lin, R. S., Lee, H. S., Han, Y., & Wang, X. Y. (2021). Experimental studies on hydration–-strength–durability of limestone-cement-calcined Hwangtoh clay ternary composite. *Construction and Building Materials*, *269*.

Liu, H., Liu, C., Wu, Y., Bai, G., He, C., Zhang, R., & Wang, Y. (2022). Hardened properties of 3D printed concrete with recycled coarse aggregate. *Cement and Concrete Research*, *159*, 106868. Available from https://doi.org/10.1016/j.cemconres.2022.106868.

Luukkonen, T., Abdollahnejad, Z., Yliniemi, J., Kinnunen, P., & Illikainen, M. (2018). One-part alkali-activated materials: A review. *Cement and Concrete Research*, *103*, 21–34. Available from https://doi.org/10.1016/j.cemconres.2017.10.001.

Marchment, T., & Sanjayan, J. (2020). Bond properties of reinforcing bar penetrations in 3D concrete printing. *Automation in Construction*, *120*, 103394. Available from https://doi.org/10.1016/j.autcon.2020.103394.

Markin, S., & Mechtcherine, V. (2023). Quantification of plastic shrinkage and plastic shrinkage cracking of the 3D printable concretes using 2D digital image correlation. *Cement and Concrete Composites*, *139*, 105050. Available from https://doi.org/10.1016/j.cemconcomp.2023.105050.

Ma, G., Li, Z., & Wang, L. (2018). Printable properties of cementitious material containing copper tailings for extrusion based 3D printing. *Construction and Building Materials.*, *162*, 613–627. Available from https://doi.org/10.1016/j.conbuildmat.2017.12.051.

Ma, G., Li, Z., Wang, L., Wang, F., & Sanjayan, J. (2019). Mechanical anisotropy of aligned fiber reinforced composite for extrusion-based 3D printing. *Construction and Building Materials*, *202*, 770–783. Available from https://doi.org/10.1016/j.conbuildmat.2019.01.008.

Ma, G., Sun, J., Wang, L., Aslani, F., & Liu, M. (2018). Electromagnetic and microwave absorbing properties of cementitious composite for 3D printing containing waste copper solids. *Cement and Concrete Composites*, *94*, 215–225. Available from https://doi.org/10.1016/j.cemconcomp.2018.09.005.

Ma, J., Wang, H., Yu, Z., Shi, H., Wu, Q., & Shen, X. (2023). A systematic review on durability of calcium sulphoaluminate cement-based materials in chloride environment. *Journal of Sustainable Cement-Based Materials*, *12*(6), 687–698. Available from https://doi.org/10.1080/21650373.2022.2113569, http://www.tandfonline.com/toc/tscm20/current.

Mazhoud, B., Perrot, A., Picandet, V., Rangeard, D., & Courteille, E. (2019). Underwater 3D printing of cement-based mortar. *Construction and Building Materials*, *214*, 458–467. Available from https://doi.org/10.1016/j.conbuildmat.2019.04.134.

McLellan, B. C., Williams, R. P., Lay, J., Van Riessen, A., & Corder, G. D. (2011). Costs and carbon emissions for geopolymer pastes in comparison to ordinary portland cement. *Journal of Cleaner Production*, *19*(9-10), 1080–1090. Available from https://doi.org/10.1016/j.jclepro.2011.02.010.

Moelich, G. M., Kruger, J., & Combrinck, R. (2020). Plastic shrinkage cracking in 3D printed concrete. *Composites Part B: Engineering*, *200*, 108313. Available from https://doi.org/10.1016/j.compositesb.2020.108313.

Mohan, M. K., Rahul, A. V., De Schutter, G., & Van Tittelboom, K. (2021a). Early age hydration, rheology and pumping characteristics of CSA cement-based 3D printable concrete. *Construction and Building Materials*, *275*, 122136. Available from https://doi.org/10.1016/j.conbuildmat.2020.122136.

Mohan, Manu K., Rahul, A. V., De Schutter, Geert, & Van Tittelboom, Kim (2021b). Extrusion-based concrete 3D printing from a material perspective: A state-of-the-art review. *Cement and Concrete Composites*, *115*, 103855. Available from https://doi.org/10.1016/j.cemconcomp.2020.103855.

Mohan, Manu K., Rahul, A. V., Van Tittelboom, Kim, & De Schutter, Geert (2021). Rheological and pumping behaviour of 3D printable cementitious materials with varying aggregate content. *Cement and Concrete Research*, *139*, 106258. Available from https://doi.org/10.1016/j.cemconres.2020.106258.

Muthukrishnan, S., Ramakrishnan, S., & Sanjayan, J. (2021). Effect of alkali reactions on the rheology of one-part 3D printable geopolymer concrete. *Cement and Concrete Composites*, *116*, 103899. Available from https://doi.org/10.1016/j.cemconcomp.2020.103899.

Nath, P., & Sarker, P. K. (2014). Effect of GGBFS on setting, workability and early strength properties of fly ash geopolymer concrete cured in ambient condition. *Construction and Building Materials*, *66*, 163–171. Available from https://doi.org/10.1016/j.conbuildmat.2014.05.080.

Nematollahi, B., Sanjayan, J., & Shaikh, F. U. A. (2015). Synthesis of heat and ambient cured one-part geopolymer mixes with different grades of sodium silicate. *Ceramics International*, *41*(4), 5696–5704. Available from https://doi.org/10.1016/j.ceramint.2014.12.154.

Nerella, V. N., Hempel, S., & Mechtcherine, V. (2019). Effects of layer-interface properties on mechanical performance of concrete elements produced by extrusion-based 3D-printing. *Construction and Building Materials*, *205*, 586–601. Available from https://doi.org/10.1016/j.conbuildmat.2019.01.235.

Olivier, J., Janssens-Maenhout, G., Muntean, M., & Peters, J. (2016). *Trends in global CO_2 emissions: 2016 report*. PBL Netherlands Environmental Assessment Agency.

Palomo, A., Grutzeck, M. W., & Blanco, M. T. (1999). Alkali-activated fly ashes: A cement for the future. *Cement and Concrete Research*, *29*(8), 1323–1329. Available from https://doi.org/10.1016/S0008-8846(98)00243-9.

Panda, B., Singh, G. B., Unluer, C., & Tan, M. J. (2019). Synthesis and characterization of one-part geopolymers for extrusion based 3D concrete printing. *Journal of Cleaner Production*, *220*, 610–619. Available from https://doi.org/10.1016/j.jclepro.2019.02.185.

Panda, B., & Tan, M. J. (2018). Experimental study on mix proportion and fresh properties of fly ash based geopolymer for 3D concrete printing. *Ceramics International*, *44*(9), 10258–10265. Available from https://doi.org/10.1016/j.ceramint.2018.03.031.

Panda, B., & Tan, M. J. (2019). Rheological behavior of high volume fly ash mixtures containing micro silica for digital construction application. *Materials Letters*, *237*, 348–351. Available from https://doi.org/10.1016/j.matlet.2018.11.131.

Panda, B., Unluer, C., & Tan, M. J. (2018). Investigation of the rheology and strength of geopolymer mixtures for extrusion-based 3D printing. *Cement and Concrete Composites*, *94*, 307–314. Available from https://doi.org/10.1016/j.cemconcomp.2018.10.002.

Panda, B., Unluer, C., & Tan, M. J. (2019). Extrusion and rheology characterization of geopolymer nanocomposites used in 3D printing. *Composites Part B: Engineering*, *176*, 107290. Available from https://doi.org/10.1016/j.compositesb.2019.107290.

Paul, S. C., Tay, Y. W. D., Panda, B., & Tan, M. J. (2018). Fresh and hardened properties of 3D printable cementitious materials for building and construction. *Archives of Civil and Mechanical Engineering*, *18*(1), 311–319. Available from https://doi.org/10.1016/j.acme.2017.02.008.

Phair, J. W. (2006). Green chemistry for sustainable cement production and use. *Green Chemistry*, *8*(9), 763–780. Available from https://doi.org/10.1039/b603997a.

Pimraksa, K., & Chindaprasirt, P. (2018). Sulfoaluminate cement-based concrete,. In F. Pacheco-Torgal, R. E. Melchers, X. Shi, N. De Belie, K. V. Tittelboom, & A. Sáez (Eds.), *Eco-efficient repair and rehabilitation of concrete infrastructures* (pp. 355–385). Elsevier Inc., Thailand Elsevier Inc. Available from http://doi.org/10.1016/B978-0-08-102181-1.00014-9.

Provis, J. L. (2009). Activating solution chemistry for geopolymers. In J. L. Provis, & J. S. J. van Deventer (Eds.), *Geopolymers: Structures, processing, properties and industrial applications* (pp. 50–71). Elsevier Ltd. Available from http://doi.org/10.1533/9781845696382.1.50.

Provis, J. L. (2014). Geopolymers and other alkali activated materials: Why, how, and what? *Materials and Structures*, *47*(1-2), 11–25. Available from https://doi.org/10.1617/s11527-013-0211-5.

Provis, J. L., & Van Deventer, J. S. J. (2009). *Geopolymers: Structures, processing, properties and industrial applications* (pp. 1–454). Elsevier Ltd, Australia Elsevier Ltd. Available from 10.1533/9781845696382.

Qian, H., Hua, S., Gao, Y., Qian, L., & Ren, X. (2021). Synergistic effect of EVA copolymer and sodium desulfurization ash on the printing performance of high volume blast furnace slag mixtures. *Additive Manufacturing*, *46*, 102183. Available from https://doi.org/10.1016/j.addma.2021.102183.

Quillin, K. (2001). Performance of belite-sulfoaluminate cements. *Cement and Concrete Research*, *31*(9), 1341–1349. Available from https://doi.org/10.1016/S0008-8846(01)00543-9.

Rahul, A. V., Santhanam, Manu, Meena, Hitesh, & Ghani, Zimam (2019). Mechanical characterization of 3D printable concrete. *Construction and Building Materials*, *227*, 116710. Available from https://doi.org/10.1016/j.conbuildmat.2019.116710.

Raza, M. H., Zhong, R. Y., & Khan, M. (2022). Recent advances and productivity analysis of 3D printed geopolymers. *Additive Manufacturing*, *52*. Available from https://doi.org/10.1016/j.addma.2022.102685.

Rehman, A. U., Lee, S. M., & Kim, J. H. (2020). Use of municipal solid waste incineration ash in 3D printable concrete. *Process Safety and Environmental Protection*, *142*, 219–228. Available from https://doi.org/10.1016/j.psep.2020.06.018.

Roussel, N. (2018). Rheological requirements for printable concretes. *Cement and Concrete Research*, *112*, 76–85. Available from https://doi.org/10.1016/j.cemconres.2018.04.005.

Ruan, S., & Unluer, C. (2016). Comparative life cycle assessment of reactive MgO and Portland cement production. *Journal of Cleaner Production*, *137*, 258–273. Available from https://doi.org/10.1016/j.jclepro.2016.07.071.

Sahu, S., & Majling, J. (1994). Preparation of sulphoaluminate belite cement from fly ash. *Cement and Concrete Research*, *24*(6), 1065–1072. Available from https://doi.org/10.1016/0008-8846(94)90030-2.

Sanjayan, J. G., Nematollahi, B., Xia, M., & Marchment, T. (2018). Effect of surface moisture on inter-layer strength of 3D printed concrete. *Construction and Building Materials*, *172*, 468–475. Available from https://doi.org/10.1016/j.conbuildmat.2018.03.232.

Sarker, P. K., & McBeath, S. (2015). Fire endurance of steel reinforced fly ash geopolymer concrete elements. *Construction and Building Materials*, *90*, 91–98. Available from https://doi.org/10.1016/j.conbuildmat.2015.04.054.

Seo, E. A., Kim, W. W., Kim, S. W., Kwon, H. K., & Lee, H. J. (2023). Mechanical properties of 3D printed concrete with coarse aggregates and polypropylene fiber in the air and underwater environment. *Construction and Building Materials*, *378*. Available from https://doi.org/10.1016/j.conbuildmat.2023.131184.

Shahmirzadi, M. R., Gholampour, A., Kashani, A., & Ngo, T. D. (2021). Shrinkage behavior of cementitious 3D printing materials: Effect of temperature and relative humidity. *Cement and Concrete Composites*, *124*. Available from https://doi.org/10.1016/j.cemconcomp.2021.104238.

Shand, M. A. (2006). *The chemistry and technology of magnesia* (pp. 1–266). John Wiley and Sons. Available from http://doi.org/10.1002/0471980579.

Sharma, M., Bishnoi, S., Martirena, F., & Scrivener, K. (2021). Limestone calcined clay cement and concrete: A state-of-the-art review. *Cement and Concrete Research*, *149*, 106564. Available from https://doi.org/10.1016/j.cemconres.2021.106564.

Sharp, J. H., Lawrence, C. D., & Yang, R. (1999). Calcium sulfoaluminate cements – low-energy cements, special cements or what? *Advances in Cement Research*, *11*(1), 3–13. Available from https://doi.org/10.1680/adcr.1999.11.1.3.

Sideris, K., Justnes, H., Soutsos, M., & Sui, T. (2018). Fly Ash. In: De Belie, N., Soutsos, M., Gruyaert, E. (eds). Properties of fresh and hardened concrete containing supplementary cementitious materials. *RILEM state-of-the-art reports* (25). Springer. Available from http://doi.org/10.1007/978-3-319-70606-1_2.

Singh, B., Ishwarya, G., Gupta, M., & Bhattacharyya, S. K. (2015). Geopolymer concrete: A review of some recent developments. *Construction and Building Materials*, *85*, 78–90. Available from https://doi.org/10.1016/j.conbuildmat.2015.03.036.

Sun, B., Zeng, Q., Wang, D., & Zhao, W. (2022). Sustainable 3D printed mortar with CO_2 pretreated recycled fine aggregates. *Cement and Concrete Composites*, *134*, 104800. Available from https://doi.org/10.1016/j.cemconcomp.2022.104800.

Tao, Y., Mohan, M. K., Rahul, A. V., De Schutter, G., & Van Tittelboom, K. (2023). Development of a calcium sulfoaluminate-Portland cement binary system for twin-pipe 3D concrete printing. *Cement and Concrete Composites*, *138*, 104960. Available from https://doi.org/10.1016/j.cemconcomp.2023.104960.

Turner, L. K., & Collins, F. G. (2013). Carbon dioxide equivalent (CO_2-e) emissions: A comparison between geopolymer and OPC cement concrete. *Construction and Building Materials*, *43*, 125–130. Available from https://doi.org/10.1016/j.conbuildmat.2013.01.023.

Unluer, C. (2018). Carbon dioxide sequestration in magnesium-based binders. In F. Pacheco-Torgal, C. Shi, & A. P. Sanchez (Eds.), *Carbon dioxide sequestration in cementitious construction materials* (pp. 129–173). Elsevier. Available from http://doi.org/10.1016/B978-0-08-102444-7.00007-1.

Ur Rehman, A., & Kim, J.-H. (2021). 3D concrete printing: A systematic review of rheology, mix designs, mechanical, microstructural, and durability characteristics. *Materials*, *14*(14), 3800. Available from https://doi.org/10.3390/ma14143800.

Van Damme, H. (2018). Concrete material science: Past, present, and future innovations. *Cement and Concrete Research*, *112*, 5–24. Available from https://doi.org/10.1016/j.cemconres.2018.05.002.

Vantyghem, G., De Corte, W., Shakour, E., & Amir, O. (2020). 3D printing of a post-tensioned concrete girder designed by topology optimization. *Automation in Construction*, *112*, 103084. Available from https://doi.org/10.1016/j.autcon.2020.103084.

Walling, S. A., & Provis, J. L. (2016). Magnesia-based cements: A journey of 150 years, and cements for the future? *Chemical Reviews*, *116*(7), 4170–4204. Available from https://doi.org/10.1021/acs.chemrev.5b00463.

Wang, D., Yue, Y., Xie, Z., Mi, T., Yang, S., McCague, C., Qian, J., & Bai, Y. (2022). Chloride-induced depassivation and corrosion of mild steel in magnesium potassium phosphate cement. *Corrosion Science*, *206*, 110482. Available from https://doi.org/10.1016/j.corsci.2022.110482.

Weng, Y., Li, M., Ruan, S., Wong, T. N., Tan, M. J., Ow Yeong, K. L., & Qian, S. (2020). Comparative economic, environmental and productivity assessment of a concrete bathroom unit fabricated through 3D printing and a precast approach. *Journal of Cleaner Production*, *261*. Available from https://doi.org/10.1016/j.jclepro.2020.121245.

Weng, Y., Li, M., Tan, M. J., & Qian, S. (2018). Design 3D printing cementitious materials via Fuller Thompson theory and Marson-Percy model. *Construction and Building Materials*, *163*, 600–610. Available from https://doi.org/10.1016/j.conbuildmat.2017.12.112.

Weng, Y., Ruan, S., Li, M., Mo, L., Unluer, C., Tan, M. J., & Qian, S. (2019). Feasibility study on sustainable magnesium potassium phosphate cement paste for 3D printing. *Construction and Building Materials*, *221*, 595–603. Available from https://doi.org/10.1016/j.conbuildmat.2019.05.053.

Winnefeld, F., & Lothenbach, B. (2010). Hydration of calcium sulfoaluminate cements — experimental findings and thermodynamic modelling. *Cement and Concrete Research*, *40*(8), 1239–1247. Available from https://doi.org/10.1016/j.cemconres.2009.08.014.

Xiao, J., Ji, G., Zhang, Y., Ma, G., Mechtcherine, V., Pan, J., Wang, Li, Ding, T., Duan, Z., & Du, S. (2021). Large-scale 3D printing concrete technology: Current status and future opportunities. *Cement and Concrete Composites*, *122*, 104115. Available from https://doi.org/10.1016/j.cemconcomp.2021.104115.

Yang, K. H., Song, J. K., & Song, K. I. (2013). Assessment of CO_2 reduction of alkali-activated concrete. *Journal of Cleaner Production*, *39*, 265–272. Available from https://doi.org/10.1016/j.jclepro.2012.08.001.

Yu, Q., Zhu, B., Li, X., Meng, L., Cai, J., Zhang, Y., & Pan, J. (2023). Investigation of the rheological and mechanical properties of 3D printed eco-friendly concrete with steel slag. *Journal of Building Engineering*, *72*, 106621. Available from https://doi.org/10.1016/j.jobe.2023.106621.

Yuan, Q., Zhou, D., Li, B., Huang, H., & Shi, C. (2018). Effect of mineral admixtures on the structural build-up of cement paste. *Construction and Building Materials*, *160*, 117–126. Available from https://doi.org/10.1016/j.conbuildmat.2017.11.050.

Zahabizadeh, B., Pereira, J., Gonçalves, C., Pereira, E. N. B., & Cunha, Vítor M. C. F. (2021). Influence of the printing direction and age on the mechanical properties of 3D printed concrete. *Materials and Structures*, *54*(2). Available from https://doi.org/10.1617/s11527-021-01660-7, 1359-5997.

Zhang, C., Jia, Z., Wang, X., Jia, L., Deng, Z., Wang, Z., Zhang, Y., & Mechtcherine, V. (2022). A two-phase design strategy based on the composite of mortar and coarse aggregate for 3D printable concrete with coarse aggregate. *Journal of Building Engineering*, *54*, 104672. Available from https://doi.org/10.1016/j.jobe.2022.104672.

Zhang, D. W., Sun, X. M., Xu, Z. Y., Xia, C. L., & Li, H. (2022). Stability of superplasticizer on NaOH activators and influence on the rheology of alkali-activated fly ash fresh pastes. *Construction and Building Materials*, *341*. Available from https://doi.org/10.1016/j.conbuildmat.2022.127864.

Zhang, J., Wang, J., Dong, S., Yu, X., & Han, B. (2019). A review of the current progress and application of 3D printed concrete. *Composites Part A: Applied Science and Manufacturing*, *125*, 105533. Available from https://doi.org/10.1016/j.compositesa.2019.105533.

Zhang, Y., Zhang, Y., She, W., Yang, L., Liu, G., & Yang, Y. (2019). Rheological and harden properties of the high-thixotropy 3D printing concrete. *Construction and Building Materials.*, *201*, 278–285. Available from https://doi.org/10.1016/j.conbuildmat.2018.12.061.

Zhao, Z., Chen, M., Jin, Y., Lu, L., & Li, L. (2022). Rheology control towards 3D printed magnesium potassium phosphate cement composites. *Composites Part B: Engineering*, *239*, 109963. Available from https://doi.org/10.1016/j.compositesb.2022.109963.

Zhao, Z., Chen, M., Xu, J., Li, L., Huang, Y., Yang, L., Zhao, P., & Lu, L. (2021). Mix design and rheological properties of magnesium potassium phosphate cement composites based on the 3D printing extrusion system. *Construction and Building Materials*, *284*, 122797. Available from https://doi.org/10.1016/j.conbuildmat.2021.122797.

Zhao, Y., Gao, Y., Chen, G., Li, S., Singh, A., Luo, X., Liu, C., Gao, J., & Du, H. (2023). Development of low-carbon materials from GGBS and clay brick powder for 3D concrete printing. *Construction and Building Materials*, *383*, 131232. Available from https://doi.org/10.1016/j.conbuildmat.2023.131232.

Zhou, Q., & Glasser, F. P. (2000). Kinetics and mechanism of the carbonation of ettringite. *Advances in Cement Research*, *12*(3), 131–136. Available from https://doi.org/10.1680/adcr.2000.12.3.131.

Zhou, Q., Milestone, N. B., & Hayes, M. (2006). An alternative to Portland Cement for waste encapsulation-The calcium sulfoaluminate cement system. *Journal of Hazardous Materials*, *136*(1), 120–129. Available from https://doi.org/10.1016/j.jhazmat.2005.11.038.

Zhu, B., Pan, J., Nematollahi, B., Zhou, Z., Zhang, Y,, & Sanjayan, J. (2019). Development of 3D printable engineered cementitious composites with ultra-high tensile ductility for digital construction. *Materials & Design, 181*, 108088. Available from https://doi.org/10.1016/j.matdes.2019.108088.

Zunino, F., Dhandapani, Y., Ben Haha, M., Skibsted, J., Joseph, S., Krishnan, S., Parashar, A., Juenger, M. C. G., Hanein, T., Bernal, S. A., Scrivener, K. L., & Avet, F. (2022). Hydration and mixture design of calcined clay blended cements: Review by the RILEM TC 282-CCL. *Materials and Structures*, *55*(9). Available from https://doi.org/10.1617/s11527-022-02060-1.

Sustainable three-dimensional printing concrete: advances, challenges, and future direction

12

Mostafa Seifan
The University of Waikato, School of Engineering, Hamilton, New Zealand

12.1 Introduction

Concrete is the most extensively used building material globally due to its characteristics such as strength, durability, reliability, serviceability, availability, fire resistance, and relatively cheap price. In addition, its raw materials are easily obtainable almost anywhere in the world, and this is the main reason that concrete is consumed at approximately 25 Gt per annum (Gursel & Ostertag, 2017). Concrete is made of four main ingredients including cement as a binder, coarse and fine aggregate, water, and admixtures. Despite the unique properties of concrete, is responsible for 8%–9% of the total global anthropogenic CO_2 emissions (Seifan et al., 2016). Additionally, conventional construction practices usually result in producing waste materials during different construction phases. Apart from the concrete material, the construction industry as a whole is known to release approximately 38% of greenhouse gas, responsible for 40% of solid waste and 12% of portable water use (Khan et al., 2021). Considering the built environment projects by the global population increase rate of 1% per year, it is expected that the footprint by the construction industry significantly increase by 2050.

The conventional construction process of a concrete structure is labor intensive and there are concerns about labor safety (Tay et al., 2017). Digital fabrication techniques to introduce more sustainable practices can be a solution to the huge market requirements to address the fundamental issues in conventional construction practices. The construction industry can benefit from the deployment of new technologies such as automation to enhance the quality of the construction element, reduce construction times, and reduce the environmental impact. In this regard, the utilization of new technologies such as additive manufacturing is becoming more relevant than ever. Over the recent decade, developments in concrete engineering have enabled the introduction of automated fabrication methods for construction and building materials (Bos et al., 2016). Among the advances in the field, three-dimensional printing concrete (3DPC) is an innovative construction technology as an alternative to conventional construction practices. This method employs an additive manufacturing technique that builds on materials layer by layer (Sandeep & Rao, 2017). The first 3D printer was invented by Charles Hull in the year 1986,

Sustainable Concrete Materials and Structures. DOI: https://doi.org/10.1016/B978-0-443-15672-4.00012-7

which was then applied in a wide range of domains and industries such as medical, automotive, and aerospace fields (Gibson et al., 2010). In the construction field, Pegna (1997) is one of the pioneers who proposed the utilization of construction robotics to build a masonry structure that cannot be made by casting.

Using this technology, the manufacture of complex shapes with construction materials is made possible. The two major categories of concrete printing techniques, namely: (1) extrusion-based and (2) powder-based, exhibit the potential of using 3DPC as a large-scale fabrication method (Lim et al., 2016). Material deposition method or extrusion-based method is based on successively layering concrete paste according to the computer-aided design (CAD) model (Panda, Bahubalendruni, et al., 2017). There are a number of automated systems that use this technique. One is contour crafting whose key feature is the use of trowels attached to the nozzle, guiding the extruded material and creating smooth surfaces. This approach allows the deposition of thicker layers of material without sacrificing the quality of the surface finish (B. Khoshnevis, 2004). Similarly, concrete printing is another construction process that constructs objects by continuous deposition of cement mortars, but with a smaller resolution of deposition. This results in higher control over complex geometries (Ma & Wang, 2018). However, a powder-based technique follows a different principle. Binder jetting (D-shape) is one example of technology that follows this 3D printing process which creates structures by extruding binder layer by layer over a powder bed. This binds 2D cross sections of the intended material to layers of material powder (Perkins & Skitmore, 2015). Features of this method include the ability to design overhanging features and more complex geometries with relatively high resolution and surface finish quality. However, the majority of this chapter is based on extrusion-based techniques since it is the most common and well documented.

Typically, 3DPC machine comprises a mixing unit, robotic arm, peristaltic pump, printing nozzle, and controller. However, other printing arrangements might be adopted based on the scale of the printed elements as illustrated in Fig. 12.1. Moreover, in some cases, additional peristaltic pump to spray water or additives can be used. The process begins with a 3D CAD model divided into several 2D layers for printing (Tay et al., 2017). Cartesian coordinates from 2D layers coupled with printing settings like the speed of the arm, pump speed, and extrusion rate are sent to the 3D printer as a machine-readable language (Paul et al., 2018). Then, under pressure from a pumping system, concrete is extruded from the tank toward a nozzle under desired speed, angle, and direction based on the control system.

Despite the recent advances and fabrication of printed construction, 3DPC is still a relatively new concept in the construction and building industry. In 2012, researchers at the University of Loughborough (Le, Austin, Lim, Buswell, Gibb, & Thorpe, 2012; Lim et al., 2012) investigated the construction scale of 3DPC. As shown in Fig. 12.2, Zareiyan and Khoshnevis (2017) investigated the influence of material and printing parameters such as extrusion and layer size and aggregate size on the properties of printed layer. They found that a higher cement to aggregate ratio and a smaller maximum aggregate size have positive impact on the bonding and the strength of printed structure. The authors also found that a short setting

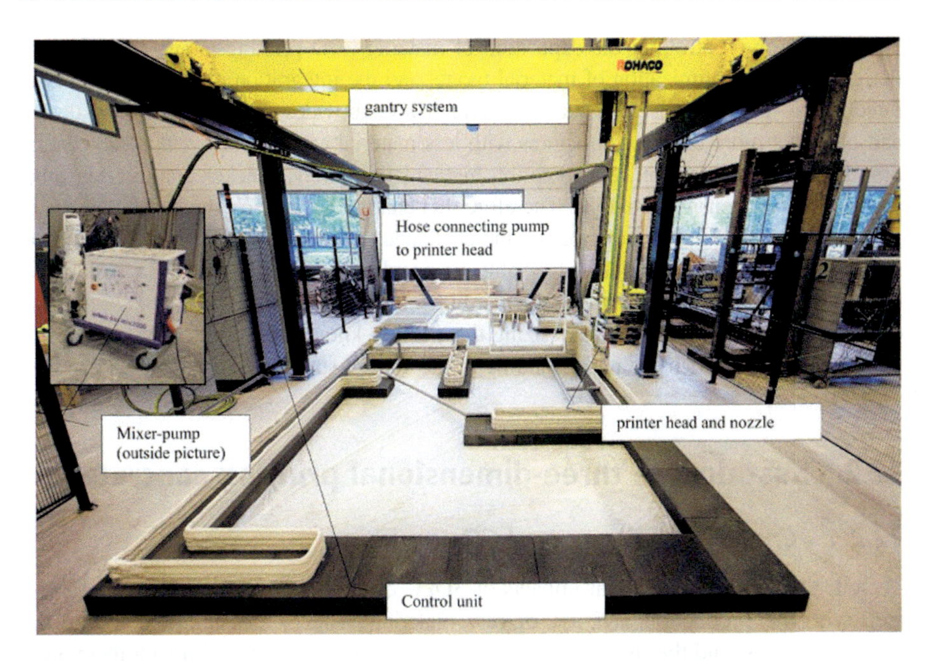

Figure 12.1 Gantry robot 3D printer.
Source: From Bos, F., Wolfs, R., Ahmed, Z., & Salet, T. (2016). Additive manufacturing of concrete in construction: Potentials and challenges of 3D concrete printing. *Virtual and Physical Prototyping, 11*(3), 209–225. https://doi.org/10.1080/17452759.2016.1209867

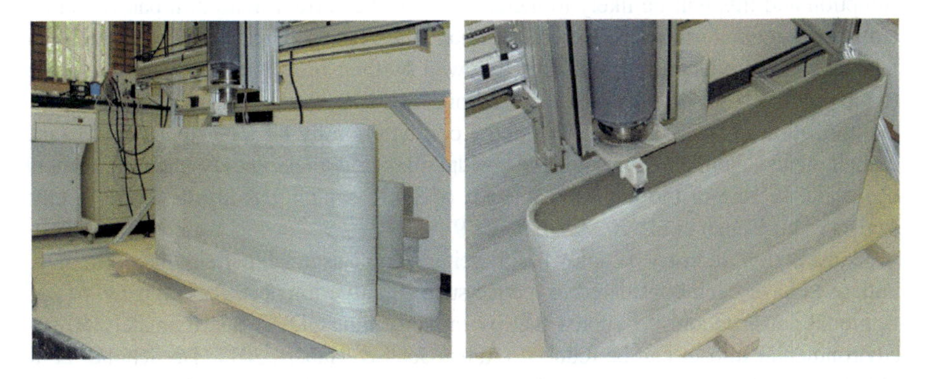

Figure 12.2 Contour crafting of 3DPC element.
Source: From Zareiyan, B., & Khoshnevis, B. (2017). Interlayer adhesion and strength of structures in Contour Crafting – effects of aggregate size, extrusion rate, and layer thickness. *Automation in Construction, 81*, 112–121. https://doi.org/10.1016/j.autcon.2017.06.013.

time can increase the likelihood of cold joints between printed layers. In other investigations, researchers have focused on other aspects of 3DPC which will be discussed in the following sections.

By precisely controlling the deposition of materials, 3D printing allows for enhanced design freedom, optimization of material usage, and the integration of functional features directly into the printed elements. Therefore 3DPC provides the opportunity to design and construct more complex structures with less material consumption and waste generation, higher recycling rate, less footprint, higher labor safety, quicker construction time, and less labor-intensive activities. Therefore this chapter focuses on the current advances in the field of 3DPC and elaborates on the challenges to be addressed for wider industrial applications. To become more acceptable to people, the 3DPC must be a sustainable technology. Hence, sustainability and environmental impacts of such technology are also discussed. By analyzing and synthesizing the existing knowledge in the field, it is aimed to provide a comprehensive understanding of the current state of sustainable 3DPC.

12.2 Sustainable three-dimensional printing concrete

12.2.1 Energy efficiency and CO_2 emission

Technically, the environmental impacts of 3DPC comprise the carbon footprint associated with the production and use of raw materials, energy consumption during the printing process, and the overall life cycle of the printed elements. Although there are a decent number of research around 3DPC trend and its applications, there are a few studies that focus on the environmental effects of such new technology. Therefore, from an environmental perspective, the exact contribution of 3DPC to the environment is still unknown. Construction industry contributes to about 40% of the energy consumption and this will be likely to increase due to the current trend in urbanization and ambitious mega-construction projects (Mohan et al., 2021). This energy consumption is related to the production of materials followed by their transportation from the factory to the construction site. In a study carried out by Weng et al. (2020), it was found that the utilization of 3DPC can reduce energy consumption by 87.1% as compared to precast elements. From an environmental standpoint, being a less energy-intensive technology could be one of the major advantages of 3DPC which was also reported in other studies (Abu-Ennab et al., 2022; Dixit, 2019).

Construction industry is known for being a nonsustainable field therefore many initiatives have been considered to reduce such negative impacts. As compared to conventional construction, it is noted that by implementing 3DPC, approximately 50% of environmental impact can be reduced (Agustí-Juan & Habert, 2017; De Schutter et al., 2018). In another study, Weng et al. (2020) reported an 85.9% reduction in CO_2 emission when 3DPC was used in the construction of bathroom structure as compared to conventional construction practices. However, from a technical point of view, at this stage of development, this value seems to be quite ambitious. This is mainly due to the fact that the most available printable cementitious materials used for printing are grout and mortar rather than concrete (Chen et al., 2022). As a result, the proportion of cement used in 3DPC of mortar ($\sim$20 wt.%) is significantly higher than concrete. The potential solutions to mitigate such effects can be partially replacing cement with pozzolans or by-products as discussed in the following sections.

12.2.2 Resource consumption and waste generation

Construction industry is largely dependent on nonrenewable resources. Over recent years, construction industry is under massive pressure to adopt sustainable practices to minimize the environmental impacts and so meet global sustainable goals. In this context, sustainable 3DPC technology aims to minimize resource consumption, waste generation, and footprint emissions as well as improve the overall life cycle performance of construction elements.

A higher embodied energy and CO_2 emissions of cement along with an increasing shortfall of natural aggregates have urged the construction industry to use various forms of waste materials as fully or partially replacing the concrete ingredients to make the final product more sustainable. As shown in Fig. 12.3, there are currently different waste or by-product materials that are used as fillers and supplemented cementitious material in concrete. Among all cement replacement materials, fly ash (FA) has received much attention (Ma, Li, & Wang, 2018; Ma et al., 2020; Nerella & Mechtcherine, 2019; Panda et al., 2019; Rahul & Santhanam, 2020). This is mainly due to its physiochemical properties and morphology. In terms of rheological properties, the spherical morphology of FA improves the extrudability and flowability of the printed mixture while the printed concrete will have a higher long-term strength as compared to the concrete made by ordinary Portland cement (PC). Similar to FA, other waste materials such as slag, silica fumes (SFs), other types of pozzolans, vegetable ashes, and quarry rock dust have the potential to be used to partially replace the cement. Although 3DPC is known for its

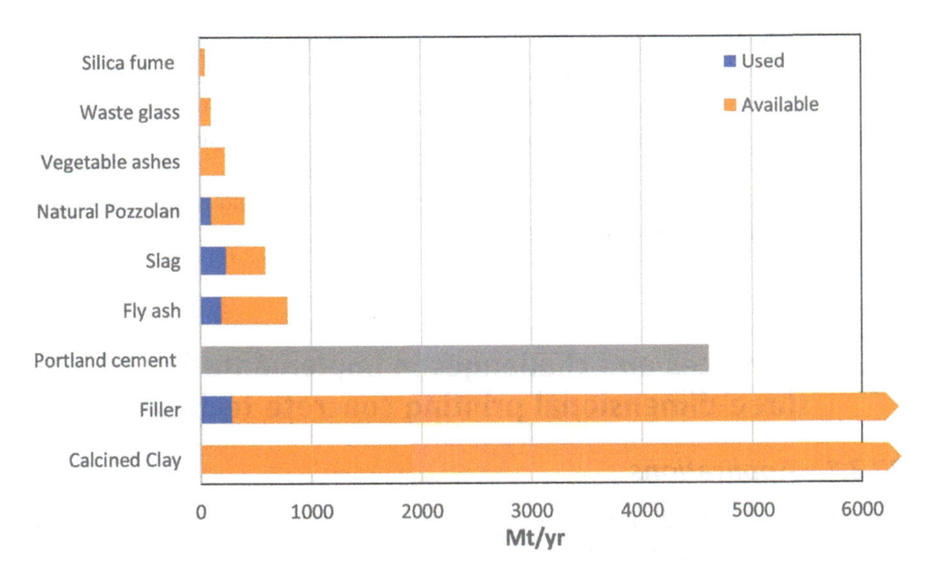

Figure 12.3 Commonly used fillers and supplementary cementitious materials as a potential replacement for cement.
Source: From Scrivener, K. L., John, V. M., & Gartner, E. M. (2018). Eco-efficient cements: Potential economically viable solutions for a low-CO_2 cement-based materials industry. *Cement and Concrete Research, 114*, 2–26. https://doi.org/10.1016/j.cemconres.2018.03.015.

sustainability as compared to conventional construction practices, it requires more cement as a binder. Therefore partial or full replacing cement with waste materials would significantly help to offset the 3DPC. Similar to cement, aggregate which is mostly obtained from natural resources can be replaced by waste materials such as crushed glass (Harrison et al., 2020), scrap tire rubber, and quarry rock dust.

In terms of waste reduction, 3DPC have the potential to significantly reduce waste generation during construction and postconstruction. For example, unlike conventional concrete casting, 3DPC does not require formwork. Additionally, the lower spillage of concrete can result in less wasting of nonrenewable sourced materials. Bhattacherjee et al. (2021) proposed the life cycle of 3DPC by considering all operations involved from mining the raw materials to transportation to demolition and dismantling (Fig. 12.4). Based on the proposed life cycle, different boundary systems can be considered namely: (1) cradle-to-cradle, (2) gate-to-gate, (3) cradle-to-grave, and finally (4) cradle-to-gate. From all possible boundaries, cradle-to-gate seems to be most relevant to 3DPC as the process covers the mining of raw materials to the printing structure on a factory or site. However, the ideal system for the construction industry would be cradle-to-cradle which covers all steps from mining to dismantling and separation of concrete from other demolition materials. Since the circular economy of construction and demolition waste has become an important aspect of life cycle, the boundary system should be considered in a way that covers the reusability aspect as well (Purchase et al., 2022). In terms of demolition and recycling, 3DPC has advantages over conventional reinforced concrete and is associated with less energy and labor-intensive activities. This is mainly due to the lack of embedded reinforcement and a lower material strength which makes mechanical, chemical, and thermal recycling easier. On the other hand, 3DPC construction can be designed and built modular and therefore it allows for easy disassembly and even the dismantled elements can be reused in different configurations or structures. From a service point of view, the modularity of 3DPC is also an advantage as it reduces the need for a complete replacement and makes the maintenance less costly and labor involved.

12.3 Advances and challenges in implementation of three-dimensional printing concrete technology

12.3.1 Applications

The traditional method of concrete construction is constructed by using formworks which are high priced and time consuming to construct, and also are restricted to certain geometries. The construction industry is facing problems such as low labor efficiency, high rate of injuries at the site, and the effect of formworks waste on the environment. In terms of labor work, precasting concrete is one of the intensive jobs where the degree of hazard is high. To build concrete construction with better geometrical freedom, higher quality, and competitive cost, construction scale

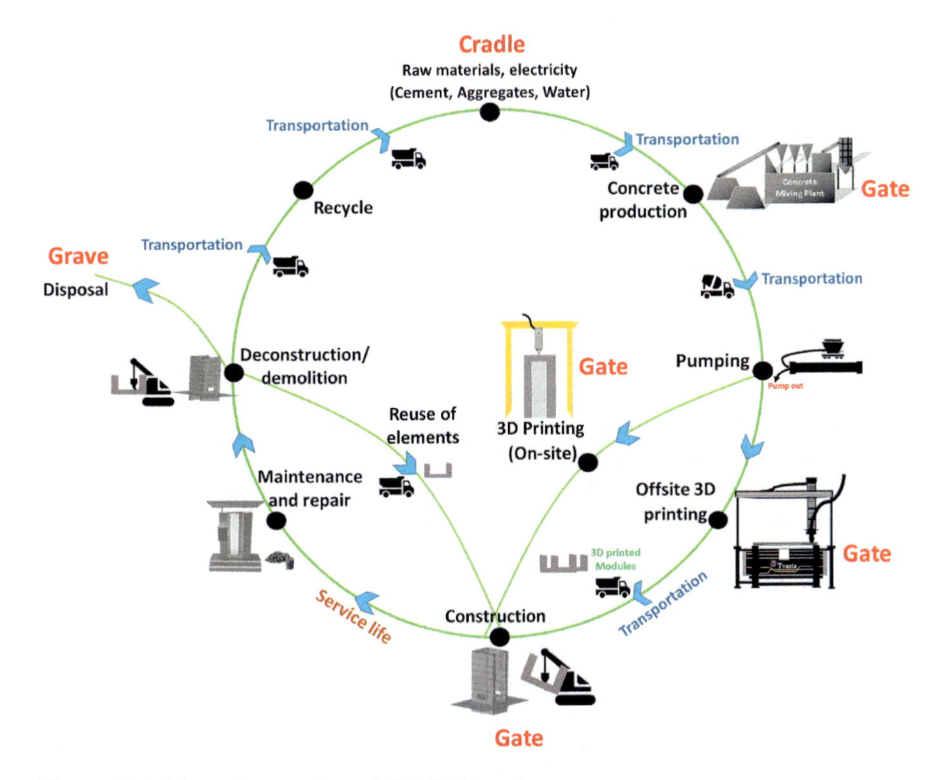

Figure 12.4 Schematic overview of 3DPC life cycle.
Source: From Bhattacherjee, S., Basavaraj, A. S., Rahul, A. V., Santhanam, M., Gettu, R., Panda, B., Schlangen, E., Chen, Y., Copuroglu, O., Ma, G., Wang, L., Basit Beigh, M. A., & Mechtcherine, V. (2021). Sustainable materials for 3D concrete printing. *Cement and Concrete Composites, 122*, 104156. https://doi.org/10.1016/j.cemconcomp.2021.104156.

additive manufacturing techniques are a feasible alternative to conventional construction methods. The 3DPC technology is successfully adopted for affordable housing construction, complex architectural designs, infrastructure and urban development, and a quick and efficient way to recover the consequences of natural hazards such as flooding, earthquake, land sliding, etc. Despite the potential applications and the recent advances in making the technology available and cost-effective, there are some challenges to be addressed before it can be widely used.

12.3.2 Challenges

12.3.2.1 Material formulation

Concrete composition is another important aspect influencing the final characteristics of 3DPC element. In this regard, discontinuity, cracking, and lack of extrudability are the main consequences if the concrete mixture is not designed as intended or the materials are not wisely selected. Discontinuity and cracking are the other challenges to control in a 3DPC element. Cracking happens during the hardening and curing stage when the concrete starts transitioning to a solid-state phase while discontinuity happens in the liquid state when printing is in progress. To avoid discontinuity flow in the element, the nozzle motion and rotation should be set at the optimum speed. It is worth noting that the optimum motion and rotational speed vary from concrete mix to concrete mix. On the other hand, the material properties along with solidification condition and duration are among the factors that control cracking. From a technical point of view a quicker solidification results in the generation of microcracks. Therefore selecting appropriate admixtures at optimum dosage to minimize such effects is recommended. The physical properties of conventional concrete are not completely ideal for this process, hence existing material for 3DPC is narrowly based on the technicality of being able to build layers on top of one another (De Schutter et al., 2018). The extruded material should support its own weight and the load of each layer added on top of it without deformation (Flynn et al., 2016). Nonetheless, several researchers have developed enhanced concrete mixtures for 3DPC. It is extremely beneficial to incorporate additives in printable concrete. Among widely used mineral admixtures in place of PC are FA, SF, blast furnace slag (BFS), limestone filler (LF), and nanosilica (NS) (Ma & Wang, 2018). In addition, chemical additives can be used such as accelerators, retarders, and superplasticizers. Utilizing such materials allow the enhancement of concrete properties and has a direct effect on physical properties such as rheology, flowability, and setting time. Not only that, incorporating these greatly influences the mechanical properties and durability of concrete. Hence, the full realization of the potential of 3DPC lies on the fine-tuning of materials suited for this process (Tay et al., 2017).

12.3.2.2 Reinforcement

Concrete is weak in tensile strength, brittle, and prone to failure under shear forces (Soltan & Li, 2018). Steel reinforcement is the typical solution to this problem. However, the challenge with 3DPC comes with adding steel reinforcement directly and automatically (Tay et al., 2017; Wangler et al., 2016). The properties of conventional reinforced concrete as a load-bearing structure simply cannot be matched by printed cementitious material. Since structural safety is a significant concern for the emerging technology of 3DPC materials, attention has been given to the possible ways to apply reinforcement in 3DPC. Currently, 3DPC could be used on the modern methods of construction technology of precast concrete formwork. The method begins with printing a concrete mold, embedding steel rebars, and printing fresh cement paste in situ shown in Figs. 12.5 and 12.6. Another source describes contour crafting that uses additional robotic arms to place steel reinforcements (Bos et al., 2016; Lim et al., 2016). These are the most generic and applicable methods to produce reinforced concrete in 3DPC; however, this does not tackle the limits of 3DPC as an automated process (Wolfs et al., 2018). Thus numerous types of reinforcements have been explored.

Figure 12.5 An example of automatic cable entrainment device for 3DPC.
Source: From Asprone, D., Menna, C., Bos, F. P., Salet, T. A. M., Mata-Falcón, J., & Kaufmann, W. (2018). Rethinking reinforcement for digital fabrication with concrete. Cement and Concrete Research, *112*, 111–121. https://doi.org/10.1016/j.cemconres.2018.05.020.

As discussed by many researchers, a possible way is to print fiber-reinforced concrete. Several research groups described the use of polypropylene fibers as reinforcement such as TU Delft, TU Eindhoven, and Loughborough University (Chen et al., 2017). Fiber-reinforced concrete could be developed for additive manufacturing which has sufficient tensile strength and ductility according to Bos et al. (2016). Moreover, the authors believed fiber reinforcements could reduce the shrinkage of concrete. This was also mentioned in Le, Austin, Lim, Buswell, Law, et al. (2012). Meanwhile, Wolfs et al. (2018) introduced the advanced concept of using cable reinforcement. According to their initial results, postcrack failure strength and deformation capacity was achieved obtained with such material. Christ et al. (2015) utilized 1 vol.% polyacrylonitrile fiber (1 cm). Their results showed a 180% increase in bending strength and 10 times increase in fracture strength compared to samples with no reinforcement. The use of carbon, basalt, glass, and steel fibers was also described by authors as potential reinforcement solutions for printable concrete (Hambach & Volkmer, 2017; Tassew & Lubell, 2014). Glass fiber–reinforced concrete studied in Tassew and Lubell (2014) which showed strain softening curves but its structural application is still not a part of the main load-bearing structure. Another method found is Hambach and Volkmer (2017), utilizing 3 vol.% carbon fibers to print concrete specimens oriented in the stress direction resulted the highest flexural strength of 120 MPa. This showed that flexural strength

Figure 12.6 An example of passive reinforcement of steel bars for 3DPC.
Source: From Asprone, D., Menna, C., Bos, F. P., Salet, T. A. M., Mata-Falcón, J., & Kaufmann, W. (2018). Rethinking reinforcement for digital fabrication with concrete. *Cement and Concrete Research, 112*, 111–121. https://doi.org/10.1016/j.cemconres.2018.05.020.

with the use of fiber reinforcements in concrete is highly dependent on its alignment to the stress direction.

However, the challenge is that existing types of the nozzle in 3DPC are much larger than those used at an experimental scale, which in this case was 2 mm (Hambach & Volkmer, 2017). In addition, the problem with blockage, segregation, and inconsistent material distribution are present even with small amounts of 1–3 vol.% fiber reinforcements. There is also difficulty in ensuring that the fibers are well embedded across the interface boundaries (Bos et al., 2016). In the study performed by Bos et al. (2018), a device was developed that directly embeds a reinforcement medium into printed concrete filaments. They have found that even stronger C1 cables resulted in bond loss between the cable and concrete whereas with weaker A1 and B1 cables complete cable breakage occurred under flexural tests. The authors have concluded that future research aimed at characterizing the cable behavior and optimizing the configuration of reinforcement can expand this concept. Therefore it is evident that more research needs to be conducted to overcome plenty of challenges and limits in producing reinforced 3DPC as a large-scale construction method.

12.3.2.3 Control settings and printing process

Moreover, there are various control settings in 3D printing of cementitious material that can affect eh ultimate properties of printed concrete. These are the motion speed of the nozzle, the extrusion speed, the nozzle size, and the height of the nozzle (Jo et al., 2020). For example, Panda et al. (2018) reported that the tensile bond strength in printed cementitious material decreases when the printing speed increases or the nozzle standoff is set to further away from the printing layer.

12.3.2.4 Extrusion nozzle

Extruded properties of concrete are greatly affected by the extrusion nozzle. Located at the tip of a printer head, the nozzle is essential in forming the intended shape and size of each concrete layer (Bos et al., 2016). Thus the nozzle shape and diameter have a direct relationship with flowability and buildability. In general, as the diameter of the nozzle increases, the flowability of the fresh paste should be decreased and vice versa. Moreover, in order to achieve good buildability on the layers, side and top trowels can be employed. Side trowels straighten the outer sides of the concrete being poured out of the nozzle while the top trowel smooths the upper surface to ensure proper bonding with the next layer.

Malaeb et al. (2015) experimented with syringes of different sizes with opening diameters between 1 and 2 cm. They determined that the optimal diameter for the nozzle was 2 cm as greater than this size caused problems with buildability wherein each layer could not hold itself. On the other hand, diameters less than 2 cm caused segregation problems in the components of the concrete mix. However, in a study by Khalil et al. (2017), an extrusion nozzle with a 5 cm opening and 1 cm tip allowed for the extrusion of mortar with sand particles less than 2 mm. In fact, it has been mentioned in previous research that maintaining a diameter ratio of D_{nozzle}/D_{max} more than 5 ensures there is no blocking during extrusion (El Cheikh et al., 2017). In terms of the shape of the nozzle, a variety has been described in literature such as circular, ellipse, square, and rectangular. Generally, circular nozzles allow printing without needing adjustments on the printing angle to achieve shape continuity or angle changes on the printed product. However, circular nozzles may create more voids in the printed structure, lowering the strength. This problem may not be as significant for square and rectangular shaped nozzles (Paul et al., 2018). Khoshnevis et al. (2001) concluded that using a square shaped nozzle is better compared with an ellipse-shaped nozzle regarding the creation of a surface finish. A square tip also had greater ease in manufacture than an elliptic one as reported by Anell (2015). The downside with square/rectangular shaped nozzles, though, is that it does not have the same ease in alignment as a circular nozzle (Paul et al., 2018). Currently being used for 3D printing of concrete, a range in size of 9 6–38 15 mm on rectangular orifices is described in previous publications (Le, Austin, Lim, Buswell, Law, et al., 2012; Nerella et al., 2016). Diameters of 4–22 mm were described in Lim et al. (2011) for circular nozzles in large-scale 3D concrete printing. Hence, it can be postulated that the mechanical properties can be affected by printing parameters such as the shape and size of the extrusion nozzle, which are the major factors to consider when designing a 3DPC process.

12.3.2.5 Pumping system or pumping speed

The pump is essential in transporting concrete paste from the tank to the extrusion nozzle. It is important that the pump is able to handle the concrete mixture and able to account for aspects such as aggregate size and flowability. In addition, pumping of concrete must be delivered without segregation of the particles as segregation may lead to loss of material on the concrete layer and blockage within the pipes. The parameters of the pump system must be selected appropriately. Pumping system or the pumping speed has no direct relationship with the physical or mechanical properties of the printed concrete; however, it does influence the implementation of the printing process. The pump pressure has a direct effect on the speed at which the concrete is poured.

For large-scale printing, a relatively high pumping pressure is demanded which is typically in the range 1–4 MPa, in order for fresh concrete to have shape retention once printed (Paul et al., 2018). A high pumping pressure is also required for mixtures with high viscosity (Ma & Wang, 2018). On the other hand, Malaeb et al. (2015) conducted a small-scale experiment wherein it was described that the use of cement a screw pump or shotcrete pump were not feasible. The high pressures exerted by this type of pump was not suitable for the scale of the project and even the lowest possible pressure was found to be inapplicable. Thus the authors described the design of a piston pump to study the concrete pouring mechanism, particularly the association between the pressure exerted by the pump and the rate at which the concrete mix is poured. They determined that their pumping speed was 5.4 L/s for a machine speed of 187.6 mm/s. In a study conducted by Paul et al. (2018), a pump speed of 3 L/min was used with a printing speed of 150 mm/s and a pump pressure of 0.4–1.0 MPa. With these settings, it was found that it was difficult to pump low slump (1–3 mm) mortar mixtures after an open time of 30–60 minutes. The study concluded by recommending that having a thixotropy value of 10,000 N mm rpm ensures better pumpability for 3DPC. In addition, Chaves et al. (2019) suggested that rheology modifiers must be employed to develop pumpable concrete in the fresh state. Meanwhile, the pumping speed generally varies depending on the dimensions and shape of the printed concrete. The speed also needs to be controlled at certain points in the direction of the printing path or wherever higher deposition of material is to be avoided (i.e., sharp curves and corners). Therefore parameters such as pumping speed, pressure, and extrusion rate must be coordinated by an appropriate control system (Paul et al., 2018).

12.3.2.6 Speed of arm (printing speed)

Determining the printing speed is highly dependent on the rheology of fresh concrete, which is a function of the material selection and mixture proportions. Inappropriate selection of arm speed can affect the quality of the print and may even cause catastrophic failure (Chen et al., 2017; Tay et al., 2017). The speed of printing is linked to several factors. First, the rate at which the concrete mixture is poured influences the layer effects. Layer effects, also called staircasing effects, are common in layer-deposited fabrication. These produce voids between the layers that weaken the interlayer bond strength of printed concrete (Paul et al., 2018). According to literature, controlling the printing speed allows control of the layer thickness which can reduce the occurrence of layer effects (Nerella et al., 2016). Too fast movement of the printing arm causes small amounts of concrete to be deposited and vice versa. A smaller amount of material poured, even though it prolongs the process, means there is finer control on the detail and finish of the printed structure (Lim et al., 2011). In addition, for extrusion-based concrete printing, the flow rate of concrete should match the speed at which it is printed otherwise inconsistencies with the layer thickness can occur which reduces the buildability of the printed product (Tay et al., 2017). Thus the speed of the arm must account for the desired printing resolution and thickness.

Another factor connected to the printing speed is the extrusion rate. The material must be extruded at a rate in which the open time of concrete is accounted for. Open time is how long the fresh printing mixture can maintain its workability before it sets and also the window it allows for each layer to bond with the next one (Paul et al., 2018). Incompatible speed of the arm with open time and workability properties results in clogging in the hose (Tay et al., 2017). Hence, in order to maintain the freshness of the material and maximize the resulting mechanical properties, the delivery through the system must be short and the material should be fed in small batches (Lim et al., 2011). In a study by Nerella et al. (2016), a printing speed of 75 mm/s and 30 seconds interlayer time gap were utilized for concrete with an open time of 90 minutes. Typical concrete mixtures take several hours to set but the addition of accelerators reduces the initial setting time of the printing material, which in this case was 3 minutes. Bos et al. (2016) used a print speed of 100 mm/s which was deliberately reduced on certain curves or angles. The print speed was deemed appropriate for the no-slump character of their concrete mixture which contained siliceous aggregates, LFs, and chemical additives. However, the resulting mechanical properties were discussed based on the physical properties of the fresh mixture rather than the printing parameters. Printing speeds of analogous concrete printing process are around 15–125 mm/s, typically for road pavements and hollow core slabs. According to Zijl and Visser (2007), the upper limit of the extrusion rate is about 200 mm/s in order to maintain the quality of the surface texture, although this may still vary depending on the materials. In conclusion, the speed of the arm affects the implementation of the 3D printing process and is determined by the physical properties of the prepared mixture.

12.3.2.7 Rheological properties

The rheology of fresh concrete material relates to the pumpability, extrudability, and buildability of 3DPC (Paul et al., 2018). Ideally, the rheology of the fresh mixture will allow it to move easily through the delivery system and allow it to be printable (Lim et al., 2011). Overall, it was observed that the main challenge with obtaining adequate rheology throughout the whole printing process was finding the correct balance between the mentioned properties, as most of them are conflicting.

For instance, the concrete paste must maintain a certain flowability to pass through the system yet maintain sufficient rigidity once it is poured from the nozzle. Second, the mixture must be set fast enough to be buildable but not fast enough that it would harden before adequately bonding with the subsequent layer (Malaeb et al., 2015). Thus obtaining the optimal rheological behavior requires a lot of consideration. Measurable properties to determine the rheology of cementitious pastes include viscosity, shear stress, open time, and green strength. Viscosity and shear stress can be measured by a rheometer. In some cases, the shear stress of fresh printing material can also be determined indirectly through a standard slump flow test (Pierre et al., 2013). Further, open time is the length of time in which the fresh material can be formed before the initial setting, while green strength is the fracture that takes place when the top of a printed layer is subjected to load. Printed concrete must have sufficient green strength to support subsequent layers (Perrot et al., 2016). These properties related to the rheology of 3DPC were searched for in relevant literature, along with the factors that influence them.

One of the main factors affecting the rheology of fresh concrete paste is particle grading (Ma, Wang, & Ju, 2018). Generally, a high packing density leads to a better rheology and fluidity which can be achieved by introducing fine mineral admixtures. In the work of Gosselin et al. (2016), they aimed to control the rheology of concrete mixture by employing materials with a fine particle size distribution. Crystalline silica (40 – 50 wt.%), SF (10 wt.%) and LF (10 wt.%) were used alongside PC. This resulted in an appropriate low critical shear and a long open time. However, it should be noted that excessive amounts of fine powders in cement pastes can increase the viscosity which can lead to adverse effects (Kwan & Wong, 2008). Ferraris et al. (2001) determined a maximum viscosity 0.06 Pa s of with the use of average particle sizes of around 11 mm, while a maximum yield stress of 30 Pa s was obtained when particles of around 5.7 mm in diameter were used. In addition, Park et al. (2005) found that the presence of high amounts of BFS together with SF results is 0.5 Pa s velocity, which is relatively low compared with the 1.2 Pa s viscosity obtained with a sample with only 15% SF. With respect to green time, there was not a lot of available literature that reported its results in connection with 3D-printed concrete and its rheology. However, theoretical research on fresh concrete material and the rheological properties has been demonstrated by Perrot et al. (2016) and Kazemian et al. (2017) that proposed a cylinder stability test to determine the buildability of concrete.

There are a few other ways to enhance the rheology of 3D-printed cementitious composites. The addition of chemical additives such as superplasticizers was reported to increase the rheology fluidity of the fresh cementitious paste (Kong et al., 2003; Mardani-Aghabaglou et al., 2013). This is due to the ability of superplasticizers to disperse flocculated cement particles which releases water molecules held inside the pores in the paste (Kong et al., 2003). It is also worth mentioning that a study by Petit et al. (2007) observed a connection between the rheological behavior of concrete to the ambient temperature and exposure time which is consistent with concrete behavior in a casting process (De Schutter et al., 2018). Furthermore, it has been well reported that a thixotropic behavior is desired in a printing material. Thixotropic behavior is a rheological property whereby material of high viscosity can become less viscous when agitated but becomes highly viscous once it has been extruded (Barnes, 1997; Mitsoulis, 2007; Petit et al., 2007). Using rheological tests, it was determined that adjustments to the mixture in terms of particle grading, chemical additives, as well as water to binder ratio (w/b) are key to achieving the desired thixotropic tendency of fresh concrete (Petit et al., 2007). Therefore several interrelated factors must be taken into account to obtain the ideal rheological properties of concrete for printing.

12.3.2.8 Flowability

Flowability is one of the key aspects in evaluating the printability of concrete mixtures whereby printed layers must possess a certain flowability that it would not stiffen during transfer. At the same time, it should have the ability to hold itself upon pouring and at the same time remain workable to bond with the next layer (Malaeb et al., 2015). Controlling the flowability ensures that the concrete paste can be pumped through the system and is easily extrudable (Ma & Wang, 2018). One of the ways to manage flowability is by selecting the proper concrete material and the particle size distribution. In general, a wider particle size distribution correlates to an increased packing density (Claisse et al., 2001; Lee et al., 2003). With proper dosage, fine powder additives have the ability to fill the voids and pores in the microstructure as well as provide lubrication by preventing the bonding of cement particles, thus improving the flowability of fresh mortar (Grzeszczyk & Lipowski, 1997; Park et al., 2005). However, it must be noted that excessive amounts of fine additives can increase particle friction, negatively affecting the fluidity of the paste (Kwan & Wong, 2008).

Several authors have studied the use of different mineral admixtures in concrete. According to Mastali and Dalvand (2016), substituting cement with 14 wt.% SF increased the slump by about 40% via a flow table test compared with the fresh paste with 100 wt.% cement. In another study (Grzeszczyk & Lipowski, 1997), the incorporation of less than 1.5 wt.% FA also improves flowability. Moreover, it was reported that ground FA is more effective at increasing fluidity compared with coarse FA. On the other hand, in a study by Güneyisi et al. (2015) which looked at a self-compacting concrete mixture, the substitution of cement by 50 wt.%. FA reduced the flowability by 43.2% and 15.6%, respectively. This could be due to the excessive amount of additive present in, which it starts to compete with the free water in the mixture instead of improving the flowability. More publications on this topic include a study by Zhang et al. (2016) which compared retarders citric acid (CA) and sodium gluconate (SG). The results revealed that the flowability of the fresh paste increased with CA addition of 0.03 wt.%; however, the fluidity started to decline with higher dosages of 0.06–0.15 wt.%. The same behavior was observed with SG wherein the optimum dosage was 0.12 wt.% SG. Studies have also shown that superplasticizers have the same ability to increase the workability of fresh pastes. Superplasticizers work by dispersing flocculated particles of cement causing the release of trapped moisture inside the pores (Kong et al., 2003;

Mardani-Aghabaglou et al., 2013). The use of 7% calcium sulfoaluminate (CSA) was also found to improve the workability by conducting a flow table test (Khalil et al., 2017). Overall, it was a common trend that flowability could be improved by appropriate material selection and its correct mixture proportion. It is a well-studied property of concrete and there are a wide variety of material that can be used.

12.3.2.9 Setting time

Printed concrete must have a fast setting time in order to develop early strength after deposition from the nozzle in order to support the subsequent layers, but not quick enough that it would set before bonding with the next layer (Malaeb et al., 2015). Hence, modifying the mixture to obtain an optimal setting time is important. Controlling the setting property of concrete is mainly brought about by the use of accelerators and retarders. Paglia et al. (2001) investigated setting time by utilizing two types of accelerators. According to the results, the alkalinity affects the setting time wherein samples with 4.5 wt.% alkaline accelerators had set 57% faster compared with samples containing 8 wt.% alkaline-free accelerators. On the other hand, incorporating common retarders were effective in prolonging the setting time. Six retarders were tested in Zhu et al. (2018) where the optimum retarder was found to be sodium tetraborate of mix ratio 0.1% to 0.3%. The final setting time was increased by 99 minutes. Other retarders such as sodium gluconate and tartaric acid also exhibited a successful retarding effect. Furthermore, studies that used different materials such as CSA and BFS also have the tendency to affect the setting time. Khalil et al. (2017) incorporated 7 wt.% CSA where it decreased the setting time compared with the ordinary PC mix. In fact, the mortar had good workability during the first hour of mixing followed by a rapid stiffening during the superposition of layers. Robeyst et al. (2008) and Gesoğlu and Özbay (2007) showed the opposite wherein the setting process is prolonged if the cement is replaced by BFS in amounts greater than 30 wt.%. However, Malaeb et al. (2015) used ordinary cement with chemical additives. According to the results of their trial and error, 1 mL of accelerator provided optimal setting time for their printable concrete without affecting other criteria. Thus achieving the desired setting time lies in determining the optimal mixture proportions of mineral admixtures, chemical additives, and other concrete components for 3D printing.

12.3.3 Experimental evaluation

12.3.3.1 Compressive strength

Compressive strength of concrete is usually taken as a priority amongst most of the other properties. This is because the compressive strength is a practical indication of concrete quality (Erdoğdu et al., 2018). This becomes even more important in 3DPC where structural safety is a significant concern (Chen et al., 2017). The compressive strength is affected by several factors such as the manufacturing process, material, curing procedure, and more (Erdoğdu et al., 2018). Generally, fine powder admixtures that have a particle size smaller than cement tend to have a filling effect. This means that the fine particles can fill the voids and pores between the cement particles and other constituents, resulting in a higher packing density and overall strength of the concrete (Gesoğlu & Özbay, 2007; Li, 2004). Materials such as SF and NS are also known for their high pozzolanic activity which equates to the improvement of the strength development (Li, 2004). This is only true for the optimal amount of these additives because excessive amounts can only result in a strength reduction (Gesoğlu & Özbay, 2007). Although the compressive strength of concrete could be improved by utilizing such materials, obtaining high strength for 3D-printed materials is still a challenge. The problem is that with the nature of 3DPC as an additive manufacturing process, there is a drawback of having a low interlayer bond strength, which in turn means lower compressive strength.

According to Lee et al. (2019) which used mineral admixtures FA and SF, the measured compressive strength of these multilayered printed mortar was 30% less than that of traditionally casted specimens. They concluded that as the layers increased, the quality of interlayer bonding decreased with it. In the contrary, Lim et al. (2011) observed a high compressive strength of 100–110 MPa with the development of their high-strength material (cement- and gypsum-based) for printing. Khalil et al. (2017) also reported good compressive strength results where they used 7 wt.% CSA and the strength is almost identical to that of normal concrete mix. However, several factors could have been at work here that affected the results. In other study, Qasimi et al. (2020) investigated the effect of nanosilica, microsilica, and microfibrillated cellulose on printability and mechanical properties of 3D printing concrete. They found that the addition of 0.4% of microfibrillated cellulose (total solid matter) and 1% of nanosilica (wt.%) can significantly enhance the mechanical properties of 3D-printed material.

Several researchers have described tests in order to identify the mechanisms that affect compressive strength in 3D-printed concrete other than the choice of material. Such literary works have reported the influence of the printing process on compressive strength. Tay et al. (2017) demonstrated that the compressive strength varies with the printing direction wherein Y direction produced the highest compressive strength. Le, Austin, Lim, Buswell, Law, et al. (2012) and Panda, Chandra Paul, and Jen Tan (2017) reported similar behavior. According to the results in Panda, Chandra Paul, and Jen Tan (2017), the highest average compressive strength was found in the longitudinal direction which is likely due to the high pressure exerted in this direction during the extrusion of the material. Consistently, the lowest compressive strength was found in the lateral direction which has the least amount of pressure during the printing process (Le, Austin, Lim, Buswell, Law, et al., 2012; Panda, Chandra Paul, & Jen Tan, 2017). Regarding the effect of time delay of printing between each layer, Tay et al. (2017) reported the highest compressive strength and interlayer bond strength when the time gap is short. In Sanjayan et al. (2018), the mean compressive strength of printed samples with a 20 minutes delay time was higher among the specimens printed within a range of 10–30 minutes time delay. The reasons were not apparent at that stage, though they may be connected to the open time of the materials. In conclusion, printed specimen typically exhibits lower compressive strengths compared to casted ones due to the higher porosity caused by the process. There remains to be a wide range of materials that could be used for printing process and studied for their mechanical properties. In addition, there are several parameters yet to be studied that affect the compressive strength of 3D-printed concrete, and understand the principles behind them.

12.3.3.2 Tensile strength

Tensile strength of concrete is usually weak and just about 10% of the compressive strength. This drove many recent researches that aimed in improving the overall strengths of concrete (Bamigboye et al., 2015). In conventional concrete production, the most common way to fix this is adding steel reinforcement; however, this is difficult to integrate automatically into a 3DPC process (Tay et al., 2017; Wangler et al., 2016).

A few publications about tensile strength of printed concrete were identified to understand the factors that affect this property. Le, Austin, Lim, Buswell, Law, et al. (2012) performed tensile strength tests on concrete with various printing time gaps in between each layer. The results determined that the time gaps for each printed layer have a direct effect on the tensile strength. The tensile strength decreased with increasing printing gap and the authors concluded that for better tensile strength of printed concrete, the printing time should be shortened. However, it was not clear whether the chemical or mechanical interfacial bonding mechanisms governed the variation in tensile test results with the interlayer printing time gap (Le, Austin, Lim, Buswell, Law, et al., 2012). Panda, Bahubalendruni, et al. (2017) found the presence of macropores in low tensile strength concrete samples, which explains the weak interfacial bond strength. This becomes more pronounced for longer time gaps between layers and mixtures with higher thixotropy. In conclusion, there is a huge gap in the literature about the tensile strength of printed concrete. More driving mechanisms on the increase or decrease of tensile strength of unreinforced 3DPC need to be understood.

12.3.3.3 Flexural strength

Flexural strength of concrete is an important parameter in the application of concrete. Loss of this property is mainly responsible for the failure observed in concrete structures (Ekwulo & Eme, 2017). Enhancing the flexural strength typically requires reinforcement; however, only unreinforced concrete is discussed in this section. The flexural or bending strength is often varied with respect to the printing direction or the printing time delays between each layer. A study by Lim et al. (2011) demonstrated the performance of a printed concrete mix comprised of 54 wt.% sand, 36 wt.% reactive cementitious material, and 10 wt.% water. The flexural strength of the samples turned out to be close to a standard casted concrete. A different result was reported by Nerella et al. (2016) wherein different directions of printing were investigated, it is unknown which direction was used in Lim et al. (2011). With printing direction of D1 (*Z* axis) and D3 (*X* axis) reached about 16% and 14% higher flexural strength in 3DPC specimens, respectively, than in a typical cast sample (Nerella et al., 2016). However, the behavior of flexural strength variation with regard to the print direction was not discussed by the authors. Paul et al. (2018) further confirmed this finding when their printed concrete had higher flexural strength in certain directions compared to the control specimen. Sanjayan et al. (2018) also performed three-point bending strength tests on specimens with different printing directions and found the same behavior. It is believed that flexural strength, as well as compressive strength, has a directional dependency due to the inherent feature of a layer-by-layer printing process (Oxman et al., 2012). Furthermore, Sanjayan et al. (2018) investigated the flexural strength of concrete with different interlayer time delays. It was found that the three-point bending strength of samples with 20 minutes time delay was highest within the testing range of 10–30 minutes. The consistently observed phenomenon of flexural strength as a function of printing direction and time delay requires further research to understand the mechanisms responsible for this.

12.3.3.4 Water absorption

Water absorption is one of the main indicators of the durability of concrete. The susceptibility of concrete to the penetration of fluids determines its serviceability (Zhang & Zong, 2014). To date, limited information is present on the water absorption of 3D-printed concrete. Most of the reviewed literature focused on the fresh properties of concrete for 3D printing as well as the mechanical properties. Unfortunately, less attention was given to the durability of the finished products since 3DPC is still an emerging technology. Nevertheless, water absorption is a function of the porosity of concrete and understanding this relationship may help predict the water absorption of 3DPC.

Pores and voids in the microstructure of concrete are expected. This network of gaps in the microstructure provides passage for the absorption of fluid into the concrete material. Nevertheless, the porosity of concrete can be controlled by the components in the fresh mixture as well as the curing condition and duration (Ramli & Akhavan Tabassi, 2012; Shafiq & Cabrera, 2004). In an experiment conducted by Zhang and Zong (2014), the curing condition (20 ± 3°C, RH 90% ± 5%) greatly affected the water absorption, in which the samples exhibited low water uptake. Concrete admixture in the form of polycarboxylate water reducers was utilized, as well as finely crushed stone aggregates and quartz sand with a fineness modulus of 2.4. Moreover, the curing condition becomes more critical when mineral admixtures are used such as FA. Numerous sources reported that mineral admixtures need a longer curing time to enable the admixture to perform its pozzolanic effect (Ekwulo & Eme, 2017; Tasdemir, 2003; Zhang & Zong, 2014). The utilization of other particles like nanoparticles on concrete has been extensively studied. Since nanoparticles are finer than ordinary PC, they can fill the voids and minimize the pores in the cement structure (Gesoğlu & Özbay, 2007). This leads to an improvement in water permeability resistance. Conversely, the addition of FA tends to increase the water uptake, since FA particles are more challenging to disperse within the structure compared with PC (Madhavi et al., 2015). Overall, the correct material choice and appropriate curing conditions can allow control of the water resistance of concrete. Since 3DPC typically exhibits more voids and pores in its structure, it can be postulated that reducing this characteristic may improve the water permeability resistance, and hence durability of printed concrete. This is yet to be seen on concrete produced by 3D printing.

12.3.3.5 Shrinkage

Shrinkage is an important parameter to be considered with printed cementitious material. Fresh concrete for 3D printing typically has a higher water content in order to be extrudable and pumpable which means that the amount of excess water is beyond the amount required for cement hydration. The occurrence of shrinkage is due to the evaporation of this free water during the setting and hardening process (Lee & Won, 2016). Shrinkage is a concern as it affects the stability of printed structures and the accuracy of their dimensions. Thus shrinkage cracking has to be controlled by identifying the correct mix proportions as well as effective curing procedures. One feasible solution to avoid shrinkage is controlling the content of mineral admixtures. Generally, the smaller the particle size of the aggregate, the less shrinkage strain on the composite (Zhang et al., 2009).

Employing 5 and 15 wt.% SF was demonstrated by Güneyisi et al. (2012) where the shrinkage of concrete was around 29%–35% less compared with ordinary concrete. Similarly, Al-Khaja (1994) proved that adding SF reduced the drying shrinkage by around 35%. Li and Yao (2001) utilized SF with ultrafine ground granulated BFS and concluded that both could significantly lessen the shrinkage. The effect of mineral admixtures on the shrinkage control of concrete can be explained by their ability to accelerate the hydration process. In other studies (Khatib, 2008; Rongbing & Jian, 2005), the effect of FA in the concrete mixture was found to decrease the shrinkage in a linear relationship. Rongbing and Jian (2005) demonstrated a reduction in shrinkage by 80% due to the combined effects of FA and CSA in concrete, while Khatib (2008) studied the effects of increasing FA content in self-compacting concrete wherein test results showed a reduction in drying shrinkage with increasing FA dosage. Up to 80 wt.% FA reduced the shrinkage by approximately 67%. This can be attributed to the ability of FA to significantly reduce shrinkage cracking as it has a lower release of hydration heat. Generally, the utilization of shrinkage reducing admixtures can significantly decrease the shrinkage of the outer part of the concrete by lowering the surface tension due to water evaporation (Ma & Wang, 2018; Shh & Sarigaphuti, 1992). Therefore it appears there are plenty of ways to avoid shrinkage which is just a matter of finding the appropriate additive and mixture proportion.

12.4 Conclusion

This chapter provides a comprehensive overview of 3DPC potential in addressing the construction challenges. In this regard, reformulating the concrete mix design to be more sustainable and environmentally friendly, will significantly help the construction industry to use this technology in wider applications. It was determined that 3D printing for a full-scale construction as an alternative to conventional techniques is still a developing technology. 3D printing technology still has a lot of challenges to overcome particularly those associated with printing material, mechanical properties, reinforcement, and durability. Discussions and future research areas drawn from this review are summarized as follows. Material for 3DPC has a very specific rheological requirement. The main challenge is developing a mixture that has sufficient flowability for extrusion yet stiff enough to support subsequently printed layers, and on top of that allows sufficient time between layers to bond. Obtaining this required material at the lowest possible cost is a key for the future of 3DPC. Mechanical properties and durability of concrete are important aspects of its application. In order for 3DPC to be viable as an industrial method, a significant understanding of its strength and durability is important. An identified research area is gaining more knowledge of the factors that affect compressive, tensile, and flexural strength, shrinkage, and water permeability resistance. Such factors that need more evidence and reasoning are the printing direction, interlayer printing time delay, and interfacial bond strength. More research in this area could overcome the problems associated with this process. Quality of printing also needs improvement which is often weakened by layer effects, macropores, and deformation during printing. The printing quality can be enhanced by considering printing parameters such as nozzle design, pump speed and pressure, and printing speed. Structural safety is of paramount importance and is the primary concern for 3DPC in large-scale implementation. Currently, there is a lack of suitable reinforcing techniques for 3DPC. Further investigations on fiber reinforcement, steel reinforcement, and cable reinforcement could be what unlock the full potential of 3D printing. The future of construction is heading toward low-cost and high-efficiency methods offering architectural complexity, reduced labor, and environmental impact. Overall, there are many challenges for 3DPC but it is a promising method that will continue to be developed in the near future. It is most likely that 3DPC will be an integrated process that allows the construction and building industry to utilize both conventional and advanced methods.

The literature studies and field trials have revealed that a few aspects to be further developed before 3DPC becomes the preferred technique in the construction industry. These areas are

- formulating a stronger while more sustainable and cheaper printable cementitious mixture;
- reducing the operation costs;
- providing a common language, guideline, and specification; and
- integration with current sustainable technology such as renewable energy.

References

Abu-Ennab, L., Dixit, M. K., Birgisson, B., & Pradeep Kumar, P. (2022). Comparative life cycle assessment of large-scale 3D printing utilizing kaolinite-based calcium sulfoaluminate cement concrete and conventional construction. *Cleaner Environmental Systems*, *5*, 100078. Available from https://doi.org/10.1016/j.cesys.2022.100078.

Agustí-Juan, I., & Habert, G. (2017). Environmental design guidelines for digital fabrication. *Journal of Cleaner Production*, *142*, 2780–2791. Available from https://doi.org/10.1016/j.jclepro.2016.10.190.

Al-Khaja, W. A. (1994). Strength and time-dependent deformations of silica fume concrete for use in Bahrain. *Construction and Building Materials*, *8*(3), 169–172. Available from https://doi.org/10.1016/S0950-0618(09)90030-7.

Anell L.H. (2015), Concrete 3d printer. MSc Thesis.

Bamigboye, G., Ede, A., Egwuatu, C., Jolayemi, J., Olowu, O., & Odewumi, T. (2015). Assessment of compressive strength of concrete produced from different brands of Portland cement. *Civil and Environmental Research*, *7*, 31–38.

Barnes, H. A. (1997). Thixotropy – a review. *Journal of Non-Newtonian Fluid Mechanics*, *70*(1–2), 1–33. Available from https://doi.org/10.1016/S0377-0257(97)00004-9.

Bhattacherjee, S., Basavaraj, A. S., Rahul, A. V., Santhanam, M., Gettu, R., Panda, B., Schlangen, E., Chen, Y., Copuroglu, O., Ma, G., Wang, L., Basit Beigh, M. A., & Mechtcherine, V. (2021). Sustainable materials for 3D concrete printing. *Cement and Concrete Composites*, *122*, 104156. Available from https://doi.org/10.1016/j.cemconcomp.2021.104156.

Bos, F., Wolfs, R., Ahmed, Z., & Salct, T. (2016). Additive manufacturing of concrete in construction: Potentials and challenges of 3D concrete printing. *Virtual and Physical Prototyping*, *11* (3), 209–225. Available from https://doi.org/10.1080/17452759.2016.1209867.

Bos, F. P., Ahmed, Z. Y., Wolfs, R. J. M., & Salet, T. A. M. (2018). 3D printing concrete with reinforcement. In D. Hordijk, & M. Luković (Eds.), *High tech concrete: Where technology and engineering meet* (pp. 2484–2493). Springer. Available from 10.1007/978-3-319-59471-2_283.

Chaves, S. F., Rodriguez, Z., Ahmed, D. H., Bos, Y., Xu, T., Salet, O., Çopuroğlu, E., Schlangen, F., & Bos. (2019). An approach to develop printable strain hardening cementitious composites. *Materials & Design*, *169*, 107651.

Chen, Y., Veer, F., & Çopuroğlu, O. (2017). A critical review of 3D concrete printing as a low CO_2 concrete approach. *Heron*, *62*(3), 167–194. Available from http://heronjournal.nl/62-3/2.pdf.

Chen, Y., He, S., Gan, Y., Çopuroğlu, O., Veer, F., & Schlangen, E. (2022). A review of printing strategies, sustainable cementitious materials and characterization methods in the context of extrusion-based 3D concrete printing. *Journal of Building Engineering*, *45*, 103599. Available from https://doi.org/10.1016/j.jobe.2021.103599.

Christ, S., Schnabel, M., Vorndran, E., Groll, J., & Gbureck, U. (2015). Fiber reinforcement during 3D printing. *Materials Letters*, *139*, 165–168. Available from https://doi.org/10.1016/j.matlet.2014.10.065.

Claisse, P. A., Lorimer, P., & Al Omari, M. (2001). Workability of cement pastes. *ACI Materials Journal*, *98*(6), 476–482.

De Schutter, G., Lesage, K., Mechtcherine, V., Nerella, V. N., Habert, G., & Agusti-Juan, I. (2018). Vision of 3D printing with concrete—technical, economic and environmental potentials. *Cement and Concrete Research*, *112*, 25–36. Available from https://doi.org/10.1016/j.cemconres.2018.06.001.

Dixit, M. K. (2019). 3-D printing in building construction: A literature review of opportunities and challenges of reducing life cycle energy and carbon of buildings. *IOP Conference Series: Earth and Environmental Science*, *290*(1), 012012. Available from https://doi.org/10.1088/1755-1315/290/1/012012.

Ekwulo, E. O., & Eme, D. B. (2017). Flexural strength and water absorption of concrete made from uniform size and graded coarse aggregates. *American Journal of Engineering Research*, *6*(10), 172–177.

El Cheikh, K., Rémond, S., Khalil, N., & Aouad, G. (2017). Numerical and experimental studies of aggregate blocking in mortar extrusion. *Construction and Building Materials*, *145*, 452–463. Available from https://doi.org/10.1016/j.conbuildmat.2017.04.032.

Erdoğdu Ş., Kurbetci Ş., Kandil U., Nas M., Nayir S., Evaluation of the compressive strength of concrete by means of cores taken from different casting direction. 13th international congress on advances in civil engineering, İzmir, Turkey, 12–14 September, pp. 1–10 (2018).

Ferraris, C. F., Obla, K. H., & Hill, R. (2001). The influence of mineral admixtures on the rheology of cement paste and concrete. *Cement and Concrete Research*, *31*(2), 245–255. Available from https://doi.org/10.1016/S0008-8846(00)00454-3.

Flynn, J. M., Shokrani, A., Newman, S. T., & Dhokia, V. (2016). Hybrid additive and subtractive machine tools – research and industrial developments. *International Journal of Machine Tools and Manufacture*, *101*, 79–101. Available from https://doi.org/10.1016/j.ijmachtools.2015.11.007.

Gesoğlu, Mehmet, & Özbay, Erdoğan (2007). Effects of mineral admixtures on fresh and hardened properties of self-compacting concretes: Binary, ternary and quaternary systems. *Materials and Structures*, *40*(9), 923–937. Available from https://doi.org/10.1617/s11527-007-9242-0.

Gibson, I., Rosen, D., & Stucker, B. (2010). *Additive manufacturing technologies – rapid prototyping to direct digital manufacturing*. Springer. Available from https://doi.org/10.1007/978-1-4419-1120-9.

Gosselin, C., Duballet, R., Roux, P., Gaudillière, N., Dirrenberger, J., & Morel, P. (2016). Large-scale 3D printing of ultra-high performance concrete – a new processing route for architects and builders. *Materials and Design*, *100*, 102–109. Available from https://doi.org/10.1016/j.matdes.2016.03.097.

Grzeszczyk, S., & Lipowski, G. (1997). Effect of content and particle size distribution of high-calcium fly ash on the rheological properties of cement pastes. *Cement and Concrete Research*, *27*(6), 907–916. Available from https://doi.org/10.1016/S0008-8846(97)00073-2.

Güneyisi, E., Gesoglu, M., Al-Goody, A., & Ipek, S. (2015). Fresh and rheological behavior of nano-silica and fly ash blended self-compacting concrete. *Construction and Building Materials*, *95*, 29–44. Available from https://doi.org/10.1016/j.conbuildmat.2015.07.142.

Güneyisi, Erhan, Gesoğlu, Mehmet, Karaoğlu, Seda, & Mermerdaş, Kasım (2012). Strength, permeability and shrinkage cracking of silica fume and metakaolin concretes. *Construction and Building Materials*, *34*, 120–130. Available from https://doi.org/10.1016/j.conbuildmat.2012.02.017.

Gursel, A. P., & Ostertag, C. (2017). Comparative life-cycle impact assessment of concrete manufacturing in Singapore. *International Journal of Life Cycle Assessment*, *22*(2), 237–255. Available from https://doi.org/10.1007/s11367-016-1149-y.

Hambach, M., & Volkmer, D. (2017). Properties of 3D-printed fiber-reinforced Portland cement paste. *Cement and Concrete Composites*, *79*, 62–70. Available from https://doi.org/10.1016/j.cemconcomp.2017.02.001.

Harrison, Edward, Berenjian, Aydin, & Seifan, Mostafa (2020). Recycling of waste glass as aggregate in cement-based materials. *Environmental Science and Ecotechnology*, *4*, 100064. Available from https://doi.org/10.1016/j.ese.2020.100064.

Jo, J. H., Jo, B. W., Cho, W., & Kim, J. H. (2020). Development of a 3D printer for concrete structures: Laboratory testing of cementitious materials. *International Journal of*

Concrete Structures and Materials, *14*(1). Available from https://doi.org/10.1186/s40069-019-0388-2.

Kazemian, A., Yuan, X., Cochran, E., & Khoshnevis, B. (2017). Cementitious materials for construction-scale 3D printing: Laboratory testing of fresh printing mixture. *Construction and Building Materials*, *145*, 639–647. Available from https://doi.org/10.1016/j.conbuildmat.2017.04.015.

Khalil, N., Aouad, G., El Cheikh, K., & Rémond, S. (2017). Use of calcium sulfoaluminate cements for setting control of 3D-printing mortars. *Construction and Building Materials*, *157*, 382–391. Available from https://doi.org/10.1016/j.conbuildmat.2017.09.109.

Khan, S. A., Koç, M., & Al-Ghamdi, S. G. (2021). Sustainability assessment, potentials and challenges of 3D printed concrete structures: A systematic review for built environmental applications. *Journal of Cleaner Production*, *303*. Available from https://doi.org/10.1016/j.jclepro.2021.127027.

Khatib, J. M. (2008). Performance of self-compacting concrete containing fly ash. *Construction and Building Materials*, *22*(9), 1963–1971. Available from https://doi.org/10.1016/j.conbuildmat.2007.07.011.

Khoshnevis, B. (2004). Automated construction by contour crafting – related robotics and information technologies. *Automation in Construction*, *13*(1), 5–19. Available from https://doi.org/10.1016/j.autcon.2003.08.012.

Khoshnevis, B., Bukkapatnam, S., Kwon, H., & Saito, J. (2001). Experimental investigation of contour crafting using ceramics materials. *Rapid Prototyping Journal*, *7*(1), 32–42. Available from https://doi.org/10.1108/13552540110365144.

Kong, H. J., Bike, S. G., & Li, V. C. (2003). Development of a self-consolidating engineered cementitious composite employing electrosteric dispersion/stabilization. *Cement and Concrete Composites*, *25*(3), 301–309. Available from https://doi.org/10.1016/S0958-9465(02)00057-4.

Kwan, A. K. H., & Wong, H. H. C. (2008). Effects of packing density, excess water and solid surface area on flowability of cement paste. *Advances in Cement Research*, *20*(1), 1–11. Available from https://doi.org/10.1680/adcr.2008.20.1.1HongKong.

Le, T. T., Austin, S. A., Lim, S., Buswell, R. A., Gibb, A. G. F., & Thorpe, T. (2012). Mix design and fresh properties for high-performance printing concrete. *Materials and Structures*, *45*(8), 1221–1232. Available from https://doi.org/10.1617/s11527-012-9828-z.

Le, T. T., Austin, S. A., Lim, S., Buswell, R. A., Law, R., Gibb, A. G. F., & Thorpe, T. (2012). Hardened properties of high-performance printing concrete. *Cement and Concrete Research*, *42*(3), 558–566. Available from https://doi.org/10.1016/j.cemconres.2011.12.003.

Lee, H., Kim, J. H. J., Moon, J. H., Kim, W. W., & Seo, E. A. (2019). Evaluation of the mechanical properties of a 3D-printed mortar. *Materials*, *12*(24). Available from https://doi.org/10.3390/ma1224104.

Lee, S. H., Kim, H. J., Sakai, E., & Daimon, M. (2003). Effect of particle size distribution of fly ash-cement system on the fluidity of cement pastes. *Cement and Concrete Research*, *33*(5), 763–768. Available from https://doi.org/10.1016/S0008-8846(02)01054-2.

Lee, S. J., & Won, J. P. (2016). Shrinkage characteristics of structural nano-synthetic fibre-reinforced cementitious composites. *Composite Structures*, *157*, 236–243. Available from https://doi.org/10.1016/j.compstruct.2016.09.001.

Li, G. (2004). Properties of high-volume fly ash concrete incorporating nano-SiO_2. *Cement and Concrete Research*, *34*(6), 1043–1049. Available from https://doi.org/10.1016/j.cemconres.2003.11.013.

Li, J., & Yao, Y. (2001). A study on creep and drying shrinkage of high performance concrete. *Cement and Concrete Research*, *31*(8), 1203–1206. Available from https://doi.org/10.1016/S0008-8846(01)00539-7.

Lim S. Buswell R. Le T. Wackrow R. Austin S. Gibb A. Thorpe T. 2011 Development of a viable concrete printing process, in: Proceedings of the 28th international symposium on automation and robotics in construction ISARC 2011, International Association for Automation and Robotics in Construction (I.A.A.R.C), United Kingdom, pp. 665–670. Available from https://doi.org/10.22260/isarc2011/0124.

Lim, S., Buswell, R. A., Le, T. T., Austin, S. A., Gibb, A. G. F., & Thorpe, T. (2012). Developments in construction-scale additive manufacturing processes. *Automation in Construction*, *21*(1), 262–268. Available from https://doi.org/10.1016/j.autcon.2011.06.010.

Lim, S., Buswell, R. A., Valentine, P. J., Piker, D., Austin, S. A., & De Kestelier, X. (2016). Modelling curved-layered printing paths for fabricating large-scale construction components. *Additive Manufacturing*, *12*, 216–230. Available from https://doi.org/10.1016/j.addma.2016.06.004.

Ma, G., Li, Z., & Wang, L. (2018). Printable properties of cementitious material containing copper tailings for extrusion based 3D printing. *Construction and Building Materials*, *162*, 613–627. Available from https://doi.org/10.1016/j.conbuildmat.2017.12.051.

Ma, G., & Wang, L. (2018). A critical review of preparation design and workability measurement of concrete material for largescale 3D printing. *Frontiers of Structural and Civil Engineering*, *12*(3), 382–400. Available from https://doi.org/10.1007/s11709-017-0430-x.

Ma, G. W., Wang, L., & Ju, Y. (2018). State-of-the-art of 3D printing technology of cementitious material—An emerging technique for construction. *Science China Technological Sciences*, *61*(4), 475–495. Available from https://doi.org/10.1007/s11431-016-9077-7.

Ma, Guowei, Li, Yanfeng, Wang, Li, Zhang, Junfei, & Li, Zhijian (2020). Real-time quantification of fresh and hardened mechanical property for 3D printing material by intellectualization with piezoelectric transducers. *Construction and Building Materials*, *241*, 117982. Available from https://doi.org/10.1016/j.conbuildmat.2019.117982.

Madhavi, T. C., Vardhan Reddy, T., Singh, B., & Lal, B. (2015). Water absorption in fly ash concrete. *International Journal of Applied Engineering Research*, *10*(9), 24587–24596.

Malaeb, Z., Hachem, H., Tourbah, A., Maalouf, T., Zarwi, N., & Hamzeh, F. (2015). 3D concrete printing: Machine and mix design. *International Journal of Civil Engineering and Technology*, *6*, 14–22.

Mardani-Aghabaglou, A., Tuyan, M., Yılmaz, G., A., Ömer, & Ramyar, K. (2013). Effect of different types of superplasticizer on fresh, rheological and strength properties of self-consolidating concrete. *Construction and Building Materials*, *47*, 1020–1025. Available from https://doi.org/10.1016/j.conbuildmat.2013.05.105.

Mastali, M., & Dalvand, A. (2016). Use of silica fume and recycled steel fibers in self-compacting concrete (SCC). *Construction and Building Materials*, *125*, 196–209. Available from https://doi.org/10.1016/j.conbuildmat.2016.08.046.

Mitsoulis, E. (2007). Flows of viscoplastic materials: Models and computations. In D. M. Binding, N. E. Hudson, & R. Keunings (Eds.), *Rheology reviews* (pp. 135–178). British Society of Rheology.

Mohan, M. K., Rahul, A. V., De Schutter, G., & Van Tittelboom, K. (2021). Extrusion-based concrete 3D printing from a material perspective: A state-of-the-art review. *Cement and Concrete Composites*, *115*, 103855. Available from https://doi.org/10.1016/j.cemconcomp.2020.103855.

Nerella V.N., Krause M., Näther M., Mechtcherine V., Studying printability of fresh concrete for formwork free concrete on-site 3D printing technology, in: 25th conference on rheology of building materials, Regensburg, 02–03 March (2016).

Nerella, V. N., & Mechtcherine, V. (2019). Studying the printability of fresh concrete for formwork-free concrete onsite 3D printing technology (CONPrint3D). In J. G. Sanjayan, A. Nazari, & B. Nematollahi (Eds.), *3D concrete printing technology* (pp. 333–347). Butterworth-Heinemann.

Oxman, Neri, Tsai, Elizabeth, & Firstenberg, Michal (2012). Digital anisotropy: A variable elasticity rapid prototyping platform. *Virtual and Physical Prototyping*, *7*(4), 261–274. Available from https://doi.org/10.1080/17452759.2012.731369.

Paglia, C., Wombacher, F., & Böhni, H. (2001). The influence of alkali-free and alkaline shotcrete accelerators within cement systems – I. Characterization of the setting behavior. *Cement and Concrete Research*, *31*(6), 913–918. Available from https://doi.org/10.1016/S0008-8846(01)00509-9.

Panda, B., Chandra Paul, S., & Jen Tan, M. (2017). Anisotropic mechanical performance of 3D printed fiber reinforced sustainable construction material. *Materials Letters*, *209*, 146–149. Available from https://doi.org/10.1016/j.matlet.2017.07.123.

Panda, B., Paul, S. C., Mohamed, N. A. N., Tay, Y. W. D., & Tan, M. J. (2018). Measurement of tensile bond strength of 3D printed geopolymer mortar. *Measurement*, *113*, 108–116. Available from https://doi.org/10.1016/j.measurement.2017.08.051.

Panda, B., Ruan, S., Unluer, C., & Tan, M. J. (2019). Improving the 3D printability of high volume fly ash mixtures via the use of nano attapulgite clay. *Composites Part B: Engineering*, *165*, 75–83. Available from https://doi.org/10.1016/j.compositesb.2018.11.109.

Panda, B. N., Bahubalendruni, R. M. V. A., Biswal, B. B., & Leite, M. (2017). A CAD-based approach for measuring volumetric error in layered manufacturing. *Proceedings of the Institution of Mechanical Engineers, Part C: Journal of Mechanical Engineering Science*, *231*(13), 2398–2406. Available from https://doi.org/10.1177/0954406216634746.

Park, C. K., Noh, M. H., & Park, T. H. (2005). Rheological properties of cementitious materials containing mineral admixtures. *Cement and Concrete Research*, *35*(5), 842–849. Available from https://doi.org/10.1016/j.cemconres.2004.11.002.

Paul, S. C., Tay, Y. W. D., Panda, B., & Tan, M. J. (2018). Fresh and hardened properties of 3D printable cementitious materials for building and construction. *Archives of Civil and Mechanical Engineering*, *18*(1), 311–319. Available from https://doi.org/10.1016/j.acme.2017.02.008.

Paul, S. C., van Zijl, G. P. A. G., Tan, M. J., & Gibson, I. (2018). A review of 3D concrete printing systems and materials properties: Current status and future research prospects. *Rapid Prototyping Journal*, *24*(4), 784–798. Available from https://doi.org/10.1108/rpj-09-2016-0154.

Pegna, J. (1997). Exploratory investigation of solid freeform construction. *Automation in Construction*, *5*(5), 427–437. Available from https://doi.org/10.1016/S0926-5805(96)00166-5.

Perkins, I., & Skitmore, M. (2015). Three-dimensional printing in the construction industry: A review. *International Journal of Construction Management*, *15*(1), 1–9. Available from https://doi.org/10.1080/15623599.2015.1012136.

Perrot, A., Rangeard, D., & Pierre, A. (2016). Structural built-up of cement-based materials used for 3D-printing extrusion techniques. *Materials and Structures*, *49*(4), 1213–1220. Available from https://doi.org/10.1617/s11527-015-0571-0.

Petit, J. Y., Wirquin, E., Vanhove, Y., & Khayat, K. (2007). Yield stress and viscosity equations for mortars and self-consolidating concrete. *Cement and Concrete Research*, *37*(5), 655–670. Available from https://doi.org/10.1016/j.cemconres.2007.02.009.

Pierre, A., Lanos, C., & Estellé, P. (2013). Extension of spread-slump formulae for yield stress evaluation. *Applied Rheology*, *23*(6). Available from https://doi.org/10.3933/ApplRheol-23-63849France.

Purchase, C. K., Al Zulayq, D. M., O'brien, B. T., Kowalewski, M. J., Berenjian, A., Tarighaleslami, A. H., & Seifan, M. (2022). Circular economy of construction and demolition waste: A literature review on lessons, challenges, and benefits. *Materials*, *15* (1). Available from https://doi.org/10.3390/ma15010076.

Qasimi, M. Al, Al Zulayq, D. M., & Seifan, M. (2020). Mechanical and rheological properties of 3D printable cement composites. *Recent Progress in Materials*, *2*(4). Available from https://doi.org/10.21926/rpm.2004022.

Rahul, A. V., & Santhanam, Manu (2020). Evaluating the printability of concretes containing lightweight coarse aggregates. *Cement and Concrete Composites*, *109*, 103570. Available from https://doi.org/10.1016/j.cemconcomp.2020.103570.

Ramli, M., & Akhavan Tabassi, A. (2012). Effects of polymer modification on the permeability of cement mortars under different curing conditions: A correlational study that includes pore distributions, water absorption and compressive strength. *Construction and Building Materials*, *28*(1), 561–570. Available from https://doi.org/10.1016/j.conbuildmat.2011.09.004.

Robeyst, N., Gruyaert, E., Grosse, C. U., & De Belie, N. (2008). Monitoring the setting of concrete containing blast-furnace slag by measuring the ultrasonic p-wave velocity. *Cement and Concrete Research*, *38*(10), 1169–1176. Available from https://doi.org/10.1016/j.cemconres.2008.04.006.

Rongbing, B., & Jian, S. (2005). Synthesis and evaluation of shrinkage-reducing admixture for cementitious materials. *Cement and Concrete Research*, *35*(3), 445–448. Available from https://doi.org/10.1016/j.cemconres.2004.07.009.

Sandeep, U., & Rao, T. (2017). A review on 3D printing of concrete-the future of sustainable construction. *i-Manager's Journal on Civil Engineering.*, *7*, 49–62.

Sanjayan, J. G., Nematollahi, B., Xia, M., & Marchment, T. (2018). Effect of surface moisture on inter-layer strength of 3D printed concrete. *Construction and Building Materials*, *172*, 468–475. Available from https://doi.org/10.1016/j.conbuildmat.2018.03.232.

Seifan, M., Samani, A. K., & Berenjian, A. (2016). Bioconcrete: next generation of self-healing concrete. *Applied Microbiology and Biotechnology*, *100*(6), 2591–2602. Available from https://doi.org/10.1007/s00253-016-7316-z.

Shafiq, N., & Cabrera, J. G. (2004). Effects of initial curing condition on the fluid transport properties in OPC and fly ash blended cement concrete. *Cement and Concrete Composites*, *26*(4), 381–387. Available from https://doi.org/10.1016/S0958-9465(03)00033-7.

Shh, S. P., & Sarigaphuti, M. E. K. (1992). Effects of shrinkage-reducing admixtures on restrained shrinkage cracking of concrete. *ACI Materials Journal*, *89*(3), 291–295.

Soltan, D. G., & Li, V. C. (2018). A self-reinforced cementitious composite for building-scale 3D printing. *Cement and Concrete Composites*, *90*, 1–13. Available from https://doi.org/10.1016/j.cemconcomp.2018.03.017.

Tasdemir, C. (2003). Combined effects of mineral admixtures and curing conditions on the sorptivity coefficient of concrete. *Cement and Concrete Research*, *33*(10), 1637–1642. Available from https://doi.org/10.1016/S0008-8846(03)00112-1.

Tassew, S. T., & Lubell, A. S. (2014). Mechanical properties of glass fiber reinforced ceramic concrete. *Construction and Building Materials*, *51*, 215–224. Available from https://doi.org/10.1016/j.conbuildmat.2013.10.046.

Tay, Y. W. D., Panda, B., Paul, S. C., Noor Mohamed, N. A., Tan, M. J., & Leong, K. F. (2017). 3D printing trends in building and construction industry: A review. *Virtual and Physical Prototyping*, *12*(3), 261–276. Available from https://doi.org/10.1080/17452759.2017.1326724.

Wangler, T., Lloret, E., Reiter, L., Hack, N., Gramazio, F., Kohler, M., Bernhard, M., Dillenburger, B., Buchli, J., Roussel, N., & Flatt, R. (2016). Digital concrete: Opportunities and challenges. *RILEM Technical Letters*, *1*, 67–75. Available from https://doi.org/10.21809/rilemtechlett.2016.16.

Weng, Y., Li, M., Ruan, S., Wong, T. N., Tan, M. J., Ow Yeong, K. L., & Qian, S. (2020). Comparative economic, environmental and productivity assessment of a concrete bathroom unit fabricated through 3D printing and a precast approach. *Journal of Cleaner Production*, *261*. Available from https://doi.org/10.1016/j.jclepro.2020.121245.

Wolfs, R. J. M., Bos, F. P., Van Strien, E. C. F., & Salet, T. A. M. (2018). A real-time height measurement and feedback system for 3d concrete printing. In D. Hordijk, & M. Luković (Eds.), *High tech concrete: Where technology and engineering meet*. Springer. Available from https://doi.org/10.1007/978-3-319-59471-2_282.

Zareiyan, B., & Khoshnevis, B. (2017). Interlayer adhesion and strength of structures in Contour Crafting – effects of aggregate size, extrusion rate, and layer thickness. *Automation in Construction*, *81*, 112–121. Available from https://doi.org/10.1016/j.autcon.2017.06.013.

Zhang, G., Li, G., & Li, Y. (2016). Effects of superplasticizers and retarders on the fluidity and strength of sulphoaluminate cement. *Construction and Building Materials*, *126*, 44–54. Available from https://doi.org/10.1016/j.conbuildmat.2016.09.019.

Zhang, J., Gong, C., Guo, Z., & Zhang, M. (2009). Engineered cementitious composite with characteristic of low drying shrinkage. *Cement and Concrete Research*, *39*(4), 303–312. Available from https://doi.org/10.1016/j.cemconres.2008.11.012.

Zhang, S. P., & Zong, L. (2014). Evaluation of relationship between water absorption and durability of concrete materials. *Advances in Materials Science and Engineering*, *2014*. Available from https://doi.org/10.1155/2014/650373.

Zhu, Y., Wen, C., Xu, G., Liu, D., & Chen, J. (2018). The preparation and performance of the cement-based concrete 3d printing materials. *Materials Science Forum*, *932*, 131–135.

Zijl, & Visser C.R. (2007). *Mechanical characteristics of extruded SHCC*. Available from http://hdl.handle.net/10019.1/44048.

Emerging resources for the development of low-carbon cementitious composites for 3D printing applications

Seyed Hamidreza Ghaffar[1], *Yazeed Al-Noaimat*[2], *Mehdi Chougan*[2] *and Mazen Al-Kheetan*[3]
[1]Department of Civil Engineering, University of Birmingham, Dubai International Academic City, Dubai, United Arab Emirates, [2]Department of Civil and Environmental Engineering, Brunel University, London, Uxbridge, Middlesex, United Kingdom, [3]Civil and Environmental Engineering Department, College of Engineering, Mutah University, Mutah, Jordan

13.1 Introduction

Concrete is known as one of the most predominant used materials on earth. As the most critical and energy-intensive component of concrete, the traditional cement production method involves heating ordinary clay and limestone materials at 1450°C (Danish & Ozbakkaloglu, 2022). In addition, fuel combustion generates an immense quantity of energy throughout the thermochemical process of calcining limestone. A ton of cement is estimated to release between 800 and 900 kg of carbon dioxide, which accounts for about 9% of total anthropogenic carbon dioxide and 2%–3% of the total energy consumed worldwide (Jim et al., 2018). Furthermore, given the exponential growth in cement production over the last 30 years, investigations have anticipated a 50% increase in annual cement production by 2050 (Danish & Ozbakkaloglu, 2022). The existing trends of energy consumption and emission factor will lead to an extra 85–105 Gt of equivalent carbon dioxide emissions and 420–505 Tj demand for energy. The existing pattern results in an additional increase of 420–505 Tj and 85–105 Gt in energy demand and carbon dioxide equivalent emissions, respectively (Monteiro et al., 2017). Research into the sustainable development of low-carbon cementitious composites is vital as the building sector and its associated processes generate 35% of worldwide CO_2 emissions and up to 65% of waste materials disposed of in landfills. However, this sector contributes significantly to addressing society's needs by improving the quality of people's living standards (Ahmad, 2021). On the other hand, customers are also looking for socially responsible and environmentally friendly building companies. Consequently, the building sector is becoming increasingly conscious that

Sustainable Concrete Materials and Structures. DOI: https://doi.org/10.1016/B978-0-443-15672-4.00013-9

establishing a competitive edge implies more than just customer satisfaction based on affordable prices and superior quality of customer service. Therefore it is essential to investigate in-depth the sustainability of construction processes and materials (Alyousef, 2021).

The construction industry is a prominent contributor to the depletion of natural aggregates such as sand and gravel. This is due to extensive mining and excavation activities, which cause soil and riverbed erosion, alter river courses, diminish subsurface aquifers, and disrupt ecosystems (Valente, 2022). The European Environmental Agency (2020) has set crucial objectives to decrease the detrimental impact of waste disposal and preserve natural resources in the construction industry. Their target is to slash up to 60% of greenhouse emissions by 2050 and promote the utilization of recycled or end-of-life materials, as well as industrial by-products, as secondary raw materials to protect natural resources.

The construction industry plays a vital role in the global economy, contributing almost 13% to the world's gross domestic product. A report by the World Economic Forum states that over 100 million people are employed in the construction industry worldwide (Forum, 2018). The construction industry, similar to the manufacturing sector, strives to improve production efficiency while maintaining quality and cost (Tafazzoli, 2022). Yet, controlling construction costs is challenging due to rising labor costs primarily due to the lack of automation and adoption of novel technologies, which leads to excess labor and reduced production efficiency (García de Soto, 2018). The construction industry has embraced digital technologies with the advent of the fourth industrial revolution, also referred to as "Industry 4.0." The latest tools and techniques, such as the Internet of Things, artificial intelligence, building information modeling, modular integrated construction, smart production, and additive manufacturing (AM)—also known as three-dimensional (3D) printing—are being used to design, construct, and maintain buildings (Maskuriy, 2019; Raza & Zhong, 2022). Over the past few years, 3D printing has gained widespread adoption due to its ability to reduce construction time, eliminate the need for highly skilled labor, enable complex structural design, and ease the burden of physically demanding work (Furet et al., 2019). Over the past 10 years, the construction sector has increasingly employed AM processes. The method used depends on the working principles of the printing process. The two main types of construction-based additive manufacturing are powder-based 3D printing and extrusion-based printing, similar to the widely-used fused deposition modeling process (Sanjayan & Nematollahi, 2019). The use of ordinary Portland cement (OPC) in AM techniques is limited due to sustainability challenges, as the production process requires high energy demands, leading to intensive emissions. Recent investigations have addressed this issue by exploring alternative low-carbon construction materials in AM technology to mitigate environmental impact. These materials include limestone-calcined clay cement (LC3) (Al-Noaimat & Hamidreza, 2023) and alkali-activated materials (AAMs)/geopolymers (Chougan, 2020, 2021, 2022; El-seidy, 2022, 2023; Valente, 2022), which have been proven to emit significantly lower levels of carbon during production. Extensive research and developments have taken place in AAM/geopolymers over the past decade. The primary focus has been on discovering low-carbon binders and integrating recycled or end-of-life aggregates to enhance the sustainability of printing materials (Chougan, 2022). However, research on the 3D printing of LC3 materials is limited to the optimization of binder materials, and more research is required to understand the feasibility of this category of materials as a feedstock for AM technology. Utilizing geopolymer 3D printing technology offers triple-dimensional sustainability benefits. This revolutionary approach not only diminishes the need for formwork but also significantly decreases waste generation during the printing process. Additionally, it eliminates the use of OPC, which plays a crucial role in reducing carbon footprints. Lastly, it substitutes natural sand with recycled aggregates, a significant step toward preserving natural resources (Yao, 2020). This chapter will thoroughly summarize the resources available for producing low-carbon cementitious composites tailored explicitly for 3D printing applications.

13.2 Available low-carbon binders

Various techniques and strategies have been explored in the creation of low-carbon concrete, including the partial and complete substitution of OPC. These binders could reduce carbon emissions while maintaining the durability and strength of the concrete. Some of those binder systems will be briefly discussed in this section.

13.2.1 Alkali-activated materials

Joseph Davidovits revolutionized the cement industry in 1979 when he introduced the concept of geopolymer, a highly sustainable and environmentally friendly material that emits significantly lower levels of carbon dioxide than traditional cement (Turner & Collins, 2013). Geopolymer, also referred to as AAM, is an inorganic polymer created through a two-stage reaction between alkaline activators and aluminosilicate source materials, namely dissolution and polycondensation. The source material and the alkaline activator are the two fundamental building blocks in the synthetic process. The precursor should have high amorphous alumina and silica concentration (Elmesalami & Celik, 2022). AAM hardening takes place through the geopolymerization reaction after combining alumina or silica-rich materials with an alkaline activator (Nodehi & Taghvaee, 2022a), in which activators break the Si−O−Si and Al−O−Al bonds in the precursor by providing alkali cations to form the strength-giving binding phases (Elzeadani et al., 2022; Provis & Van Deventer, 2013). The alkaline activators can be used in either a liquid form (known as two-part geopolymer) or a solid form (known as one-part geopolymer or just add water), with the two-part system being the widely adopted system for 3D printing applications.

3D printable AAMs are known as a sustainable substitute for OPC printable mixtures. To evaluate the sustainability of these mixtures, two key parameters, embodied carbon, and embodied energy, should be taken into account (Bhattacherjee, 2021). Fig. 13.1 displays the sustainability assessment of various

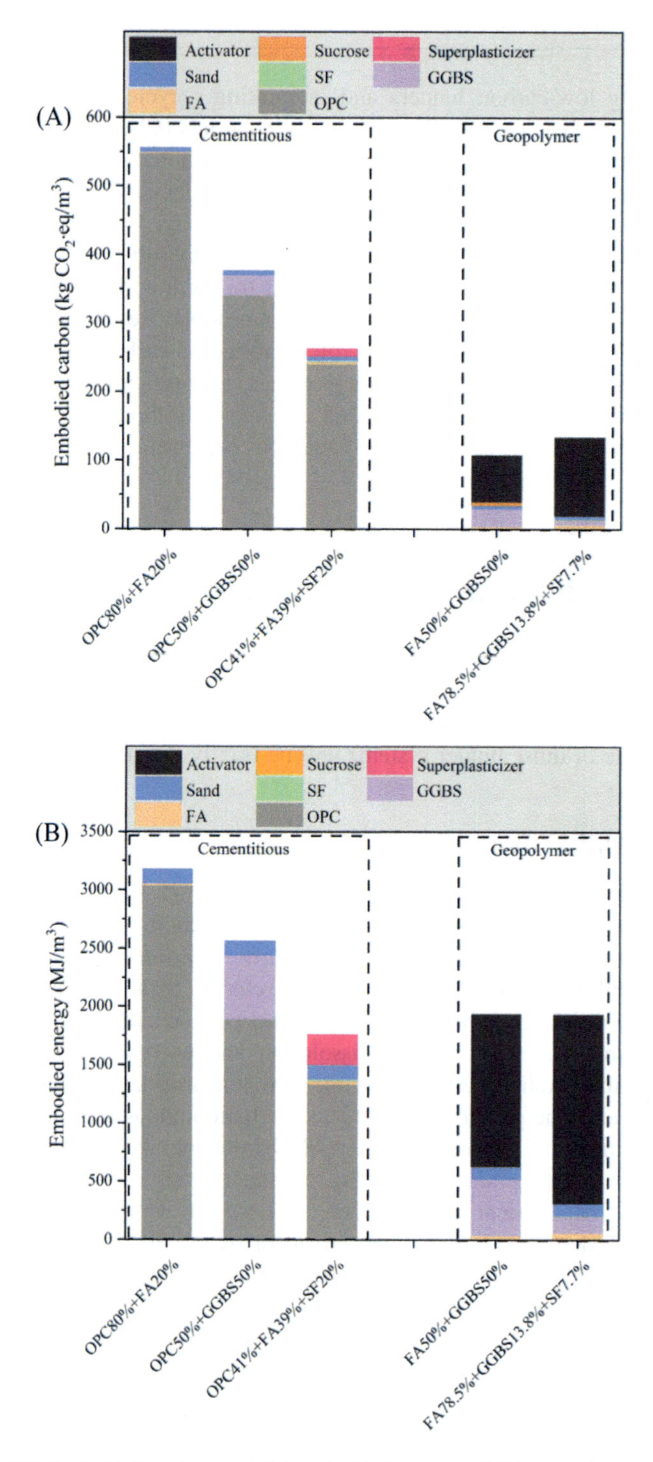

Figure 13.1 (A) Embodied carbon and (B) embodied energy of 3D printed cement–based versus alkali-activated materials (Zhong & Zhang, 2022).
Source: From https://doi.org/10.1016/j.cemconcomp.2022.104455.

3D printable mixtures that possess comparable initial static yield stress and a similar range of compressive strength after 28 days (Zhong & Zhang, 2022). The results indicated that the 3D printable OPC-based mixture (80% OPC and 20% fly ash [FA]) exhibited the highest embodied carbon and embodied energy of around 555.51 kg CO_2 eq/m^3 and 3173.6 MJ/m^3, respectively (Rahul et al., 2020), which is much higher than that of the embodied carbon ($\sim$107.48 CO_2 eq/m^3) and embodied energy ($\sim$2000 MJ/m^3) registered for 3D printable AAM (50% FA + 50% granulated blast-furnace slag [GGBS]) mixture (Bong, 2021).

13.2.1.1 Two-part alkali-activated materials

Two-part AAM (conventional AAM) is produced by dry mixing the solid ingredients of the mixture (i.e., binder and aggregate) followed by adding a liquid alkaline activator. FA, metallurgical slag, and metakaolin are highly prevalent precursors that are extensively employed in the production of two-part geopolymers. FA and slag are by-products generated from various industries, such as power plants and metal manufacturing, while metakaolin is created by the kaolinite clay calcination process at elevated temperatures. Throughout the last few years, an array of emerging feasible binders have been identified. These include red mud, rice husk ash, palm oil fuel ash, mineral processing tailings, coal bottom ash, recycled ceramic, incineration products of sludge, waste glass, laterite soil, and biomass ash (Athira, 2021; Elizabeth et al., 2024; Kaze, 2021; Mounika, 2023; Provis et al., 2015).

To produce two-part geopolymers, alkali hydroxides and alkali metal silicate solutions (i.e., sodium and potassium) are adopted as activators (Provis & Bernal, 2014). Alternatively, alkali metal carbonates or sulfates can also be employed to activate the precursor materials. Several alternative activators, including alkali carbonates, hydroxides, and aluminates derived from biomass and industrial residues, have also been introduced to activate the precursors to produce two-part geopolymers. Alkali silicates derived from the noncrystalline form of silicon dioxide also hold immense potential to be employed for this purpose (Adesanya, 2021). It is essential to mention that the efficacy of an activator solution is primarily determined by the chemical composition and reactivity of the precursor materials. It is widely acknowledged that waste-based activators exhibit significantly lower CO_2 emissions compared to commercially available ones. Yet, the global implementation of such activators could pose certain challenges due to potential limitations in the supply chain associated with obtaining the required raw materials. For instance, according to the European Glass Container Federation, Europe recycled more than 75% of its waste glasses in 2017 (Nair, 2017), indicating a potential scarcity of silica sources to produce alternative sodium silicate. However, it is crucial to note that Europe has several other silica sources, such as silica fume, diatomaceous earth, chlorosilane, and other amorphous silica materials, which are equally suitable for this purpose (Adesanya, 2021).

Numerous studies have demonstrated that using two-part AAM offers remarkable performance for a range of applications. Specifically, for 3D printing applications, it is crucial to opt for two-part AAMs due to their exceptional mechanical performance, rheological characteristics, and rapid setting time, making them suitable candidates (Alghamdi et al., 2019; Chougan, 2020). The mixtures used in AAM 3D printing are similar to the ones used in conventionally cast samples. It is imperative to note that employing a combination of NaOH and Na_2SiO_3 is an effective measure toward achieving success with the AAM additive manufacturing process (Chougan, 2021; Zhang, 2018). Research findings have verified that an activator modulus (SiO_2/Na_2O or K_2O) ranging between 0.5 and 2.0 is required for optimal effectiveness. However, it is recommended that 3D printable mixtures should only consist of fine aggregates that are less than 2 mm in size since coarse aggregates can cause complications when passing through the hoses and nozzles of most 3D printing systems (Mechtcherine, 2019). For optimal 3D printing, it is recommended to maintain a specific aggregate/binder ratio in 3D printable mixtures, ranging from 1.2 to 1.9 (Panda & Tan, 2018; Zhong & Zhang, 2022).

AAM 3D printing involves carefully considering various properties to ensure high-quality feedstock is prepared for 3D printing applications. These properties include fresh properties (i.e., setting time, extrusion open time, flowability, rheology, and thixotropy), printing quality (i.e., shape retention, printability, and buildability), bulk properties (i.e., density, porosity, drying shrinkage), and hardened properties (i.e., compressive, flexural, tensile, and interlayer bond strengths) (Zhong & Zhang, 2022). Various types of additives, including Halloysite clay minerals (Chougan, 2022), graphene-based materials (Chougan, 2020), and attapulgite nanoclay (Chougan, 2021), have been proposed to tailor the aforementioned properties of 3D printed AAMs. Moreover, to address the challenges associated with employing traditional steel rebar reinforcement in extrusion-based 3D printing, various fiber types, including polypropylene (Nematollahi, 2018), glass (Panda et al., 2017), polyvinyl alcohol (Chougan, 2021), and steel fibers (Chen et al., 2023), have been suggested for inclusion in 3D printable AAM mixtures.

Choosing alternative low-carbon construction materials relies on various factors such as sustainability, mechanical performance, and cost. It has been proven that AAMs meet the first two requirements and deliver satisfactory results in sustainability and mechanical properties. However, the cost of using AAMs is also a crucial consideration when selecting materials. A study found that AAMs cost more than twice as much as OPC-based concrete (Fernando, 2021). This is due to the high cost of alkali activators, which make up 80% of the material cost of AAMs (Abdollahnejad, 2020). In addition, to achieve the desired fresh properties of 3D printed mixtures, it is necessary to include chemical admixtures such as minerals, retarders, viscosity modifiers, and superplasticizers, which makes the cost of 3D printable mixtures considerably higher than conventional mixtures (Raza & Zhong, 2022; Saedi et al., 2021). Nevertheless, the total cost of cementitious materials' 3D printing method has been reported to be comparatively lower than conventional methods due to the elimination of formwork and workforce costs (Han, 2021).

13.2.1.2 One-part alkali-activated materials

Although two-part AAMs have shown great potential to be utilized in the construction industry, the transporting operation of hazardous activator solutions limited its application. In this regard, one-part AAM was developed to overcome the handling challenges of the viscous and hazardous liquid alkaline activator on-site for large-scale applications (Al-Noaimat & Hamidreza, 2023; van Deventer, 2007). One-part AAM, also known as "just add water," is produced by adding water to a dry mix that includes all the solid ingredients, such as precursor, aggregates, and powder activators. Up-to-date, the most used solid activators are grade sodium silicate and anhydrous sodium silicate (Zhong & Zhang, 2022). Over the past few years, there have been numerous studies conducted to explore the effectiveness and efficiency of utilizing one-part geopolymer for 3D printing applications, and the findings have shown promising results from several perspectives, which will be further discussed in this section.

Similar to two-part AAM, the properties of the resultant one-part mixtures are mainly influenced by the activator content and materials used in the precursor, which can have a major impact on the workability, setting time, and mechanical properties of the mix. For instance, increasing the activator content resulted in the rapid evolution of yield stress, which increased the required pumping energy and limited the setting and open time of the mixtures, and decreased the flowability. Nevertheless, it is beneficial for the shape retention of the mixture after extrusion due to the rapid reflocculation of the mix which enables more layers to be printed in the buildability test (Muthukrishnan et al., 2021). Similarly, increasing GGBS content in the precursor decreased the flowability of the mixture. This may be because the GGBS contains more calcium that creates more nucleation sites at the early stage, leading to the mixtures' hardening rapidly. This, in turn, improved the compression and flexural strength performance of the mixtures. However, the setting time was shortened due to the rapid hardening of the mixture (Shah, 2020). Particle morphology of the precursor materials also plays a role in affecting the properties of the mixtures. For example, it was reported that the angular shape of GGBS particles plays a role in decreasing the workability of the mixture compared to the spherical shape to that of FA (Yousefi Oderji, 2019).

One of the main advantages of AAM is its ability to synthesize any aluminosilicate sources to be used as a binder, which could further increase the environmental benefits it provides by decreasing the amount of waste being landfilled through value-adding to be used as a precursor (Wong, 2020). Hence, some researchers' interest has been shifted to employ waste generated at the end of construction life, such as brick, in one-part AAM and 3D printing applications (Al-Noaimat & Ghaffar, 2023a, 2023b). However, these materials possess less reactivity and geopolymerization than FA used in the precursor (Rovnaník, 2018). Promising results were obtained when brick powder partially replaced FA by 30% in FA- and GGBS-based one-part geopolymer, possessing similar compressive strength at 7 and 28 days and exhibiting 20 layers in the buildability test without showing any printing issues, as shown in Fig. 13.2 (Pasupathy et al., 2023). This indicates the feasibility of reusing waste brick powder and promoting waste recycling in AAMs for 3D printing applications.

Although one-part AAM does not exhibit similar strength performance to two-part AAM, it is still higher than OPC (Al-Noaimat & Hamidreza, 2023).

Figure 13.2 Buildability test of one-part geopolymer containing 30% brick powder (Pasupathy et al., 2023).
Source: From https://doi.org/10.1016/j.cemconcomp.2023.104943.

Nevertheless, one of the main challenges that face one-part AAM is the rapid strength development when the water is added to the dry mixture, which evolves the mixture temperature, leading to faster hardening and setting time (Al-Noaimat & Hamidreza, 2023). Many authors have considered the addition of retards and superplasticizers to tackle this problem. From an environmental perspective, one-part AAM is better than both two-part AAM and OPC systems, in which it was recorded that the production of a 3D printable one-part geopolymer mixture can reduce carbon emissions by up to 70% and embodied energy by 15% when compared to 3D printable OPC with a similar compressive strength value (Paul & Tay, 2018). The production of alkali activators is a key contributor to the environmental impact of geopolymer production (Petrillo, 2016; Turner & Collins, 2013). Moreover, it was reported that one-part AAM exhibits lower environmental impact than two-part AAM (Luukkonen, 2018) that is due to the elimination of the use of two types of activators in two-part AAM into one type of solid powder activator in one-part AAM, which resulted in the production of a more environmentally friendly mix. In addition to its environmental benefits, it was found that one-part AAM could reduce the cost by 80% compared to OPC (Habert & Ouellet-Plamondon, 2016). It also allows for further reduction in cost compared to conventional (two-part) AAM. For instance, it was found that two-part AAM had the highest cost, followed by Portland cement concrete and one-part AAM to be the cheapest. This is due to the lower price of the powder activator compared to the liquid activator by four times (Vinai & Soutsos, 2019). However, it should be noted that the production cost of AAM is significantly influenced by the materials used in the precursor.

13.2.2 Limestone-calcined clay cement

The partial replacement of OPC with supplementary cementitious materials (SCMs) has received significant attention from researchers over the past few decades. However, common SCMs share the problem of limited availability, leading researchers, and industry to seek alternatives (Chen et al., 2020; Du & Pang, 2020). The suitable chemical composition, high availability, and reactivity of calcined clay and limestone have shown the potential to be alternative SCMs (Al-Noaimat & Chougan, 2023). The combination of limestone-calcined clay attracted the researchers' focus in the past few years due to its potential to replace a high amount of OPC (more than 45%) (Al-Noaimat & Chougan, 2023), produce mixtures with comparable strength performance to traditional cement after 7 days, and improve resistance to chemical attacks (Antoni, 2012; Avet & Scrivener, 2018; Avet et al., 2019). However, its implementation in 3D printing applications is still new, and research on it is limited.

There are different classifications of clay depending on the arrangement of the octahedral (O) and tetrahedra (T) sheets and their ratio in clay structure, which can be divided into 1:1-type clay, such as kaolinite, and 2:1-type clay, such as montmorillonite and illite, which are the most widely known clay minerals. Most studies have utilized kaolinite because of its superior pozzolanic reactivity compared to other clay minerals, as well as the simpler removal of water molecules in kaolinitic clays due to its simple structure formation (Al-Noaimat & Akis, 2023; Fernandez et al., 2011). However, clays should be activated prior to being used to partially replace cement, which can be done in several ways, including thermally (Samet, 2007), mechanically (Vizcayno, 2010), or chemically activating (Amin, 2012). One of the most commonly employed methods for clay processing is thermal activation. This involves subjecting the clay to high temperatures ranging from 600°C to 900°C for a sufficient duration in order to eliminate water molecules from its structure (Fernandez et al., 2011; Muzenda, 2020). Depending on the type of clay, the optimum calcination temperature differs. However, increasing the calcination temperature above 900°C could decrease the reactivity for most of the calcined clays due to sintering and recrystallization phenomena (Alujas, 2015).

Some recent studies have been conducted to develop a 3D printable LC3 and investigate the properties of the 3D printed mixture in order to reduce reliance on the OPC system. These studies investigated the effect of different replacement levels, additive incorporation, different mix ratios, and the addition of other SCMs to the mix on the printability and other properties of the mixture. For example, Long (2021) investigated the effect of replacing 40% and 50% of OPC with limestone-calcined clay and found that it increased the static yield stress of the resultant mixture by 7 and 15 times, respectively. However, increasing replacement levels resulted in difficulties in extrusion and produced defects on the surface of the extruded layers and decreased the compressive strength performance of the resultant mixture. Similarly, Chen et al. (2021) investigated the impact of replacing up to 90% of OPC with a limestone-calcined clay combination and reported a workability loss, and the extrudability window decreased with increasing the substitution ratio.

It was also reported that increasing the replacement level significantly decreased the compressive strength. The decrement in compressive strength is due to the fewer hydration products formed due to less OPC in the mix, in which the combination of limestone-calcined clay could compensate for the high replacement of OPC (Zhou, 2022). Although increasing the replacement level deteriorates the strength performance of the mixture, it enhances the buildability and shape retention of the mixtures (Chen et al., 2021; Long, 2021).

Other than the replacement level, the purity and grade of the used clay were reported to play a significant role in the printability and properties of the resultant mixtures (Chen et al., 2019, 2020). According to the results obtained by Chen et al. (2020), increasing metakaolin content in clay enhanced the shear yield stress and buildability of the LC3 mixture while decreasing the open time of the mixture (Fig. 13.3). Compressive strength of cast and different directions of 3D printed LC3 mixtures with different grades of calcined clay: low-grade calcined clay (LCC), medium-grade calcined clay (MCC); and high-grade calcined clay (HCC). Chen et al. (2020) presented the impact of different calcined clay grades on the compressive strength of the cast and 3D printed LC3 mixtures. While the cast LC3 exhibited higher compressive strength performance when using calcined clay with higher kaolinite content (high-grade clay), 3D printed LC3 mixtures had the opposite trend and exhibited higher strength performance when using MCC (around 70% kaolinite content) (Chen et al., 2020). That is because using HCC in 3D printing could

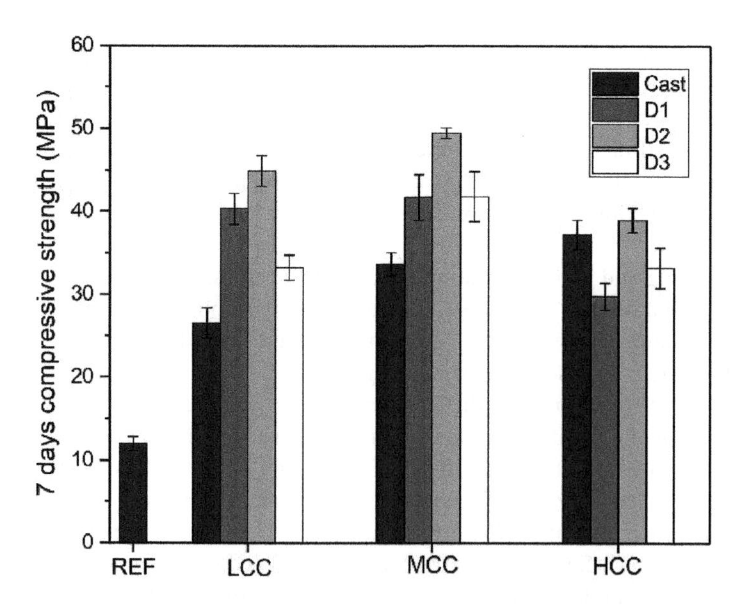

Figure 13.3 Compressive strength of the cast and different directions of 3D printed LC3 mixtures with different grades of calcined clay. *LCC*, low grade calcined clay; *MCC*, medium-grade calcined clay; *HCC*, high-grade calcined clay (Chen, 2020).
Source: From https://doi.org/10.1016/j.cemconcomp.2020.103708.

increase the thixotropy that could allow the formation of cold joints and weaker interfaces (Roussel, 2018; Wangler, 2016).

LC3 mixtures had lower porosity with finer microstructure than OPC, which is induced by the additional products formed by the pozzolanic reaction (Antoni, 2012; Skibsted & Snellings, 2019). The filler effect of limestone, due to its high fineness, plays a role in refining the pore microstructure and reduces the porosity of the mixture (Environment, 2018). The grade of calcined clay also plays a role in lowering the porosity of the mixture at early ages. While at later ages, even LCC (less than 50% kaolinite content) significantly refines the pore microstructure (Avet & Scrivener, 2018).

Calcined clay is an environmentally friendly material that needs lower energy and releases fewer CO_2 emissions than OPC. This helps decrease the environmental impact when replacing OPC to produce blended cement. The clay calcination process is reported to require around 60% of the energy needed to produce OPC (Gettu, 2019) and releases around 30% CO_2 of that from OPC production. Utilizing LC3 in the casting method could be responsible for lowering CO_2 emissions from 930 to 610 kgCO_2/ton and energy consumption from 5945 to 4850 MJ/ton of cement (Sánchez-Berriel, 2016).

13.3 Recycled aggregate to replace natural sand

13.3.1 Construction and demolition waste

The construction industry is one of the largest consumers of natural resources, including OPC and natural sand, and its activities generate enormous waste. Every year around 5–8 million tons from South Africa (Bester et al., 2017) and 820 million tons from the European Union of construction and demolition waste (CDW) are generated (Styles et al., 2018). The U.S. Environmental Protection Agency defines CDW as waste typically produced during the renovation and demolition of existing structures or new construction projects (United States Environmental Protection Agency, 2019). Approximately 45% of the waste materials consist of brick and ceramic waste (Fatta, 2003; Oikonomou, 2005; Reig, 2013). Using CDW as a substitute for natural aggregates has become an increasingly popular solution to reduce the construction industry's carbon footprint. This approach not only minimizes landfill waste but also conserves natural resources and reduces the overall environmental impact of construction activities, which gives the potential to contribute to sustainable development in the construction industry significantly. The yearly generated CDW accounts for half of the solid waste, threatening the environment since less than 5% is recycled and reused while the rest is landfilled (Bai, 2021).

The use of recycled concrete aggregate (RCA) has been investigated and has some limitations that can impact the performance of concrete compared to natural aggregate. RCA consists of two parts, which are the natural sand and the cementitious paste attached to their surface. Hence, in recycled aggregate concrete (RAC), two interfacial transition zones (ITZs) are present. The first one is the old ITZ between natural aggregate and the old cementitious mortar, while the second one is the new ITZ between recycled aggregate and the new mortar, as shown in Fig. 13.4, which presents an illustration of RCA in concrete containing RCA. That makes RAC has a larger volume of ITZ compared to natural aggregate concrete (NAC) which has only one new ITZ (Wang, 2021). Therefore undesirable properties may arise, such as increased crushability, water absorption, dust content, reduced abrasion resistance, and lower specific gravity (Mohammed et al., 2021). This, in turn, could result in a noticeable decrease in mechanical strength and increase dry shrinkage (Loukili, 2018). Thus RCA partially replaced natural sand in studies to limit the deterioration impact on the mixture properties (Loukili, 2018). For instance, Ding (2020) investigated the impact of replacing up to 50% of natural sand with recycled sand obtained from processing waste concrete on the properties of 3D printed concrete. The compressive and flexural strength of 3D printed mixtures containing recycled sand showed slightly lower performance than the mixture with natural sand, which is associated with the unhydrated paste on the surface of the recycled sand particles and their porous structure. Moreover, compressive, splitting, and flexural strength results showed anisotropic behaviour, as shown in Fig. 13.5.

Another CDW product that could be recycled and used as aggregate is waste brick. Numerous studies have examined the use of recycled brick aggregate (RBA) in concrete, but it is difficult to predict the impact on concrete strength and water absorption properties from replacing natural aggregates with RBA. This is due to the variation of RBA from different origins and different countries (Christen et al., 2022). Although incorporating RBA increases the mixtures' transport properties (Dang, 2018; Paul & Babafemi, 2018), it has been found to improve thermal and acoustic insulation (Dang, 2018) and reduce drying shrinkage due to the high water absorption of brick aggregate (BA). As the binder loses water during hydration and

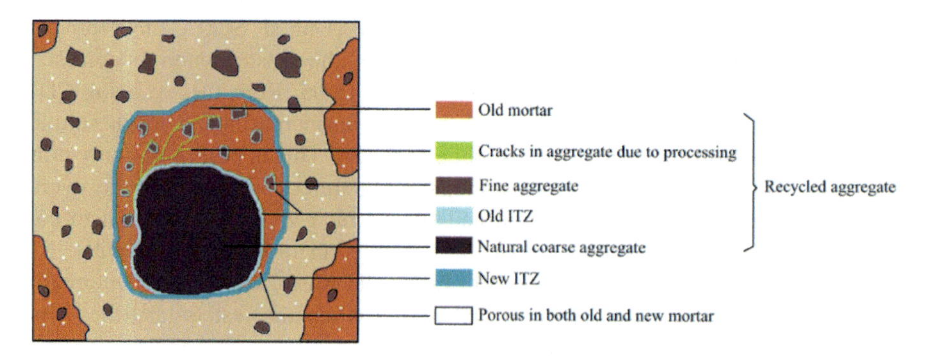

Figure 13.4 Illustration of interfacial transition zones and principal components of recycled aggregate concrete(Wang, 2021).
Source: From https://doi.org/10.1016/j.resconrec.2021.105565.

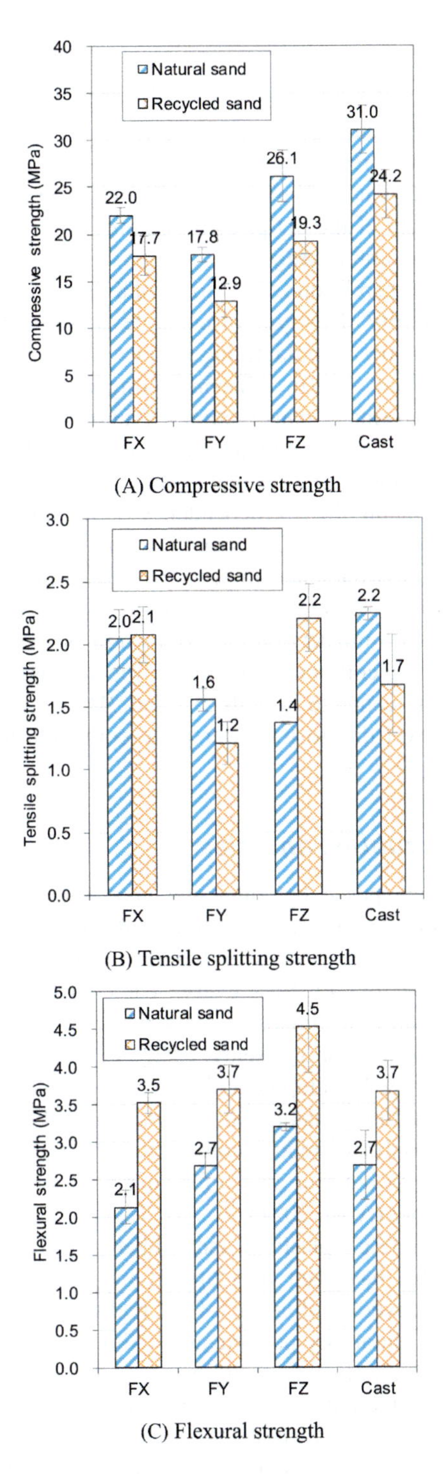

Figure 13.5 Anisotropic mechanical behavior of 3D printed concrete: (A) compressive strength; (B) tensile strength; and (C) flexural strength (Ding, 2020).
Source: From https://doi.org/10.1016/j.cemconcomp.2020.103724.

moisture exchange with the environment, the BA will release water to the binder matrix, providing internal curing (Wong, 2018; Zhang, 2018). Christen et al. (2022) investigated the impact of replacing 64% of natural sand using two different types of RBA separately and their combination on the properties of 3D printed concrete. The compressive strength of the 3D printed mixtures with RBA was lower by 25% than the mixture with natural aggregate.

Applications and research on concrete containing recycled aggregates have reached a level of maturity. This type of concrete is commonly utilized in various industries, such as floor tile manufacturing (López Gayarre, 2017), structural component production (Pranav, 2020), self-compacting concrete preparation (Revilla-Cuesta, 2020), and pavement construction (Maduabuchukwu Nwakaire, 2020). However, there is limited up-to-date research on utilizing those wastes in 3D printing applications. Their incorporation and reuse in 3D printing technology could bring the benefit of reducing the overdependency on natural resources and lowering the environmental impact. Han (2021) found that increasing the recycled aggregate content decreased the environmental impacts, as shown in Fig. 13.6. However, the transport distance for recycled aggregate was short in their case study, and it could be the main for decreased impacts on the production of the materials according to the authors. Moreover, incorporating CDW to replace aggregate also allows for reducing the cost of the concrete in both cast and 3D printing methods by increasing the replacement level due to its lower price compared to the natural sand, as shown in Fig. 13.7.

13.3.2 Polymers, rubber, and glass

The mounting challenge posed by municipal and industrial waste pollution and landfill overflow demands immediate action and innovation toward sustainable practices. The improper disposal of over 1 billion used tyres, 353 million tons of plastic, and 130 million tons of glass every year contributes significantly to this environmental crisis—they decompose slowly while inflicting severe damage on nature through pollution release (El-seidy, 2022; Ferdous, 2021; Organisation for Economic Co-operation & Development, 2022). A possible remedy to address this issue could lie in substituting natural aggregate with solid waste in cementitious composites, which not only can reduce construction-related ecological harm (depletion of natural resources) but can also offer an innovative solution for managing waste disposal effectively (Alwi, 2021).

The use of polymer (e.g., plastics) and rubber (e.g., crumb rubber) waste in cementitious composites has been investigated by many researchers, with most results indicating the adverse influence of its replacement to aggregate on concrete's mechanical and durability performance (Ahmed & Rana, 2023; Xu, 2020). However, the incorporation of glass waste in cementitious composites showed varying results depending on the replaced component of concrete, where the use of glass waste as sand exhibited more favourable mechanical and durability characteristics than its incorporation as coarse aggregate (Ahmed & Rana, 2023; Al-awabdeh, 2022; Nodehi & Taghvaee, 2022b). This is associated with the high reactivity of

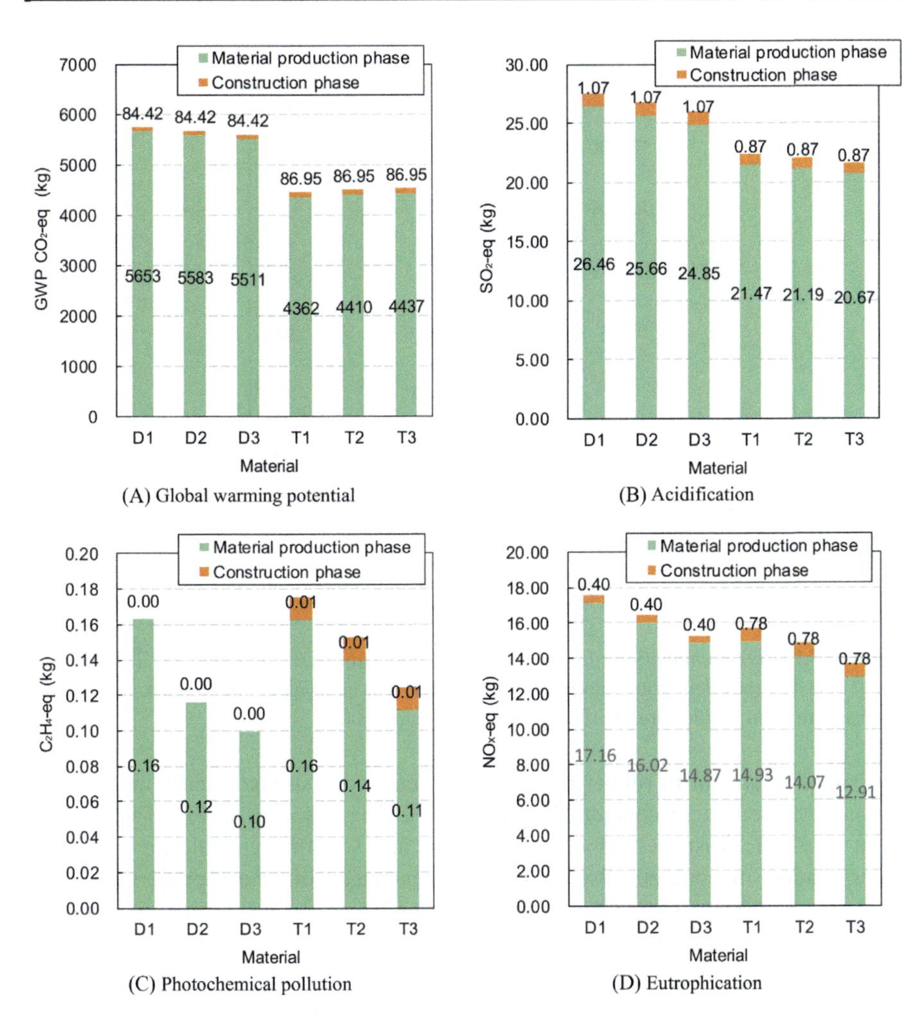

Figure 13.6 Environmental impact of 3D printed (D) and cast (T) concrete with natural aggregate (D1,T1), 50% recycled aggregate (D2,T2) and 100% recycled aggregate (D3,T3) on (A) global warming potential, (B) acidification, (C) photochemical pollution, and (D) eutrophication (Han, 2021).
Source: From https://doi.org/10.1016/j.jclepro.2020.123884.

silica with cement when using fine glass aggregate. To combat the negative impact of replacing aggregates with different solid waste materials, physical and chemical treatments of waste particles prior to their inclusion in cementitious composites were used. For instance, the pretreatment of rubber aggregate with sodium hydroxide (NaOH) proved its efficacy in increasing the compressive strength of rubberized concrete by 40% (Challa & Das, 2019) and enhancing the adhesion between the cementitious matrix and rubber particles leading to a reduced drying shrinkage (Si et al., 2017). Similar results were obtained when applying plasma polymerization

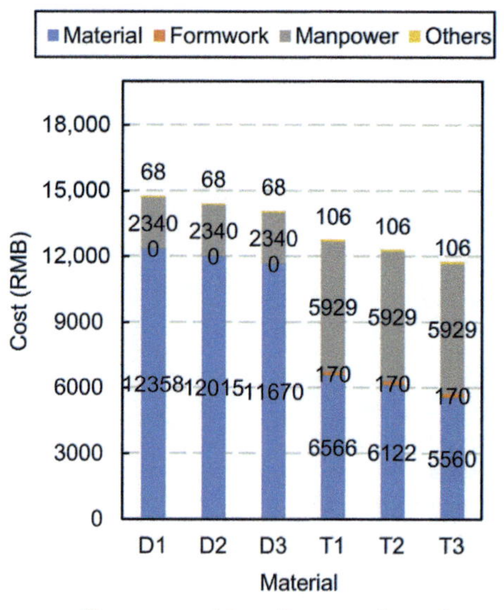

Figure 13.7 Cost composition of concrete with natural sand (D1,T1), 50% (D2,T2) and 100% (D3,T3) recycled aggregate in 3D printing (D) and cast (T) methods (Han, 2021). *Source*: From https://doi.org/10.1016/j.jclepro.2020.123884.

and silane treatment to rubber particles, with enhanced strength and long-term performance observed in concrete (Dong et al., 2013; Xiaowei, 2017). The pretreatment of plastic waste with sodium hydroxide and sodium hypochlorite prior to its incorporation in concrete also managed to enhance concrete's properties, which refers to the effect of such treatment processes in increasing the hydrophilicity of plastic particles and their bond to the cementitious matrix (Kaur & Pavia, 2021; Sharma & Pal, 2016).

Researchers took one step toward sustainability by integrating rubber, polymer, and glass waste particles within low-carbon cementitious composites to mitigate the negative environmental effect of ordinary concrete, teamed up with the solid waste problem. Replacing sand with rubber particles in AAMs significantly reduced greenhouse gas (GHG) emissions, whereas a total replacement of sand with rubber particles in AAM mixtures resulted in a drop of 40% in CO_2 emission (El-seidy, 2023; Valente, 2022; Valente & Sambucci, 2021). Moreover, replacing sand with polymer particles (e.g., polyvinyl chloride [PVC]) in AAM composites managed to reduce CO_2 emission by 93% (for a 50% replacement ratio by volume), whereas 45% reduction in CO_2 emission was achieved when replacing sand with 100% waste glass (El-seidy, 2023). Furthermore, those environmental gains are associated with enhancements in other properties like the acoustic insulation of composites, where LC3 mixtures incorporated with rubber particles have shown superior sound insulation properties (Tanaçan, 2022). The use of PVC aggregate in geopolymer was also beneficial in enhancing its thermal insulation properties and reducing its density to be used as a lightweight mixture (El-seidy, 2023; Kunthawatwong, 2023). The waste glass was found to be a feasible alternative to sand in AAM composites due to its ability to enhance AAMs' thermal insulation and energy absorption (El-seidy, 2023). However, the strength and durability of composites were significantly compromised due to the incorporation of those waste materials (Tanaçan, 2022; Valente & Sambucci, 2021). Accordingly, the treatment of rubber, glass, and polymer wastes with sulfuric acid, sodium hydroxide, calcium hydroxide, and other alkaline solutions prior to their integration into low-carbon cementitious composites exhibited promising results in terms of mechanical, physical, and durability properties (Saloni, 2021; Zhang, 2022). For instance, the pretreatment of rubber with NaOH improved its bonding with the geopolymer binder, reduced the water absorption of geopolymer, improved its compressive strength, and enhanced its resistance to acid attack (Saloni, 2021).

The incorporation of municipal solid waste in 3D printable low-carbon cementitious composites is a novel approach that has the ability to maximize the reduction of carbon emissions and minimize the consumption of virgin materials (Chougan, 2021). Despite the bright pathway of this approach, only one study to this date dealt with replacing natural sand with municipal solid waste in the 3D printing of low-carbon cementitious composites. Promising results were reported by the researchers indicating that rubberized AAMs exhibited higher flexural strength, enhanced thermal and acoustic insulation, and better mechanical isotropy than rubberized OPC used in 3D printing (Valente, 2023). However, this approach is still in its infancy, and more research is still needed to investigate the practicality of replacing sand with other municipal wastes like polymers and glass in 3D printed low-carbon cementitious composites.

13.4 Limitations and future directions

Despite the environmental gains of replacing sand with waste materials in the 3D printing of low-carbon cementitious composites, there are many challenges facing the development and adoption of this approach. Firstly, the chemical composition of most solid waste aggregates (e.g., polymers, rubber) differs from virgin aggregates, limiting the compatibility between the binder and aggregates and imposing difficulties in the buildability and extrudability of the final printed composite (Bai, 2021). Also, replacing virgin aggregates with recycled aggregates was seen to compromise the compressive strength of 3D printed low-carbon cementitious composites depending on the mix design and the replacement ratio, which should be considered with considerable care (Cagatay, 2022; Ilcan, 2022). Secondly, the 3D printing of low-carbon cementitious composites like one-part geopolymers has a narrow open time emerging from the quick development of yield stress, which hinders its implementation in large-scale applications. Furthermore, the presence of solid waste aggregates like CDW in the composition of the printed mixture is expected to significantly reduce the printing window as they might increase the geopolymerization reaction (Cuevas, 2021; Fonseca & Matos, 2023; Peng & Unluer, 2023). Thirdly, the use of 3D printing in construction generally lacks the regulations and testing standards to facilitate its use and adequately evaluate the properties of the manufactured composite (Martínez-García et al., 2021). Given that the implementation of waste materials in 3D printed low-carbon cementitious composites is still in its infancy and it is not a well-established area like the 3D printing of OPC, standardizing its testing and regulating its usage are still far away from being developed. Finally, the 3D printing of cementitious composites with low embodied carbon and containing recycled/waste aggregates may impose some risks to the environment springing from its high energy consumption and its operation that depends mainly on lithium batteries (Agustí-Juan, 2017).

On the other hand, the unique capabilities of low-carbon cementitious composites in improving the sustainability of the built environment and confining waste materials within their composition coupled with the economic and environmental benefits of the 3D printing technology would result in emerging a more efficient construction method than conventional subtractive one. This approach, if adopted by decision-makers, regulators, industrialists, and environmentalists, will guarantee the development of a nourished construction industry with lower manpower and materials costs, higher productivity, and lower environmental impact (Peng & Unluer, 2023). To achieve this goal, technical issues related to the narrow open time of printing, buildability, extrudability, and mechanical properties of composites need to be enhanced by employing different environmentally friendly admixtures in the mixtures and exploring new preparation parameters and precursors (Al-Noaimat & Hamidreza, 2023). In addition, the use of renewable energy sources to run the 3D printing units would definitely reduce their carbon footprint and further improve their environmental performance.

13.5 Conclusions

This chapter presents some of the many ways to reduce the cement industry's carbon footprint by producing more sustainable concrete by incorporating low-carbon binders, such as limestone-calcined clay cement or AAMs. Recycling CDW and end-of-life products and reusing them in concrete by replacing natural sand could also reduce the environmental impact from the disposal process of those landfills and allow for preserving the limited available natural resources. From an engineering perspective, it is also beneficial since the developed low-carbon concrete has similar or better quality to that of traditional concrete. Combining the benefits obtained from developing those low-carbon concretes with 3D printing technology could be beneficial in meeting the demand for sustainable strategies since it allows the production of complex shapes while producing almost zero waste and reduces GHG emission due to the less steel reinforcement and removal of formwork. That would also enable the reduction of the overall cost of the concrete mixture compared to the conventional concrete since the developed mixtures would contain industrial by-products to be used as binders or recycled materials to replace the natural sand.

References

Abdollahnejad, Z., et al. (2020). Mix design properties and cost analysis of fly ash-based geopolymer foam. *Construction and Building Materials*, *80*(May 2010), 18–30. Available from https://doi.org/10.1016/j.conbuildmat.2015.01.063.

Adesanya, E., et al. (2021). Opportunities to improve sustainability of alkali-activated materials: A review of side-stream based activators'. *Journal of Cleaner Production*, *286*, 125558. Available from https://doi.org/10.1016/j.jclepro.2020.125558.

Agustí-Juan, I., et al. (2017). Potential benefits of digital fabrication for complex structures: Environmental assessment of a robotically fabricated concrete wall. *Journal of Cleaner Production*, *154*, 330–340. Available from https://doi.org/10.1016/j.jclepro.2017.04.002.

Ahmad, W., et al. (2021). A scientometric review of waste material utilization in concrete for sustainable construction. *Case Studies in Construction Materials*, *15*(August), e00683. Available from https://doi.org/10.1016/j.cscm.2021.e00683.

Ahmed, K. S., & Rana, L. R. (2023). Fresh and hardened properties of concrete containing recycled waste glass: A review'. *Journal of Building Engineering*, *70*(March), 106327. Available from https://doi.org/10.1016/j.jobe.2023.106327.

Al-awabdeh, F. W., et al. (2022). Comprehensive investigation of recycled waste glass in concrete using silane treatment for performance improvement. *Results in Engineering*, *16*(October), 100790. Available from https://doi.org/10.1016/j.rineng.2022.100790.

Alghamdi, H., Nair, S. A. O., & Neithalath, N. (2019). Insights into material design, extrusion rheology, and properties of 3D-printable alkali-activated fly ash-based binders. *Materials and Design*, *167*, 107634. Available from https://doi.org/10.1016/j.matdes.2019.107634.

Al-Noaimat, Y. A., & Akis, T. (2023). Influence of cement replacement by calcinated kaolinitic and montmorillonite clays on the properties of mortars. *Arabian Journal for Science and Engineering*, *48*, 14043–14057. Available from https://doi.org/10.1007/s13369-023-08041-y.

Al-Noaimat, Y. A., Chougan, M., et al. (2023). 3D printing of limestone-calcined clay cement: A review of its potential implementation in the construction industry'. *Results in Engineering*, *18*(March), 101115. Available from https://doi.org/10.1016/j.rineng.2023.101115.

Al-Noaimat, Y. A., Ghaffar, S. H., et al. (2023a). Recycled brick aggregates in one-part alkali-activated materials: Impact on 3D printing performance and material properties. *Developments in the Built Environment*, *16*, 100248. Available from https://doi.org/10.1016/j.dibe.2023.100248.

Al-Noaimat, Y. A., Ghaffar, S. H., et al. (2023b). Upcycling end-of-life bricks in high-performance one-part alkali-activated materials. *Developments in the Built Environment*, *16*, 100231. Available from https://doi.org/10.1016/j.dibe.2023.100231.

Al-Noaimat, Y. A., Hamidreza, S., et al. (2023). A review of 3D printing low-carbon concrete with one-part geopolymer: Engineering, environmental and economic feasibility'. *Case Studies in Construction Materials*, *18*(October 2022), e01818. Available from https://doi.org/10.1016/j.cscm.2022.e01818.

Alujas, A., et al. (2015). Pozzolanic reactivity of low grade kaolinitic clays: Influence of calcination temperature and impact of calcination products on OPC hydration. *Applied Clay Science*, *108*, 94–101. Available from https://doi.org/10.1016/j.clay.2015.01.028.

Alwi, R., et al. (2021). Properties of concrete with untreated and treated crumb rubber: A review. *Journal of Materials Research and Technology*, *11*, 1753–1798. Available from https://doi.org/10.1016/j.jmrt.2021.02.019.

Alyousef, R., et al. (2021). Potential use of recycled plastic and rubber aggregate in cementitious materials for sustainable construction: A review'. *Journal of Cleaner Production*, *329*(November), 129736. Available from https://doi.org/10.1016/j.jclepro.2021.129736.

Amin, N. U., et al. (2012). Activation of clay in cement mortar applying mechanical, chemical and thermal techniques. *Advances in Cement Research*, *24*(6), 319–324. Available from https://doi.org/10.1680/adcr.11.00020.

Antoni, M., et al. (2012). Cement substitution by a combination of metakaolin and limestone. *Cement and Concrete Research*, *42*(12), 1579–1589. Available from https://doi.org/10.1016/j.cemconres.2012.09.006.

Athira, V. S., et al. (2021). Agro-waste ash based alkali-activated binder: Cleaner production of zero cement concrete for construction'. *Journal of Cleaner Production*, *286*, 125429. Available from https://doi.org/10.1016/j.jclepro.2020.125429.

Avet, F., & Scrivener, K. (2018). Investigation of the calcined kaolinite content on the hydration of limestone calcined clay cement (LC3). *Cement and Concrete Research*, *107* (January), 124–135. Available from https://doi.org/10.1016/j.cemconres.2018.02.016.

Avet, F., Boehm-Courjault, E., & Scrivener, K. (2019). Investigation of C-A-S-H composition, morphology and density in limestone calcined clay cement (LC3). *Cement and Concrete Research*, *115*(October 2018), 70–79. Available from https://doi.org/10.1016/j.cemconres.2018.10.011.

Bai, G., et al. (2021). 3D printing eco-friendly concrete containing under-utilised and waste solids as aggregates. *Cement and Concrete Composites*, *120*(March), 104037. Available from https://doi.org/10.1016/j.cemconcomp.2021.104037.

Bester, J., Kruger, D., & Miller, B. (2017). South African construction and demolition waste procedure and its sourced material effects on concrete. *MATEC Web of Conferences*, *120*, 02008.

Bhattacherjee, S., et al. (2021). Sustainable materials for 3D concrete printing. *Cement and Concrete Composites*, *122*(June), 104156. Available from https://doi.org/10.1016/j.cemconcomp.2021.104156.

Bong, S. H., et al. (2021). Ambient temperature cured "just-add-water" geopolymer for 3D concrete printing applications. *Cement and Concrete Composites*, *121*(April), 104060. Available from https://doi.org/10.1016/j.cemconcomp.2021.104060.

Cagatay, N., et al. (2022). Mechanical anisotropy evaluation and bonding properties of 3D-printable construction and demolition waste-based geopolymer mortars. *Cement and Concrete Composites*, *134*(October), 104814. Available from https://doi.org/10.1016/j.cemconcomp.2022.104814.

Challa, P. R., & Das, B. B. (2019). Methods to monitor resources and logistic planning at project sites. In B. B. Das, & N. Neithalath (Eds.), *Sustainable Construction and Building Materials. Lecture Notes in Civil Engineering* (vol. 25, pp. 793–802). Springer.

Chen., et al. (2019). Limestone and calcined clay-based sustainable cementitious materials for 3D concrete printing: A fundamental study of extrudability and early-age strength development. *Applied Sciences*, *9*(9), 1809. Available from https://doi.org/10.3390/app9091809.

Chen, W., et al. (2023). Improving mechanical properties of 3D printable "one-part" geopolymer concrete with steel fiber reinforcement. *Journal of Building Engineering*, *75*(June), 107077. Available from https://doi.org/10.1016/j.jobe.2023.107077.

Chen, Y., et al. (2020). Effect of different grade levels of calcined clays on fresh and hardened properties of ternary-blended cementitious materials for 3D printing. *Cement and Concrete Composites, 114*(March). Available from https://doi.org/10.1016/j.cemconcomp.2020.103708.

Chen, Y., et al. (2021). 3D printing of calcined clay-limestone-based cementitious materials. *Cement and Concrete Research, 149*, 106553. Available from https://doi.org/10.1016/j.cemconres.2021.106553.

Chougan, M., et al. (2020). The influence of nano-additives in strengthening mechanical performance of 3D printed multi-binder geopolymer composites. *Construction and Building Materials, 250*, 118928. Available from https://doi.org/10.1016/j.conbuildmat.2020.118928.

Chougan, M., et al. (2021). Investigation of additive incorporation on rheological, microstructural and mechanical properties of 3D printable alkali-activated materials. *Materials and Design, 202*, 109574. Available from https://doi.org/10.1016/j.matdes.2021.109574.

Chougan, M., et al. (2022). Effect of natural and calcined halloysite clay minerals as low-cost additives on the performance of 3D-printed alkali-activated materials. *Materials & Design, 223*, 111183. Available from https://doi.org/10.1016/j.matdes.2022.111183.

Christen, H., van Zijl, G., & de Villiers, W. (2022). The incorporation of recycled brick aggregate in 3D printed concrete. *Cleaner Materials, 4*(March), 100090. Available from https://doi.org/10.1016/j.clema.2022.100090.

Cuevas, K., et al. (2021). 3D printable lightweight cementitious composites with incorporated waste glass aggregates and expanded microspheres – Rheological, thermal and mechanical properties. *Journal of Building Engineering, 44*(April), 102718. Available from https://doi.org/10.1016/j.jobe.2021.102718.

Dang, J., et al. (2018). Properties of mortar with waste clay bricks as fine aggregate. *Construction and Building Materials, 166*, 898–907. Available from https://doi.org/10.1016/j.conbuildmat.2018.01.109.

Danish, A., & Ozbakkaloglu, T. (2022). Greener cementitious composites incorporating sewage sludge ash as cement replacement: A review of progress, potentials, and future prospects'. *Journal of Cleaner Production, 371*(August), 133364. Available from https://doi.org/10.1016/j.jclepro.2022.133364.

Ding, T., et al. (2020). Hardened properties of layered 3D printed concrete with recycled sand. *Cement and Concrete Composites, 113*(February), 103724. Available from https://doi.org/10.1016/j.cemconcomp.2020.103724.

Dong, Q., Huang, B., & Shu, X. (2013). Rubber modified concrete improved by chemically active coating and silane coupling agent. *Construction and Building Materials, 48*, 116–123. Available from https://doi.org/10.1016/j.conbuildmat.2013.06.072.

Du, H., & Pang, S. D. (2020). High-performance concrete incorporating calcined kaolin clay and limestone as cement substitute. *Construction and Building Materials, 264*, 120152. Available from https://doi.org/10.1016/j.conbuildmat.2020.120152.

Elizabeth, S., Ramaswamy, K. P., & Skariah, B. (2024). A review on rheological behaviour of alkali activated materials and the influence of composition factors. *Materials Today: Proceedings [Preprint]*. Available from https://doi.org/10.1016/j.matpr.2023.03.277.

Elmesalami, N., & Celik, K. (2022). A critical review of engineered geopolymer composite: A low-carbon ultra-high-performance concrete. *Construction and Building Materials, 346*(July), 128491. Available from https://doi.org/10.1016/j.conbuildmat.2022.128491.

El-seidy, E., et al. (2022). Mechanical and physical characteristics of alkali- activated mortars incorporated with recycled polyvinyl chloride and rubber aggregates. *Journal of Building Engineering, 60*(June), 105043. Available from https://doi.org/10.1016/j.jobe.2022.105043.

El-seidy, E., et al. (2023). Lightweight alkali-activated materials and ordinary Portland cement composites using recycled polyvinyl chloride and waste glass aggregates to fully replace natural sand. *Construction and Building Materials*, *368*(October 2022), 130399. Available from https://doi.org/10.1016/j.conbuildmat.2023.130399.

Elzeadani, M., Bompa, D. V., & Elghazouli, A. Y. (2022). One part alkali activated materials: A state-of-the-art review Ordinary Portland cement'. *Journal of Building Engineering*, *57*(April), 104871. Available from https://doi.org/10.1016/j.jobe.2022.104871.

Environment, U. N., et al. (2018). Eco-efficient cements: Potential economically viable solutions for a low-CO_2 cement-based materials industry'. *Cement and Concrete Research*, *114*(June), 2–26. Available from https://doi.org/10.1016/j.cemconres.2018.03.015.

Fatta, D. *et al.* (2003) 'Generation and management of construction and demolition waste in Greece — An existing challenge', Resources, Conservation and Recycling, 40, 81–91.

Ferdous, W., et al. (2021). Recycling of landfill wastes (tyres, plastics and glass) in construction – A review on global waste generation, performance, application and future opportunities. *Resources, Conservation & Recycling*, *173*(June), 105745. Available from https://doi.org/10.1016/j.resconrec.2021.105745.

Fernandez, R., Martirena, F., & Scrivener, K. L. (2011). The origin of the pozzolanic activity of calcined clay minerals: A comparison between kaolinite, illite and montmorillonite. *Cement and Concrete Research*, *41*(1), 113–122. Available from https://doi.org/10.1016/j.cemconres.2010.09.013.

Fernando, S., et al. (2021). Life cycle assessment and cost analysis of fly ash – Rice husk ash blended alkali-activated concrete. *Journal of Environmental Management*, *295* (June), 113140. Available from https://doi.org/10.1016/j.jenvman.2021.113140.

Fonseca, M., & Matos, A. M. (2023). 3D construction printing standing for sustainability and circularity: Material-Level Opportunities. '. *Materials*, *16*(6), 2458. Available from https://doi.org/10.3390/ma16062458.

Forum, W.E. (2018). *6 ways the construction industry can build for the future*. World Economic Forum. https://www.weforum.org/agenda/2018/03/how-construction-industry-can-build-its-future/

Furet, B., Poullain, P., & Garnier, S. (2019). 3D printing for construction based on a complex wall of polymer-foam and concrete. *Additive Manufacturing*, *28*(April), 58–64. Available from https://doi.org/10.1016/j.addma.2019.04.002.

García de Soto, B., et al. (2018). Productivity of digital fabrication in construction: Cost and time analysis of a robotically built wall. *Automation in Construction*, *92*(May), 297–311. Available from https://doi.org/10.1016/j.autcon.2018.04.004.

Gettu, R., et al. (2019). Influence of supplementary cementitious materials on the sustainability parameters of cements and concretes in the Indian context. *Materials and Structures*, *52*(1), 1–11. Available from https://doi.org/10.1617/s11527-019-1321-5.

Habert, G., & Ouellet-Plamondon, C. (2016). Recent update on the environmental impact of geopolymers. *RILEM Technical Letters*, *1*, 17. Available from https://doi.org/10.21809/rilemtechlett.v1.6.

Han, Y., et al. (2021). Environmental and economic assessment on 3D printed buildings with recycled concrete. *Journal of Cleaner Production*, *278*, 123884. Available from https://doi.org/10.1016/j.jclepro.2020.123884.

Ilcan, H., et al. (2022). Rheological properties and compressive strength of construction and demolition waste-based geopolymer mortars for 3D-Printing. *Construction and Building Materials*, *328*(February), 127114. Available from https://doi.org/10.1016/j.conbuildmat.2022.127114.

Jim, L. F., Dom, A., & Vega-azamar, R. E. (2018). Carbon footprint of recycled aggregate concrete. *Advances in Civil Engineering, 2018*. Available from https://doi.org/10.1155/2018/7949741, Article ID 7949741.

Kaur, G., & Pavia, S. (2021). Chemically treated plastic aggregates for eco-friendly cement mortars. *Journal of Material Cycles and Waste Management*, *23*(4), 1531–1543. Available from https://doi.org/10.1007/s10163-021-01235-2.

Kaze, C. R., et al. (2021). Alkali-activated laterite binders: Influence of silica modulus on setting time, Rheological behaviour and strength development. *Cleaner Engineering and Technology*, *4*(January), 100175. Available from https://doi.org/10.1016/j.clet.2021.100175.

Kunthawatwong, R., et al. (2023). Performance of geopolymer mortar containing PVC plastic waste from bottle labels at normal and elevated temperatures. *Buildings*, *13*(4), 1031. Available from https://doi.org/10.3390/buildings13041031.

Long, W. J., et al. (2021). Printability and particle packing of 3D-printable limestone calcined clay cement composites. *Construction and Building Materials*, *282*, 122647. Available from https://doi.org/10.1016/j.conbuildmat.2021.122647.

López Gayarre, F., et al. (2017). Use of recycled mixed aggregates in floor blocks manufacturing. *Journal of Cleaner Production*, *167*(2017), 713–722. Available from https://doi.org/10.1016/j.jclepro.2017.08.193.

Loukili, A., et al. (2018). How do recycled concrete aggregates modify the shrinkage and self-healing properties ? *Cement and Concrete Composites*, *86*, 72–86. Available from https://doi.org/10.1016/j.cemconcomp.2017.11.003.

Luukkonen, T., et al. (2018). One-part alkali-activated materials: A review. *Cement and Concrete Research*, *103*(September 2017), 21–34. Available from https://doi.org/10.1016/j.cemconres.2017.10.001.

Maduabuchukwu Nwakaire, C., et al. (2020). Utilisation of recycled concrete aggregates for sustainable highway pavement applications: A review. *Construction and Building Materials*, *235*, 117444. Available from https://doi.org/10.1016/j.conbuildmat.2019.117444.

Martínez-García, A., Monzón, M. and Paz, R. (2021) 'Standards for additive manufacturing technologies: Structure and impact', in J. Pou, A. Riveiro, and J.P.B.T.-A.M. Davim (eds) Handbooks in advanced manufacturing. Elsevier, pp. 395–408.

Maskuriy, R., et al. (2019). Industry 4.0 for the construction industry: Review of management perspective. *Economies*, *7*(3), 68. Available from https://doi.org/10.3390/economies7030068.

Mechtcherine, V., et al. (2019). Large-scale digital concrete construction – CONPrint3D concept for on-site, monolithic 3D-printing. *Automation in Construction*, *107*(April), 102933. Available from https://doi.org/10.1016/j.autcon.2019.102933.

Mohammed, M. S., Elkady, H., & Gawwad, H. A. A.- (2021). Utilization of construction and demolition waste and synthetic aggregates. *Journal of Building Engineering*, *43* (August), 103207. Available from https://doi.org/10.1016/j.jobe.2021.103207.

Monteiro, P. J. M., Miller, S. A., & Horvath, A. (2017). Produce and use with care. *Nature Publishing Group*, *16*(7), 698–699. Available from https://doi.org/10.1038/nmat4930.

Mounika, G., et al. (2023). A review on effect of red mud on properties of alkali activated materials (AAMs) and geopolymers (GPs). *Materials Today: Proceedings*. Available from https://doi.org/10.1016/j.matpr.2023.03.446.

Muthukrishnan, S., Ramakrishnan, S., & Sanjayan, J. (2021). Effect of alkali reactions on the rheology of one-part 3D printable geopolymer concrete. *Cement and Concrete Composites*, *116*(August 2020), 103899. Available from https://doi.org/10.1016/j.cemconcomp.2020.103899.

Muzenda, T. R., et al. (2020). The role of limestone and calcined clay on the rheological properties of LC3. *Cement and Concrete Composites*, *107*(May 2019), 103516. Available from https://doi.org/10.1016/j.cemconcomp.2020.103516.

Nair, A. (2017). *Recycled glass market: Europe exhibits continual growth.* Inkwood Research. https://inkwoodresearch.com/recycled-glass-market-europe-exhibits-continual-growth/

Nematollahi, B., et al. (2018). Effect of polypropylene fibre addition on properties of geopolymers made by 3D printing for digital construction. *Materials*, *11*(12). Available from https://doi.org/10.3390/ma11122352.

Nodehi, M., & Taghvaee, V. M. (2022a). Alkali-activated materials and geopolymer: A review of common precursors and activators addressing circular economy. *Circular Economy and Sustainability*, *2*(1), 165–196. Available from https://doi.org/10.1007/s43615-021-00029-w.

Nodehi, M., & Taghvaee, V. M. (2022b). Sustainable concrete for circular economy: A review on use of waste glass'. *Glass Structures & Engineering*, *7*(1), 3–22. Available from https://doi.org/10.1007/s40940-021-00155-9.

Oikonomou, N. D. (2005). Recycled concrete aggregates. *Cement and Concrete Composites*, *27*(2), 315–318. Available from https://doi.org/10.1016/j.cemconcomp.2004.02.020.

Organisation for Economic Co-operation and Development. (2022). *Plastic pollution is growing relentlessly as waste management and recycling fall short, says OECD.* https://www.oecd.org/newsroom/plastic-pollution-is-growing-relentlessly-as-waste-management-and-recycling-fall-short.htm

Panda, B., & Tan, M. J. (2018). Experimental study on mix proportion and fresh properties of fly ash based geopolymer for 3D concrete printing. *Ceramics International*, *44*(9), 10258–10265. Available from https://doi.org/10.1016/j.ceramint.2018.03.031.

Panda, B., Chandra Paul, S., & Jen Tan, M. (2017). Anisotropic mechanical performance of 3D printed fiber reinforced sustainable construction material. *Materials Letters*, *209*, 146–149. Available from https://doi.org/10.1016/j.matlet.2017.07.123.

Pasupathy, K., Ramakrishnan, S., & Sanjayan, J. (2023). 3D concrete printing of eco-friendly geopolymer containing brick waste. *Cement and Concrete Composites*, *138*(January), 104943. Available from https://doi.org/10.1016/j.cemconcomp.2023.104943.

Paul, S. C., Babafemi, A. J., et al. (2018). Properties of normal and recycled brick aggregates for production of medium range (25–30 MPa) structural strength concrete. *Buildings*, *8* (5), 72. Available from https://doi.org/10.3390/BUILDINGS8050072.

Paul, S. C., Tay, Y. W. D., et al. (2018). Fresh and hardened properties of 3D printable cementitious materials for building and construction. *Archives of Civil and Mechanical Engineering*, *18*(1), 311–319. Available from https://doi.org/10.1016/j.acme.2017.02.008.

Peng, Y., & Unluer, C. (2023). Development of alternative cementitious binders for 3D printing applications: A critical review of progress, advantages and challenges'. *Composites Part B*, *252*(May 2022), 110492. Available from https://doi.org/10.1016/j.compositesb.2022.110492.

Petrillo, A., et al. (2016). Eco-sustainable geopolymer concrete blocks production process. *Agriculture and Agricultural Science Procedia*, *8*, 408–418. Available from https://doi.org/10.1016/j.aaspro.2016.02.037.

Pranav, S., et al. (2020). Alternative materials for wearing course of concrete pavements: A critical review. *Construction and Building Materials*, *236*, 117609. Available from https://doi.org/10.1016/j.conbuildmat.2019.117609.

Provis, J. L., & Van Deventer, J. S. (2013). *Alkali activated materials: State-of-the-art report, RILEM TC 224-AAM*. Springer. Available from https://doi.org/10.1007/978-94-007-7672-2.

Provis, J. L., & Bernal, S. A. (2014). Geopolymers and related alkali-activated materials. *Annual Review of Materials Research*, *44*, 299–327. Available from https://doi.org/10.1146/annurev-matsci-070813-113515.

Provis, J. L., Palomo, A., & Shi, C. (2015). Advances in understanding alkali-activated materials. *Cement and Concrete Research*, *78*, 110–125. Available from https://doi.org/10.1016/j.cemconres.2015.04.013.

Rahul, A. V., Sharma, A., & Santhanam, M. (2020). A desorptivity-based approach for the assessment of phase separation during extrusion of cementitious materials. *Cement and Concrete Composites*, *108*(July 2019), 103546. Available from https://doi.org/10.1016/j.cemconcomp.2020.103546.

Raza, M. H., & Zhong, R. Y. (2022). A sustainable roadmap for additive manufacturing using geopolymers in construction industry. *Resources, Conservation & Recycling*, *186* (August), 106592. Available from https://doi.org/10.1016/j.resconrec.2022.106592.

Reig, L., et al. (2013). Properties and microstructure of alkali-activated red clay brick waste. *Construction and Building Materials*, *43*, 98–106. Available from https://doi.org/10.1016/j.conbuildmat.2013.01.031.

Revilla-Cuesta, V., et al. (2020). Self-compacting concrete manufactured with recycled concrete aggregate: An overview. *Journal of Cleaner Production*, *262*, 121362. Available from https://doi.org/10.1016/j.jclepro.2020.121362.

Roussel, N. (2018). Rheological requirements for printable concretes. *Cement and Concrete Research*, *112*(January), 76–85. Available from https://doi.org/10.1016/j.cemconres.2018.04.005.

Rovnaník, P., et al. (2018). Rheological properties and microstructure of binary waste red brick powder/metakaolin geopolymer. *Construction and Building Materials*, *188*, 924–933. Available from https://doi.org/10.1016/j.conbuildmat.2018.08.150.

Saedi, A., Jamshidi-zanjani, A., & Khodadadi, A. (2021). A review of additives used in the cemented paste tailings: Environmental aspects and application. *Journal of Environmental Management*, *289*(March), 112501. Available from https://doi.org/10.1016/j.jenvman.2021.112501.

Saloni., et al. (2021). Effect of pre-treatment methods of crumb rubber on strength, permeability and acid attack resistance of rubberised geopolymer concrete. *Journal of Building Engineering*, *41*, 102448. Available from https://doi.org/10.1016/j.jobe.2021.102448.

Samet, B. (2007). Use of a kaolinitic clay as a pozzolanic material for cements: Formulation of blended cement. ', *Cement and Concrete Composites*, *29*, 741–749. Available from https://doi.org/10.1016/j.cemconcomp.2007.04.012.

Sánchez-Berriel, S., et al. (2016). Assessing the environmental and economic potential of limestone calcined clay cement in Cuba. *Journal of Cleaner Production*, *124*, 361–369. Available from https://doi.org/10.1016/j.jclepro.2016.02.125.

Sanjayan, J.G. and Nematollahi, B. (2019) 3D concrete printing for construction applications, in: J.G. Sanjayan, A. Nazari, B. Nematollahi, Eds., 3D concrete printing technology. Elsevier Inc., pp. 1–11.

Shah, S. F. A., et al. (2020). Improvement of early strength of fly ash-slag based one-part alkali activated mortar. *Construction and Building Materials*, *246*, 118533. Available from https://doi.org/10.1016/j.conbuildmat.2020.118533.

Sharma, R., & Pal, P. (2016). Use of different forms of waste plastic in concrete – A review. *Journal of Cleaner Production, 112*, 473–482. Available from https://doi.org/10.1016/j.jclepro.2015.08.042.

Si, R., Guo, S., & Dai, Q. (2017). Durability performance of rubberized mortar and concrete with NaOH-Solution treated rubber particles. *Construction and Building Materials, 153*, 496–505. Available from https://doi.org/10.1016/j.conbuildmat.2017.07.085.

Skibsted, J., & Snellings, R. (2019). Reactivity of supplementary cementitious materials (SCMs) in cement blends. *Cement and Concrete Research, 124*(July), 105799. Available from https://doi.org/10.1016/j.cemconres.2019.105799.

Styles, D., Schoenberger, H., & Zeschmar-lahl, B. (2018). Construction and demolition waste best management practice in Europe. *Resources, Conservation & Recycling, 136* (December 2017), 166–178. Available from https://doi.org/10.1016/j.resconrec.2018.04.016.

Tafazzoli, M. (2022) 'Construction automation and sustainable development', in H. Jebelli et al. (eds). Automation and robotics in the architecture, engineering, and construction industry. Springer International Publishing, pp. 73–95.

Tanaçan, L. (2022). Sound transmission loss of LC3-based mortars with barite and waste rubber. *Building Acousticss, 29*(4), 503–528. Available from https://doi.org/10.1177/1351010X221129264.

Turner, L. K., & Collins, F. G. (2013). Carbon dioxide equivalent (CO2-e) emissions: A comparison between geopolymer and OPC cement concrete. *Construction and Building Materials, 43*, 125–130. Available from https://doi.org/10.1016/j.conbuildmat.2013.01.023.

United States Environmental Protection Agency. (2019). *Sustainable management of construction and demolition materials*. Accessed June 23, 2023 from https://www.epa.gov/smm/sustainable-management-construction-and-demolition-materials#what

Valente, M., & Sambucci, M. (2021). Geopolymers vs. cement matrix materials: How nanofiller can help a sustainability approach for smart construction applications — A review. *Nanomaterials, 11*(8), 2007. Available from https://doi.org/10.3390/nano11082007.

Valente, M., et al. (2022). Reducing the emission of climate-altering substances in cementitious materials: A comparison between alkali-activated materials and Portland cement-based composites incorporating recycled tire rubber'. *Journal of Cleaner Production, 333*(August 2021), 130013. Available from https://doi.org/10.1016/j.jclepro.2021.130013.

Valente, M., et al. (2023). Composite alkali-activated materials with waste tire rubber designed for additive manufacturing: An eco-sustainable and energy saving approach'. *Journal of Materials Research and Technology, 24*, 3098–3117. Available from https://doi.org/10.1016/j.jmrt.2023.03.213.

van Deventer, J. S. J., et al. (2007). Reaction mechanisms in the geopolymeric conversion of inorganic waste to useful products. *Journal of Hazardous Materials, 139*(3), 506–513. Available from https://doi.org/10.1016/j.jhazmat.2006.02.044.

Vinai, R., & Soutsos, M. (2019). Production of sodium silicate powder from waste glass cullet for alkali activation of alternative binders. *Cement and Concrete Research, 116*(June 2018), 45–56. Available from https://doi.org/10.1016/j.cemconres.2018.11.008.

Vizcayno, C., et al. (2010). Pozzolan obtained by mechanochemical and thermal treatments of kaolin. *Applied Clay Science, 49*(4), 405–413. Available from https://doi.org/10.1016/j.clay.2009.09.008.

Wang, B., et al. (2021). A comprehensive review on recycled aggregate and recycled aggregate concrete. *Resources, Conservation and Recycling, 171*(September 2020), 105565. Available from https://doi.org/10.1016/j.resconrec.2021.105565.

Wangler, T., et al. (2016). Digital concrete: Opportunities and challenges. *RILEM Technical Letters*, *1*, 67. Available from https://doi.org/10.21809/rilemtechlett.2016.16.

Wong, C. L., et al. (2018). Potential use of brick waste as alternate concrete-making materials: A review. *Journal of Cleaner Production*, *195*, 226–239. Available from https://doi.org/10.1016/j.jclepro.2018.05.193.

Wong, C. L., et al. (2020). Mechanical strength and permeation properties of high calcium fly ash-based geopolymer containing recycled brick powder. *Journal of Building Engineering*, *32*(July), 101655. Available from https://doi.org/10.1016/j.jobe.2020.101655.

Xiaowei, C., et al. (2017). Crumb waste tire rubber surface modification by plasma polymerization of ethanol and its application on oil-well cement. *Applied Surface Science*, *409*, 325–342. Available from https://doi.org/10.1016/j.apsusc.2017.03.072.

Xu, J., et al. (2020). Research on crumb rubber concrete: From a multi-scale review'. *Construction and Building Materials*, *232*, 117282. Available from https://doi.org/10.1016/j.conbuildmat.2019.117282.

Yao, Y., et al. (2020). Life cycle assessment of 3D printing geo-polymer concrete: An ex-ante study. *Journal of Industrial Ecology*, *24*(1), 116–127. Available from https://doi.org/10.1111/jiec.12930.

Yousefi Oderji, S., et al. (2019). Fresh and hardened properties of one-part fly ash-based geopolymer binders cured at room temperature: Effect of slag and alkali activators. *Journal of Cleaner Production*, *225*, 1–10. Available from https://doi.org/10.1016/j.jclepro.2019.03.290.

Zhang, D., et al. (2018). The study of the structure rebuilding and yield stress of 3D printing geopolymer pastes. *Construction and Building Materials*, *184*, 575–580. Available from https://doi.org/10.1016/j.conbuildmat.2018.06.233.

Zhang, T., et al. (2022). Modification of glass powder and its effect on the compressive strength of hardened alkali-activated slag-glass powder paste. *Journal of Building Engineering*, *58*(April), 105030. Available from https://doi.org/10.1016/j.jobe.2022.105030.

Zhang, Z., et al. (2018). A review of studies on bricks using alternative materials and approaches. *Construction and Building Materials*, *188*, 1101–1118. Available from https://doi.org/10.1016/j.conbuildmat.2018.08.152.

Zhong, H., & Zhang, M. (2022). 3D printing geopolymers: A review. *Cement and Concrete Composites*, *128*(November 2021), 104455. Available from https://doi.org/10.1016/j.cemconcomp.2022.104455.

Zhou, Y., et al. (2022). Sustainable lightweight engineered cementitious composites using limestone calcined clay cement (LC 3). *Composites Part B*, *243*(July), 110183. Available from https://doi.org/10.1016/j.compositesb.2022.110183.

Sustainable 3D printed concrete structures using high-quality secondary raw materials

14

Farhad Aslani and Yifan Zhang
Materials and Structures Innovation Group, School of Engineering, The University of Western Australia, WA, Australia

14.1 Introduction

3D concrete printing (3DCP), also known simply as concrete printing, stands as a prominent exemplar of revolutionary automation technologies that challenge traditional construction methodologies. Within the realm of additive manufacturing (AM), 3DCP is recognized as a pivotal process, as per the definition provided by the American Society of Testing and Materials (ASTM), which characterizes AM as "a process of joining materials to create objects from 3D model data, typically layer by layer"(ASTM F2792−12a,92−12a, 2015). Its advantages encompass greater flexibility in shaping elements, cost savings through reduced reliance on formwork for casting, decreased labor expenses, minimized waste generation, and expedited production timelines when juxtaposed with conventional cast-in-situ methods (Ambily et al., 2023; Bakar et al., 2023; Hamidi & Aslani, 2019). Fig. 14.1 shows a summary of advantages and disadvantages of 3DCP over conventional cast-in-situ methods (Hamidi & Aslani, 2019). In addition to academic research, there exist numerous instances of successful 3DCP implementations within the industry. An illustrative example is the collaboration between ICON Build and Lennar, a prominent homebuilder in the United States. This partnership is focused on the development of the largest community of 3D-printed homes, encompassing the construction of 100 printed homes in Georgetown, Texas (Bauguess Brooke, 2022). In August of 2021, the world's largest pedestrian 3D printed concrete bridge, spanning 29 m, was inaugurated in the city of Nijmegen, Dutch (Everett, 2021). Furthermore, in November 2022, ICON Build disclosed that it had entered into a contract with the National Aeronautics and Space Administration (NASA) for a project valued at $60 million. This project is aimed at the development of a versatile system that utilizes Lunar or Martian resources for construction projects in space (ICON Team, 2022).

Fig. 14.2 presents a demonstration of printed engineered cementitious composites incorporating polyvinyl alcohol (PVA) fiber (Zhang & Aslani, 2021). Since 3DCP constructs structures layer by layer, with materials deposited in a predefined pattern, it inherently gives rise to an anisotropic effect. This effect manifests as

Sustainable Concrete Materials and Structures. DOI: https://doi.org/10.1016/B978-0-443-15672-4.00014-0

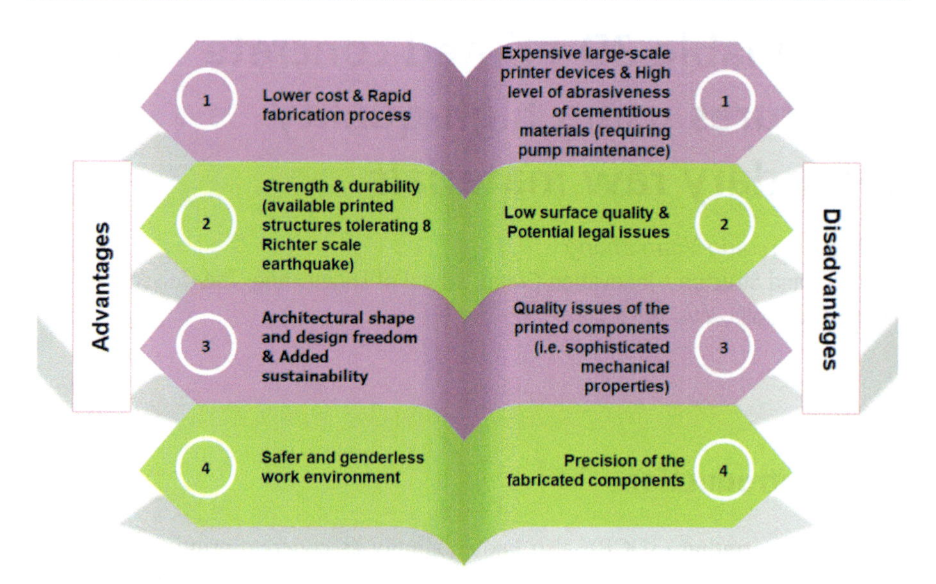

Figure 14.1 A summary of advantages and disadvantages of 3DCP (Hamidi & Aslani, 2019).

variations in strength, porosity, and other mechanical properties along distinct directions. Consequently, researchers have been diligently exploring the phenomenon of anisotropy within the realm of 3DCP. Fig. 14.3 shows commonly used directions to define loading directions (Sun et al., 2021).

Concrete, being the most commonly utilized building material worldwide, is a versatile and durable substance extensively employed in diverse construction applications, encompassing buildings, bridges, pavements, and various infrastructure projects. According to data published by the Global Cement and Concrete Association (GCCA), the global production of concrete in 2020 amounted to a volumetric total of 14.0 billion cubic meters, with annual global cement production reaching a substantial 4.2 billion tons in the same year (Global Cement & Concrete Association (GCCA), n.d.). Meanwhile, data from the report by Global Environmental Alert Service (GEAS) of the United Nations Environment Program (UNEP) states that the cement industry is the second only to power generation in the CO_2 production. Producing 1 ton of Portland cement releases 1 ton of CO_2 into the atmosphere, resulting in approximately 7%–8% of overall human-produced CO_2 (Sheikh & Ahmad, 2023; UNEP, 2010).

Furthermore, aggregates, encompassing both fine and coarse materials derived from sources such as sand, gravel, and crushed rocks, play a pivotal role in curtailing the demand for cement and water while simultaneously augmenting the mechanical robustness of concrete formulations. Data compiled by the United States Geological Survey (USGS) reveals that between 2017 and 2021, the annual consumption of sand and gravel in the United States, accounting for both domestic production and imports, increased from 888 million metric tons to 1000 million

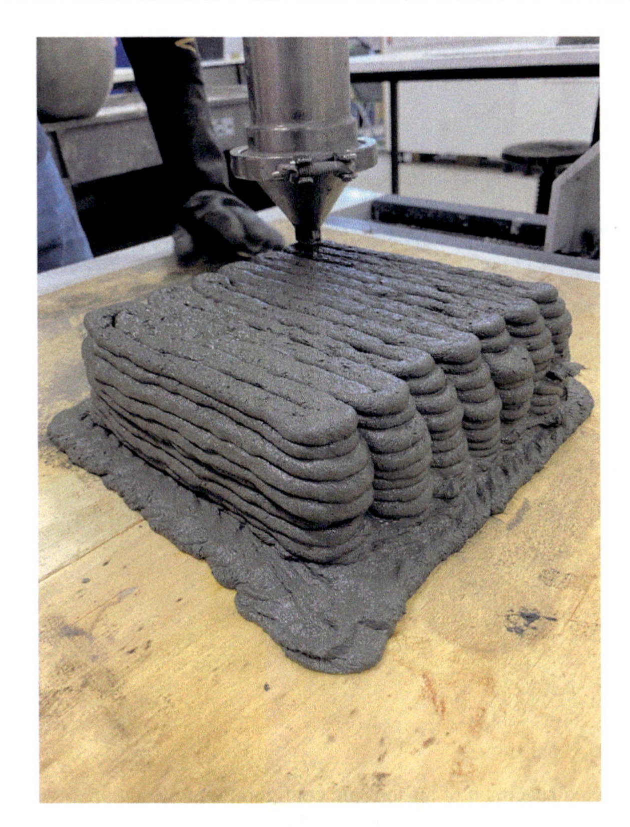

Figure 14.2 A typical example of 3DCP incorporating fibers (Zhang & Aslani, 2021).

metric tons (U.S. Geological Survey, 2022). The report further posits that approximately 46% of construction sand and gravel served as aggregates for Portland cement concrete, while the remaining 54% found application in various purposes including road base coverage and stabilization, construction fill, production of asphaltic concrete, bituminous mixtures, miscellaneous uses, and infrastructure maintenance (U.S. Geological Survey, 2022). USGS asserts in its annual reports that global resources of construction aggregates, encompassing both sand and gravel as well as crushed stone, are abundant across the globe. Nonetheless, in light of the increasing emphasis on sustainable development, coupled with concerns about the depletion of naturally occurring nonrenewable resources, the impact of climate change, adverse consequences associated with overexploitation and degradation, the substantial demand for primary and secondary raw materials to support rapid development in emerging economies, and varying environmental regulations and legislation governing resource extraction, the availability of construction aggregates resources has begun to exhibit signs of scarcity (Ren et al., 2022). UNEP underscored and highlighted this scarcity of resources in the aggregates industry in

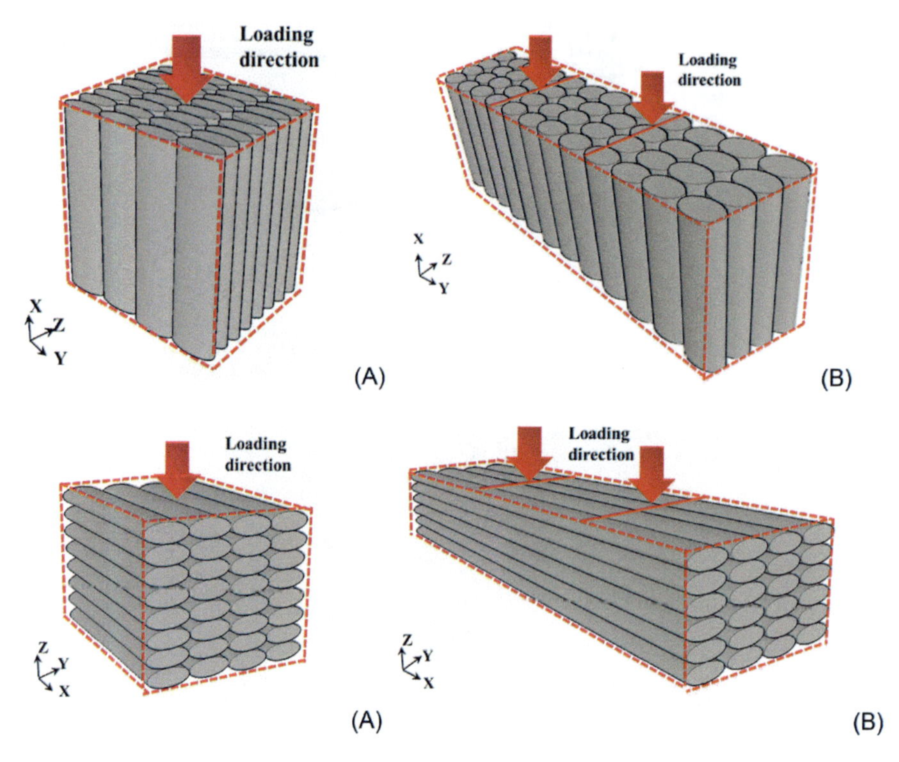

Figure 14.3 Printed specimen of anisotropic configuration (A) compressive strength and (B) flexural strength. This figure illustrates the loading directions used to evaluate anisotropic effects (J. Sun et al., 2021).

its 2019 report on sustainable development of aggregate resources, referring to it as the "sand crisis."

The construction sector is currently undergoing a transformation toward sustainability and resource efficiency. Consequently, the incorporation of high-quality secondary raw materials, including recycled aggregates, industrial by-products, and waste materials, holds the potential to make substantial contributions to sustainable construction practices. Furthermore, the synergy between 3DCP and secondary raw materials in the construction of concrete structures not only confers environmental advantages but also yields economic benefits. These benefits encompass reductions in the consumption of natural resources, mitigation of environmental concerns linked to the processing and manufacturing of virgin materials, minimization of waste generation, and optimization of resource efficiency, among others. Additionally, using secondary raw material can help divert waste materials, which always involves heavy cost for treatment, stocking, and addressing environmental and sanitation issues resulting from the land disposal of certain wastes.

Within this context, this chapter delves into the potential of sustainable 3D printed concrete structures through the utilization of high-quality secondary raw

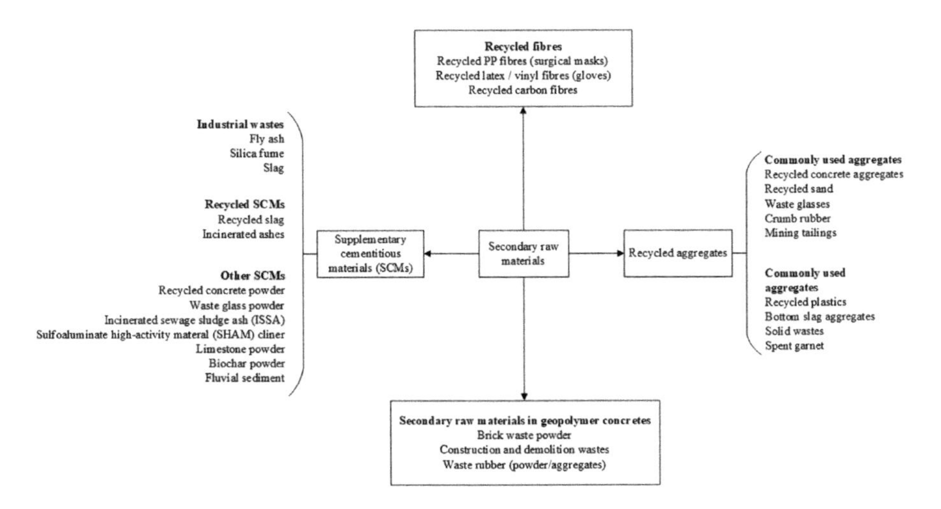

Figure 14.4 Categories of secondary raw materials.

materials. It examines these materials from multiple angles, considering their properties and characteristics that can be integrated into 3DCP. Furthermore, it assesses the impact of these secondary raw materials on the durability and mechanical properties of the printed concrete structures. Besides, this chapter discusses the opportunities and challenges associated with this approach, as well as the current progress in this field. Within the confines of this chapter, the secondary raw materials employed in 3DCP are classified into four distinct categories: secondary supplementary cementitious materials (SCMs), recycled aggregates, recycled fibers, and secondary raw materials designated for use in geopolymer concretes. Fig. 14.4 shows the information aggregation with categories.

14.2 Secondary supplementary cementitious materials

The mix design for 3DCP is contingent not solely on mechanical strengths but also on three critical parameters: pumpability, extrudability, and buildability. These parameters delineate the material's capacity to be pumped, extruded through nozzles, and retain adequate strength to support subsequent layers (Bhattacherjee et al., 2021; Chen et al., 2022; Hamidi & Aslani, 2019; Nodehi et al., 2022; Panda & Tan, 2018). Pumpability, extrudability, and buildability are primarily influenced by the rheological properties of the fresh paste, including yield stress, viscosity, and thixotropy (Bhattacherjee et al., 2021; Rahul et al., 2019; Roussel, 2018). These properties are further influenced by the degree of hydration over time (Suiker, 2018). To attain the targeted rheological properties for the printing process, a range of chemical admixtures is typically employed. These include superplasticizers (SP), viscosity-modifying agents (VMA), accelerators/retarders, and water-reducing agents, which are commonly used to fine-tune the fresh properties of the material.

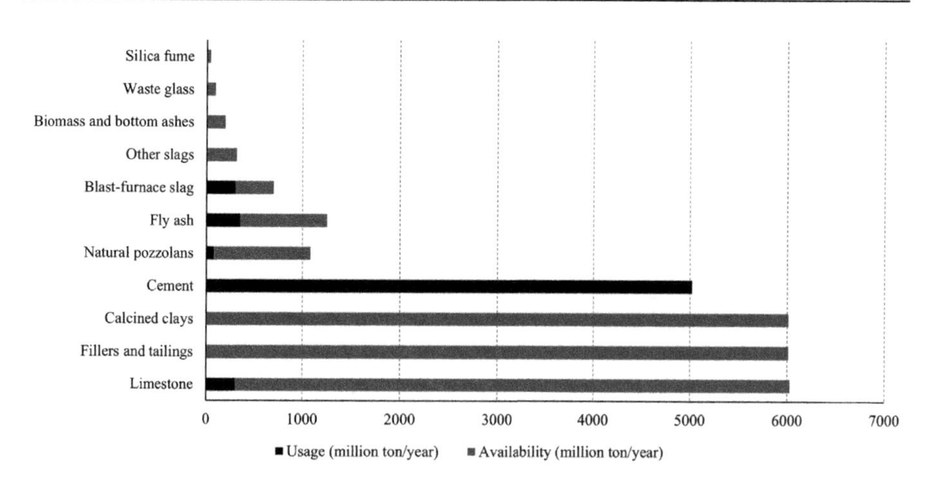

Figure 14.5 Availability of commonly used SCMs (Scrivener et al., 2018; Suraneni, 2021). *Source:* Replotted from Dey, D., Srinivas, D., Boddepalli, U., Panda, B., Gandhi, I. S. R., & Sitharam, T. G. (2022). 3D printability of ternary Portland cement mixes containing fly ash and limestone. *Materials Today: Proceedings*, *70*, 195–200; Scrivener, K. L., John, V. M., Gartner, E. M. (2018). Eco-efficient cements: Potential economically viable solutions for a low-CO2 cement-based materials industry. *Cement and Concrete Research. 114*, 2–26; Suraneni, P. (2021). Recent developments in reactivity testing of supplementary cementitious materials. *RILEM Technical Letters*, *6*, 131–139.

Furthermore, SCMs such as fly ash (FA), silica fume (SF), and ground-granulated blast-furnace slag (GGBFS) are frequently integrated into the concrete mixture. This incorporation serves the dual purpose of achieving the desired rheological properties and enhancing the mechanical strength and durability of the printed concrete.

Fig. 14.5 depicts the availability of commonly used SCMs. The figure illustrates that FA boasts the highest annual production rate, totaling approximately 1200 million tons, while also recording an annual consumption of 352 million tons. This elevated consumption can be attributed to the cost-effectiveness of FA. However, a study by Omran et al. (2018) suggests that the production rate and cost of FA may change over time due to the closure of traditional coal-fired power plants. On the contrary, SF, another frequently used SCM, exhibits a relatively lower annual production rate when compared to FA, and its specific utilization remains less well-documented. However, it is widely acknowledged that SF plays a pivotal role in enhancing concrete properties and is a prevalent choice in applications involving high-performance concrete (Dey, Srinivas, Panda, et al., 2022).

FA, SF, and GGBFS stand out as some of the primary secondary SCMs frequently employed in concrete technology. These substances are renowned for their capacity to enhance the long-term mechanical strength and durability of concrete by means of pozzolanic reactions and filling effects (Bhattacherjee et al., 2021). FA is a fine powder ranging in color from tan to dark gray, and it is a by-product of coal combustion in power plants (International Energy Agency, 2018).

ASTM categorizes FA into Class C and Class F based on its CaO content (ASTM C618−19, 2019). The particles of FA have a diameter ranging from submicron to 300 μm (Abualrous et al., 2016), and their specific surface area typically falls within the range of 300−450 m^2/kg (Feng et al., 2018; Weng et al., 2019). When incorporated as a SCM in concrete, FA enhances workability, diminishes the heat generated during hydration, augments long-term strength and durability, and serves as a mitigating factor against the risk of alkali-silica reaction (ASR) (Abualrous et al., 2016).

SF, also known as micro silica, is a by-product of the production of silicon and ferrosilicon alloys (Dey, Srinivas, Panda, et al., 2022). SF is a fine, amorphous material composed of highly reactive silicon dioxide (SiO_2) particles with an average size of 0.1−0.3 μm (Abualrous et al., 2016). When used in concrete, it reacts with calcium hydroxide (CH) produced during hydration to form additional calcium silicate hydrate (C-S-H) gel. This results in denser, stronger, and more durable concrete. Benefits of using SF in concrete include (1) increasing mechanical strengths of concrete; (2) reducing concrete permeability through densification, hence improving concrete durability; (3) further improving concrete durability in terms of chemical attack, sulfate attack, ASR, chloride iron corrosion; (4) improving concrete workability; and (5) enhancing thermal resistance against thermal cracking induced at high temperature.

GGBFS, a by-product of the iron and steel industry, is a commonly used SCM in concrete production. GGBFS is a reactive pozzolan obtained by rapidly quenching molten slag from a blast furnace with water or steam, resulting in a glassy, granular material (Dey, Srinivas, Panda, et al., 2022). It is predominantly composed of silicates and aluminosilicates. Slag has a particle size less than 45 μm with a specific surface area of 400−600 m^2/kg (El-Chabib, 2020). Similar to FA and SF, GGBFS's pozzolanic properties involve reaction with CH produced during hydration to form additional C-S-H gel. This contributes to improving strength and durability of concrete. Apart from the benefits in properties, inclusion of GGBFS in concrete can also help with reducing amount of heat generated from hydration reaction, which is particularly beneficial in massive concrete structures.

The study by Ferraris et al. (2001) concluded that properties of SCMs, including specific gravity, particle morphology, surface roughness, porosity, particle size distribution, and fineness, have significant effects on mixture's early-age properties. For instance, the inclusion of SCMs in mixture will increase the water demand. Nevertheless, changes in mixture's properties can vary for different SCMs. Assem et al. (n.d.) and Ferraris et al. (2001) reported that FA improves mixture's workability, while others reduces it because FA reduces the yield stress, while slag or SF increases it. Besides, adding SCMs also changes the setting time of the mixture, and generally, a finer SCM accelerates hydration reaction, thus showing early strength and reducing setting time, while a coarser SCM delays the setting time (Dey, Srinivas, Panda, et al., 2022; Khayat et al., n.d.). Moreover, SCMs usually reduce early-age strength due to limited reactivity, expected for highly reactive materials, such as SF and calcined clay, which can enhance early-age strength through filling effect (Oey et al., 2013; Ramanathan et al., 2020). However, at later ages and reasonable replacement ratios, SCMs improve both strength and concrete durability, including chloride penetration resistance, sulfate attack resistance, alkali-silica expansion, and calcium oxychloride damage (Nath & Sarker, 2011; Suraneni et al., 2016).

14.2.1 Fly ash

Dey, Srinivas, Panda, et al. (2022) summarized the influential available literatures regarding the usage of FA in 3D printed concretes from 2012 to 2021 in their review study. This study reported that increasing the content of Class F FA in an ordinary Portland cement (OPC)-based mixture reduces static yield stress and viscosity, improving extrudability but causing higher deformation of bottom layers in 3D printable mixtures. The impact of FA on rheology depends on its physical and chemical properties, with ultrafine FA being more effective (Panda & Tan, 2019). Another study by Yuan et al. (2018) found that replacing 20 wt.% of OPC with FA negatively affected the growth rate of static yield stress and structural built-up, potentially due to a slower overall reaction, reducing the buildability of 3D structures. However, Panda and Tan (2019) improved buildability by incorporating nanoclay admixtures to enhance flocculation strength and shear-yield stress. Additionally, Zahabizadeh et al. (2021) observed that higher FA content in OPC-based mixtures increased flowability, reduced drying shrinkage, and mitigated early-age defects in 3D printed components, such as high porosity and weak interfacial bond strength.

Xu et al. (2022) investigated the effect of including FA and GGBFS into the blend of OPC and sulfoaluminate on compressive strength, extrudability, and buildability of concretes. The study found that as the amount of FA/GGBFS replacing cement increased, the slump and fluidity of fresh pastes initially increased and then decreased, while the setting time increased gradually. The optimal conditions were observed at 20% FA content, with maximum slump and expansion and the highest flexural and compressive strengths at 28 days. Moeini et al. (2022) investigated the thixotropic behavior of cement-based materials used in large-scale 3D printing. Different mortar mixtures with optimized admixtures, including a high-range water reducing agent, attapulgite nanoclay, a VMA, FA, and SF, were examined through rheological measurements and uniaxial compression tests. The results showed that incorporating attapulgite nanoclay and modified starch-based VMA improved thixotropic behavior, while the proposed evaluation methods effectively captured the evolution of thixotropy and evaluated the impact of admixtures.

14.2.2 Silica fume

Zhang et al. (2018) reported in their study that the addition of approximately 2 wt.% of SF increased the buildability of mixture by about 117% compared to the control mix. This was attributed to the high specific surface area, early-age hydration reaction, and accelerated cement hydration, leading to the formation of flocculated microstructures, resulting in early strength and higher buildability. Results and observations from other studies also reported similar observations (Kazemian et al., 2017; Panda & Tan, 2019; Soltan & Li, 2018).

14.2.3 Slag

Zhao et al. (2023) conducted a study to investigate the mechanical properties and printability of printed mortars incorporating recycled GGBFS and clay brick powder (CBP). In their study, 65 wt.% of cement was replaced with recycled GGBFS, CBP, or a combination of both. The results of the study revealed several important findings. Firstly, replacing 65% of the cement with GGBFS led to a 45.5% increase in the flow diameter of the mortar. When CBP was added to replace GGBFS at levels ranging from 10% to 30%, there was a slight reduction in the flow diameter of the mortar. In terms of opening time, the mix containing 65% GGBFS exhibited a significantly longer opening time compared to the control mix, with the opening time being approximately 150% higher than the control mix (which had an opening time of 30 minutes). This extended opening time was mainly attributed to the lower hydration rate resulting from the reduced cement content.

To assess the buildability of the mortars, the authors measured the deformation of 20-layer printed structures. The results showed that the control mix had the minimum height loss, while the mixes containing 65% GGBFS and 30% CBP experienced significant collapses when 20 layers were printed. However, mixes containing GGBFS with 10% and 20% CBP (relative to the GGBFS content) exhibited desired height loss, with reductions of 6.2% and 5.9%, respectively. The authors concluded that the buildability of the mortar was correlated with the static yield stress of the paste. The replacement of cement with 65% GGBFS led to a 60% reduction in static yield stress, while replacing GGBFS with CBP at 10%, 20%, and 30% increased the static yield stress by 58.6%, 48.2%, and 23.9%, respectively, which was consistent with the findings from the 20-layer structures.

The study also evaluated the 28-day compressive strength of the mortars and found that reducing the cement content with GGBFS and CBP led to a decrease in both early- and late-age strength. This was attributed to the fact that the strength contribution from the secondary hydration reaction of GGBFS and CBP could not compensate for the strength loss resulting from the reduction in cement content. Furthermore, the increase in porosity due to the presence of capillary pores during printing resulted in a 10%–50% reduction in compressive strength compared to cast-in-situ specimens. Additionally, the slower reaction kinetics of CBP also had a negative effect on the strength of the mortars. In terms of chloride ion penetration, the inclusion of GGBFS and CBP in the mortars led to a significant inhibition of chloride ion migration. This inhibition was attributed to the filling effect of GGBFS and CBP, which resulted in a denser and more impermeable microstructure. Overall, the study highlighted the influence of GGBFS and CBP on the mechanical properties, buildability, and chloride ion penetration of printed mortars. It provided valuable insights into the use of these recycled materials in 3D printed construction applications.

Ma, Li, et al. (2018) and Ma, Sun, et al. (2018) conducted a study to fabricate cementitious composites with the ability to absorb electromagnetic waves (EMWs) using copper slag (CS). The study involved seven different mix designs, ranging from 0 to 30 wt.% of CS, to investigate the printability and buildability of the mixtures. Additionally, five different mass ratios of copper powder, ranging from 2 to 10 wt.%, were incorporated to optimize the properties of the composites. The results of the study showed that the inclusion of CS and copper powder significantly improved the buildability of the mixtures, as indicated by the height to width ratio. This suggests that the mixtures had better structural integrity and were more suitable for 3D printing. Furthermore, the study also examined the mechanical properties of these mixtures in separate investigations (Sun, Huang, Aslani, Ma Huang, Aslani, Ma, 2020a, 2020b).

Wang et al. (2022) conducted a study using an orthogonal experimental design approach to investigate the effects of industrial wastes, namely GGBFS, FA, and SF, on the printability and mechanical performance of printed mortar. The study concluded that FA had a more pronounced influence on the fluidity and slump of the mortar compared to GGBFS and SF. Specifically, FA had a positive effect on the fluidity and slump properties of the mortar. The optimal mix design identified in the study consisted of 20 wt.% FA, 15 wt.% SF, and 10 wt.% GGBFS. This mix design resulted in the desired rheological properties for printing and achieved a compressive strength of 56.3 MPa, which was higher than the compressive strength achieved by cast specimens. Furthermore, the literature review mentioned that several other studies have reported their observations and results on the incorporation of slag into 3D printed cementitious composites (Chaves Figueiredo et al., 2019; Rahul et al., 2020).

14.2.4 Other supplementary cementitious materials

Besides the commonly used secondary SCMs, researchers have also attempted other secondary SCMs in 3DCP. Table 14.1 summarizes the available previous literature on the attempts of other secondary SCMs for 3DCP.

14.2.4.1 Recycled concrete powder

Qian et al. (2022) conducted a study on the use of recycled concrete powder (RCP) as a replacement for GGBFS and the incorporation of recycled concrete aggregate (RCA) as a substitute for natural aggregates in 3D printed mortar (3DPM). The study found that replacing conventional sand with RCA improved the yield stress of 3DPM due to the self-consolidation properties of unhydrated cement on the RCA and increased water absorption of the RCA. Furthermore, the compressive strength of 3DPM was enhanced with the addition of RCA, making it more suitable for 3D printing in the short term when 100% natural sand was substituted with RCA. On the other hand, substituting GGBFS with recycled ceramic powder reduced the hydration heat of 3DPM. By optimizing the gradation resulting from the

Table 14.1 Summary of previous literature of using other SCMs in 3DCP.

References	Waste materials	Source	Remarks
Qian et al. (2022)	Recycled construction powder	Construction and demolition wastes	Investigated the possibility of introducing recycled construction powder into 3DCP through particle packing optimization and analyzed its influences on fresh and mechanical properties
Li et al. (2023)	Incinerated sewage sludge ash (ISSA)	Local sewage sludge incineration plant	Proposed a low-pH mix design for 3DCP using solid wastes including waste glass powder, slag, ISSA, properties including workability and printability were assessed
Deng et al. (2023)	Waste glass powder	Waste glass bottles	Attempted replacing OPC with waste glass powder at ratios of 0–60 wt.% and compared rheological, mechanical properties, and drying shrinkage with cast-in-situ specimens
Rehman et al. (2020)	Incineration ash (FA and bottom ash)	Municipal solid wastes	Studied buildability and mechanical strengths of printed mortars incorporating two incineration ashes, FA and bottom ash
Shahzad et al. (2020)	Clinker	Waste solids	Studied the rheological and mechanical properties of printed mortar containing sulfoaluminate high-activity material (SHAM), which was made by glue gas desulfurization gypsum, calcium carbide slag, aluminum ash, and red mud
Vergara et al. (2023)	Biochar powder	Wood biomass	Reported the attempts of using biochar powder recycled from wood biomass gasification into printable mortar development
Daher et al. (2022)	Fluvial sediment	Flash calcination of crushed calcareous sand	Used recycled flash-calcined dredge sediment to replace cement in printable mortar development; extrudability, buildability, and compressive strength were assessed
Dey, Srinivas, Boddepalli, et al. (2022)	Limestone powder	Milled limestone	Studied the printability and anisotropic bonding strength of printed ternary Portland cement incorporating limestone powder

substitution of 10% GGBFS with RCP, the harmful macropore (>1000 nm) content in 3DPM decreased from 10.7% to 7.1%, and the yield stress of 3DPM increased.

The mechanical test results indicated that replacing 10% of GGBFS with RCP only slightly reduced the compressive strength of 3DPM from 49.5 to 47.2 MPa. However, increasing the volume of RCP to 20% further decreased the compressive strength to 44.1 MPa. To maintain the desired printing properties of 3DPM and maximize the volume of RCP, the study controlled the RCP content at 10%. Despite the drawbacks of high porosity and inadequate mineral composition of RCP, the study suggested that these issues could be partially compensated by the packing model (Qian et al., 2022).

14.2.4.2 *Recycled solid waste powder*

Li et al. (2023) conducted a study on the potential use of waste glass powder (WGP) and incinerated sewage sludge ash (ISSA) as replacements for SCMs in low-pH concrete for 3D printing. The study focused on assessing workability, printability, mechanical strength, and shrinkage resistance. The researchers found that the inclusion of WGP and ISSA in the conventional low-pH recipe, as substitutes for OPC and SF, improved printability by reducing yield stress and enhancing shape maintenance. This substitution helped mitigate the excessive yield stress caused by high SF dosage in low-pH recipes, resulting in improved extrudability and buildability. By partially replacing OPC and SF with WGP, slag, and ISSA, the study achieved a lower surface pH in the concrete. However, this replacement also led to a reduction in compressive strength. The study quantified the contribution of all binders to the surface pH, demonstrating the significant control of pH achieved through the inclusion of these materials. The researchers identified that reducing OPC content and increasing SF content is the most effective approach for developing low-pH concrete with desirable properties for 3D printing. This approach allows for better control of surface pH and improved printability while still maintaining acceptable levels of mechanical strength.

In a study by Deng et al. (2023), a mortar mix incorporating WGP as a partial replacement for cement in 3DCP was proposed. The researchers investigated the effects of different weight ratios of WGP (0%, 20%, 40%, and 60%) on various properties, including flowability, rheological properties, buildability, compressive strength, anisotropic effects, and drying shrinkage. The results of the study showed that when the WGP content was below 40%, its influence on properties such as fluidity, initial static yield stress, buildability, and drying shrinkage was limited. However, a noticeable reduction in compressive strength was still observed. On the other hand, when the WGP content reached 60%, it significantly reduced drying shrinkage by 50% compared to the control mix. However, the large quantity of WGP had detrimental effects on the initial fluidity, initial static yield stress, buildability, and compressive strength of the mortar. Based on their findings, the study recommended using a mix with 40% WGP as the optimum composition. This proportion provided a balance between achieving reduced drying shrinkage and maintaining acceptable levels of fluidity, static yield stress, buildability, and compressive strength (Deng et al., 2023).

In a study conducted by Rehman et al. (2020), the rheological and mechanical performances of cementitious composites incorporating two types of municipal solid waste (MSW) incineration ash, namely FA and bottom ash, were investigated for 3D printing applications. The researchers replaced cement with FA and bottom ash at different ratios (0%, 5%, 7.5%, 10%, and 15% by weight of cement) in the mixtures. The results showed that both FA and bottom ash had an impact on the setting time of the mixtures. They reduced the setting time, which is beneficial for maintaining the shape of the composites after extrusion. FA was found to be more effective than bottom ash in this regard. Furthermore, the inclusion of waste ashes contributed to the yield stress of the mixtures, which improved the shape stability and buildability of the composites. However, it negatively affected the opening time required for printing. Regarding the mechanical properties, the addition of incinerated ashes led to a reduction in the compressive strength of the composites at 28 days. The maximum reduction was observed when FA was replaced by 15%, resulting in a decrease of approximately 36% in compressive strength. Similarly, the inclusion of FA also had a negative influence on the bonding strength among layers. Overall, the study indicated that while the incorporation of MSW incineration ashes in cementitious composites for 3D printing had positive effects on flowability, yield stress, and buildability, it had a detrimental impact on setting time, compressive strength, and interlayer bonding strength.

Shahzad et al. (2020) conducted a study on the development of the construction of 3D printing material using sulfoaluminate high-activity material (SHAM) derived from solid wastes. The SHAM clinker was produced from a mixture consisting of 24% flue gas desulfurization gypsum, 22% calcium carbide slag, 25% aluminum ash, and 29% red mud. The production process involved a pilot rotary kiln with a capacity of 2 t/d, and the final calcination temperature was reported to be 1250ºC. The researchers optimized the mix design by adding 0.1 wt.% boric acid and 0.05 wt.% lithium carbonate to adjust the setting time of the mixtures. The water-to-solid ratio was set at 0.28, resulting in an initial setting time of 42 minutes and a final setting time of 62 minutes. At 28 days, the compressive strength of the material reached 97 MPa. The cost of the SHAM clinker was reported to be around 28–42 US$/metric ton, which was lower than the cost of OPC, which ranged from 56 to 71 US$/metric ton. In another study by Shahzad et al. (2021), the performance of the SHAM-based material was further investigated by incorporating a phase change material (PCM) with a chemical formula of C31H64 and a melting point of 52ºC–54ºC. The results showed that the inclusion of PCM not only achieved sufficient 28-day compressive strength but also addressed the problem of clogging during extrusion in the printing process.

14.2.4.3 Limestone powder

A study conducted by Dey, Srinivas, Boddepalli, et al. (2022) focused on the use of limestone as a partial replacement for FA in high-volume fly ash mortar for 3D printing. Ternary binder formulations were prepared with limestone at replacement levels of up to 15 wt.%. The study investigated the effects of limestone on the fresh properties, printability, and interlayer bond strength of the printed mortar. The inclusion of limestone in the ternary mixes led to changes in the fresh properties of the mortar, including slump, slump flow, and yield stress. It was observed that the addition of limestone affected these properties. In terms of printability, the mixes containing limestone exhibited improved buildability during 3D printing. However, the study also revealed that the interlayer bond strength decreased as the dosage of limestone increased. Notably, the bond strength was lower in the bottom layers compared to the middle and top layers. This reduction in bond strength in the bottom layers could be attributed to X-directional porosity resulting from an increased nozzle stand-off distance caused by the suppression of the bottom layer during the printing process.

14.3 Recycled aggregates

The United States Environmental Protection Agency (EPA) defines construction and demolition (C&D) materials as a combination of the debris generated during the construction, renovation, and demolition of buildings and infrastructures (U.S. Environmental Protection Agency, 2020). According to the factsheet provided by EPA released in 2020, in 2018, the recycling ratio reached 75.8% (U.S. Environmental Protection Agency, 2020). According to data from Ren et al. (2022), in 2018, China emerged a recycling rate of C&D debris less than 10% (approximately 3.6% of total waste generated). With regard to European Union, the EU also generates large quantity of C&D wastes, where France and Germany contributed 240 and 225 million tons to the total amount, respectively (UEPG European Aggregates Association, 2021). UEPG's annual review for 2019–2020 shows that the production of secondary aggregates comprising recycled plus re-used aggregates was 327 million tons, representing 10.6% of total production from EU27 (UEPG European Aggregates Association, 2020). These numbers and studies indicate that currently, there is a significant amount of untreated construction waste generated globally, and this number continues to increase at an alarming rate. From the perspective of sustainable development, especially considering the natural resource depletion and low recycling rate, the academic and industrial sectors of civil engineering have been stepping into the field of using recycled materials in construction materials. This section will discuss the studies, findings, and outcomes of previous studies on using high quality secondary raw materials in printable cementitious composites by pioneer researchers.

14.3.1 Recycled concrete aggregates

RCA refers to the crushed concrete that is obtained from the demolition of old concrete structures or from leftover concrete after construction. RCA is a sustainable alternative to natural aggregates and can be used in various construction applications. It consists of crushed concrete particles that retain the original properties of the parent concrete, such as strength and durability. RCA offers several benefits, including reduced waste disposal, conservation of natural resources, and lower environmental impact compared to traditional aggregates. Table 14.2 summarizes the previous literatures of using RCA or recycled sand (RS) in 3DCP.

Liu, Liu, Wu, Bai, He, Zhang et al. (2022) conducted a study on the mechanical properties of engineered cementitious composites (ECC) incorporating RCA. In their research, natural coarse aggregates (NCA) were partially replaced with FA in the control mix, and then RCA was used to replace NCA at ratios of 50 and 100 wt. %. The CA used in their study had particle sizes of 5–7, 7–10, and 10–12 mm with a mass ratio of 8:15:7. The results showed that the compressive and flexural strength of cast specimens decreased significantly with higher replacement ratios. For printed specimens, the compressive strength decreased by 30% regardless of anisotropy. The reduction in flexural strength was not as pronounced as the compressive strength at 28 days.

In a subsequent investigation, Liu, Liu, Wu, Bai, He, Yao, et al. (2022) explored the potential of printed ECC incorporating RCA at different replacement ratios and assessed the interlayer bonding properties of these mixes. The presence of RCA increased the surface roughness of the filaments, leading to the formation of large voids at the interfaces between layers. These large-sized pores at the interlayer interfaces played a crucial role in crack initiation and propagation, thereby degrading the bonding strength among the layers. However, the inclusion of RCA actually improved the interlayer bonding strength compared to printed mixtures, mainly due to the increased bond surface area resulting from the use of coarser aggregates. Nevertheless, as the replacement ratio of RCA increased to 100 wt.%, the interlayer bonding strength decreased, mainly attributed to the inherent weaknesses in the mechanical properties of the old cement matrix and poor adhesion at the Interface Transition Zone (ITZ).

Xiao et al. (2022) conducted a study on the mechanical properties of printed concrete using different aggregate combinations, including RCA and RFA. The results indicated that the inclusion of RCA in the concrete mixture improved buildability due to its higher water absorption and rough particle shapes. The higher water absorption and specific surface area of RCA also positively influenced the hydration process, particularly during air curing. The study found that the fully replaced concrete with recycled aggregates exhibited a minimum anisotropic compressive strength of 16.36 MPa, which was approximately 67.15% of the strength observed in the cast-in-situ specimen. When comparing the full replacement of RFA with the full replacement of RCA, the former showed an improvement in mechanical properties, with a 4.1 MPa increase in compressive strength at 28 days. However, this improvement was not observed under standard curing conditions. Additionally, using a combination of fully RCA and RFA resulted in a reduction in mechanical properties that was less than the sum of the reductions observed with each individual aggregate, regardless of the maintenance or pouring method employed (Xiao et al., 2022).

Table 14.2 Review of RCA in 3DCP.

References	Concrete type	Recycled aggregate	Replacement object[a]	Content[b]	w/b[c]	Fiber	Remarks
Liu, Liu, et al. (2022)	ECC	RCA	CA	0, 50, 100 wt.%	0.3	PVA	Investigated hardened properties of both cast and printed specimens
Liu, Liu, et al. (2022)	ECC	RFA	FA	0, 50, 100 wt.%	0.3	PVA	Studied the pore structure and bonding strength of cast and printed specimens
Xiao et al. (2022)	Concrete	RCA and RFA	CA, FA, CA + FA	100 vol.%	0.36		Studied mechanical properties of printed concrete with FA, CA, and CA + FA being replaced by RCA
Wu et al. (2021)	ECC	RCA	CA	0, 50, 100 wt.%	0.3	PVA	Mainly focused on the rheological properties and buildability of printed ECC
Xiao et al. (2020)	Mortar	RS	FA	0, 12.5, 25, 50 wt.%	0.35		Analyzed effects of using RS on mechanical properties and anisotropy of 3D printed mortar, also investigated the effects of curing age and nozzle height
Ding, Xiao, Qin, et al. (2020)	Mortar	RS	FA	0, 25, 50 wt. %	0.35		Studied the unconfined uniaxial compressive strength, load-displacement, lateral deformation, and stress-strain behavior of specimens after 2.5 hours of printing were measured and analyzed

Zou et al. (2021)	Mortar	RS	FA	0, 50, 100 wt.%	0.35		Focused on the rheological properties of printed mortar incorporating RS
Ding et al. (2021)	ECC	RS	FA	0 and 50 wt.%	0.35	PE	Studied the flexural behavior of ECC incorporating RS and PE fibers as well as the anisotropic effects
Xiao et al. (2021)	ECC	RS	FA	0 and 100 wt.%	0.35	PE	Studied the mechanical properties including compressive strength, direct and splitting tensile strength, and flexural strength of printed ECC with fully RS
Liu et al. (2023)	ECC	RS	FA	0, 50, 100 wt.%	0.3	PVA	Investigated the mechanical properties of printed ECC incorporating RS and correlation between strength and macroscopic constitution
De Vlieger et al. (2023)	ECC	RS	FA	25, 50, 100 vol.%		Not specified	Focused on the rheological properties and buildability of printed ECC containing RS

[a]Refer to fine aggregate (FA) or coarse aggregate (CA).
[b]With respect to the content of replacement object.
[c]w/b: Water-to-binder ratio.

In a separate study, Wu et al. (2021) investigated the rheological properties and buildability of printed ECC incorporating RCA as a replacement for NCA. The findings showed that as the content of recycled aggregates increased, the flowability of the fresh paste decreased, and the yield stress increased due to the higher specific area, water absorption, and internal friction induced by the recycled aggregates. During the initial 15 minutes of the printing process, the buildability of the recycled aggregate concrete improved with an increase in the replacement rate. However, this improvement came at the cost of a decreased available open time for printing (Wu et al., 2021).

14.3.2 Recycled sand

The topic of RS has received less attention compared to the extensive studies conducted on RCA for cementitious composites (Ding, Xiao, Zou, et al., 2020). This lack of discussion on the use of RS is primarily due to the perception that it is not preferred as fine aggregate in cementitious composites. One of the main reasons for this perception is that RS has a lower density and higher water absorption, which can be attributed to its porous structures and the presence of old cement paste attached to the surface (Dapena et al., 2011; Jiménez et al., 2013). In recent years, researchers have started to address the negative impacts associated with the inclusion of RS by adjusting the composition and have focused on developing strategies to overcome the challenges posed by RS in printed cementitious composites. This section aims to provide a summary of some of the studies conducted on printed cementitious composites that incorporate RS.

Xiao et al. (2020) and Ding, Xiao, Zou, et al. (2020) have been publishing a series of research articles on the use of recycled fine aggregate, specifically RS. Their initial investigations reported that increasing the replacement ratio from 0 to 50 wt.% decreased the 28-day compressive strength of cast-in-situ samples, with the reduction reaching approximately 25% at a 50% replacement ratio. This deterioration was attributed to two factors: (1) the porous microstructures of the old cement paste attached to the surface of RS, which are naturally weaker than the newly formed cement matrix, and (2) the presence of primary microcracks detected at the ITZ. Consequently, as the replacement ratio increased, the compressive strength of the concrete decreased. For printed specimens, they generally had lower compressive strength than the cast-in-situ specimens, regardless of the anisotropic effect. The minimum reduction in compressive strength was observed with mixes containing 12.5% RS and when the loading direction was perpendicular to the filament (Ding, Xiao, Zou, et al., 2020).

Regarding splitting tensile strength, the inclusion of RS decreased the tensile strength of cast mixtures from 2.2 to 1.4 MPa at a 50% replacement ratio. For printed specimens, the splitting tensile strength initially increased at 12.5% replacement and then decreased. However, it reached 2.8 MPa at a 50% replacement ratio in the Z-direction. However, this study did not provide a detailed explanation for this behavior. Similarly, the flexural strength slightly increased and peaked at a replacement ratio of 25%, and then decreased for cast specimens. For printed specimens, due to the existence of layered filament, they generally had higher flexural strength in the Z-direction compared to the cast-in-situ specimens. However, the improvement was less significant with higher RS content (Ding, Xiao, Zou, et al., 2020).

Additionally, Ding, Xiao, Qin, et al. (2020) measured the mechanical properties at early ages, specifically 2.5 hours after printing, using the same mixtures. Uniaxial unconfined compressive tests were conducted in this study. According to their results, at 2.5 hours of curing, the addition of RS in 3DPM shifted the failure pattern from significant lateral deformation and plastic behavior to minimal lateral deformation and a prominent shear failure plane at early ages. The addition of RS in the mortar transformed the material from a deformable state to a relatively stiff state, increasing peak load and reducing lateral deformation. The development of green strength and Young's modulus followed linear and quadratic trends, respectively, with a faster development observed with higher amounts of RS, especially in mature specimens (Ding, Xiao, Qin, et al., 2020).

Zou et al. (2021) conducted a study on the rheological properties of mixes incorporating RS. The study aimed to assess the rheology, failure pattern, and buildability of such mixes. Their findings revealed that the inclusion of RS increased the static yield stress, viscosity, and area of thixotropy hoop due to the higher specific area and water absorption of RS. To improve the open time for printing and maintain sufficient buildability, the researchers used HMPC and sodium gluconate to retard the setting time. In a separate study, Zhang et al. (2022) investigated the drying shrinkage of 3DPM with RS. The study concluded that the addition of recycled fine aggregates in mortar resulted in a decrease in flowability over time, which affected the open time for 3D printing processes. The 3D printing process itself accelerated the early-stage drying shrinkage of the mortar while delaying the later-stage development, leading to lower shrinkage strains at 60 days. Furthermore, the use of transverse and oblique printing paths in 3D printing mortar reduced the drying shrinkage strains compared to the longitudinal printing path (Zhang et al., 2022).

Following the results obtained with mortars, Ding et al. (2021) and Xiao et al. (2021) conducted trials involving the addition of PE into mixes incorporating RS, up to 100% replacement. Their findings revealed that the inclusion of PE fibers in the mixtures to form ECCs significantly improved the flexural strength of ECCs. This improvement was attributed to the bridging effect of fibers and the strong bonding between the fiber and cement matrix. However, the addition of PE fibers did not lead to a significant increase in flexural stiffness. These conclusions were supported by their scanning electron microscopy (SEM) results. Furthermore, the authors claimed that at a 100% replacement ratio, the specimens achieved higher compressive, flexural, and splitting tensile strength, as well as improved flexural capacity, demonstrating the potential of RS (Xiao et al., 2021). As part of their research, they also attempted to construct a small room using 3DCP with RS (Xiao et al., 2020).

Liu et al. (2023) conducted a study to investigate the mechanism of the formation and distribution of extruded pore defects on the compressive strength and constitutive model of printed ECC incorporating RS. Their results revealed that the porosity of ECCs increased by 2.28% and 4.19% at the 50% and 100% replacement ratios, respectively. Consequently, the inclusion of RS led to a deterioration in compressive strength. Furthermore, the shape of the extruded pore defects resembled an ellipsoid, and their distribution varied based on different loading directions. The occurrence and propagation of cracks were primarily influenced by the arrangement of these extruded pore defects (Liu et al., 2023).

De Vlieger et al. (2023) conducted a systematic investigation into the buildability of ECC incorporating RS for 3DCP. The study examined properties such as reflocculation rate (R_{thix}), flowability, rheology, and initial setting time using a rheometer and Vicat test. In their experiments, natural sand was replaced with RS at ratios of 25%, 50%, and 100% by volume. The results indicated that as the replacement ratio increased, the slump of the mortar decreased from 21.1 to 13.9 cm at 100%. This decrease was attributed to two factors:

- The RS had lower sphericity and roundness compared to natural sand, resulting in increased internal friction between particles and reduced flowability.
- The particle size distribution of RS was finer than that of natural sand, leading to a higher specific surface area, thus having higher water absorption.

Furthermore, the stress-growth test confirmed these observations. The static yield stress increased by 22%, 57%, and 201% at the replacement ratios of 25%, 50%, and 100% respectively. Similarly, the dynamic yield stress increased by 11%, 25%, and 96% compared to the control mix. These increases in yield stress were due to the heightened internal friction caused by the presence of RS, which improved the microstructure of the mortar. However, this also resulted in insufficient extrudability and pumpability of the ECCs. Additionally, the inclusion of RS led to a decrease in the initial setting time of ECCs. This was attributed to the presence of nonhydrated cement particles attached to the RS, which lowered the water-to-cement ratio and accelerated the hydration process (De Vlieger et al., 2023).

14.3.3 Waste glass

The EPA released a factsheet in 2022 that provides an overview of waste glass generation and recycling rates in the United States (U.S. Environmental Protection Agency, 2022). According to the factsheet, in 2018, a total of 12.3 million US tons of waste glass was generated in the United States. Out of this amount, 3.06 million tons (approximately 25%) were recycled, 1.64 million tons were burned for energy recovery, and a significant portion of 7.55 million tons (about 61%) was disposed of in landfills. On the other hand, data from the China National Resources Recycling Association (CRRA) for the same year, 2018, indicate that a total of 18.8 million tons of waste glass was generated in China (China National Resources Recycling Association CRRA, 2019). Out of this amount, approximately 10.4 million tons (around 55%) were recycled. These figures demonstrate that while the recycling rate for waste glass is relatively higher in China (55%), there is still significant room for improvement. In the United States, the recycling rate stands at around 25%, indicating a greater reliance on landfill disposal for waste glass. These statistics emphasize the importance of promoting recycling initiatives and implementing sustainable waste management strategies to reduce the environmental impact associated with waste glass and maximize its potential as a valuable resource. Table 14.3 summarizes the available studies on the usage of waste glass as aggregates for 3DCP. Fig. 14.6 presents the waste glass generated and recycled in the United States and China in 2018.

Ting et al. (2019) investigated the rheological and mechanical properties of mixes incorporating recycled glass as a FA for 3DCP. The mix design employed a cement-FA-SF ratio of 7:2:1 and an aggregate-to-binder ratio of 1.2, while maintaining an overall water-to-binder ratio of 0.46. The aggregates consisted of conventional river sand and recycled glass. Materials deposition was accomplished through a rectangular nozzle with dimensions of 15 × 30 mm. The study assessed both the fresh properties (such as static yield stress and dynamic yield stress) and hardened properties (including compressive strength, splitting tensile strength, and flexural strength) of the concrete mixes. The results indicated a reduction in mechanical strengths for the hardened specimens and decreased buildability of the fresh paste. This reduction was attributed to weak adhesion between the recycled glass and the

Table 14.3 Review summary of composites incorporating waste glass.

References	Glass aggregate fraction	w/b	Remarks
Ting et al. (2019)	100%	0.46	Studied the rheological and mechanical properties of printed mortar containing recycled glass aggregate as replacement of FA
Ting et al. (2021)	0, 25, 50, 75, 100 wt.%	0.35	Investigated rheological and mechanical properties of concrete incorporating waste glass aggregate at varied replacement ratios
Ting et al. (2021)	Medium: up to 32% Fine: up to 25% Superfine: up to 95%	0.31	Studied the effects of combinations of glass aggregate with different size, glass to binder ratio on rheological and mechanical properties of printed concrete, thus proposed optimum mix design with maximum buildability and extrudability
Liu, Li, et al. (2022)	Fine glass: 0%, 17%, 33%, 50% Coarse glass: 50%, 33%, 17%, 0%	0.24	Studied the effects of glass gradation on flexural strength of printed concrete and analyzed the microstructures of such mixtures; this study also involved coarse recycled glass aggregate
Cuevas et al. (2021)	0 and 50 vol.%	0.35	Studied effects of glass aggregates on mechanical, rheological, and thermal properties of lightweight concrete for 3DCP

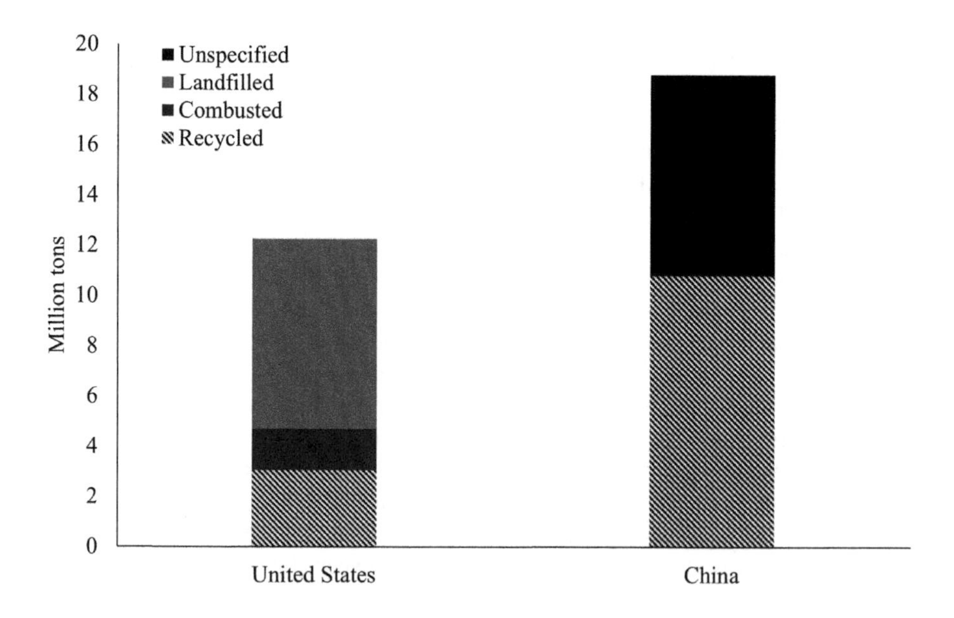

Figure 14.6 Glass waste generated in the United States and China in 2018.
Source: This figure was generated based on the data obtained from EPA and CRRA (China National Resources Recycling Association CRRA, 2019; U.S. Environmental Protection Agency, 2022).

cement matrix at the ITZ, as well as the particle packing density influenced by the gradation of fine glass particles (Mohajerani et al., 2017).

On the other hand, the inclusion of recycled glass aggregate resulted in better flowing properties compared to the sand mix, mainly due to its lower water absorption capacity than river sand (Tan & Du, 2013). However, the presence of recycled glass particles still led to a reduction in the mix's buildability, as reported in previous literature (Marchon et al., 2018; Panda et al., 2019; Qian & De Schutter, 2018).

Building upon these findings, Ting et al. (2021) further investigated the effects of using recycled glass cullet as a partial replacement for FA (river sand) at replacement ratios of 25%, 50%, 75%, and 100% by weight, specifically studying the potential for ASR expansion. The results showed that higher glass content increased ASR expansion but remained below the innocuous limit of 0.1% specified by ASTM C1260 (2022) (ASTM,60–22, 2022). The inclusion of glass particles also acted to reduce the alkalinity in the paste, thus mitigating ASR. However, experimental results still indicated that the presence of recycled glass, at any replacement ratio, reduced the pumpability and buildability of the fresh paste.

To address the challenges associated with the fresh properties of the cement paste when using recycled glass aggregate in 3DPC, Ting et al. (2021) explored the use of nanoclay at ratios of 0.2%, 0.9%, and 2.4% to optimize the mix design based

Table 14.4 Optimized mix design for mixes containing recycled glass aggregates (Ting et al., 2021).

	w/b	Glass/binder ratio	Aggregate ratio, %			Nanoclay content, %
			Medium	Fine	Superfine	
NC04	0.31	0.8	32.0	25.0	43.0	0.4
GB10	0.31	1.0	32.0	25.0	43.0	0.0
FM24	0.31	0.8	11.4	25.0	63.6	0.0

on mechanical and rheological properties for 3DCP. The proposed mix designs are summarized in Table 14.4.

In addition to investigating rheological and mechanical properties, Liu, Setunge, et al. (2022) conducted further research on the microstructures of 3D printed concrete containing recycled glass and its correlation with flexural properties. They performed experiments using X-ray micro-computed tomography, SEM, and three-point bending tests on a mix that incorporated 50 wt.% of recycled glass as both fine and coarse aggregates, with a water-to-binder ratio of 0.45. The results showed that the inclusion of coarse recycled glass increased the overall porosity of the specimen, while the impact of fine recycled glass on porosity was less significant. The presence of voids influenced crack propagation. Contrary to the hypothesis of weaker adhesion between recycled glass particles and the cement matrix, SEM micrographs revealed a well-bonded interface between the glass surface and the cement matrix at the ITZ. Furthermore, the bending test demonstrated that the incorporation of recycled glass aggregate reduced the flexural strength of the concrete in both loading directions, regardless of the printing direction (Liu, Li, et al., 2022).

Cuevas et al. (2021) investigated the rheological, thermal, and mechanical properties of lightweight concrete with incorporated waste glass aggregates as a replacement for basalt fine aggregate at 50 and 100 vol.% for 3DCP. The results indicated that replacing 50 and 100 vol.% of basalt aggregate with waste glass led to a reduction in the thermal conductivity of the composites by 11% and 17%, respectively. The incorporation of waste glass also resulted in a shorter initial setting time (50 vol.%: 2.4 hours, 100 vol.%: 3.1 hours, control: 3.5 hours) and final setting time (50 vol.%: 4.1 hours, 100 vol.%: 4.6 hours, control: 5.2 hours) of the mixes. The authors noted that minor adjustments to the mix composition were required to achieve sufficient flowability for the printing process. At 28 days of curing, incorporating 50 vol.% of waste glass showed a slight increase of approximately 0.2 MPa in flexural strength, while full replacement indicated a reduction in strength of approximately 0.4 MPa. The same trend was observed for compressive strength.

14.3.4 Recycled rubber

Fig. 14.7 provides a summary of global waste rubber generation by countries or regions, as reported by the World Business Council for Sustainable Development

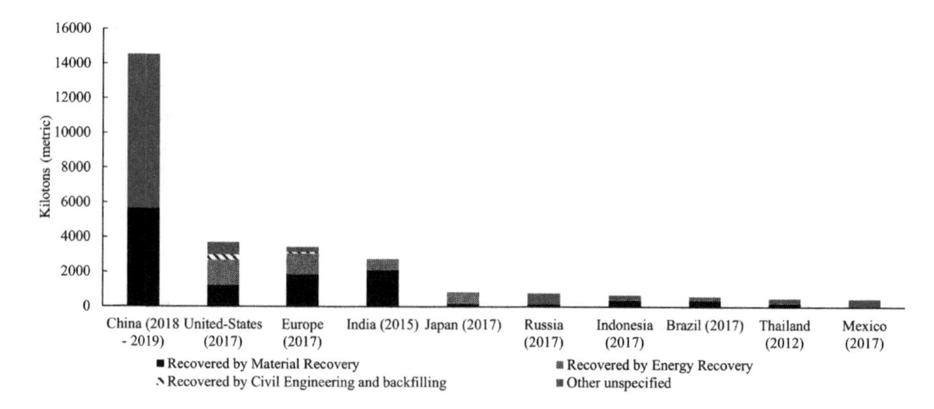

Figure 14.7 ELT generation and recovery by countries.
Source: This figure is re-generated based on data from WBCSD report (World Business Council for Sustainable Development).

WBCSD (2019); details can be found in Table 14.5. The figure presents a visual representation of the waste rubber generation trends across different locations. According to the Australia Department of the Environment, in the period of 2009–2010, Australia produced 48 million equivalent passenger units of end-of-life tires (ELTs). Out of this amount, 66% was disposed of in landfills, stockpiled, or illegally dumped, 16% was recycled, and 18% was exported (Australian Government Department of the Environment, 2014). The European Tyre and Rubber Manufacturers Association (ETRMA) reported that in 2018, the European Union (EU) generated 3.57 million tons of ELTs. Among this total, 50% of the ELTs were recycled, and 30% were used for energy consumption (ETRMA, 2018). The table provided in the source summarizes the percentages of rubber that were recovered, used for energy generation, utilized in civil engineering or backfilling applications, and includes other unspecified uses. These figures demonstrate the considerable amount of waste rubber being recovered and indicate the potential for incorporating waste rubber in civil engineering projects as a means of consuming recycled rubber materials. Table 14.6 summarizes the available studies into the use of crumb rubber (CR) in printable cementitious composites.

In the study conducted by Aslani et al. (2022), the focus was on understanding the properties and behavior of 3D printed rubberized cementitious composites incorporating PVA fibers. CR was used at various ratios (5, 10, 15, and 20 vol.%) relative to the fine aggregate, while the PVA content was controlled at 1.75 vol.% of the binder. As the rubber content increased, the flowability of the fresh paste decreased due to higher interparticle friction caused by the presence of rubber (Aslani et al., 2022; Siddika et al., 2019; Wanasinghe et al., 2022). To maintain the desired flowability for printing, the dosage of chemical admixtures needed to be adjusted accordingly. The compressive strength of the rubberized composites was reduced compared to the control mix. This reduction can be attributed to the weak adhesion at the ITZ between the rubber particles and the cement matrix (Aslani, 2016; Aslani et al., 2018; Zhang, Aslani, & Lehane, 2021). The minimum

Table 14.5 Percentages of waste rubber recovered (World Business Council for Sustainable Development, WBCSD, 2019).

	China (2017)	**United States (2017)**	**Europe (2017)**	**India (2015)**	**Japan (2017)**	**Russia (2017)**	**Indonesia (2017)**	**Brazil (2017)**	**Thailand (2012)**	**Mexico (2017)**
Subtotal material recovery	39%	33%	54%	76%	19%	19%	55%	64%	39%	6%
Subtotal Energy	0	39%	34%	22%	73%	1%	20%	35%	15%	14%
Subtotal civil engineering and backfilling	0	9%	3%	0	0	0	0	0	0	0
Other unspecified	61%	19%	8%	2%	8%	80%	25%	0	46%	80%

Table 14.6 Summary of previous literature on crumb rubber for 3DCP.

References	Replaced material	Rubber replacement ratio	Remarks
Aslani et al. (2022)	FA	5, 10, 15, and 20 vol.%	Studied the effects of fine crumb rubber on mechanical and shrinkage properties of PVA reinforced composites incorporating rubber
Liu and Tran (2022); Liu, Setunge, et al. (2022)	FA	15 wt.%	Studied the mechanical behaviors of printed rubberized mortar and analyzed the effects of surface treatment on mechanical and rheological properties of mixtures
Ye et al. (2021)	FA	0, 20, 40, 60 vol.%	Investigated the fresh and mechanical properties of printed ultrahigh ductile concrete with crumb rubber and PE fiber

reduction in compressive strength was observed at a rubber replacing ratio of 5 vol.% as shown in Fig. 14.8. The inclusion of CR resulted in an improvement in the 7-day flexural strength of the printed specimens. However, the flexural strength declined at 28 days as shown in Fig. 14.9. Among the different rubber ratios, the mix incorporating 5 vol.% of CR showed the least deterioration, with a reduction of 5.5 MPa compared to the control mix. The 3D printed specimens incorporating CR exhibited significantly less shrinkage strain compared to cast-in-situ specimens. At 56 days of curing, the perpendicular and parallel printing directions showed shrinkage strain reductions equivalent to 64% and 74%, respectively. This reduction in shrinkage strain can be attributed to the presence of more voids induced by the CR, which enhanced the water-blocking capacity. These findings provide valuable insights for the development and optimization of rubberized concrete mixtures for 3D printing applications.

As aforementioned, the weak adhesion at ITZ of rubberized concrete limits the use of rubber into concrete or cementitious composite development. To overcome the deterioration in strength due to rubber particles, researchers attempted using rubber surface treatment to achieve such aim. Liu and Tran (2022) and Liu, Setunge, et al. (2022) attempted coating CR particles with fresh cement paste for mixes incorporating CR at 15 wt.% with respect to fine aggregate. Their studies investigated the effect of factor of cement-to-rubber (C/R) ratios on mechanical performance of printed mixes. C/R ratios ranged from 0.25 to 0.55. In terms of coating process, five ceramic balls (weighted 11 g, diameter 15 mm) were placed into the container and the container was then rotated for 10 minutes at a speed of 400 rpm. Coated particles were air-dried for 24 hours and then air-cured for another 6 hours in a sealed plastic bag before mixing and printing. Their results show that with higher C/R ratio, compressive strength was

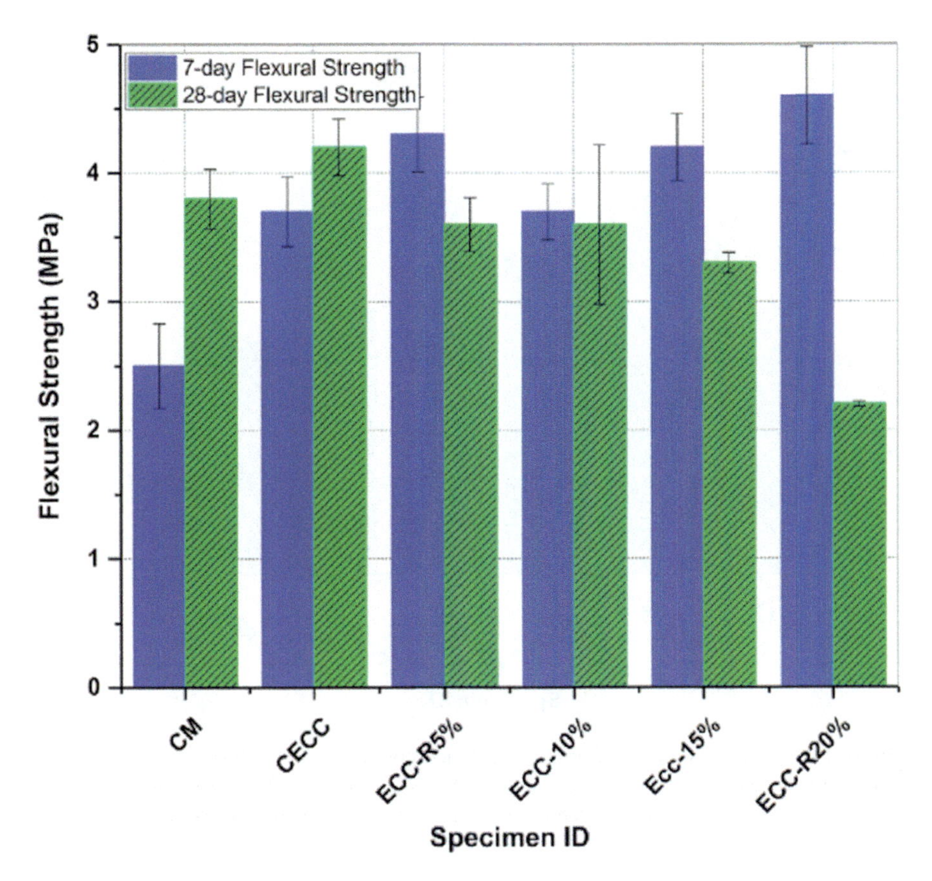

Figure 14.8 Compressive strength measurements.
Source: From Aslani, F., Dale, R., Hamidi, F., & Valizadeh, A. (2022). Mechanical and shrinkage performance of 3D-printed rubberised engineered cementitious composites. *Construction and Building Materials*, *339*, 127665.

improved up to 16.1% for loading direction in parallel to deposition direction and for loading direction parallel to printing direction, C/R ratio of 0.4 gave the highest improvement of 25.7% while C/R of 0.55 gave 23.9%. SEM results also proved the observations of increase in compressive strength due to the stronger bond between coating surface and cement matrix, thus densified ITZ. Besides, μCT images show that elongated pores were observed in printed specimens while cast-in-situ specimens presented spherical voids. Meanwhile, results also tells that the cast specimens had a constant higher percentage of large pores (>1 mm) compared to the printed, thus showing less compressive strength. Furthermore, their μCT images of posttest specimens with failure patterns suggested that with high C/R ratio (0.4 and 0.55), compressive strength was affected by the pore morphology and alignment relative to loading direction, while for low C/R ratio, weak bonding at ITZ plays an important role in compressive strength (Liu, Setunge, et al., 2022).

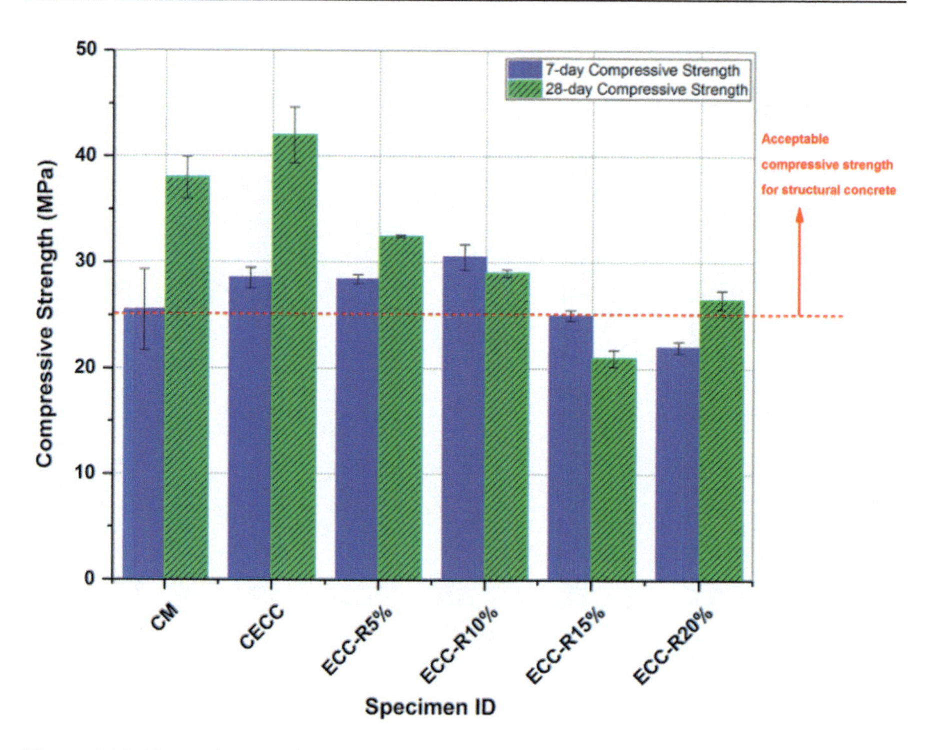

Figure 14.9 Flexural strength of printed ECC incorporating crumb rubber.
Source: From Aslani, F., Dale, R., Hamidi, F., & Valizadeh, A. (2022). Mechanical and shrinkage performance of 3D-printed rubberised engineered cementitious composites. *Construction and Building Materials, 339*, 127665.

In an effort to overcome the issue of weak adhesion at the ITZ in rubberized concrete, Liu and Tran (2022) and Liu, Setunge, et al. (2022) conducted studies on coating CR particles with fresh cement paste for mixes containing 15 wt.% of CR with respect to the fine aggregate. The research aimed to investigate the effect of the C/R ratios on the mechanical performance of the printed mixes. The C/R ratios ranged from 0.25 to 0.55. The coating process involved rotating the container with five ceramic balls and the CR particles for 10 minutes at a speed of 400 rpm. The coated particles were then air-dried for 24 hours and air-cured for another 6 hours in a sealed plastic bag before mixing and printing. Results revealed that higher C/R ratios led to an improvement in compressive strength. For the loading direction parallel to the deposition direction, the compressive strength improved up to 16.1%. In the case of the loading direction parallel to the printing direction, the C/R ratio of 0.4 resulted in the highest improvement of 25.7%, while a C/R ratio of 0.55 showed an improvement of 23.9%. The SEM results confirmed these observations, showing a stronger bond between the coating surface and the cement matrix, leading to densification of the ITZ. μCT images revealed that elongated pores were observed in the printed specimens, whereas cast-in-situ specimens exhibited spherical voids.

The cast specimens consistently had a higher percentage of large pores (>1 mm) compared to the printed specimens, resulting in lower compressive strength. The posttest μCT images of specimens with different failure patterns suggested that, with high C/R ratios (0.4 and 0.55), the compressive strength was influenced by the morphology and alignment of pores relative to the loading direction. In contrast, for low C/R ratios, the weak bonding at the ITZ played a significant role in determining the compressive strength. Overall, the research demonstrated that coating CR particles with fresh cement paste can improve the compressive strength of rubberized concrete. Higher C/R ratios and the resulting stronger bond between the coating surface and the cement matrix led to densification of the ITZ and enhanced mechanical performance. These findings provide insights into optimizing the mix design and surface treatment techniques for incorporating rubber particles into 3D printed concrete.

A study by Ye et al. (2021) then investigated the effects of including CR in ultrahigh ductile concrete (UHDC) that is desired for 3DCP. This research focused on evaluating the performance of printable UHDC in its fresh and hardened states. The findings revealed that 40% CR resulted in the optimal printable UHDC, with a tensile strength of 4.67 MPa and a strain of 7.50% after printing. The study also highlighted the importance of open time on flowability and printability, with an ideal range of 5–35 minutes. The printed UHDC exhibited enhanced flexural deformability and energy dissipation capacity compared to the cast but lower tensile strength and strain due to twisting fiber orientation. The material demonstrated minimal anisotropy in flexural strength but significant anisotropy in flexural deformability and compressive and flexural energy dissipation capacity. Additionally, the authors propose the incorporation of bionics in 3D printed structures, designing bio-inspired components such as bamboo-inspired columns, nacre-inspired beams, and honeycomb-inspired slabs.

14.3.5 Mining tailings

Álvarez-Fernández et al. (2021) conducted a study to investigate the inclusion of mining tailings with natural sand as an aggregate combination for printable concretes suitable for building purposes. They evaluated the properties of the concrete mix, including compressive strength, workability, and buildability. The tailings were characterized as fine sand, with 99% of the particles being smaller than 0.25 mm and an average size of 0.1 mm. The tailings consisted primarily of silica (80%), calcite (15%), and feldspars (5%). Mixes without natural sand exhibited extremely low compressive and tensile strength due to the weak structure of the tailings and high water-to-binder (w/b) ratio. However, by reducing the w/b and incorporating natural sand, both compressive and tensile strength were significantly improved. The optimum replacing ratio was found to be 20% by weight of the overall fine aggregate at a w/b ratio of 0.625. With this composition, the compressive strength reached 23.5 MPa, and the tensile strength reached 2.9 MPa. The experiment results indicated that the mortar incorporating tailings exhibited insufficient buildability for printing, as the bottom layers of the printed structures were significantly deformed with the deposition of more layers. The authors suggested that the use of SP and accelerators would be necessary to improve the strength and early strength development of the mortar, thereby enhancing buildability. Additional measures, such as adjusting the w/b ratio and including natural sand, as well as the use of SP and accelerators, are needed to optimize the workability and buildability of the mix for 3D printing applications. These findings provide insights into the potential utilization of mining tailings as a sustainable alternative in concrete production for construction purposes.

In a study by Ma, Li, et al. (2018), the potential of 3D printed ECC incorporating copper tailings as a replacement for fine aggregates was investigated. The researchers evaluated various properties, including extrudability, buildability, flowability, open time, fresh rheological properties, and hardened properties. The open time, which represents the duration during which the mixture maintains suitable flow properties for printing, was controlled at 70 minutes for all the mixtures. The printability of the mixtures was reported to be adequate for the printing process. River sand was replaced with copper tailings at different ratios ranging from 0 to 50 wt.%, with increments of 10%. However, mixtures with replacement ratios of 40% and 50% exhibited insufficient stiffness to resist deformation during deposition and showed high flowability. Additionally, polypropylene fibers (PP) were incorporated into the mixture. In terms of bending behavior, the average flexural strength and maximum deflection of the printed specimens were found to be reduced by 31.4% and 36.3%, respectively, compared to the cast-in-situ specimens. This deterioration in bending behavior was attributed to the combined effects of weak interfaces, a reduction in adhesion, and cold joint problems caused by the time gap between the deposition of two layers.

Li et al. (2020) conducted a study to examine the microstructures of printed mortar incorporating different contents of copper and iron tailings as a replacement for fine aggregate. The researchers used X-ray diffraction, Fourier-transform infrared spectroscopy, and SEM techniques to characterize the microstructures and gelation activities of the printed mortar. Based on the rheological and mechanical properties obtained, an optimum mix design was identified. It consisted of 10 wt.% copper tailings, 40% iron tailings, 19% FA, 30% belite cement, and 1% water reducing agent. This mixture exhibited an overall fluidity of 197.5 mm and an initial and final setting time of 50 and 82 minutes, respectively. The 28-day compressive strength of the printed mortar reached 45.2 MPa, with a flexural strength of 8.2 MPa. Microstructural analysis revealed that the dominant hydration products in the mortar incorporating tailings were C-S-H gel and ettringite. These products were closely interconnected, forming a web-like 3D structure that densified the entire mortar structure.

14.3.6 Other aggregates

Indeed, besides the commonly discussed recycled aggregates, there are several other types of recycled materials that hold potential for incorporation into materials for

3D concrete 3DCP. Some of these materials include plastics, bottom ashes, solid wastes, and various types of minerals.

14.3.6.1 Recycled plastics

Skibicki et al. (2021) conducted a study to evaluate the potential of using recycled polyethylene terephthalate (PET) as an aggregate in 3DPM. The researchers focused on assessing the compressive and flexural strength of the mortar. In their study, they replaced fine aggregate (sand) with 30 vol.% of PET. The results of the study showed that the incorporation of PET in the mortar led to a decrease in both compressive and flexural strength compared to the control mix. Specifically, at 7 days, the compressive and flexural strength of the PET-containing mortar were approximately 15% and 24% lower, respectively, than that of the control mix. The authors suggested that using PET in the development of concrete or mortar mixes for 3DCP could be a potential direction. However, they emphasized the need for further investigation and analysis. In particular, they highlighted the importance of assessing the mechanical properties at later ages, such as 28 days, to gain a more comprehensive understanding of the long-term performance of the PET-containing mortar.

14.3.6.2 Bottom slag aggregate

In a study by Butkute et al. (2023), the influence of carbonated bottom slag granules as a replacement for natural aggregates in mixtures containing CEM I 42.5R cement, CH, and burnt shale ash was investigated. Carbonated bottom slag, which is an artificial aggregate, was created using granulated and carbonated construction materials. The granules were made by combining bottom slag, burnt shale ash, hydrated lime, and cement. Bottom slag is a waste material generated after the municipal waste burning process in a cogeneration power plant. The bottom slag aggregates produced were characterized as fine-grained (<4 mm), porous, and lightweight, with a particle density ranging from 1610 to 1670 kg/m^3. The printed specimens incorporating carbonated bottom slag granules exhibited a slight increase in compressive strength, ranging from 3% to 9% compared to the control mix.

14.3.6.3 Solid wastes

In a study conducted by Bai et al. (2021), the use of underutilized solids and solid wastes as aggregates in concrete for 3D printing was investigated. Materials including desert sand (DS), river-sediment ceramsite sand (ceramsite: CPs), and RCA were tested. The study focused on evaluating the effects of particle gradations of these aggregates on rheological properties, mechanical properties, and shrinkage resistance of the printed concrete. The results revealed that the use of gradated particles in the concrete mixture led to a reduction in flowability. However, it also improved the structural build-up and interlayer bonding of the printed specimens. Furthermore, the addition of these underutilized aggregates helped to reduce shrinkage in the printed concrete. This indicates that the incorporation of such aggregates effectively utilized waste particles and contributed to the sustainability of 3D printing technology.

14.3.6.4 Spent garnet

In a study conducted by Skibicki et al. (2022), the use of spent garnet as a replacement for fine aggregate in printable mortar was investigated. The study involved replacing natural sand with spent garnet at varying ratios, ranging from 0% to 100% with an increment of 25%. The focus of the study was on evaluating the early age mechanical properties of the mortar mixtures incorporating spent garnet. The results indicated that as the replacement ratio of spent garnet increased, there was a decrease in the green strength of the mortar. Green strength refers to the ability of the material to support its own weight and resist deformation at an early age. Furthermore, the study found that the Young's modulus of the mortar decreased by up to 70% and 80% with increasing replacement ratios of spent garnet. This suggests a reduction in the stiffness and overall mechanical performance of the mortar. Based on their findings, the authors recommended a replacement ratio of 50% for maintaining sufficient printability and buildability of the mortar.

14.4 Recycled fibers

In a study conducted by Rajeev et al. (2023) from Swinburne University of Technology and RMIT University, the potential use of fiber recovered from face mask waste as reinforcement in 3DCP was explored. With the significant increase in the usage of medical surgical masks during the COVID-19 pandemic, the generation of related waste has become a concern in terms of waste management and sustainability. The study aimed to assess the extrudability, buildability, compressive strength, and flexural strength of 3DCP composites incorporating face mask fibers and compare them to cast-in-situ specimens. Additionally, the performance of the face mask fibers was compared to PP fibers commonly used as reinforcement. The results showed that the inclusion of face mask fibers significantly improved the rheological properties of the mixture, demonstrating good extrudability and buildability at all dosage levels. On the other hand, the mixes containing PP fibers exhibited poorer extrudability, particularly at higher fiber dosages. In terms of mechanical properties, the addition of 1% dosage of face mask fibers resulted in a substantial increase of 41% in compressive strength compared to the unreinforced concrete. The flexural strength along the weaker interface also showed remarkable improvements of 74% and 82% with the inclusion of 1% and 2% face mask fibers, respectively. Furthermore, the interlayer bond strength experienced a 21% enhancement with 1% face mask fiber content, which was also associated with the highest surface moisture content. The mechanical performance of the face mask fibers appeared to be comparable to that of PP fibers at a dosage of 1%. Additionally, the study found that the mechanical properties of both the printed and mold-cast specimens incorporating face mask fibers were similar, indicating that 3DCP can achieve comparable performance to traditional casting methods. These findings suggest that the utilization of face mask waste fibers as reinforcement in 3DCP has the potential to enhance the mechanical properties of the printed structures while addressing the issue of waste generated by the widespread use of medical surgical masks. Further research is needed to assess the long-term durability and other relevant properties of the printed structures incorporating face mask fibers.

In a study conducted by Mousavi et al. (2022), the potential of using recycled latex and vinyl gloves as fibers in mortar or concrete for 3D printing was explored. With the increased usage of gloves during the COVID-19 pandemic, the generation of waste gloves has become a significant concern. The study aimed to assess the effects of incorporating recycled latex and vinyl glove fibers, as well as nano graphene oxide (GO), on the flowability, buildability, and mechanical properties of the mixtures. Latex and vinyl gloves were recycled and cut into slices to be used as fibers in the mixtures at volumetric ratios of 0.1% and 0.2%. The average thickness and length of the recycled fibers were 0.2 and 16 mm, respectively. Nano GO was also included in the mixtures at a content of 0.03% by weight of cement to compensate for any undesired deterioration caused by the recycled fibers. The presence of the recycled fibers and nano GO increased the average flow diameter of the mixtures by approximately 10% on average, attributed to the increased internal friction caused by the fibers and nanoparticles. The buildability of the mixtures was significantly improved with the inclusion of fibers and nano GO, with latex fibers showing a more significant effect than vinyl fibers. However, the inclusion of latex and vinyl fibers resulted in a reduction in compressive strength, with higher fiber content leading to more substantial strength reduction. The maximum strength reduction was observed in the mix containing 0.2% vinyl fiber, with a decrease of 37.9%. Nevertheless, when nano GO was added to the mix containing 0.2% vinyl fiber, significant improvements of 60.5% and 61.3% were observed in direct tensile strength and flexural strength, respectively, compared to the control mix. Based on these findings, the authors concluded that the mixtures incorporating 0.1% latex fiber and GO, or 0.2% vinyl fiber and GO, could be potential alternatives. The addition of nano GO helped to enhance the mechanical properties and compensate for the strength reduction caused by the recycled fibers. Further research is needed to assess the long-term durability and other relevant properties of the printed structures incorporating recycled glove fibers and nano GO.

In a study conducted by Liu et al. (2023), the impact of layer offset on the mechanical properties of printed mortar incorporating recycled carbon fiber (rCF) was investigated. The study aimed to determine the optimal fiber fraction for printing and assess the effect of increasing fiber content on the fluidity, compressive strength, and flexural strength of the printed mortar. The researchers found that increasing the fiber content from 0 to 2.0 vol.% resulted in a decrease in the fluidity of the fresh mixture from 139 to 123 mm. This reduction in fluidity can be attributed to the increased presence of fibers, which tend to hinder the flow of the mixture. In terms of mechanical properties, the maximum compressive strength of 69.89 ± 2.86 MPa was observed in the mix containing 1.5 vol.% of rCF. This compressive strength value was 2.5% and 4% higher than the mix containing 1.0 and 0 vol.% of rCF, respectively. The flexural strength of the different mixes was maintained at a similar level of approximately 7.5 ± 0.5 MPa, indicating that the addition of rCF did not have a significant effect on the flexural strength of the printed mortar. Based on tests conducted on printability and buildability, the mix containing 1.5 vol.% of rCF exhibited adequate extrudability and buildability for printing purposes.

14.5 Geopolymer with secondary raw materials

In the study by Pasupathy et al. (2023), the utilization of alkali-activated brick waste powder as a binder in 3D printable geopolymer mixes is investigated. The incorporation of brick waste improved the fresh properties of the printable mixes, including flow, setting time, and rheological properties. However, higher brick waste content had a negative impact on the hardened properties such as compressive strength and interlayer strength, although incorporating up to 10% brick waste showed improvements in these properties. Furthermore, a sustainability assessment revealed that the proposed geopolymer concrete using brick waste has the potential to significantly reduce embodied energy and carbon emissions, achieving reductions of approximately 60%–80% compared to OPC concrete.

Demiral et al. (2022) explored the possibility of using demolition wastes in printable geopolymer mortar in their study of assessing the mechanical anisotropy and bonding properties of geopolymer incorporating recycled wastes. In this study, demolition wastes including hollow brick, red clay brick, root tile, concrete wastes, and glass wastes were incorporated into sodium hydroxide and CH geopolymerizations. These demolition wastes were used as fine aggregates (Demiral et al., 2022; Şahin et al., 2021). Their results reported that increasing molarity of NaOH had positive effect on compressive and flexural strength while splitting and direct strength were reduced due to the increase in viscosity and, hence, resulting in loss of adhesion in the interfacial bonding among layers. The inclusion of $Ca(OH)_2$ did not show expected enhancement due to the excessive increase in the viscosity. After 90 days of curing, both cast- in-situ and printed specimens reached a compressive strength of 20 MPa, which was deemed to be sufficient for masonry walls (Demiral et al., 2022). Another study from the same research group by Ilcan et al. (2022) explored the rheological properties of geopolymer composites incorporating these demolition wastes. According to their observations, flowability of mixture was heavily correlated to the molarity of NaOH solution, and the optimum molarity was found to be 10 M. And the inclusion of $Ca(OH)_2$ increased the viscosity and buildability but resulted in a loss of flowability on the contract. However, the presence of Na_2SiO_3 increased the initial yield strength and improved the flowability of the mixture but shortened the open time to less than 30 minutes, on the other hand. The paper also claimed that through adjusting the mix composition, the optimized mix could attain sufficient rheological properties for printing without any chemical admixtures and achieve a 28-day compressive strength of 36 MPa (Ilcan et al., 2022).

Valente et al. (2023) attempted using two types of waste rubber (rubber powder: 0–1 mm, CR: 1–3 mm) to replace natural sand used in geopolymer materials. In this study, natural sand was replaced with rubber at 50 and 100 vol.%, and properties including static and dynamic mechanical properties, thermos-acoustic insulation characteristics, and fresh properties were examined. Experiment results show that the inclusion of waste rubber significantly reduce the fluidity due to (1) the interlocking structure of rubber particles resisting the flowability of mixtures; and (2) larger surface area resulting in higher internal friction (Valente et al., 2023; Youssf et al., 2014). The results also indicated that mixtures incorporating waste rubber exhibited superior flexural performance when compared to OPC composites, primarily due to the enhanced compatibility of waste rubber with geopolymers. The optimal mixture consisted of 25% rubber powder and 75% coarser rubber, yielding a flexural strength of 2 MPa, thereby meeting the requirements outlined in the ASTM standard for structural lightweight concrete. Additionally, a thermal-acoustic assessment unveiled that the incorporation of rubber aggregates into geopolymers could enhance thermal insulation performance, owing to lower thermal conductivity, and improve low-frequency noise insulation compared to OPC materials (Valente et al., 2023).

14.6 Conclusion

This chapter introduces and examines the context of high-quality secondary raw materials suitable for application in 3DCP. The secondary raw materials covered in this chapter encompass secondary supplementary cementitious composites such as FA, SF, and slag, along with recycled aggregates, including RCAs, RS, waste glass, recycled rubber, tailings, and various other types of recycled aggregates. Furthermore, the chapter explores the incorporation of recycled fibers and geopolymer concrete utilizing secondary raw materials.

- The studies investigating the utilization of FA in 3DCP have demonstrated its influence on rheology, extrudability, buildability, and the potential for incorporating hybrid materials to further enhance the properties of printed structures. The incorporation of SF results in an increase in early-age strength by accelerating cement hydration, thereby enhancing the material's buildability. Previous research has also indicated that the use of GGBFS or other types of slag delays the cement hydration reaction, improving pumpability and extrudability. However, it is important to note that mixtures with a high slag content may experience strength degradation at both early and late stages, as the secondary hydration induced by slag may not fully compensate for the loss of strength due to the lower cement content.
- Research on the potential of various SCMs, such as RCP, ISSA, WGP, recycled FA, bottom ash, and SHAM clinker, has shown promise in replacing conventional cementitious materials in 3DCP. These studies have shed light on their effects on workability, printability, extrudability, and mechanical strength, indicating the potential to harness these waste materials for developing sustainable and cost-effective alternatives in 3DCP. It is crucial to recognize that the inclusion of these SCMs significantly impacts the rheological

properties of cementitious composites, and the extent of this influence varies among different types of SCMs and their respective proportions. Additionally, given that SCMs profoundly influence the fresh properties of composites, the incorporation of admixtures, including chemical admixtures and nanomaterials, becomes pivotal. The dosages of these admixtures should be meticulously tailored on a case-by-case basis to uphold the desired fresh properties for the printing process.

- Research findings regarding recycled aggregates indicate that their incorporation typically results in a decrease in mechanical properties due to their inherent weak structures and weaker bonding or adhesion to the cement matrix when compared to natural aggregates. Most recycled aggregates diminish the flowability of the mixture due to their higher internal friction or water absorption rates. However, this can actually enhance buildability. Conversely, recycled glass aggregate tends to improve flowability due to its lower water absorption rate but may reduce buildability. Studies have suggested that the use of recycled aggregates can be a sustainable option. Still, the mix design should be individually optimized, and adjusting rheological properties for printing may necessitate the use of chemical admixtures or nanomaterials.
- This chapter also encompasses three types of recycled fibers obtained from surgical masks, gloves, and rCF. A review of the literature indicates that the inclusion of fibers generally leads to reduced flowability and extrudability in fresh mixtures due to increased internal friction, but it notably enhances buildability. Fiber-reinforced composites tend to exhibit higher flexural strength due to the bridging effects of the fibers. The anisotropic effect is more pronounced in printed mixtures that incorporate fibers compared to other materials. However, the impact of fibers on compressive strength is subject to uncertainty and is influenced by factors such as fiber content and void structures within the printed specimens.
- This chapter also provides summaries of several studies on geopolymer concrete that incorporate waste or recycled materials for printing. The findings suggest that the decrease in mechanical properties of the mixtures may be attributed to the weak structure of recycled materials and the poor adhesion between particles. The rheological properties of the mixtures are influenced by the molarity of alkali solutions. However, experimental observations have also indicated that the inclusion of waste materials may increase internal friction, resulting in reduced rheological properties.
- Building upon these findings, the utilization of high-quality secondary raw materials offers not only sustainability benefits but also substantial contributions to improving rheological properties, attaining specific desired characteristics and enhancing the durability of composites, among other advantages. However, considering factors such as shape, size, and inherent properties like reactivity, surface condition, and strength, the incorporation of secondary materials in 3DCP concretes may yield contrasting outcomes. To mitigate any potential deterioration in rheological and mechanical properties, it is imperative to conduct thorough research and optimize the mix design. Additionally, the use of chemical admixtures, with carefully calibrated dosages, should be tailored to each specific case and rigorously studied.

References

Abualrous, Y., Panesar, D.K. Hooton, R.D., & Singh, B. (2016). Particle size analysis as a means to better understand the influence of fly ash variability in concrete (pp. 1424–1431). *Proceedings, annual conference*. Canada: Canadian Society for Civil Engineering.

Ambily, P. S., Kaliyavaradhan, S. K., & Rajendran, N. (2023). Top challenges to widespread 3D concrete printing (3DCP) adoption – A review. *European Journal of Environmental and Civil Engineering*, 1–29. Available from https://doi.org/10.1080/19648189.2023.2213294.

Álvarez-Fernández, M. I., Prendes-Gero, M. B., González-Nicieza, C., Guerrero-Miguel, D. J., & Martínez-Martínez, J. E. (2021). Optimum mix design for 3D concrete printing using mining tailings: A case study in Spain. *Sustainability*, *13*(3), 1–14. Available from https://doi.org/10.3390/su13031568, https://www.mdpi.com/2071-1050/13/3/1568/pdf.

Aslani, F., Dale, R., Hamidi, F., & Valizadeh, A. (2022). Mechanical and shrinkage performance of 3D-printed rubberised engineered cementitious composites. *Construction and Building Materials*, *339*, 127665. Available from https://doi.org/10.1016/j.conbuildmat.2022.127665.

Aslani, F., Ma, G., Yim Wan, D. L., & Muselin, G. (2018). Development of high-performance self-compacting concrete using waste recycled concrete aggregates and rubber granules. *Journal of Cleaner Production*, *182*, 553–566. Available from https://doi.org/10.1016/j.jclepro.2018.02.074.

Aslani, F. (2016). Mechanical properties of waste tire rubber concrete. *Journal of Materials in Civil Engineering*, *28*(3). Available from https://doi.org/10.1061/(ASCE)MT.1943-5533.0001429, http://ascelibrary.org/mto/resource/1/jmcee7/.

Assem, A. A., Hassan, M. A., Khandaker, M. L., & Hossain. (2002). Effect of metakaolin and silica fume on rheology of self-consolidating concrete. *ACI Materials Journal*, *109*(6).

ASTMC1260-22. (2022). Unpublished content standard test method for potential alkali reactivity of aggregates (Mortar-Bar Method).

ASTM C618-19. (2019). Unpublished content standard specification for coal fly ash and raw or calcined natural pozzolan for use in concrete.

ASTM F2792-12a. (2015). Unpublished content standard terminology for additive manufacturing technology ASTM F2792 – 12a (withdrawn).

Australian Government Department of the Environment. (2014). Unpublished content fact-sheet – Product stewardship for end-of-life tyres.

Bai, G., Wang, L., Ma, G., Sanjayan, J., & Bai, M. (2021). 3D printing eco-friendly concrete containing under-utilised and waste solids as aggregates. *Cement and Concrete Composites*, *120*, 104037. Available from https://doi.org/10.1016/j.cemconcomp.2021.104037.

Bakar, M.S. A., Sa'ude, N., Ibrahim, M., & Ismail, N.A. N. (2023). AIP Conference Proceedings American Institute of Physics Inc. Malaysia Additive Manufacturing (3D Printing): A review of current 3D concrete printing on materials, methods, applications, properties and challenges. Available from http://scitation.aip.org/content/aip/proceeding/aipcp 10.1063/5.0120868

Bauguess Brooke. (2022). ICON and Lennar announce community of 3D-printed homes is now underway in Georgetown, TX. Available from https://www.iconbuild.com/newsroom/icon-and-lennar-announce-community-of-3d-printed-homes-is-now-underway-in-georgetown-tx.

Bhattacherjee, S., Basavaraj, A. S., Rahul, A. V., Santhanam, M., Gettu, R., Panda, B., Schlangen, E., Chen, Y., Copuroglu, O., Ma, G., Wang, L., Basit Beigh, M. A., & Mechtcherine, V. (2021). Sustainable materials for 3D concrete printing. *Cement and Concrete Composites*, *122*, 104156. Available from https://doi.org/10.1016/j.cemconcomp.2021.104156.

Butkute, K., Vaitkevicius, V., Sinka, M., Augonis, A., & Korjakins, A. (2023). Influence of carbonated bottom slag granules in 3D concrete printing. *Materials*, *16*(11). Available from https://doi.org/10.3390/ma16114045, http://www.mdpi.com/journal/materials.

Chaves Figueiredo, S., Romero Rodríguez, C., Ahmed, Z. Y., Bos, D. H., Xu, Y., Salet, T. M., Çopuroğlu, O., Schlangen, E., & Bos, F. P. (2019). An approach to develop printable strain hardening cementitious composites. *Materials and Design*, *169*. Available from https://doi.org/10.1016/j.matdes.2019.107651, https://www.journals.elsevier.com/materials-and-design.

Chen, Y., He, S., Gan, Y., Çopuroğlu, O., Veer, F., & Schlangen, E. (2022). A review of printing strategies, sustainable cementitious materials and characterization methods in the context of extrusion-based 3D concrete printing. *Journal of Building Engineering*, *45*. Available from https://doi.org/10.1016/j.jobe.2021.103599, http://www.journals.elsevier.com/journal-of-building-engineering/.

China National Resources Recycling Association (CRRA). (2019). Unpublished content China's renewable resource recycling industry development report.

Cuevas, K., Chougan, M., Martin, F., Ghaffar, S. H., Stephan, D., & Sikora, P. (2021). 3D printable lightweight cementitious composites with incorporated waste glass aggregates and expanded microspheres – Rheological, thermal and mechanical properties. *Journal of Building Engineering*, *44*. Available from https://doi.org/10.1016/j.jobe.2021.102718, http://www.journals.elsevier.com/journal-of-building-engineering/.

Daher, J., Kleib, J., Benzerzour, M., Abriak, N. E., & Aouad, G. (2022). Recycling of flash-calcined dredged sediment for concrete 3D printing. *Buildings*, *12*(9). Available from https://doi.org/10.3390/buildings12091400, http://www.mdpi.com/journal/buildings.

Dapena, E., Alaejos, P., Lobet, A., & Pérez, D. (2011). Effect of recycled sand content on characteristics of mortars and concretes. *Journal of Materials in Civil Engineering*, *23* (4), 414–422. Available from https://doi.org/10.1061/(ASCE)MT.1943-5533.0000183.

Demiral, N. C., Ozkan Ekinci, M., Sahin, O., Ilcan, H., Kul, A., Yildirim, G., & Sahmaran, M. (2022). Mechanical anisotropy evaluation and bonding properties of 3D-printable construction and demolition waste-based geopolymer mortars. *Turkey Cement and Concrete Composites*, *134*. Available from https://doi.org/10.1016/j.cemconcomp.2022.104814, http://www.sciencedirect.com/science/journal/09589465.

Deng, Q., Zou, S., Xi, Y., & Singh, A. (2023). Development and characteristic of 3D-printable mortar with waste glass powder. *Buildings*, *13*(6). Available from https://doi.org/10.3390/buildings13061476, http://www.mdpi.com/journal/buildings.

Dey, D., Srinivas, D., Boddepalli, U., Panda, B., Gandhi, I. S. R., & Sitharam, T. G. (2022). 3D printability of ternary Portland cement mixes containing fly ash and limestone. *Materials Today: Proceedings*, *70*, 195–200. Available from https://doi.org/10.1016/j.matpr.2022.09.020, https://www.sciencedirect.com/journal/materials-today-proceedings.

Dey, D., Srinivas, D., Panda, B., Suraneni, P., & Sitharam, T. G. (2022). Use of industrial waste materials for 3D printing of sustainable concrete: A review. *Journal of Cleaner Production*, *340*, 130749. Available from https://doi.org/10.1016/j.jclepro.2022.130749.

Ding, T., Xiao, J., Qin, F., & Duan, Z. (2020). Mechanical behavior of 3D printed mortar with recycled sand at early ages. *Construction and Building Materials*, *248*, 118654. Available from https://doi.org/10.1016/j.conbuildmat.2020.118654.

Ding, T., Xiao, J., Zou, S., & Wang, Y. (2020). Hardened properties of layered 3D printed concrete with recycled sand. *Cement and Concrete Composites*, *113*, 103724. Available from https://doi.org/10.1016/j.cemconcomp.2020.103724.

Ding, T., Xiao, J., Zou, S., & Yu, J. (2021). Flexural properties of 3D printed fibre-reinforced concrete with recycled sand. *Construction and Building Materials*, *288*, 123077. Available from https://doi.org/10.1016/j.conbuildmat.2021.123077.

El-Chabib, H. (2020). *Properties of SCC with supplementary cementing materials* (pp. 283–308). Elsevier BV. Available from http://doi.org/10.1016/b978-0-12-817369-5.00011-8.

ETRMA. (2018). Unpublished content end of life tyres management – Europe – Status.

Everett H. (2021). World's longest 3D printed concrete pedestrian bridge unveiled in Nijmegen. Available from https://3dprintingindustry.com/news/worlds-longest-3d-printed-concrete-pedestrian-bridge-unveiled-in-nijmegen-195951/.

Feng, J., Sun, J., & Yan, P. (2018). The influence of ground fly ash on cement hydration and mechanical property of mortar. *Advances in Civil Engineering, 2018*. Available from https://doi.org/10.1155/2018/4023178, http://www.hindawi.com/journals/ace/.

Ferraris, C. F., Obla, K. H., & Hill, R. (2001). The influence of mineral admixtures on the rheology of cement paste and concrete. *Cement and Concrete Research, 31*(2), 245–255. Available from https://doi.org/10.1016/S0008-8846(00)00454-3, http://www.sciencedirect.com/science/journal/00088846.

Global Cement and Concrete Association (GCCA). (2023). Cement and concrete around the world. Available from https://gccassociation.org/concretefuture/cement-concrete-around-the-world/.

Hamidi, F., & Aslani, F. (2019). Additive manufacturing of cementitious composites: Materials, methods, potentials, and challenges. *Construction and Building Materials, 218*, 582–609. Available from https://doi.org/10.1016/j.conbuildmat.2019.05.140.

ICON Team. (2022). ICON to develop lunar surface construction system with $57.2 million NASA award. Available from https://www.iconbuild.com/newsroom/icon-to-develop-lunar-surface-construction-system-with-57-2-million-nasa-award.

Ilcan, H., Sahin, O., Kul, A., Yildirim, G., & Sahmaran, M. (2022). Rheological properties and compressive strength of construction and demolition waste-based geopolymer mortars for 3D-Printing. *Construction and Building Materials, 328*, 127114. Available from https://doi.org/10.1016/j.conbuildmat.2022.127114.

International Energy Agency. (2023). Unpublished content coal 2018: Analysis and forecasts.

Jiménez, J. R., Ayuso, J., López, M., Fernández, J. M., & De Brito, J. (2013). Use of fine recycled aggregates from ceramic waste in masonry mortar manufacturing. *Construction and Building Materials, 40*, 679–690. Available from https://doi.org/10.1016/j.conbuildmat.2012.11.036.

Kazemian, A., Yuan, X., Cochran, E., & Khoshnevis, B. (2017). Cementitious materials for construction-scale 3D printing: Laboratory testing of fresh printing mixture. *Construction and Building Materials, 145*, 639–647. Available from https://doi.org/10.1016/j.conbuildmat.2017.04.015.

Khayat, K. H., Sayed, M., & Yahia, A. (2008). Effect of supplementary cementitious materials on rheological properties, bleeding, and strength of structural grout. *ACI Materials Journal, 105*(6).

Liu, H., Liu, C., Wu, Y., Bai, G., He, C., Yao, Y., Zhang, R., & Wang, Y. (2022). 3D printing concrete with recycled coarse aggregates: The influence of pore structure on interlayer adhesion. *Cement and Concrete Composites, 134*. Available from https://doi.org/10.1016/j.cemconcomp.2022.104742, http://www.sciencedirect.com/science/journal/09589465.

Liu, H., Liu, C., Wu, Y., Bai, G., He, C., Zhang, R., & Wang, Y. (2022). Hardened properties of 3D printed concrete with recycled coarse aggregate. *Cement and Concrete Research, 159*, 106868. Available from https://doi.org/10.1016/j.cemconres.2022.106868.

Liu, J., Li, S., Gunasekara, C., Fox, K., & Tran, P. (2022). 3D-printed concrete with recycled glass: Effect of glass gradation on flexural strength and microstructure. *Construction and Building Materials, 314*, 125561. Available from https://doi.org/10.1016/j.conbuildmat.2021.125561.

Liu, J., Setunge, S., & Tran, P. (2022). 3D concrete printing with cement-coated recycled crumb rubber: Compressive and microstructural properties. *Construction and Building*

Materials, 347, 128507. Available from https://doi.org/10.1016/j.conbuildmat.2022.128507.

Liu, J., Tran, P., Nguyen Van, V., Gunasekara, C., & Setunge, S. (2023). 3D printing of cementitious mortar with milled recycled carbon fibres: Influences of filament offset on mechanical properties. *Cement and Concrete Composites, 142*, 105169. Available from https://doi.org/10.1016/j.cemconcomp.2023.105169.

Liu, J., & Tran, P. (2022). Experimental study on 3D-printed cementitious materials containing surface-modified recycled crumb rubber. *Materials Today: Proceedings, 70*, 90–94. Available from https://doi.org/10.1016/j.matpr.2022.08.550, https://www.sciencedirect.com/journal/materials-today-proceedings.

Liu, C., Wang, Z., Wu, Y., Liu, H., Zhang, T., Wang, X., & Zhang, W. (2023). 3D printing concrete with recycled sand: The influence mechanism of extruded pore defects on constitutive relationship. *Journal of Building Engineering, 68*. Available from https://doi.org/10.1016/j.jobe.2023.106169, http://www.journals.elsevier.com/journal-of-building-engineering/.

Li, X.-S., Li, L., & Zou, S. (2023). Developing low-pH 3D printing concrete using solid wastes. *Buildings, 13*(2), 454. Available from https://doi.org/10.3390/buildings13020454.

Li, X., Zhang, N., Yuan, J., Wang, X., Zhang, Y., Chen, F., & Zhang, Y. (2020). Preparation and microstructural characterization of a novel 3D printable building material composed of copper tailings and iron tailings. *Construction and Building Materials, 249*, 118779. Available from https://doi.org/10.1016/j.conbuildmat.2020.118779.

Marchon, D., Kawashima, S., Bessaies-Bey, H., Mantellato, S., & Ng, S. (2018). Hydration and rheology control of concrete for digital fabrication: Potential admixtures and cement chemistry. *Cement and Concrete Research, 112*, 96–110. Available from https://doi.org/10.1016/j.cemconres.2018.05.014, http://www.sciencedirect.com/science/journal/00088846.

Ma, G., Li, Z., & Wang, L. (2018). Printable properties of cementitious material containing copper tailings for extrusion based 3D printing. *Construction and Building Materials, 162*, 613–627. Available from https://doi.org/10.1016/j.conbuildmat.2017.12.051.

Ma, G., Sun, J., Wang, L., Aslani, F., & Liu, M. (2018). *Electromagnetic and microwave absorbing properties of cementitious composite for 3D printing containing waste copper solids, Cement and Concrete Composites* (94, pp. 215–225). Available from http://www.sciencedirect.com/science/journal/09589465, http://doi.org/10.1016/j.cemconcomp.2018.09.005.

Moeini, M. A., Hosseinpoor, M., & Yahia, A. (2022). 3D printing of cement-based materials with adapted buildability. *Construction and Building Materials, 337*, 127614. Available from https://doi.org/10.1016/j.conbuildmat.2022.127614.

Mohajerani, A., Vajna, J., Cheung, T. H. H., Kurmus, H., Arulrajah, A., & Horpibulsuk, S. (2017). Practical recycling applications of crushed waste glass in construction materials: A review. *Construction and Building Materials, 156*, 443–467. Available from https://doi.org/10.1016/j.conbuildmat.2017.09.005.

Mousavi, S. S., & Dehestani, M. (2022). Influence of latex and vinyl disposable gloves as recycled fibers in 3D printing sustainable mortars. *Sustainability, 14*(16). Available from https://doi.org/10.3390/su14169908, http://www.mdpi.com/journal/sustainability/.

Nath, P., & Sarker, P. (2011). Procedia Engineering Australia. Effect of fly ash on the durability properties of high strength concrete (pp. 1149–1156). Available from http://doi.org/10.1016/j.proeng.2011.07.144.

Nodehi, M., Ozbakkaloglu, T., & Gholampour, A. (2022). Effect of supplementary cementitious materials on properties of 3D printed conventional and alkali-activated concrete:

A review. *Automation in Construction*, *138*, 104215. Available from https://doi.org/10.1016/j.autcon.2022.104215.

Oey, T., Kumar, A., Bullard, J. W., Neithalath, N., & Sant, G. (2013). The filler effect: The influence of filler content and surface area on cementitious reaction rates. *Journal of the American Ceramic Society*, *96*(6), 1978–1990. Available from https://doi.org/10.1111/jace.12264.

Omran, A., Soliman, N., Xie, A., Davidenko, T., & Tagnit-Hamou, A. (2018). Field trials with concrete incorporating biomass-fly ash. *Construction and Building Materials*, *186*, 660–669. Available from https://doi.org/10.1016/j.conbuildmat.2018.07.084.

Panda, B., Ruan, S., Unluer, C., & Tan, M. J. (2019). Improving the 3D printability of high volume fly ash mixtures via the use of nano attapulgite clay. *Composites Part B: Engineering*, *165*, 75–83. Available from https://doi.org/10.1016/j.compositesb.2018.11.109.

Panda, B., & Tan, M. J. (2018). Experimental study on mix proportion and fresh properties of fly ash based geopolymer for 3D concrete printing. *Ceramics International*, *44*(9), 10258–10265. Available from https://doi.org/10.1016/j.ceramint.2018.03.031.

Panda, B., & Tan, M. J. (2019). Rheological behavior of high volume fly ash mixtures containing micro silica for digital construction application. *Materials Letters*, *237*, 348–351. Available from https://doi.org/10.1016/j.matlet.2018.11.131, http://www.journals.elsevier.com/materials-letters/.

Pasupathy, K., Ramakrishnan, S., & Sanjayan, J. (2023). 3D concrete printing of eco-friendly geopolymer containing brick waste. *Cement and Concrete Composites*, *138*, 104943. Available from https://doi.org/10.1016/j.cemconcomp.2023.104943.

Qian, Y., & De Schutter, G. (2018). Enhancing thixotropy of fresh cement pastes with nanoclay in presence of polycarboxylate ether superplasticizer (PCE). *Cement and Concrete Research*, *111*, 15–22. Available from https://doi.org/10.1016/j.cemconres.2018.06.013, http://www.sciencedirect.com/science/journal/00088846.

Qian, H., Hua, S., Yue, H., Feng, G., Qian, L., Jiang, W., & Zhang, L. (2022). Utilization of recycled construction powder in 3D concrete printable materials through particle packing optimization. *Journal of Building Engineering*, *61*, 105236. Available from https://doi.org/10.1016/j.jobe.2022.105236.

Rahul, A. V., Santhanam, M., Meena, H., & Ghani, Z. (2019). 3D printable concrete: Mixture design and test methods. *Cement and Concrete Composites*, *97*, 13–23. Available from https://doi.org/10.1016/j.cemconcomp.2018.12.014, http://www.sciencedirect.com/science/journal/09589465.

Rahul, A. V., Sharma, A., & Santhanam, M. (2020). A desorptivity-based approach for the assessment of phase separation during extrusion of cementitious materials. *Cement and Concrete Composites*, *108*, 103546. Available from https://doi.org/10.1016/j.cemconcomp.2020.103546.

Rajeev, P., Ramesh, A., Navaratnam, S., & Sanjayan, J. (2023). Using fibre recovered from face mask waste to improve printability in 3D concrete printing. *Cement and Concrete Composites*, *139*, 105047. Available from https://doi.org/10.1016/j.cemconcomp.2023.105047.

Ramanathan, S., Croly, M., & Suraneni, P. (2020). Comparison of the effects that supplementary cementitious materials replacement levels have on cementitious paste properties. *Cement and Concrete Composites*, *112*, 103678. Available from https://doi.org/10.1016/j.cemconcomp.2020.103678.

Rehman, A. U., Lee, S. M., & Kim, J. H. (2020). Use of municipal solid waste incineration ash in 3D printable concrete. *Process Safety and Environmental Protection*, *142*,

219–228. Available from https://doi.org/10.1016/j.psep.2020.06.018, http://www.elsevier.com/wps/find/journaldescription.cws_home/713889/description#description.

Ren, Z., Jiang, M., Chen, D., Yu, Y., Li, F., Xu, M., Bringezu, S., & Zhu, B. (2022). Stocks and flows of sand, gravel, and crushed stone in China (1978–2018): Evidence of the peaking and structural transformation of supply and demand. *Resources, Conservation and Recycling, 180*. Available from https://doi.org/10.1016/j.resconrec.2022.106173, http://www.elsevier.com/locate/resconrec.

Roussel, N. (2018). Rheological requirements for printable concretes. *Cement and Concrete Research, 112*, 76–85. Available from https://doi.org/10.1016/j.cemconres.2018.04.005, http://www.sciencedirect.com/science/journal/00088846.

Şahin, O., İlcan, H., Ateşli, A. T., Kul, A., Yıldırım, G., & Şahmaran, M. (2021). Construction and demolition waste-based geopolymers suited for use in 3-dimensional additive manufacturing. *Cement and Concrete Composites, 121*. Available from https://doi.org/10.1016/j.cemconcomp.2021.104088, http://www.sciencedirect.com/science/journal/09589465.

Scrivener, K. L., John, V. M., & Gartner, E. M. (2018). Eco-efficient cements: Potential economically viable solutions for a low-CO_2 cement-based materials industry. *Cement and Concrete Research, 114*, 2–26. Available from https://doi.org/10.1016/j.cemconres.2018.03.015, http://www.sciencedirect.com/science/journal/00088846.

Shahzad, Q., Shen, J., Naseem, R., Yao, Y., Waqar, S., & Liu, W. (2021). Influence of phase change material on concrete behavior for construction 3D printing. *Construction and Building Materials, 309*, 125121. Available from https://doi.org/10.1016/j.conbuildmat.2021.125121.

Shahzad, Q., Wang, X., Wang, W., Wan, Y., Li, G., Ren, C., & Mao, Y. (2020). Coordinated adjustment and optimization of setting time, flowability, and mechanical strength for construction 3D printing material derived from solid waste. *Construction and Building Materials, 259*, 119854. Available from https://doi.org/10.1016/j.conbuildmat.2020.119854.

Sheikh, T., & Ahmad, H. (2023). Concrete, CO_2, and catalysis: Merging industry and research goals for sustainable development.

Siddika, A., Mamun, M. A. A., Alyousef, R., Amran, Y. H. M., Aslani, F., & Alabduljabbar, H. (2019). Properties and utilizations of waste tire rubber in concrete: A review. *Construction and Building Materials, 224*, 711–731. Available from https://doi.org/10.1016/j.conbuildmat.2019.07.108, https://www.journals.elsevier.com/construction-and-building-materials.

Skibicki, S., Jakubowska, P., Kaszyńska, M., Sibera, D., Cendrowski, K., & Hoffmann, M. (2022). Early-age mechanical properties of 3d-printed mortar with spent garnet. *Materials, 15*(1). Available from https://www.mdpi.com/1996-1944/15/1/100/pdf, http://doi.org/10.3390/ma15010100.

Skibicki, S., Pultorak, M., & Kaszynska, M. (2021). Evaluation of material modification using PET in 3D concrete printing technology. *IOP Conference Series: Materials Science and Engineering, 1044*(1), 012002. Available from https://doi.org/10.1088/1757-899x/1044/1/012002.

Soltan, D. G., & Li, V. C. (2018). A self-reinforced cementitious composite for building-scale 3D printing. *Cement and Concrete Composites, 90*, 1–13. Available from https://doi.org/10.1016/j.cemconcomp.2018.03.017, http://www.sciencedirect.com/science/journal/09589465.

Suiker, A. S. J. (2018). Mechanical performance of wall structures in 3D printing processes: Theory, design tools and experiments. *International Journal of Mechanical Sciences, 137*, 145–170. Available from https://doi.org/10.1016/j.ijmecsci.2018.01.010.

Sun, J., Aslani, F., Lu, J., Wang, L., Huang, Y., & Ma, G. (2021). Fibre-reinforced lightweight engineered cementitious composites for 3D concrete printing. *Ceramics International*, *47*(19), 27107–27121. Available from https://doi.org/10.1016/j.ceramint.2021.06.124, https://www.journals.elsevier.com/ceramics-international.

Sun, J., Huang, Y., Aslani, F., & Ma, G. (2020). Properties of a double-layer EMW-absorbing structure containing a graded nano-sized absorbent combing extruded and sprayed 3D printing. *Construction and Building Materials*, *261*, 120031. Available from https://doi.org/10.1016/j.conbuildmat.2020.120031.

Sun, J., Huang, Y., Aslani, F., & Ma, G. (2020). Electromagnetic wave absorbing performance of 3D printed wave-shape copper solid cementitious element. *Cement and Concrete Composites*, *114*, 103789. Available from https://doi.org/10.1016/j.cemconcomp.2020.103789.

Suraneni, P., Azad, V. J., Isgor, O. B., & Weiss, W. J. (2016). Calcium oxychloride formation in pastes containing supplementary cementitious materials: Thoughts on the role of cement and supplementary cementitious materials reactivity. *RILEM Technical Letters*, *1*, 24–30. Available from https://doi.org/10.21809/rilemtechlett.2016.7, http://letters.rilem.net/index.php/rilem/article/download/7/10.

Suraneni, P. (2021). Recent developments in reactivity testing of supplementary cementitious materials. *RILEM Technical Letters*, *6*, 131–139. Available from https://doi.org/10.21809/rilemtechlett.2021.150, https://letters.rilem.net/2f8ebe8a-dc06-42a2-8086-becae0520e35.

Tan, K. H., & Du, H. (2013). Use of waste glass as sand in mortar: Part I – Fresh, mechanical and durability properties. *Cement and Concrete Composites*, *35*(1), 109–117. Available from https://doi.org/10.1016/j.cemconcomp.2012.08.028.

Ting, G. H. A., Tay, Y. W. D., Qian, Y., & Tan, M. J. (2019). Utilization of recycled glass for 3D concrete printing: Rheological and mechanical properties. *Journal of Material Cycles and Waste Management*, *21*(4), 994–1003. Available from https://doi.org/10.1007/s10163-019-00857-x.

Ting, G. H. A., Tay, Y. W. D., & Tan, M. J. (2021). Experimental measurement on the effects of recycled glass cullets as aggregates for construction 3D printing. *Journal of Cleaner Production*, *300*. Available from https://doi.org/10.1016/j.jclepro.2021.126919, https://www.journals.elsevier.com/journal-of-cleaner-production.

U.S. Environmental Protection Agency. (2022). Glass: Material – Specific data. Available from https://www.epa.gov/facts-and-figures-about-materials-waste-and-recycling/glass-material-specific-data.

U.S. Environmental Protection Agency. (2020). Unpublished content. Advancing sustainable materials management: 2018 fact sheet.

U.S. Geological Survey. (2022). Unpublished content mineral commodity summaries 2022 (sand and gravel).

UEPG European Aggregates Association. (2020). Unpublished content UEPG annual review 2019–2020.

UEPG European Aggregates Association. (2021). Unpublished content UEPG annual review 2020–2021.

UNEP. (2010). Unpublished content thematic focus: Resource efficiency, harmful substances and hazardous waste, and climate change greening cement production has a big role to play in reducing greenhouse gas emissions.

Valente, M., Sambucci, M., Chougan, M., & Ghaffar, S. H. (2023). Composite alkali-activated materials with waste tire rubber designed for additive manufacturing: an eco-sustainable and energy saving approach. *Journal of Materials Research and Technology*,

24, 3098–3117. Available from https://doi.org/10.1016/j.jmrt.2023.03.213, http://www.elsevier.com/journals/journal-of-materials-research-and-technology/2238-7854.

Vergara, L. A., Perez, J. F., & Colorado, H. A. (2023). 3D printing of ordinary Portland cement with waste wood derived biochar obtained from gasification. *Case Studies in Construction Materials*, *18*. Available from https://doi.org/10.1016/j.cscm.2023.e02117, http://www.journals.elsevier.com/case-studies-in-construction-materials/.

De Vlieger, J., Boehme, L., Blaakmeer, J., & Li, J. (2023). Buildability assessment of mortar with fine recycled aggregates for 3D printing. *Construction and Building Materials*, *367*, 130313. Available from https://doi.org/10.1016/j.conbuildmat.2023.130313.

Wanasinghe, D., Aslani, F., & Dai, K. (2022). Effect of age and waste crumb rubber aggregate proportions on flexural characteristics of self-compacting rubberized concrete. *Structural Concrete*, *23*(4), 2041–2060. Available from https://doi.org/10.1002/suco.202000597, http://onlinelibrary.wiley.com/journal/10.1002/(ISSN)1751-7648.

Wang, B., Zhai, M., Yao, X., Wu, Q., Yang, M., Wang, X., Huang, J., & Zhao, H. (2022). Printable and mechanical performance of 3D printed concrete employing multiple industrial wastes. *MDPI, China Buildings*, *12*(3). Available from https://doi.org/10.3390/buildings12030374, https://www.mdpi.com/2075-5309/12/3/374/pdf.

Weng, Y., Ruan, S., Li, M., Mo, L., Unluer, C., Tan, M. J., & Qian, S. (2019). Feasibility study on sustainable magnesium potassium phosphate cement paste for 3D printing. *Construction and Building Materials*, *221*, 595–603. Available from https://doi.org/10.1016/j.conbuildmat.2019.05.053.

World Business Council for Sustainable Development (WBCSD). (2019). Unpublished content Global ELT Management – A global state of knowledge on regulation, management systems, impacts of recovery and technologies.

Wu, Y., Liu, C., Liu, H., Zhang, Z., He, C., Liu, S., Zhang, R., Wang, Y., & Bai, G. (2021). Study on the rheology and buildability of 3D printed concrete with recycled coarse aggregates. *Journal of Building Engineering*, *42*, 103030. Available from https://doi.org/10.1016/j.jobe.2021.103030.

Xiao, J., Lv, Z., Duan, Z., & Hou, S. (2022). Study on preparation and mechanical properties of 3D printed concrete with different aggregate combinations. *Journal of Building Engineering*, *51*, 104282. Available from https://doi.org/10.1016/j.jobe.2022.104282.

Xiao, J., Zou, S., Ding, T., Duan, Z., & Liu, Q. (2021). Fiber-reinforced mortar with 100% recycled fine aggregates: A cleaner perspective on 3D printing. *Journal of Cleaner Production*, *319*, 128720. Available from https://doi.org/10.1016/j.jclepro.2021.128720.

Xiao, J., Zou, S., Yu, Y., Wang, Y., Ding, T., Zhu, Y., Yu, J., Li, S., Duan, Z., Wu, Y., & Li, L. (2020). 3D recycled mortar printing: System development, process design, material properties and on-site printing. *Journal of Building Engineering*, *32*, 101779. Available from https://doi.org/10.1016/j.jobe.2020.101779.

Xu, Z., Zhang, D., Li, H., Sun, X., Zhao, K., & Wang, Y. (2022). Effect of FA and GGBFS on compressive strength, rheology, and printing properties of cement-based 3D printing material. *Construction and Building Materials*, *339*, 127685. Available from https://doi.org/10.1016/j.conbuildmat.2022.127685.

Ye, J., Cui, C., Yu, J., Yu, K., & Xiao, J. (2021). Fresh and anisotropic-mechanical properties of 3D printable ultra-high ductile concrete with crumb rubber. *Composites Part B: Engineering*, *211*, 108639. Available from https://doi.org/10.1016/j.compositesb.2021.108639.

Youssf, O., Elgawady, M. A., Mills, J. E., & Ma, X. (2014). An experimental investigation of crumb rubber concrete confined by fibre reinforced polymer tubes. *Construction and*

Building Materials, *53*, 522–532. Available from https://doi.org/10.1016/j.conbuildmat.2013.12.007.

Yuan, Q., Zhou, D., Li, B., Huang, H., & Shi, C. (2018). Effect of mineral admixtures on the structural build-up of cement paste. *Construction and Building Materials*, *160*, 117–126. Available from https://doi.org/10.1016/j.conbuildmat.2017.11.050.

Zahabizadeh, B., Pereira, J., Gonçalves, C., Pereira, E. N. B., & Cunha, V. M. C. F. (2021). Influence of the printing direction and age on the mechanical properties of 3D printed concrete. *Materials and Structures/Materiaux et Constructions*, *54*(2). Available from https://doi.org/10.1617/s11527-021-01660-7, https://www.springer.com/journal/11527.

Zhang, Y., & Aslani, F. (2021). Development of fibre reinforced engineered cementitious composite using polyvinyl alcohol fibre and activated carbon powder for 3D concrete printing. *Construction and Building Materials*, *303*. Available from https://doi.org/10.1016/j.conbuildmat.2021.124453, https://www.journals.elsevier.com/construction-and-building-materials.

Zhang, Y., Aslani, F., & Lehane, B. (2021). Compressive strength of rubberized concrete: Regression and GA-BPNN approaches using ultrasonic pulse velocity. *Construction and Building Materials*, *307*, 124951. Available from https://doi.org/10.1016/j.conbuildmat.2021.124951.

Zhang, H., Xiao, J., Duan, Z., Zou, S., & Xia, B. (2022). Effects of printing paths and recycled fines on drying shrinkage of 3D printed mortar. *Construction and Building Materials*, *342*, 128007. Available from https://doi.org/10.1016/j.conbuildmat.2022.128007.

Zhang, Y., Zhang, Y., Liu, G., Yang, Y., Wu, M., & Pang, B. (2018). Fresh properties of a novel 3D printing concrete ink. *Construction and Building Materials*, *174*, 263–271. Available from https://doi.org/10.1016/j.conbuildmat.2018.04.115.

Zhao, Y., Gao, Y., Chen, G., Li, S., Singh, A., Luo, X., Liu, C., Gao, J., & Du, H. (2023). Development of low-carbon materials from GGBS and clay brick powder for 3D concrete printing. *Construction and Building Materials*, *383*. Available from https://doi.org/10.1016/j.conbuildmat.2023.131232, https://www.journals.elsevier.com/construction-and-building-materials.

Zou, S., Xiao, J., Duan, Z., Ding, T., & Hou, S. (2021). On rheology of mortar with recycled fine aggregate for 3D printing. *Construction and Building Materials*, *311*, 125312. Available from https://doi.org/10.1016/j.conbuildmat.2021.125312.

Sustainable seawater sea-sand concrete materials and structures

Feng Yu[1], *Siqi Ding*[2], *Ashraf Ashour*[3], *Sufen Dong*[1] *and Baoguo Han*[1]
[1]School of Infrastructure Engineering, Dalian University of Technology, Dalian, P.R. China, [2]Department of Civil and Environmental Engineering, The Hong Kong Polytechnic University, Hong Kong, P.R. China, [3]Faculty of Engineering and Digital Technologies, University of Bradford, Bradford, United Kingdom

15.1 Introduction

With the rapid development of modern infrastructures, the global production of concrete has reached 30 billion tons per year, resulting in an increasing demand for natural resources, including fresh water and river-sand (Miller et al., 2018; Monteiro et al., 2017). This situation has become one of the critical challenges for the sustainable development of concrete. Global freshwater resources only account for 2.5% of the total water resources, and less than 1% of freshwater resources are accessible to human beings directly, leading to water scarcity for 2–3 billion people (Ovink et al., 2023), becoming a crucial factor limiting economic and social development. Moreover, the global production of aggregates reaches 40–50 billion tons per year (Tabsh & Alhoubi, 2022). The excessive exploitation of river-sand resources has seriously threatened flood control, river stability, and navigation safety in inland regions. Additionally, constructing infrastructures in coastal and marine areas requires the long-distance transportation of freshwater and river-sand resources from inland regions, substantially increasing project costs, construction periods, and carbon emissions (Xiao et al., 2017). Considering that the Earth can provide nearly infinite seawater resources and abundant sea-sand resources, seawater sea-sand concrete (SSC) manufactured by seawater and sea-sand instead of fresh water and river-sand, respectively, can contribute to alleviating resource scarcity, reducing construction costs, improving construction efficiency, reducing carbon footprints of long-distance transportation, conserving inland resources together with protecting river ecology (Teng et al., 2019). Therefore sustainable SSC has broad application prospects in coastal and marine infrastructures.

Nevertheless, using seawater and sea-sand inevitably introduces large quantities of chloride ions into concrete, leading to steel corrosion of reinforced concrete structures (Dhondy et al., 2019). In order to ensure the durability of building structures, current specifications generally stipulate that only desalinated sea-sand can be used in concrete production. Additionally, these regulations impose strict limits on the chloride ion content of raw materials. However, the desalination of seawater and sea-sand is expected to result in high fresh water consumption and costs, posing

Sustainable Concrete Materials and Structures. DOI: https://doi.org/10.1016/B978-0-443-15672-4.00015-2

a challenge to the sustainable development of modern concrete practices (Xiao et al., 2017). To address the steel corrosion of SSC, researchers have offered a series of solutions from the perspectives of both SSC matrix and steel bars. It is well known that the free chloride ions can migrate to steel surface and damage the passivation film, whereas bound chloride ions are harmless (Yuan et al., 2009). Incorporating alumina-rich cementitious materials can promote the reaction of chloride binding, thus decreasing the content of free chlorides to prevent steel corrosion (Zhao et al., 2021). Moreover, corrosion inhibitors can inhibit or retard chloride corrosion by forming a passivation film on the steel surface (Soeylev & Richardson, 2008). Additionally, corrosion-resistant bars, including stainless steel and fiber-reinforced polymer (FRP) bars, have great potential for reinforcing SSC in various structures due to their remarkable corrosion resistance (J. Teng, 2018). Therefore adopting the strategies of adding corrosion inhibitors, incorporating alumina-rich cementitious materials, and substituting conventional bars with corrosion-resistant ones are expected to address the problem of steel corrosion in SSC.

This chapter presents a critical review of hydration, microstructures, and performances of SSC materials, focusing on the contribution of seawater and sea-sand together with exploring the approaches employed to address steel corrosion in SSC. Novel forms of sustainable concrete, namely coral aggregate concrete (CAC), recycled aggregate concrete, alkali-activated concrete, and ultra-high performance concrete, produced from seawater and sea-sand are, then, presented, aiming to further improve the sustainability of SSC materials. Additionally, structural members manufactured with the combination of SSC and corrosion-resistant materials are described, including SSC structures reinforced with stainless steel/FRP bars, SSC-filled stainless steel/FRP tubes, and SSC wrapped with FRP. The future challenges facing SSC materials and structures are also discussed.

15.2 Microstructures of sustainable seawater sea-sand concrete

15.2.1 Hydration process

Seawater and sea-sand contain around 3% soluble inorganic salts. Cl^-, Na^+, SO_4^{2-}, Mg^{2+}, Ca^{2+}, and K^+ are main ions. It has been proven in numerous studies that the hydration process of SSC is faster than that of normal concrete (NC) prepared by fresh water and river-sand (Guo et al., 2020; Han, Zhong, Ding, et al., 2021; Han, Zhong, Yu, et al., 2021; Yaseen et al., 2023; Zhang, Sun, et al., 2021). This is attributed to the promotion of cement hydration by chlorides. It is well known that $CaCl_2$ is usually used as an early strength agent for concrete (Riding et al., 2010). Firstly, NaCl introduced by seawater and sea-sand can react with portlandite ($Ca(OH)_2$) to form $CaCl_2$, as shown in Eq. (15.1). Subsequently, $CaCl_2$ can react with $Ca(OH)_2$ to produce calcium oxychloride (CAOXY) as the insoluble solid phase, as shown in Eq. (15.2). Meanwhile, the consumption of Ca^{2+} reduces the alkalinity and concentration of calcium in solution $(OH)_2$, promoting the dissolution of the minerals in the cement, and, thus, further increasing the rate of hydration.

$$Ca(OH)_2 + 2NaCl \rightarrow CaCl_2 + 2Na^+ + 2OH^- \tag{15.1}$$

$$CaCl_2 + 3Ca(OH)_2 + 12H_2O \rightarrow 3Ca(OH)_2 \cdot CaCl_2 \cdot 12H_2O \tag{15.2}$$

The effects of chlorides in seawater and sea-sand on specific hydration periods are as follows. Li, Li, et al. (2020) found that the effect of seawater on the initial hydration induction period is not significant, as shown in Fig. 15.1. Subsequently, as chlorides promote the hydration of C_3S, the arrival time of the first exothermic peak is advanced by 36%, and the exothermic heat is increased by 44% in the accelerated period of cement hydration. Furthermore, an exotherm with acromion is formed due to the promotion of the hydration of C_3A and C_4AF by Friedel's salt after the complete depletion of gypsum. As illustrated in Fig. 15.1, Younis et al. (2018) demonstrated that the cumulative heat release of seawater cement paste is 7.7% higher than that of freshwater cement paste at the curing age of 3 days. The difference between the two gradually disappears until the curing age of 7 days. It can be concluded that the effect of chlorides on the hydration process of SSC is significant during the early ages. However, as time progresses, their impact becomes less significant in the later stages.

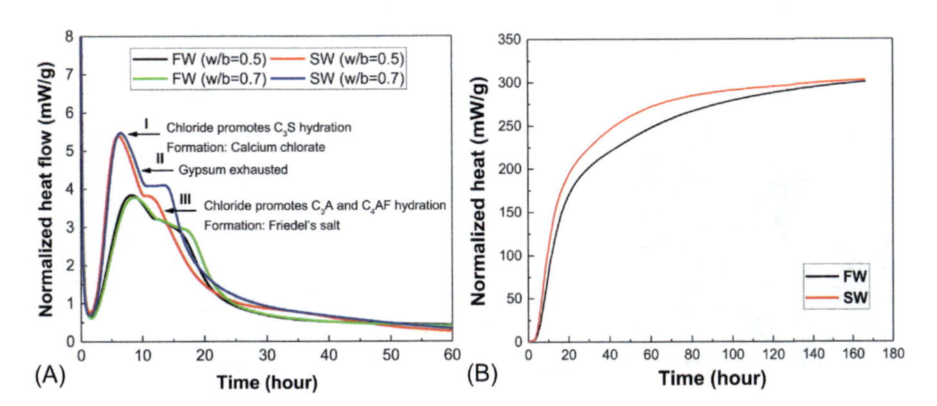

Figure 15.1 Heat evolution of seawater cement paste and freshwater cement paste: (A) exothermic rate; and (B) cumulative heat. The isothermal calorimetry tests were conducted on seawater cement paste and freshwater cement paste. The test results indicated that the effect of seawater on the hydration process of cement paste was significant in the early ages, but it was not significant in the later stages.
Source: From Li, P., Li, W., Sun, Z., Shen, L., & Sheng, D. (2021). Development of sustainable concrete incorporating seawater: A critical review on cement hydration, microstructure and mechanical strength. *Cement and Concrete Composites*, *121*. https://doi.org/10.1016/j.cemconcomp.2021.104100.

15.2.2 Hydration products

Fig. 15.2 illustrates typical hydration products formed by the action of Cl^-, SO_4^{2-}, and Mg^{2+} ions (Ragab et al., 2016). Ettringite (AFt) is formed by the action of SO_4^{2-} (Cai et al., 2023), as shown in Eq. (15.3). Li, Li, et al. (2020) proved that the amount of AFt of seawater cement paste is higher than that of freshwater paste by 67%. Thaumasite ($CaSiO_3 \cdot CaCO_3 \cdot CaSO_4 \cdot 15H_2O$) sulfate attack (TSA) is more destructive than the formation of AFt (Rahman & Bassuoni, 2014). The occurrence of TSA in concrete needs a range of conditions involving enough calcium silicate, carbonate, sulfate, and water. Compared to the volume expansion triggered by AFt, TSA can directly disrupt calcium silicate hydrate (C-S-H) gels as the primary source of cement strength (Sotiriadis et al., 2020). Furthermore, chloride ions can form Friedel's salt by instantly reacting with C_3A or anion exchange with AFm, respectively (Bachtiar et al., 2022), as illustrated in Eqs. (15.4) and (15.5). This is the significant difference between NC and SSC. Friedel's salt can immobilize some chloride ions to reduce the concentration of free chloride ions and

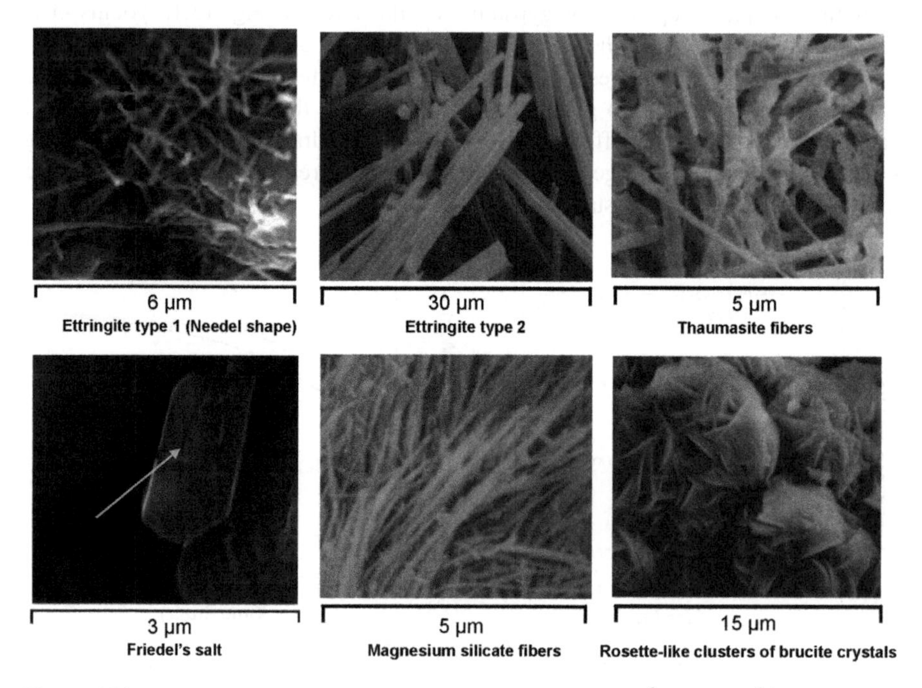

Figure 15.2 Hydration products formed by the action of Cl^-, SO_4^{2-}, and Mg^{2+} ions SEM images of ettringite and thaumasite formed by the action of SO_4^{2-} ion, magnesium silicate hydrates and brucite formed by the action of Mg^{2+} ion, and Friedel's salt formed by the action of Cl^- ion.

Source: From Ragab, A. M., Elgammal, M. A., Hodhod, O. A. G., & Ahmed, T. E. S. (2016). Evaluation of field concrete deterioration under real conditions of seawater attack. *Construction and Building Materials*, *119*, 130–144. https://doi.org/10.1016/j.conbuildmat.2016.05.014.

promote further densification of the microstructure due to the shape of the hexagonal slice (Yuan et al., 2009). Additionally, Mg^{2+} can replace Ca^{2+} of $Ca(OH)_2$ to generate brucite ($Mg(OH)_2$), as shown in Eq. (15.6), attaching to the cement surface and thus reducing the hydration rate of cement (Zhao et al., 2019). Meanwhile, $Mg(OH)_2$ can also convert C-S-H gels into noncementitious magnesium silicate hydrates (M-S-H), as shown in Eq. (15.7), resulting in the destruction of the cementitious structures and the decrease of mechanical properties.

$$\begin{aligned} &4CaO \cdot Al_2O_3 \cdot 13H_2O + 3(CaSO_4 \cdot 2H_2O) \\ &\quad + 14H_2O \rightarrow 3CaO \cdot Al_2O_3 \cdot 3CaSO_4 \cdot 32H_2O + Ca(OH)_2 \end{aligned} \tag{15.3}$$

$$\begin{aligned} &3CaO \cdot Al_2O_3 + 2NaCl + Ca(OH)_2 \\ &\quad + 10H_2O \rightarrow 3CaO \cdot Al_2O_3 \cdot CaCl_2 \cdot 10H_2O + 2NaOH \end{aligned} \tag{15.4}$$

$$\left[Ca_2Al(OH)_6\right]^+ - R + Cl^- \rightarrow \left[Ca_2Al(OH)_6\right]^+ - Cl^- + R \tag{15.5}$$

$$Ca(OH)_2 + Mg^{2+} \rightarrow Ca^{2+} + Mg(OH)_2 \tag{15.6}$$

$$4Mg(OH)_2 + SiO_2 \cdot nH_2O \rightarrow 4MgO \cdot SiO_2 \cdot 8.5H_2O + (n - 4.5)H_2O \tag{15.7}$$

Liu, An, et al. (2022) reported that incorporating seawater and sea-sand decreased the content of $Ca(OH)_2$ crystals and increased the content of C-S-H gels. Sun, Yu, et al. (2022) demonstrated that owing to seawater's accelerating effect on cement hydration, the hydrated degree of cement and the content of hydration products increase by 4.67% and 12.25%, respectively. The microscopic appearance of C-S-H gels converts the long and thick fibers to short and thin fibers owing to the incorporation of seawater (Sun, Zhang, et al., 2021). The polymerization degree and average molecular chain length of C-S-H gels are reduced, and the indentation modulus and packing density of C-S-H gels are improved. This can be attributed to the electrostatic attraction between the alkali cations and the negatively charged C-S-H gels (Lam et al., 2022; Sun, Lu, et al., 2022).

Based on the above analysis, it can be inferred that the addition of abundant ions from seawater and sea-sand can substantially influence the structure and morphology of cement hydration products and even generate a range of harmful hydration products, leading to the deterioration of microstructures. This mechanism could potentially play a crucial role in determining the properties and characteristics of SSC.

15.2.3 Pore structures

The formation of SSC pore structure is closely related to the hydration process and hydration products. It is reported that the porosity of SSC at a curing age of 28 days is higher than NC by 14.07%, and the volumes of middle capillary pores

(50–100 nm) and large capillary pores (> 100 nm) are increased by 43.62% and 49.59%, respectively (Han, Zhong, Ding, et al., 2021; Han, Zhong, Yu, et al., 2021). But the mesopores (5–50 nm) are increased by 9.85%, in agreement with the increase in gel pores below 10 nm of C-S-H gels in seawater cement paste, according to Zhang, Sun, et al. (2022). This may be due to the adsorption of Na^+ and Cl^- by C-S-H gels, leading to the structural reorganization of C-S-H gels and the decrease of the mean chain lengths. Liu, Liu, Jin, et al. (2022) found that incorporating seawater and sea-sand reduces the volume of total pores and harmful capillary pores by 16.42% and 49.91%, respectively (Liu, Liu, Jin, et al., 2022). In conclusion, due to the faster hydration process of SSC compared to NC, more hydration products such as C-S-H gels and Friedel's salts are formed. These products effectively fill the capillary pores, thus reducing the porosity.

15.3 Properties of sustainable seawater sea-sand concrete

15.3.1 Fresh properties

The fresh properties of concrete play a crucial role in the transportation and cast-in-place construction. Numerous studies have shown that the fluidity of SSC is reduced due to the addition of seawater and sea-sand, as shown in Table 15.1. The decreasing rate of fluidity ranges from 2.56% to 45.40%. This can be explained in two aspects. On the one hand, due to the accelerating effect of NaCl in seawater and sea-sand on cement hydration, a large amount of solid hydration products is formed, and the flocculation between suspended particles is

Table 15.1 Fluidity of SSC compared to NC.

Water to binder ratio	Slump flow (mm)		Increase/decrease of slump flow		References
	NC	SSC	Absolute value (mm)	Relative value (%)	
0.42	198	170	−28	−14.14	Xiao et al. (2023)
0.48	113	88	25	−21.92	Liu, An, et al. (2022)
0.38	195	190	−5	−2.56	Vafaei et al. (2021)
0.48	174	95	−79	−45.40	Chen et al. (2020)
0.34	600	480	−120	−20.00	Younis et al. (2018)

strengthened, driving SSC to set and harden rapidly (Liu, An, et al., 2022). On the other hand, the small particle size and smooth surface of sea-sand particles result in a larger specific surface area than the river-sand particles, allowing more cement paste to wrap around the sea-sand surface, consequently, reducing cement paste available for achieving fluidity. Additionally, sea-sand contains shells with $CaCO_3$ as the main component (Dhondy et al., 2019). These shells significantly increase the time-dependent loss of fluidity, slightly decrease the apparent density, and slightly increase the air content of SSC (Vafaei et al., 2021). This can be explained by an increase in friction between sand particles and a decrease in the compactness of the matrix.

Notably, Xiao et al. (2023) reported that compared to NC, the setting times of sea-sand concrete, seawater-mixed concrete, and SSC decrease by 9.73%, 16.79%, and 26.53%, respectively, while the fluidities decrease by 4.04%, 9.09%, and 14.14%, respectively, in agreement with the findings reported by Chen et al. (2020), confirming the two main factors controlling the reduction in fresh properties of SSC, with seawater playing a more significant role than sea-sand.

15.3.2 Mechanical properties

15.3.2.1 Static mechanical properties

Numerous studies have been carried out on the static mechanical properties of SSC, as shown in Table 15.2. However, conflicting results are reported on the variation in the compressive, flexural, and tensile strengths between SSC and NC at different ages. Fig. 15.3 illustrates that the compressive strength of SSC is significantly higher than NC at the ages of 3 days and 7 days, with maximum increases of 27.64% and 30.53%, respectively, signifying the rapid hydration reaction of SSC in the early ages and the pore-filling effect of Friedel's salt (Xie et al., 2021). However, the difference in compressive strength at the age of 28 days ranges from −18.97% to 17.25%, possibly related to seawater's physical and chemical properties and sea-sand from different sea regions. On the hand, Cui et al. (2022) reported that as the salinity value of seawater doubles, the 28 days compressive strength of SSC with a W/B ratio of 0.6 and 0.65 increases by 11.34% and 10.16%, respectively, confirming the remarkable contribution of the inorganic salt content of seawater and sea-sand to the strength development. On the other hand, Geng et al. (2022) concluded that there are significant differences in the gradation, surface texture, and salinity of sea-sand from different regions. The sea-sand with smooth surface texture and high shell content can reduce the interfacial bonding strength with the cement paste, thereby reducing the compressive strength of SSC. Moreover, the addition of seawater and sea-sand causes a reduction in long-term compressive strength in the range of 2.12%−13.18%. This reduction can be explained by the increased crystallization pressure within the pores and the disruption of the C-S-H gel caused by the presence of SO_4^{2-} and Mg^{2+} ions in SSC, as explained in the studies conducted by Li, Li, et al. (2020) and Vafaei et al. (2021).

Table 15.2 Recent studies on the compressive, flexural, and tensile strengths of SSC compared to NC.

W/B ratio	Increase/decrease of compressive strength		Increase/decrease of flexural strength		Increase/decrease of tensile strength		References
	Absolute value (MPa)	Relative value (%)	Absolute value (MPa)	Relative value (%)	Absolute value (MPa)	Relative value (%)	
0.35	5.00 (28d)	9.80	–	–	–	–	Iqbal et al. (2023)
0.45	0.42 (3d)	1.56	–	–	–	–	Vidhyadharan and Johny (2023)
	0.62 (7d)	1.48	–	–	–	–	
	−1.34 (28d)	−2.84	–	–	–	–	
0.51	3.49 (7d)	16.41	–	–	–	–	Cui et al. (2022)
	2.77 (28d)	9.64	–	–	–	–	
0.6	2.74 (7d)	18.85	–	–	–	–	
	−0.45 (28d)	−1.85	–	–	–	–	
0.65	−1.73 (7d)	−11.89	–	–	–	–	
	−4.38 (28d)	−18.97	–	–	–	–	
0.35	−4.30 (28d)	−7.92	0.09 (28d)	1.36	−0.42 (28d)	−8.71	Vafaei et al. (2022a)
0.48	4.92 (3d)	18.00	–	–	−0.18 (3d)	−1.00	Liu, An, et al. (2022)
	2.39 (7d)	7.00	–	–	−0.57 (7d)	−3.00	
	2.90 (28d)	7.00	–	–	0.22 (28d)	1.00	
	−5.56 (90d)	−11.00	–	–	0.24 (90d)	1.00	
	−7.13 (360d)	−12.00	–	–	−0.52 (360d)	−2.00	
0.50	2.90 (3d)	13.06	–	–	–		Geng et al. (2022)
	1.20 (7d)	4.24	–	–	–	–	
	−4.10 (28d)	−10.07	0.30 (28d)	6.12	0.40 (28d)	12.12	
	−3.90 (56d)	−8.63	–	–	–	–	
	−4.40 (90d)	−9.24	–	–	−0.10 (90d)	−2.33	
	−5.30 (180d)	−9.78	–	–	–	–	
	−7.50 (360d)	−13.18	–	–	−0.30 (360d)	−6.25	

0.48	2.63 (28d)	6.37	–	–	–	–	Liu et al. (2021)
0.38	7.74 (3d)	23.30	–	–	–	–	Pan et al. (2021)
	−0.92 (7d)	−1.90	–	–	–	–	
	−4.89 (28d)	−7.60	1.14 (28d)	18.20	–	–	
0.28	7.81 (3d)	15.00	–	–	–	–	
	2.23 (7d)	3.20	–	–	–	–	
	−1.23 (28d)	−1.40	0.94 (28d)	12.00	–	–	
0.38	5.76 (3d)	14.81	–	–	–	–	Vafaei et al. (2021)
	4.62 (7d)	10.69	–	–	–	–	
	−4.34 (28d)	−7.99	0.09 (28d)	1.36	−0.42 (28d)	−8.71	
	−6.86 (90d)	−10.89	–	–	–	–	
0.60	6.68 (7d)	30.53	–	–	–	–	Dhondy et al. (2020)
	4.40 (28d)	17.25	–	–	–	–	

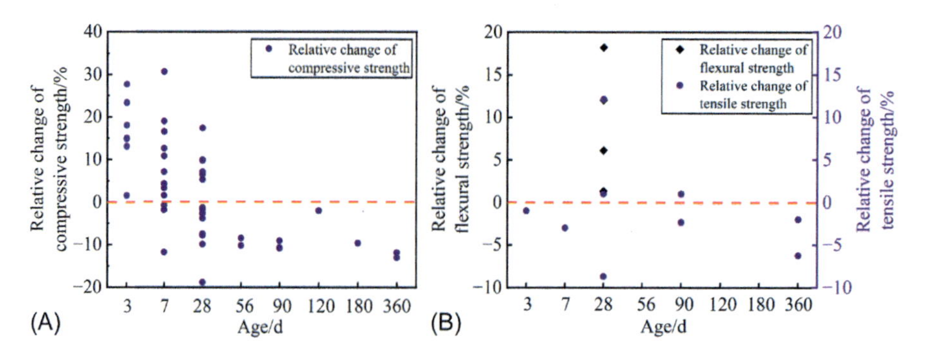

Figure 15.3 Time-dependent difference in the strength between SSC and NC: (A) relative change of compressive strength; and (B) relative change of flexural and tensile strengths. The difference in compressive strength, flexural and tensile strengths between SSC and NC at different ages were investigated, indicating that using seawater and sea-sand can improve the early mechanical properties of concrete, and result in the slow development of long-term mechanical properties.

Fig. 15.3 shows that the flexural strength of SSC is significantly higher than OC at the age of 28 days, with maximum increases of 18.2%. Nevertheless, there are significant variation in the results of tensile strength, ranging from −8.71% to 12.12%, and a considerable reduction in the long-term tensile strength of SSC can be concluded. It is well known that concrete's flexural and tensile failure generally starts at the interfacial transition zone (ITZ) between the aggregate and cement paste. Due to the accelerated cement hydration and the formation of Friedel's salt in SSC, the compactness of ITZ at the early ages can be improved, thus increasing flexural and tensile strengths (Guan et al., 2023). As the age increases, the formation of harmful hydration products causes a reduction in the compactness, resulting in a slow strength gain of SSC (D. Pan et al., 2021).

15.3.2.2 Dynamic mechanical properties

Investigating the dynamic mechanical properties of SSC is of great significance for the structural design of impact and blast resistance. Zhang et al. (2021) concluded that as the strain rate increases from 10^{-5} to $10^{-1}\,s^{-1}$, the peak stress and modulus of elasticity of SSC increased, but the deformability of SSC exhibits no significant change. Vafaei, Hassanli, et al. (2022) conducted a drop-weight impact test to study the impact resistance of SSC and found that the average number of blows and the impact energy of SSC are higher than that of NC by 25.93% and 28.00%, respectively. This impact resistance improvement can be attributed to the denser microstructure of SSC, absorbing more fracture energy. Xu et al. (2022) and Xu, Zhao et al. (2023) employed the split Hopkinson pressure bar (SHPB) testing to investigate the dynamic compressive properties of SSC. Their study showed that due to the dynamic compressive strengths of SSC corresponding to air pressures of 0.3, 0.6, and 0.9 MPa are improved by 1.55%, 12.83%, and 3.88%, respectively.

However, the energy absorption capacity is slightly reduced. The general laws of SSC under static loads still apply to dynamic loads.

In summary, the mechanical properties of SSC are strongly influenced by the physical and chemical properties of seawater and sea-sand. Incorporating seawater and sea-sand can improve the early mechanical properties but result in the slow development of long-term mechanical properties.

15.3.3 Shrinkage

Shrinkage has a significant effect on the strength development and durability of concrete. It has been proven that the numbers of micropores (pore diameter less than 2.5 nm) and mesopores (pore diameter ranging from 2.5 to 50 nm) determine the shrinkage deformation of concrete (Liu, An, et al., 2022; Vafaei et al., 2022b). Liu, An, et al. (2022) illustrated that the drying shrinkage of SSC is essentially unchanged after reaching the age of 180 days. The cumulative drying shrinkage of SSC at the age of 360 days is 8.61% higher than that of NC owing to the increasing number of pores with diameters less than 50 nm, leading to increased water drainage and pore tensile stress. Vafaei et al. (2022b) reported that the drying shrinkage of SSC is more significant than the autogenous shrinkage. Due to the incorporation of seawater and sea-sand, the drying and autogenous shrinkage strains of SSC at the age of 90 days were increased by 29.79% and 68.75%, respectively, in agreement with the observed increases in the total porosity and average pore diameter of SSC (Vafaei et al., 2022b). Owing to the accelerating effect of seawater and sea-sand on cement hydration, the internal humidity of SSC is rapidly reduced. Meanwhile, the formation of large amounts of C-S-H gels refines the pore structure and increases the number of micropores and mesopores, promoting water drainage and increasing pore tensile stresses. Consequently, this underlying mechanism highlights the more pronounced shrinkage observed in SSC.

15.3.4 Durability

15.3.4.1 Steel corrosion

The steel corrosion induced by chloride ions poses a significant challenge to the durability of reinforced concrete structures. Chloride ions can diffuse and penetrate the surface of steel bars from the outside of concrete, destroying the passivation film on the surface of the reinforcement, subsequently leading to the corrosion reaction of steel bars (Yuan et al., 2009). Therefore the corrosion risk of endogenous chloride ions introduced by seawater and sea-sand has become the primary challenge in the widespread adoption of SSC (Xiao et al., 2017). Wu et al. (2019) found that under the actions of dry-wet cycles and seawater immersion, the weight loss rates of steel bars inside SSC are 0.186% and 0.111% at the age of 90 days, respectively, while no steel corrosion occurs in NC. Wang et al. (2019) studied the corrosion products of mild steel bars embedded in SSC and observed both lepidocrocite (γ-FeOOH) and goethite (α-FeOOH) are formed in the internal rust layer of SSC and NC. However, unlike maghemite (γ-Fe_2O_3) formation in NC, magnetite (Fe_3O_4) is found in SSC, indicating a higher corrosion rate of steel bars inside SSC. Some corrosion products (FeOOH) continue promoting corrosion reactions even without oxygen in seawater immersion. Yu, Al-Saadi, Zhao, et al. (2021) and Yu, Al-Saadi, Kohli, et al. (2021) illustrated that the open circuit potential, current density, and impedance of SSC containing 304 and 316 L stainless steel are higher than those of SSC containing mild steel, indicating a lower corrosion risk of the stainless steel due to the protection of Cr and Mo elements. However, increasing the alkalinity of SSC is beneficial to improve the corrosion resistance of the mild steel while also affecting the formation of Mo- and Cr-enriched passivation film, thereby decreasing the corrosion resistance of stainless steel reinforcement.

The approaches employed to address steel corrosion of SSC include various methods, such as adding corrosion inhibitors, improving chloride-binding capacity, and substituting conventional steel bars with corrosion-resistant bars (Bazli et al., 2021; Xiao et al., 2017; Zhao et al., 2021). The review of improving chloride-binding capacity and using corrosion-resistant bars will be described in Sections 15.3.4.2 and 15.5.1, respectively, whereas the findings on the incorporation of corrosion inhibitors are presented below. Adding corrosion inhibitors has been widely used in many practical projects, including organic and inorganic corrosion inhibitors. Due to the severe pollution and high dosage of inorganic corrosion inhibitors, applying organic corrosion inhibitors to SSC has gained significant consideration (Elshami et al., 2020). Xu et al. (2011) reported that a compound corrosion inhibitor composed of 50% of triethanolamine, 20% of dimethylethanolamine, 25% of triethoxysilane, and 5% of lithium nitrate successfully prevented the corrosion of steel bars embedded in SSC even at 420d, indicating the effectiveness of compounding corrosion inhibitors in retarding the corrosion reaction. Xu et al. (2019) found that a 2% triethanolamine corrosion inhibitor can decrease the passive film damage and the weight loss of steel rebar inside SSC to a similar level as NC, while Pan et al. (2020) demonstrated that the corrosion risk of steel bars inside SSC with a 0.75% imidazoline corrosion inhibitor is lower than that of a 3% triethylenetetramine corrosion inhibitor. The common mechanism of corrosion inhibitors involves the formation of a protective film layer on the surface of steel bars through adsorption and passivation, effectively acting as a barrier against chloride ion penetration and inhibiting electrochemical reactions.

15.3.4.2 Chloride binding

Chloride binding is a crucial factor, significantly influencing the durability of reinforced concrete structures. The chloride ions in concrete are divided into free chloride ions and bound chloride ions (Yuan et al., 2009). The free chloride ions can move freely in the pores and migrate to the surface of steel bars. As the chloride ions content accumulates to a critical level, the passivation film on steel surface is gradually penetrated and damaged, thus losing the function of protecting steel bars (Zhao et al., 2021). Furthermore, the chloride-binding capacity of concrete can delay the time of free chloride ions arriving at the surface of steel bars and reduce the critical chloride ion concentration, thus extending the service life of reinforced concrete structures (J. Xiao et al., 2017).

The bound chloride ions can be physically adsorbed by C-S-H gels and be chemically bound by Friedel's salt (Wang et al., 2020). On the one hand, the chloride ions can exist in the chemisorbed layer, penetrate the interlayer, or be adsorbed into the lattice of the C-S-H gel (Li, Li, et al., 2020). Considerable evidence suggests that a high Ca/Si ratio can enhance the adsorption capacity of C-S-H gels, resulting from the positive charge of Ca^{2+} facilitating the adsorption of chloride ions (Li, Li, et al., 2021). Nevertheless, due to the reversible nature of physical adsorption, the desorption of chloride ions may occur in C-S-H gels under low chloride ion content (Wang et al., 2013). This may not be a concern for SSC with a high chloride content. On the other hand, Friedel's salt is a mineral as the AFm-phase consisting of the main layers $[Ca_2Al(OH)_6]^+$ and the interlayers $X^- \cdot nH_2O$ where X is chloride in the case of Friedel's salt (Qu et al., 2021). However, AFm-phases can be interconverted under certain conditions (Ebead et al., 2022). As shown in Fig. 15.4, C_3A firstly reacts with SO_4^{2-} to form AFt until SO_4^{2-} is depleted. Subsequently, Friedel's salt is formed until Cl^- is exhausted. It is noticeable that chloride ions can replace other anions to create AFm-phases, including Friedel's salt and Kuzer's salt. Nevertheless, since Kuzer's salts generally are formed at low chloride ion content, the chemical binding products in SSC are dominated by Friedel's salts. After the two anions are entirely depleted, AFm is formed by the reaction of the remaining C_3A and AFt. Additionally, CO_3^{2-} significantly influences the chloride-binding capacity because carbonation decreases the alkalinity of the solution and reduces the stability of Friedel's salt (Geng et al., 2016; Sun, Sun, et al., 2021), as illustrated in Eq. (15.8). CO_3^{2-} has the same ability as SO_4^{2-} to replace chloride ions in AFm-phases, driving the decomposition of Friedel's salt, as depicted in

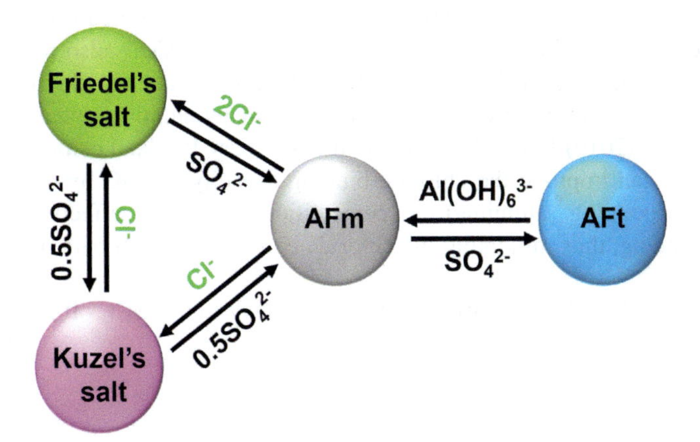

Figure 15.4 The transformation relationship of AFm-phases and AFt. There is a competitive mechanism between Cl^- and SO_4^{2-} ions in the process of chloride binding, indicating AFm phases and AFt can be transformed into each other under different Cl^- and SO_4^{2-} ions concentrations.

Eq. (15.9). According to the stability of hydrated products formed between anions and AFm, the order of reactions is SO_4^{2-}, CO_3^{2-}, and Cl^- in turn (Wang et al., 2013), causing more severe chloride corrosion of steel bars under the carbonation and sulfate attack due to decreased chloride binding capacity.

$$Ca(OH)_2 + CO_2 \rightarrow CaCO_3 + H_2O \tag{15.8}$$

$$3CaO \cdot Al_2O_3 \cdot CaCl_2 \cdot 10H_2O + 3CO_2 \rightarrow 3CaCO_3 + Al_2O_3 + CaCl_2 + 10H_2O \tag{15.9}$$

In recent years, incorporating alumina-rich cementitious materials to improve the chloride-binding capacity of SSC has gained considerable attention. The content of alumina is closely related to the formation of Friedel's salt (Talero et al., 2011), as shown in Eq. (15.10). Li, Jin, et al. (2021) reported that metakaolin (MK), fly ash (FA), and calcium aluminate cement can improve the chloride-binding capacity that is the ratio of bound to free chloride ions content of seawater and sea-sand mortar by 65%, 35%, and 200%, respectively. Calcium aluminate cement improves the chloride-binding capacity by enhancing C-S-H gels' physical adsorption. The difference between MK and FA in the chemical binding is that most of the alumina in MK is available, while most of the alumina in FA is crystallized in mullite ($3Al_2O_3 \cdot 2SiO_2$). Yao et al. (2021) observed that incorporating MK can enhance the chloride-binding capacity of SSC. Saleh, Mahmood, et al. (2023) reported that 25% slag (SG) can increase the chloride-binding capacity by 13.99% due to the high alumina content. On the other hand, silica fume has a slight negative effect on the chloride-binding capacity of SSC because the pozzolanic effect of silica fume reduces the alkalinity of SSC and promotes the solubility of Friedel's salts, thus decreasing the chemical binding effect (Saleh, Mahmood, et al., 2023). Meanwhile, a large number of C-S-H gels with a low Ca/Si ratio are formed owing to the secondary reaction of silica fume, thus promoting the physical adsorption of C-S-H gels. Zhou, Zhou, et al. (2022) reported that limestone calcined clay cement (LC^3) is beneficial for forming Friedel's salt due to the many aluminate phases in calcined clay. Xu et al. (2019) found that compared to ordinary Portland cement, sulfoaluminate cement can improve the chloride-binding capacity of SSC by 42.2%. In summary, incorporating alumina-rich cementitious material effectively enhances chloride binding, thus mitigating the chloride corrosion of steel bars in SSC.

$$Al_2O_3^{r-} + 2NaCl + 4Ca(OH)_2 + 7H_2O \rightarrow 3CaO \cdot Al_2O_3 \cdot CaCl_2 \cdot 10H_2O + 2NaOH \tag{15.10}$$

15.3.4.3 Chloride penetration resistance and impermeability

The chloride penetration resistance and impermeability of concrete represent the abilities of concrete to resist the intrusion of external chloride ions and moisture, which are essential factors in evaluating the service life of structures.

Iqbal et al. (2023) conducted a rapid chloride permeability (RCP) test and water penetration test (WPT) to investigate the durability performance of SSC. They found that due to the incorporation of seawater and sea-sand, the depth of water penetration and the RCP value are reduced by 7.90% and 9.13%, respectively. Saleh, Mahmood, et al. (2023) reported that the chloride migration coefficients of SSC at the age of 7 days, 28 days, and 90 days are lower than those of NC by 42.84%, 44.69%, and 42.52%, respectively. Ting et al. (2020) illustrated that the passed charges at the age of 28 days and 90 days decreased by 28.71% and 24.54%, respectively, owing to the incorporation of seawater and sea-sand. Liu, Liu, Fan, et al. (2022) found that the depths of water penetration of SSC at the age of 28 days, 56 days, and 90 days are lower than those of NC by 20.96%, 12.86, and 30.18%, respectively. The improvement of chloride penetration resistance and impermeability of SSC can be attributed to the formation of dense microstructure that effectively hinders the intrusion of aggressive external media. Moreover, the low salinity content gradient between the SSC pore solution and external seawater contributes to SSC's high chloride penetration resistance (Zhou & He, 2021).

15.3.4.4 Additional relevant durability properties

Other relevant durability features of SSC, including sulfate, frost, and carbonation resistance, are partially addressed in the existing studies. Han, Zhong, Yu, et al. (2021) found that the mass change and strength loss of SSC under the sulfate attack are lower than that of NC by 37.3% and 37.4%, implying the utilization of seawater and sea-sand can increase the sulfate resistance due to the filling and compaction effects against sulfate attack products. Li, Sun, et al. (2022) and Sun, Li, et al. (2022) concluded that the sulfate attack induced less severe damage to SSC compared to NC. This observation can be attributed to the increase of Vickers hardness and the decrease in porosity.

Li et al. (2013) found no significant difference in mass loss and compressive strength between SSC and NC after 50 freeze-thaw cycles, agreeing with the observations of Li, Liu, et al. (2020) and Li, Sun, et al. (2022). Zhou and Li (2022) concluded that the strength and energy dissipation capacity of SSCs is weakened under the action of seawater freeze-thaw cycles, compared to freshwater freeze-thaw cycles. Due to the high salinity of SSC, the decrease in the freezing point of pore solution increases the number of capillary pores and cracks.

He et al. (2022) illustrated that SSC, after 56d accelerated carbonation shows a slight increase in carbonation depth compared to NC. Zhang, Xiao, et al. (2022) found that the carbonation depths of SSC at the ages of 7 days, 14 days, and 28 days were 8.24 times, 8.37 times, and 6.23 times that of NC, respectively, largely attributed to the deterioration of pore size distribution and porosity under carbonation. Liu et al. (2021) observed that the carbonation depths of SSC after the accelerated carbonization for 7 days, 14 days, 28 days, and 112 days are higher than that of NC by 436.78%, 82.14%, 11.81%, and 4.67%, respectively, but compressive strengths are reduced by −7.38%, 1.89%, 5.10%, and 9.43%, respectively. This finding suggests that carbonation of SSC is more remarkable than NC during the early ages, but the difference in carbonation between the two becomes less pronounced at later ages. Furthermore, with the growth of carbonization time, the calcium carbonate ($CaCO_3$) generated gradually fills the pores and hinders further carbon penetration. The decrease in the long-term strength of SSC may be related to the decomposition of Friedel's salts due to the anion exchange of CO_3^{2-}.

In conclusion, SSC has comparable chloride penetration resistance, impermeability, sulfate resistance, frost resistance, and carbonation resistance to NC, indicating the impact of seawater and sea-sand on the durability is acceptable, except for steel corrosion.

15.4 New types of sustainable concrete based on seawater and sea-sand

15.4.1 Seawater sea-sand-based coral aggregate concrete

The use of coral aggregate (CA) to produce concrete has been widely used in the construction of island projects (A. Wang et al., 2018). Coarse and fine aggregates account for about 70%–80% of concrete volume, and the engineering demand is enormous. Utilizing seawater and sea-sand in the production of CAC can alleviate the shortage of freshwater and aggregate resources in island areas, thus further improving the sustainability of concrete.

CA is characterized by high porosity, high permeability, low crushing strength, and high salinity (Huang et al., 2018). Therefore mechanical properties of seawater sea-sand-based coral aggregate concrete (SSCAC) have become the research focus of many investigations. Huang et al. (2019) reported that seawater positively affects SSCAC's strength and modulus of elasticity, but the effect of sea-sand is negative. SSCAC exhibited a more brittle fracture mode than conventional CAC due to the accelerated crack propagation by seawater and sea-sand. Furthermore, extensive evidence suggests that due to CA's low crushing strength, cracks occur firstly in CA of SSCAC under compression, bending, and tensile actions. In contrast, the damage pattern of SSC composed of gravel is dominated by cracking in the ITZ between aggregate and matrix. Thus the mechanical strengths of SSCAC are significantly lower than that of SSC based on conventional gravel. Moreover, Xu et al. (2022) and Xu, Zhao, et al. (2023) found that the strain rate effect of SSCAC under the impact loading is more significant than SSC composed of gravel due to the increase in dynamic peak strain and energy absorption density. Yang, Zhu, et al. (2023) concluded that with the rise of load duration and applied load level, both the total creep and residual strain of SSCAC are increased, implying that part of the recoverable strain is transformed into an irrecoverable strain. Other properties of SSCAC, in addition to mechanical properties, have been partially reported. Sun et al. (2023) reported that the prewetting of CA has a significant and little effect on the workability and strength of SSCAC, respectively. Zhou, Guo, et al. (2022) found that incorporating 20% saturated CA can eliminate auto shrinkage of SSC due to the increase in internal humidity and the workability, compressive strength, and microstructure of SSC with CA content ranging from 10% to 20% were the best.

Some researchers adopt modified methods to improve the mechanical properties of SSCAC. Wang et al. (2023) modified CA by filling pores with superfine cement and found that the increasing rates of compressive and tensile strengths of SSCAC by the modification measures are within 30%. Qi et al. (2020) incorporated stainless steel and polypropylene (PP) fibers into SSCAC. They concluded that stainless steel fibers effectively increase the compressive strength and elastic modulus, hindering the crack development of SSCAC.

In summary, using seawater and sea-sand in combination with CA can produce SSCAC with a satisfactory performance by the appropriate mix design. CA modification and fiber reinforcement can overcome the weak mechanical properties of SSCAC caused by the natural properties of CA.

15.4.2 Seawater sea-sand-based recycled aggregate concrete

The technology of using recycled coarse aggregates (RCA) to replace natural aggregates in concrete production has been widely studied. The utilization of RCA not only alleviates the environmental damage caused by haphazard disposal of waste concrete and excessive-mining of natural coarse aggregates (NCA) but also serves as an effective means of recycling and utilizing waste resources (J. Xiao et al., 2012). In recent years, the development of recycled aggregate concrete (RAC) using seawater and sea-sand has gradually attracted increasing attention from researchers, aiming to enhance sustainability of concrete industry.

RCA is characterized by higher porosity, higher water absorption, and lower crushing strength than NCA (Rabadia & Aslani, 2023). Most importantly, SSC has only one ITZ between NCA and cement paste. However, seawater sea-sand-based recycled aggregate concrete (SSRAC) has an old ITZ between primary aggregate and residual old cement paste together with a new ITZ between residual old cement paste and fresh cement paste (Etxeberria et al., 2007). Therefore the new and old interfaces are closely influencing the mechanical properties and durability of SSRAC. The incorporation of seawater and sea-sand has a positive impact on the mechanical properties and a negative effect on the workability and early-cracking resistance, according to Xiao et al. (2019). This finding can be attributed to the improved ITZ and increased early shrinkage caused by the formation of Friedel's salt and rapid cement hydration. Furthermore, numerous studies proved that the increased RCA replacement rate significantly reduces the static and dynamic mechanical properties (Xiao et al., 2019, 2021; Zhang et al., 2021a). Additionally, the fluidity of SSRAC is decreased slightly owing to the water absorption of RCA pores. Zhang, Xiao, et al. (2022) found that the optimum replacement ratio of RCA is 50% to achieve the highest level of carbonation resistance for SSRAC. The pores in RCA can provide space for the growth of harmful hydration products, thus alleviating the leaching and high crystallization pressure of those hydration products, as confirmed by the reduced porosity of SSRAC. Lu et al. (2023) observed that with the increase of seawater immersion time of SSRAC, there is a transition in the flexural damage mode changes from ITZ to the aggregate due to the increase in RCA defects.

Several modification techniques have been employed to improve the performance of SSRAC. Notably, incorporating supplementary cementitious materials (SCMs), including silica fume (SF) and FA, can exert a pozzolanic effect which improves the interfacial strength and fill the RCA pores, thus enhancing the workability, mechanical properties, and durability of SSRAC (Lu et al., 2023). Due to the high aluminate content of calcined clay, LC^3 promotes the formation of Friedel's salt in SSRAC, thus improving the microstructure and mechanical properties (Zhou, Zhou, et al., 2022). Furthermore, basalt, PP, and stainless steel fibers are used to reinforce SSRAC, effectively restricting the crack propagation of SSRAC as highlighted by Huang, Wang, et al. (2022). It is worth stating that the combined use of seawater, sea-sand, RCA, and basalt fibers can counteract the adverse effects of RCA on the mechanical properties due to the improved ITZ by eco-friendly basalt fibers. Zhou, Zhou, et al. (2022) found that incorporating polyethylene (PE) fiber-strengthened RCA can improve the compressive strength of SSC at the age of 28 days and 90 days by 19.34% and 15.13%, respectively, owing to the impeding effect of the reinforcement layer on the crack propagation.

In general, using seawater and sea-sand in combination with SCMs can improve the performance of SSRAC by enhancing the interfacial strength of RCA. Incorporating fibers to hinder crack propagation is an effective way to improve the mechanical properties of SSRAC.

15.4.3 Seawater sea-sand-based alkali-activated concrete

The cement industry accounts for about 7% of the world's total carbon emissions (Ali et al., 2011). Alkali-activated materials (AAMs) with low carbon emissions, high strength, and corrosion resistance offer a potential solution to address environmental issues induced by cement production and industrial waste emission (Provis, 2014). Developing seawater sea-sand-based alkali-activated concrete (SSAAC) significantly reduces natural resource consumption and carbon emissions.

AAMs are made by the reactions between SCMs with pozzolanic activity or potential water hardness and alkaline activators (Provis, 2018). According to the calcium content of precursors, AAMs can be classified as no calcium ($<1\%$, such as MK), low calcium ($<10\%$, such as FA), and high calcium ($>10\%$, such as SG). Some studies have compared the performance differences between SSAAC and alkali-activated concrete (AAC). Zhao et al. (2022) reported that the chloride-binding mechanisms of alkali-activated slag (AAS) include the physical adsorption by C-A-S-H gel and the chemical binding by Friedel's salt. Nevertheless, seawater reduces the hydration rate of AAS, increasing the fluidity and setting time. AAS's slow strength development is associated with the deterioration of pore distribution. This results from the reduced silica content and impeded dissolution of precursors due to the interaction between seawater's $MgCl_2$ and alkali activators, as suggested by Lao et al. (2023). Yang et al. (2019) observed that incorporating seawater and sea-sand to slag-based AAC results in slightly increased mechanical properties, drying shrinkage, and chloride penetration resistance of slag-based AAC. Li, Zhao, et al. (2018) concluded that the strength degradation of slag-based SSAAC at high temperatures is due to the thermal expansion incompatibility between the paste matrix's contraction and the aggregates' expansion.

Lyu et al. (2022) found that NaOH and Na_2SiO_3 as liquid activators are superior to anhydrous and pentahydrate metasilicate as solid activators concerning the compressive strength of SSAAC. Owing to the dense microstructure formed by calcium-rich C-A-S-H gels, the combination of FA and SG in AAC performs better than that of FA and MK. Yang, Wang, et al. (2022) concluded that increasing the mass ratio of SG to FA benefits the mechanical properties of sea-sand AAC exposed a temperature below 400°C, due to the formation of tobermorite generated from the interaction of FA and SG at high temperatures. As the external temperature rises to 600°C, the decomposition of C-S-A-H and the formation of calcite and gypsum lead to a decrease in the mechanical properties of AAC with high SG content.

Overall, using seawater, sea-sand, AAMs, and activators can produce green and low-carbon SSAAC. The working mechanism of seawater and sea-sand on the hydration process of the AAMs system is different from that of the cement system due to the interaction between $MgCl_2$ and activators. The calcium content of SCMs plays a crucial role in the mechanical properties of SSAAC, with SG outperforming FA and MK.

15.4.4 Seawater sea-sand-based ultra-high performance concrete

With the increasing complexity of modern infrastructures and service environments, the shortcomings of concrete brittleness, susceptibility to cracking, and high self-weight are highlighted. Furthermore, the enormous consumption of concrete significantly impacts resources, energy usage, and the environment. To address these challenges, ultra-high performance concrete (UHPC) with excellent strength, toughness, and durability has become a research hotspot (D. Wang et al., 2015). UHPC offers a range of benefits, including significant weight reduction, excellent impact resistance, long service life, and low life cycle cost, making it very promising for critical infrastructures such as large-span bridges, thin-walled structures, blast-resistant structures, and offshore platforms (Xue et al., 2020). Seawater sea-sand-based ultra-high performance concrete (SSUHPC) is expected to extend the advantages of UHPC in resource-saving and carbon emission reduction, thus further promoting the sustainable development of concrete.

The critical preparation principle of UHPC matrix is using ground quartz sand and SCMs to improve the fineness and activity of concrete components, minimizing the defects of composites to obtain high homogeneity and compactness (Shi et al., 2015). Furthermore, steel fibers are used to prepare UHPC to improve the natural deficiencies of brittleness and easy cracking of cement-based materials. Numerous studies have proved the feasibility of seawater and sea-sand instead of freshwater and ground quartz sand to produce UHPC with reinforcing fibers, as shown in Table 15.3. It can be concluded that SF and FA are widely used in SSUHPC matrix, and various types of corrosion-resistant fibers are used in SSUHPC to address the chloride corrosion risk of steel fibers caused by seawater and sea-sand. In terms of the effect of seawater and sea-sand on the UHPC matrix, Saleh, Li, et al. (2023) and Saleh, Mahmood, et al. (2023) demonstrated that the incorporation of seawater and sea-sand endows UHPC with high early strength, resistivity, shrinkage, and chloride penetration resistance. Sun, Yu, et al. (2022) found that seawater accelerates the hydration process of cement and SF and promotes the transformation of capillary pores into gel pores, thus decreasing the workability but increasing the early strength in addition to aggravating the shrinkage of UHPC matrix. Zhang, Chang, et al. (2022) demonstrated that 10% SF significantly improves the mechanical properties and corrosion resistance of seawater-mixed UHPC. When the SF content increases to 30%, the long-term strength decreases due to the decrease of unhydrated clinkers.

Introducing various types of SCMs into SSUHPC can further reduce cement consumption and improve performance. Li et al. (2018) reported that seawater promotes the formation of C-S-H gels with a low Ca/Si ratio in the UHPC matrix containing SF, while C-S-H gels of UHPC matrix containing SG have a high Ca/Si ratio. The positive effect of aluminum-rich SG on strength development and chloride binding of seawater-mixed UHPC is more significant than that of SF. Wang and Huang (2023) concluded that due to the pozzolanic reaction and synergistic effect between the limestone (LS), MK, and cement, the combination of 20% MK and 10% LS makes SSUHPC achieving the highest long-term strength. This can be proved by the increase of chemically bound water and the decrease of Ca/Al and Ca/Si ratios in C-S-H gels. Saleh, Li, et al. (2023) recommended the combination of 12.5% SF and 37.5% SG to achieve optimum workability, long-term strength, and dimensional stability properties of the SSUHPC matrix.

Concerning the corrosion of steel fibers in SSUHPC, Sun, Li, et al. (2022) found rust spots on the surface of SSUHPC after 6 months of sulfate attack. Li, Liu, et al. (2020) found that rusty steel fibers are located on the surface of SSUHPC in the marine environment after only 1 year, but there is no corrosion of the internal steel fibers of SSUHPC. To eliminate the corrosion risk of steel fibers, ultra-high-molecular-weight polyethylene (UHMWPE) fibers are incorporated into SSUHPC, endowing it remarkable multiple-cracking and tensile strain-hardening performance (Huang, Wang, et al., 2021; Huang, Wu, et al., 2021). But the adverse effect of UHMWPE fibers on the compressive strength of SSUHPC is also reported. Zhu et al. (2022) incorporated PP, polyvinyl alcohol (PVA), UHMWPE, basalt, and alkali-resistant glass fibers into SSUHPC to study the mechanical properties. They found that owing to the addition of basalt and alkali-resistant glass fibers, the compressive strengths of SSUHPC are reduced by 14.28% and 9.5%, respectively. PP, PVA, and UHMWPE fibers all belong to organic fibers, susceptible to weak interface bonding strength. Additionally, due to the influence of external environmental factors such as moisture, temperature, and light, the degradation of organic fibers

Table 15.3 Recent studies on mechanical properties of seawater sea-sand-based ultra-high performance concrete.

W/B ratio	SCM type and content (%)	Fiber type and content (%)	Compressive strength (MPa)	Flexural strength (MPa)	Tensile strength (MPa)	References
0.14	SF (15); FA (25)	Steel fiber (2.4)	151.5	–	–	Li, Liu, et al. (2020)
			159.7	–	–	Sun, Li, et al. (2022)
			162.1	–	–	Li, Sun, et al. (2022)
0.16	SF (20); FA (10)	PP, PVA, basalt, alkali-resistant glass, and UHMWPE fibers (0–0.3)	135.2–155.8	15.5–19.3	–	Zhu et al. (2022)
0.18	SF (25)	UHMWPE fiber (0–1.5)	131.9–141.2	–	4.1–8.2	Huang, Wang, et al. (2021)
0.18	SF (20)	UHMWPE fiber (1.0–2.0)	131.6–137.7		6.1–8.2	Huang, Wu, et al. (2021)
0.24	SF (25); FA (20)	SSW (0–1.5)	116.1–138.6	8.8–13.8	–	Yu et al. (2023)

makes it difficult for SSUHPC to maintain long-term performance in the extreme service environment. Yu et al. (2023) proposed a new type of SSUHPC using seawater and ground sea-sand in combination with superfine stainless wires (SSWs), as shown in Fig. 15.5. The aim of using ground sea-sand is to increase the fineness of sea-sand to fully play its activity together with avoid the waste of sea-sand resources caused by the direct removal of large particles. They concluded that SSWs do not rust after immersion in seawater. The flexural and compressive strengths of SSUHPC incorporating 1.5% SSWs are 13.8 and 138.6 MPa, respectively, and the flexural toughness of UHPSSC is increased by 428.9%. The high specific surface area of SSW and enrichment of SF on its surface enhance the interfacial bond between SSWs and matrix, further promoting the full play of the SSWs' reinforcing mechanisms, as proved by the decrease of the Ca/Si ratio at the SSW surface. The C-S-H gels with a high Ca/Si ratio within the ITZ as well as Friedel's salt, are conducive to immobilize chlorides, blocking the migration of chlorides through the matrix and further mitigating the risk of long-term chloride corrosion of SSWs.

The corrosion behavior of steel bars embedded in SSUHPC has also been partly reported. Wang, Wang, et al. (2021) reported that steel bars embedded in uncracked SSUHPC do not rust after 40 dry-wet cycles of NaCl. The potential differences between steel bars and $Cu/CuSO_4$ reference electrodes (CSE) are more than -200 mV, indicating a low corrosion risk of steel bars according to ASTM C876-2015. Increasing the cover thickness can facilitate the protection effect of SSUHPC on steel bars. Wei et al. (2020) reported that as the chloride ion content of sea-sand increases, the corrosion rate of reinforcement embedded in sea-sand UHPC gradually increases during the early stage. However, the thick passivation film can be

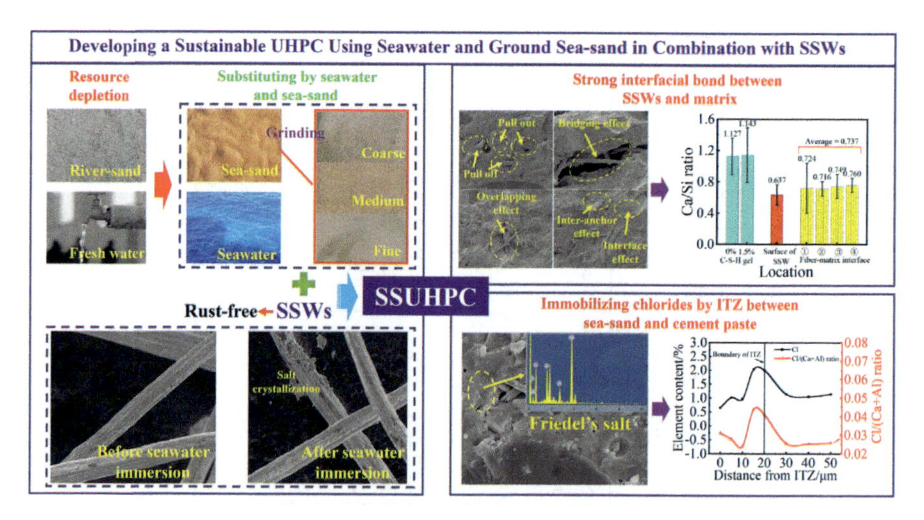

Figure 15.5 Sustainable UHPC using seawater and ground sea-sand in combination with SSWs. Using seawater and ground sea-sand in combination with SSWs can produce UHPC with improved sustainability, interfacial bond, and chloride-binding capacity.

formed within 10 days, indicating that the chloride content of the sea-sand has no significant effect on the corrosion rate after the passivation of steel bars (Wei et al., 2020). Huang, Pan, et al. (2022) found that if the endogenous chloride ion content is below 1.2%, the steel bars embedded in SSUHPC can form the passivation film. Zheng et al. (2022) concluded that the steel corrosion rate in seawater-mixed UHPC is high after casting due to the high Cl^-/OH^- ratio and abundant oxygen and water. With the rapid reduction of the internal humidity and densification of microstructure, the corrosion rate can be reduced to a negligible level within the first three days. The lack of water rather than oxygen inhibits the corrosion process in seawater-mixed UHPC. As shown in Fig. 15.6, the corrosion rate of mild steel in seawater-mixed UHPC is lower than that in seawater-mixed NC, indicating that seawater-mixed UHPC can be employed as a high-resistivity material to prevent steel corrosion.

In brief, it is feasible to develop UHPC with excellent mechanical properties and durability properties using seawater and sea-sand in combination with corrosion-resistant fibers. The utilization of SCMs in SSUHPC can bring a significant improvement in performance and sustainability. The corrosion reaction of steel bars is concentrated during the early ages of SSUHPC. The low W/B ratio and dense microstructure of SSUHPC contribute to the inhibition of corrosion reactions, resulting in a low long-term corrosion risk of steel bars embedded in SSUHPC.

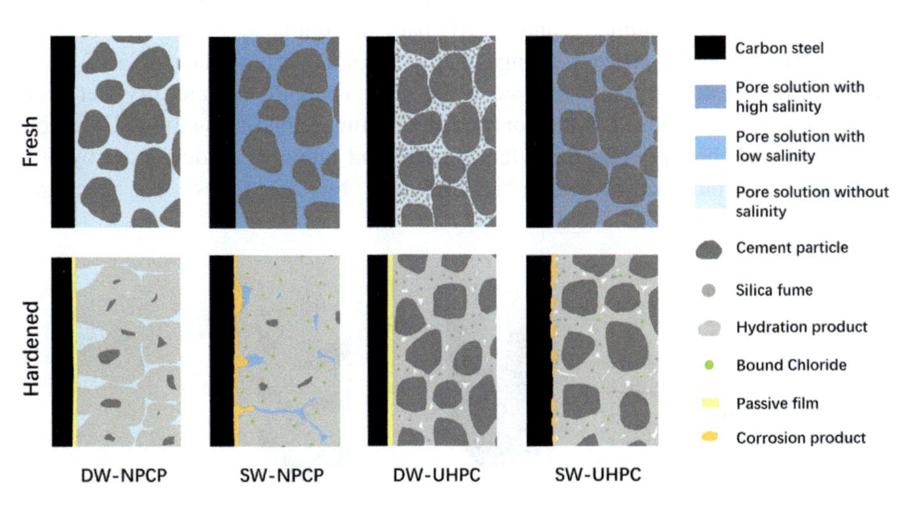

Figure 15.6 Model of the steel–paste interface in fresh and hardened cement pastes was showed in the figure. DW, SW, NPCP, and UHPC represent deionized water, seawater, normal Portland cement paste with a W/B ratio of 0.4, ultra-performance cement paste with a W/B ratio of 0.14, respectively, indicating the seawater-mixed UHPC has a lower long-term corrosion risk than that of seawater-mixed NC.
Source: From Zheng, H., Lu, J., Shen, P., Sun, L., Poon, C. S., & Li, W. (2022). Corrosion behavior of carbon steel in chloride-contaminated ultra-high-performance cement pastes. *Cement and Concrete Composites*, *128*. https://doi.org/10.1016/j.cemconcomp.2022.104443.

15.5 Structural SSC members reinforced with stainless steel or fiber-reinforced polymer

15.5.1 Seawater sea-sand concrete reinforced with stainless steel/fiber-reinforced polymer bars

Substituting conventional steel bars with corrosion-resistant bars is an effective way to overcome the steel corrosion in SSC. Two different types of corrosion-resistant reinforcement were explored in the literature, namely stainless steel bars and FRP bars. Owing to the remarkable corrosion resistance, using stainless steel bars is one of the solutions to deal with the corrosion of conventional steel bars (García-Alonso et al., 2007). Nevertheless, the high cost of stainless steels bar limits their application to critical parts of the structure. Furthermore, because of the excellent corrosion resistance, high strength-weight ratio, and good designability, FRP materials, including glass-fiber, carbon-fiber, and basalt-fiber FRP (GFRP, CFRP, and BFRP) are regarded as the ideal choice for developing innovative high-performance SSC structure systems (Bazli et al., 2021), as shown in Fig. 15.7.

With the same reinforcement ratio and strength grade, the load capacity of stainless steel bars reinforced SSC columns is higher than that of FRP bars reinforced SSC columns (J. Xu, Wu, et al., 2023). Using stainless steel bars to replace a proportion of FRP bars can significantly improve ductility. Dong et al. (2020) compared the flexural performance and durability of SSC beams reinforced with HRB400 steel, 304 stainless, and CFRP bars in simulated ocean environments. They observed no significant change in the flexural performance of SSC beams reinforced with steel and stainless steel bars after 9-month seawater dry-wet cycles and seawater immersion. On the contrary, the ultimate load capacity of CFRP bars reinforced SSC beams under high temperature and humidity is decreased by 30%, and the maximum crack width is increased by 2.2 times. The long-term

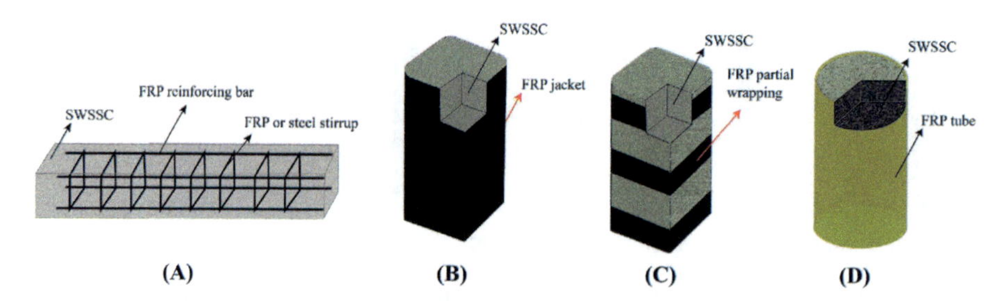

Figure 15.7 Typical FRP-SSC structural members: (A) SSC member reinforced with FRP bars; (B) fully FRP-wrapped SSC member; (C) partially FRP-wrapped SWSSC member; (D) SSC-filled FRP tube.
From Bazli, M., Heitzmann, M., & Hernandez, B. V. (2021). Hybrid fibre reinforced polymer and seawater sea sand concrete structures: A systematic review on short-term and long-term structural performance. *Construction and Building Materials, 301*. https://doi.org/10.1016/j.conbuildmat.2021.124335.

performances of FRP bars are easily susceptible to temperature, humidity, and alkalinity, resulting in strength degradation in alkaline environments and salt-alkali complex corrosive environments (Guo et al., 2022; Zhang, Zhang, et al., 2022). Wang, Gong, et al. (2021) concluded that under the action of water molecules, hydroxide ions, and chloride ions, the strength retentions of FRP bars were 71%–77%, 26%–98%, and 49%–77%, respectively. Generally, CFRP bars show the best durability and bonding strength in SSC, followed by GFRP and BFRP bars (Guo et al., 2018; Guo et al., 2022). Reducing the alkalinity and permeability of SSC by incorporating SCMs is beneficial for alleviating the strength degradation of FRP bars (Feng et al., 2023; Yi et al., 2021; Yi, Zhu, Guo, et al., 2022; Yi, Zhu, Rahman, et al., 2022). The degradation of resin, basalt fiber, and fiber-resin interface of FRP bars embedded in the low-alkalinity SSC can be significantly mitigated by the appropriate design of ternary components containing cement, silica fume, and FA.

Overall, stainless steel bars and FRP bars can, both, be used to develop SSC structural members due to their excellent corrosion resistance. However, stainless steel bars can endow SSC with better mechanical and durability performance than FRP bars. CFRP bars show the best strength retention than GFRP and BFRP bars. The strength degradation of FRP bars embedded in SSC can be mitigated by reducing the alkalinity and permeability of SSC.

15.5.2 Seawater sea-sand concrete –filled stainless steel/fiber-reinforced polymer tubes

Concrete-filled steel tubes can make full use of the properties of steel and concrete materials, offering several advantages, including excellent load-bearing capacity, ease of construction, and high seismic performance. SSC-filled stainless steel tubes can significantly reduce the consumption of expensive stainless steel, thus reducing the overall project cost (Y. Cai & Kwan, 2021). Moreover, SSC-filled FRP tubes can effectively restrain SSC by improving its strength and deformation capacity, with FRP tubes acting as a permanent formwork to improve construction efficiency (Y. L. Li, Teng, et al., 2018).

Considerable studies have been conducted on the mechanical and durability performances of SSC-filled stainless steel/FRP tubes. Due to the different material properties between stainless steel and FRP, SSC-filled FRP tubes exhibit a more distinct restraint effect than SSC-filled stainless steel tubes, but the latter has a higher ductility. When the lateral expansion of SSC exceeds that of FRP tubes, the linear stress–strain curve of FRP leads to an increasing confining pressure on SSC, which is called as passive confinement effect (Y. L. Li, Zhao, Raman Singh, et al., 2016; Y. L. Li, Zhao, Raman, et al., 2016). Whereas, the confining pressure provided by stainless steel tubes slows down after stainless steel reaches yield strength, implying that stainless steel with the rounded stress–stress curve can provide active confinement effect. Li, Li, et al. (2021) found that SSC-filled stainless steel tubes in indoor environment or immersed in NaCl solution showed no reduction in the loading capacity. However, the hydrothermal environment can significantly aggravate the strength degradation of SSC-filled FRP cubes (Y.-L. Li, Zhao, et al., 2021). Li, Teng, et al. (2018) found that the hoop strengths of SSC-filled GFRP/BFRP/CFRP tubular stub columns in artificial seawater for six months are reduced by 23%, 39%, and 8%, respectively. Bazli et al. (2019) reported that the bond strength of CFRP tubes is greater than that of GFRP and BFRP tubes due to the strong frictional resistance. Fiber orientation, tube surface roughness, and tube aspect ratio significantly impact the bonding strength between FRP tubes and SSC. Regarding strength degradation in seawater, SSC-filled CFRP tubes exhibited the best long-term strength compared to SSC-filled GFRP and BFRP tubes (Bazli, Li, et al., 2020; Bazli, Zhao, et al., 2020; Guo et al., 2022).

It can be summarized that SSC-filled stainless steel tubes have a less constraining effect and higher ductility and durability than SSC-filled FRP tubes. The bonding and long-term strength between CFRP tubes and SSC are higher than that of GFRP and BFRP tubes.

15.5.3 Seawater sea-sand concrete wrapped with fiber-reinforced polymer

Applying FRP as the transverse confinement in the ring direction of SSC columns can substantially increase the ultimate strength and deformation capacity (Li, Yang, et al., 2021). To protect the internal SSC from seawater corrosion, the fully FRP wrapped SSC columns are applied to structural parts in submerged zones. However, in terms of SSC members in tidal and atmospheric zones, adopting the partially FRP wrapped strategy rather than the fully wrapped strategy is beneficial for accelerating the escape of internal moisture through the unconfined region between two adjacent FRP strips, thus mitigating the strength degradation of FRP, improving the durability of SSC members and the utilization efficiency of FRP (Yang, Wang, et al., 2022).

Due to the remarkable strength retention, CFRP is the ideal option for confining SSC columns. It has been proved that the ultimate strength of CFRP partially wrapped SSC is lower than that of CFRP fully wrapped SSC. Nevertheless, the former exhibits superior strength retention under wet-dry cycles in the seawater environment (Yang et al., 2020a, Yang et al., 2020b, Yang, Lu, et al., 2023). This is because the CFRP partially scheme's small installation area helps mitigate the CFRP degradation and FRP-concrete interface damage. The axial stress−strain relationship for partially CFRP-wrapped SSC columns has a more noticeable transition segment and strain-softening behavior than that of fully FRP-wrapped SSC columns (Yang et al., 2020a, 2020b). This can be solved by decreasing the clear spacing ratio. The lateral stress−axial strain relationship is determined by the clear spacing ratio and confinement stiffness of CFRP jackets. The SSC column's ultimate strength and deformation capacity can be significantly improved with the decrease in clear spacing ratio and the increase in confinement stiffness. Additionally, FRP confinement has been applied to SSC-filled stainless steel tubes to enhance the strength and ductility of members (Chen et al., 2023; Zeng et al., 2019; Zeng et al., 2022). FRP confinement can improve the combined effect between the stainless steel tube and SSC under the axial compression action together with retarding the outward expansion of the stainless steel tube. The constraining effect of FRP jacket is more significant than that of the stainless steel tube.

It can be concluded that CFRP utilization efficiency and durability of CFRP partially wrapped SSC columns are higher than that of CFRP fully wrapped SSC columns. FRP confinement can further improve the constraining effect of SSC-filled stainless steel tubes.

15.6 Conclusions and future needs

The massive production and consumption of concrete heavily burden resources, energy, and the environment. In constructing coastal and marine infrastructures, utilizing seawater and sea-sand in place of fresh water and river-sand gains remarkable economic and ecologic benefits, such as alleviating resource depletion, protecting river ecology, reducing project costs and lower carbon emissions due to long-distance transportation. This is of great significance in promoting sustainable development of concrete. In this chapter, the research findings on hydration, microstructures, and performances of SSC as well as new types of sustainable concrete based on seawater and sea-sand are introduced; structural members fabricated with SSC and corrosion-resistant reinforcement (stainless steel and FRP) are also presented. The conclusions are summarized as follows.

1. Cl^- in seawater and sea-sand effectively accelerate cement hydration and promotes the formation of C-S-H gels and Friedel's salts, improving microstructures and pore structures, reducing the workability, and increasing the early strength of SSC. SSC exhibits lower long-term strength and more significant shrinkage compared to NC. This is associated with the formation of harmful hydration products from SO_4^{2-} and Mg^{2+}. The effect of seawater and sea-sand on the durability, except for steel corrosion of concrete, is acceptable. The solutions for steel corrosion include adding corrosion inhibitors to protect the steel passivation film, incorporating alumina-rich cementitious materials to improve chloride-binding capacity, and substituting conventional steel bars with corrosion-resistant bars.
2. Utilizing seawater and sea-sand in combination with CAC, RAC, AAC, and UHPC can produce new sustainable concrete with satisfactory performance. The incorporation of SCMs brings a significant improvement in microstructures and performances, whereas the incorporation of corrosion-resistant fibers is an effective way to improve strength and toughness. Notably, the hydration process of AAMs system is negatively affected by seawater and sea-sand due to the interaction of $MgCl_2$ and activators. Owing to the low W/B ratio and dense microstructure, steel bars embedded in SSUHPC have a low long-term corrosion risk even in the presence of high chloride content.
3. Stainless steel bars endow better mechanical and durability properties of SSC members compared to FRP bars. The bond and strength retention of CFRP is more remarkable than that of GFRP and BFRP. The strength degradation of FRP bars can be mitigated by

reducing the alkalinity and permeability of SSC. SSC-filled stainless steel tubes have a lower constraining effect and higher ductility and durability in comparison with SSC-filled FRP tubes. The FRP utilization efficiency and durability of CFRP partially wrapped SSC columns are higher than that of CFRP fully wrapped SSC columns. CFRP confinement can further improve the constraining effect of SSC-filled stainless-steel tubes.

Although a number of studies have been conducted in this field, many challenges still need to be addressed. Firstly, the salinity of seawater and sea-sand as well as the shape of sea-sand particles, exhibit significant variations among different regions. The effect of seawater sea-sand from different sea areas on SSC performance needs to be further investigated. Moreover, the effects of SO_4^{2-}, CO_3^{2-}, Mg^{2+}, and Ca^{2+} ions on chloride-binding capacity and steel corrosion of SSC deserve further research and exploration. Additionally, it is surprising that SSUHPC produces a low long-term corrosion risk for conventional steel bars. Further systematic studies on the durability and long-term corrosion behavior of SSUHPC are required to control steel corrosion within an acceptable range. Lastly, with the development of concrete technology, seawater and sea-sand are expected to be used in more new types of concrete, further promoting the sustainable improvement of concrete. SSC combined with stainless steel or FRP can form innovative high-performance structural systems, holding great promise in coastal and marine infrastructures.

Acknowledgments

The authors thank the funding support from the National Science Foundation of China (51978127, 52178188, and 51908103) and the China Postdoctoral Science Foundation (2022M710973 and 2022M720648).

References

Ali, M. B., Saidur, R., & Hossain, M. S. (2011). A review on emission analysis in cement industries. *Renewable and Sustainable Energy Reviews*, *15*(5), 2252–2261. Available from https://doi.org/10.1016/j.rser.2011.02.014.

Bachtiar, E., Rachim, F., Makbul, R., Tata, A., Irfan-Ul-Hassan, M., Kırgız, M. S., Syarif, M., de Sousa Galdino, A. G., Khitab, A., Benjeddou, O., Kolovos, K. G., Ledesma, E. F., Yusri, A., & Papatzani, S. (2022). Monitoring of chloride and Friedel's salt, hydration components, and porosity in high-performance concrete. *Case Studies in Construction Materials*, *17*. Available from https://doi.org/10.1016/j.cscm.2022.e01208, http://www.journals.elsevier.com/case-studies-in-construction-materials/.

Bazli, M., Heitzmann, M., & Hernandez, B. V. (2021). Hybrid fibre reinforced polymer and seawater sea sand concrete structures: A systematic review on short-term and long-term structural performance. *Construction and Building Materials*, *301*. Available from https://doi.org/10.1016/j.conbuildmat.2021.124335, https://www.scopus.com/inward/

record.uri?eid = 2-s2.0-85111246941&doi = 10.1016/2fj.conbuildmat.2021.124335&partnerID = 40&md5 = b5f64a413892b53d2edbdc7a5c2b47ac.

Bazli, M., Li, Y.-L., Zhao, X.-L., Raman, R. K. S., Bai, Y., Al-Saadi, S., & Haque, A. (2020). Durability of seawater and sea sand concrete filled filament wound FRP tubes under seawater environments. *Composites Part B: Engineering*, *202*. Available from https://doi.org/10.1016/j.compositesb.2020.108409, https://www.scopus.com/inward/record.uri?eid = 2-s2.0-85090992038&doi = 10.1016/2fj.compositesb.2020.108409&partnerID = 40&md5 = 7c94f3181c7ab28f5f402837cca5cd0b.

Bazli, M., Zhao, X. L., Bai, Y., Singh Raman, R. K., & Al-Saadi, S. (2019). Bond-slip behaviour between FRP tubes and seawater sea sand concrete. *Engineering Structures*, *197*109421. Available from https://doi.org/10.1016/j.engstruct.2019.109421.

Bazli, M., Zhao, X. L., Singh Raman, R. K., Bai, Y., & Al-Saadi, S. (2020). Bond performance between FRP tubes and seawater sea sand concrete after exposure to seawater condition. *Construction and Building Materials*, *265*. Available from https://doi.org/10.1016/j.conbuildmat.2020.120342, https://www.journals.elsevier.com/construction-and-building-materials.

Cai, Y., & Kwan, A. K. H. (2021). Behaviour and design of cold-formed austenitic stainless steel circular tubes infilled with seawater sea-sand concrete. *Engineering Structures*, *241*. Available from https://doi.org/10.1016/j.engstruct.2021.112435, https://www.scopus.com/inward/record.uri?eid = 2-s2.0-85105304865&doi = 10.1016/2fj.engstruct.2021.112435&partnerID = 40&md5 = 850b2f8a564cbc1f2a952a4def5a09c8.

Cai, Y. M., Tao, Y., Xuan, D. X., Sun, Y. J., & Poon, C. S. (2023). Effect of seawater on the morphology, structure, and properties of synthetic ettringite. *Cement and Concrete Research*, *163*. Available from https://doi.org/10.1016/j.cemconres.2022.107034, https://www.scopus.com/inward/record.uri?eid = 2-s2.0-85142205720&doi = 10.1016/2fj.cemconres.2022.107034&partnerID = 40&md5 = 2d7d5fa0add1ae329b72484f5329243d.

Chen, G., Liu, P., Jiang, T., He, Z., Wang, X., Lam, L., & Chen, J. F. (2020). Effects of natural seawater and sea sand on the compressive behaviour of unconfined and carbon fibre-reinforced polymer-confined concrete. *Advances in Structural Engineering*, *23*(14), 3102–3116. Available from https://doi.org/10.1177/1369433220920459, https://uk.sagepub.com/en-gb/eur/advances-in-structural-engineering/journal202466.

Chen, Z., Qin, W., Liang, Y., & Zhou, J. (2023). Axial compressive performance of seawater sea sand concrete-filled CFRP-stainless steel tube short columns. *Construction and Building Materials*, *369*. Available from https://doi.org/10.1016/j.conbuildmat.2023.130501, https://www.journals.elsevier.com/construction-and-building-materials.

Cui, Y. X., Jiang, J. F., Fu, T. F., & Liu, S. F. (2022). Feasibility of using waste brine/seawater and sea sand for the production of concrete: An experimental investigation from mechanical properties and durability perspectives. *Sustainability*, *14*(20). Available from https://doi.org/10.3390/su142013340, https://www.scopus.com/inward/record.uri?eid = 2-s2.0-85141064970&doi = 10.3390/2fsu142013340&partnerID = 40&md5 = b9b4247845e1e4bfdad47c1c3fb2ee68.

Dhondy, T., Remennikov, A., & Neaz Sheikh, M. (2020). Properties and application of sea sand in sea sand-seawater concrete. *Journal of Materials in Civil Engineering*, *32*(12). Available from https://doi.org/10.1061/(ASCE)MT.1943-5533.0003475, http://ascelibrary.org/mto/resource/1/jmcee7/.

Dhondy, T., Remennikov, A., & Shiekh, M. N. (2019). Benefits of using sea sand and seawater in concrete: a comprehensive review. *Australian Journal of Structural Engineering*, *20*(4), 280–289. Available from https://doi.org/10.1080/13287982.2019.1659213, http://www.tandfonline.com/toc/TSEN20/current.

Dong, Z., Wu, G., Zhao, X.-L., Zhu, H., & Lian, J.-L. (2020). The durability of seawater sea-sand concrete beams reinforced with metal bars or non-metal bars in the ocean environment. *Advances in Structural Engineering*, *23*(2), 334–347. Available from https://doi.org/10.1177/1369433219870580, https://www.scopus.com/inward/record.uri?eid = 2-s2.0-85071543823&doi = 10.1177/2f1369433219870580&partnerID = 40&md5 = eb3c03b8cf50d1b131e78afa0847fe20.

Ebead, U., Lau, D., Lollini, F., Nanni, A., Suraneni, P., & Yu, T. (2022). A review of recent advances in the science and technology of seawater-mixed concrete. *Cement and Concrete Research*, *152*. Available from https://doi.org/10.1016/j.cemconres.2021.106666, http://www.sciencedirect.com/science/journal/00088846.

Elshami, A., Bonnet, S., Khelidj, A., & Sail, L. (2020). Effectiveness of corrosion inhibitors in simulated concrete pore solution. *European Journal of Environmental and Civil Engineering*, *24*(13), 2130–2150. Available from https://doi.org/10.1080/19648189.2018.1500309, http://www.tandfonline.com/loi/tece20.

Etxeberria, M., Vázquez, E., Marí, A., & Barra, M. (2007). Influence of amount of recycled coarse aggregates and production process on properties of recycled aggregate concrete. *Cement and Concrete Research*, *37*(5), 735–742. Available from https://doi.org/10.1016/j.cemconres.2007.02.002.

Feng, G., Zhu, D., Guo, S., Rahman, M. Z., Ma, W., Yi, Y., Jin, Z., & Shi, C. (2023). A comparative study of bare and seawater sea sand concrete wrapped basalt fiber-reinforced polymer bars exposed to laboratory and real marine environments. *Construction and Building Materials*, *371*. Available from https://doi.org/10.1016/j.conbuildmat.2023.130764, https://www.scopus.com/inward/record.uri?eid = 2-s2.0-85148545439&doi = 10.1016/2fj.conbuildmat.2023.130764&partnerID = 40&md5 = 5c80ab154c67c3350e8ec7aa0a5552c3.

García-Alonso, M. C., Escudero, M. L., Miranda, J. M., Vega, M. I., Capilla, F., Correia, M. J., Salta, M., Bennani, A., & González, J. A. (2007). Corrosion behaviour of new stainless steels reinforcing bars embedded in concrete. *Cement and Concrete Research*, *37*(10), 1463–1471. Available from https://doi.org/10.1016/j.cemconres.2007.06.003.

Geng, J., Easterbrook, D., Liu, Q. F., & Li, L. Y. (2016). Effect of carbonation on release of bound chlorides in chloride-contaminated concrete. *Magazine of Concrete Research*, *68*(7), 353–363. Available from https://doi.org/10.1680/jmacr.15.00234, http://www.ice-virtuallibrary.com/content/serial/macr.

Geng, J., Zhu, D., Guo, S., Yi, Y., & Zhou, L. (2022). Experimental study on mechanical properties of seawater sea-sand concrete with sea-sands from different regions. *Materials Reports*, *36*(3). Available from https://doi.org/10.11896/cldb.21010189, https://www.scopus.com/inward/record.uri?eid = 2-s2.0-85125414909&doi = 10.11896/2fcldb.21010189&partnerID = 40&md5 = 9dd534a5dcd284b1e98bfc211f4d2334.

Guan, H., Hao, B., & Zhang, G. (2023). Mechanical properties of concrete prepared using seawater, sea sand and spontaneous combustion coal gangue. *Structures*, *48*, 172–181. Available from https://doi.org/10.1016/j.istruc.2022.12.089, https://www.journals.elsevier.com/structures.

Guo, F., Al-Saadi, S., Singh Raman, R. K., & Zhao, X. L. (2018). Durability of fiber reinforced polymer (FRP) in simulated seawater sea sand concrete (SWSSC) environment. *Corrosion Science*, *141*, 1–13. Available from https://doi.org/10.1016/j.corsci.2018.06.022, https://www.scopus.com/inward/record.uri?eid = 2-s2.0-85049012135&doi = 10.1016/2fj.corsci.2018.06.022&partnerID = 40&md5 = d0312847c975c448a1ae5743dcb89fe4.

Guo, M., Hu, B., Xing, F., Zhou, X., Sun, M., Sui, L., & Zhou, Y. (2020). Characterization of the mechanical properties of eco-friendly concrete made with untreated sea sand and seawater based on statistical analysis. *Construction and Building Materials, 234*. Available from https://doi.org/10.1016/j.conbuildmat.2019.117339, https://www.scopus.com/inward/record.uri?eid = 2-s2.0-85074348426&doi = 10.1016/2fj.conbuildmat.2019.117339&partnerID = 40&md5 = e08018ba79e268b01ed2c2efdc58399b.

Guo, X., Xiong, C., Jin, Z., & Pang, B. (2022). A review on mechanical properties of FRP bars subjected to seawater sea sand concrete environmental effects. *Journal of Building Engineering, 58*. Available from https://doi.org/10.1016/j.jobe.2022.105038, https://www.scopus.com/inward/record.uri?eid = 2-s2.0-85135719916&doi = 10.1016/2fj.jobe.2022.105038&partnerID = 40&md5 = 576d2da7a6dde9caa3b2c24808fa2ad1.

Han, S., Zhong, J., Ding, W., & Ou, J. (2021). Strength, hydration, and microstructure of seawater sea-sand concrete using high-ferrite Portland cement. *Construction and Building Materials, 295*. Available from https://doi.org/10.1016/j.conbuildmat.2021.123703, https://www.scopus.com/inward/record.uri?eid = 2-s2.0-85106622404&doi = 10.1016/2fj.conbuildmat.2021.123703&partnerID = 40&md5 = 7b519a240ec6fe2282399aba73803a97.

Han, S., Zhong, J., Yu, Q., Yan, L., & Ou, J. (2021). Sulfate resistance of eco-friendly and sulfate-resistant concrete using seawater sea-sand and high-ferrite Portland cement. *Construction and Building Materials, 305*. Available from https://doi.org/10.1016/j.conbuildmat.2021.124753, https://www.scopus.com/inward/record.uri?eid = 2-s2.0-85114129431&doi = 10.1016/2fj.conbuildmat.2021.124753&partnerID = 40&md5 = 6ae101754f10e9fa7a312b61937cf0a6.

He, S., Jiao, C., Niu, Y., & Li, S. (2022). Utilizing of coral/sea sand as aggregates in environment-friendly marine mortar: Physical properties, carbonation resistance and microstructure. *Case Studies in Construction Materials, 16*. Available from https://doi.org/10.1016/j.cscm.2022.e00981.

Huang, Y., He, X., Sun, H., Sun, Y., & Wang, Q. (2018). Effects of coral, recycled and natural coarse aggregates on the mechanical properties of concrete. *Construction and Building Materials, 192*, 330–347. Available from https://doi.org/10.1016/j.conbuildmat.2018.10.111.

Huang, Y., Li, X., Lu, Y., Wang, H., Wang, Q., Sun, H., & Li, D. (2019). Effect of mix component on the mechanical properties of coral concrete under axial compression. *Construction and Building Materials, 223*, 736–754. Available from https://doi.org/10.1016/j.conbuildmat.2019.07.015, https://www.journals.elsevier.com/construction-and-building-materials.

Huang, W., Pan, A., Wei, J., Chen, B., Chen, R., Bian, X., Chen, Z., & Zhang, H. (2022). Critical content of endogenous chloride ion and service life prediction of simulated marine materials UHPC. Southeast University, China. *Journal of Southeast University, 52*(5), 890–898. Available from https://doi.org/10.3969/j.issn.1001-0505.2022.05.009, http://journal.seu.edu.cn/oa/oascriptissuelsit.aspx?kind = upissue&issuenolist = 2022/C4/EA/B5/DA5/C6/DA.

Huang, Y. J., Wang, T. C., Sun, H. L., Li, C. X., Yin, L., & Wang, Q. (2022). Mechanical properties of fibre reinforced seawater sea-sand recycled aggregate concrete under axial compression. *Construction and Building Materials, 331*. Available from https://doi.org/10.1016/j.conbuildmat.2022.127338.

Huang, B. T., Wang, Y. T., Wu, J. Q., Yu, J., Dai, J. G., & Leung, C. K. Y. (2021). Effect of fiber content on mechanical performance and cracking characteristics of ultra-high-performance seawater sea-sand concrete (UHP-SSC). *Advances in Structural Engineering, 24*(6), 1182–1195. Available from https://doi.org/10.1177/1369433220972452, https://uk.sagepub.com/en-gb/eur/advances-in-structural-engineering/journal202466.

Huang, B. T., Wu, J. Q., Yu, J., Dai, J. G., Leung, C. K. Y., & Li, V. C. (2021). Seawater sea-sand engineered/strain-hardening cementitious composites (ECC/SHCC): Assessment and modeling of crack characteristics. *Cement and Concrete Research, 140*. Available from https://doi.org/10.1016/j.cemconres.2020.106292, http://www.sciencedirect.com/science/journal/00088846.

Iqbal, M., Zhang, D., Khan, K., Amin, M. N., Ibrahim, M., & Salami, B. A. (2023). Evaluating mechanical, microstructural and durability performance of seawater sea sand concrete modified with silica fume. *Journal of Building Engineering, 72*. Available from https://doi.org/10.1016/j.jobe.2023.106583, https://www.scopus.com/inward/record.uri?eid = 2-s2.0-85154054509&doi = 10.1016/2fj.jobe.2023.106583&partnerID = 40&md5 = ebfd7db8ab3dd8849a0f8f128d6b2f73.

Lam, W., Shen, P., Cai, Y., Sun, Y., Zhang, Y., & Sun Poon, C. (2022). Effects of seawater on UHPC: Macro and microstructure properties. *Construction and Building Materials, 340*. Available from https://doi.org/10.1016/j.conbuildmat.2022.127767, https://www.scopus.com/inward/record.uri?eid = 2-s2.0-85130912469&doi = 10.1016/2fj.conbuildmat.2022.127767&partnerID = 40&md5 = 4713b62fa2d398d4a4c84698d45bcb2f.

Lao, J. C., Huang, B. T., Xu, L. Y., Khan, M., Fang, Y., & Dai, J. G. (2023). Seawater sea-sand Engineered Geopolymer Composites (EGC) with high strength and high ductility. *Cement and Concrete Composites, 138*. Available from https://doi.org/10.1016/j.cemconcomp.2023.104998, http://www.sciencedirect.com/science/journal/09589465.

Liu, J., An, R., Jiang, Z., Jin, H., Zhu, J., Liu, W., Huang, Z., Xing, F., Liu, J., Fan, X., & Sui, T. (2022). Effects of w/b ratio, fly ash, limestone calcined clay, seawater and sea-sand on workability, mechanical properties, drying shrinkage behavior and micro-structural characteristics of concrete. *Construction and Building Materials, 321*. Available from https://doi.org/10.1016/j.conbuildmat.2022.126333, https://www.scopus.com/inward/record.uri?eid = 2-s2.0-85122517889&doi = 10.1016/2fj.conbuildmat.2022.126333&partnerID = 40&md5 = d61768dc6dfb34480ab5ec679ca6f906.

Liu, J., Fan, X., Liu, J., Jin, H., Zhu, J., & Liu, W. (2021). Investigation on mechanical and micro properties of concrete incorporating seawater and sea sand in carbonized environment. *Construction and Building Materials, 307*. Available from https://doi.org/10.1016/j.conbuildmat.2021.124986, https://www.scopus.com/inward/record.uri?eid = 2-s2.0-85115628248&doi = 10.1016/2fj.conbuildmat.2021.124986&partnerID = 40&md5 = 417573ee106645e973711fdee67fd3e7.

Liu, J., Liu, J., Fan, X., Jin, H., Zhu, J., Huang, Z., Xing, F., & Sui, T. (2022). Experimental analysis on water penetration resistance and micro properties of concrete: Effect of supplementary cementitious materials, seawater, sea-sand and water-binder ratio. *Journal of Building Engineering, 50*. Available from https://doi.org/10.1016/j.jobe.2022.104153, https://www.scopus.com/inward/record.uri?eid = 2-s2.0-85124622315&doi = 10.1016/2fj.jobe.2022.104153&partnerID = 40&md5 = 1d0145edc953e6e857fdfe73099cc04a.

Liu, J., Liu, J., Jin, H., Fan, X., Jiang, Z., Zhu, J., & Liu, W. (2022). Evaluation of the pore characteristics and microstructures of concrete with fly ash, limestone-calcined clay, seawater, and sea sand. *Arabian Journal for Science and Engineering, 47*(10), 13603−13622. Available from https://doi.org/10.1007/s13369-022-06809-2, https://www.scopus.com/inward/record.uri?eid = 2-s2.0-85141121473&doi = 10.1007/2fs13369-022-06809-2&partnerID = 40&md5 = f79c3e8387043216b30c6503451c63d1.

Li, H., Farzadnia, N., & Shi, C. (2018). The role of seawater in interaction of slag and silica fume with cement in low water-to-binder ratio pastes at the early age of hydration. *Construction and Building Materials, 185*, 508−518. Available from https://doi.org/10.1016/j.conbuildmat.2018.07.091.

Li, S., Jin, Z., & Yu, Y. (2021). Chloride binding by calcined layered double hydroxides and alumina-rich cementitious materials in mortar mixed with seawater and sea sand. *Construction and Building Materials, 293*. Available from https://doi.org/10.1016/j.conbuildmat.2021.123493, https://www.scopus.com/inward/record.uri?eid = 2-s2.0-85107622894&doi = 10.1016/2fj.conbuildmat.2021.123493&partnerID = 40&md5 = baf14435efe9fc14904a5f7f71c05c55.

Li, T., Liu, X., Zhang, Y., Yang, H., Zhi, Z., Liu, L., Ma, W., Shah, S. P., & Li, W. (2020). Preparation of sea water sea sand high performance concrete (SHPC) and serving performance study in marine environment. *Construction and Building Materials, 254*. Available from https://doi.org/10.1016/j.conbuildmat.2020.119114, https://www.journals.elsevier.com/construction-and-building-materials.

Li, P., Li, W., Sun, Z., Shen, L., & Sheng, D. (2021). Development of sustainable concrete incorporating seawater: A critical review on cement hydration, microstructure and mechanical strength. *Cement and Concrete Composites, 121*. Available from https://doi.org/10.1016/j.cemconcomp.2021.104100, http://www.sciencedirect.com/science/journal/09589465.

Li, P., Li, W., Yu, T., Qu, F., & Tam, V. W. Y. (2020). Investigation on early-age hydration, mechanical properties and microstructure of seawater sea sand cement mortar. *Construction and Building Materials, 249*. Available from https://doi.org/10.1016/j.conbuildmat.2020.118776, https://www.journals.elsevier.com/construction-and-building-materials.

Li, T., Sun, X., Shi, F., Zhu, Z., Wang, D., Tian, H., Liu, X., Lian, X., Bao, T., & Hou, B. (2022). The mechanism of anticorrosion performance and mechanical property differences between seawater sea-sand and freshwater river-sand ultra-high-performance polymer cement mortar (UHPC). *Polymers, 14*(15), 3105. Available from https://doi.org/10.3390/polym14153105.

Li, Y. L., Teng, J. G., Zhao, X. L., & Singh Raman, R. K. (2018). Theoretical model for seawater and sea sand concrete-filled circular FRP tubular stub columns under axial compression. *Engineering Structures, 160*, 71–84. Available from https://doi.org/10.1016/j.engstruct.2018.01.017, https://www.scopus.com/inward/record.uri?eid = 2-s2.0-85044964309&doi = 10.1016/2fj.engstruct.2018.01.017&partnerID = 40&md5 = 399ba1a7adaa190af3b83534dc95f3bb.

Li, P., Yang, T., Zeng, Q., Xing, F., & Zhou, Y. (2021). Axial stress–strain behavior of carbon FRP-confined seawater sea-sand recycled aggregate concrete square columns with different corner radii. *Composite Structures, 262*. Available from https://doi.org/10.1016/j.compstruct.2021.113589, https://www.scopus.com/inward/record.uri?eid = 2-s2.0-85100507158&doi = 10.1016/2fj.compstruct.2021.113589&partnerID = 40&md5 = 13e8ad722f6f1daf6cc90e973e795436.

Li, Y. L., Zhao, X. L., Raman Singh, R. K., & Al-Saadi, S. (2016). Tests on seawater and sea sand concrete-filled CFRP, BFRP and stainless steel tubular stub columns. *Thin-Walled Structures, 108*, 163–184. Available from https://doi.org/10.1016/j.tws.2016.08.016, https://www.scopus.com/inward/record.uri?eid = 2-s2.0-84984633420&doi = 10.1016/2fj.tws.2016.08.016&partnerID = 40&md5 = 2fade6bdf6658d6210f1158eff890bf8.

Li, Y. L., Zhao, X. L., Raman, R. K. S., & Al-Saadi, S. (2016). Compression tests on seawater and sea sand concrete (SWSSC) filled stainless steel, CFRP and BFRP tubular stub columns. *Thin-Walled Structures*, 270–276.

Li, Y.-L., Zhao, X.-L., & Raman, R. K. S. (2021). Durability of seawater and sea sand concrete and seawater and sea sand concrete–filled fibre-reinforced polymer/stainless steel tubular stub columns. *Advances in Structural Engineering, 24*(6), 1074–1089. Available from https://doi.org/10.1177/1369433220944509, https://www.scopus.com/inward/record.uri?eid = 2-s2.0-85088583826&doi = 10.1177/2f1369433220944509&partnerID = 40&md5 = 23f96e7505e51b6e5d4f0d47e05c2df0.

Li, Y. L., Zhao, X. L., Singh Raman, R. K., & Al-Saadi, S. (2018). Thermal and mechanical properties of alkali-activated slag paste, mortar and concrete utilising seawater and sea sand. *Construction and Building Materials*, *159*, 704–724. Available from https://doi.org/10.1016/j.conbuildmat.2017.10.104.

Li, Y. T., Zhou, L., Zhang, Y., Cui, J. W., & Shao, J. (2013). Study on long-term performance of concrete based on seawater, sea sand and coral sand. *Advanced Materials Research*, *706*, 512–515.

Lu, Z., Liu, G., Wu, Y., Dai, M., Jiang, M., & Xie, J. (2023). Recycled aggregate seawater–-sea sand concrete and its durability after immersion in seawater. *Journal of Building Engineering*, *65*. Available from https://doi.org/10.1016/j.jobe.2022.105780, https://www.scopus.com/inward/record.uri?eid = 2-s2.0-85145353360&doi = 10.1016/2fj.jobe.2022.105780&partnerID = 40&md5 = 198b39b31e600c3992643b059a53f39f.

Lyu, X., Robinson, N., Elchalakani, M., Johns, M. L., Dong, M., & Nie, S. (2022). Sea sand seawater geopolymer concrete. *Journal of Building Engineering*, *50*. Available from https://doi.org/10.1016/j.jobe.2022.104141, http://www.journals.elsevier.com/journal-of-building-engineering/.

Miller, S. A., Horvath, A., & Monteiro, P. J. M. (2018). Impacts of booming concrete production on water resources worldwide. *Nature Sustainability*, *1*(1), 69–76. Available from https://doi.org/10.1038/s41893-017-0009-5, https://doi.org/10.1038/s41893-017-0009-5.

Monteiro, P. J. M., Miller, S. A., & Horvath, A. (2017). Towards sustainable concrete. *Nature Materials*, *16*(7), 698–699. Available from https://doi.org/10.1038/nmat4930, https://doi.org/10.1038/nmat4930.

Ovink, H., Rahimzoda, S., Cullman, J., & Imperiale, A. J. (2023). The UN 2023 Water Conference and pathways towards sustainability transformation for a water-secure world. *Nature Water*, *1*(3), 212–215. Available from https://doi.org/10.1038/s44221-023-00052-1, https://doi.org/10.1038/s44221-023-00052-1.

Pan, C., Li, X., & Mao, J. (2020). The effect of a corrosion inhibitor on the rehabilitation of reinforced concrete containing sea sand and seawater. *Materials*, *13*(6). Available from https://doi.org/10.3390/ma13061480, https://www.scopus.com/inward/record.uri?eid = 2-s2.0-85082618376&doi = 10.3390/2fma13061480&partnerID = 40&md5 = 913cfd8d369626e2877fc54e61522fca.

Pan, D., Yaseen, S. A., Chen, K., Niu, D., Ying Leung, C. K., & Li, Z. (2021). Study of the influence of seawater and sea sand on the mechanical and microstructural properties of concrete. *Journal of Building Engineering*, *42*. Available from https://doi.org/10.1016/j.jobe.2021.103006, https://www.scopus.com/inward/record.uri?eid = 2-s2.0-85111885359&doi = 10.1016/2fj.jobe.2021.103006&partnerID = 40&md5 = 281c322a74af1144b6b2dfe45fa12ee9.

Provis, J. L. (2018). Alkali-activated materials. *Cement and Concrete Research*, *114*, 40–48. Available from https://doi.org/10.1016/j.cemconres.2017.02.009, http://www.sciencedirect.com/science/journal/00088846.

Provis, J. L. (2014). Geopolymers and other alkali activated materials: Why, how, and what? *Materials and Structures*, *47*(1-2), 11–25. Available from https://doi.org/10.1617/s11527-013-0211-5.

Qi, X., Huang, Y., Li, X., Hu, Z., Ying, J., & Li, D. (2020). Mechanical properties of sea water sea sand coral concrete modified with different cement and fiber types. *Journal of*

Renewable Materials, *8*(8), 914–937. Available from https://doi.org/10.32604/jrm.2020.010991.

Qu, F., Li, W., Wang, K., Tam, V. W. Y., & Zhang, S. (2021). Effects of seawater and undesalted sea sand on the hydration products, mechanical properties and microstructures of cement mortar. *Construction and Building Materials*, *310*125229. Available from https://doi.org/10.1016/j.conbuildmat.2021.125229.

Rabadia, K., & Aslani, F. (2023). Development of creep and shrinkage prediction models for recycled aggregate concrete. *Structural Concrete*. Available from https://doi.org/10.1002/suco.202300072, http://onlinelibrary.wiley.com/journal/10.1002/(ISSN)1751-7648.

Ragab, A. M., Elgammal, M. A., Hodhod, O. A. G., & Ahmed, T. E. S. (2016). Evaluation of field concrete deterioration under real conditions of seawater attack. *Construction and Building Materials*, *119*, 130–144. Available from https://doi.org/10.1016/j.conbuildmat.2016.05.014.

Rahman, M. M., & Bassuoni, M. T. (2014). Thaumasite sulfate attack on concrete: Mechanisms, influential factors and mitigation. *Construction and Building Materials*, *73*, 652–662. Available from https://doi.org/10.1016/j.conbuildmat.2014.09.034.

Riding, K., Silva, D. A., & Scrivener, K. (2010). Early age strength enhancement of blended cement systems by CaCl2 and diethanol-isopropanolamine. *Cement and Concrete Research*, *40*(6), 935–946. Available from https://doi.org/10.1016/j.cemconres.2010.01.008, https://www.sciencedirect.com/science/article/pii/S0008884610000232.

Saleh, S., Li, Y.-L., Hamed, E., Mahmood, A. H., & Zhao, X.-L. (2023). Workability, strength, and shrinkage of ultra-high-performance seawater, sea sand concrete with different OPC replacement ratios. *Journal of Sustainable Cement-Based Materials*, *12*(3), 271–291. Available from https://doi.org/10.1080/21650373.2022.2050831, https://www.scopus.com/inward/record.uri?eid = 2-s2.0-85126815683&doi = 10.1080/2f21650373.2022.2050831&partnerID = 40&md5 = 8c1ed48a6f536d028995ae386ff9571d.

Saleh, S., Mahmood, A. H., Hamed, E., & Zhao, X.-L. (2023). The mechanical, transport and chloride binding characteristics of ultra-high-performance concrete utilising seawater, sea sand and SCMs. *Construction and Building Materials*, *372*. Available from https://doi.org/10.1016/j.conbuildmat.2023.130815, https://www.scopus.com/inward/record.uri?eid = 2-s2.0-85149178437&doi = 10.1016/2fj.conbuildmat.2023.130815&partnerID = 40&md5 = 3477886d345d30f22cac13272d14c25c.

Shi, C., Wu, Z., Xiao, J., Wang, D., Huang, Z., & Fang, Z. (2015). A review on ultra high performance concrete: Part I. Raw materials and mixture design. *Construction and Building Materials*, *101*, 741–751. Available from https://doi.org/10.1016/j.conbuildmat.2015.10.088.

Soeylev, T. A., & Richardson, M. G. (2008). Corrosion inhibitors for steel in concrete: State-of-the-art report. *Construction and Building Materials*, *22*(4), 609–622. Available from https://doi.org/10.1016/j.conbuildmat.2006.10.013.

Sotiriadis, K., Mácová, P., Mazur, A. S., Viani, A., Tolstoy, P. M., & Tsivilis, S. (2020). Long-term thaumasite sulfate attack on Portland-limestone cement concrete: A multi-technique analytical approach for assessing phase assemblage. *Cement and Concrete Research*, *130*. Available from https://doi.org/10.1016/j.cemconres.2020.105995, http://www.sciencedirect.com/science/journal/00088846.

Sun, X., Li, T., Shi, F., Liu, X., Zong, Y., Hou, B., & Tian, H. (2022). Sulphate corrosion mechanism of ultra-high-performance concrete (UHPC) prepared with seawater and sea sand. *Polymers*, *14*(5). Available from https://doi.org/10.3390/polym14050971, https://

www.scopus.com/inward/record.uri?eid = 2-s2.0-85126271810&doi = 10.3390/2fpolym14050971&partnerID = 40&md5 = 02a11245061a1b5e81afd78fb93e21a0.

Sun, Y., Lu, J. X., & Poon, C. S. (2022). Strength degradation of seawater-mixed alite pastes: an explanation from statistical nanoindentation perspective. *Cement and Concrete Research, 152*. Available from https://doi.org/10.1016/j.cemconres.2021.106669, http://www.sciencedirect.com/science/journal/00088846.

Sun, M., Sun, C., Zhang, P., Liu, N., Li, Y., Duan, J., & Hou, B. (2021). Influence of carbonation on chloride binding of mortars made with simulated marine sand. *Construction and Building Materials, 303*. Available from https://doi.org/10.1016/j.conbuildmat.2021.124455, https://www.journals.elsevier.com/construction-and-building-materials.

Sun, L., Yang, Z. Y., Qin, R. Y., & Wang, C. (2023). Mix design optimization of seawater sea sand coral aggregate concrete. *Science China Technological Sciences, 66*(2), 378–389. Available from https://doi.org/10.1007/s11431-022-2242-3, https://www.scopus.com/inward/record.uri?eid = 2-s2.0-85145578279&doi = 10.1007/2fs11431-022-2242-3&partnerID = 40&md5 = 8fda9b6b4f7dbb93c470e360d1dc4647.

Sun, M., Yu, R., Jiang, C., Fan, D., & Shui, Z. (2022). Quantitative effect of seawater on the hydration kinetics and microstructure development of ultra high performance concrete (UHPC). *Construction and Building Materials, 340*. Available from https://doi.org/10.1016/j.conbuildmat.2022.127733, https://www.journals.elsevier.com/construction-and-building-materials.

Sun, Y., Zhang, Y., Cai, Y., Lam, W. L., Lu, J.-X., Shen, P., & Poon, C. S. (2021). Mechanisms on accelerating hydration of alite mixed with inorganic salts in seawater and characteristics of hydration products. *ACS Sustainable Chemistry and Engineering, 9*(31), 10479–10490. Available from https://doi.org/10.1021/acssuschemeng.1c01730, https://www.scopus.com/inward/record.uri?eid = 2-s2.0-85112521711&doi = 10.1021/2facssuschemeng.1c01730&partnerID = 40&md5 = 99ed59d1f4cfd43b4fd68534aa27b62e.

Tabsh, S. W., & Alhoubi, Y. (2022). Experimental investigation of recycled fine aggregate from demolition waste in concrete. *Sustainability, 14*(17), 10787. Available from https://doi.org/10.3390/su141710787.

Talero, R., Trusilewicz, L., Delgado, A., Pedrajas, C., Lannegrand, R., Rahhal, V., Mejía, R., Delvasto, S., & Ramírez, F. A. (2011). Comparative and semi-quantitative XRD analysis of Friedel's salt originating from pozzolan and Portland cement. *Construction and Building Materials, 25*(5), 2370–2380. Available from https://doi.org/10.1016/j.conbuildmat.2010.11.037.

Teng, J. (2018). New-material hybrid structures. *China Civil Engineering Journal, 51*(12), 1–11.

Teng, J.-G., Xiang, Y., Yu, T., & Fang, Z. (2019). Development and mechanical behaviour of ultra-high-performance seawater sea-sand concrete. *Advances in Structural Engineering, 22*(14), 3100–3120. Available from https://doi.org/10.1177/1369433219858291, https://www.scopus.com/inward/record.uri?eid = 2-s2.0-85068606163&doi = 10.1177/2f1369433219858291&partnerID = 40&md5 = a752f6a587f7c33684b60ef3bb6882be.

Ting, M. Z. Y., Wong, K. S., Rahman, M. E., & Selowara Joo, M. (2020). Mechanical and durability performance of marine sand and seawater concrete incorporating silicomanganese slag as coarse aggregate. *Construction and Building Materials, 254*. Available from https://doi.org/10.1016/j.conbuildmat.2020.119195, https://www.journals.elsevier.com/construction-and-building-materials.

Vafaei, D., Hassanli, R., Ma, X., Duan, J., & Zhuge, Y. (2022). Fracture toughness and impact resistance of fiber-reinforced seawater sea-sand concrete. *Journal of Materials in*

Civil Engineering, *34*(5). Available from https://doi.org/10.1061/(ASCE)MT.1943-5533.0004179, http://ascelibrary.org/mto/resource/1/jmcee7/.

Vafaei, D., Hassanli, R., Ma, X., Duan, J., & Zhuge, Y. (2021). Sorptivity and mechanical properties of fiber-reinforced concrete made with seawater and dredged sea-sand. *Construction and Building Materials*, *270*. Available from https://doi.org/10.1016/j.conbuildmat.2020.121436, https://www.journals.elsevier.com/construction-and-building-materials.

Vafaei, D., Ma, X., Hassanli, R., Duan, J., & Zhuge, Y. (2022a). Microstructural behaviour and shrinkage properties of high-strength fiber-reinforced seawater sea-sand concrete. *Construction and Building Materials*, *320*. Available from https://doi.org/10.1016/j.conbuildmat.2021.126222, https://www.journals.elsevier.com/construction-and-building-materials.

Vafaei, D., Ma, X., Hassanli, R., Duan, J., & Zhuge, Y. (2022b). Microstructural and mechanical properties of fiber-reinforced seawater sea-sand concrete under elevated temperatures. *Journal of Building Engineering*, *50*. Available from https://doi.org/10.1016/j.jobe.2022.104140, http://www.journals.elsevier.com/journal-of-building-engineering/.

Vidhyadharan, V. P., & Johny, A. (2023). Utilization of seawater and sea sand in concrete for the sustainability of natural resources. *Materials Today: Proceedings*. Available from https://doi.org/10.1016/j.matpr.2023.03.559, https://www.scopus.com/inward/record.uri?eid = 2-s2.0-85152300788&doi = 10.1016/2fj.matpr.2023.03.559&partnerID = 40&md5 = 7202b29940e07a3ac1f484865e88751c.

Wang, D., Gong, Q., Yuan, Q., & Luo, S. (2021). Review of the properties of fiber-reinforced polymer-reinforced seawater-sea sand concrete. *Journal of Materials in Civil Engineering*, *33*(10). Available from https://doi.org/10.1061/(ASCE)MT.1943-5533.0003894, http://ascelibrary.org/mto/resource/1/jmcee7/.

Wang, J., & Huang, Y. (2023). Mechanical properties and hydration of ultra-high-performance seawater sea-sand concrete (UHPSSC) with limestone calcined clay cement (LC3). *Construction and Building Materials*, *376*. Available from https://doi.org/10.1016/j.conbuildmat.2023.130950, https://www.journals.elsevier.com/construction-and-building-materials.

Wang, A., Lyu, B., Zhang, Z., Liu, K., Xu, H., & Sun, D. (2018). The development of coral concretes and their upgrading technologies: A critical review. *Construction and Building Materials*, *187*, 1004–1019. Available from https://doi.org/10.1016/j.conbuildmat.2018.07.202.

Wang, X., Shi, C., He, F., Yuan, Q., Wang, D., Huang, Y., & Li, Q. (2013). Chloride binding and its effects on microstructure of cement-based materials. *Journal of the Chinese Ceramic Society*, *41*(2), 187–198. Available from https://doi.org/10.7521/j.issn.0454-5648.2013.02.11.

Wang, D., Shi, C., Wu, Z., Xiao, J., Huang, Z., & Fang, Z. (2015). A review on ultra high performance concrete: Part II. Hydration, microstructure and properties. *Construction and Building Materials*, *96*, 368–377. Available from https://doi.org/10.1016/j.conbuildmat.2015.08.095.

Wang, F., Sun, Y., Xue, X., Wang, N., Zhou, J., & Hua, J. (2023). Mechanical properties of modified coral aggregate seawater sea-sand concrete: Experimental study and constitutive model. *Case Studies in Construction Materials*, *18*. Available from https://doi.org/10.1016/j.cscm.2023.e02095, http://www.journals.elsevier.com/case-studies-in-construction-materials/.

Wang, Z., Wang, B., Yang, D., & Han, J. (2020). Research progress on the chloride binding capability of cement-based composites. *Journal of the Ceramic Society of Japan*, *128*

(5), 238–253. Available from https://doi.org/10.2109/jcersj2.19146, https://www.jstage.jst.go.jp/article/jcersj2/128/5/128_19146/_pdf/-char/en.

Wang, Z., Wang, J., Zhou, Y., Song, P., Guan, Q., & Zhou, Z. (2021). The ability of UHPC prepared with unpurified sea sand and seawater to protect embedded steel against corrosion. *E3S Web of Conferences*, *293*. Available from https://doi.org/10.1051/e3sconf/202129301024, http://www.e3s-conferences.org/.

Wang, G., Wu, Q., Li, X. Z., Xu, J., Xu, Y., Shi, W. H., & Wang, S. L. (2019). Microscopic analysis of steel corrosion products in seawater and sea-sand concrete. *Materials*, *12*(20). Available from https://doi.org/10.3390/ma12203330, https://res.mdpi.com/d_attachment/materials/materials-12-03330/article_deploy/materials-12-03330.pdf.

Wei, J., Bian, X., Huang, W., Chen, R., Chen, B., Huang, Q., Zhou, J., & Wu, Y. (2020). Passivation behavior of steel bar in sea sand ultra-high performance cement evaluated using electrochemical method. *Journal of the Chinese Ceramic Society*, *48*(8), 1223–1232. Available from https://doi.org/10.14062/j.issn.0454-5648.20200205, https://www.scopus.com/inward/record.uri?eid = 2-s2.0-85090212836&doi = 10.14062/2fj.issn.0454-5648.20200205&partnerID = 40&md5 = ebdaf10dfee315f1760d66b5cda8a467.

Wu, Q., Li, X., Xu, J., Wang, G., Shi, W., & Wang, S. (2019). Size distribution model and development characteristics of corrosion pits in concrete under two curing methods. *Materials*, *12*(11). Available from https://doi.org/10.3390/ma12111846, https://res.mdpi.com/materials/materials-12-01846/article_deploy/materials-12-01846.pdf?filename = &attachment = 1.

Xiao, J., Li, W., Fan, Y., & Huang, X. (2012). An overview of study on recycled aggregate concrete in China (1996-2011). *Construction and Building Materials*, *31*, 364–383. Available from https://doi.org/10.1016/j.conbuildmat.2011.12.074.

Xiao, J., Qiang, C., Nanni, A., & Zhang, K. (2017). Use of sea-sand and seawater in concrete construction: Current status and future opportunities. *Construction and Building Materials*, *155*, 1101–1111. Available from https://doi.org/10.1016/j.conbuildmat.2017.08.130, https://www.scopus.com/inward/record.uri?eid = 2-s2.0-85028698935&doi = 10.1016/2fj.conbuildmat.2017.08.130&partnerID = 40&md5 = ef0febc81f6f7b5bebd77a374e69f464.

Xiao, J., Zhang, Q., Zhang, P., Shen, L., & Qiang, C. (2019). Mechanical behavior of concrete using seawater and sea-sand with recycled coarse aggregates. *Structural Concrete*, *20*(5), 1631–1643. Available from https://doi.org/10.1002/suco.201900071, http://onlinelibrary.wiley.com/journal/10.1002/(ISSN)1751-7648.

Xiao, J., Zhang, K., & Zhang, Q. (2021). Strain rate effect on compressive stress–strain curves of recycled aggregate concrete with seawater and sea sand. *Construction and Building Materials*, *300*. Available from https://doi.org/10.1016/j.conbuildmat.2021.124014, https://www.journals.elsevier.com/construction-and-building-materials.

Xiao, S., Zhang, M., Zou, D., Liu, T., Zhou, A., & Li, Y. (2023). Influence of seawater and sea sand on the performance of Anti-washout underwater concrete: The overlooked significance of Mg2 + . *Construction and Building Materials*, *374*. Available from https://doi.org/10.1016/j.conbuildmat.2023.130932, https://www.journals.elsevier.com/construction-and-building-materials.

Xie, Q., Xiao, J., Zhang, K., & Zong, Z. (2021). Compressive behavior and microstructure of concrete mixed with natural seawater and sea sand. *Frontiers of Structural and Civil Engineering*, *15*(6), 1347–1357. Available from https://doi.org/10.1007/s11709-021-0780-2, https://www.scopus.com/inward/record.uri?eid = 2-s2.0-85119680232&doi = 10.1007/2fs11709-021-0780-2&partnerID = 40&md5 = db09e59bba6d5353448352d38cf52b25.

Xue, J., Briseghella, B., Huang, F., Nuti, C., Tabatabai, H., & Chen, B. (2020). Review of ultra-high performance concrete and its application in bridge engineering. *Construction and Building Materials, 260*. Available from https://doi.org/10.1016/j.conbuildmat.2020.119844, https://www.journals.elsevier.com/construction-and-building-materials.

Xu, Q., Ji, T., Yang, Z., & Ye, Y. (2019). Steel rebar corrosion in artificial reef concrete with sulphoaluminate cement, sea water and marine sand. *Construction and Building Materials, 227*. Available from https://doi.org/10.1016/j.conbuildmat.2019.116685, https://www.journals.elsevier.com/construction-and-building-materials.

Xu, C., Ou, Z., Zhou, J., Chen, H., & Qiao, C. (2011). Investigation on protectional ability on steel bar of compound corrosion inhibitor applied in seawater-and-sea sand concrete. *Applied Mechanics and Materials, 71-78*, 864–870. Available from https://doi.org/10.4028/http://www.scientific.net/AMM.71-78.864.

Xu, J., Tang, Y., Chen, Y., & Chen, Z. (2022). Split-Hopkinson pressure bar tests on dynamic properties of concrete made of seawater and marine aggregates. *Journal of Vibration and Shock., 41*(14), 233–242. Available from https://doi.org/10.13465/j.cnki.jvs.2022.14.031, http://jvs.sjtu.edu.cn/CN/volumn/home.shtml.

Xu, J., Wu, Z., Jia, H., Yu, R. C., & Cao, Q. (2023). Axial compression of seawater sea sand concrete columns reinforced with hybrid FRP-stainless steel bars. *Magazine of Concrete Research, 75*(13), 685–702. Available from https://doi.org/10.1680/jmacr.22.00249, http://www.icevirtuallibrary.com/content/serial/macr.

Xu, J., Zhao, X., Tang, Y., Liu, T., & Chen, L. (2023). Compressive stress-strain constitutive relationship of seawater and marine aggregates fabricated concrete under dynamic response. *Journal of Vibration Engineering, 36*(1), 207–216. Available from https://doi.org/10.16385/j.cnki.issn.1004-4523.2023.01.022.

Yang, J., Lu, S., Zeng, J. J., Wang, J., & Wang, Z. (2023). Durability of CFRP-confined seawater sea-sand concrete (SSC) columns under wet-dry cycles in seawater environment. *Engineering Structures, 282*. Available from https://doi.org/10.1016/j.engstruct.2023.115774, http://www.journals.elsevier.com/engineering-structures/.

Yang, S., Wang, J., Dong, K., Zhang, X., & Sun, Z. (2022). A predictive solution for fracture modeling of alkali-activated slag and fly ash blended sea sand concrete after exposure to elevated temperature. *Construction and Building Materials, 329*. Available from https://doi.org/10.1016/j.conbuildmat.2022.127111, https://www.journals.elsevier.com/construction-and-building-materials.

Yang, J., Wang, J., Lu, S., Zhang, L., & Wang, Z. (2022). Axial compressive behavior of FRP nonuniformly wrapped seawater sea-sand concrete in square columns. Beijing University of Aeronautics and Astronautics (BUAA). *China Acta Materiae Compositae Sinica., 39*(6), 2801–2809. Available from https://doi.org/10.13801/j.cnki.fhclxb.20210708.004, https://fhclxb.buaa.edu.cn/.

Yang, J., Wang, J., & Wang, Z. (2020a). Axial compressive behavior of partially CFRP confined seawater sea-sand concrete in circular columns – part II: A new analysis-oriented model. *Composite Structures, 246*. Available from https://doi.org/10.1016/j.compstruct.2020.112368, http://www.elsevier.com/inca/publications/store/4/0/5/9/2/8.

Yang, J., Wang, J., & Wang, Z. (2020b). Axial compressive behavior of partially CFRP confined seawater sea-sand concrete in circular columns – part I: Experimental study. *Composite Structures, 246*. Available from https://doi.org/10.1016/j.compstruct.2020.112373, http://www.elsevier.com/inca/publications/store/4/0/5/9/2/8.

Yang, S., Xu, J., Zang, C., Li, R., Yang, Q., & Sun, S. (2019). Mechanical properties of alkali-activated slag concrete mixed by seawater and sea sand. *Construction and*

Building Materials, *196*, 395–410. Available from https://doi.org/10.1016/j.conbuildmat.2018.11.113.

Yang, Z., Zhu, H., Zhang, B., Dong, Z., & Wu, P. (2023). Short-term creep behaviors of seawater sea-sand coral aggregate concrete: An experimental study with Rheological model and neural network. *Construction and Building Materials*, *363*. Available from https://doi.org/10.1016/j.conbuildmat.2022.129786, https://www.journals.elsevier.com/construction-and-building-materials.

Yao, Q., Li, Z., Lu, C., Peng, L., Luo, Y., & Teng, X. (2021). Development of engineered cementitious composites using sea sand and metakaolin. *Frontiers in Materials*, *8*. Available from https://doi.org/10.3389/fmats.2021.711872, http://journal.frontiersin.org/journal/materials.

Yaseen, S. A., Yiseen, G. A., Poon, C. S., Leung, C. K., & Li, Z. (2023). The effectuation of seawater on the microstructural features and the compressive strength of fly ash blended cement at early and later ages. *Journal of the American Ceramic Society*, *106*(8), 4967–4986. Available from https://doi.org/10.1111/jace.19109, https://www.scopus.com/inward/record.uri?eid = 2-s2.0-85152359984&doi = 10.1111/2fjace.19109&partnerID = 40&md5 = ba0ac4ca7e55303dd952efa14112bb33.

Yi, Y., Guo, S., Li, S., Zillur Rahman, M., Zhou, L., Shi, C., & Zhu, D. (2021). Effect of alkalinity on the shear performance degradation of basalt fiber-reinforced polymer bars in simulated seawater sea sand concrete environment. *Construction and Building Materials*, *299*. Available from https://doi.org/10.1016/j.conbuildmat.2021.123957, https://www.journals.elsevier.com/construction-and-building-materials.

Yi, Y., Zhu, D., Guo, S., Li, S., Feng, G., Liu, Z., Zhou, L., & Shi, C. (2022). Development of a low-alkalinity seawater sea sand concrete for enhanced compatibility with BFRP bar in the marine environment. *Cement and Concrete Composites*, *134*. Available from https://doi.org/10.1016/j.cemconcomp.2022.104778, http://www.sciencedirect.com/science/journal/09589465.

Yi, Y., Zhu, D., Rahman, M. Z., Shuaicheng, G., Li, S., Liu, Z., & Shi, C. (2022). Tensile properties deterioration of BFRP bars in simulated pore solution and real seawater sea sand concrete environment with varying alkalinities. *Composites Part B: Engineering*, *243*. Available from https://doi.org/10.1016/j.compositesb.2022.110115, https://www.journals.elsevier.com/composites-part-b-engineering.

Younis, A., Ebead, U., Suraneni, P., & Nanni, A. (2018). Fresh and hardened properties of seawater-mixed concrete. *Construction and Building Materials*, *190*, 276–286. Available from https://doi.org/10.1016/j.conbuildmat.2018.09.126.

Yuan, Q., Shi, C., De Schutter, G., Audenaert, K., & Deng, D. (2009). Chloride binding of cement-based materials subjected to external chloride environment - a review. *Construction and Building Materials*, *23*(1), 1–13. Available from https://doi.org/10.1016/j.conbuildmat.2008.02.004.

Yu, X., Al-Saadi, S., Kohli, I., Zhao, X. L., & Singh Raman, R. K. (2021). Austenitic stainless-steel reinforcement for seawater sea sand concrete: Investigation of stress corrosion cracking. *Metals*, *11*(3), 1–10. Available from https://doi.org/10.3390/met11030500, https://www.mdpi.com/2075-4701/11/3/500/pdf.

Yu, X., Al-Saadi, S., Zhao, X. L., & Raman, R. K. S. (2021). Electrochemical investigations of steels in seawater sea sand concrete environments. *Materials*, *14*(19). Available from https://doi.org/10.3390/ma14195713, https://www.mdpi.com/1996-1944/14/19/5713/pdf.

Yu, F., Dong, S., Li, L., Ashour, A., Ding, S., Han, B., & Ou, J. (2023). Developing a sustainable ultrahigh-performance concrete using seawater and sea sand in combination with superfine stainless wires. *Journal of Materials in Civil Engineering*, *35*(10) 04023368. Available from https://doi.org/10.1061/JMCEE7.MTENG-16072.

Zeng, J. J., Liao, J. J., Liang, W. F., Guo, Y. C., Zhou, J. K., Lin, J. X., & Yan, K. (2022). Cyclic axial compression behavior of FRP-confined seawater sea-sand concrete-filled stainless steel tube stub columns. *Frontiers in Materials*, *9*. Available from https://doi.org/10.3389/fmats.2022.872055, http://journal.frontiersin.org/journal/materials.

Zeng, L., Yu, W.L., Lu, R.Q., Yuan, H., Li, L.J. (2019). APFIS 2019 Proceedings – 7th Asia-Pacific Conference on FRP in structures. APFIS Conference Series China. Axial compressive behavior of seawater and sea-sand concrete (SSC) filled stainless steel tubular columns with FRP confined ssc core.

Zhang, Y., Chang, J., Zhao, Q., Lam, W. L., Shen, P., Sun, Y., Zhao, D., & Poon, C. S. (2022). Effect of dosage of silica fume on the macro-performance and micro/nanostructure of seawater Portland cement pastes prepared with an ultra-low water-to-binder ratio. *Cement and Concrete Composites*, *133*. Available from https://doi.org/10.1016/j.cemconcomp.2022.104700, http://www.sciencedirect.com/science/journal/09589465.

Zhang, Y., Sun, Y., Shen, P., Lu, J., Cai, Y., & Poon, C. S. (2022). Physicochemical investigation of Portland cement pastes prepared and cured with seawater. *Materials and Structures/Materiaux et Constructions.*, *55*(6). Available from https://doi.org/10.1617/s11527-022-01991-z, https://link.springer.com/journal/11527.

Zhang, Y., Sun, Y., Zheng, H., Cai, Y., Lam, W. L., & Poon, C. S. (2021). Mechanism of strength evolution of seawater OPC pastes. *Advances in Structural Engineering*, *24*(6), 1256–1266. Available from https://doi.org/10.1177/1369433221993299, https://www.scopus.com/inward/record.uri?eid = 2-s2.0-85101082778&doi = 10.1177/2f1369433221993299&partnerID = 40&md5 = 9214ddafbe8508eab62e338a5f1cb5dd.

Zhang, K., Xiao, J., Hou, Y., & Zhang, Q. (2022). Experimental study on carbonation behavior of seawater sea sand recycled aggregate concrete. *Advances in Structural Engineering*, *25*(5), 927–938. Available from https://doi.org/10.1177/13694332211026221, https://journals.sagepub.com/home/ASE.

Zhang, K., Xiao, J., & Zhang, Q. (2021a). Experimental study on stress-strain curves of seawater sea-sand concrete under uniaxial compression with different strain rates. *Advances in Structural Engineering*, *24*(6), 1124–1137. Available from https://doi.org/10.1177/1369433220958765, https://uk.sagepub.com/en-gb/eur/advances-in-structural-engineering/journal202466.

Zhang, K., Xiao, J., & Zhang, Q. (2021b). Complete stress-strain curves of seawater sea sand recycled aggregate concrete under uniaxial compression. *Journal of Tongji University.*, *49*(12), 1738–1745. Available from https://doi.org/10.11908/j.issn.0253-374x.21085, http://tjxb.cnjournals.cn/ch/index.aspx.

Zhang, K., Zhang, Q., & Xiao, J. (2022). Durability of FRP bars and FRP bar reinforced seawater sea sand concrete structures in marine environments. *Construction and Building Materials*, *350*. Available from https://doi.org/10.1016/j.conbuildmat.2022.128898, https://www.journals.elsevier.com/construction-and-building-materials.

Zhao, Y., Hu, X., Shi, C., Zhang, Z., & Zhu, D. (2021). A review on seawater sea-sand concrete: Mixture proportion, hydration, microstructure and properties. *Construction and Building Materials*, *295*. Available from https://doi.org/10.1016/j.conbuildmat.2021.123602, https://www.journals.elsevier.com/construction-and-building-materials.

Zhao, G., Li, J., Han, F., Shi, M., & Fan, H. (2019). Sulfate-induced degradation of cast-in-situ concrete influenced by magnesium. *Construction and Building Materials*, *199*, 194–206. Available from https://doi.org/10.1016/j.conbuildmat.2018.12.022.

Zhao, J., Long, B., Yang, G., Cheng, Z., & Liu, Q. (2022). Characteristics of alkali-activated slag powder mixing with seawater: Workability, hydration reaction kinetics and mechanism. *Case Studies in Construction Materials, 17.* Available from https://doi.org/10.1016/j.cscm.2022.e01381, http://www.journals.elsevier.com/case-studies-in-construction-materials/.

Zheng, H., Lu, J., Shen, P., Sun, L., Poon, C. S., & Li, W. (2022). Corrosion behavior of carbon steel in chloride-contaminated ultra-high-performance cement pastes. *Cement and Concrete Composites, 128.* Available from https://doi.org/10.1016/j.cemconcomp.2022.104443, http://www.sciencedirect.com/science/journal/09589465.

Zhou, L., Guo, S., Ma, W., Xu, F., Shi, C., Yi, Y., & Zhu, D. (2022). Internal curing effect of saturated coral coarse aggregate in high-strength seawater sea sand concrete. *Construction and Building Materials, 331.* Available from https://doi.org/10.1016/j.conbuildmat.2022.127280, https://www.journals.elsevier.com/construction-and-building-materials.

Zhou, J., & He, X. (2021). Experimental study on internal and external salt attack from seawater and sea-sand to mortars. *KSCE Journal of Civil Engineering*, *25*(8), 2951–2961. Available from https://doi.org/10.1007/s12205-021-0375-4, http://www.springer.com/engineering/journal/12205.

Zhou, J., & Li, D. (2022). Cyclic compressive behavior of seawater sea-sand concrete after seawater freeze–thaw cycles: Experimental investigation and analytical model. *Construction and Building Materials, 345.* Available from https://doi.org/10.1016/j.conbuildmat.2022.128227, https://www.journals.elsevier.com/construction-and-building-materials.

Zhou, Q., Zhou, Y., Guan, Z., Xing, F., Guo, M., & Hu, B. (2022). Mechanical performance and constitutive model analysis of concrete using PE fiber-strengthened recycled coarse aggregate. *Polymers, 14*(19). Available from https://doi.org/10.3390/polym14193964, http://www.mdpi.com/journal/polymers.

Zhu, D., Li, L., & Guo, S. (2022). Research on influence factors of performance of ultra-high performance seawater sea-sand concrete. *Journal of Hunan University Natural Sciences.*, *49*(3), 187–195. Available from https://doi.org/10.16339/j.cnki.hdxbzkb.2022039, http://hdxbzkb.cnjournals.net/ch/index.aspx.

Sustainable ultra-high-performance concrete materials and structures

16

Tong Sun[1], *Xinyue Wang*[2], *Ashraf Ashour*[3] *and Baoguo Han*[1]
[1]School of Civil Engineering, Dalian University of Technology, Dalian, Liaoning, P.R. China, [2]School of Civil Engineering, Tianjin University, Tianjin, P.R. China, [3]Faculty of Engineering and Digital Technologies, University of Bradford, Bradford, United Kingdom

16.1 Introduction

Ultra-high-performance concrete (UHPC) is a fiber-reinforced cementitious material with remarkable mechanical properties and durability. However, conventional UHPC heavily relies on energy-intensive materials and natural resources, notably cement. Compared to normal concrete (NC), UHPC employs about three to four times more cement, consuming significant amount of energy, and emitting large amounts of carbon dioxide during its production and use (Ferdosian et al., 2017). The cement industry accounts for 5%−8% of the total global carbon emissions and is significantly responsible for the emergence of the global greenhouse effect (Mohamad et al., 2021). Furthermore, UHPC uses quartz sand (QS) as fine aggregates, which is a nonrenewable resource. The extensive mining required to extract QS can lead to ecological damage and contribute to potential energy shortage (Wang, Yan, et al., 2021). In addition, civil engineering infrastructures are characterized by complex and changing service environments, large heights and spans, diverse systems, and changing functions, which urgently require the use of UHPC materials that exhibit superior strength, toughness, durability, and multifunctionality to meet these challenges effectively (Li, Wu, et al., 2020).

To improve the properties of UHPC and promote further development of UHPC and their related structures in a sustainable direction, various research investigations have been conducted to modify UHPC properties, including refining the preparation process, optimizing the mix proportion, and carefully selecting raw materials for UHPC. By optimizing the preparation process and mix proportion design of UHPC, the close particle packing structure of UHPC can be achieved to reduce internal defects of UHPC. Only 30%−40% of cement in UHPC undergoes hydration, and the rest of the cement only plays a filling role. Therefore cement can partly be replaced with supplementary cementitious materials (SCMs). Up to now, several new cementitious material systems are developed and tested. Some industrial, agricultural, and municipal solid wastes have similar properties to UHPC raw materials and can be used as SCMs or aggregates for the preparation of UHPC (Wang et al., 2019). In addition, traditional materials exhibit single performance and show slow progress in terms of performance improvement. Meanwhile, the resource dependence, preparation cost,

Sustainable Concrete Materials and Structures. DOI: https://doi.org/10.1016/B978-0-443-15672-4.00016-4

and recycling difficulty are increasing. As a result, the full exploitation or expansion of material performance, such as multifunctionalization, structure-function integration, function-intelligence integration, has become an important direction for the sustainable development of materials. The strength and durability of UHPC can be further improved by adding high-performance fibers or nanomaterials while giving UHPC materials multifunctional and intelligent features (Li et al., 2022).

Sustainable ultrahigh performance concrete (SUHPC) combines environmental, social, and economic sustainability aspects to create durable components that can withstand marine or extremely harsh environments. It is particularly suitable for constructing lightweight bridge elements and structures with special seismic needs (Shao et al., 2022). The application of SUHPC can minimize carbon emissions and reduce costs during the construction stages, as well as reduce the self-weight of structures, extend their service life, and cut their whole life-cycle costs. In addition, incorporating high-performance fibers and nanomaterials into SUHPC can endow SUHPC and its structures with multifunctional and intelligent performance (Ding et al., 2023; Han et al., 2017a; Lefever et al., 2022). This chapter first provides an introduction to the relevant properties of UHPC and its environmental impact. It, then, focuses on the optimization of UHPC mix proportion and the realization of UHPC sustainable development by using solid wastes as SCMs or aggregates, developing new cementitious material systems, and adding high-performance fibers and nanomaterials. Finally, the structural applications of SUHPC are presented.

16.2 Ultra-high-performance concrete and its impact on the recourse, energy, and environment

16.2.1 Material composition and properties of UHPC

UHPC is a high-performance fiber-reinforced cementitious composite with ultra-high strength, ductility, and durability (Huang, Wang, et al., 2023; Yoo et al., 2022; Zhang et al., 2023). As shown in Fig. 16.1, UHPC is formulated by incorporating finely ground QS and mineral admixtures to increase the activity of concrete ingredients, employing superplasticizer to guarantee a low water-binder ratio, and employing thermal curing to promote hydration reaction. The most central features are increasing the quantity of hydrated calcium silicate gels and reducing the amount of calcium hydroxide, leading to an increased proportion of silicon-oxygen bonds in the hydration products, while reducing the proportion of calcium-oxygen bonds. By minimizing the defects, such as pores and microcracks within the composites, UHPC achieves a high level of homogeneity and compactness. In addition, it is equally important to improve the natural deficiencies of cementitious materials, such as brittleness, susceptibility to cracking, low tensile properties, and poor impact resistance by incorporating steel fibers in the production of UHPC to increase the proportion of metallic bonds in the composites (Hung et al., 2019).

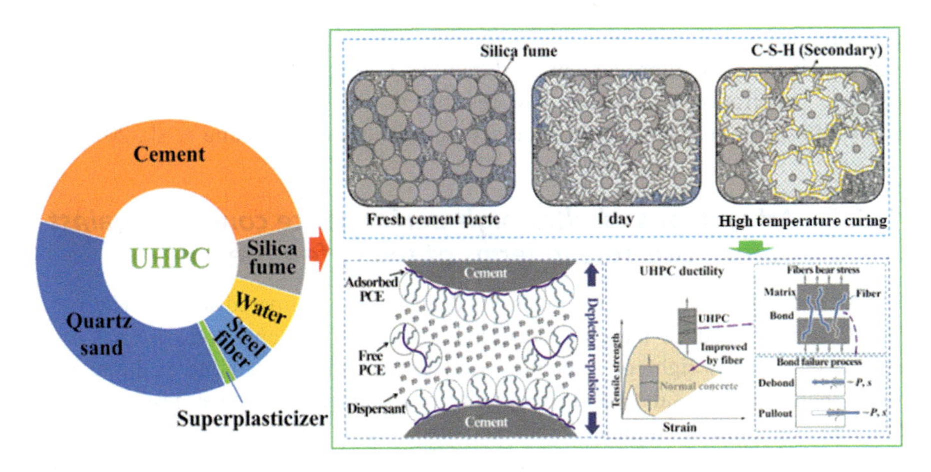

Figure 16.1 Formulation principle of UHPC. In the preparation process of UHPC, finely ground quartz sand and mineral admixtures are used to improve the activity, thermal curing is used to promote hydration reaction, and superplasticizers are used to ensure low water-binder ratio.

Table 16.1 Comparison of the main properties of ultra-high-performance concrete (UHPC) and normal concrete (NC).

Performance indicators	NC	UHPC	Comparison of UHPC and NC
Compressive strength (MPa)	20–50	120–230	~5 times
Flexural strength (MPa)	2–5	30–60	~10 times
Modulus of elasticity (GPa)	30–40	40–60	~1.2 times
Fracture toughness (kJ/m^2)	0.12	20–40	~200 times
Creep coefficient	1.3 to 2.1	0.29–0.31	~1/5
Chloride diffusion coefficient (10^{-12} m^2/s)	1.1	0.02	~1/50
Freeze-thaw spalling (g/cm)^2	>1000	7	~1/140
Water absorption characteristics (kg/m^3)	2.7	0.2	~1/13
Abrasion factor	4	1.3	~1/3
Compressive strength/material weight	1.7	6.5	~4 times

The basic properties of UHPC are shown in Table 16.1, compared to NC. The compressive strength of UHPC is generally 150–200 N/mm^2, about five times more than that of NC; UHPC's toughness is over 300 times higher than NC's; while the self-weight of UHPC for a given bearing capacity is only 1/3 or 1/2 of that of NC (Tahwia et al., 2023; Yin et al., 2023; Zhong et al., 2023). UHPC has a large improvement in permeability due to its extremely low water-binder ratio, low porosity, and small most probable pore size (Alkaysi et al., 2016). UHPC exhibits varying degrees of improvement in durability indexes, such as sulfate resistance, chloride penetration resistance, frost resistance, and carbonation resistance, when compared to NC, particularly in uncracked conditions (Magureanu et al., 2012).

16.2.2 Advantages of ultra-high-performance concrete against normal concrete in terms of total life-cycle cost

UHPC has higher strength, toughness, and durability compared to NC. The use of UHPC helps to reduce the cross-sectional size of components and material usage, reduce the dead weight of structures, and release more available building space while increasing the safety and service life of structures. Likewise, less waste after building demolition reduces transportation needs and environmental impact (Sameer et al., 2019). UHPC structures possess excellent performance, economic efficiency, and recyclability, contributing to their enhanced sustainability. However, the sustainable development of UHPC and its associated structures faces challenges due to the high carbon footprint, high demand for resources and energy, and high preparation costs involved. Firstly, the amount of cement used in UHPC is large, which requires significant energy consumption and emits a substantial amount of carbon dioxide during production and use. The carbon emission of UHPC is about twice as much as that of NC, and the energy consumption of UHPC is 20%–30% higher than that of NC (Table 16.2). Secondly, the production of UHPC requires a considerable reliance on natural resources, particularly QS, which constitutes approximately 50%–60% of the mixture. At the same time, to improve the toughness and mitigate brittle failure in UHPC, steel fibers are often incorporated at a certain percentage. However, the cost of steel fibers, comprising only 1% of the total material content, may exceed the cost of the UHPC matrix, greatly increasing the initial cost of UHPC. In addition, civil engineering structures present complex and variable service environments, large heights and spans, diverse systems, and versatile functions. The performance of conventional UHPC is prone to degradation in extreme, harsh, and multifactor coupled environments, while it cannot meet the requirements of the multifunctionality of building structures. Therefore it is especially important to reduce the amount of cement and QS in UHPC, reduce the dependence of UHPC on the environment and energy, and improve the performance related to UHPC, enhancing their multifunctionality and intelligence as well as sustainability.

Table 16.2 UHPC energy consumption and carbon footprint (/m^3).

kg CO_2 equivalent	Total primary energy (MJ)	References
1049	7520	Bouhaya et al. (2009)
950 with coated fibers	11488	Behloul and Guise (2008)
1350	–	Stengel and Schießl (2008)
974	9148	UHPC for the precast industry

16.3 Optimum design of ultra-high-performance concrete mix proportion

Mix proportion design methods for UHPC encompass various approaches, including experimental optimization (Stovall et al., 1986), statistical design (Soliman & Tagnit-Hamou, 2018), and methods based on the principle of compact stacking (Mehdipour & Khayat, 2018). These methods enable the formulation of UHPC mixes by either conducting iterative experiments to refine the proportions, utilizing statistical models to analyze the effects of variables, or focusing on achieving dense packing of particles. The particle continuous stacking method is by far the most widely used mix proportion design method for UHPC, which can achieve compact stacking of UHPC, significantly reduce the porosity, and enhance the strength of UHPC. However, most particle stacking models only consider the stacking density of the solid skeleton and do not reflect the interaction of liquids (e.g., water and superplasticizer) with solid particles. As a result, the measured stacking density is lower than that in the actual wet state, leading to an inaccurate prediction of UHPC strength.

In recent years, two mathematical modeling methods, the design of quadratic saturated D-optimal mixture design (DOMD) and the genetic algorithm-based artificial neural network (GA-ANN) approach, have been widely used in mix proportion design of UHPC (Ghafari et al., 2015; Hammoudi et al., 2019). The essence of the quadratic saturated DOMD is to transform the mixture ratio design into a mathematical model by determining the relationship between variables and responses. This method enables the establishment of the relationship between the components of UHPC (including the liquid phase) and the stacking system. It is shown that by using the quadratic saturated DOMD to achieve the maximum wet stack density, the effects of aggregate proportions, admixtures, and fibers on the performance of UHPC can be accurately predicted, providing theoretical support for the design of environmentally friendly UHPC (Fan et al., 2020). GA-ANN fully combines the advantages of genetic algorithm and artificial neural networks, which is widely used for the prediction of concrete-related properties. Akkurt et al. (2003) showed that GA-ANN can reduce experimental tests, improve the cost-effectiveness of UHPC, and provide accurate predictions by continuously varying the input values and their weights. GA-ANN methods are simpler and more straightforward than conventional statistical methods through linear and nonlinear programming. The prediction and optimization of the workability, packing density, mechanical properties, porosity, and durability of UHPC with different mix proportions are achieved by the DOMD method or GA-ANN, as shown in Fig. 16.2, which is an effective way to further realize the sustainable development of UHPC.

16.4 Sustainable ultra-high-performance concrete fabricated with solid waste

UHPC preparation should reduce energy consumption and pollution, reduce the requirement for natural materials, reduce greenhouse gas emissions and consumption of

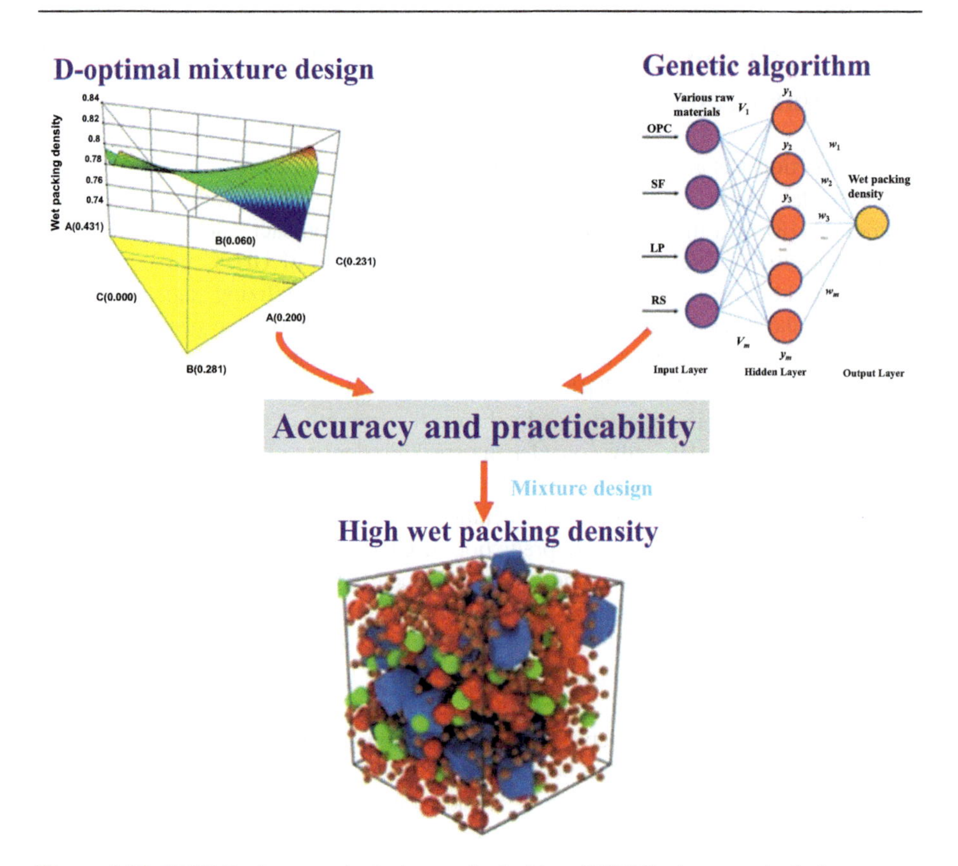

Figure 16.2 SUHPC mixture ratio design methods. Two SUHPC mixture ratio design methods, the design of quadratic saturated D-optimal mixture design (DOMD) and the genetic algorithm-based artificial neural network (GA-ANN) approach, can accurately predict the effects of aggregate ratio, admixtures, and fibers on the performance of UHPC. *Source:* From Yuan, Z., Wang, L. N., & Ji, X. (2014). Prediction of concrete compressive strength: Research on hybrid models genetic based algorithms and ANFIS. *Advances in Engineering Software*, *67*, 156–163. https://doi.org/10.1016/j.advengsoft.2013.09.004.

resources by reducing the amount of cement, and using recycled materials (Wang et al., 2017) or solid waste (Hamada et al., 2023). Most of cement in UHPC only plays a filling role, so inactive fillers can be used as SCMs to partially replace cement (Wu et al., 2017). SCMs can not only reduce the amount of cement but improve the workability, mechanical properties, volume stability, and durability of UHPC. However, conventional SCMs (such as fly ash (Kim et al., 2013), slag (Biskri et al., 2017), and silica fume (Shi et al., 2015) are more expensive and from unstable sources, and their stability and quality issues have been a major concern for the concrete industry. Studies have shown that some industrial wastes (tailings (Gu et al., 2022), steel slag (Guo et al., 2018), agricultural wastes (rice husk ash (Hasan et al., 2022), sugarcane bagasse ash (Jahanzaib Khalil et al., 2021), and municipal wastes (e.g., glass waste) have similar properties and stable sources to UHPC raw materials and can be used as SCMs or aggregates in UHPC (Jiao et al., 2020). The use of solid waste for the preparation of UHPC can reduce energy consumption and environmental pollution and reduce the cost of UHPC based on mechanical properties and durability, which is an effective way to achieve further development of UHPC in a sustainable direction (Shafieifar et al., 2017; Shi et al., 2021).

16.5 Sustainable ultra-high-performance concrete fabricated with industrial waste

16.5.1 Sustainable ultra-high-performance concrete with iron ore tailing powder/iron ore tailing sand

Tailings are globally recognized as the industrial waste with the largest volume and lowest utilization rate. Among various tailings, iron ore tailings (IOTs), which account for about one-third of all types of tailings reserves, are mining wastes obtained during the beneficiation process for concentrating iron ore (Zhang, Gu, et al., 2020). Magalhães et al. (2020) reported that more than 80% of IOTs are nonmetallic minerals, with chemical composition mainly of calcium, magnesium, silicon, and aluminum oxides and small amounts of sodium, potassium, sulfur, and iron oxides, among which, the higher content of silicon and aluminum and the larger size distribution of IOT, can be used as SCMs or aggregates for the preparation of SUHPC, as shown in Fig. 16.3.

IOT has low volcanic ash activity, and grinding IOT sand into iron ore tailings powder (ITP) can improve its volcanic ash activity as SCMs. The microstructure and elementary composition of ITP are similar to those of cement (Fig. 16.5), but the particles are larger than those of cement, corresponding to a smaller specific surface area, with a crystal structure and poorer water absorption. The use of ITP as SCM released more free water from the UHPC matrix and could improve the workability of UHPC (Table 16.3).

ITP can reduce the early hydration products within UHPC and decrease the compressive strength. However, due to the angular structural appearance of ITP that enhances the adhesion between ITP and UHPC matrix, the long-term compressive strength of UHPC was not significantly affected when the ITP substitution ratio did not exceed 30% (Gu et al., 2022). Ling et al. (2021) showed that the 3-day compressive strength of the ITP substitution ratio of 30% was reduced by 23.17% compared to the control group, which was due to less early hydration products within UHPC, less dense interfacial transition zone (ITZ) between ITP and cement paste, and reduced early compressive strength. With the increase of curing age, the ITZ between ITP and cement paste was denser and the degree of hydration increased, and a large amount of C-S-H gels filling the pores between particles could be observed on 28-day, which promotes the development of compressive strength, and the 28-day strength of UHPC only decreased by 2.5% (Fig. 16.3).

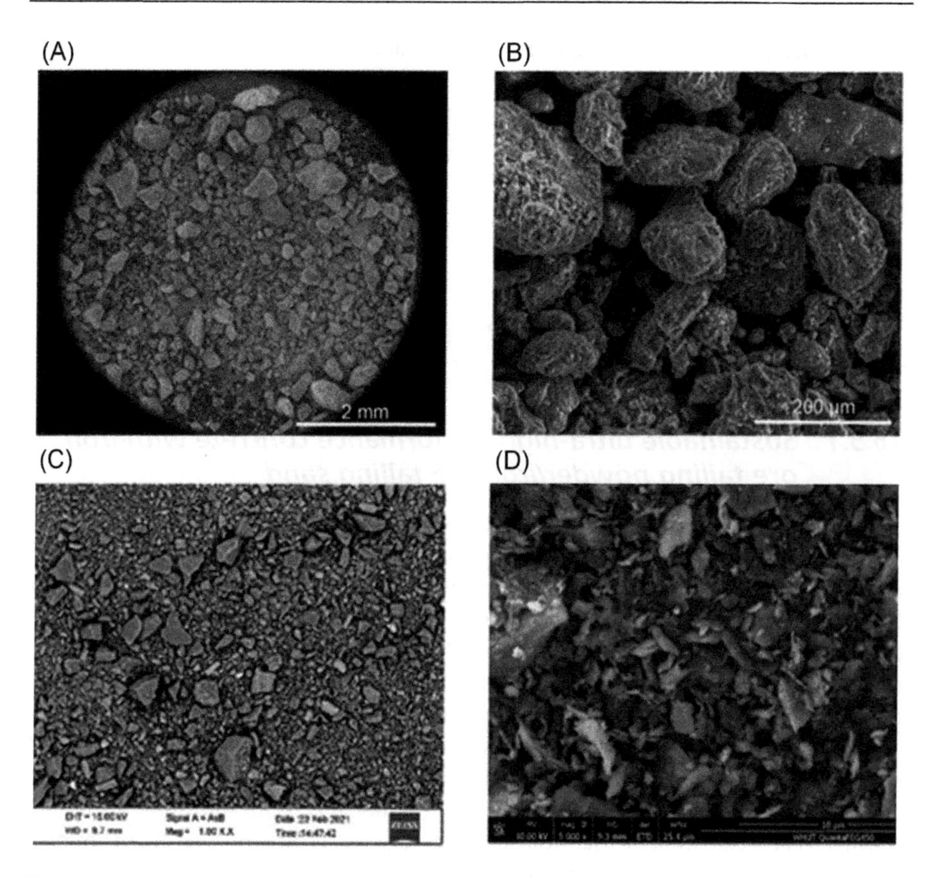

Figure 16.3 Microstructure of iron ore tailings powder and sand. Hydration products and pore structure of iron ore tailings powder and sand under scanning electron microscope. (A, B) iron ore tailings powder (C, D) iron ore tailings sand.

Due to lower ITP activity resulting in fewer hydration products within the matrix, UHPC microstructure is significantly deteriorated and durability is reduced when the ITP substitution ratio is high but still has greater advantages compared to NC. Ruidong et al. (2021) studied the shrinkage performance of UHPC incorporating ITP, and the results showed that partial substitution of ITP for cement reduced early hydration and increased water content, slowing down UHPC's early autogenous shrinkage, and the volume stability of UHPC could be improved by 20% when the ITP substitution ratio was 30% (Li et al., 2010).

IOTs exhibit relatively low volcanic ash activity, used as SCMs in UHPC. Their activity requires activation by processes, such as grinding or heat treatment. However, these activation methods lead to increased energy consumption in UHPC preparation. The particle size range of IOT is large, and some of the IOT sand with larger particle sizes can be used to replace natural fine aggregates in UHPC. Compared to natural sand, the specific surface area of IOTs sand is much higher and can absorb more water. At the same time, the IOT sand is angular and irregular

Table 16.3 Effect of adding ITP on SUHPC performance (relative values).

The contents of ITP (%)	Workability (%)	Compressive strength (%)			Volume stability (%)	Chloride ion permeability (%)	References
		3d	7d	28d			
10	7.70	−3.67	0	16.30	22.30	4.10	Ling et al., (2021)
20	19.2	−7.32	−13.13	1.20	29.50	13.60	
30	53.8	−23.17	−21.21	−2.50	44.60	32.10	
10	–	4.76	3.92	2.50	–	–	Gu et al. (2022)
20	–	0.14	−0.72	−3.33	–	–	
30	–	−7.14	−11.76	−9.17	–	–	
5	–	–	9.09	18.87	–	–	Ruidong et al. (2021)
10	–	–	11.36	20.75	−22.22	–	
15	–	–	16.25	41.18	−4.44	–	
20	–	–	12.50	35.29	6.67	–	
25			10.00	33.33	–	–	
30			3.75	17.65	20.00	–	

in shape, which increases the friction between the particles and reduces the workability of the wet mixture, and the water-binder ratio needs to be increased to maintain the compatibility of UHPC.

The particle size of IOT sand is larger than cement but smaller than fine aggregates, so the fine particles of IOT sand can fill the pores, optimize the pore structure, and improve the compact packing density of UHPC. Zhang, Gu, et al. (2020) and Mu (2021) showed that with IOT sand substitution between 20% and 50%, some of the finer IOT sand particles have volcanic ash activity, which can consume the remaining $Ca(OH)_2$ in the UHPC matrix, producing more C-S-H gels, which contributes to the development of compressive strength of UHPC, and the 28-day compressive strength of UHPC can increase up to 14.29%. However, due to the irregular shape of the IOT sand surface, when its substitution ratio exceeds 50%, it is not conducive to the particle stack of UHPC raw materials, and the reduction of UHPC compactness hinders the development of strength (Table 16.4).

IOT sand can fill the pores inside the UHPC matrix within a reasonable content range, which can effectively reduce and offset the stress caused by the absence of free water in the capillaries and reduce the drying shrinkage of UHPC. In addition, with the increase of curing time, some IOT sand particles can consume excess $Ca(OH)_2$, producing more C-S-H gels, making the UHPC structure more compact and substantially improving the UHPC resistance to chloride ion permeability (Fig. 16.4).

The CO_2 emissions of different mix proportions can be effectively reduced when IOT is used as SCM or aggregate for UHPC, as shown in Table 16.5. Since the CO_2 emissions of UHPC mainly come from cement, the CO_2 emissions can be reduced by about 29% with conventional UHPC when IOT is used as SCM to replace cement, when the replacement ratio is 30%.

16.5.2 Sustainable ultra-high-performance concrete fabricated with steel slag

Steel slag is a typical steelmaking byproduct produced when steel is separated from impurities in the furnace. It is reported that the total global steel slag volume is about 230–290 million tons in 2022, but the recovery ratio is less than 30%, leading to a continuous increase in the amount of steel slag waste (Guo et al., 2018). The current disposal methods of steel slag are mainly open-air toppling and landfills, causing many environmental problems such as soil pollution, water pollution, and waste of land resources (Wang et al., 2012; Du, 2022). The chemical composition of steel slag is mainly Fe_2O_3, SiO_2, Al_2O_3, CaO, and MnO, and the mineral composition is mainly C_2S, C_3S, C_4AF, and C_2F. It has volcanic ash activity and cementing performance (Fig. 16.5) and can be used as SCM in the preparation of UHPC (Jiang, Shan, et al., 2018; Shi, 2004).

When steel slag is mixed into UHPC, the mixing water in the flocculation structure is released, and the steel slag can be well filled into the interstices of cement particles, reducing the surface energy of UHPC mix proportion. The steel slag powder is also microscopically spherical, which can produce the ball effect. Many

Table 16.4 Effect of adding iron tailings sand on SUHPC performance (relative values).

The contents of IOT (%)	Workability (%)	Compressive strength (%)		Volume stability (%)	Chloride ion permeability (%)	References
		7d	28d			
20	−5.71	4.55	6.67	–	−5.00	Zhang, Gu, et al. (2020)
40	2.86	7.95	14.29	–	10.00	
60	–	2.27	5.71	–	−15.00%	
80	–	−1.14	−0.95	–	−25.00%	
100	−14.29	−5.68	−9.52	–	−35.00%	
25	−22.73	9.26	5.71	7.89	−7.73	Mu (2021)
50	−47.27	12.96	12.86	15.79	−17.27	
75	−56.36	−1.85	1.43	9.56	−21.36	
100	−58.18	−7.41	−0.71	10.53	−23.64	
25	−5.67	–	2.05	0.63	–	Zhao et al. (2021)
50	−14.54	–	5.61	1.75	–	
75	−24.82	–	9.50	3.50	–	
100	−31.56	–	−16.17	7.96	–	

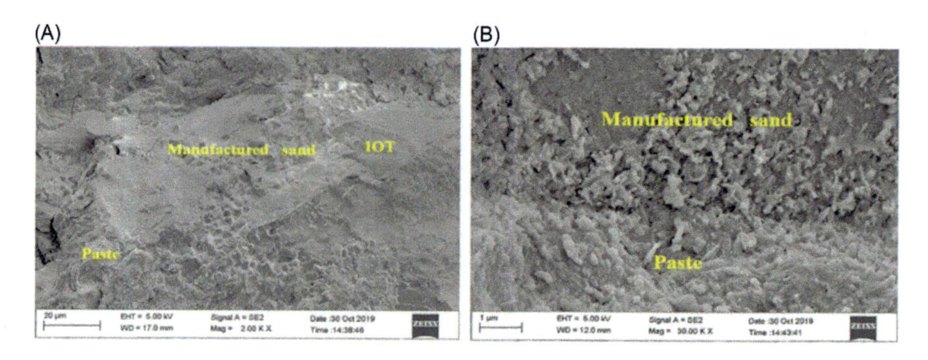

Figure 16.4 SEM images of IOT sand. Research on the characteristics of UHPC hydration products from iron tailings shows that IOT particles can consume excessive $Ca(OH)_2$ and produce more C-S-H, making the UHPC structure more uniform and dense. (A) ITZ between paste and mortars of UHPC without IOT, (B) ITZ between paste and mortars of UHPC with IOT.

Table 16.5 SUHPC carbon emissions from iron-bearing tailings fines.

ITP (%)	Emission of CO_2 (kg/m^3)	References
0	672.06	Zhang, Gu, et al. (2020)
0	690.96	
0	659.42	Ling et al. (2021)
10	596.52	
20	533.43	
30	470.42	

investigations showed that the workability of UHPC gradually increases with the increase of the steel slag substitution ratio, and when the steel slag substitution ratio is raised to 50%, the workability of UHPC can be increased to 33.94% (Xu et al., 2023; Du, 2022).

The volcanic ash activity of steel slag is relatively low compared to that of cement, which reduces the early compressive strength of UHPC. However, the hydration of steel slag gradually increases with the increase of curing age, and the smaller particle size of steel slag can better improve the compact packing density of UHPC and compensate for part of the loss of compressive strength (Table 16.6). When steel slag is used as SCM, the ratio of hydration reaction inside UHPC becomes slower, and at the same time, steel slag fills the pores inside the hydration products, which has a more obvious inhibiting effect on the shrinkage of UHPC. However, the reduction of C-S-H gels of hydration products leads to an increase in chloride ion transport channels and a decrease in the resistance to chloride ion permeability of UHPC. Despite the impact of steel slag on the durability of UHPC, UHPC prepared using steel slag as SCM within reasonable limits still has greater advantages compared to NC.

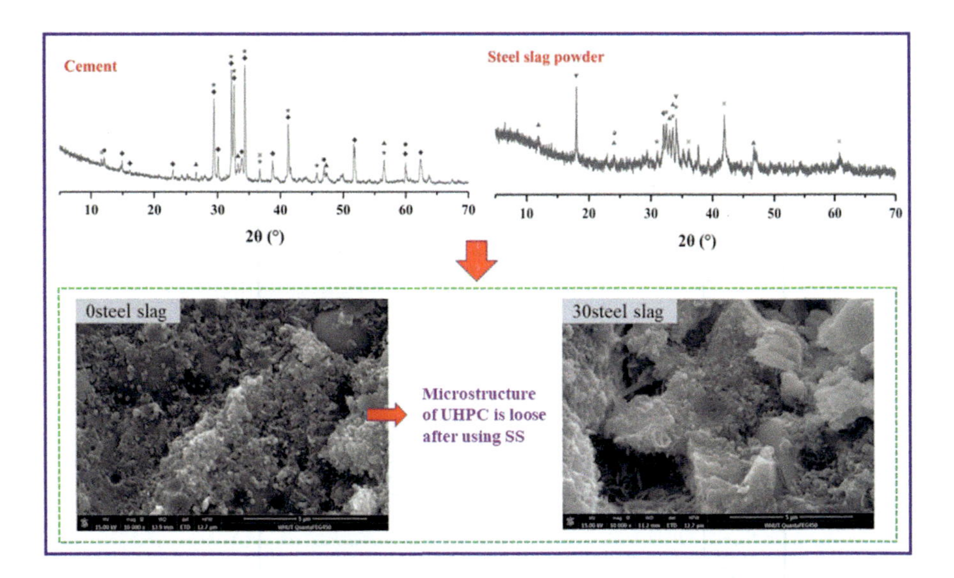

Figure 16.5 SUHPC chemical composition and microstructure containing steel slag. Chemical composition, hydration products, and microstructure of SUHPC containing steel slag.
Source: From Fan, D., Yu, R., Shui, Z., Liu, K., Feng, Y., Wang, S., Li, K., Tan, J., & He, Y. (2021). A new development of eco-friendly ultra-high performance concrete (UHPC): Towards efficient steel slag application and multi-objective optimization. *Construction and Building Materials*, *306*, 124913. https://doi.org/10.1016/j.conbuildmat.2021.124913.

The effective use of steel slag in UHPC would also mitigate soil acidification by reducing steel slag landfilling. Furthermore, when the steel slag content in UHPC reaches 30%, the negative impact of UHPC preparation on terrestrial ecosystems is reduced by 33.5% (Jiang, Ling, et al., 2018). Thus increasing the steel slag content in UHPC not only helps in reducing energy consumption and emissions but also enhances the ecological benefits of UHPC containing steel slag (Li, Cheng, et al., 2020).

16.6 Sustainable ultra-high-performance concrete fabricated with agricultural waste

16.6.1 *Sustainable ultra-high-performance concrete fabricated with rice husk ash*

Rice husk is a residue produced during rice processing, accounting for about 20%–23% of the total weight of rice, of which about 83% is toppled as waste near water streams or disposed of in landfills, seriously affecting water, soil, and the overall environment (Kang et al., 2019; Prasara-A & Gheewala, 2017). When burned at temperatures below

Table 16.6 Effect of adding steel slag on SUHPC performance (relative values).

The contents of SS (%)	Workability (%)	Compressive strength (%)			Volume stability (%)	Chloride ion permeability (%)	References
		3d	7d	28d			
5	3.04	–	−4.43	–	6.00%	–	(Du, 2022)
10	13.04	–	−11.39	–	17.00%	–	
15	15.65	–	−14.56	–	27.00%	–	
20	19.13	–	−17.72	–	31.00%	–	
25	20.87	–	−20.25	–	34.00%	–	
10	8.20	−5.43	−3.98	−1.98	–	48.96	Fan et al. (2021)
20	11.47	−9.44	−4.57	−1.69	–	88.54	
30	13.76	−12.15	−6.72	−0.58	–	200.00	
40	33.94	−20.48	−11.41	0	–	343.75	
5	3.57	−4.51	−4.43	−1.10	1.86	100.00	Li, Cheng, et al. (2020)
10	6.25	−10.79	−8.42	−3.37	4.60	236.67	
15	8.93	−21.12	−18.19	−8.13	19.25	483.33	
20	10.71	−24.79	−22.81	−6.20	25.47	450.00	
25	11.61	−28.35	−25.07	−7.02	32.92	423.33	

700°C, rice husk can be effectively converted into rice husk ash (RHA), containing 90%−96% amorphous silicon dioxide, which is comparable to the volcanic ash activity of silica fume (SF). The high price of SF and the relatively complex recovery process greatly increase the cost of UHPC incorporating SF. Due to the similar chemical composition of similar RHA and SF, RHA would be highly effective in the hydration process and improvement of cement paste microstructure. As a result, RHA serves as a superior SCM due to its abundant availability and cost-effectiveness, making it a favorable choice in comparison to SF (Alex et al., 2016).

Table 16.7 shows the impact of adding RHA on SUHPC. RHA has a porous structure and high water absorption characteristics, which would reduce the workability of UHPC incorporating RHA as SCM (Alex et al., 2016). The improvement of the compressive strength of UHPC by RHA can be attributed to several factors. Firstly, RHA can fill the internal pores of UHPC to further improve the degree of compact stacking of the matrix, leading to improved compressive strength. Another factor contributing to the improvement of compressive strength in UHPC with RHA is its large specific surface area, containing a large amount of amorphous silicon dioxide, which exhibits high volcanic ash activity (Cheah et al., 2016). This allows RHA to react with the calcium hydroxide present in the cement paste, resulting in the formation of additional C-S-H gels (Kang et al., 2019). This reaction improves the microstructure of the interface between the cement paste and the ITZ, leading to increased compressive strength (Fig. 16.6). In addition, the porous structure of RHA can absorb water in the early stage and gradually release the absorbed water as hydration proceeds, reducing the decrease of relative humidity inside UHPC during hydration and improving the phenomenon of large autogenous shrinkage in the early stage of UHPC (Mosaberpanah & Umar, 2020; Mostafa et al., 2022).

The use of SCMs can reduce the total CO_2 emissions of concrete, but CO_2 emissions will vary for different materials (Table 16.8). Quispe et al. (2019) showed that when using the same mass of cement or RHA for the preparation of UHPC, the CO_2 emissions of RHA were only 29% of that of cement, and RHA required less energy consumption than cement for recycling as waste. RHA has a beneficial impact on both reducing the environmental load of UHPC and improving the environmental compatibility of UHPC.

Compared to conventional UHPC, incorporation of RHA can reduce the cost of UHPC with certain life-cycle economic benefits, considering that the cost of RHA is only 1/200 of that of cement (Table 16.9). With regard to mechanical properties, durability, environmental factors, and the cost of UHPC, the adoption of RHA can further promote the sustainable development of UHPC and contribute positively to its overall sustainability.

16.6.2 Sustainable ultra-high-performance concrete fabricated with sugarcane bagasse ash

Sugarcane is the world's major sugar crop. Global sugarcane production is estimated to be more than 1.5 billion tons per year, and 26% of fibrous bagasse waste

Table 16.7 Effect of adding RHA on SUHPC performance (relative values).

The contents of RHA (%)	Workability (%)	Compressive strength (%)			Chloride ion permeability (%)	Permeability (%)	References
		3d	7d	28d			
10	−2.67		9.07	8.76	−14.92	−12.46	Mostafa et al. (2022)
20	−4.00		13.76	13.21	−26.44	−20.76	
30	−5.87		10.87	10.03	−35.25	−28.72	
40	−8.27		6.57	5.59	−43.39	−36.68	
50	−10.40		0.86	0	−48.81	−43.25	
10	−16.07	5.45	12.59	1.22	—	—	Ozturk et al. (2020)
20	−9.52	−0.25	2.22	5.49	—	—	
1/6	−2.27	9.10	—	4.17	−56.45	−3.33	Mosaberpanah and Umar (2020)
1/3	−9.09	9.23	—	6.67	−80.65	−54.84	
1/2	−11.36	9.04	—	12.5	−66.94	−22.58	
2/3	−15.91	8.76	—	16.67	−61.29	−4.84	
5/6	−22.73	8.23	—	10.0	−58.06	−27.42	
1	−54.55	8.10	—	—	—	—	
5	−8.23	—	1.65	6.34	—	—	Hasan et al. (2022)
10	−31.76	—	2.33	6.01	—	—	
15	-64.71	—	5.20	1.63	—	—	
20	−88.24	—	9.83	−6.09	—	—	

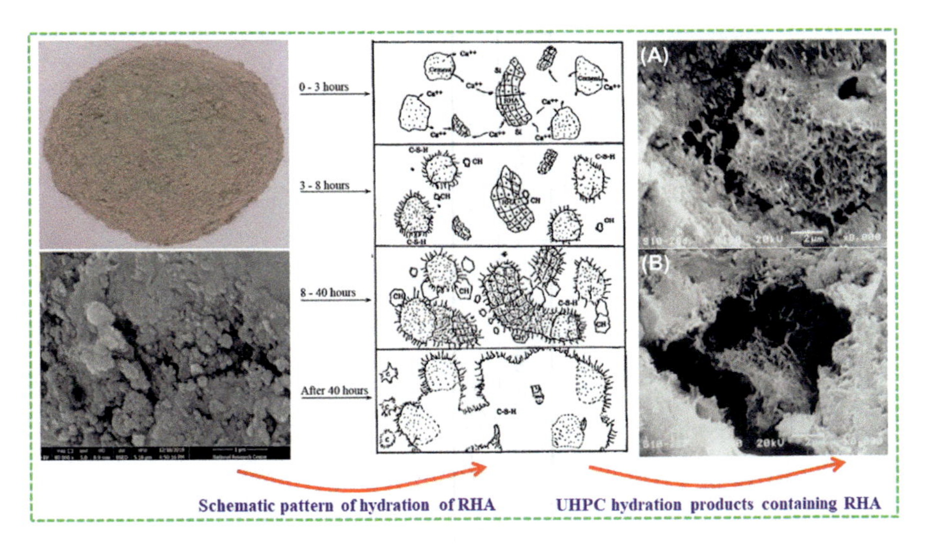

Figure 16.6 Effect of RHA on the ecological properties of SUHPC. The effect of RHA on the hydration products of SUHPC, thus changing the internal pore structure of SUHPC.
Source: From Mosaberpanah, M. A., & Umar, S. A. (2020). Utilizing rice husk ash as supplement to cementitious materials on performance of ultra-high-performance concrete: A review. *Materials Today Sustainability*, *7–8*, 100030. https://doi.org/10.1016/j.mtsust.2019.100030.

Table 16.8 CO_2 emissions from concrete components.

Materials	Emission of CO_2 (kg/m^3)	References
Concrete	821	Flower and Sanjayan (2007)
Sand	13.9	Turner and Collins (2013)
Stone chips	40.9	Turner and Collins (2013)
Water	0.196	Long et al. (2015)
RHA	157	Quispe et al. (2019)
Superplasticizer	720	Yang et al. (2013)

is generated per ton of sugarcane. When fibrous bagasse is burned at about 600°C–800°C, sugarcane bagasse ash (SCBA) containing large amounts of amorphous silica is produced (Inbasekar et al., 2016), as shown in Fig. 16.7. SCBA has a porous structure and excellent volcanic ash activities, making it an outstanding SCM, with excellent performance (Venkatesh et al., 2018). Also, the use of SCBA to prepare UHPC can realize the rational utilization of bagasse and reduce the pollution of water bodies and soil by bagasse.

The surface texture of SCBA, coupled with its low loss on ignition, contributes to improving the workability of UHPC due to the reduced friction between particles. The higher volcanic ash activity of SCBA makes the matrix of UHPC denser, and SCBA can react with calcium hydroxide to form additional C-S-H gels in the late stage of the hydration reaction, thus improving the strength of UHPC (Table 16.10).

Table 16.9 Prices of materials.

Materials	Cement	RHA	Sand	Stone Chips	Water	Superplasticizer	References
Cost (USD/kg)	0.14118	0.00059	0.02353	0.07059	0.00059	2.35294	Hasan et al. (2022)

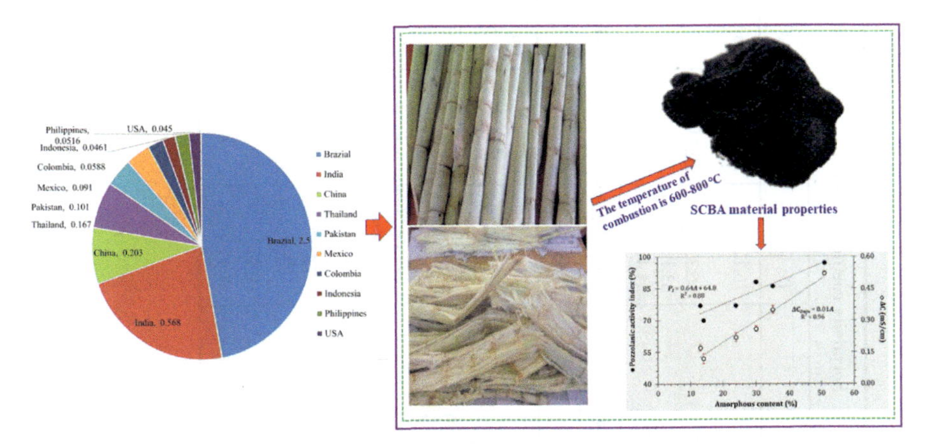

Figure 16.7 Preparation technology and material properties of SCBA. The yield distribution of sugarcane in the world and the preparation process and related properties of RHA. *Source:* From Jahanzaib Khalil, M., Aslam, M., & Ahmad, S. (2021). Utilization of sugarcane bagasse ash as cement replacement for the production of sustainable concrete – A review. *Construction and Building Materials, 270.* https://doi.org/10.1016/j.conbuildmat.2020.121371.

Huang, Huang, et al. (2023) investigated the compressive and flexural strengths of UHPC containing SCBA under different curing conditions and showed that there is an optimal threshold between 40% and 80% SCBA to maximize the compressive and flexural strengths of UHPC. Thermal curing is more favorable for the development of UHPC strength compared to normal temperature curing. The volcanic ash activity of SCBA was fully developed after thermal curing, and long calcium silica hydrate chains were formed within the UHPC matrix, which further improved the strength of UHPC.

Rajasekar et al. (2018) investigated the durability of UHPC and showed that SCBA improved the chloride ion permeability resistance of UHPC, but the effect of curing conditions on the chloride ion permeability resistance of UHPC was not consistent with the effect on strength. UHPC after thermal curing exhibited better chloride ion permeation resistance, and the chloride ion permeation resistance of UHPC could be improved by 63.33% when the SCBA substitution ratio was 15%. Steam curing led to uneven distribution of hydration products and more porous microstructure, resulting in lower chloride ion permeation resistance of UHPC, but ambient temperature curing produced UHPC with properties between those cured by these two methods. However, regardless of the curing method, the chloride ion permeation resistance of UHPC is better than that of NC.

16.7 Sustainable ultra-high-performance concrete fabricated with municipal waste

Glass has become one of the most versatile materials worldwide due to its excellent light transmittance, chemically inert, high strength, and low permeability properties.

Table 16.10 Effect of adding SCBA on SUHPC performance (relative values).

The contents of SCBA (%)	Workability (%)	Compressive strength (%)			Chloride ion permeability (%)			Sorptivity coefficient (%)			References
		N	T	S	N	T	S	N	T	S	
5	−0.53	0.40	5.8	7.7	−8.87	−46.67	−8.51	14.12	16.12	18.22	Rajasekar et al. (2018)
10	4.55	5.88	26.32	22.22	−31.43	−56.68	−17.02	25.21	16.33	20.31	
15	10.20	13.50	50	32	−40	−63.33	−36.17	41.34	33.12	38.45	
20%	13.10	5.10	36.84	19.44	−38.57	−33.33	−40.43	42.55	34.43	38.55	
The contents of SCBA (%)	**Workability (%)**	**Compressive strength (%)**			**Flexural strength (%)**			**Porosity (%)**	**$Ca(OH)_2$ contents (%)**		**Reference**
		3d	**28d**				**60d**	**3d**	**28d**	**60d**	
20	6.5	−3.57	−3.45	−5.74	−10.14	−4.17	5.71	1.59	2.94		Huang, Huang, et al. (2023)
40	10	−2.38	6.90	1.64	−5.80	8.33	2.86	3.51	−2.94		
60	4	3.42	14.66	12.30	−5.23	14.58	13.33	−26.79	−8.82		
80	0	2.13	13.79	9.02	2.17	16.67	12.38	−54.23	−2.35		
The contents of SCBA (%)	**Workability (%)**	**Compressive strength (%)**			**Flexural strength (%)**			**Porosity (%)**	**Volume stability (%)**		**Reference**
		7d	**28d**				**91d**	**7d**	**28d**	**91d**	
20	4.44	–	4.45	3.05	–	−1.55	5.34	−4.23	20.15		Wu et al. (2022)
40	11.11	–	0.17	7.74	–	−5.70	−3.88	8.94	21.33		
60	0	–	−11.08	0	–	−20.21	−20.87	4.13	37.33		

Note: N, T, and S stand for different concrete curing systems; N is normal temperature curing, T is thermal curing, and S is steam curing.

With the increase of industrialization and the improvement of living standards, the total amount of waste glass continues to increase, causing serious environmental pollution, for example, air, water, and soil contamination, due to the nonbiodegradable nature of glass (Qaidi et al., 2022).

Glass sand (GS) made from waste glass has a high silica content, similar density, and water absorption to natural fine aggregates. It is considered as a cost-effective alternative to natural aggregates, offering environmentally friendly, easy accessibility, and chemically stable solution (Topçu & Canbaz, 2004). GS has a smooth surface with fewer edges, its water absorption is lower than that of QS, and UHPC prepared using GS exhibits good workability. Due to the sharp edges of GS and smooth particle surface, the adhesion between cement and glass is lower than that between cement and natural sand, and the bond between cement mortar and GS in the ITZ is poor. In addition, the alkali-silica reaction (ASR) between the high alkali pore water in the cement paste and the reactive silica in the waste glass leads to the destructive expansion of concrete. Therefore GS adversely affects the strength of UHPC (Table 16.11). Although the strength of UHPC appears to experience different degrees of reduction, its indices such as compressive strength and flexural strength still meet the minimum requirements for UHPC classification, ensuring that UHPC remains capable of satisfying the demands of most practical projects.

Waste glass can be finely ground and used as SCM, as the silica in the glass can be converted into hydrated calcium silicate by the reaction of calcium hydrate ($Ca(OH)_2$). When the glass powder (GP) particle size is less than 150 μm, it starts to show volcanic ash activity, making it suitable as SCM for preparing SUHPC. GP can improve the workability of UHPC for two main reasons. Firstly, the dilution effect of cement reduces the hydration products in the early stage of cement paste and diminishes the number of bridgings between hydration products, leading to decreased resistance of cement paste flow. Secondly, the smooth surface of GP particles along with their low water absorption reduces the frictional resistance between GP particles and cement paste.

The early volcanic ash activity of GP is weak and only plays the role of microaggregate filling in the UHPC matrix, which reduces the early hydration products and decreases the strength of UHPC. The increase in the compressive strength of UHPC is mainly attributed to the further excitation of the volcanic ash activity of GP, which consumes the calcium hydroxide in the matrix through the "secondary hydration" reaction, resulting in additional C-S-H gels and partially hydrated calcium aluminate (C-A-H), leading to improved microstructure of UHPC. However, when the GP content was too much, the cement dilution effect became more significant and the formed hydration products were significantly reduced, causing deterioration of the microstructure of UHPC.

GP improves durability more than cement, and the volcanic ash reaction of GP refines the pores, reducing the porosity and pore connectivity for chloride ion transport (Rakshvir et al., 2006). The addition of 20%–30% GP reduced the chloride ion permeability of UHPC by 40%–90% (Table 16.12), and GP was more effective in enhancing chloride ion resistance compared to other SCMs (Ali-Boucetta et al., 2021).

Table 16.11 Effect of GS addition on SUHPC performance (relative values).

Contents of GP (%)	Workability (%)	Compressive strength (%)			Tensile strength (%)			Volume stability (%)	Chloride ion permeability (%)	References
7d	14d	28d	7 d	14 d	28 d					
25	57.89	8.11	5.56	9.09	−10.71	−8.57	−20.00	–	–	Jiao et al. (2020)
50	105.26	28.38	17.78	23.64	−1.43	−2.86	10.23	–	–	
75	321.05	35.16	27.78	30.91	−0.71	−7.43	−0.27	–	–	
100	557.89	18.92	24.44	24.45	2.14	−1.71	1.010	–	–	
25	17.78	14.29	–	3.45	−8.75	–	−13.90	−14.28	−15.06	Tahwia et al. (2022)
50	25.14	34.78	–	5.69	−9.56	–	−13.70	−25.71	−27.55	
100	32.22	38.41	–	10.25	10.35	–	−14.10	−34.29	29.77	
25	3.45	−4.05	–	−2.90	–	–	−4.51	–	–	Wei et al. (2021)
50	4.93	−7.43	–	−8.72	–	–	−9.33	–	–	
75	5.91	−10.81	–	−14.54	–	–	−17.75	–	–	
100	5.42	−12.81	–	17.88	–	–	−18.68	–	–	

Table 16.12 Effect of adding GP on SUHPC performance (relative values).

The contents of GP (%)	Workability (%)	Compressive strength (%)		Tensile strength		Water permeability (%)	References
7d	28d	7d	28d				
10	2.63	−5.08	−5.88	−12.5	−10.91	−4.05	Amin et al. (2023)
20	7.89	−9.32	5.29	−18.75	16.36	−9.80	
30	10.53	−13.56	8.23	−20	30.91	−7.79	
40	13.16	−19.49	11.76	–	–	−12.32	
50	15.79	−27.97	17.65	–	–	−15.22	
10	8.33	26.15	−17.47	4.04	−13.73	−5.71	Tahwia et al. (2022)
30	13.89	−7.69	4.37	−1.52	3.92	−30.12	
50	30.56	−9.23	8.20	−16.16	12.55	−57.14	
10	3.78	−5.12	−4.23	−12.11	8.33	–	Hamada et al. (2022)
20	5.95	−12.36	2.11	−19.05	5.67	–	
30	10.81	−27.65	4.93	−25.64	15.12	–	

Overall, the application of waste glass in UHPC reduces the cost of product preparation and has a value-added effect on solid waste. Rakshvir and Barai (2006) studied the cost and CO_2 emissions of different raw materials in UHPC, where 1 ton of cement and GP produce 846 and 63 kg of CO_2, respectively. By using waste glass as an SCM instead of cement, the CO_2 emission of UHPC can be greatly reduced, which has good economic and environmental benefits (Soliman & Tagnit-Hamou, 2016).

16.8 Sustainable ultra-high-performance concrete fabricated with new cementitious system

Limestone calcined clay cement (LC3) is recognized as a sustainable cementitious material system, benefiting from the wide global distribution and abundant availability of limestone and clay. Calcined clay is the product of clay calcination at 600°C–800°C. During the calcination process, kaolinite, the main active component of clay, is transformed into metakaolin, which can react with calcium hydroxide formed by cement hydration in a volcanic ash reaction. The amorphous alumina formed after calcination can also react with calcium carbonate to form aluminous carbonate phase material (Scrivener et al., 2018). These additional hydration products contribute to filling the pores in cementitious system, thus playing a positive role in strength and toughness.

Several studies (Dong, Liu, et al., 2023; Mo et al., 2020; Xiong & Liew, 2020) have shown that high content of calcined clay and limestone powder reduces the workability of UHPC. The degree of saturation of the high-efficiency superplasticizer adsorbed on the particle surface increases with the increase of particle fineness. Hence, the calcined clay, with a large specific surface area and a great adsorption efficiency for the high-efficiency superplasticizer, requires more superplasticizer to disperse the calcined clay particles and ensure good workability of UHPC (Dong, Liu, et al., 2023). In this way, UHPC prepared by LC3 exhibits acceptable workability that meets most engineering applications (Xuan et al., 2022).

The mechanical properties of UHPC are influenced by the microstructure of the internal pores of UHPC matrix. The appropriate calcined clay can fill the tiny pores in the UHPC matrix, refine the pore size in the UHPC matrix, and improve the compressive strength of the matrix (Mo et al., 2021). However, the decrease in the ratio of calcined clay to cement clinker decreases the volcanic ash activity of UHPC, leading to the reduction of hydration products in the UHPC matrix, a decrease in the ability to fill internal pores, and a decrease in the overall mechanical properties of UHPC. In addition, limestone powder can exert a nucleation effect, that is, providing additional nucleation sites and promoting the formation of C-S-H gels on its surface (Wang & Huang, 2023). Therefore with the increase of limestone powder content and the decrease of calcined clay-to-limestone ratio, the degree of cement hydration is enhanced. It has also been shown (Long et al., 2022) that the lack of large pores in a denser structure limits the diffusion and migration of ions in

Table 16.13 LC3 impact of SUHPC performance (relative values).

Sample	Workability (%)	Compressive strength (%)			References
		1d	3d	28d	
LC3 20/40	–	−68.97	−46.75	–	Dong, Liu, et al. (2023)
LC3 30/30	–	−67.23	−45.12	–	
LC3 40/20	–	−62.07	−35.06	–	
LC3 35/10	−20.50	–	–	13.78	Guo et al. (2023)
LC3 40/10	−26.70	–	–	−5.19	
LC3 45/10	−29.20	–	–	−22.96	
LC3 5/30	−3.04	−9.52	1.35	25.12	Mo et al. (2020)
LC3 10/30	−7.43	−2.38	8.20	30.56	
LC3 15/30	−9.80	−7.14	11.48	30.88	
LC3 20/30	−16.22	−14.29	−18.03	13.89	

Note: LC_3 x/y, x/y denotes the mass fraction of total cementitious material accounted for by calcined clay and limestone powder, respectively.

the pore solution and slows down the hydration of the cement, which explains the lower degree of cement hydration at high calcined clay and limestone ratios (Table 16.13).

LC3 cementitious material system can significantly reduce the use of cement and contribute to energy conservation and environmental protection (Maraghechi et al., 2018). Meanwhile, the aluminum phase in calcined clay reacts with limestone powder in an alkaline environment to form monocarboaluminate (Mc), which can refine the pore structure and compensate for the strength loss caused by cement dilution in UHPC (Dong, Liu, et al., 2023). In addition, the chloride ion permeability, water absorption, and drying shrinkage of UHPC containing LC3 cementitious system were reduced, and the durability and volume stability of UHPC were improved (Dhandapani et al., 2018). And LC3 system is beneficial to reduce the preparation cost and carbon footprint of UHPC, which can further promote the sustainable development of UHPC.

16.9 Sustainable ultra-high-performance concrete fabricated with high-performance fibers and nanomaterials

UHPC requires the addition of steel fibers in the preparation process to improve the strength and toughness as well as the brittle damage of UHPC. Many research investigations have been conducted on the basic properties of UHPC, confirming that increasing the volume content of steel fibers within a certain range can improve the tensile and flexural strengths of UHPC (Dong et al., 2022) and its toughness is also improved. However, the increase in steel fiber content significantly increases the cost of UHPC, for example, the cost of steel fibers at 1% content exceeds the cost of the UHPC matrix (Heinz et al., 2004). Moreover, a large volume of steel fibers not only increases the strength-to-density of UHPC, but the UHPC mixing process is prone to fiber agglomeration that impacts the workability and mechanical properties of UHPC (Abdul-Rahman et al., 2020).

The development history of UHPC is shown in Fig. 16.8. Conventional steel fibers with single characteristics tend to improve the performance of UHPC as adopted at the early stages of development. However, this approach leads to resource dependence, cost, and recycling difficulty of the material. Recent developments have further achieved high performance, sustainability, and intelligence of UHPC structures. Naaman and Wille (2012) demonstrated that compounding steel fibers and other high-performance fibers (e.g., polyvinyl alcohol fibers, polypropylene fibers, and polyester fibers) in SUHPC can reduce the construction difficulty and enhance the SUHPC bending toughness and crack resistance, while the economic problems caused by a single fiber can be tackled. Some key properties of hybrid fiber-reinforced SUHPC are summarized in Tables 16.14 and 16.15. Owing to their unique nanosize, nanomaterials can exhibit filling and micro agglomeration effects in cementitious materials (Wang et al., 2022a). By these effects, the presence of nanomaterials can lead to a dense and homogeneous cement matrix at a low content level. Moreover, nanomaterials with pozzolanic activity, for example, nanosilica, can undergo additional reactions during hydration to produce more C-S-H gels (Wang et al., 2022c, 2024), effectively improving the cement matrix and matrix-aggregate interface. In addition, the incorporation of nanomaterials offers an effective approach to enhance the SUHPC matrix and the interface and cobearing capacity between the steel fibers and the matrix (Wang et al., 2020a, 2020b, 2021, 2022a). Such improvements result

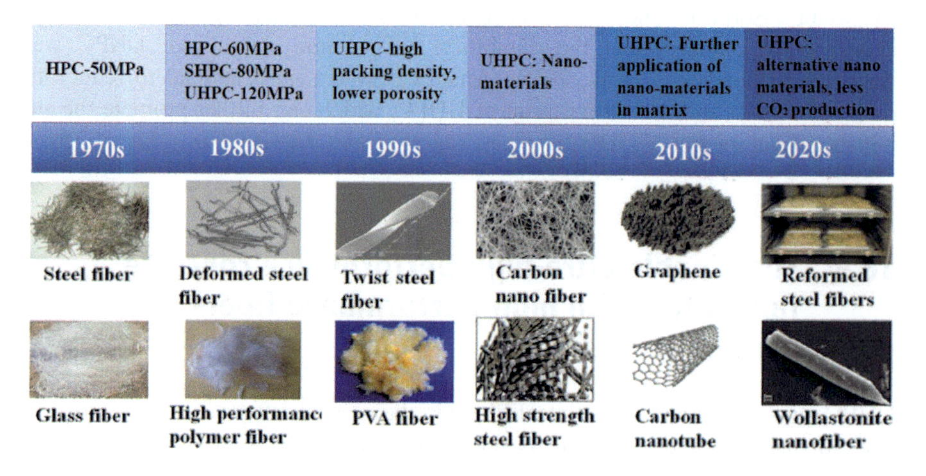

Figure 16.8 Development of high-performance concrete (HPC) and UHPC since 1960. Since 1960, the performance of UHPC has been improved by the addition of high-performance fibers and nanomaterials.
Source: From Gong, J., Ma, Y., Fu, J., Hu, J., Ouyang, X., Zhang, Z., & Wang, H. (2022). Utilization of fibers in ultra-high performance concrete: A review. *Composites Part B: Engineering*, *241*, 109995. https://doi.org/10.1016/j.compositesb.2022.109995.

Table 16.14 Effect of different types of fibers and nanomaterials on the mechanical properties of SUHPC.

Hybrid fibers		Compressive strength	Flexural strength	Tensile strength	Other properties	References
Name	\\V_f (%)					
Steel fiber + PVA	1.5 + 0.5	160.0 MPa (+4.58%)	23.42 MPa (+9.86%)	–	15.22 MPa (+4.83%) (first cracking strength)	(Meng & Khayat, 2018)
Steel fiber + CF	2.5 + 0.011	193.9 MPa (+1.61%)	41.82 MPa (+36.31%)	–		(Ma et al., 2019)
Steel fiber + PP	0.75 + 0.25	144.7 MPa (−6.58%)	–		13.5 MPa (−2.17%) (splitting tensile strength)	(Smarzewski & Barnat-Hunek, 2018)
Steel fiber + basalt	1.0 + 0.5	128.2 MPa (−14.09%)	–	14.74 MPa (+3.08%)	13.42 MPa (+36.94%) (first cracking strength)	(Kang et al., 2016)
Steel fiber + PVA	1.0 + 0.5	143.0 MPa (−4.03%)	–	11.84 MPa (−17.20%)	8.71 MPa (−11.12%) (first cracking strength)	
Steel fiber + PE	1.0 + 0.5	142.0 MPa (−4.70%)	–	16.21 MPa (+13.36%)	11.13 MPa (+13.57%) (first cracking strength)	
Steel fiber + nanosilica	2.5 + 4.0 (wt.)	+15.70%	+34.6%	–	–	(Yu et al., 2014)
Steel fiber + nano-TiO_2	2.5 + 2.0 (wt.)	+5.60%	+17.57%	–	–	Ghafari et al. (2015)
Steel fiber + MWCNTs	2.0 + 0.09	+2.2	+20.4%	+92.9%	–	(Aayisha & Mohan, 2020)

Table 16.15 Effect of SSWs and different types of nanomaterials on the flexural toughness, interfacial bond strength, and antisoftening coefficient of SUHPC.

Hybrid fibers		**Flexural toughness ($J = m^2$)**	**Interfacial bond strength (MPa)**	**Antisoftening coefficient (%)**	**Reference**
Name	V_f (%)	+4.75%	+3.10%	+21.03	Dong, Meng, et al. (2023)
SSWs + MWCNTs	1.2 + 0.25 (wt.)	+3.39%	+1.71%	+7.62	
SSWs + nano-TiO_2	0.6 + 0.25 (wt.)	+24.89%	+24.81%	+37.71	
SSWs + nano-TiO_2	1.2 + 0.25 (wt.)	+32.90%	+33.33%	+55.15	

in enhanced strength, impermeability, toughness, and durability of UHPC composites. The use of steel fibers with high-performance fibers or nanomaterials in SUHPC provides a new way to develop SUHPC with low steel fiber content and high strength-to-density ratios, with the potential to reduce the environmental footprint of SUHPC and improve structural safety (Dong, Meng, et al., 2023).

Apart from bearing materials for infrastructures, SUHPC is expected to possess more functional properties to fulfill the requirements of sustainable infrastructures in harsh, complex, and variable service environments. As steel fiber is an indispensable raw material in the production of UHPC, its electrically and thermally conductive properties are first considered to endow SUHPC with related functional properties. However, owing to the low content level, the steel fibers are difficult to reach the percolation threshold at commonly used dosage, exhibiting no obvious multifunctional/smart performance of SUHPC (Han et al., 2017). Many investigations (Dong et al., 2022; Dong et al., 2021; Li, Zheng, et al., 2020) showed that incorporating high-performance fibers or nanomaterials can not only improve the structural performance of SUHPC but endow SUHPC with functional/smart and additional properties (self-sensing, self-healing, electrically conductive, thermal, electromagnetic properties, light transmittance/luminescence, photocatalysis, energy harvesting, antimicrobial/viral, and the ability to respond to external stimuli). For example, the use of carbon fibers and steel fibers can form electrically and thermally conductive networks within SUHPC, giving SUHPC superior electrical and thermal conductivity as well as excellent electro-thermal functionality (Liu et al., 2020). Super-fine stainless wires (SSWs), owing to their microscale diameter and high aspect ratios, are effective in improving the weak interfacial zones and microstructure compactness of SUHPC (Dong et al., 2020, 2021). Meanwhile, SSWs can form widely distributed enhancing, toughening, electrical, and thermal networks within SUHPC at low dosage levels (Sun et al., 2024). The bending load-displacement curve of SUHPC compounding 0.2% stainless steel wire and 1.8% steel fiber had multiple cracking stages, and the flexural strength/toughness index of SUHPC can be increased by 87%/86% (Wang et al., 2022d). In addition, incorporating SSWs can reduce the electrical resistivity of SUHPC from 12900 to 1310 Ω/cm (Table 16.16). Therefore the use of SSWs not only gives SUHPC remarkable mechanical properties, durability, and resistance to early hydration-thermal cracking but also endows SUHPC with functional/smart properties, for example, electrically conductive, thermal, self-sensing, and electromagnetic property (Sufen Dong et al., 2020).

In addition, carbon nanotubes (Han et al., 2016; Wang et al., 2022b), graphene (Li, Ding, et al., 2021; Li, Zheng, et al., 2021), and other nanoscale functional fillers compounding with steel fibers can form long- and short-range overlap networks of the electrically and thermally conductive, enabling SUHPC to exhibit superior electrically and thermal conductivity and showing stable response under repeated load and impact load. While nano-TiO_2 (NT) not only enables SUHPC to possess remarkable mechanical properties and electromagnetic shielding/absorption properties but also antibacterial properties, strong redox properties, and acid-base corrosion resistance (Li et al., 2019). Thanks to the antimicrobial properties and acid-base corrosion

Table 16.16 Effect of different types of steel fibers and nanomaterials on the functionality of SUHPC.

Name	Content (wt.%)	Electrical property		Stress sensitivity (%/MPa)	References
DC resistivity (W/cm)	AC resistivity (W/cm)				
MWCNTs + SSWs	0.25 + 0.6	23.22	22.92	0.032	Zhang et al. (2019)
	0.25 + 1.2	17.02	16.72	0.02	
	0.5 + 0.6	126.36	123.44	0.046	
	0.5 + 1.2	33.06	32.72	0.016	
NT + SSWs	1.5 + 0.6	8.72	8.92	0.01	Li et al. (2019)
	1.5 + 1.2	5.32	5.60	0.02	
	3.0 + 0.6	19.31	19.68	0.03	
	3.0 + 1.2	10.08	10.34	0.02	
SSWs + SF	0.4 + 1.4	–	1310	0.006	Wang et al., 2022d)
	0.4 + 1.6	–	–	0.013	
	0.4 + 1.8	–	–	0.007	

resistance, NT can improve the erosion resistance of SUHPC in sewage and seawater. NT composite cementitious materials showed 96.81% and 80.93% inhibition of surface microorganisms in marine and sewage, respectively, improving the safety and extending the service life of infrastructure (Li et al., 2019; Ruan et al., 2018). With the development of nanocomposites technology, nanomaterials with multicomponent and multiscale features such as helical carbon nanotubes (Han et al., 2015), electrostatic self-assembled TiO_2/CNT (Zhang, Li, et al., 2020), electrostatically self-assembled NCB/CNT (Ding et al., 2019), nickel-plated carbon nanotubes (Wang et al., 2022d), silica-coated TiO_2 (Han et al., 2017), and in-situ-grown carbon nanotubes on cement (Ding et al., 2022) have been gradually developed and applied to SUHPC. The nanomaterials (Wang et al., 2020) with multicomponent and multiscale features not only alleviate the dispersion problem of traditional nanomaterials in cementitious material but improve the mechanical, electrical, pressure-sensitive, temperature-sensitive, and electromagnetic wave-absorbing properties of SUHPC. Incorporating functional fillers are expected to achieve the structural/functional integration of SUHPC (Ding et al., 2019; Jiang, Shan, et al., 2018).

16.10 Structural applications of sustainable ultra-high-performance concrete

UHPC has excellent mechanical properties and durability performance and is widely used in architectural and engineering applications such as bridges, tunnels, building elevation and exterior decoration, industrial structures, and architectural landscapes (Chen et al., 2019). In addition to the above-mentioned conventional architectural and engineering structures, UHPC plays a key role in various application-specific scenarios such as structural connection systems, complex components, support systems, and reinforcement and rehabilitation of large infrastructures such as bridges, highways, and airport runways. SUHPC has a wider range of applications as a new construction material due to its excellent mechanical properties and durability, as well as its green and low-cost characteristics, such as durable components exposed to marine or harsh environments, the construction of lightweight and sustainable bridge elements, and structures with seismic requirements. In addition, SUHPC compounding steel fiber and high-performance fibers or nanomaterials can make the UHPC structure develop toward multifunctionality and intelligence.

Reinforced concrete structures near marine environments are susceptible to corrosion caused by chloride ions, and, consequently, the service life of such structures is greatly reduced, consuming a lot of human and financial resources to repair and strengthen deteriorated structures at a later stage, greatly increasing the life-cycle cost of such structures. SUHPC usually has a higher resistance to ion penetration and can be used in extremely exposed environmental conditions to reduce reinforcement corrosion and improve the durability and service life of the structure (Lv et al., 2021).

SUHPC with high-performance fibers usually has higher compressive strength, tensile strength, corrosion resistance, denser microstructure, and superior crack control performance compared to conventional UHPC. The application of SUHPC in bridge construction can effectively reduce the cross-sectional area and the amount of reinforcement in the tensile zone, achieving the lightweight sustainability of bridge structures. It can also enhance structural life-cycle performance, reduce post-maintenance costs, and significantly reduce carbon emissions. In addition, SUHPC can be applied in bridge construction for bridge rehabilitation, deck overlays, bridge columns, seismic columns, and connections of prefabricated elements. SUHPC is also promising for applications such as strengthening and retrofitting old bridges due to the increasing number of vehicles, increased vehicle loads, freeze-thaw cycles, deck cracking, and steel corrosion (Abdal et al., 2023).

The higher strength of SUHPC can effectively reduce the weight of structures, thus reducing the effect of earthquakes on structures. The higher ductility and toughness characteristics of SUHPC can also improve the deformation capacity of structural components. The energy dissipation capacity of the structural members can be improved, effectively consuming the earthquake energy and enhancing the seismic performance of structures (Hu et al., 2021). The use of SUHPC in critical areas of conventional concrete structures can reinforce and modify the structure to withstand greater seismic loads, while greatly reducing the construction cost of structures and improving its safety and service life.

SUHPC with electrical, thermal, acoustic, optical, and germicidal functions can realize self-sensing, self-healing, and self-regulation of structures, making UHPC structures both structural-functional integration and functional-intelligent integration. For one thing, it can meet the functional requirements of infrastructure construction, and for another, it can improve the safety and service life of related infrastructure. Many studies have shown that SUHPC with excellent electrical and thermal conductivity can be used for self-curing, deicing and snow melting (Ding et al., 2023), energy storage (Zhang et al., 2012), and electromagnetic shielding of large and complex infrastructures such as roads, bridges, dams, and nuclear power plants (Han et al., 2015). SUHPC with photocatalytic properties degrades harmful gases in the atmosphere and organic pollutants in water into nontoxic and harmless products that do not cause secondary pollution through photocatalytic reactions under light conditions to achieve air purification inside buildings and self-cleaning of buildings under rainwater wash (Li et al., 2022).

At the present stage, civil engineering structures are characterized by complex and changing service environments, large heights and spans, diverse systems, and changing functions. Within this context, SUHPC materials offer a range of advantageous attributes, including environmental friendliness, increased strength, enhanced toughness, improved durability, and multifunctionality. SUHPC materials have the potential to reduce carbon emissions and costs during structure construction, minimize the overall weight of buildings and structures, prolong the service life of engineering structures, and decrease the overall life-cycle costs. Therefore using SUHPC is beneficial for largely promoting the sustainable development of engineering structures.

16.11 Conclusion

Optimizing the mix proportion of UHPC by reducing the amount of cement and natural aggregates and incorporating steel fibers and other high-performance fibers or nanomaterials are effective ways to promote the sustainable development of UHPC. SUHPC can reduce the demand for resources and energy, reduce environmental pollution, and lower the preparation cost, while further improving the properties of SUHPC and its associated structures. This chapter describes the mix proportion design methodology of SUHPC, the impact of using various types of solid waste as SCMs and aggregate substitutes on the performance and environmental impact of UHPC, and the use of high-performance fibers and nanomaterials to further improve the related properties of UHPC. Finally, it describes the structural applications of SUHPC. The following main conclusions may be drawn:

1. UHPC has excellent mechanical properties and durability, but it is extremely dependent on cement materials and natural aggregates, which are costly to use and cause a negative environmental impact. SUHPC is optimized based on traditional UHPC to reduce its dependence on natural materials and energy-consuming materials and to reduce the cost of UHPC preparation and CO_2 emissions. The DOMD or GA-ANN can be utilized to achieve optimum SUHPC mix proportion.
2. The development of new cementitious material systems and the use of industrial, municipal, and agricultural wastes as SCMs or aggregate substitutes are effective ways to produce SUHPC, which can simultaneously meet the requirements of strength and durability of UHPC and reduce the carbon footprint of UHPC. In addition, the incorporation of high-performance fibers and nanomaterials in UHPC can improve the safety and service life of UHPC structures, achieving multifunctional and intelligent UHPC.

3. SUHPC as a new construction material with excellent mechanical properties and durability, as well as environmentally friendly features, is widely used in durable components exposed to marine or harsh environments, the construction of lightweight sustainable bridge elements, and structures with seismic needs. Incorporating high-performance fibers or nanomaterials into SUHPC enables the construction of UHPC structures that are multi-functional and intelligent.
4. The utilization of SUHPC and its related structures has been gradually increasing, but due to the extensive scope covered by SUHPC, there is a lack of more systematic and in-depth research on SUHPC. As a result, subsequent research and evaluation system of SUHPC properties needs to be gradually improved so that they can be better utilized in practical engineering projects.

Acknowledgements

The authors thank the funding supported from the National Science Foundation of China (52308236, 52368031, 52178188 and 52308243), National Building Materials Industry Major Science and Technology Tackling Project (Intrinsic Self-Sensing Concrete for Full-Time and Full-Area Intelligent Monitoring of Major Civil Infrastructures), Natural Science Joint Foundation of Liaoning Province (2023-BSBA-077), and Natural Science Foundation of Heilongjiang Province (LH2023E069).

References

Aayisha, A., & Mohan, R. P. (2020). *An experimental study on addition of carbon nanotubes to improve the performance of UHPC* (*3*, pp. 811–817). Springer International Publishing. Available from https://doi.org/10.1007/978-3-030-26365-2_74.

Abdal, S., Mansour, W., Agwa, I., Nasr, M., Abadel, A., Onuralp Özkılıç, Y., & Akeed, M. H. (2023). Application of ultra-high-performance concrete in bridge engineering: Current status, limitations, challenges, and future prospects. *Buildings*, *13*(1). Available from https://doi.org/10.3390/buildings13010185, http://www.mdpi.com/journal/buildings.

Abdul-Rahman, M., Al-Attar, A. A., Hamada, H. M., & Tayeh, B. (2020). Microstructure and structural analysis of polypropylene fibre reinforced reactive powder concrete beams exposed to elevated temperature. *Journal of Building Engineering*, *29*. Available from https://doi.org/10.1016/j.jobe.2019.101167, http://www.journals.elsevier.com/journal-of-building-engineering/.

Akkurt, S., Ozdemir, S., Tayfur, G., & Akyol, B. (2003). The use of GA-ANNs in the modelling of compressive strength of cement mortar. *Cement and Concrete Research*, *33*(7), 973–979. Available from https://doi.org/10.1016/S0008-8846(03)00006-1.

Alex, J., Dhanalakshmi, J., & Ambedkar, B. (2016). Experimental investigation on rice husk ash as cement replacement on concrete production. *Construction and Building Materials*, *127*, 353–362. Available from https://doi.org/10.1016/j.conbuildmat.2016.09.150.

Ali-Boucetta, T., Behim, M., Cassagnabere, F., Mouret, M., Ayat, A., & Laifa, W. (2021). Durability of self-compacting concrete containing waste bottle glass and granulated slag. *Construction and Building Materials*, *270*121133. Available from https://doi.org/10.1016/j.conbuildmat.2020.121133.

Alkaysi, M., El-Tawil, S., Liu, Z., & Hansen, W. (2016). Effects of silica powder and cement type on durability of ultra high performance concrete (UHPC). *Cement and Concrete Composites*, *66*, 47–56. Available from https://doi.org/10.1016/j.cemconcomp.2015.11.005, http://www.sciencedirect.com/science/journal/09589465.

Amin, M., Agwa, I. S., Mashaan, N., Mahmood, S., & Abd-Elrahman, M. H. (2023). Investigation of the physical mechanical properties and durability of sustainable ultra-high performance concrete with recycled waste glass. *Sustainability*, *15*(4), 3085. Available from https://doi.org/10.3390/su15043085.

Behloul, M., & Guise, O. (2008). Ultra high performance fiber reinforced concrete: A material for green buildings. *Proceedings of 7th international congress concrete*. Construction's Sustainable Option.

Biskri, Y., Achoura, D., Chelghoum, N., & Mouret, M. (2017). Mechanical and durability characteristics of high performance concrete containing steel slag and crystalized slag as aggregates. *Construction and Building Materials*, *150*, 167–178. Available from https://doi.org/10.1016/j.conbuildmat.2017.05.083.

Bouhaya, L., Le Roy, R., & Feraille-Fresnet, A. (2009). Simplified environmental study on innovative bridge structure. *Environmental Science and Technology*, *43*(6), 2066–2071. Available from https://doi.org/10.1021/es801351gFrance, http://pubs.acs.org/doi/pdfplus/10.1021/es801351g.

Cheah, W.-K., Ooi, C.-H., & Yeoh, F.-Y. (2016). Rice husk and rice husk ash reutilization into nanoporous materials for adsorptive biomedical applications: A review. *Open Material Sciences*, *3*(1). Available from https://doi.org/10.1515/mesbi-2016-0004.

Chen, B. C., Wei, J. G., Su, J. Z., Huang, W., Chen, Y. C., Huang, Q. W., & Chen, Z. H. (2019). State-of-the-art progress on application of ultra-high performance concrete. *Journal of Architecture and Civil Engineering*, *36*(2), 10–20. Available from http://jace.chd.edu.cn.

Dhandapani, Y., Sakthivel, T., Santhanam, M., Gettu, R., & Pillai, R. G. (2018). Mechanical properties and durability performance of concretes with limestone calcined clay cement (LC3). *Cement and Concrete Research*, *107*, 136–151. Available from https://doi.org/10.1016/j.cemconres.2018.02.005, http://www.sciencedirect.com/science/journal/00088846.

Ding, S., Dong, S., Wang, X., Ding, S., Han, B., & Ou, J. (2023). Self-heating ultra-high performance concrete with stainless steel wires for active deicing and snow-melting of transportation infrastructures. *Cement and Concrete Composites*, *138*, 105005. Available from \//doi.org/10.1016/j.cemconcomp.2023.105005.

Ding, S., Ruan, Y., Yu, X., Han, B., & Ni, Y. Q. (2019). Self-monitoring of smart concrete column incorporating CNT/NCB composite fillers modified cementitious sensors. *Construction and Building Materials*, *201*, 127–137. Available from https://doi.org/10.1016/j.conbuildmat.2018.12.203.

Ding, S., Wang, X., Qiu, L., Ni, Y. Q., Dong, X., Cui, Y., Ashour, A., Han, B., & Ou, J. (2023). Self-sensing cementitious composites with hierarchical carbon fiber-carbon nanotube composite fillers for crack development monitoring of a maglev girder. *Small*, *19*(9), 2206258. Available from https://doi.org/10.1002/smll.202206258.

Ding, S., Xiang, Y., Ni, Y. Q., Thakur, V. K., Wang, X., Han, B., & Ou, J. (2022). In-situ synthesizing carbon nanotubes on cement to develop self-sensing cementitious composites for smart high-speed rail infrastructures. *Nano Today*, *43*. Available from https://doi.org/10.1016/j.nantod.2022.101438, http://www.elsevier.com/wps/find/journaldescription.cws_home/706735/description#description.

Dong, S., Dong, X., Ashour, A., Han, B., & Ou, J. (2020). Fracture and self-sensing characteristics of super-fine stainless wire reinforced reactive powder concrete. *Cement and Concrete Composites*, *105*, 103427. Available from https://doi.org/10.1016/j.cemconcomp.2019.103427.

Dong, Y., Liu, Y., & Hu, C. (2023). Towards greener ultra-high performance concrete based on highly-efficient utilization of calcined clay and limestone powder. *Journal of Building Engineering*, *66*, 105836. Available from https://doi.org/10.1016/j.jobe.2023.105836.

Dong, S., Meng, W., Wang, D., Zhang, W., Wang, X., Han, B., & Ou, J. (2023). Principle and implementation of incorporating nanomaterials to develop ultrahigh-performance concrete with low content of steel fibers. *Journal of Materials in Civil Engineering*, *35* (6). Available from https://doi.org/10.1061/JMCEE7.MTENG-14849, http://ascelibrary.org/mto/resource/1/jmcee7/.

Dong, S., Wang, X., Ashour, A., Han, B., & Ou, J. (2022). Enhancement and underlying mechanisms of stainless steel wires to fatigue properties of concrete under flexure. *Cement and Concrete Composites*, *126*. Available from https://doi.org/10.1016/j.cemconcomp.2021.104372, http://www.sciencedirect.com/science/journal/09589465.

Dong, S., Wang, Y., Ashour, A., Han, B., & Ou, J. (2021). Uniaxial compressive fatigue behavior of ultra-high performance concrete reinforced with super-fine stainless wires. *International Journal of Fatigue*, *142*, 105959. Available from https://doi.org/10.1016/j.ijfatigue.2020.105959.

Du, H. (2022). Preparation and Properties of Green Steel Slag UHPC. Master thesis. Wuhan Institute of Technology, Wuhan, China (in Chinese).

Fan, D., Yu, R., Shui, Z., Liu, K., Feng, Y., Wang, S., Li, K., Tan, J., & He, Y. (2021). A new development of eco-friendly Ultra-High performance concrete (UHPC): Towards efficient steel slag application and multi-objective optimization. *Construction and Building Materials*, *306*, 124913. Available from https://doi.org/10.1016/j.conbuildmat.2021.124913.

Fan, D., Yu, R., Shui, Z., Wu, C., Wang, J., & Su, Q. (2020). A novel approach for developing a green ultra-high performance concrete (UHPC) with advanced particles packing meso-structure. *Construction and Building Materials*, *265*, 120339.

Ferdosian, I., Camões, A., & Ribeiro, M. (2017). High-volume fly ash paste for developing ultra-high performance concrete (UHPC). *Tecnologia dos Materiais*, *29*(1), e157–e161. Available from https://doi.org/10.1016/j.ctmat.2016.10.001, http://www.journals.elsevier.com/ciencia-and-tecnologia-dos-materiais/.

Flower, D. J. M., & Sanjayan, J. G. (2007). Green house gas emissions due to concrete manufacture. *International Journal of Life Cycle Assessment*, *12*(5), 282–288. Available from https://doi.org/10.1065/lca2007.05.327, http://www.springerlink.com/content/0948-3349.

Ghafari, E., Bandarabadi, M., Costa, H., & Júlio, E. (2015). Prediction of fresh and hardened state properties of UHPC: Comparative study of statistical mixture design and an artificial neural network model. *Journal of Materials in Civil Engineering*, *27*(11). Available from https://doi.org/10.1061/(ASCE)MT.1943-5533.0001270, http://ascelibrary.org/mto/resource/1/jmcee7/.

Guo, J., Bao, Y., & Wang, M. (2018). Steel slag in China: Treatment, recycling, and management. *Waste Management*, *78*, 318–330. Available from https://doi.org/10.1016/j.wasman.2018.04.045, http://www.elsevier.com/locate/wasman.

Guo, D., Guo, M., Xing, F., Zhou, Y., Huang, Z., & Cao, W. (2023). Using limestone calcined clay cement and recycled fine aggregate to make ultra-high-performance concrete: Properties and environmental impact. *Construction and Building Materials*, *394*, 132026. Available from https://doi.org/10.1016/j.conbuildmat.2023.132026.

Gu, X., Zhang, W., Zhang, X., Li, X., & Qiu, J. (2022). Hydration characteristics investigation of iron tailings blended ultra high performance concrete: The effects of mechanical activation and iron tailings content. *Journal of Building Engineering*, *45*. Available from https://doi.org/10.1016/j.jobe.2021.103459, http://www.journals.elsevier.com/journal-of-building-engineering/.

Hamada, H., Alattar, A., Tayeh, B., Yahaya, F., & Thomas, B. (2022). Effect of recycled waste glass on the properties of high-performance concrete: A critical review. *Case Studies in Construction Materials*, *17*, e01149. Available from https://doi.org/10.1016/j.cscm.2022.e01149.

Hamada, H. M., Shi, J., Abed, F., Al Jawahery, M. S., Majdi, A., & Yousif, S. T. (2023). Recycling solid waste to produce eco-friendly ultra-high performance concrete: A review of durability, microstructure and environment characteristics. *Science of the Total Environment*, *876*. Available from https://doi.org/10.1016/j.scitotenv.2023.162804, http://www.elsevier.com/locate/scitotenv.

Hammoudi, A., Moussaceb, K., Belebchouche, C., & Dahmoune, F. (2019). Comparison of artificial neural network (ANN) and response surface methodology (RSM) prediction in compressive strength of recycled concrete aggregates. *Construction and Building Materials*, *209*, 425–436. Available from https://doi.org/10.1016/j.conbuildmat.2019.03.119.

Han, B., Ding, S., & Yu, X. (2015). Intrinsic self-sensing concrete and structures: A review. *Measurement*, *59*, 110–128. Available from https://doi.org/10.1016/j.measurement.2014.09.048.

Han, B., Li, Z., Zhang, L., Zeng, S., Yu, X., Han, B., & Ou, J. (2017). Reactive powder concrete reinforced with nano SiO_2-coated TiO_2. *Construction and Building Materials*, *148*, 104–112. Available from https://doi.org/10.1016/j.conbuildmat.2017.05.065.

Han, B., Zhang, L., Zeng, S., Dong, S., Yu, X., Yang, R., & Ou, J. (2017a). Nano-core effect in nano-engineered cementitious composites. *Composites Part A: Applied Science and Manufacturing*, *95*, 100–109. Available from https://doi.org/10.1016/j.compositesa.2017.01.008.

Han, B., Zhang, L., Zhang, C., Wang, Y., Yu, X., & Ou, J. (2016). Reinforcement effect and mechanism of carbon fibers to mechanical and electrically conductive properties of cement-based materials. *Construction and Building Materials*, *125*, 479–489. Available from https://doi.org/10.1016/j.conbuildmat.2016.08.063.

Hasan, N. M. S., Sobuz, M. H. R., Khan, M. M. H., Mim, N. J., Meraz, M. M., Datta, S. D., Rana, M. J., Saha, A., Akid, A. S. M., Mehedi, M. T., Houda, M., & Sutan, N. M. (2022). Integration of rice husk ash as supplementary cementitious material in the production of sustainable high-strength concrete. *Materials*, *15*(22). Available from https://doi.org/10.3390/ma15228171, http://www.mdpi.com/journal/materials.

Heinz, D., Dehn, F., & Urbonas, L. (2004). Fire resistance of ultra high performance concrete (UHPC)-Testing of laboratory samples and columns under load. *International Symposium on Ultra High Performance Concrete*, 703–715.

Huang, P., Huang, B., Li, J., Wu, N., & Xu, Q. (2023). Application of sugar cane bagasse ash as filler in ultra-high performance concrete. *Journal of Building Engineering*, *71*, 106447. Available from https://doi.org/10.1016/j.jobe.2023.106447.

Huang, Y., Wang, J., Wei, Q. 'an, Shang, H., & Liu, X. (2023). Creep behaviour of ultra-high-performance concrete (UHPC): A review. *Journal of Building Engineering*, *69*, 106187. Available from https://doi.org/10.1016/j.jobe.2023.106187.

Hung, C. C., Lee, H. S., & Chan, S. N. (2019). Tension-stiffening effect in steel-reinforced UHPC composites: Constitutive model and effects of steel fibers, loading patterns, and rebar sizes. *Composites Part B: Engineering*, *158*, 269–278. Available from https://doi.org/10.1016/j.compositesb.2018.09.091.

Hu, R., Fang, Z., Shi, C., Benmokrane, B., & Su, J. (2021). A review on seismic behavior of ultra-high performance concrete members. *Advances in Structural Engineering*, *24*(5), 1054–1069. Available from https://doi.org/10.1177/1369433220968451, https://uk.sagepub.com/en-gb/eur/advances-in-structural-engineering/journal202466.

Inbasekar, M., Hariprasath, P., & Senthilkumar, D. (2016). Study on potential utilization of sugarcane bagasse ash in steel fiber reinforced concrete. *International Journal of Engineering Sciences & Research Technology*, *5*(4), 43–50.

Jahanzaib Khalil, M., Aslam, M., & Ahmad, S. (2021). Utilization of sugarcane bagasse ash as cement replacement for the production of sustainable concrete – A review. *Construction and Building Materials*, *270*. Available from https://doi.org/10.1016/j.conbuildmat.2020.121371, https://www.journals.elsevier.com/construction-and-building-materials.

Jiang, Y., Ling, T. C., Shi, C., & Pan, S. Y. (2018). Characteristics of steel slags and their use in cement and concrete—A review. *Resources, Conservation and Recycling*, *136*, 187–197. Available from https://doi.org/10.1016/j.resconrec.2018.04.023, http://www.elsevier.com/locate/resconrec.

Jiang, S., Shan, B., Ouyang, J., Zhang, W., Yu, X., Li, P., & Han, B. (2018). Rheological properties of cementitious composites with nano/fiber fillers. *Construction and Building Materials*, *158*, 786–800. Available from https://doi.org/10.1016/j.conbuildmat.2017.10.072.

Jiao, Y., Zhang, Y., Guo, M., Zhang, L., Ning, H., & Liu, S. (2020). Mechanical and fracture properties of ultra-high performance concrete (UHPC) containing waste glass sand as partial replacement material. *Journal of Cleaner Production*, *277*, 123501. Available from https://doi.org/10.1016/j.jclepro.2020.123501.

Kang, S. T., Choi, J. I., Koh, K. T., Lee, K. S., & Lee, B. Y. (2016). Hybrid effects of steel fiber and microfiber on the tensile behavior of ultra-high performance concrete. *Composite Structures*, *145*, 37–42. Available from https://doi.org/10.1016/j.compstruct.2016.02.075.

Kang, S. H., Hong, S. G., & Moon, J. (2019). The use of rice husk ash as reactive filler in ultra-high performance concrete. *Cement and Concrete Research*, *115*, 389–400. Available from https://doi.org/10.1016/j.cemconres.2018.09.004, http://www.sciencedirect.com/science/journal/00088846.

Kim, K., Shin, M., & Cha, S. (2013). Combined effects of recycled aggregate and fly ash towards concrete sustainability. *Construction and Building Materials*, *48*, 499–507. Available from https://doi.org/10.1016/j.conbuildmat.2013.07.014.

Lefever, G., De Belie, N., Snoeck, D., Aggelis, D. G., & Van Hemelrijck, D. (2022). Nanomaterials in self-healing cementitious composites. *Recent Advances in Nano-Tailored Multi-Functional Cementitious Composites*, 141–159. Available from https://doi.org/10.1016/B978-0-323-85229-6.00013-5, https://www.sciencedirect.com/book/9780323852296.

Ling, G., Shui, Z., Gao, X., Sun, T., Yu, R., & Li, X. (2021). Utilizing iron ore tailing as cementitious material for eco-friendly design of ultra-high performance concrete (UHPC). *Materials*, *14*(8), 1829. Available from https://doi.org/10.3390/ma14081829.

Liu, Y., Tian, W., Wang, M., Qi, B., & Wang, W. (2020). Rapid strength formation of on-site carbon fiber reinforced high-performance concrete cured by ohmic heating. *Construction and Building Materials*, *244*, 118344. Available from https://doi.org/10.1016/j.conbuildmat.2020.118344.

Li, S., Cheng, S., Mo, L., & Deng, M. (2020). Effects of steel slag powder and expansive agent on the properties of ultra-high performance concrete (UHPC): Based on a case study. *Materials*, *13*(3), 683. Available from https://doi.org/10.3390/ma13030683.

Li, J., Wu, Z., Shi, C., Yuan, Q., & Zhang, Z. (2020). Durability of ultra-high performance concrete – A review. *Construction and Building Materials*, *255*, 119296. Available from https://doi.org/10.1016/j.conbuildmat.2020.119296.

Li, L., Zheng, Q., Dong, S., Wang, X., & Han, B. (2020). The reinforcing effects and mechanisms of multi-layer graphenes on mechanical properties of reactive powder concrete. *Construction and Building Materials*, *251*, 118995. Available from https://doi.org/10.1016/j.conbuildmat.2020.118995.

Li, Z., Ding, S., Kong, L., Wang, X., Ashour, A., Han, B., & Ou, J. (2022). Nano TiO_2-engineered anti-corrosion concrete for sewage system. *Journal of Cleaner Production*, *337*, 130508. Available from https://doi.org/10.1016/j.jclepro.2022.130508.

Li, H., Ding, S., Zhang, L., Ouyang, J., & Han, B. (2021). Rheological behaviors of cement pastes with multi-layer graphene. *Construction and Building Materials*, *269*, 121327. Available from https://doi.org/10.1016/j.conbuildmat.2020.121327.

Li, L., Zheng, Q., Han, B., & Ou, J. (2021). Fatigue behaviors of graphene reinforcing concrete composites under compression. *International Journal of Fatigue*, *151*, 106354. Available from https://doi.org/10.1016/j.ijfatigue.2021.106354.

Li, Z., Han, B., Yu, X., Zheng, Q., & Wang, Y. (2019). Comparison of the mechanical property and microstructures of cementitious composites with nano- and micro-rutile phase TiO_2. *Archives of Civil and Mechanical Engineering*, *19*(3), 615–626. Available from https://doi.org/10.1016/j.acme.2019.02.002, http://www.sciencedirect.com/science/journal/16449665.

Li, C., Sun, H., Bai, J., & Li, L. (2010). Innovative methodology for comprehensive utilization of iron ore tailings. Part 1. The recovery of iron from iron ore tailings using magnetic separation after magnetizing roasting. *Journal of Hazardous Materials*, *174*(1-3), 71–77. Available from https://doi.org/10.1016/j.jhazmat.2009.09.018.

Long, G., Gao, Y., & Xie, Y. (2015). Designing more sustainable and greener self-compacting concrete. *Construction and Building Materials*, *84*, 301–306. Available from https://doi.org/10.1016/j.conbuildmat.2015.02.072.

Long, W. J., Wu, Z., Khayat, K. H., Wei, J., Dong, B., Xing, F., & Zhang, J. (2022). Design, dynamic performance and ecological efficiency of fiber-reinforced mortars with different binder systems: Ordinary Portland cement, limestone calcined clay cement and alkali-activated slag. *Journal of Cleaner Production*, *337*. Available from https://doi.org/10.1016/j.jclepro.2022.130478, https://www.journals.elsevier.com/journal-of-cleaner-production.

Lv, L. S., Wang, J. Y., Xiao, R. C., Fang, M. S., & Tan, Y. (2021). Chloride ion transport properties in microcracked ultra-high performance concrete in the marine environment. *Construction and Building Materials*, *291*. Available from https://doi.org/10.1016/j.conbuildmat.2021.123310, https://www.journals.elsevier.com/construction-and-building-materials.

Ma, R., Guo, L., Ye, S., Sun, W., & Liu, J. (2019). Influence of hybrid fiber reinforcement on mechanical properties and autogenous shrinkage of an ecological UHPFRCC. *Journal of Materials in Civil Engineering*, *31*(5), 04019032. Available from https://doi.org/10.1061/(ASCE)MT.1943-5533.0002650.

Magalhães, L. F. D., França, S., Oliveira, Md. S., Peixoto, R. A. F., Bessa, S. A. L., & Bezerra, A. Cd. S. (2020). Iron ore tailings as a supplementary cementitious material in the production of pigmented cements. *Journal of Cleaner Production*, *274*. Available from https://doi.org/10.1016/j.jclepro.2020.123260, https://www.journals.elsevier.com/journal-of-cleaner-production.

Magureanu, C., Sosa, I., Negrutiu, C., & Heghes, B. (2012). Mechanical properties and durability of ultra-high-performance concrete. *ACI Materials Journal*, *109*(2), 177–184.

Maraghechi, H., Avet, F., Wong, H., Kamyab, H., & Scrivener, K. (2018). Performance of limestone calcined clay cement (LC3) with various kaolinite contents with respect to chloride transport. *Materials and Structures*, *51*(5). Available from https://doi.org/10.1617/s11527-018-1255-3.

Mehdipour, I., & Khayat, K. H. (2018). Understanding the role of particle packing characteristics in rheo-physical properties of cementitious suspensions: A literature review. *Construction and Building Materials*, *161*, 340–353. Available from https://doi.org/10.1016/j.conbuildmat.2017.11.147.

Mohamad, N., Muthusamy, K., Embong, R., Kusbiantoro, A., & Hashim, M. H. (2021). Environmental impact of cement production and Solutions: A review. *Materials Today: Proceedings*, *48*, 741–746. Available from https://doi.org/10.1016/j.matpr.2021.02.212, https://www.sciencedirect.com/journal/materials-today-proceedings.

Mosaberpanah, M. A., & Umar, S. A. (2020). Utilizing rice husk ash as supplement to cementitious materials on performance of ultra high performance concrete: A review. *Materials Today Sustainability*, *7-8*, 100030. Available from https://doi.org/10.1016/j.mtsust.2019.100030.

Mostafa, S. A., Ahmed, N., Almeshal, I., Tayeh, B. A., & Elgamal, M. S. (2022). Experimental study and theoretical prediction of mechanical properties of ultra-high-performance concrete incorporated with nanorice husk ash burning at different temperature treatments. *Environmental Science and Pollution Research*, *29*(50), 75380–75401. Available from https://doi.org/10.1007/s11356-022-20779-w, https://www.springer.com/journal/11356.

Meng, W., & Khayat, K. H. (2018). Effect of hybrid fibers on fresh properties, mechanical properties, and autogenous shrinkage of cost-effective UHPC. *Journal of Materials in Civil Engineering*, *30*(4), 04018030. Available from https://doi.org/10.1061/(ASCE)MT.1943-5533.0002212.

Mo, Z., Wang, R., & Gao, X. (2020). Hydration and mechanical properties of UHPC matrix containing limestone and different levels of metakaolin. *Construction and Building Materials*, *256*, 119454. Available from https://doi.org/10.1016/j.conbuildmat.2020.119454.

Mo, Z., Zhao, H., Jiang, L., Jiang, X., & Gao, X. (2021). Rehydration of ultra-high performance concrete matrix incorporating metakaolin under long-term water curing. *Construction and Building Materials*, *306*, 124875. Available from https://doi.org/10.1016/j.conbuildmat.2021.124875.

Mu, C. G. (2021). Study on strength and permeability of iron tailings modified ultra-high performance concrete. *Comprehensive Utilization of Fly Ash*, *35*(06), 68–72.

Naaman, A. E., & Wille, K. (2012). The path to ultra-high performance fiber reinforced concrete (UHP-FRC): five decades of progress. *Proceedings of Hipermat*, 3–15.

Ozturk, M., Karaaslan, M., Akgol, O., & Sevim, U. K. (2020). Mechanical and electromagnetic performance of cement based composites containing different replacement levels of ground granulated blast furnace slag, fly ash, silica fume and rice husk ash. *Cement and Concrete Research*, *136*. Available from https://doi.org/10.1016/j.cemconres.2020.106177, http://www.sciencedirect.com/science/journal/00088846.

Prasara-A, J., & Gheewala, S. H. (2017). Sustainable utilization of rice husk ash from power plants: A review. *Journal of Cleaner Production*, *167*, 1020–1028. Available from https://doi.org/10.1016/j.jclepro.2016.11.042, https://www.journals.elsevier.com/journal-of-cleaner-production.

Qaidi, S., Najm, H. M., Abed, S. M., Özkılıç, Y. O., Al Dughaishi, H., Alosta, M., Sabri, M. M. S., Alkhatib, F., & Milad, A. (2022). Concrete containing waste glass as an environmentally friendly aggregate: A review on fresh and mechanical characteristics. *Materials*, *15*(18). Available from https://doi.org/10.3390/ma15186222, http://www.mdpi.com/journal/materials.

Quispe, I., Navia, R., & Kahhat, R. (2019). Life cycle assessment of rice husk as an energy source: A Peruvian case study. *Journal of Cleaner Production*, *209*, 1235–1244. Available from https://doi.org/10.1016/j.jclepro.2018.10.312, https://www.journals.elsevier.com/journal-of-cleaner-production.

Rajasekar, A., Arunachalam, K., Kottaisamy, M., & Saraswathy, V. (2018). Durability characteristics of ultra high strength concrete with treated sugarcane bagasse ash. *Construction and Building Materials*, *171*, 350–356. Available from https://doi.org/10.1016/j.conbuildmat.2018.03.140.

Rakshvir, M., & Barai, S. V. (2006). Studies on recycled aggregates-based concrete. *Waste Management and Research*, *24*(3), 225–233. Available from https://doi.org/10.1177/0734242X06064820.

Ruan, Y., Han, B., Yu, X., Li, Z., Wang, J., Dong, S., & Ou, J. (2018). Mechanical behaviors of nano-zirconia reinforced reactive powder concrete under compression and flexure. *Construction and Building Materials*, *162*, 663–673. Available from https://doi.org/10.1016/j.conbuildmat.2017.12.063.

Ruidong, W., Yu, S., Juanhong, L., Linian, C., Guangtian, Z., & Yueyue, Z. (2021). Effect of iron tailings and slag powders on workability and mechanical properties of concrete. *Frontiers in Materials*, *8*. Available from https://doi.org/10.3389/fmats.2021.723119, http://journal.frontiersin.org/journal/materials.

Sameer, H., Weber, V., Mostert, C., Bringezu, S., Fehling, E., & Wetzel, A. (2019). Environmental assessment of ultra-high-performance concrete using carbon, material, and water footprint. *Materials*, *12*(6), 851. Available from https://doi.org/10.3390/ma12060851.

Scrivener, K., Martirena, F., Bishnoi, S., & Maity, S. (2018). Calcined clay limestone cements (LC3). *Cement and Concrete Research*, *114*, 49–56. Available from https://doi.org/10.1016/j.cemconres.2017.08.017, http://www.sciencedirect.com/science/journal/00088846.

Shafieifar, M., Farzad, M., & Azizinamini, A. (2017). Experimental and numerical study on mechanical properties of ultra high performance concrete (UHPC). *Construction and Building Materials*, *156*, 402–411. Available from https://doi.org/10.1016/j.conbuildmat.2017.08.170.

Shao, Y., Nguyen, W., Bandelt, M. J., Ostertag, C. P., & Billington, S. L. (2022). Seismic performance of high-performance fiber-reinforced cement-based composite structural members: A review. *Journal of Structural Engineering*, *148*(10). Available from https://doi.org/10.1061/(asce)st.1943-541x.0003428.

Shi, Y., Long, G., Zeng, X., Xie, Y., & Wang, H. (2021). Green ultra-high performance concrete with very low cement content. *Construction and Building Materials*, *303*, 124482. Available from https://doi.org/10.1016/j.conbuildmat.2021.124482.

Shi, C., Wang, D., Wu, L., & Wu, Z. (2015). The hydration and microstructure of ultra high-strength concrete with cement-silica fume-slag binder. *Cement and Concrete Composites*, *61*, 44–52. Available from https://doi.org/10.1016/j.cemconcomp.2015.04.013, http://www.sciencedirect.com/science/journal/09589465.

Shi, C. (2004). Steel slag – Its production, processing, characteristics, and cementitious properties. *Journal of Materials in Civil Engineering*, *16*(3), 230–236. Available from https://doi.org/10.1061/(ASCE)0899-1561(2004)16:3(230).

Smarzewski, P., & Barnat-Hunek, D. (2018). Property assessment of hybrid fiber-reinforced ultra-high-performance concrete. *International Journal of Civil Engineering*, *16*, 593–606.

Soliman, N. A., & Tagnit-Hamou, A. (2016). Development of ultra-high-performance concrete using glass powder – Towards ecofriendly concrete. *Construction and Building Materials, 125*, 600–612. Available from https://doi.org/10.1016/j.conbuildmat.2016.08.073.

Soliman, N. A., & Tagnit-Hamou, A. (2018). Using particle packing and statistical approach to optimize eco-efficient ultra-high-performance concrete. *ACI Materials Journal, 115* (5), 795–797.

Stengel, T., & Schießl. (2008). Sustainable construction with UHPC-from life cycle inventory data collection to environmental impact assessment. *Proceedings of the 2nd international symposium on ultra high performance concrete* (pp. 461–468).

Stovall, T., de Larrard, F., & Buil, M. (1986). Linear packing density model of grain mixtures. *Powder Technology, 48*(1), 1–12. Available from https://doi.org/10.1016/0032-5910(86)80058-4.

Sun, T., Wang, X., Maimaitituersun, N., Dong, S., Li, L., & Han, B. (2024). Synergistic effects of steel fibers and steel wires on uniaxial tensile mechanical and self-sensing properties of UHPC. *Construction and Building Materials, 416*, 134991. Available from https://doi.org/10.1016/j.conbuildmat.2024.134991.

Tahwia, A. M., Essam, A., Tayeh, B. A., & Elrahman. (2022). Enhancing sustainability of ultra-high performance concrete utilizing high-volume waste glass powder. *Case Studies in Construction Materials, 17*, e01807.

Tahwia, A. M., Hamido, M. A., & Elemam, W. E. (2023). Using mixture design method for developing and optimizing eco-friendly ultra-high performance concrete characteristics. *Case Studies in Construction Materials, 18*. Available from https://doi.org/10.1016/j.cscm.2022.e01807, http://www.journals.elsevier.com/case-studies-in-construction-materials/.

Topçu, I. B., & Canbaz, M. (2004). Properties of concrete containing waste glass. *Cement and Concrete Research, 34*(2), 267–274. Available from https://doi.org/10.1016/j.cemconres.2003.07.003.

Turner, L. K., & Collins, F. G. (2013). Carbon dioxide equivalent (CO_2-e) emissions: A comparison between geopolymer and OPC cement concrete. *Construction and Building Materials, 43*, 125–130. Available from https://doi.org/10.1016/j.conbuildmat.2013.01.023.

Venkatesh, K. R., Rani, R., Thamilselvi, M., & Rajahariharasudhan, R. (2018). Experimental study on partial replacement of cement with sugaecane bagasse ash in concrete. *International Journal of Advance Engineering and Research Development* (04), 778–785.

Wang, X., Ding, S., Ashour, A., Ye, H., Thakur, V. K., Zhang, L., & Han, B. (2024). Back to basics: Nanomodulating calcium silicate hydrate gels to mitigate CO_2 footprint of concrete industry. *Journal of Cleaner Production, 434*, 139921. Available from https://doi.org/10.1016/j.jclepro.2023.139921.

Wang, X., Ding, S., Qiu, L., Ashour, A., Wang, Y., Han, B., & Ou, J. (2022a). Improving bond of fiber-reinforced polymer bars with concrete through incorporating nanomaterials. *Composites Part B: Engineering, 239*, 109960. Available from https://doi.org/10.1016/j.compositesb.2022.109960.

Wang, J., Dong, S., Dai Pang, S., Yu, X., Han, B., & Ou, J. (2022b). Tailoring anti-impact properties of ultra-high performance concrete by incorporating functionalized carbon nanotubes. *Engineering, 8*, 232–245. Available from https://doi.org/10.1016/j.eng.2021.04.030.

Wang, X., Dong, S., Li, Z., Han, B., & Ou, J. (2022c). Nanomechanical characteristics of interfacial transition zone in nano-engineered concrete. *Engineering, 17*, 99–109. Available from https://doi.org/10.1016/j.eng.2020.08.025.

Wang, X., Dong, S., Ashour, A., Ding, S., & Han, B. (2021). Bond behaviors between nano-engineered concrete and steel bars. *Construction and Building Materials*, *299*, 124261. Available from https://doi.org/10.1016/j.conbuildmat.2021.124261.

Wang, B., Yan, L., Fu, Q., & Kasal, B. (2021). A comprehensive review on recycled aggregate and recycled aggregate concrete. *Resources, Conservation and Recycling*, *171*, 105565. Available from https://doi.org/10.1016/j.resconrec.2021.105565.

Wang, X., Dong, S., Ashour, A., Zhang, W., & Han, B. (2020). Effect and mechanisms of nanomaterials on interface between aggregates and cement mortars. *Construction and Building Materials*, *240*, 117942. Available from https://doi.org/10.1016/j.conbuildmat.2019.117942.

Wang, X., Zheng, Q., Dong, S., Ashour, A., & Han, B. (2020). Interfacial characteristics of nano-engineered concrete composites. *Construction and Building Materials*, *259*, 119803. Available from https://doi.org/10.1016/j.conbuildmat.2020.119803.

Wang, D., Dong, S., Wang, X., Maimaitituersun, N., Shao, S., Yang, W., & Han, B. (2022d). Sensing performances of hybrid steel wires and fibers reinforced ultra-high performance concrete for in-situ monitoring of infrastructures. *Journal of Building Engineering*, *58*, 105022. Available from https://doi.org/10.1016/j.jobe.2022.105022.

Wang, J., & Huang, Y. (2023). Mechanical properties and hydration of ultra-high-performance seawater sea-sand concrete (UHPSSC) with limestone calcined clay cement (LC3). *Construction and Building Materials*, *376*, 130950. Available from https://doi.org/10.1016/j.conbuildmat.2023.130950.

Wang, Q., Yan, P., & Mi, G. (2012). Effect of blended steel slag-GBFS mineral admixture on hydration and strength of cement. *Construction and Building Materials*, *35*, 8–14. Available from https://doi.org/10.1016/j.conbuildmat.2012.02.085.

Wang, X., Yu, R., Shui, Z., Song, Q., Liu, Z., bao, M., Liu, Z., & Wu, S. (2019). Optimized treatment of recycled construction and demolition waste in developing sustainable ultra-high performance concrete. *Journal of Cleaner Production*, *221*, 805–816. Available from https://doi.org/10.1016/j.jclepro.2019.02.201, https://www.journals.elsevier.com/journal-of-cleaner-production.

Wang, X., Yu, R., Shui, Z., Song, Q., & Zhang, Z. (2017). Mix design and characteristics evaluation of an eco-friendly ultra-high performance concrete incorporating recycled coral based materials. *Journal of Cleaner Production*, *165*, 70–80. Available from https://doi.org/10.1016/j.jclepro.2017.07.096.

Wei, H., Liu, T., Zou, D., & Zhou, A. (2021). Preparation and properties of green ultra-high performance concrete containing waste glass. *Journal of Building Materials*, *24*(3), 492–498. Available from https://doi.org/10.3969/j.issn.1007-9629.2021.03.007, http://jzclxb.allmaga.net/ch/index.aspx.

Wu, N., Ji, T., Huang, P., Fu, T., Zheng, X., & Xu, Q. (2022). Use of sugar cane bagasse ash in ultra-high performance concrete (UHPC) as cement replacement. *Construction and Building Materials*, *317*, 125881. Available from https://doi.org/10.1016/j.conbuildmat.2021.125881.

Wu, Z., Shi, C., & He, W. (2017). Comparative study on flexural properties of ultra-high performance concrete with supplementary cementitious materials under different curing regimes. *Construction and Building Materials*, *136*, 307–313. Available from https://doi.org/10.1016/j.conbuildmat.2017.01.052.

Xiong, M. X., & Liew, J. Y. R. (2020). Buckling behavior of circular steel tubes infilled with C170/185 ultra-high-strength concrete under fire. *Engineering Structures*, *212*. Available from https://doi.org/10.1016/j.engstruct.2020.110523, http://www.journals.elsevier.com/engineering-structures/.

Xuan, M. Y., Bae, S. C., Kwon, S. J., & Wang, X. Y. (2022). Sustainability enhancement of calcined clay and limestone powder hybrid ultra-high-performance concrete using belite-rich Portland cement. *Construction and Building Materials*, *351*. Available from https://doi.org/10.1016/j.conbuildmat.2022.128932, https://www.journals.elsevier.com/construction-and-building-materials.

Xu, J., Zhan, P., Zhou, W., Zuo, J., Shah, S. P., & He, Z. (2023). Design and assessment of eco-friendly ultra-high performance concrete with steel slag powder and recycled glass powder. *Powder Technology*, *419*. Available from https://doi.org/10.1016/j.powtec.2023.118356, http://www.elsevier.com/locate/powtec.

Yang, K. H., Song, J. K., & Song, K. I. (2013). Assessment of CO_2 reduction of alkali-activated concrete. *Journal of Cleaner Production*, *39*, 265–272. Available from https://doi.org/10.1016/j.jclepro.2012.08.001.

Yin, T., Liu, K., Fan, D., & Yu, R. (2023). Derivation and verification of multilevel particle packing model for ultra-high performance concrete (UHPC): Modelling and experiments. *Cement and Concrete Composites*, *136*, 104889. Available from https://doi.org/10.1016/j.cemconcomp.2022.104889.

Yoo, D. Y., Oh, T., & Banthia, N. (2022). Nano-materials in ultra-high-performance concrete (UHPC) – A review. *Cement and Concrete Composites*, 104730.

Yu, R., Spiesz, P., & Brouwers, H. J. H. (2014). Effect of nano-silica on the hydration and microstructure development of Ultra-High Performance Concrete (UHPC) with a low binder amount. *Construction and Building Materials*, *65*, 140–150. Available from https://doi.org/10.1016/j.conbuildmat.2014.04.063.

Zhang, W., Gu, X., Qiu, J., Liu, J., Zhao, Y., & Li, X. (2020). Effects of iron ore tailings on the compressive strength and permeability of ultra-high performance concrete. *Construction and Building Materials*, *260*, 119917. Available from https://doi.org/10.1016/j.conbuildmat.2020.119917.

Zhang, L., Li, L., Wang, Y., Yu, X., & Han, B. (2020). Multifunctional cement-based materials modified with electrostatic self-assembled CNT/TiO_2 composite filler. *Construction and Building Materials*, *238*, 117787. Available from https://doi.org/10.1016/j.conbuildmat.2019.117787.

Zhang, K., Han, B., & Yu, X. (2012). Electrically conductive carbon nanofiber/paraffin wax composites for electric thermal storage. *Energy Conversion and Management*, *64*, 62–67. Available from https://doi.org/10.1016/j.enconman.2012.06.021.

Zhang, S. S., Wang, J. J., Lin, G., Yu, T., & Fernando, D. (2023). Stress-strain models for ultra-high performance concrete (UHPC) and ultra-high performance fiber-reinforced concrete (UHPFRC) under triaxial compression. *Construction and Building Materials*, *370*, 130658. Available from https://doi.org/10.1016/j.conbuildmat.2023.130658.

Zhang, W., Zheng, Q., Wang, D., Yu, X., & Han, B. (2019). Electromagnetic properties and mechanisms of multiwalled carbon nanotubes modified cementitious composites. *Construction and Building Materials*, *208*, 427–443. Available from https://doi.org/10.1016/j.conbuildmat.2019.03.029.

Zhao, Y., Gu, X., Qiu, J., Zhang, W., & Li, X. (2021). Study on the utilization of iron tailings in ultra-high-performance concrete: Fresh properties and compressive behaviors. *Materials*, *14*(17), 4807. Available from https://doi.org/10.3390/ma14174807.

Zhong, H., Chen, M., & Zhang, M. (2023). Effect of hybrid industrial and recycled steel fibres on static and dynamic mechanical properties of ultra-high performance concrete. *Construction and Building Materials*, *370*, 130691. Available from https://doi.org/10.1016/j.conbuildmat.2023.130691.

Sustainable nano concrete materials and structures

17

Dong Lu[1,2] *and Jing Zhong*[1,2]
[1]School of Civil Engineering, Harbin Institute of Technology, Harbin, P.R. China, [2]Key Lab of Structures Dynamic Behavior and Control of the Ministry of Education (Harbin Institute of Technology), Harbin, P.R. China

17.1 Introduction

Cement-based composite, including cement paste, cement mortar, and concrete, is one of the most visible manifestations of mankind's physical footprint on the planet (Jiang et al., 2023; Monteiro et al., 2017; Rong, Li, Huang, et al., 2023; Rong, Li, Zheng, et al., 2023; Schneider et al., 2011). To date, it has become the largest man-made engineering material (e.g., accounting for approximately 40% of artificial objects, an annual amount of ~1.4 billion cubic meters, equivalent to more than 30 billion tons) due to the characteristics of low cost, high compressive strength, outstanding durability, and long service life (Rong, Li, Huang, et al., 2023; Rong, Li, Zheng, et al., 2023; Schneider et al., 2011; Shi et al., 2011). However, the inherently quasibrittle behavior of cement composites has limited their structural design and applications (Huo, Liu, et al., 2023; Huo, Lu, et al., 2023; Lu et al., 2022a). Additionally, the cement and concrete industry has high energy consumption and a large environmental footprint, placing a large burden on the environment (John et al., 2019; Liu et al., 2022; Shi et al., 2011; Tian et al., 2020).

To meet the development direction of modern concrete infrastructures in terms of green and sustainability, new-generation cement-based composites not only have excellent mechanical and durability properties but also have the characteristics of low-carbon and environmentally friendly (John et al., 2019; Ouyang et al., 2020; Schneider et al., 2011). In recent years, a lot of attempts have been adopted to develop low-carbon cement composites, for example, applying supplementary cementitious materials (SCMs) to partially replace cement to prepare sustainable cement composites (Althoey & Farnam, 2019; Gan et al., 2022; Huang et al., 2021; Lothenbach et al., 2011; Lu, Tang, et al., 2020; Lu, Wang, et al., 2021; Lu, Zhong, et al., 2021), developing new types of cementitious binders to replace cement for concrete production (Li et al., 2020; Xu, Zhong, et al., 2018), and applying nanomaterials to modify cement composite to enhance its performance and thus reducing cement consumption (Lu et al., 2022b, 2022c, 2022d; Lu, Wang, et al., 2023). Among them, nanomaterials demonstrate great potential to significantly improve the microstructure of cement-based composites, which was first proposed in the

Sustainable Concrete Materials and Structures. DOI: https://doi.org/10.1016/B978-0-443-15672-4.00017-6

2010s (Chintalapudi & Pannem, 2019; Ding et al., 2023; Dong et al., 2023), as shown in Fig. 17.1.

Typically, nanomaterial is a material that consists of a group of substances, at least one of which has dimensions smaller than ~100 nm, that exhibit unique physical and chemical properties (Brittain et al., 2023; Budde et al., 2016; Chen et al., 2023). The development of nanotechnology and nanomaterials over the past century has provided valuable opportunities to enhance the microstructure of cement composites at the nanoscale (Hou et al., 2020; Singh et al., 2017). In general, the addition of very small amounts of nanomaterials can significantly improve the properties of cement composites (Chintalapudi & Pannem, 2020b; Gao et al., 2019; Horszczaruk et al., 2015; Jing, Ye, et al., 2020; Xu et al., 2019). In particular, carbon-based nanomaterials, including carbon nanotubes (CNT) (Azeem & Azhar Saleem, 2020; Du et al., 2020; Ghaharpour et al., 2016; Wu et al., 2019), carbon nanofibers (CNF) (He & Yang, 2021; Meng & Khayat, 2018; Shi et al., 2019), graphene (Baomin & Shuang, 2019; Chougan et al., 2019; Lin & Du, 2020), and graphene oxide (GO) (Imanian Ghazanlou et al., 2020; Jing, Ye, et al., 2020; Long et al., 2018), have numerous potential to enhance the performance of cement composites. What is more, carbon-based nanomaterials are abundant in nature and have been industrially produced on a large scale (Dharmasiri et al., 2022; Ghahremani et al., 2022; Guo, Zhang, et al., 2022; Lu & Zhong, 2022; Lu, Leng, et al., 2023). According to published studies (Liu et al., 2023; Long et al., 2019; Lu, Hanif, et al., 2018; Lu, Hou, et al., 2018), adding 0.01 wt.% GO can enhance the 28-d compressive strength of cement-based materials by approximately 20%–30%. As a result, the reduction in cement consumption in the concrete industry could trigger substantial environmental and economic benefits (Ding et al., 2023; Liu et al., 2023).

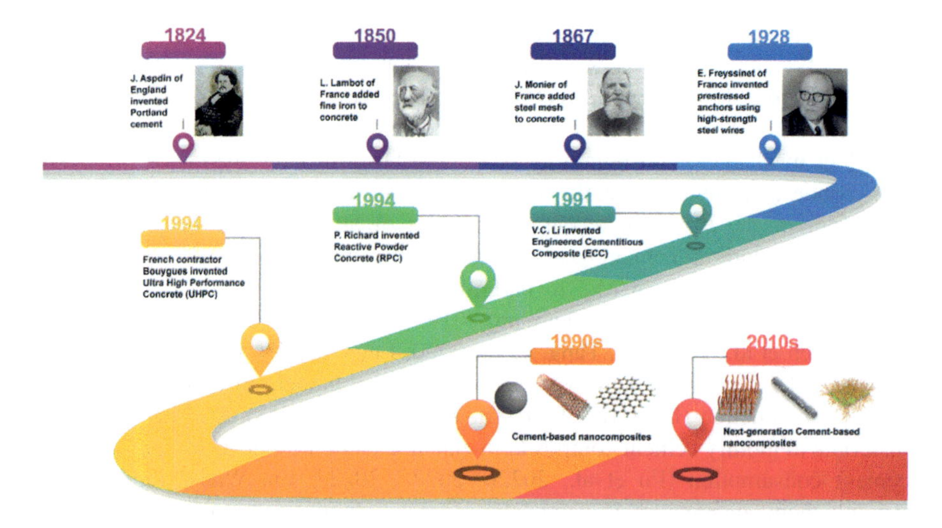

Figure 17.1 The development of cement composites: from conventional cement composites to multifunctional and smart cement composites.

The application of nanomaterials contributes to a deeper understanding of the behavior of cement composites, enables the manipulation of structures at the nanoscale, and reduces the production and ecological costs of cementitious composites, thus prolonging the service life of concrete infrastructures, which is of great and far-reaching significance in guiding the development and application of concrete structures (Han et al., 2016; Qiu et al., 2020; Sun et al., 2013). As shown in Fig. 17.2, the general design principle for cement composites is to customize the composition, fabrication, and structure, and concrete structures based on the inherent properties of the cement materials and the needs of the specific application, with due consideration to effectiveness, economy, and sustainability.

This chapter will give an overall introduction to the main properties of carbon-based nanomaterials used in cement composite modification, the different dispersion methods used to prepare cement composites, the hydration and rheological properties of fresh cement composites, the influence of admixed carbon-based nanomaterials in cement hydration system, the effects of carbon-based nanomaterials on the mechanical and durability properties of hardened cement composites, and the future development and challenges of carbon-based nanomaterials engineered cement composites and structures are discussed.

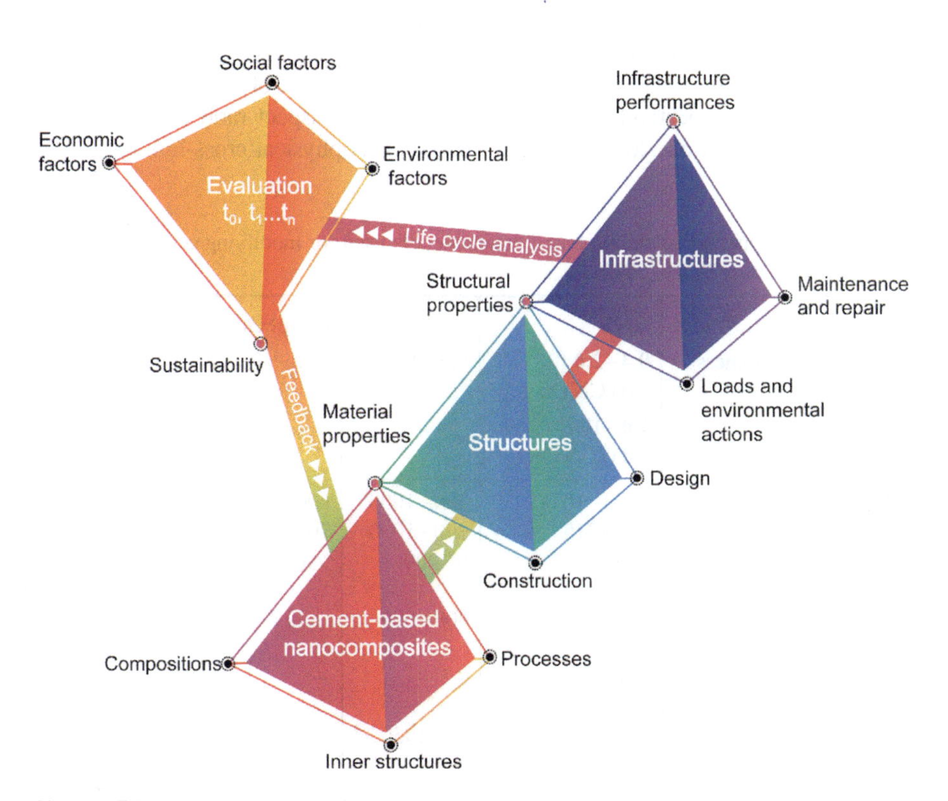

Figure 17.2 Design principles of cement composites.

17.2 Fundamentals of carbon-based nanomaterials

Nanomaterials are low-dimensional materials consisting of building blocks of submicron or nanoscale dimensions (i.e., 1–100 nm) with size effects in at least one direction (Han et al., 2023; Jeong et al., 2022; Lee et al., 2023). In bulk materials, physical properties are independent of size, whereas, in nanomaterials, different physical properties may depend on the size and shape of the nanomaterial (Guo, Wang, et al., 2022; Han et al., 2023; Ruoff et al., 2020). Nanomaterials used in modifying cement composites are typically categorized into four kinds: (1) inorganic-based nanomaterials; (2) carbon-based nanomaterials; (3) organic-based nanomaterials; and (4) composite-based nanomaterials. In this chapter, we only discussed one-dimensional (1D) nanofibers (i.e., CNT and CNF) and two-dimensional (2D) nanosheets (i.e., graphene and GO), as presented in Table 17.1, because they are the most investigated carbon-based nanomaterials in improving the performance of cement composites (Lin & Du, 2020; Shi et al., 2019; Zhao, Guo, et al., 2020).

Carbon-based nanomaterials engineered cement composites are designed using a predominantly bottom-up approach, so it is important to first understand what carbon-based nanomaterials are and how they can be utilized to develop sustainable cement composites (Fig. 17.3). In particular, CNT has a 1D concentric tubular structure with a hexagonal arrangement of carbon atoms and it was first reported by Iijima (1991), and its quality, property, and production have been continuously improving since then. Generally, the diameter and length of CNT are 1–100 nm and 1–100 μm, respectively, resulting in a very high aspect ratio and thus severe entanglement (Li et al., 2005; Liu et al., 2003). This physical cross-linking, as well

Table 17.1 Properties of carbon-based nanomaterials used for modifying cement composites (Lu & Zhong, 2022).

Component	**CNT**	**CNF**	**Graphene**	**GO**
Diameter/Thickness (nm)	**0.4–2.0 (SWCNT)a** **1.0–100 (MWCNT)a**	**0.5–100**	**~1**	**~0.67**
Aspect ratio	1000–10,000	100–1000	600–600,000	1500–45,000
SSAa (m^2/g)	20–1315	100–1000	700–1500	2000–2600
Elastic modulus (GPa)	>1000	6–200	>1100	>300
Tensile strength (GPa)	50–200	400–600	~125	>112
Electrical resistivity ($\Omega \cdot cm$)	$\sim 10^{-2}$	0.1	–	–
Thermal conductivity W/(m · K)	~6600	20–1950	–	–

[a]SWCNT and MWCNT are single-wall CNT and multiwalled CNT, respectively. SSA is a specific surface area.

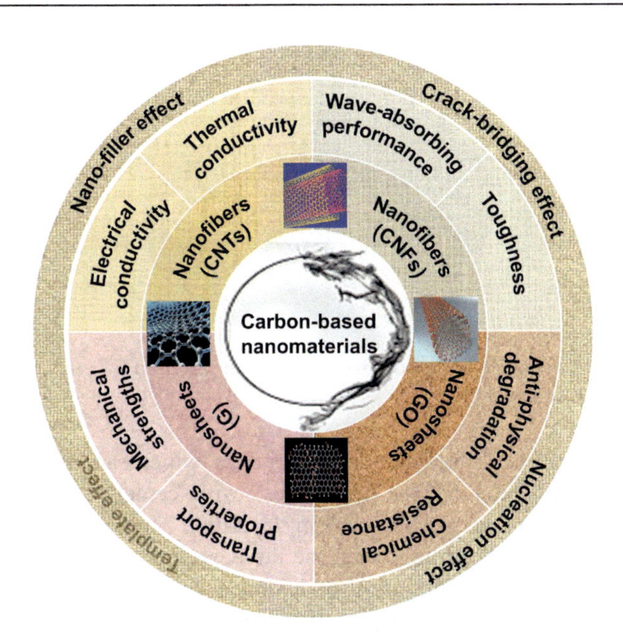

Figure 17.3 Carbon-based nanomaterials used for modification of cement composites.

as van der Waals interactions between the CNT, makes its dispersion in aqueous solutions and cement matrices extremely challenging. CNF is a cylindrical nanostructure, the graphene layer arranged in it, and it can be regarded as a fiber-shaped nanomaterial with a stacked tapered shape, its length is typically 50–200 nm (De Jong & Geus, 2007; Wang & Aslani, 2019). CNF has also been widely used as an additive for cement modification because it has geometric and physical characteristics similar to CNT, but the price is much lower (usually one-fifth of CNT) (Shi et al., 2019; Sun et al., 2013). In addition, graphene, CNT, and CNF all show great potential for the future development of smart cement composites, for instance, carbon-based nanomaterials with excellent electrical conductivity are directly mixed with cement and then bound it with aggregate to develop conductive cement composites (Gwon et al., 2023; Wang & Aslani, 2019); additionally, the carbon-based nanomaterials can be coated onto the surface of cement or aggregate and then mixed with other components to develop conductive cement composites (Lu et al., 2022; Lu, Huo, et al., 2023; Lu, Ma, et al., 2023; Lu, Wang, et al., 2022), which can be used for structural health monitoring, electromagnetic shielding, and snow-melting and deicing in the pavement, but this is beyond the scope of this chapter.

Graphene is a one-atom-thick fat nanosheet of SP^2 bonded carbon atoms arranged in a cellular lattice (Marchesini et al., 2020; Mypati et al., 2021). It is the parent of all graphite materials (Pei & Cheng, 2012; Shang et al., 2019). According to the previous studies (Pei & Cheng, 2012; Shang et al., 2019; Yan et al., 2020), the bending stiffness of GO is similar to that of graphene (assuming similar thickness), which is about 2 kT, meaning that these single-layer 2D materials bend easily

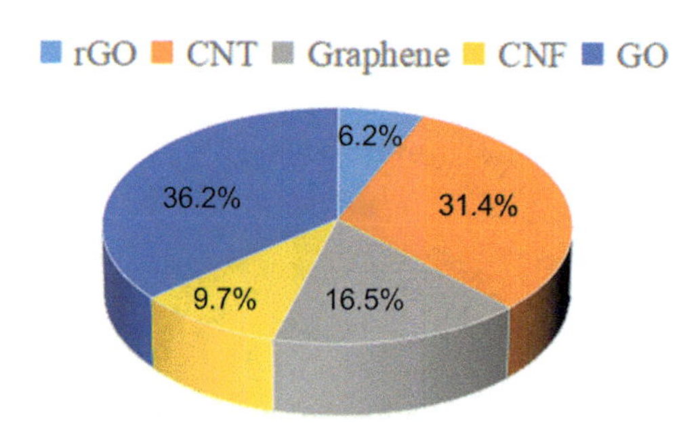

Figure 17.4 The percentage of carbon-based nanomaterials used for preparing cement composites. rGO is reduced GO.

and will naturally form a large number of wrinkles, which can influence the interaction between graphene (and its derivatives) with cement matrix. It is widely believed that GO can accelerate cement hydration since its surface is rich in oxygen-containing energy groups, increase the degree of polymerization of cement hydrates, and regulate the formation of hydrated crystals (Xu et al., 2019; Zhao et al., 2018). With these advantages, GO has become one of the most studied carbon-based nanomaterials for modifying cement materials in the past decade, as indicated in Fig. 17.4.

17.3 Dispersion and characterization of carbon-based nanomaterials in water and cement matrix

17.3.1 Challenges in the dispersion of carbon-based nanomaterials

As mentioned above, the possible agglomeration of carbon-based nanomaterials in the cement matrix is defective and can negatively affect the performance of cement composites, that is, the agglomeration of nanomaterials in cement composite can affect the workability of fresh mixtures (increasing viscosity and decreasing slump), decreasing mechanical properties, and increasing transport properties (Jung et al., 2020; Lin & Du, 2020; Xu, Zeng, et al., 2018; Zhao, Guo, et al., 2020). Namely, a high dispersion quality of admixing carbon-based nanomaterials in cement composites is a prerequisite for their effective enhancement (Dong et al., 2021; Du et al., 2020). However, it is still very challenging to realize an acceptable dispersion of admixing carbon-based nanomaterials in cement composites, considering the strong Van der Waal's force and the complex electrolytic environment provided by cement pore solution (i.e., divalent ions).

Table 17.2 collects the dispersion methods of carbon-based nanomaterials in water and cement matrix. In Table 17.2, acid treatment and surfactant wrapping have been widely adopted to overcome attractions between carbon-based nanomaterials (Lu et al., 2022d; Wang et al., 2020). Specifically, acids can impart a charge

Table 17.2 Dispersion methods of carbon-based nanomaterials in water and cement matrix.

Dispersion method		System	Description	References
Ultrasonic dispersion		Water	Ultrasonic energy can cavitate and exfoliate the carbon-based nanomaterials	
Acid treatment		Water	The strong steric hindrance effects can separate carbon-based nanomaterials from charged ions	Lu et al. (2022d), Wang et al. (2020)
Surface modification		Water/cement pore solution		Zhao, Guo, et al. (2020), Zhao et al. (2018)
Mechanical dispersion	Mechanical agitation	Cement matrix	Dispersion of carbon-based nanomaterials using mechanical shearing force	Jing, Xu, et al. (2020)
	Ball milling			Jing et al. (2019, 2020)
Integrated dispersion method		Cement matrix	The combined application of the aforementioned methods	Baomin and Shuang (2019), Chintalapudi and Pannem (2020a)
Silica fume		Cement matrix	Using silica fume to prevent agglomeration of carbon-based nanomaterials in cement composites	Lu et al. (2022b), Lu, Sheng, et al. (2023)

to the surface of carbon-based nanomaterials at the cost of potentially destroying their structure (Lu et al., 2022d; Wang et al., 2020), whereas surfactant encapsulation can make carbon-based nanomaterials more compatible with water, either by a mechanism of charge repulsion or steric repulsion, depending on the molecular structure of the surfactants (Zhao et al., 2018; Zhao, Zhu, et al., 2020). Since carbon-based nanomaterials are usually dispersed in water prior to mixing with cement, this chapter summarizes the dispersion of carbon-based nanomaterials in water, cement pore solution, and cement matrix, respectively, to elucidate the main mechanisms and related colloidal behaviors in each step.

17.3.2 Dispersion of carbon-based nanomaterials in solution

As summarized in Table 17.2, the typical dispersion methods of carbon-based nanomaterials in solution mainly include mechanical shearing, ultrasonication, and surface physiochemical modification (Konsta-Gdoutos et al., 2010; Silvestro & Jean Paul Gleize, 2020). Among them, ultrasonication is a technique that relies on cavitation with high local energy, and the effectiveness of this approach can be optimized by properly selecting sonication power and time (Lu, Leng, et al., 2023; Rahman et al., 2022; Wang & Aslani, 2019). However, the crystal structure of carbon-based nanomaterials may be destroyed due to the high cavitation energy. To further improve the dispersion quality of mixed carbon-based nanomaterials, hydrophilic functional groups can be grafted onto the material (Lu et al., 2022d). For instance, GO can be regarded as graphene chemically grafted with a large number of oxygen-containing functional groups (Li et al., 2019; Lin et al., 2020). Surface active chemicals, such as superplasticizers, shrinkage-reducing admixtures, and air-entraining surfactants, are prepared in the form of an aqueous solution before mixing with cement (Li et al., 2017; Lv et al., 2014; Zhao, Guo, Liu, et al., 2017; Zhao, Zhu, et al., 2020). As such, preparing uniform carbon-based nanomaterial solutions can facilitate the practical applications of carbon-based nanomaterials engineered cement composites (Li et al., 2020; Liu et al., 2020). Polycarboxylate superplasticizer (SP) is the most used surfactant and has been extensively used to disperse carbon-based nanomaterials (Zhao, Guo, et al., 2020; Zhao, Zhu, et al., 2020). As suggested in Fig. 17.5, applying SP in GO-modified cement composites can simultaneously disperse GO and cement. Similarly, with the help of SP, it can achieve good dispersion of the graphene in the solution (Fig. 17.5). However, the absorbed surfactant itself may be stripped off when the ionic strength in cement composites is high enough (Zhao et al., 2018).

More than two decades ago, the environmental science community began to systematically study the stability of surfactants encapsulating carbon-based nanomaterials in various electrolytes (Singh et al., 2017). Since then, the agglomeration of particles has been one of the key research topics. However, the exact chemical structures of most kinds of SP are generally kept secret, which makes the optimization of the dispersion of carbon-based nanomaterials ambiguous. Since the purpose of synthesizing SP is to modify the properties of cement through the interaction of SP with cement grains, a new surfactant should be designed and prepared

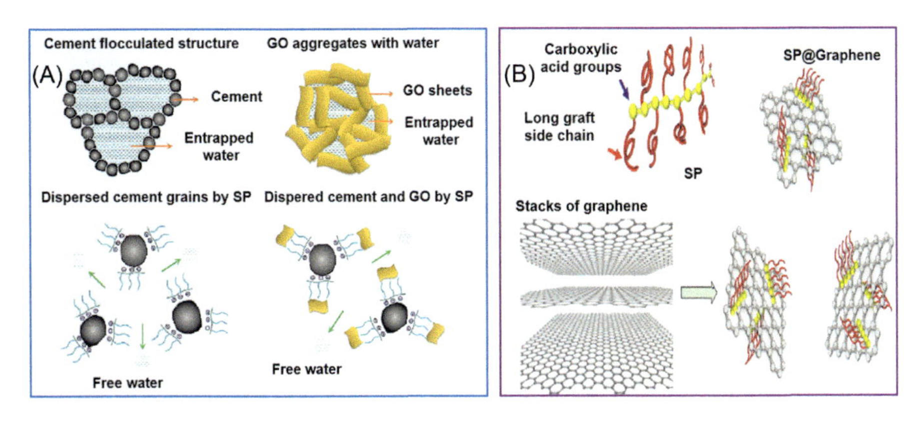

Figure 17.5 Illustration of (A) SP dispersing GO and cement and (B) the dispersion of graphene by SP.

specifically for the dispersion of carbon-based nanomaterials. In addition, the stability of surfactants on the dispersion of carbon-based nanomaterials remains to be studied and clarified.

17.3.3 Dispersion of carbon-based nanomaterials in the cement matrix

Carbon-based nanomaterials can also be used as an admixture and then mixed with dry powders, and then mixed with water and other raw materials to prepare cement composites, similar to SCMs (Ghosh et al., 2019; Jing, Xu, et al., 2020; Jing, Ye, et al., 2020). As reported by Mohammed et al. (2015), 0.01 wt.% GO was mixed with cement for 5 minutes to achieve a uniform mixture, while they failed to achieve acceptable dispersion as the GO dosage increased to 0.06 wt.%. However, it is virtually impossible to unwind carbon-based nanomaterial agglomerates with ordinary shear mixing equipment alone. Recently, silica fume has been used to disperse carbon-based nanomaterials in cement matrices through a premixing process (Lu et al., 2022b; Lu, Sheng, et al., 2023), as presented in Fig. 17.6. Due to the high specific surface area (SSA) of the carbon-based nanomaterials, the effectiveness of using silica fume as a carrier for dispersing carbon-based nanomaterials depends largely on the available surface area of the carrier.

17.3.4 Evaluation of dispersion quality of carbon-based nanomaterials

Table 17.3 shows the techniques adopted to evaluate the dispersion quality of carbon-based nanomaterials in water or cement matrix. The SEM equipped with an energy dispersive spectrometer (EDS) has been widely used to characterize the dispersion of carbon-based nanomaterials in the solution and cement matrix (Lu, Hou, et al., 2018). However, SEM-EDS cannot achieve quantitative characterization of

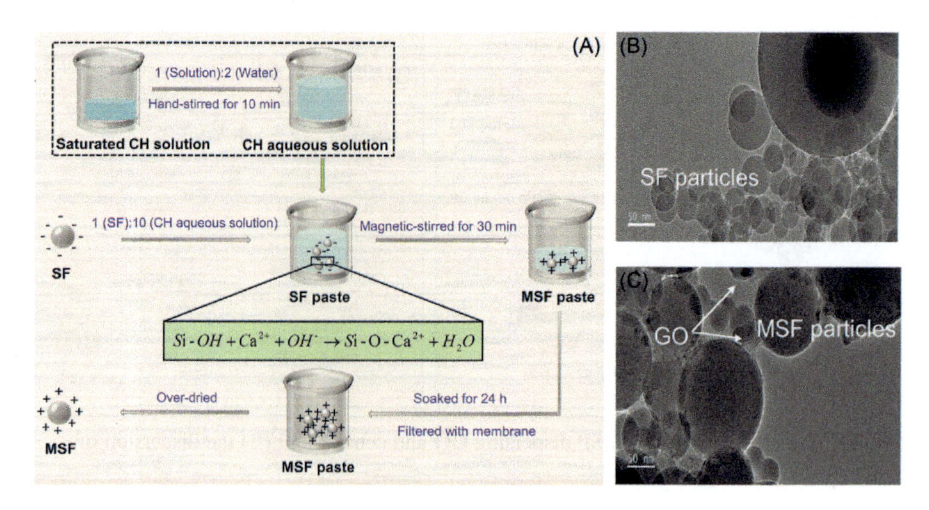

Figure 17.6 (A) Illustration of GO coated on the silica fume (SF) surface; and SEM images of (B) SF particles and (C) GO-coated SF. MSF is modified SF.

carbon-based nanomaterial dispersion in the cement matrix. Again, some cement hydrates, such as ettringite (AFt), monosulfate (AFm), and calcium hydroxide (CH), all have their shape, making it very difficult to distinguish them from the admixing carbon-based nanomaterials. Overall, the current evaluation of the dispersibility of carbon-based nanomaterials is mainly for aqueous solutions, and there are few effective means of evaluating the dispersion quality of carbon-based nanomaterials in cement matrices. As such, more research should be needed in the future to identify and evaluate the dispersion quality of carbon-based nanomaterials in cement matrix.

17.4 Effect of carbon-based nanomaterials on the performance of cement composites

17.4.1 Fresh properties of cement composites

The fresh properties of cement composites including hydration kinetics and rheology properties, both of which can influence the following mechanical and durability properties of hardened cement composites.

Although there are a large number of studies committed to revealing the nuclear effect of CNT, previous reports of CNTs accelerate cement hydration are somewhat contradictory (Azeem & Azhar Saleem, 2020; Ghaharpour et al., 2016; Makar & Chan, 2009; Silvestro & Jean Paul Gleize, 2020), as indicated in Fig. 17.7. For instance, Sobolkina et al. (2016) indicated that adding 0.25 wt.% HNO_3-oxidized CNT hardly affects C_3S hydration, mainly due to the limited adsorption sites present on CNT surface (Fig. 17.7). Tafesse and Kim (2019) suggested that the admixing pure CNT cannot accelerate or delay cement hydration, according to the

Table 17.3 Evaluation of dispersion quality of carbon-based nanomaterials in water or cement matrix.

Methods	System	Description	References
UV-vis spectroscopy	Suspension	Using Beer–Lambert law to calculate the concentration of carbon-based nanomaterials	Lu, Hou, et al. (2018)
Laser particle size		Smaller particle size suggests higher dispersion quality	Li et al. (2020), Lu, Hou, et al. (2018)
Dynamic light scattering			Du et al. (2020), Muthu et al. (2021)
Zeta potential		A higher zeta potential value (absolute value) indicates higher dispersion quality	Du and Pang (2018), Zhao et al. (2018)
Optical microscope	Suspension/ cement matrix	Evaluation of dispersion quality of carbon-based nanomaterials according to observed size	Zhao et al. (2018)
SEM			Lv et al. (2013, 2014), Zhao, Guo, Liu, et al. (2017)
Transmission electron microscope			Zhao et al. (2018)
Raman spectrum			Lu et al. (2019), Lu, Hanif, et al. (2018), Lu, Hou, et al. (2018)

findings of nonevaporated water in cement composites (Fig. 17.7), they only provide a nanofiller effect. We believe that this is mainly because the surface of the raw CNT is relatively smooth, and it presents inertia (without oxygen-containing group and chemical affinity), resulting in almost no chemical interactions between CNT and cement grains. Interestingly, previous studies have proven that admixing GO can accelerate cement hydration (G. Xu et al., 2019). Indeed, a recent study indicated that the admixing 0.025 wt.% GO can lead to an earlier heat flow peak (Ghazizadeh et al., 2018), as presented in Fig. 17.7. Unfortunately, there is still a lack of direct evidence for the seeding effect of carbon-based nanomaterials, which is described in more detail below.

In general, the addition of carbon-based nanomaterials to cement composite could decrease the workability of the mixture, this is mainly due to the ultra-high

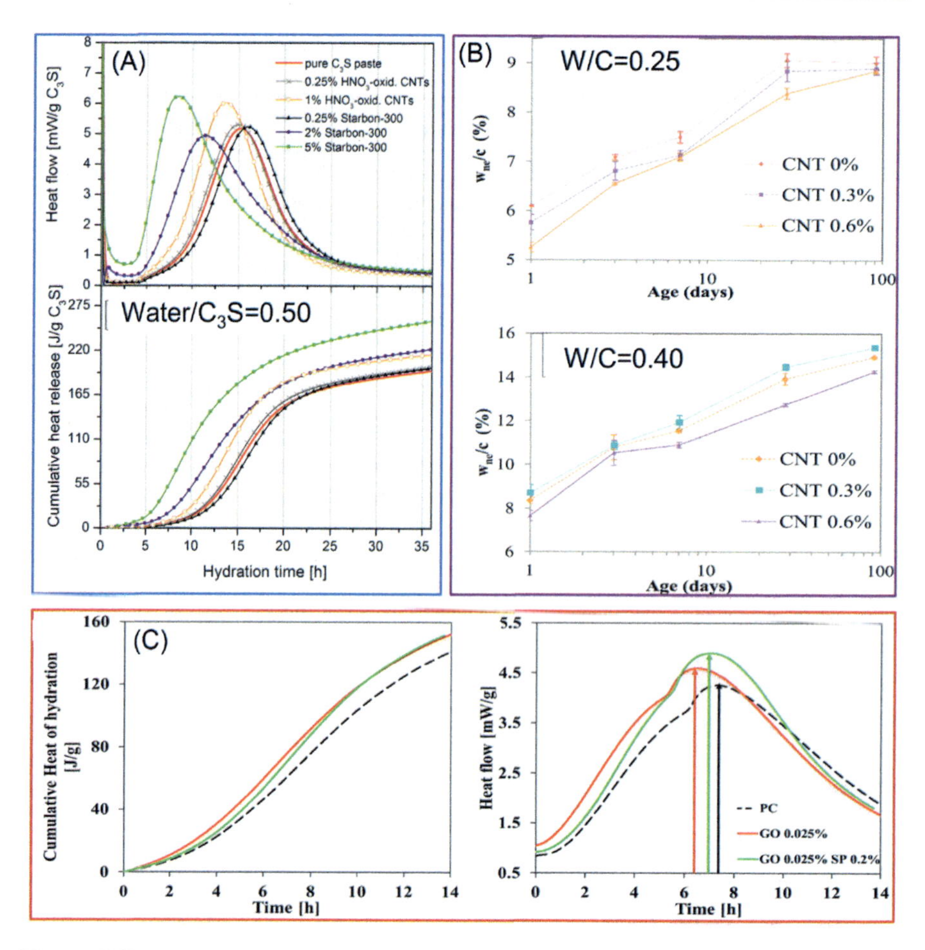

Figure 17.7 (A) Isothermal calorimetry of C3S samples. (B) Nonvaporated water. (C) Isothermal calorimeter of cement paste.

SSA of nanoparticles requiring more water for wetting their surface and the unique aspect ratio of carbon-based nanomaterials, such as CNT, CNF, and graphene (Long et al., 2018; Song et al., 2020; Sun et al., 2016). The reduced slump value of fresh mixtures negatively affects their casting and transportation processes, which in turn affects the final mechanical and durability properties of cement composites (Li et al., 2020; Singh et al., 2017). Surfactant is a promising candidate for addressing this issue. As provided in Fig. 17.8, Li, Ding, et al. (2021) showed that introducing SP can reduce the yield stress and minimum viscosity of cement pastes (w/c = 0.24) by 80% and 70%, respectively, when admixing 0.75 wt.% graphene. However, in the presence of SP, especially when the SP content exceeds 0.75 wt.%, the incorporation of graphene hardly affects the yield stress and viscosity of the cement paste (Leonavičius et al., 2020; Meng & Khayat, 2018).

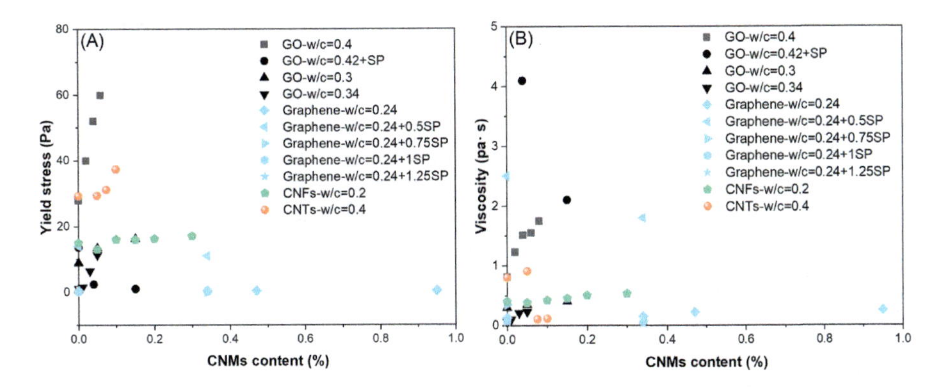

Figure 17.8 Influence of admixing carbon-based nanomaterials on the rheology property of cement composites: (A) yield stress and (B) viscosity. CNMs are carbon-based nanomaterials.

17.4.2 Mechanical properties of hardened cement composites

The mechanical properties of cement composites are often considered to be the most important parameter for their practical applications (Cui et al., 2018; Li, Fei, et al., 2021; Lu, Jiang, Leng, et al., 2023; Lu, Jiang, Tan, et al., 2023). In the last decades, it has been widely recognized that the incorporation of small amounts of carbon-based nanomaterials at the optimum dosage can significantly increase the mechanical strength of cement composites (Li et al., 2021; Paul et al., 2018; Rong et al., 2015).

Table 17.4 illustrates the experimental results collected from the published studies related to the enhancement of carbon-based nanomaterials to mechanical properties (e.g., compressive, flexural, and tensile strength) of cement composites. In Table 17.4, 0.01–0.10 wt.% of GO is usually used for the manufacture of cement paste or mortar, whereas CNF and CNT are usually used for the preparation of cement concrete due to their excellent aspect ratios, which can effectively enhance the toughness and cracking resistance of cementitious composites, especially for the preparation of ultra-high-performance concrete.

In terms of the compressive strength of cement composites, adding GO is more effective than that of CNT or CNF, while CNT and CNF seem more effective for enhancing the flexural and tensile strengths of cement composites (Lu et al., 2022d), as suggested in Fig. 17.9. Recently, our group used GO to coat aggregates surface (fine, coarse, and recycled aggregates), and found that the introduction of GO-coated aggregates can target enhance the interfacial transition zone of cement composites (Lu et al., 2022c; Lu, Wang, et al., 2023), which significantly reduces the amount of GO since the specific surface area of the aggregates is much smaller than that of the cement particles, which has significant economic and environmental advantages, and provides a candidate for the realization of sustainable concrete structures (Fig. 17.10).

Table 17.4 Enhancement of carbon-based nanomaterials to mechanical strengths of cement composites at 28 days.

Matrix	Types of carbon-based nanomaterials (by weight of cement)	w/c	Increment in mechanical strengths			References
			Compressive	Flexural	Tensile	
Paste	Graphene (0.05 wt.%)	0.35	3%–8%	15%–24%	–	Wang et al. (2016)
Paste	Graphene (0.025 wt.%)	0.40	14.9%	23.6%	15.2	Liu et al. (2019)
Paste	Graphene (2 wt.%)	0.38	−33.3%			Xu and Zhang (2017)
Paste	GO (0.05 wt.%)	0.35	29.0%	–	–	Xu et al. (2019)
Paste	GO (0.01 wt. %)	0.30	10.0%	15.6%	–	Lv et al. (2014)
Paste	GO (0.03 wt. %)	0.30	20.1%	27.3%	–	
Paste	GO (0.05 wt. %)	0.30	27.5%	30.7%	–	
Paste	CNFs (0.1 wt.%)	0.485	–	20%	–	Gao et al. (2019)
Paste	CNTs (0.15 wt.%)	0.40	14.3%	3%	–	Qin et al. (2021)
Mortar	GO (0.022 wt.%)	0.42	34.10%	30.37%	33.0%	Zhao, Guo, Ge, et al. (2017)
Mortar	GO (0.1 wt.%)	0.48	27.7%	–	–	Ebrahimizadeh Abrishami and Zahabi (2016)
Mortar	Graphene (0.05 wt.%)	N.A.	8.3%	15.6%		Matalkah and Soroushian (2020)
Mortar	GO (0.15 wt.%)	0.35	13.7%	14.5%		Akarsh et al. (2021)
Mortar	GO (0.01 wt.%)	0.37	13.4%	51.7%	47.0%	Lv et al. (2013)
Mortar	GO (0.03 wt.%)	0.37	38.9%	60.7%	78.6%	
Mortar	GO (0.05 wt. %)	0.37	47.9%	30.2%	35.8%	
Mortar	CNTs (0.3 wt.%)	0.40	−2.7%	–	–	Kim et al. (2014)
Mortar	CNTs (0.5 wt.%)	0.60	−1.8%	–	–	
Mortar	CNFs (0.1 wt.%)	0.485	6.15	–	–	Gao et al. (2019)
Concrete	CNFs (0.30 wt.%)	0.20	60.0%	10.1%	55.0%	Meng and Khayat (2018)

Concrete	CNTs (0.002 wt.%)	0.23	−1.1%	–	–	Jung et al. (2020)
Concrete	CNTs (0.005 wt.%)	0.23	5.5%	–	–	
Concrete	CNTs (0.008 wt.%)	0.23	−0.8%	–	–	
Concrete	CNTs (0.01 wt.%)	0.23	−4.9%	–	–	
Concrete	CNFs (0.1 wt.%)	0.51	8.13%	–	–	Gao et al. (2019)
Concrete	CNFs (0.10 wt.%)	0.51	6.16%	–	–	Zhu et al. (2018)
Concrete	Graphene (0.30 wt.%)	0.20	40%	59%	–	Meng and Khayat (2016)
Concrete	Graphene (0.10 wt.%)	0.16	62.25	9.25	–	Guo et al. (2020)

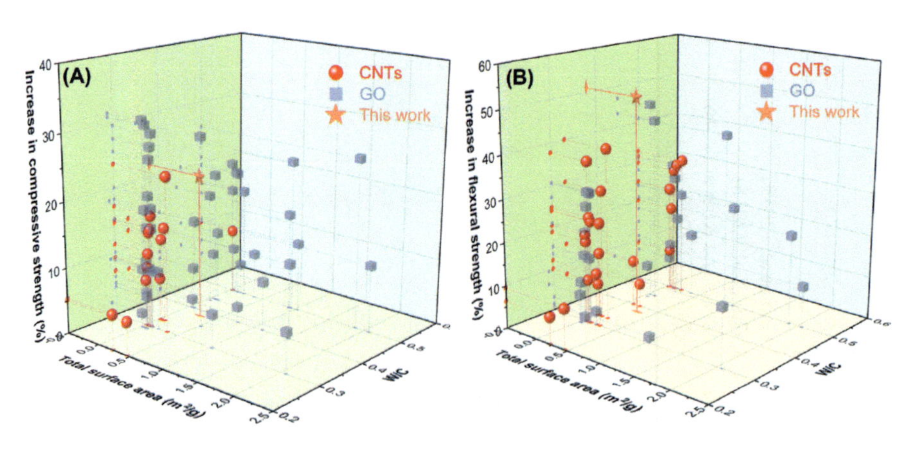

Figure 17.9 The reinforcement efficiency of CNT and GO for cement composites: (A) increase in compressive strength and (B) increase in flexural strength.

17.4.3 Durability of hardened cement composites

Durability is essential to ensure the reliability and stability of cement composites in their long-term service life (Li et al., 2021). In this section, the durability of cement composites is evaluated in terms of transportation characteristics and chemical resistance (Shi et al., 2019; Zhao, Guo, et al., 2020). This section mainly discussed the findings brought out by carbon-based nanomaterials on the transport property and chemical degradation of cement composites.

The transport properties of cement-based materials reflect their resistance to the passage of aggressive media (e.g., water and ions) through the interior, which can be improved by the introduction of well-dispersed carbon-based nanomaterials (Cao et al., 2023; Li et al., 2021). For example, Qureshi and Panesar (2019) suggested that the maximum water vapor adsorption of 0.06% GO-modified cement composites increased to 2%, since additional C-S-H was formed and it can induce a denser microstructure (Fig. 17.11). Similar to the role of GO in the cement composites, admixing CNT or CNF exerts the nanofiller effect and bridging effect to improve the microstructure of cement materials (Cao et al., 2023; Liew et al., 2016; Wu et al., 2019), as shown in Fig. 17.11.

Chemical degradation of cement composites typically involves sulfate attack, carbonation and acid attack (Cao et al., 2023; Konstantopoulos et al., 2020; Lee et al., 2018; Lu, Cao, et al., 2020). As discussed above, adding carbon-based nanomaterials also showed great potential to enhance the chemical resistance of cement composites. For instance, Long et al. (2018) found that adding 0.2 wt.% GO can reduce the carbonation depth of cement composites by about 40%, mainly ascribed to the refined pore structure, as shown in Fig. 17.11. Mohammed et al. (2018) reported that the carbonation depth of the cement composites could be reduced from 11 to 2 mm with the addition of 0.06 wt.% GO, which was mainly due to the benefit of the reduction in mesopore volume (Fig. 17.11). Other studies have

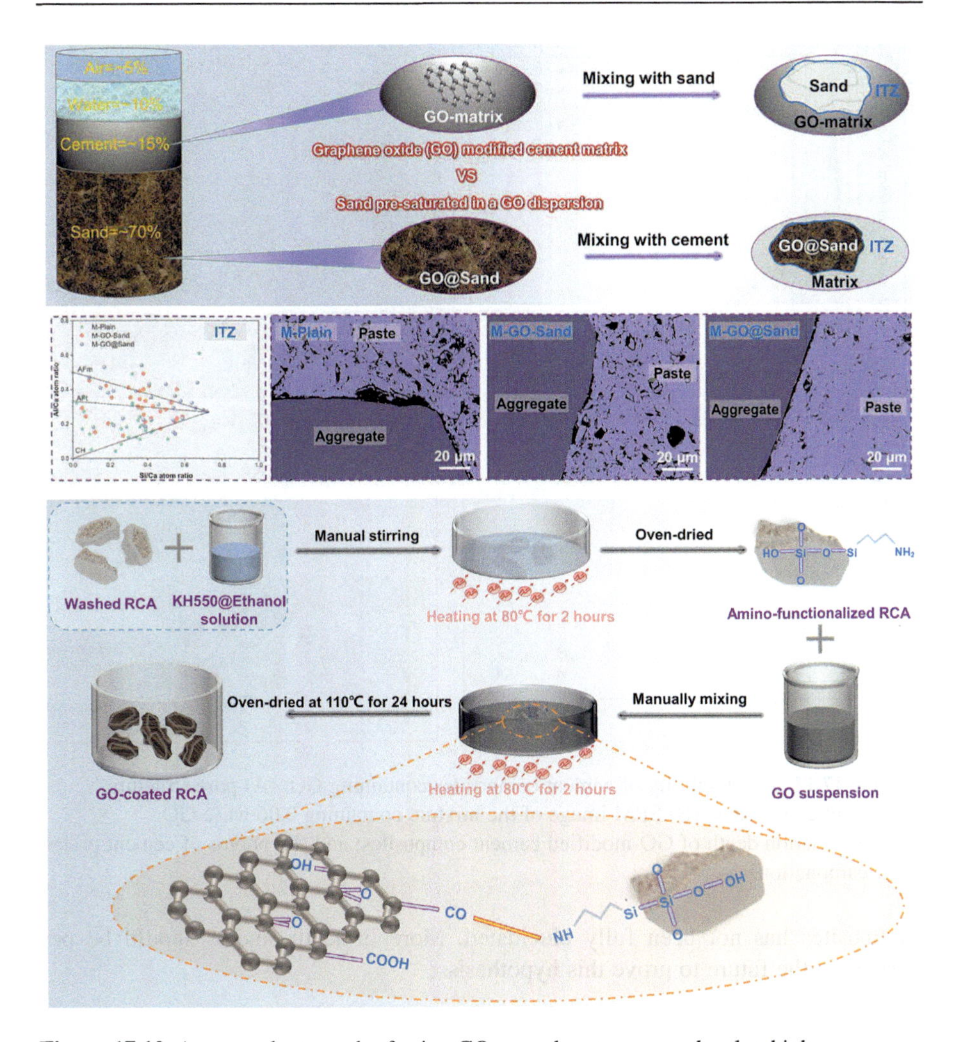

Figure 17.10 A targeted approach of using GO-coated aggregate to develop high-performance and sustainable cement composites.

observed similar benefits of incorporating CNT or CNF on the chemical resistance of cement composites and attributed this benefit to the refinement of the microstructure (Jung et al., 2020; Lee et al., 2018).

In conclusion, adding carbon-based nanomaterials to cement composites has demonstrated good durability. Compared with CNT or CNF, GO has a higher SSA and abundant hydrophilic functional groups, so more hydration products can be generated, resulting in a denser microstructure of the composites, which is better for improving the durability of cementitious materials. At present, although the durability of cement composites mainly depends on their microstructure, the mechanism of admixing carbon-based nanomaterials to enhance the durability of cement

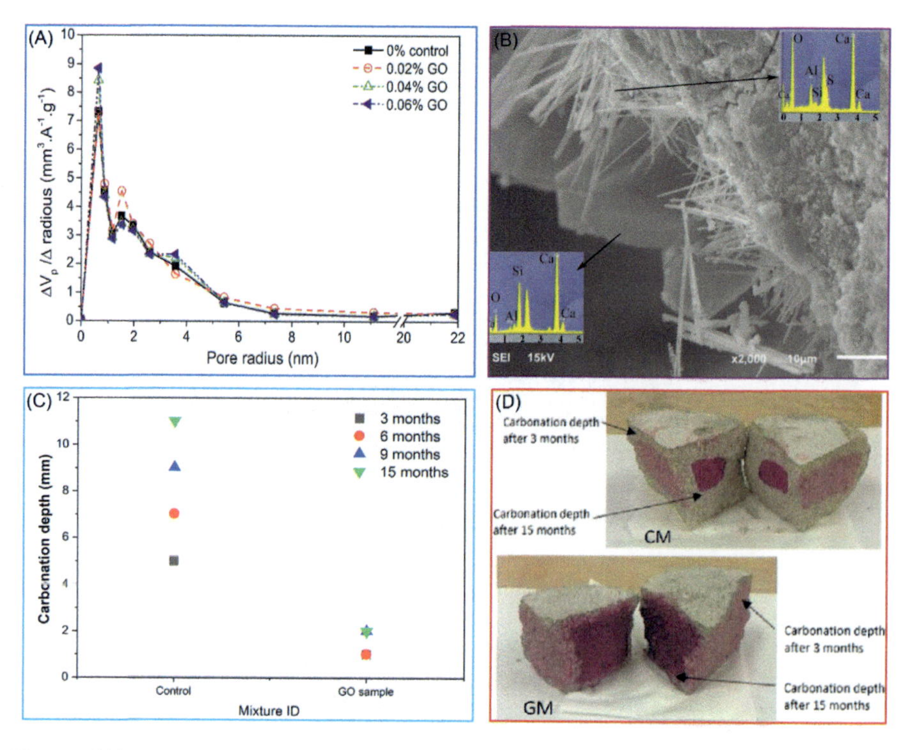

Figure 17.11 The durability of cement composites containing GO: (A) pore structure of cement composites; (B) SEM image of the mixture containing 0.06 wt.% GO. (C) Carbonation depth of GO-modified cement composites; and (D) photos of cement pastes after carbonation.

composites has not been fully elucidated. More in-depth studies should be performed in the future to prove this hypothesis.

17.5 Future development and challenges of carbon-based nanomaterials engineered cement composites and structures

In the past decades, the development of nanoscience and nanotechnology has brought cement composites into a new era. The development and application of new-generation and sustainable cement composites are still a hot topic and research frontier in the field of civil engineering. Many of the potential advantages of this technology over conventional cement composites have been demonstrated and widely validated, such as increased reliability and service life of concrete infrastructures, enhanced concrete structural performance and durability, increased safety against natural disasters and vibrations, reduced life cycle costs of infrastructure operation and management, and reduced impact on resources, energy, and the environment. Despite the phenomenal growth of cement composites, there are both opportunities and risks. Issues such as the mechanism of nano-strengthening/nano-modification of cement composites and the design of cement composites need to be addressed. More in-depth exploration of the new generation of cement composites is needed. In addition, more attention should be paid to the impact on global health for promoting the industrialization of cement composites.

Due to the complexity, diversity, and variability of cement composites, there is a need for a materials innovation infrastructure based on materials genome engineering. By deciphering the genetic code of materials and reconfiguring or modifying the structural units of materials at the nanoscale, it is possible to draw a blueprint of nanoscale properties to develop new theories and methods for a new generation of cement composites. This new integrated design continuum, which combines increased use of artificial intelligence, biotechnology, and numerical data, will also replace lengthy and costly empirical studies with mathematical models and computational simulations, providing model parameters, validating key predictions, and complementing and extending the range of model validity and reliability. In addition, it will help accelerate the discovery process and enable the scientific, informative, optimization, fabrication, and deployment of the new generation of cement composites, as shown in Fig. 17.12.

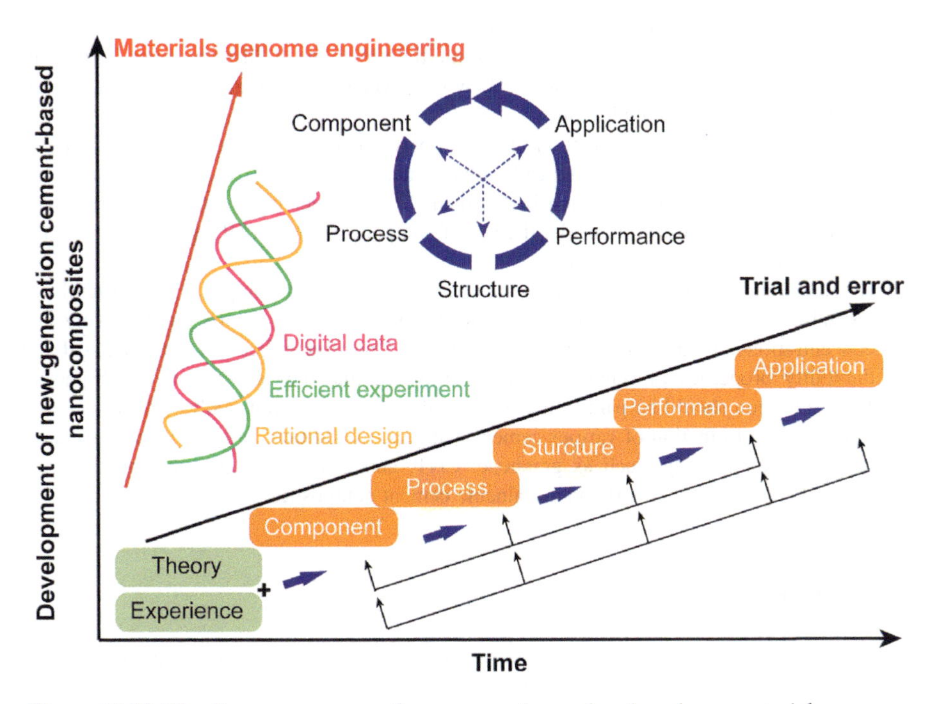

Figure 17.12 The discovery process of new-generation carbon-based nanomaterials engineered cement composites based on materials genome engineering.

17.6 Summary

This chapter first gives a glimpse of the numerous unique and interesting, mainly the different dispersion methods employed to manufacture cement composites and the current evaluation methods, the hydration and rheological properties of a fresh mixture, the influences of carbon-based nanomaterials on the mechanical and durability properties of hardened cement composites. The conclusions are summarized below:

1. Locating and evaluating the dispersion quality and distribution of admixing carbon-based nanomaterials in cement composites is key to understanding their enhancement mechanisms. Although previous studies on the incorporation of carbon-based nanomaterials have yielded some satisfactory results in terms of enhancement efficiency, the effective dispersion of carbon-based nanomaterials in the cement matrix remains a primary challenge, which undoubtedly restricts the feasibility of their widespread application in the construction industry.
2. It is believed that GO can accelerate cement hydration, in which the admixing GO can act as a nanofiller and template for cement hydration. While pure CNT or CNF generally do not promote or delay cement hydration, they play a role in crack-bridging in cement composites. Interestingly, CNT with oxygen-containing functional groups grafted on the surface was found to have the potential to promote cement hydration. Namely, no consensus incorporated CNT can accelerate cement hydration. Overall, direct evidence is still very limited and more detailed investigations are needed to study the effect of carbon-based nanomaterials on the formation of cement hydrates. In particular, due to the long duration of cement hydration and the production of multiple hydrates at different times, further in-depth studies are needed to investigate how (and when) incorporated carbon-based nanomaterials affect cement hydration.
3. For fresh mixtures, the introduction of carbon-based nanomaterials leads to an increase in viscosity, that is, the incorporated carbon-based nanomaterials impair the setting time and workability of the mixture, which is detrimental to the later transportation, pouring, and compaction processes. Fortunately, the use of some SCMs such as fly ash and silica fume can compensate for this deficiency.
4. Research on carbon-based nanomaterials-engineered cement composites is currently in ascendant, but there are still many obstacles to overcome. The incorporation of very low concentrations of carbon-based nanomaterials has a positive effect on the mechanical and durability properties of cement materials, provided that they are effectively dispersed. More attention should be paid to the study of the effect of various carbon-based nanomaterials on the formation of various cement hydrates. In addition, carbon-based nanomaterials with conductivity, such as graphene, CNT, and CNF, have great potential for the future development of smart and sustainable cement composites.

References

Akarsh, P. K., Marathe, S., & Bhat, A. K. (2021). Influence of graphene oxide on properties of concrete in the presence of silica fumes and M-sand. *Construction and Building Materials, 268*. Available from https://doi.org/10.1016/j.conbuildmat.2020.121093, https://www.journals.elsevier.com/construction-and-building-materials.

Althoey, F., & Farnam, Y. (2019). The effect of using supplementary cementitious materials on damage development due to the formation of a chemical phase change in cementitious materials exposed to sodium chloride. *Construction and Building Materials*, *210*, 685–695. Available from https://doi.org/10.1016/j.conbuildmat.2019.03.230.

Azeem, M., & Azhar Saleem, M. (2020). Hydration model for the OPC-CNT mixture: Theory and experiment. *Construction and Building Materials*, *264*. Available from https://doi.org/10.1016/j.conbuildmat.2020.120691, https://www.journals.elsevier.com/construction-and-building-materials.

Baomin, W., & Shuang, D. (2019). Effect and mechanism of graphene nanoplatelets on hydration reaction, mechanical properties and microstructure of cement composites. *Construction and Building Materials*, *228*. Available from https://doi.org/10.1016/j.conbuildmat.2019.116720, https://www.journals.elsevier.com/construction-and-building-materials.

Brittain, R., Liskiewicz, T., Morina, A., Neville, A., & Yang, L. (2023). Diamond-like carbon graphene nanoplatelet nanocomposites for lubricated environments. *Carbon*, *205*, 485–498. Available from https://doi.org/10.1016/j.carbon.2023.01.061, http://www.journals.elsevier.com/carbon/.

Budde, H., Coca-López, N., Shi, X., Ciesielski, R., Lombardo, A., Yoon, D., Ferrari, A. C., & Hartschuh, A. (2016). Raman radiation patterns of graphene. *ACS Nano*, *10*(2), 1756–1763. Available from https://doi.org/10.1021/acsnano.5b06631, http://pubs.acs.org/journal/ancac3.

Cao, R., Yang, J., Li, G., Zhou, Q., & Niu, M. (2023). Durability performance of multiwalled carbon nanotube reinforced ordinary Portland/calcium sulfoaluminate cement composites to sulfuric acid attack at early stage. *Materials Today Communications*, *35*. Available from https://doi.org/10.1016/j.mtcomm.2023.105748, http://www.journals.elsevier.com/materials-today-communications/.

Chen, Z., Zhang, Y., Wang, Z., Wu, Y., Zhao, Y., Liu, L., & Ji, G. (2023). Bioinspired moth-eye multi-mechanism composite ultra-wideband microwave absorber based on the graphite powder. *Carbon*, *201*, 542–548. Available from https://doi.org/10.1016/j.carbon.2022.09.035, http://www.journals.elsevier.com/carbon/.

Chintalapudi, K., & Pannem, R. M. R. (2020a). The effects of graphene oxide addition on hydration process, crystal shapes, and microstructural transformation of Ordinary Portland Cement. *Journal of Building Engineering*, *32*. Available from https://doi.org/10.1016/j.jobe.2020.101551, http://www.journals.elsevier.com/journal-of-building-engineering/.

Chintalapudi, K., & Pannem, R. M. R. (2020b). An intense review on the performance of graphene oxide and reduced graphene oxide in an admixed cement system. *Construction and Building Materials*, *259*. Available from https://doi.org/10.1016/j.conbuildmat.2020.120598, https://www.journals.elsevier.com/construction-and-building-materials.

Chintalapudi, K., & Pannem, R. M. R. (2019). Strength properties of graphene oxide cement composites. *Materials Today: Proceedings*, *45*, 3971–3975. Available from https://doi.org/10.1016/j.matpr.2020.08.369, https://www.sciencedirect.com/journal/materials-today-proceedings.

Chougan, M., Marotta, E., Lamastra, F. R., Vivio, F., Montesperelli, G., Ianniruberto, U., & Bianco, A. (2019). A systematic study on EN-998-2 premixed mortars modified with graphene-based materials. *Construction and Building Materials*, *227*. Available from https://doi.org/10.1016/j.conbuildmat.2019.116701, https://www.journals.elsevier.com/construction-and-building-materials.

Cui, H., Jin, Z., Zheng, D., Tang, W., Li, Y., Yun, Y., Lo, T. Y., & Xing, F. (2018). Effect of carbon fibers grafted with carbon nanotubes on mechanical properties of cement-based composites. *Construction and Building Materials*, *181*, 713–720. Available from https://doi.org/10.1016/j.conbuildmat.2018.06.049.

Dharmasiri, B., Randall, J. D., Stanfield, M. K., Ying, Y., Andersson, G. G., Nepal, D., Hayne, D. J., & Henderson, L. C. (2022). Using surface grafted poly(acrylamide) to simultaneously enhance the tensile strength, tensile modulus, and interfacial adhesion of carbon fibres in epoxy composites. *Carbon*, *186*, 367–379. Available from https://doi.org/10.1016/j.carbon.2021.10.046, http://www.journals.elsevier.com/carbon/.

Ding, S., Wang, X., & Han, B. (2023). *New-generation cement-based nanocomposites*. Singapore: Springer Nature. Available from 10.1007/978-981-99-2306-9.

Dong, S., Li, L., Ashour, A., Dong, X., & Han, B. (2021). Self-assembled 0D/2D nano carbon materials engineered smart and multifunctional cement-based composites. *Construction and Building Materials*, *272*. Available from https://doi.org/10.1016/j.conbuildmat.2020.121632, https://www.journals.elsevier.com/construction-and-building-materials.

Dong, S., Zhang, W., Wang, X., & Han, B. (2023). New-generation pavement empowered by smart and multifunctional concretes: A review. *Construction and Building Materials*, *402*. Available from https://doi.org/10.1016/j.conbuildmat.2023.132980, https://www.journals.elsevier.com/construction-and-building-materials.

Du, H., & Pang, S. D. (2018). Dispersion and stability of graphene nanoplatelet in water and its influence on cement composites. *Construction and Building Materials*, *167*, 403–413. Available from https://doi.org/10.1016/j.conbuildmat.2018.02.046.

Du, Y., Yang, J., Skariah Thomas, B., Li, L., Li, H., & Nazar, S. (2020). Hybrid graphene oxide/carbon nanotubes reinforced cement paste: An investigation on hybrid ratio. *Construction and Building Materials*, *261*. Available from https://doi.org/10.1016/j.conbuildmat.2020.119815, https://www.journals.elsevier.com/construction-and-building-materials.

Ebrahimizadeh Abrishami, M., & Zahabi, V. (2016). Reinforcing graphene oxide/cement composite with NH2 functionalizing group. *Bulletin of Materials Science*, *39*(4), 1073–1078. Available from https://doi.org/10.1007/s12034-016-1250-7.

Gan, Y., Zhang, X., Jiang, Z., Lu, D., Xu, N., Han, L., & Han, X. (2022). Cementitious fillers in cement asphalt emulsion mixtures: Long-term performance and microstructure. *Arabian Journal for Science and Engineering*, *47*(4), 4943–4953. Available from https://doi.org/10.1007/s13369-021-06266-3, https://link.springer.com/journal/13369.

Gao, Y., Zhu, X., Corr, D. J., Konsta-Gdoutos, M. S., & Shah, S. P. (2019). Characterization of the interfacial transition zone of CNF-Reinforced cementitious composites. *Cement and Concrete Composites*, *99*, 130–139. Available from https://doi.org/10.1016/j.cemconcomp.2019.03.002, http://www.sciencedirect.com/science/journal/09589465.

Ghaharpour, F., Bahari, A., Abbasi, M., & Ashkarran, A. A. (2016). Parametric investigation of CNT deposition on cement by CVD process. *Construction and Building Materials*, *113*, 523–535. Available from https://doi.org/10.1016/j.conbuildmat.2016.03.080.

Ghahremani, P., Mostafatabar, A. H., Bahlakeh, G., & Ramezanzadeh, B. (2022). Rational design of a novel multi-functional carbon-based nano-carrier based on multi-walled-CNT-oxide/polydopamine/chitosan for epoxy composite with robust pH-sensitive active anti-corrosion properties. *Carbon*, *189*, 113–141. Available from https://doi.org/10.1016/j.carbon.2021.11.067, http://www.journals.elsevier.com/carbon/.

Ghazizadeh, S., Duffour, P., Skipper, N. T., & Bai, Y. (2018). Understanding the behaviour of graphene oxide in Portland cement paste. *Cement and Concrete Research*, *111*, 169–182. Available from https://doi.org/10.1016/j.cemconres.2018.05.016, http://www.sciencedirect.com/science/journal/00088846.

Ghosh, S., Harish, S., Rocky, K. A., Ohtaki, M., & Saha, B. B. (2019). Graphene enhanced thermoelectric properties of cement based composites for building energy harvesting. *Energy and Buildings*, *202*. Available from https://doi.org/10.1016/j.enbuild.2019.109419, https://www.journals.elsevier.com/energy-and-buildings.

Guo, Y., Wang, Y., Huang, Z., Tong, X., & Yang, N. (2022). Size effect of rhodium nanoparticles supported on carbon black on the performance of hydrogen evolution reaction. *Carbon*, *194*, 303–309. Available from https://doi.org/10.1016/j.carbon.2022.04.008, http://www.journals.elsevier.com/carbon/.

Guo, L., Wu, J., & Wang, H. (2020). Mechanical and perceptual characterization of ultra-high-performance cement-based composites with silane-treated graphene nano-platelets. *Construction and Building Materials*, *240*. Available from https://doi.org/10.1016/j.conbuildmat.2019.117926, https://www.journals.elsevier.com/construction-and-building-materials.

Guo, S., Zhang, X., Shi, C., Zhao, D., He, C., & Zhao, N. (2022). Simultaneously enhanced mechanical properties and electrical property of Cu-2 wt% Ag alloy matrix composites with analogy-bicontinuous structures constructed via in-situ synthesized graphene nanoplatelets. *Carbon*, *198*, 207–218. Available from https://doi.org/10.1016/j.carbon.2022.07.025, http://www.journals.elsevier.com/carbon/.

Gwon, S., Kim, H., & Shin, M. (2023). Self-heating characteristics of electrically conductive cement composites with carbon black and carbon fiber. *Cement and Concrete Composites*, *137*. Available from https://doi.org/10.1016/j.cemconcomp.2023.104942, http://www.sciencedirect.com/science/journal/09589465.

Han, L., Li, K., Xiao, C., Yin, X., Gui, X., Song, Q., & Ye, F. (2023). Carbon nanotube-vertical edge rich graphene hybrid sponge as multifunctional reinforcements for high performance epoxy composites. *Carbon*, *201*, 871–880. Available from https://doi.org/10.1016/j.carbon.2022.09.074, http://www.journals.elsevier.com/carbon/.

Han, B., Zhang, L., Zhang, C., Wang, Y., Yu, X., & Ou, J. (2016). Reinforcement effect and mechanism of carbon fibers to mechanical and electrically conductive properties of cement-based materials. *Construction and Building Materials*, *125*, 479–489. Available from https://doi.org/10.1016/j.conbuildmat.2016.08.063.

He, S., & Yang, E. H. (2021). Strategic strengthening of the interfacial transition zone (ITZ) between microfiber and cement paste matrix with carbon nanofibers (CNFs). *Cement and Concrete Composites*, *119*. Available from https://doi.org/10.1016/j.cemconcomp.2021.104019, http://www.sciencedirect.com/science/journal/09589465.

Horszczaruk, E., Mijowska, E., Kalenczuk, R. J., Aleksandrzak, M., & Mijowska, S. (2015). Nanocomposite of cement/graphene oxide - impact on hydration kinetics and Young's modulus. *Construction and Building Materials*, *78*, 234–242. Available from https://doi.org/10.1016/j.conbuildmat.2014.12.009.

Hou, P., Shi, J., Prabakar, S., Cheng, X., Wang, K., Zhou, X., & Shah, S. P. (2020). Effects of mixing sequences of nanosilica on the hydration and hardening properties of cement-based materials. *Construction and Building Materials*, *263*. Available from https://doi.org/10.1016/j.conbuildmat.2020.120226, https://www.journals.elsevier.com/construction-and-building-materials.

Huang, K., Zhang, X., Lu, D., Xu, N., Gan, Y., Han, X., & Liu, Y. (2021). The role of iron tailing powder in ultra-high-strength concrete subjected to elevated temperatures. *Advances in Civil Engineering*, *2021*, 1–11. Available from https://doi.org/10.1155/2021/6681429.

Huo, Y., Liu, T., Lu, D., Han, X., Sun, H., Huang, J., Ye, X., Zhang, C., Chen, Z., & Yang, Y. (2023). Dynamic tensile properties of steel fiber reinforced polyethylene fiber-engineered/strain-hardening cementitious composites (PE-ECC/SHCC) at high strain rate. *Cement and Concrete Composites, 143*. Available from https://doi.org/10.1016/j.cemconcomp.2023.105234, http://www.sciencedirect.com/science/journal/09589465.

Huo, Y., Lu, D., Wang, Z., Liu, Y., Chen, Z., & Yang, Y. (2023). Bending behavior of strain hardening cementitious composites based on the combined fiber-interface constitutive model. *Computers & Structures, 281*. Available from https://doi.org/10.1016/j.compstruc.2023.107017.

Iijima, S. (1991). Helical microtubules of graphitic carbon. *Nature, 354*(6348), 56–58. Available from https://doi.org/10.1038/354056a0.

Imanian Ghazanlou, S., Jalaly, M., Sadeghzadeh, S., & Habibnejad Korayem, A. (2020). High-performance cement containing nanosized Fe3O4–decorated graphene oxide. *Construction and Building Materials, 260*. Available from https://doi.org/10.1016/j.conbuildmat.2020.120454, https://www.journals.elsevier.com/construction-and-building-materials.

Jeong, D. W., Kim, K., Lee, G., Kang, M., Chang, H., Jang, A. R., & Lee, J. O. (2022). Electrochemical transparency of graphene. *ACS Nano, 16*(6), 9278–9286. Available from https://doi.org/10.1021/acsnano.2c01786, http://pubs.acs.org/journal/ancac3.

Jiang, Z., Zheng, W., Wang, Y., Sun, P., Lu, D., & Sun, L. (2023). The use of soundless chemical demolition agents in reinforced concrete deep beam demolition: Experimental and numerical study. *Journal of Building Engineering, 69*. Available from https://doi.org/10.1016/j.jobe.2023.106260, http://www.journals.elsevier.com/journal-of-building-engineering/.

Jing, G., Xu, K., Feng, H., Wu, J., Wang, S., Li, Q., Cheng, X., & Ye, Z. (2020). The non-uniform spatial dispersion of graphene oxide: A step forward to understand the inconsistent properties of cement composites. *Construction and Building Materials, 264*. Available from https://doi.org/10.1016/j.conbuildmat.2020.120729, https://www.journals.elsevier.com/construction-and-building-materials.

Jing, G.-J., Ye, Z.-M., Li, C., Cui, J., Wang, S.-X., & Cheng, X. (2020). A ball milling strategy to disperse graphene oxide in cement composites. *Carbon, 159*. Available from https://doi.org/10.1016/j.carbon.2019.12.082.

Jing, G.-j, Ye, Z.-m, Li, C., Cui, J., Wang, S.-x, & Cheng, X. (2019). A ball milling strategy to disperse graphene oxide in cement composites. *New Carbon Materials, 34*(6), 569–577. Available from https://doi.org/10.1016/s1872-5805(19)60032-6.

John, V. M., Quattrone, M., Abrão, P. C. R. A., & Cardoso, F. A. (2019). Rethinking cement standards: Opportunities for a better future. *Cement and Concrete Research, 124*. Available from https://doi.org/10.1016/j.cemconres.2019.105832, http://www.sciencedirect.com/science/journal/00088846.

De Jong, K. P., & Geus, J. W. (2007). Carbon nanofibers: Catalytic synthesis and applications. *Catalysis Reviews, 42*(4), 481–510. Available from https://doi.org/10.1081/CR-100101954.

Jung, M., Lee, Y. S., Hong, S. G., & Moon, J. (2020). Carbon nanotubes (CNTs) in ultra-high performance concrete (UHPC): Dispersion, mechanical properties, and electromagnetic interference (EMI) shielding effectiveness (SE). *Cement and Concrete Research, 131*. Available from https://doi.org/10.1016/j.cemconres.2020.106017, http://www.sciencedirect.com/science/journal/00088846.

Kim, H. K., Park, I. S., & Lee, H. K. (2014). Improved piezoresistive sensitivity and stability of CNT/cement mortar composites with low water–binder ratio. *Composite Structures, 116*(1), 713–719. Available from https://doi.org/10.1016/j.compstruct.2014.06.007.

Konsta-Gdoutos, M. S., Metaxa, Z. S., & Shah, S. P. (2010). Highly dispersed carbon nanotube reinforced cement based materials. *Cement and Concrete Research*, *40*(7), 1052–1059. Available from https://doi.org/10.1016/j.cemconres.2010.02.015.

Konstantopoulos, G., Koumoulos, E., Karatza, A., & Charitidis, C. (2020). Pore and phase identification through nanoindentation mapping and micro-computed tomography in nanoenhanced cement. *Cement and Concrete Composites*, *114*. Available from https://doi.org/10.1016/j.cemconcomp.2020.103741, http://www.sciencedirect.com/science/journal/09589465.

Lee, H. S., Balasubramanian, B., Gopalakrishna, G. V. T., Kwon, S. J., Karthick, S. P., & Saraswathy, V. (2018). Durability performance of CNT and nanosilica admixed cement mortar. *Construction and Building Materials*, *159*, 463–472. Available from https://doi.org/10.1016/j.conbuildmat.2017.11.003.

Lee, J. E., Choi, J., Jang, D., Lee, S., Kim, T. H., & Lee, S. (2023). Processing-controlled radial heterogeneous structure of carbon fibers and primary factors determining their mechanical properties. *Carbon*, *206*, 16–25. Available from https://doi.org/10.1016/j.carbon.2023.02.002, http://www.journals.elsevier.com/carbon/.

Leonavičius, D., Pundienė, I., Pranckevičienė, J., & Kligys, M. (2020). Selection of superplasticisers for improving the rheological and mechanical properties of cement paste with CNTs. *Construction and Building Materials*, *253*. Available from https://doi.org/10.1016/j.conbuildmat.2020.119182, https://www.journals.elsevier.com/construction-and-building-materials.

Liew, K. M., Kai, M. F., & Zhang, L. W. (2016). Carbon nanotube reinforced cementitious composites: An overview. *Composites Part A: Applied Science and Manufacturing*, *91*, 301–323. Available from https://doi.org/10.1016/j.compositesa.2016.10.020.

Lin, Y., & Du, H. (2020). Graphene reinforced cement composites: A review. *Construction and Building Materials*, *265*. Available from https://doi.org/10.1016/j.conbuildmat.2020.120312, https://www.journals.elsevier.com/construction-and-building-materials.

Lin, J., Shamsaei, E., Basquiroto de Souza, F., Sagoe-Crentsil, K., & Duan, W. H. (2020). Dispersion of graphene oxide–silica nanohybrids in alkaline environment for improving ordinary Portland cement composites. *Cement and Concrete Composites*, *106*.

Liu, S., Chen, Y., Li, X., Wang, L., & Du, H. (2023). Development of bio-inspired cement-based material by magnetically aligning graphene oxide nanosheets in cement paste. *Construction and Building Materials*, *369*. Available from https://doi.org/10.1016/j.conbuildmat.2023.130545, https://www.journals.elsevier.com/construction-and-building-materials.

Liu, J., Fu, J., Yang, Y., & Gu, C. (2019). Study on dispersion, mechanical and microstructure properties of cement paste incorporating graphene sheets. *Construction and Building Materials*, *199*, 1–11. Available from https://doi.org/10.1016/j.conbuildmat.2018.12.006.

Liu, Y., Li, Q.-Q., Zhang, H., Yu, S.-P., Zhang, L., & Yang, Y.-Z. (2020). Research progress on the use of micro/nano carbon materials for antibacterial dressings. *New Carbon Materials*, *35*(4), 323–335. Available from https://doi.org/10.1016/s1872-5805(20)60492-9.

Liu, Z., Yuan, X., Zhao, Y., Chew, J. W., & Wang, H. (2022). Concrete waste-derived aggregate for concrete manufacture. *Journal of Cleaner Production*, *338*. Available from https://doi.org/10.1016/j.jclepro.2022.130637, https://www.journals.elsevier.com/journal-of-cleaner-production.

Liu, F., Zhang, X., Cheng, J., Tu, J., Kong, F., Huang, W., & Chen, C. (2003). Preparation of short carbon nanotubes by mechanical ball milling and their hydrogen adsorption behavior. *Carbon*, *41*(13), 2527–2532. Available from https://doi.org/10.1016/S0008-6223(03)00302-6.

Li, C. Y., Chen, S. J., Li, W. G., Li, X. Y., Ruan, D., & Duan, W. H. (2019). Dynamic increased reinforcing effect of graphene oxide on cementitious nanocomposite. *Construction and Building Materials*, *206*, 694–702. Available from https://doi.org/10.1016/j.conbuildmat.2019.02.001.

Li, H., Ding, S., Zhang, L., Ouyang, J., & Han, B. (2021). Rheological behaviors of cement pastes with multi-layer graphene. *Construction and Building Materials*, *269*. Available from https://doi.org/10.1016/j.conbuildmat.2020.121327, https://www.journals.elsevier.com/construction-and-building-materials.

Li, Z., Fei, M.-E., Huyan, C., & Shi, X. (2020). Fly ash-based geopolymer composites: An overview. *Resources, Conservation and Recycling*.

Li, Z., Fei, M.-E., Huyan, C., & Shi, X. (2021). Fly ash-based geopolymer composites: An overview. *Resources, Conservation and Recycling*.

Li, X., Lu, Z., Chuah, S., Li, W., Liu, Y., Duan, W. H., & Li, Z. (2017). Effects of graphene oxide aggregates on hydration degree, sorptivity, and tensile splitting strength of cement paste. *Composites Part A: Applied Science and Manufacturing*, *100*, 1–8. Available from https://doi.org/10.1016/j.compositesa.2017.05.002.

Li, G. Y., Wang, P. M., & Zhao, X. (2005). Mechanical behavior and microstructure of cement composites incorporating surface-treated multi-walled carbon nanotubes. *Carbon*, *43*(6), 1239–1245. Available from https://doi.org/10.1016/j.carbon.2004.12.017.

Long, W. J., Gu, Y. C., Xing, F., & Khayat, K. H. (2018). Microstructure development and mechanism of hardened cement paste incorporating graphene oxide during carbonation. *Cement and Concrete Composites*, *94*, 72–84. Available from https://doi.org/10.1016/j.cemconcomp.2018.08.016, http://www.sciencedirect.com/science/journal/09589465.

Long, W. J., Ye, T. H., Gu, Y. C., Li, H. D., & Xing, F. (2019). Inhibited effect of graphene oxide on calcium leaching of cement pastes. *Construction and Building Materials*, *202*, 177–188. Available from https://doi.org/10.1016/j.conbuildmat.2018.12.194.

Lothenbach, B., Scrivener, K., & Hooton, R. D. (2011). Supplementary cementitious materials. *Cement and Concrete Research*, *41*(12), 1244–1256. Available from https://doi.org/10.1016/j.cemconres.2010.12.001.

Lu, D., Cao, H., Shen, Q., Gong, Y., Zhao, C., & Yan, X. (2020). Dynamic characteristics and chloride resistance of basalt and polypropylene fibers reinforced recycled aggregate concrete. *Advances in Polymer Technology*, *2020*, 1–9. Available from https://doi.org/10.1155/2020/6029047.

Lu, Z., Hanif, A., Sun, G., Liang, R., Parthasarathy, P., & Li, Z. (2018). Highly dispersed graphene oxide electrodeposited carbon fiber reinforced cement-based materials with enhanced mechanical properties. *Cement and Concrete Composites*, *87*, 220–228. Available from https://doi.org/10.1016/j.cemconcomp.2018.01.006, http://www.sciencedirect.com/science/journal/09589465.

Lu, Z., Hou, D., Hanif, A., Hao, W., Sun, G., & Li, Z. (2018). Comparative evaluation on the dispersion and stability of graphene oxide in water and cement pore solution by incorporating silica fume. *Cement and Concrete Composites*, *94*, 33–42. Available from https://doi.org/10.1016/j.cemconcomp.2018.08.011, http://www.sciencedirect.com/science/journal/09589465.

Lu, D., Huo, Y., Jiang, Z., & Zhong, J. (2023). Carbon nanotube polymer nanocomposites coated aggregate enabled highly conductive concrete for structural health monitoring. *Carbon*, *206*, 340–350. Available from https://doi.org/10.1016/j.carbon.2023.02.043, http://www.journals.elsevier.com/carbon/.

Lu, D., Jiang, X., Leng, Z., Zhang, S., Wang, D., & Zhong, J. (2023). Dual responsive microwave heating-healing system in asphalt concrete incorporating coal gangue and functional aggregate. *Journal of Cleaner Production, 422.*

Lu, D., Jiang, X., Tan, Z., Yin, B., Leng, Z., & Zhong, J. (2023). Enhancing sustainability in pavement Engineering: A-state-of-the-art review of cement asphalt emulsion mixtures. *Cleaner Materials, 9.* Available from https://doi.org/10.1016/j.clema.2023.100204, http://www.elsevier.com/locate/issn/2772-3976.

Lu, D., Leng, Z., Lu, G., Wang, D., & Huo, Y. (2023). A critical review of carbon materials engineered electrically conductive cement concrete and its potential applications. *International Journal of Smart and Nano Materials, 14*(2), 189–215. Available from https://doi.org/10.1080/19475411.2023.2199703, http://www.tandfonline.com/loi/tsnm20.

Lu, D., Ma, L. P., Zhong, J., Tong, J., Liu, Z., Ren, W., & Cheng, H. M. (2023). Growing nanocrystalline graphene on aggregates for conductive and strong smart cement composites. *ACS Nano, 17*(4), 3587–3597. Available from https://doi.org/10.1021/acsnano.2c10141, http://pubs.acs.org/journal/ancac3.

Lu, D., Sheng, Z., Yan, B., Jiang, Z., Wang, D., & Zhong, J. (2023). Rheological behavior of fresh cement composites with graphene oxide-coated silica fume. *Journal of Materials in Civil Engineering, 35*(10)19435533. Available from https://doi.org/10.1061/JMCEE7.MTENG-15428, http://ascelibrary.org/mto/resource/1/jmcee7/.

Lu, D., Shi, X., Wong, H. S., Jiang, Z., & Zhong, J. (2022). Graphene coated sand for smart cement composites. *Construction and Building Materials, 346.* Available from https://doi.org/10.1016/j.conbuildmat.2022.128313, https://www.journals.elsevier.com/construction-and-building-materials.

Lu, D., Shi, X., & Zhong, J. (2022a). Nano-engineering the interfacial transition zone in cement composites with graphene oxide. *Construction and Building Materials, 356.* Available from https://doi.org/10.1016/j.conbuildmat.2022.129284, https://www.journals.elsevier.com/construction-and-building-materials.

Lu, D., Shi, X., & Zhong, J. (2022b). Understanding the role of unzipped carbon nanotubes in cement pastes. *Cement and Concrete Composites, 126.* Available from https://doi.org/10.1016/j.cemconcomp.2021.104366, http://www.sciencedirect.com/science/journal/09589465.

Lu, D., Shi, X., & Zhong, J. (2022c). Interfacial nano-engineering by graphene oxide to enable better utilization of silica fume in cementitious composite. *Journal of Cleaner Production, 354.* Available from https://doi.org/10.1016/j.jclepro.2022.131381, https://www.journals.elsevier.com/journal-of-cleaner-production.

Lu, D., Shi, X., & Zhong, J. (2022d). Interfacial bonding between graphene oxide coated carbon nanotube fiber and cement paste matrix. *Cement and Concrete Composites, 134.* Available from https://doi.org/10.1016/j.cemconcomp.2022.104802, http://www.sciencedirect.com/science/journal/09589465.

Lu, D., Tang, Z., Zhang, L., Zhou, J., Gong, Y., Tian, Y., & Zhong, J. (2020). Effects of combined usage of supplementary cementitious materials on the thermal properties and microstructure of high-performance concrete at high temperatures. *Materials, 13*(8) 19961944. Available from https://doi.org/10.3390/MA13081833, https://www.mdpi.com/1996-1944/13/8/1833.

Lu, D., Wang, Y., Leng, Z., & Zhong, J. (2021). Influence of ternary blended cementitious fillers in a cold mix asphalt mixture. *Journal of Cleaner Production, 318.* Available from https://doi.org/10.1016/j.jclepro.2021.128421, https://www.journals.elsevier.com/journal-of-cleaner-production.

Lu, D., Wang, D., Wang, Y., & Zhong, J. (2023). Nano-engineering the interfacial transition zone between recycled concrete aggregates and fresh paste with graphene oxide.

Construction and Building Materials, 384. Available from https://doi.org/10.1016/j.conbuildmat.2023.131244, https://www.journals.elsevier.com/construction-and-building-materials.

Lu, D., Wang, D., & Zhong, J. (2022). Highly conductive and sensitive piezoresistive cement mortar with graphene coated aggregates and carbon fiber. *Cement and Concrete Composites, 134*. Available from https://doi.org/10.1016/j.cemconcomp.2022.104731, http://www.sciencedirect.com/science/journal/09589465.

Lu, Z., Yao, J., & Leung, C. K. Y. (2019). Using graphene oxide to strengthen the bond between PE fiber and matrix to improve the strain hardening behavior of SHCC. *Cement and Concrete Research, 126*. Available from https://doi.org/10.1016/j.cemconres.2019.105899, http://www.sciencedirect.com/science/journal/00088846.

Lu, D., & Zhong, J. (2022). Carbon-based nanomaterials engineered cement composites: a review. *Journal of Infrastructure Preservation and Resilience*, *3*(1). Available from https://doi.org/10.1186/s43065-021-00045-y.

Lu, D., Zhong, J., Yan, B., Gong, J., He, Z., Zhang, G., & Song, C. (2021). Effects of curing conditions on the mechanical and microstructural properties of ultra-high-performance concrete (UHPC) incorporating iron tailing powder. *Materials*, *14*(1), 1−14. Available from https://doi.org/10.3390/ma14010215, https://www.mdpi.com/1996-1944/14/1/215/pdf.

Lv, S., Liu, J., Sun, T., Ma, Y., & Zhou, Q. (2014). Effect of GO nanosheets on shapes of cement hydration crystals and their formation process. *Construction and Building Materials*, *64*, 231−239. Available from https://doi.org/10.1016/j.conbuildmat.2014.04.061.

Lv, S., Ma, Y., Qiu, C., Sun, T., Liu, J., & Zhou, Q. (2013). Effect of graphene oxide nanosheets of microstructure and mechanical properties of cement composites. *Construction and Building Materials*, *49*, 121−127. Available from https://doi.org/10.1016/j.conbuildmat.2013.08.022.

Makar, J. M., & Chan, G. W. (2009). Growth of cement hydration products on single-walled carbon nanotubes. *Journal of the American Ceramic Society*, *92*(6), 1303−1310. Available from https://doi.org/10.1111/j.1551-2916.2009.03055.x.

Marchesini, S., Turner, P., Paton, K. R., Reed, B. P., Brennan, B., Koziol, K., & Pollard, A. J. (2020). Gas physisorption measurements as a quality control tool for the properties of graphene/graphite powders. *Carbon*, *167*, 585−595. Available from https://doi.org/10.1016/j.carbon.2020.05.083, http://www.journals.elsevier.com/carbon/.

Matalkah, F., & Soroushian, P. (2020). Graphene nanoplatelet for enhancement the mechanical properties and durability characteristics of alkali activated binder. *Construction and Building Materials*, *249*. Available from https://doi.org/10.1016/j.conbuildmat.2020.118773, https://www.journals.elsevier.com/construction-and-building-materials.

Meng, W., & Khayat, K. H. (2018). Effect of graphite nanoplatelets and carbon nanofibers on rheology, hydration, shrinkage, mechanical properties, and microstructure of UHPC. *Cement and Concrete Research*, *105*, 64−71. Available from https://doi.org/10.1016/j.cemconres.2018.01.001, http://www.sciencedirect.com/science/journal/00088846.

Meng, W., & Khayat, K. H. (2016). Mechanical properties of ultra-high-performance concrete enhanced with graphite nanoplatelets and carbon nanofibers. *Composites Part B: Engineering*, *107*, 113−122. Available from https://doi.org/10.1016/j.compositesb.2016.09.069.

Mohammed, A., Sanjayan, J. G., Duan, W. H., & Nazari, A. (2015). Incorporating graphene oxide in cement composites: A study of transport properties. *Construction and Building Materials*, *84*, 341−347. Available from https://doi.org/10.1016/j.conbuildmat.2015.01.083.

Mohammed, A., Sanjayan, J. G., Nazari, A., & Al-Saadi, N. T. K. (2018). The role of graphene oxide in limited long-term carbonation of cement-based matrix. *Construction and Building Materials*, *168*, 858−866. Available from https://doi.org/10.1016/j.conbuildmat.2018.02.082.

Monteiro, P. J. M., Miller, S. A., & Horvath, A. (2017). Towards sustainable concrete. *Nature Materials*, *16*(7), 698–699. Available from https://doi.org/10.1038/nmat4930, http://www.nature.com/nmat/.

Muthu, M., Ukrainczyk, N., & Koenders, E. (2021). Effect of graphene oxide dosage on the deterioration properties of cement pastes exposed to an intense nitric acid environment. *Construction and Building Materials*, *269*. Available from https://doi.org/10.1016/j.conbuildmat.2020.121272, https://www.journals.elsevier.com/construction-and-building-materials.

Mypati, S., Sellathurai, A., Kontopoulou, M., Docoslis, A., & Barz, D. P. J. (2021). High concentration graphene nanoplatelet dispersions in water stabilized by graphene oxide. *Carbon*, *174*, 581–593. Available from https://doi.org/10.1016/j.carbon.2020.12.068, http://www.journals.elsevier.com/carbon/.

Ouyang, K., Shi, C., Chu, H., Guo, H., Song, B., Ding, Y., Guan, X., Zhu, J., Zhang, H., Wang, Y., & Zheng, J. (2020). An overview on the efficiency of different pretreatment techniques for recycled concrete aggregate. *Journal of Cleaner Production*, *263*. Available from https://doi.org/10.1016/j.jclepro.2020.121264, https://www.journals.elsevier.com/journal-of-cleaner-production.

Paul, S. C., van Rooyen, A. S., van Zijl, G. P. A. G., & Petrik, L. F. (2018). Properties of cement-based composites using nanoparticles: A comprehensive review. *Construction and Building Materials*, *189*, 1019–1034. Available from https://doi.org/10.1016/j.conbuildmat.2018.09.062.

Pei, S., & Cheng, H. M. (2012). The reduction of graphene oxide. *Carbon*, *50*(9), 3210–3228. Available from https://doi.org/10.1016/j.carbon.2011.11.010, http://www.journals.elsevier.com/carbon/.

Qin, Renyuan, Zhou, Ao, Yu, Zechuan, Wang, Quan, & Lau, Denvid (2021). Role of carbon nanotube in reinforcing cementitious materials: An experimental and coarse-grained molecular dynamics study. *Cement and Concrete Research*, *147*. Available from https://doi.org/10.1016/j.cemconres.2021.106517.

Qiu, Liangsheng, Dong, Sufen, Ashour, Ashraf, & Han, Baoguo (2020). Antimicrobial concrete for smart and durable infrastructures: A review. *Construction and Building Materials*, *260*. Available from https://doi.org/10.1016/j.conbuildmat.2020.120456.

Qureshi, T. S., & Panesar, D. K. (2019). Impact of graphene oxide and highly reduced graphene oxide on cement based composites. *Construction and Building Materials*, *206*, 71–83. Available from https://doi.org/10.1016/j.conbuildmat.2019.01.176.

Rahman, M. L., Malakooti, A., Ceylan, H., Kim, S., & Taylor, P. C. (2022). A review of electrically conductive concrete heated pavement system technology: From the laboratory to the full-scale implementation. *Construction and Building Materials*, *329*. Available from https://doi.org/10.1016/j.conbuildmat.2022.127139, https://www.journals.elsevier.com/construction-and-building-materials.

Rong, X. L., Li, L., Huang, W. Y., Dong, L. G., Zheng, S. S., Wang, F., Lu, D., & Wang, J. Y. (2023). Experimental investigation of the seismic resistance of RC beam–column connections after freeze–thaw cycle treatment. *Engineering Structures*, *290*. Available from https://doi.org/10.1016/j.engstruct.2023.116330, http://www.journals.elsevier.com/engineering-structures/.

Rong, X. L., Li, L., Zheng, S. S., Wang, F., Huang, W. Y., Zhang, Y. X., & Lu, D. (2023). Freeze-thaw damage model for concrete considering a nonuniform temperature field. *Journal of Building Engineering*, *72*.

Rong, Z., Sun, W., Xiao, H., & Jiang, G. (2015). Effects of nano-SiO2 particles on the mechanical and microstructural properties of ultra-high performance cementitious composites.

Cement and Concrete Composites, *56*, 25–31. Available from https://doi.org/10.1016/j.cemconcomp.2014.11.001, http://www.sciencedirect.com/science/journal/09589465.

Ruoff, R. S., Lee, K. H., & Lee, S. H. (2020). Synthesis of diamond-like carbon nanofiber films. *ACS Nano*, *14*(10), 13663–13672. Available from https://doi.org/10.1021/acsnano.0c05810, http://pubs.acs.org/journal/ancac3.

Schneider, M., Romer, M., Tschudin, M., & Bolio, H. (2011). Sustainable cement production —present and future. *Cement and Concrete Research*, *41*(7), 642–650. Available from https://doi.org/10.1016/j.cemconres.2011.03.019.

Shang, H., Li, Y., Zhang, Y., Wei, X., Qi, Y., & Zhao, D. (2019). Influences of adding nano-graphite powders on microstructure and gaseous hydrogen storage properties of ball milled Mg90Al10 alloy. *Carbon*, *149*, 93–104. Available from https://doi.org/10.1016/j.carbon.2019.04.028, http://www.journals.elsevier.com/carbon/.

Shi, C., Jiménez, A. F., & Palomo, A. (2011). New cements for the 21st century: The pursuit of an alternative to Portland cement. *Cement and Concrete Research*, *41*(7), 750–763. Available from https://doi.org/10.1016/j.cemconres.2011.03.016, http://www.sciencedirect.com/science/journal/00088846.

Shi, T., Li, Z., Guo, J., Gong, H., & Gu, C. (2019). Research progress on CNTs/CNFs-modified cement-based composites – A review. *Construction and Building Materials*, *202*, 290–307. Available from https://doi.org/10.1016/j.conbuildmat.2019.01.024.

Silvestro, L., & Jean Paul Gleize, P. (2020). Effect of carbon nanotubes on compressive, flexural and tensile strengths of Portland cement-based materials: A systematic literature review. *Construction and Building Materials*, *264*. Available from https://doi.org/10.1016/j.conbuildmat.2020.120237, https://www.journals.elsevier.com/construction-and-building-materials.

Singh, N. B., Kalra, Meenu, & Saxena, S. K. (2017). Nanoscience of Cement and Concrete. *Materials Today: Proceedings*, *4*(4), 5478–5487. Available from https://doi.org/10.1016/j.matpr.2017.06.003.

Sobolkina, A., Mechtcherine, V., Bergold, S. T., Neubauer, J., Bellmann, C., Khavrus, V., Oswald, S., Leonhardt, A., & Reschetilowski, W. (2016). Effect of carbon-based materials on the early hydration of tricalcium silicate. *Journal of the American Ceramic Society*, *99*(6), 2181–2196.

Song, C., Hong, G., & Choi, S. (2020). Effect of dispersibility of carbon nanotubes by silica fume on material properties of cement mortars: Hydration, pore structure, mechanical properties, self-desiccation, and autogenous shrinkage. *Construction and Building Materials*, *265*. Available from https://doi.org/10.1016/j.conbuildmat.2020.120318, https://www.journals.elsevier.com/construction-and-building-materials.

Sun, G., Liang, R., Lu, Z., Zhang, J., & Li, Z. (2016). Mechanism of cement/carbon nanotube composites with enhanced mechanical properties achieved by interfacial strengthening. *Construction and Building Materials*, *115*, 87–92. Available from https://doi.org/10.1016/j.conbuildmat.2016.04.034.

Sun, S., Yu, X., Han, B., & Ou, J. (2013). In situ growth of carbon nanotubes/carbon nanofibers on cement/mineral admixture particles: A review. *Construction and Building Materials*, *49*, 835–840. Available from https://doi.org/10.1016/j.conbuildmat.2013.09.011.

Tafesse, M., & Kim, H. K. (2019). The role of carbon nanotube on hydration kinetics and shrinkage of cement composite. *Composites Part B: Engineering*, *169*, 55–64. Available from https://doi.org/10.1016/j.compositesb.2019.04.004.

Tian, Y., Lu, D., Ma, R., Zhang, J., Li, W., & Yan, X. (2020). Effects of cement contents on the performance of cement asphalt emulsion mixtures with rapidly developed early-age strength. *Construction and Building Materials*, *244*. Available from https://doi.org/

10.1016/j.conbuildmat.2020.118365, https://www.journals.elsevier.com/construction-and-building-materials.

Wang, L., & Aslani, F. (2019). A review on material design, performance, and practical application of electrically conductive cementitious composites. *Construction and Building Materials*, *229*. Available from https://doi.org/10.1016/j.conbuildmat.2019.116892, https://www.journals.elsevier.com/construction-and-building-materials.

Wang, B., Jiang, R., & Wu, Z. (2016). Investigation of the mechanical properties and microstructure of graphene nanoplatelet-cement composite. *Nanomaterials*, *6*(11). Available from https://doi.org/10.3390/nano6110200, http://www.mdpi.com/2079-4991/6/11/200/pdf.

Wang, Y., Lu, D., Wang, F., Zhang, D., Zhong, J., Liang, B., Gui, X., & Sun, L. (2020). A new strategy to prepare carbon nanotube thin film by the combination of top-down and bottom-up approaches. *Carbon*, *161*, 563–569. Available from https://doi.org/10.1016/j.carbon.2020.01.090, http://www.journals.elsevier.com/carbon/.

Wu, Q., Ji, Z., Xin, L., Li, D., Zhang, L., Liu, C., Yang, T., Wen, Z., Wang, H., Xin, B., Xue, H., Chen, F., Xu, Z., Cui, H., & He, M. (2019). Iron silicide-catalyzed growth of single-walled carbon nanotubes with a narrow diameter distribution. *Carbon*, *149*, 139–143. Available from https://doi.org/10.1016/j.carbon.2019.04.059, http://www.journals.elsevier.com/carbon/.

Xu, G., Du, S., He, J., & Shi, X. (2019). The role of admixed graphene oxide in a cement hydration system. *Carbon*, *148*, 141–150. Available from https://doi.org/10.1016/j.carbon.2019.03.072, http://www.journals.elsevier.com/carbon/.

Xu, Y., Zeng, J., Chen, W., Jin, R., Li, B., & Pan, Z. (2018). A holistic review of cement composites reinforced with graphene oxide. *Construction and Building Materials*, *171*, 291–302. Available from https://doi.org/10.1016/j.conbuildmat.2018.03.147.

Xu, J., & Zhang, D. (2017). Pressure-sensitive properties of emulsion modified graphene nanoplatelets/cement composites. *Cement and Concrete Composites*, *84*, 74–82. Available from https://doi.org/10.1016/j.cemconcomp.2017.07.025, http://www.sciencedirect.com/science/journal/09589465.

Xu, G., Zhong, J., & Shi, X. (2018). Influence of graphene oxide in a chemically activated fly ash. *Fuel*, *226*, 644–657. Available from https://doi.org/10.1016/j.fuel.2018.04.033, http://www.journals.elsevier.com/fuel/.

Yan, Y., Zhai, D., Liu, Y., Gong, J., Chen, J., Zan, P., Zeng, Z., Li, S., Huang, W., & Chen, P. (2020). van der Waals Heterojunction between a Bottom-Up Grown Doped Graphene Quantum Dot and Graphene for Photoelectrochemical Water Splitting. *ACS nano*, *14*(1), 1185–1195. Available from https://doi.org/10.1021/acsnano.9b09554.

Zhao, L., Guo, X., Ge, C., Li, Q., Guo, L., Shu, X., & Liu, J. (2017). Mechanical behavior and toughening mechanism of polycarboxylate superplasticizer modified graphene oxide reinforced cement composites. *Composites Part B: Engineering*, *113*, 308–316. Available from https://doi.org/10.1016/j.compositesb.2017.01.056.

Zhao, L., Guo, X., Liu, Y., Ge, C., Chen, Z., Guo, L., Shu, X., & Liu, J. (2018). Investigation of dispersion behavior of GO modified by different water reducing agents in cement pore solution. *Carbon*, *127*, 255–269. Available from https://doi.org/10.1016/j.carbon.2017.11.016, http://www.journals.elsevier.com/carbon/.

Zhao, L., Guo, X., Liu, Y., Ge, C., Guo, L., Shu, X., & Liu, J. (2017). Synergistic effects of silica nanoparticles/polycarboxylate superplasticizer modified graphene oxide on mechanical behavior and hydration process of cement composites. *RSC Advances*, *7* (27), 16688–16702. Available from https://doi.org/10.1039/c7ra01716b, http://pubs.rsc.org/en/journals/journalissues.

Zhao, L., Guo, X., Song, L., Song, Y., Dai, G., & Liu, J. (2020). An intensive review on the role of graphene oxide in cement-based materials. *Construction and Building Materials, 241*. Available from https://doi.org/10.1016/j.conbuildmat.2019.117939, https://www.journals.elsevier.com/construction-and-building-materials.

Zhao, L., Zhu, S., Wu, H., Zhang, X., Tao, Q., Song, L., Song, Y., & Guo, X. (2020). Deep research about the mechanisms of graphene oxide (GO) aggregation in alkaline cement pore solution. *Construction and Building Materials, 247*. Available from https://doi.org/10.1016/j.conbuildmat.2020.118446, https://www.journals.elsevier.com/construction-and-building-materials.

Zhu, X., Gao, Y., Dai, Z., Corr, D. J., & Shah, S. P. (2018). Effect of interfacial transition zone on the Young's modulus of carbon nanofiber reinforced cement concrete. *Cement and Concrete Research, 107*, 49–63. Available from https://doi.org/10.1016/j.cemconres.2018.02.014, http://www.sciencedirect.com/science/journal/00088846.

Sustainable thermal energy storage concrete incorporated with phase change materials

18

Yushi Liu[1,2,3], *Yunshi Pan*[1], *Kunyang Yu*[1] *and Yingzi Yang*[1,2,3]
[1]School of Civil Engineering, Harbin Institute of Technology, Harbin, P.R. China, [2]Key Lab of Structures Dynamic Behavior and Control of the Ministry of Education, Harbin Institute of Technology, Harbin, P.R. China, [3]Key Lab of Smart Prevention and Mitigation of Civil Engineering Disasters of the Ministry of Industry and Information Technology, Harbin Institute of Technology, Harbin, P.R. China

18.1 Introduction

Aggregates and cementitious elements combine to form concrete. It has numerous advantages, including a straightforward manufacturing process, simple access to raw materials, great strength, low cost, etc., and is the most commonly used building material around the world (Liu et al., 2023). Researchers are paying more attention to smart concrete that have multiple functionalities in order to make better use of concrete (Madbouly et al., 2020). On the foundation of traditional concrete, functional elements or components are combined to create smart concrete (Chung, 2021). It has adaptive abilities and can sense changes in the serving environment, such as stress level, environmental temperature, etc. (Shao et al., 2022). At present, the research on smart concrete mostly focuses on the perception and response functions of the load state (Han et al., 2015). Self-healing (Pan & Gencturk, 2023) and conductive concrete (Liu et al., 2021), for instance, have been shown to have promising futures in the monitoring of structural health and other areas (Gaumet et al., 2021). There are, however, few investigations on smart concrete that can adapt to and control the distribution of the surrounding temperature.

The world's population is now continuing its pattern of fast expansion (Lutz & Kc, 2010). With the increase in population and accelerated urbanization, environmental issues including greenhouse gas emissions and energy constraints are getting worse (Abbasi & Abbasi, 2011). The amount of natural resources used by the building sector now accounts for 40% of total resource consumption worldwide (Xia et al., 2020). Additionally, more than half of all building energy consumption has been used to regulate thermal comfort during the operational stage of buildings (Akeiber et al., 2016). Therefore designing concrete materials with thermal energy storage and temperature control properties is a very promising technique to lower building energy consumption.

Sustainable Concrete Materials and Structures. DOI: https://doi.org/10.1016/B978-0-443-15672-4.00018-8

Normally, the temperature-sensing function of smart concrete is generally realized by adding conductive materials to the concrete (Ding, 2023). The change in concrete resistance can serve as an indirect indicator of changes in the surrounding temperature, and an external power source have to be used to regulate concrete temperature in accordance with application scenario requirements (Zhang et al., 2020). However, conductive concrete tends to be expensive and needs extra equipment for measuring resistance (Arabzadeh et al., 2019), and the concrete is still unable to react in real time to changes in the ambient temperature. One of the ideal approaches to address the short board of smart concrete in temperature perception and regulation is to combine phase change materials (PCMs) and concrete.

PCM is a type of functional material that can convert and store thermal energy through the phase change process in response to changes in environmental temperature (Alehosseini & Jafari, 2019). PCMs have the unique ability to store heat energy in the form of latent heat, which distinguishes them from conventional energy storage materials (Kousksou et al., 2014). Compared to sensible heat, latent heat energy storage has a higher-energy density. This means that it can store more heat energy with less material consumption (Yu et al., 2021). Generally, PCMs can be classified into four categories, including solid–solid, solid–liquid, solid–gas, and liquid–gas (Gao et al., 2022). In recent years, PCMs have been widely used in fields including electronic equipment thermal management (Chen et al., 2012), clean energy transition and storage (Huang et al., 2022), medical care (Mohammadi et al., 2022), and textile (Mondal, 2008), as a low-carbon and environmentally friendly energy storage material. Considering the application of PCMs in the construction, solid–liquid PCMs are most commonly used in order to maximize the heat storage function in the thermal comfort temperature range of buildings. However, the volume of solid–liquid PCMs changes significantly during the phase change process (Alva et al., 2017). Encapsulation can solve the problem of leakage caused by size instability. Commonly used encapsulation strategies include physical adsorption (Chen et al., 2015), core–shell encapsulation (Peng et al., 2022), covalent grafting (Sun et al., 2020), and polymerization (Zhang et al., 2018). The physical adsorption method and core–shell encapsulation method are relatively simple to prepare as they do not involve any chemical reactions between PCMs and the encapsulation materials. The covalent grafting and polymerization method achieves the goal of encapsulating PCMs by forming a chemical bond between the PCMs and the encapsulation material. However, the preparation process is relatively complex. The selection of the encapsulation method for PCMs should be based on their properties and application scenarios.

The potential for synergistic collaboration between PCMs and concrete has been confirmed in numerous studies (Pasupathy et al., 2008). There are usually two ways to composite PCMs with concrete, one is to directly incorporation, and the other is to encapsulate PCMs to obtain form-stable PCMs before adding them to the concrete mixture. The direct incorporation method is easy to operate and the most economical, but it often leads to rapid reduction of the mechanical properties of concrete (Navarro et al., 2016). The indirect incorporation method can make concrete have more functions on the basis of energy storage through the collaborative application of encapsulation materials and PCMs, which has gained lots of attention recently (Singh et al., 2023). The combination of PCM and concrete material can make concrete have the function of temperature self-sensing and self-adaptive adjustment (Ling & Poon, 2013). When the ambient temperature changes, PCMs absorb or release energy through phase transition process, which allows for the storage of heat energy and temperature regulation without the need of external force. This type of smart thermal energy storage concrete can serve as a passive thermal energy storage unit in building structures. Based on this premise, the function of adjusting the thermal comfort of a building can be achieved without consuming additional energy. This approach offers the advantage of being both eco-friendly and more efficient when compared to other methods of thermal comfort adjustment (Sun et al., 2022).

Smart thermal energy storage concrete is an important subfield of smart concrete. As a novel type of low-carbon emission and environmentally friendly thermal energy storage material, the smart thermal energy storage concrete has become the focus of research worldwide. In this chapter, the classification and encapsulation strategies of PCMs are introduced in details. The method and effectiveness of incorporating PCMs into concrete are analyzed and evaluated retrospectively. Furthermore, this chapter delves into the impact of PCMs on both the mechanical and thermal characteristics of concrete. This analysis and discussion provide a valuable theoretical reference for the research of smart thermal energy storage concrete.

18.2 Phase change materials

18.2.1 Classification

PCMs can store thermal energy in the form of latent heat through the transition of their physical state when the external temperature changes (Samimi et al., 2016). The typical classification strategies of PCMs are shown in Fig. 18.1.

Generally, PCMs can be divided into single and eutectic. A single PCM is composed of only one component and can be further divided into organic PCM and inorganic PCM. Organic PCMs (such as paraffin, fatty acid, sugar alcohol, etc.) usually have the advantages of stable chemical properties, minor supercooling and free of being corrosion (Cabeza et al., 2011). However, they show a great limitation in thermal conductivity (Liang et al., 2023). Inorganic PCMs mainly contain hydrated salt, molten salt, and metal. The phase change temperature of molten salts is usually above 200°C, making it difficult to apply it in the field of construction engineering. The hydrated salts usually have a phase change temperature below 100°C, which has great potential for application in the construction field (Chen et al., 2014). Nevertheless, hydrated salts have the problem of phase stratification during the phase change process (Yu et al., 2021), which can be remedied through encapsulation, modification, and other treatments (Liu & Yang, 2017). The eutectic PCM consists of two or more different types of components (Sharma et al., 2009). For eutectic PCMs, the interaction between the crystallization behavior of different

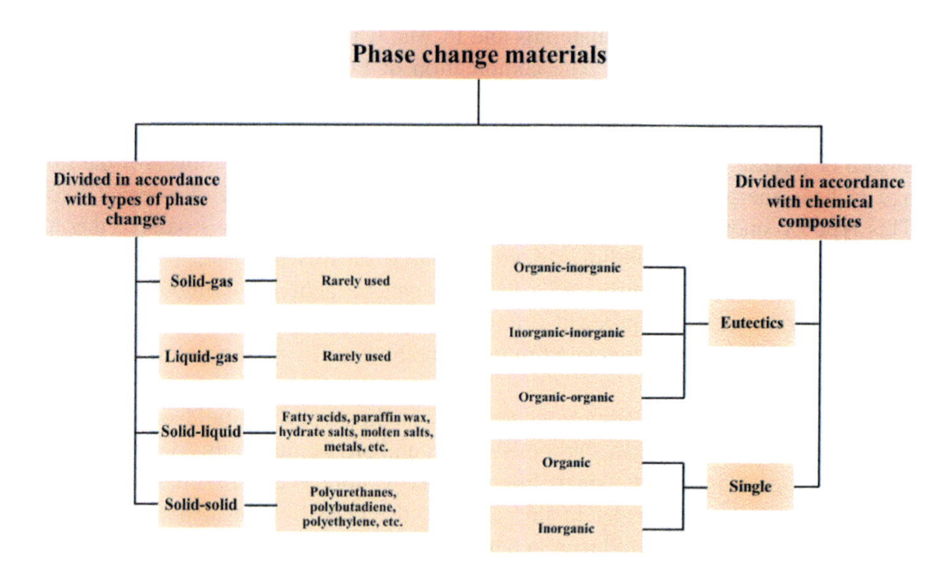

Figure 18.1 Typical classification of phase change materials.

components has a significant impact on their thermal properties. Recently, our group developed a new type of decanoic acid/polyethylene glycol eutectic PCM that was suitable for building thermal comfort adjusting (Yu et al., 2023). The crystallization behavior and evolution mechanism of the decanoic acid/polyethylene glycol eutectic PCM had been pointed out. The results showed that DA adsorbed around the long chain of polyethylene glycol (PEG) due to intermolecular hydrogen bonding, making it more orderly arranged. Consequently, the degree of crystallization increased and supercooling decreased, leading to an enhancement of thermal energy storage capacity.

On the basis of the phase change process, PCMs can be divided into solid–liquid PCMS, solid–solid PCMs, solid–gas PCMs, and liquid–gas PCMs (Alehosseini & Jafari, 2019). The phase change behavior of solid-solid PCMs mainly benefits from the local transformation of crystal structure, which has a small volume change, a relatively high phase change temperature, and a more complex preparation process (Yu et al., 2021). Solid–liquid PCMs are the most commonly used type of PCM in research on building materials due to their characteristics of appropriate phase change temperature, wide range of sources, low prices, and stable performance (Su et al., 2015). Whereas solid–gas and liquid–gas PCMs are rarely used in various fields due to the large shape deformation during the phase transition process.

18.2.2 Stabilizing strategies

The shape change of PCMs during the phase change process can affect the efficiency of their latent heat (Trigui et al., 2014). Therefore researchers are considering using supplementary materials to make PCMs formally stable (i.e., preparing form-stable PCMs) so that their physical characteristics remain relatively stable during the phase change process, such as volume and shape. There are already many different stabilizing strategies for PCMs with different physical and chemical properties. The most used stabilizing strategies can be divided into two categories: physical stabilizing strategies and chemical stabilizing strategies according to whether PCMs participate in chemical reactions directly during the synthesis process. The physical stabilizing strategy refers to the fact that supplementary materials do not undergo direct chemical reactions with PCMs. On the contrary, chemical stabilizing strategy refers to the formation of new chemical bonds through chemical reactions between PCMs and supplementary materials to achieve the purpose of stabilizing.

18.2.2.1 Physical stabilizing strategies

Physical stabilizing strategies are mainly suitable for stabilizing solid–liquid PCMs. The mainstream physical stabilizing strategies include physical absorption and core–shell encapsulation. The supplementary materials used in the physical absorption method usually have high porosity and a large specific surface area in order to load PCMs. The shaping of PCMs is mainly achieved through capillary action and surface tension (Liang et al., 2023). The commonly used loading carrier materials in physical adsorption methods include various porous carbon materials (e.g., carbon aerogel (Kashyap et al., 2020), hierarchical porous carbon (Feng et al., 2021), biochar (Muchtar et al., 2022), etc.), porous minerals (e.g., diatomite (Qian et al., 2015), expanded perlite (Fu et al., 2017), expanded vermiculite (Wang et al., 2021), etc.), and metal foam (Borhani et al., 2021). It is necessary to select the appropriate loading carrier for physical absorption according to the characteristics of PCMs. The basic requirements (Deng et al., 2018) for loading materials include:

No leakage of melted PCMs.

Having stable physical and chemical properties.

No fusion and destruction during the phase change process.

No chemical reaction between the loading materials and PCMs.

The loading carrier improves the performance such as low thermal conductivity and excessive undercooling in addition to enhancing the dimensional stability of PCMs. Moreover, the loading materials also perform a number of other tasks, including photothermal and electrothermal conversion (Wu et al., 2021).

Wang et al. (2022) used waste plastic as a raw material to prepare carbon nanotubes and used them as loading carriers to prepare form-stable PCMs through physical blending methods. Compared with pure PCMs, the form-stable PCMs achieved great leak resistance and thermal cycling stability with a latent heat retention rate exceeding 90%. Meanwhile, the thermal conductivity of form-stable PCMs had increased by 104%. Yan et al. (2023) used expanded graphite to load erythrose to prepare an erythritol/expanded graphite form-stable PCM through physical blending. The obtained form-stable PCMs had good thermal storage capacity and thermal conductivity. The prepare process of physical blending was shown in Fig. 18.2A.

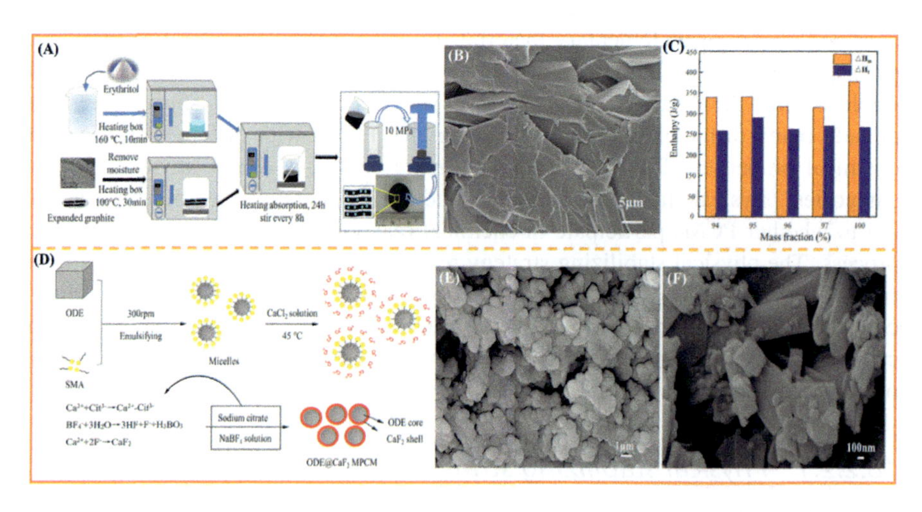

Figure 18.2 (A) Prepare process of physical blending, (B) microstructure of expanded graphite after absorption, (C) enthalpy of prepared erythritol/expanded graphite (Yan et al., 2023), (D) prepare process of core−shell encapsulation, (E) microstructure of octadecane/ CaF_2 at core−shell ratio of 3, (F) microstructure of octadecane/CaF_2 at core−shell ratio of 4 (Hu et al., 2022).

First, PCMs were melted at a high temperature and then mixed with expanded graphite to obtain powder-like form-stable PCMs. In order to test the mechanical strength of the form-stable PCMs, the powder was subjected to compression treatment. As shown in Fig. 18.2B, the PCM had been absorbed into the expanded graphite successfully. Fig. 18.2C indicated that the enthalpy of form-stable PCMs reached 290.52 J/g with erythritol contents of 95%. However, when the mass fraction of erythritol was further increased, the enthalpy nearly remained constant, indicating that the adsorption of expanded graphite had reached a saturation state. Apart from that, the supercooling of the form-stable PCMs was significantly reduced, and the thermal conductivity was increased compared with pure erythritol, confirming that the form-stable PCM obtained through physical absorption had better functionality.

In order to improve the adsorption capacity, a vacuum impregnation method was further developed by optimizing the adsorption method based on the physical blending method. The vacuum impregnation method can effectively eliminate air and water in the pores of the loading materials through the auxiliary effect of air pressure in the vacuum environment (Lee et al., 2019), significantly improving the loading ratio of PCMs and achieving higher thermal storage capacity (Nomura et al., 2009). Xiao et al. (2022) used silver nanoparticles to modify biochar to prepare form-stable PCMs loofah foam/PEG by vacuum impregnation. The enthalpy of loofah foam/PEG was 140.5 J/g, and the encapsulation efficiency was up to 95%. In addition, the thermal conductivity of loofah foam/PEG reached 0.632 W/m · K after the modification. Furthermore, the thermophysical properties of loofah foam/PEG remained stable after 100 thermal cycles.

The physical blending and vacuum impregnation methods in the physical adsorption method both had their own advantages. On the one hand, the physical blending method had the advantages of a simple process, easy operation, and low requirements for experimental equipment (Zhang et al., 2022). On the other hand, the form-stable PCM obtained by the vacuum impregnation method had a high PCM loading ratio (Mehrali et al., 2013). However, the vacuum impregnation method had some disadvantages, such as high requirements for equipment and high production costs, which limited its application. After all, the physical adsorption method cannot achieve complete encapsulation of PCMs. Due to the exposure of surface pores, the antileakage ability of form-stable PCMs obtained by the physical adsorption method was relatively poor at higher temperatures. It was also difficult to ensure long-term stability for the thermal physical properties of form-stable PCMs in working environments such as high pressure and impact. Considering a further improvement in the encapsulation level of PCMs, a core–shell encapsulation method can be adopted.

Core–shell encapsulation was to use PCMs as core materials and encapsulation materials as shell materials to realize shape stabilization. The melting point of shell materials was higher than that of PCMs, ensuring the form-stable PCM remained shape stable throughout the entire phase change process and avoiding leakage (Arshad et al., 2019). The core–shell encapsulation can synthesis form-stable PCMs with smaller sizes and uniform shapes. This method had certain applicability to PCMs with different chemical compositions and can optimize the thermal properties of form-stable PCMs (Zhang et al., 2012). Therefore core–shell encapsulation had also attracted more and more attention. In recent years, researchers around the world had successfully prepared form-stable PCMs with particle sizes ranging from micrometers to microns and even nanometers.

Recently, a research group from Nanjing, China, used melamine-urea-formaldehyde (MUF) as shell materials to encapsulate tetradecane (Wang et al, 2022). In their study, phase change nanocapsules with a core–shell structure were successfully prepared and further modified by titanium dioxide (TiO_2). The obtained nanocapsules had a melting enthalpy of 156.2 J/g and an encapsulation efficiency of 68.6%. Besides, the nanocapsules still maintained thermal stability and exhibited a regular spherical shape after 100 melting-crystallization cycles. In addition, after modification with TiO_2, the thermal conductivity of phase change nanocapsules had been significantly improved. Apart from organic polymers, many inorganic materials can also serve as shell materials to achieve collaborative work with organic PCMs through core–shell encapsulation. Hu et al. (2022) used CaF_2 as shell materials to prepare microencapsulated PCMs (MPCMs) with different core–shell ratios through core–shell encapsulation. The preparation process was shown in Fig. 18.2D. Octadecane was emulsified into microdroplets to form an O/W system under the action of surfactant. Then, the hydrophilic group of the surfactant on the surface of microdroplets adsorbed Ca^{2+} from $CaCl_2$ solution and F^- from $NaBF_4$ solution. Finally, the CaF_2 shells were formed on the surface of octadecane microdroplets through the self-assembly reaction. The microstructures of MPCMs with different core–shell ratio were shown in Fig. 18.2E and F. It can be seen from the figure that the micromorphology was spherical when the core–shell ratio was below 3:1, while the MPCMs appeared sheet-like when the core–shell ratio reached 4:1. In the experimental test, the MPCMs showed good thermal performance, with a latent heat of 131 J/g and a maximum encapsulation ratio of 54.02%. The CaF_2 shell not only made the MPCMs had good thermal cycling stability but also had high thermal conductivity. Moreover, TiO_2 (Cao et al., 2014), calcium carbonate (Zhang et al., 2023), and polyurea (Park et al., 2014) had also been used as shell materials to prepare form-stable PCMs with excellent performance.

18.2.2.2 Chemical stabilizing strategies

The PCMs and stabilizing materials are joined by chemical bonds in the chemical stabilizing techniques (Jiang et al., 2001). PCM would maintain its location fixed, guaranteeing dimensional stability on a macro scale because of the function played by the covalent link between PCMs and the stabilizing materials (Su & Liu, 2006).

Rong et al. (2022) developed a flexible shape table phase change material (FSPCM) based on paraffin, styrene-b-ethylene-co-butylene-b-styrene (SEBS) and expanded graphite (EG) for thermal management of batteries. The connection mechanism between PA, SEBS, and EG was shown in Fig. 18.3A and the preparation scheme was shown in Fig. 18.3B. Differential scanning calorimeter (DSC)

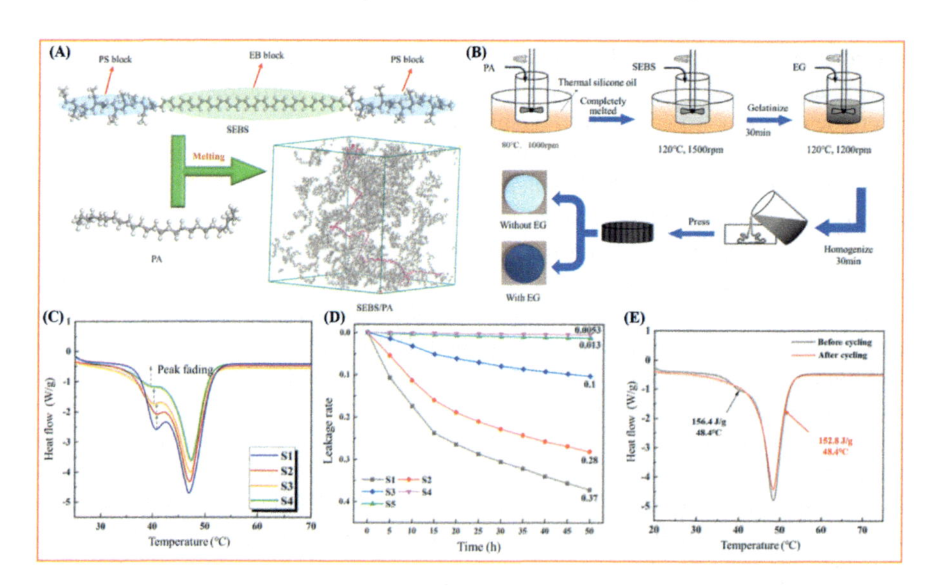

Figure 18.3 (A) Mechanism of the bonding between PA, styrene-b-ethylene-co-butylene-b-styrene, and expanded graphite, (B) scheme of the preparation process, (C) differential scanning calorimeter (DSC) curves of flexible shape table phase change material (FSPCM), (D) leakage ratio of FSPCM, (E) DSC curves of FSPCM before and after thermal cycling (Rong et al., 2022). *PA*, paraffin.

curves of FSPCM were shown in Fig. 18.3C, indicating that the phase change temperature of FSPCM was 47.01°C and the enthalpy was 193.6 J/g. It was noticed that the maximum thermal conductivity can reach 1.265 W/m · K, which was more than three times higher than that of pure PCMs. In terms of shape stability and thermal cycling stability, the leakage of PCMs was tested for 10 heating-cooling cycles. As shown in Fig. 18.3D, the leakage of PCMs showed a decreasing trend with the increase in SEBS content. The DSC curves of FSPCM before and after thermal cycling were shown in Fig. 18.3E. It was indicated that the phase change behavior of FSPCM remained almost unchanged before and after thermal cycling, demonstrating good thermal cycling stability. Sarı et al. (2017) grafted PEG onto a backbone consisted of styrene and maleic anhydride monomers (SMA) to prepare SMA-graft-PEG solid-solid form-stable PCMs. The phase change temperature of SMA-graft-PEG was 42.59°C and the enthalpy was 107.04 J/g. The overall shape of SMA-graft-PEG was stable during the heating process, and it still had good thermal reliability after 5000 heating-cooling cycles.

PCMs can also be fixed in polymers with three-dimensional (3D) network structures through chemical stabilizing strategies. The network polymer can maintain stability at high temperatures, ensuring that the shape of the form-stable PCMs remains unchanged during the phase change process. Form-stable PCMs with network structures typically exhibit excellent stability under high temperature conditions (Chen et al., 2014). Liu et al. (2020) synthesized form-stable PCMs by loading PEG onto a poly acrylic acid (PAA)/silica (SiO_2) network structure. The test results showed that the maximum enthalpy of these form-stable PCMs can reach 172.9 J/g and the form-stable PCMs can maintain shape stability at a temperature of 350°C. Besides, chemical stabilizing strategies can also help form-stable PCMs realize multifunctionality.

The solid–solid PCMs prepared by chemical stabilizing strategies had a higher phase change temperature and were mainly used for battery thermal management or large thermal storage systems. However, this type of stabilizing strategy was rarely used in the production process of concrete or other building materials because of the relatively higher phase change temperature, which was inconsistent with the physical comfort temperature.

18.3 Influence of form-stable phase change materials on properties of concrete

18.3.1 Incorporation methods of phase change materials into concrete

The incorporation methods of PCMs into concrete are usually divided into two categories: one is to directly add PCMs to the concrete through blending or immersion, and the other is to preencapsulate the PCMs and then composite them with the concrete.

18.3.1.1 Direct incorporation method

The direct incorporation method of PCMs was the most convenient and cost-effective method for preparing intelligent thermal storage concrete. As early as 1989, Hawes et al. (1989) had tried to increase the latent heat of concrete by directly incorporating PCMs during the mix process. As most kind of the concrete had a high alkaline environment, it was particularly important to choose appropriate PCMs to prevent the incompatible with concrete when using the direct incorporation method. This research showed that butyl stearate, dodecanol and PEG 600 were suitable for concrete preparation. Besides, Hawes also came up with the standards for PCMs chosen for thermal storage concrete. To be specific, PCMs used in concrete should not only have a good thermal property and chemical stability, but also not obstructing the bonding between cement slurry and aggregates. Afterward, many researchers adopted a direct incorporation method to prepare concrete with higher latent heat. The direct incorporation method can be further divided into two types: direct mixing and impregnation.

Direct mixing was to add PCMs without coating or encapsulation during the blending process of concrete (Soares et al., 2013). Cunha et al. (2020) directly added unencapsulated PCMs to cement mortars and studied the performance of the mortars after being exposed to low and high temperatures. The test results showed that the direct addition of PCMs significantly reduced the compressive and flexural strengths of the mortars. PCMs also had a serious negative impact on the mechanical properties of mortars at high temperatures, while PCMs had a certain positive effect on the frost resistance of the mortars in a relatively low-temperature environment. Since the adverse effect of direct incorporation method on the mechanical properties of concrete, it was rarely adopted in recent studies.

The impregnation method involved impregnating melted PCMs into concrete through capillary action. The quality of preparation mainly depended on the pore structure of the matrix as well as the viscosity of PCMs (Ling & Poon, 2013). Some researchers directly used PCMs to impregnate concrete specimens. The results showed that PCMs were successfully sucked into the pores of the concrete, but leakage occurred under thermal cycling (Lee et al., 2000). Lopez-Arias et al. (2023) immersed melted PCMs into the pores of hardened concrete in a vacuum environment. The process of impregnation was shown in Fig. 18.4A, and the relationship between the content of PCMs and porosity of concrete was shown in Fig. 18.4B. It was indicated that prolonging the impregnation time at different porosities can increase the content of PCMs, but the porosity had little effect on the absorption of PCMs under a short impregnation time. In this study, the compressive strength of concrete showed a slight increase with the increase in PCM content due to the filling effect of PCMs on surface pores. Apart from directly using hardened concrete as the impregnation carrier, the current mainstream impregnation method also included using porous aggregates as carriers to impregnate PCMs in a vacuum environment. Kalombe et al. (2023) prepared a low-cost antifreeze concrete by impregnating PCMs into lightweight aggregates. The prepared concrete can delay the freezing time and reduce the damage caused by freeze-thaw cycles to the concrete when used in road construction. Hao et al. (2022) impregnated paraffin into recycled fine aggregate to prepare a 3D-printed concrete with a thermal storage function. The

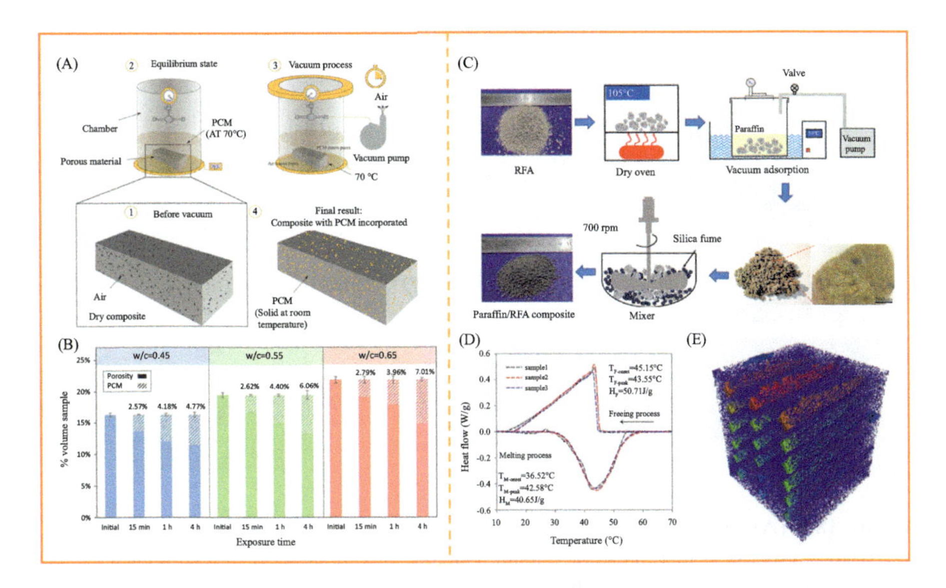

Figure 18.4 (A) Scheme of the impregnation process, (B) relationship between the content of phase change materials and porosity of concrete, scheme of the preparation process (Lopez-Arias et al., 2023), (C) scheme of impregnating paraffin into recycled fine aggregate, (D) differential scanning calorimeter curves of the impregnated fine aggregate, (E) three-dimensional scanning of inner pore structure of the prepared concrete (Hao et al., 2022).

impregnation scheme was shown in Fig. 18.4C. A layer of silica fume was wrapped on the surface of the impregnated fine aggregate in order to reduce the leakage of paraffin during the subsequent preparation process. Fig. 18.4D showed the DSC curves of the impregnated fine aggregate. The PCM leakage ratio was only 2.7 wt.% when continuously heated at 70°C for 120 hours. The compressive strength of the prepared 3D-printed concrete decreased with the increase in paraffin content due to the hydrophobic characteristics of paraffin wax, which hindered the hydration of cement. The thermal conductivity also decreased with the increase in paraffin content and exhibited anisotropy. As shown in Fig. 18.4E, the thermal conductivity in different directions was mainly affected by porosity. The use of man-made lightweight aggregates (Tuncel & Pekmezci, 2018), shale ceramsite (Ying et al., 2023), expanded vermiculite (Dora et al., 2023), and other porous materials as carriers to impregnate PCMs was also reported in some literature and the compressive strength of the prepared concrete all showed a decrease ranging from 30% to 50%.

18.3.1.2 Indirect incorporation by form-stable phase change materials

Most of the PCMs used in thermal storage concrete were solid–liquid PCMs. The adverse effects of the leakage of PCMs on the mechanical performance and durability of concrete can be prevented by encapsulating PCMs before mixing with other

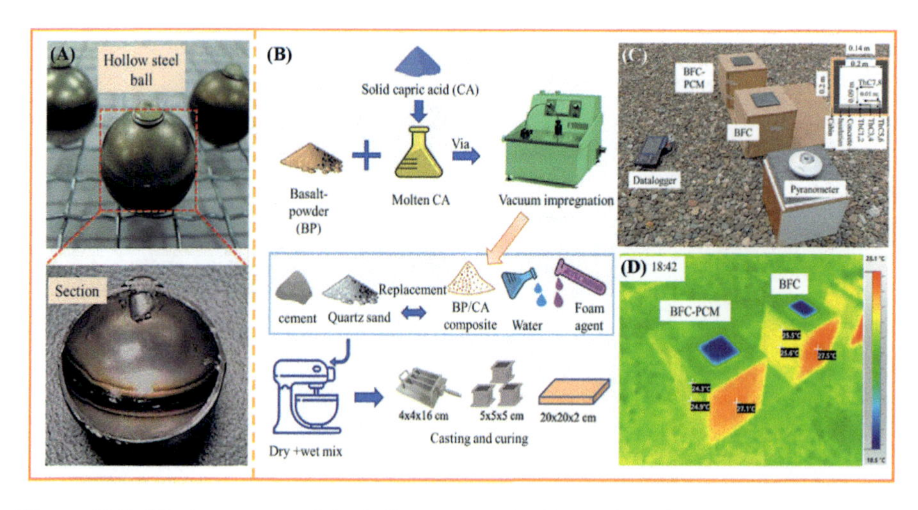

Figure 18.5 (A) Appearance of the hollow steel balls (Cui et al., 2022), (B) scheme for the preparation process of thermal storage foam concrete, (C) small-scale simulation test chamber, (D) images captured by the infrared thermal imager (Kocyigit et al., 2023).

components of concrete (Sharshir et al., 2023). This preparation method was called indirect incorporation. The indirect incorporation methods of PCMs in thermal storage concrete can be divided into macroencapsulation and microencapsulation in accordance with different encapsulation scales (Cao et al., 2022). Cui et al. (2022) used hollow steel balls to encapsulate PCMs and prepared high-performance thermal storage concrete materials for pile foundations. The appearance of the hollow steel balls was shown in Fig. 18.5A. Because of the weak bonding between PCM steel balls and cementitious materials, a decrease in the compressive strength of concrete occurred. Meanwhile, the strength loss can be effectively compensated for by incorporating steel fibers. Additionally, hollow steel balls with a smaller wall thickness should be selected as encapsulation carriers for better thermal bearing capacity of the concrete piles. The macroencapsulation carrier had a large size and limited compatibility with cement-based materials. Therefore more research has chosen to incorporate PCMs into concrete through microencapsulation to reduce adverse effects on the performance of concrete.

Microencapsulation refers to the encapsulation of PCMs on a smaller scale through physical adsorption or core–shell encapsulation to form microspheres or microcapsules, and then the microspheres or microcapsules were added to concrete. Compared with macroencapsulation, the concrete incorporating MPCMs had better integrity and was used in various components such as enclosure structures, beams, slabs, and columns. Al-Absi et al. (2023) used formaldehyde-free bio-based MPCMs to prepare heat storage foam concrete. The MPCMs were introduced into concrete by replacing sand in equal volume percentages. The prepared thermal storage concrete was applied to the exterior walls of the building. This type of thermal storage concrete exterior wall can reduce the indoor peak temperature by 5.5°C compared with ordinary concrete exterior walls. It was illustrated that the thermal storage concrete had a positive effect on reducing the internal heat gain of buildings in high temperature environments. However, MPCMs also weakened the mechanical properties of concrete. Eddhahak-Ouni et al. (2014) added MPCMs with a particle size in the range of 100–300 μm to concrete at different volume fractions. The compressive strength of concrete decreases by more than 30% when the content of the MPCMs was 5 vol.%. In the research of Hunger et al. (2009), the compressive strength of cement-based concrete mixed with 5% MPCMs decreased by more than 70%. Directly adding MPCMs to the concrete system resulted in a significant loss of strength (Figueiredo et al., 2016), as reported in many research studies. Under these circumstances, the negative impact of MPCMs on mechanical properties can be reduced by replacing ordinary components of concrete (Berardi & Gallardo, 2019). Kocyigit (2023) used the physical absorption method to prepare form-stable PCMs based on basalt grain. The form-stable PCMs were added to foam concrete and the scheme for the preparation process of thermal storage foam concrete was shown in Fig. 18.5B. For the purpose of testing the temperature regulation function of the thermal storage concrete, a small-scale simulation test chamber was designed with slabs made of the foam concrete (Fig. 18.5C). The test chambers were placed on the roof and exposed to the sun light. The images captured by the infrared thermal imager were shown in Fig. 18.5D. The thermal storage foam concrete can store a certain amount of solar energy under sun light and release it when the temperature dropped at night, so as to reduce the thermal fluctuation in the space. The infrared thermal images also confirmed that the thermal storage foam concrete can reduce the temperature change. Furthermore, the incorporation of PCMs into concrete by microencapsulation can also regulate its early hydration heat and reduce temperature cracks (Arora et al., 2017). In later applications, it can also reduce the stress failure of concrete caused by environmental temperature gradients (Pilehvar et al., 2019).

18.3.2 Mechanical strength

The mechanical properties of thermal storage concrete are not only related to the mix design of the concrete matrix but are also affected by the properties, contents, and introduction methods of PCMs. To be specific, PCMs directly mixed into concrete without encapsulation measures will have a great negative impact on mechanical properties in most cases. As for the indirect incorporation method, the influences of form-stable PCMs on mechanical properties depend on the properties of the encapsulation materials and the preparation technique. Mechanical properties of smart thermal storage concrete are summarized in Table 18.1.

18.3.2.1 Compressive strength

Generally, the compressive strength of thermal storage concrete was lower compared to ordinary concrete. A multiangle study on the different mechanical properties parameters of lightweight aggregate concrete containing MPCMs carried out by

Table 18.1 Summary of mechanical properties of smart thermal storage concrete.

Types of PCM	Types of concrete	Incorporation method	PCM content	Compressive strength (MPa)	Compressive strength change	Flexural strength (MPa)	Flexural strength change	Ref.
Attapulgite/Capric-myristic acid	Foam concretes	Indirect method	15 wt.%	13.46	−11%	2.33	35%	Gencel, Ustaoglu, et al. (2022)
Paraffin/ low-density polyethylene-ethylvinylacetate	Geopolymer concrete (GC)	Indirect method	3.2 wt.%	34.1	−51%	–	–	Cao et al. (2017)
Paraffin/polymer	Portland cement concrete (PCC)	Indirect method	5 wt.%	25.25	−43.12%	–	–	D'Alessandro et al. (2018)
Paraffin/ polymethyl methacrylate	Lightweight aggregate concrete	Indirect method	5 wt.%	25.45	−25.87%	–	–	Zhu et al. (2021)
Paraffin/ polymethyl methacrylate	Cementitious Concrete	Indirect method	5 wt.%	20	−69%	–	–	Jayalath et al. (2016)
Micronal DS 5001X	Mortar	Indirect method	20 wt.%	4.32	− 76.2%	1.91	65.02%	Rebelo et al. (2022)
Microencapsulated PCM	Self-compacting concrete	Indirect method	5 wt.%	21.36	−71.2%	–	–	Hunger et al. (2009)
Polyethylene glycol (PEG) /silica fume	Cement mortar	Indirect method	5 wt.%	25.9	−15.36%	3.64	3.7	Hattan et al. (2021)
Paraffin/ polyurethane-polyurea	Ultra-high performance concrete	Indirect method	5 wt.%	88	−27%	19.8	+24%	Ren et al. (2021)

Paraffin/ low-density polyethylene-ethylvinylacetate	GC	Indirect method	20 wt.%	42.5	−45.2%	–	–	Pilehvar et al. (2019)
Micronal DS 5001X	PCC	Indirect method	1 wt.%	21	−16%	–	–	Eddhahak-Ouni et al. (2014)
Paraffin/polymer	GC	Indirect method	5.2 wt.%	32	−60.5%	–	–	Cao et al. (2018)
Micronal DS 5001	PCC	Indirect method	–	–	−36%	–	25%	Figueiredo et al. (2016)
Capric acid	Foam concretes	Direct method	23.8 wt.%	7.96	−23.83%	2.37	2.1%	Kocyigit et al. (2023)
Hydrated salt PCMs/fly ash cenosphere	Cement mortar	Indirect method	5 vol.%	44.3	+1.1%	7.9	+5%	Yu, Liu, et al. (n.d.)
Macroencapsulated PCM	PCC	Indirect method	31.9 wt.%	42	−36.36%	–	–	Cui et al., (2022)
Paraffin	Cement mortar	Direct method	3.38 wt.%	6	−45.4%	2.5	+4.2%	Cunha et al. (2020)
Paraffin	Lightweight aggregate concrete	Direct method	6%	59.5	−18.7%	9	9.1%	Tuncel and Pekmezci (2018)
CA-ethyl alcohol	Foam concrete	Direct method	5%	6.8	−23.6%	–	–	Dora et al. (2023)
Paraffin/styrene maleic anhydride-diethylenetriamine	Cement mortar	Indirect method	10 wt.%	43.2	−33.6%	9.4	17.6%	Cui et al. (2015)

PCM, Phase change material; *CA*, capric acid.

Zhu et al. (2021) showed that the compressive strength significantly decreased with PCM content increasing. When the MPCM content is 5 wt% and 10 wt%, the compressive strength decreases by 22.93% and 33.1%, respectively. Jayalath et al. (2016) studied the effect of MPCMs on the properties of cementious mortars and concrete. The MPCMs with a mass fraction of 5% reduced the compressive strength of the mortar and concrete by 50% and 69% at 28 days, respectively. Pilehvar et al. (2017) tested the effect of MPCMs on the mechanical properties of geopolymer concrete. The data showed that with the increase of PCM content, the compressive strength of geopolymer concrete showed a decreasing trend. The compressive strength decreased more significantly at a higher temperature. D'Alessandro et al. (2018) prepared multifunctional smart concrete based on two encapsulation methods: macroencapsulation and microencapsulation. The mechanical properties of concrete contained macro- and microencapsulated PCMs with the content of 5 wt.% decreased by 33.59% and 43.12%, respectively. Many other studies had also reported that the introduction of PCMs led to the loss of compressive strength in concrete (Cao et al., 2018). The negative impact of PCMs on the compressive strength of concrete was mainly due to the low mechanical strength of PCMs and the weak interface bonding between the PCMs and the concrete matrix, which resulted in a high porosity (Kheradmand et al., 2020). Modification of the form-stable PCMs was an effective method to reduce the strength loss of heat storage concrete. Our group microencapsulated EHS (a binary eutectic hydrated salt PCM) with perforated fly ash cenospheres, and modified the surface of the phase change microspheres by secondary coating with silica (Yu et al., 2022). The microstructure of perforated fly ash cenospheres, microencapsulated EHS after impregnation, and the phase change microspheres finally synthesized were shown in Fig. 18.6A–C, respectively. DSC curves of phase change microspheres were shown in Fig. 18.6D. The enthalpy of the phase change microspheres used in concrete was 129.7 J/g. The modified phase change microspheres were used to replace the standard sand with a volume replacement ratio of 0%–20% to prepare energy storage mortars. The compressive strengths of mortars with different contents of phase change microspheres were shown in Fig. 18.6E, respectively. It can be seen from the figure that, with the increase in the content of phase change microspheres, the compressive strength of thermal storage mortar showed a trend of first increasing and then decreasing. When the content of phase change microspheres was 5%, the compressive strength at 28d of mortars was improved. This was because the silica coated on the surface of the microsphere participated in the pozzolanic reaction and enhanced the interface between the phase change microsphere and the cementitious matrix, as shown in Fig. 18.6F. This study showed that using materials with certain activity in the hydration process of cementitious materials to encapsulate PCMs can achieve better functionality of thermal storage concrete materials.

18.3.2.2 Flexural strength

Different types and incorporation methods of PCMs had different effects on the flexural strength of heat storage concrete. Ren et al. (2021) added commercially

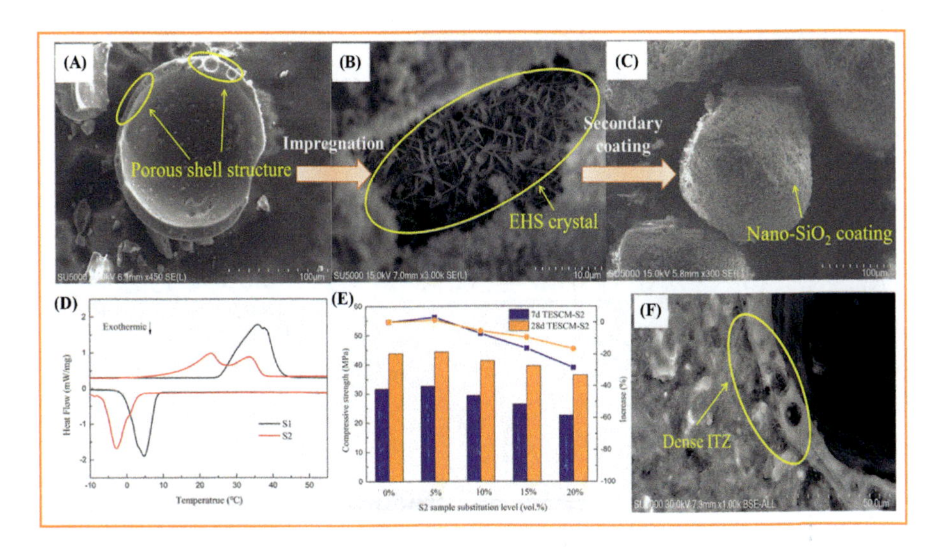

Figure 18.6 (A) Perforated structure of FAC, (B) EHS impregnated into the cavity of FAC, (C) secondary coating of nano-SiO_2, (D) differential scanning calorimeter curves of phase change microspheres, (E) compressive strength of mortars with different content of phase change microspheres, (F) dense ITZ between the phase change microsphere and the cementitious materials (Yu et al., 2022). *FAC*, Fly ash cenosphere; *EHS*, eutectic hydrated salt; *ITZ*, interface transition zone.

available microcapsules into ultra-high performance concrete (UHPC). As shown in Fig. 18.7A, the flexural strength of UHPC increased by 24% with the phase change microcapsule content of 5 wt.%. However, the flexural strength decreased by 4% as the content of phase change microcapsules increased to 10 wt.%. The load deflection curves of UHPC with a phase change microcapsules content of 5 wt.% was shown in the Fig. 18.7B. UHPC had a more obvious hardening behavior compared to ordinary UHPC (Fig. 18.7C). The load deflection curves confirmed that the interface bonding between steel fiber and matrix had been strengthened, explaining the reason for the improvement in flexural strength at the phase change microcapsule content of 5 wt.%. When the content of phase change microcapsules increased to 10 wt.%, the negative effects of agglomeration and leakage of phase change microcapsules (Fig. 18.7D) dominated, causing a decrease in flexural strength. Gencel, Ustaoglu, et al. (2022) studied the physical and mechanical properties of foamed concrete containing attapulgite/capric-myristic acid eutectic form-stable PCMs. According to the three-point bending test, the flexural strength of the heat storage foam concrete decreased by 68.61% and 57.38% at 7d and 28d, respectively. The porosity of foamed concrete increased by 56% at most after adding form-stable PCMs, and the increase of porosity was one of the main reasons for the loss of flexural strength. Hattan et al. (2021) adopted to find out the influence of form-stable PCMs on the flexural strength of cementious mortars with different water-to-cement ratios. The test results showed that the loss of flexural strength was greater

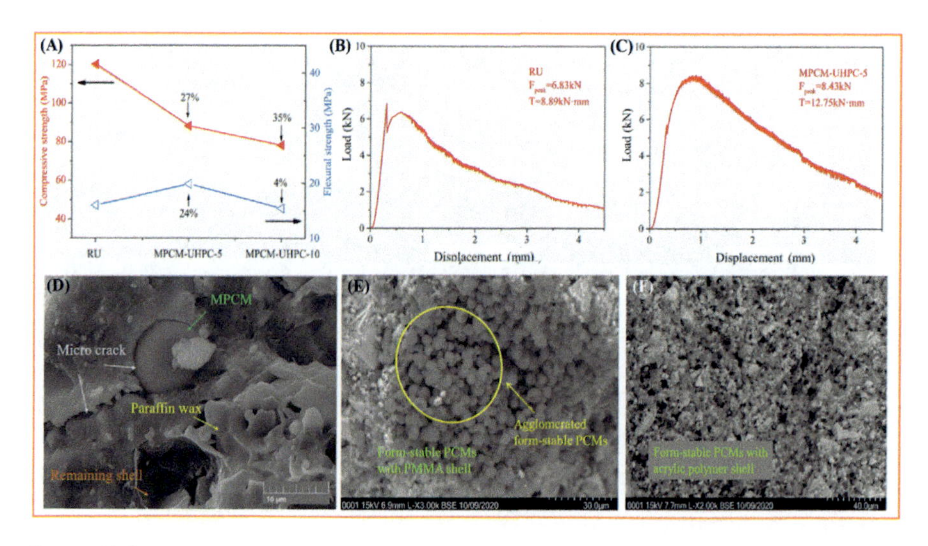

Figure 18.7 (A) Mechanical properties of ultra-high performance concrete (UHPC), (B) load deflection curves of UHPC with phase change microcapsules content of 5 wt.%, (C) load deflection curves of UHPC without phase change microcapsules, (D) SEM image of UHPC with form-stable phase change material (PCM) (Ren et al., 2021), (E) and (F) micromorphology of cement mortar mixed with the two different kinds of form-stable PCMs (Rebelo et al., 2022). *SEM*, Scanning electron microscope.

at a low water to cement ratio. Besides, the adverse effect of PCMs on flexural strength tended to increase with the extension of curing time. Rebelo et al. (2022) added two different kinds of form-stable PCMs to cement mortar. The shell materials of the two form-stable PCMs were polymethyl methacrylate (PMMA) and acrylic polymer, respectively. The core materials of both form-stable PCMs were paraffin. The flexural strength decreased by 68% and 57%, respectively, when the content of form-stable PCMs was 20 wt.%. The micromorphology of cement mortar mixed with the two different types of form-stable PCMs was shown in Fig. 18.7E and F, respectively. It can be seen that the form-stable PCMs with PMMA shells were more seriously agglomerated in the cement mortar, which was one of the reasons why they caused more flexural strength loss. The adverse effect of PCMs on flexural strength also came from the increase in porosity and microstructure degradation. It was an effective way to make up for the loss of flexural strength of heat storage concrete by strengthening the matrix with fiber reinforcement.

To sum up, no matter what kind of incorporation method is used, the mechanical properties of heat storage concrete have a certain decline compared with ordinary concrete in most studies. The main causation for the weaken of mechanical properties included the hindering of cement hydration by PCMs and the deterioration of microstructure due to a weak interface between phase PCMs and the concrete matrix. Encapsulation and modification of PCMs with appropriate materials is an effective way to make up for the lack of mechanical properties of thermal storage concrete. because of its lack of mechanical properties, thermal storage concrete is mostly used as a heat storage module in maintenance structures such as exterior walls and ceilings. And it is rarely used in important load-bearing structures such as beams and columns. In order to further expand the application scope of thermal storage concrete, form-stable PCMs suitable for working with cement-based materials are worth studying in the future.

18.3.3 Thermal properties

The perception and regulation of environmental temperature by smart thermal energy storage concrete mainly rely on its thermal properties. The parameters used to characterize the thermal properties of smart thermal storage concrete usually include thermal conductivity, specific heat capacity, and thermal storage capacity. The introduction of PCMs may cause apparent changes in the thermal performance of concrete, making the thermal properties of smart thermal storage concrete significantly different from ordinary concrete. The thermal properties of different kinds of smart thermal storage concrete were summarized in Table 18.2.

18.3.3.1 Thermal conductivity

The thermal conductivity of smart thermal storage concrete was influenced by factors such as the composition of the concrete matrix materials, microstructure, and properties of PCMs. The thermal conductivity of ordinary concrete was around 1.2 W/m · K. The thermal conductivities of concrete with large porosity, such as aerated concrete (Schackow et al., 2014) and foam concrete (Liu et al., 2014), ranged from 0.2 to 0.9 W/m · K. The influences of introducing PCMs on the thermal conductivity of the smart thermal storage concrete varied in existing research, which were mainly related to the type and incorporation method of PCMs as well as the type of concrete matrix. Gencel's group added MPCMs to foam concrete and tested its thermophysical properties (Gencel, Subasi, et al., 2022). The thermal conductivity of foam concrete matrix was only 0.09 W/m · K. When the content of MPCM was 5 wt%, 10 wt%, and 15 wt%, the thermal conductivity of heat storage foam concrete (FC-MPCM) was 0.118, 0.129, and 0.153 W/m · K, respectively. The thermal conductivity of FC-MPCM was linearly proportional to the MPCM content. The microstructure of FC-MPCM was shown in Fig. 18.8A and B. It can be seen that the incorporation of MPCM caused the collapse of pores in the foam concrete matrix. The thermal conductivity increased with the decrease in porosity, as shown in Fig. 18.8C. Although the thermal conductivity of foam concrete with MPCM was significantly increased, it still met the requirements of thermal insulation materials. Sukontasukkul et al. (2019) prepared a phase change lightweight aggregate (PCM-LWA) by adsorbing PCMs into porous lightweight aggregates through the vacuum impregnation method. Then, the prepared PCM-LWA was added to the concrete by replacing part of the aggregates. The thermal conductivity of the thermal storage lightweight aggregate concrete in the environment above and below the phase change temperature of the PCMs involved was measured by a heat flow meter apparatus as shown in Fig. 18.8D. The test results were shown in Fig. 18.8E.

Table 18.2 Summary of thermal properties of smart thermal energy storage concrete.

Types of PCM	Types of concrete	Incorporation method	PCM content	λ (W/m · K)	C (J/g · K)	Thermal storage capacity		Ref.
						ΔH_m (J/g)	Peak temperature decrease (°C)	
Attapulgite/Capric-myristic acid	Foam concretes	Indirect	15 wt.%	0.511	–	3.12	0.56	Gencel, Ustaoglu, et al. (2022)
Paraffin/ polymethyl methacrylate	Cementitious mortar	Indirect method	20 vol.%	1.3	1.49	–	–	Jayalath et al. (2016)
Micronal DS 5001X	Mortar	Indirect method	20 wt.%	0.029	–	12.256	–	Rebelo et al. (2022)
Microencapsulated PCM	Self-compacting concrete	Indirect method	5 wt.%	2.1	–	–	–	Hunger et al. (2009)
Micronal DS 5001X	Portland cement concrete (PCC)	Indirect method	1%	1.9	1.03	22.32	–	Eddhahak-Ouni et al. (2014)
Paraffin/polymer	Geopolymer concrete	Indirect method	5.2 wt.%	2.1	3	3	4.5	Cao et al. (2018)
Macroencapsulated PCM	PCC	Indirect method	31.9 wt.%	1.03	0.89	–	–	Cui et al. (2022)
Capric acid/basalt powder	Foam concretes	Indirect method	23.8 wt.%	–	–	17.4	-	Kocyigit (2023)
CA-ethyl alcohol	Foam concrete	Direct method	5 wt.%	0.11	–	86.17	–	Dora et al. (2023)

Paraffin/ polymethyl methacrylate	Lightweight aggregate concrete	Indirect method	2.5 wt.%	1.46	2.24	9	–	Zhu et al. (2023)
Bio-based PCM/ formaldehyde	Cement mortar	Indirect method	5 wt.%	1.83	1.5	–	–	Das et al. (2022)
PEG	Lightweight concrete	Direct method	7.8 wt.%	0.841	–	7.72	–	Sukontasukkul et al. (2019)
Paraffin/char	Cement mortar	Indirect method	3.33 wt.%	–	1.12	–	–	Ryms (2022)
Micronal DS5040X	Cement mortar	Indirect method	10 vol.%	0.26	–	15	–	Guardia et al. (2019)
n-Octadecane/ silica	Cement mortar	Indirect method	5 vol.%	0.89	–	0.706	0.3	Wang et al. (2022)
Paraffin/styrene maleic anhydride-diethylenetriamine	Cement mortar	Indirect method	10 wt.%	1.018		–	2.8	Cui et al. (2015)
Phase change microcapsule	Cement mortar	Indirect method	20 wt.%	–	2.22	23.13	6.3	Xie et al., (2023)
Paraffin	Lightweight aggregate concrete	Direct method	4 wt.%	0.71	1.21	–	–	Shen (2021)
PureTemp 23	Concrete	Direct method	1 wt.%	–	–	–	4	Yousefi et al. (2022)
n-Octadecane/ cenosphere	Cement mortar	Indirect method	3 wt.%	–	–	5.49	–	Halder (2022)
CA/silica aerogel granules	Cement mortar	Indirect method	17.5 wt.%	0.75	–	8.31	–	Kumar (2023)

PCM, Phase change material; *CA*, capric acid.

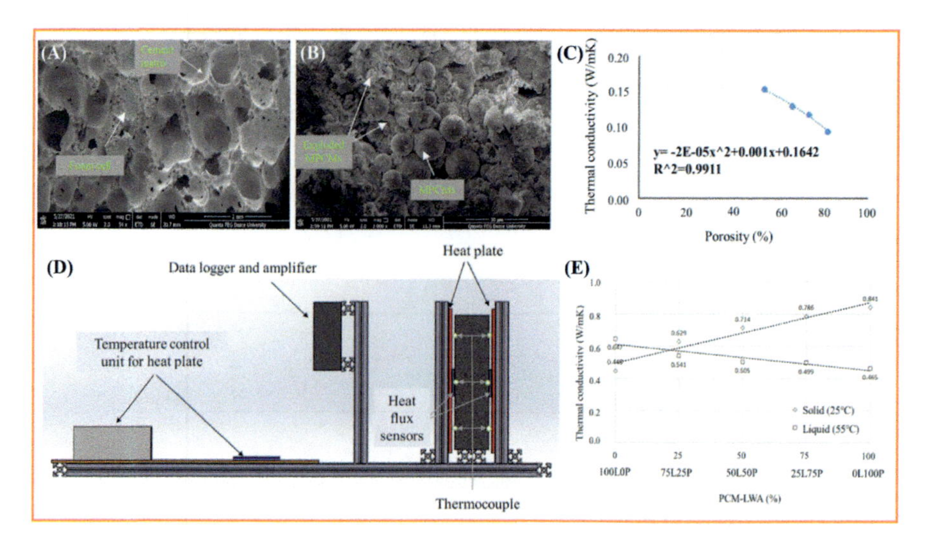

Figure 18.8 (A) and (B) Microstructure of FC-MPCM, (C) relationship between thermal conductivity and porosity of the foam concrete (Gencel, Subasi, et al., 2022), (D) heat flow meter apparatus for test of thermal conductivity, (E) thermal conductivities of the thermal storage lightweight aggregate concrete (Sukontasukkul et al., 2019). *FC-MPCM*, Foam concretes containing microencapsulated phase change material.

When the temperature was lower than the phase change temperature, the thermal conductivity of heat storage lightweight aggregate concrete increased with the increase in the PCM-LWA replacement ratio. On the contrary, when PCMs were in liquid phase, the thermal conductivity had a decreasing tendency with the increase in PCM-LWA content. It can be seen that the influence of PCMs on the thermal conductivity of concrete at different temperatures had obvious differences. This was because the solid PCM filled the voids of the lightweight aggregate, which reduced the air content and improved the thermal conductivity. As the temperature rose and the PCM melted, parts of the external heat were stored as latent heat, which improved the thermal resistance of the system and reduced the thermal conductivity. Cao et al. (2017) used commercially available paraffin-based PCM microcapsules to replace part of the sand and mixed them into different types of concrete. With the increase in the content of PCM microcapsules, the thermal conductivity of cement concrete and geopolymer concrete had a downward trend, regardless of whether the PCMs were in a solid or liquid state. There were two reasons for this phenomenon. One was that the thermal conductivity of phase change microcapsules was about 0.2 W/m · K, which was lower than that of sand. The other was that the porosity of concrete increased with the increase in phase change microcapsule content, and the existence of pores made the thermal conductivity decrease. In addition, this research had shown that under the same PCM microcapsules content, the thermal conductivity of concrete with PCMs in the solid state was higher than that in the liquid state, which was consistent with the experimental results obtained by the incorporation method of directly impregnating PCMs into aggregates. Cui et al. (2015) studied the incorporation of phase change microcapsules into cement mortar and also obtained similar results.

Above all, the different types and introduction methods of PCMs have different effects on the thermal conductivity of concrete. Therefore the thermal conductivity of thermal storage concrete can be designed from the perspective of materials and preparation processes according to the requirements of the application scenario so as to obtain smart thermal storage concrete that meets the application requirements.

18.3.3.2 Specific heat capacity

Specific heat capacity is the heat required to increase 1 K per unit mass of substance. A larger specific heat capacity means a smaller temperature change under the same heat difference. In order to achieve the function of environmental thermal comfort adjustment and reduce temperature fluctuation, smart thermal storage concrete should have a relatively large specific heat capacity.

Kehli et al. (2023) impregnated straw with bio-based palm oil to prepare low-carbon emission form-stable PCMs and used them to prepare composite gypsum materials with heat storage functions. The specific heat capacity of gypsum samples mixed with 3 wt.% form-stable PCMs was 0.858 J/g · K, which was increased by 4.4% compared with the control group. Xie et al. (2023) used carbon nanotubes to prepare a composite PCM microcapsule. The basic thermophysical properties and microstructure of this microcapsule were shown in Fig. 18.9A and B, respectively. The specific heat capacity of cement mortar containing 20% PCM microcapsules

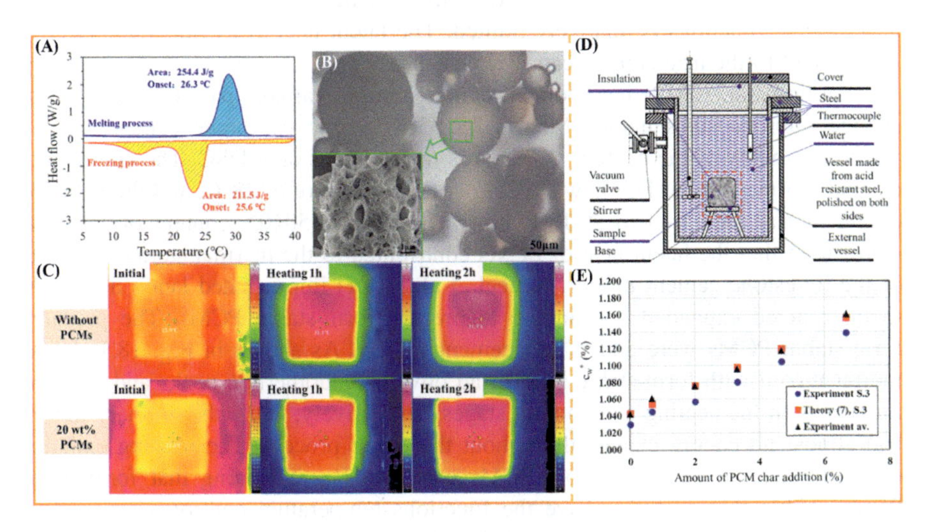

Figure 18.9 (A) Differential scanning calorimeter curves of the microcapsule, (B) microstructure of the microcapsule, (C) infrared thermal imager photos (Xie et al., 2023), (D) schematic diagram of the testing device, (E) the specific heat capacity of the thermal storage mortar (Ryms et al., 2022).

was 2.22 J/g · K, which was increased by 52.05% compared with the control group. In addition, the infrared thermal imager photos shown in Fig. 18.9C also confirmed that PCM microcapsules can improve the thermal inertia of cement mortar and further improve its temperature control performance. Ryms et al. (2022) used char made from recycled waste tires as a support skeleton to prepare paraffin-based form-stable PCMs and mixed them into cement mortar. This study improved the design of the top cover of the calorimeter. In the experiment, the sample was directly placed on an open base for measurement, and the schematic illustration of the testing device was shown in Fig. 18.9D. The test results revealed that the specific heat capacity of mortar increased with the increase in PCM content. In practical terms, the specific heat capacity of thermal storage mortar reached 1.138 J/g · K with a form-stable PCMs content of 6.67%, which was 10.5% higher than that of ordinary mortar. As shown in Fig. 18.9E, the specific heat capacity data obtained from the test was closed to the theoretical calculation value, which proved the accuracy of the experimental design. Other studies also reported that the incorporation of PCMs increased the specific heat capacity of concrete (Shen, 2021). Normally, the specific heat capacity of thermal storage concrete showed an increasing trend with the increase in PCM content, indicating the incorporation of PCMs definitely benefitted the temperature control ability of thermal storage concrete.

18.3.3.3 Heat storage capacity

The heat storage capacity of smart thermal storage concrete consists of two parts: the sensible heat of the concrete and the latent heat provided by PCMS. The sensible heat of concrete is mainly related to its own density, and its role in exerting heat storage functions is relatively limited. The latent heat of PCMs is an effective supplement to the heat storage capacity of concrete.

Guardia et al. (2019) incorporated commercially available paraffin-based phase change microcapsules named Micronal DS5040X into limestone cement mortar, and the enthalpy of the cement mortar was measured using a DSC. The enthalpy of cement mortar without PCMs was 15 J/g at a temperature of 28°C, and the enthalpy of the mortar increased to 20 J/g with a Micronal DS5040X content of 10 wt.%, confirming that the latent heat of PCMs contributed to the thermal storage capacity of the limestone cement mortar. Gencel et al. (2021) impregnated lauric acid (LA) to micronized expanded vermiculite to prepare form-stable PCMs. The prepared form-stable PCMs were added into cement mortar. As shown in Fig. 18.10A, the mortar mixed with form-stable PCMs had obvious endothermic peaks and exothermic peaks in the heating and cooling stages, respectively, while the ordinary mortar had almost no energy change throughout the entire process. The melting enthalpy of mortar containing form-stable PCMs was 18.9 J/g, indicating that form-stable PCMs enable the mortar to have the functions temperature control. Halder et al. (2022) impregnated PCM with etched fly ash microspheres and modified its surface with the biomimetic material dopamine, as shown in Fig. 18.10B. The modified form-stable PCMs did not reduce the mechanical properties of cement mortar due to the presence of a dopamine biomimetic coating. In order to obtain reliable

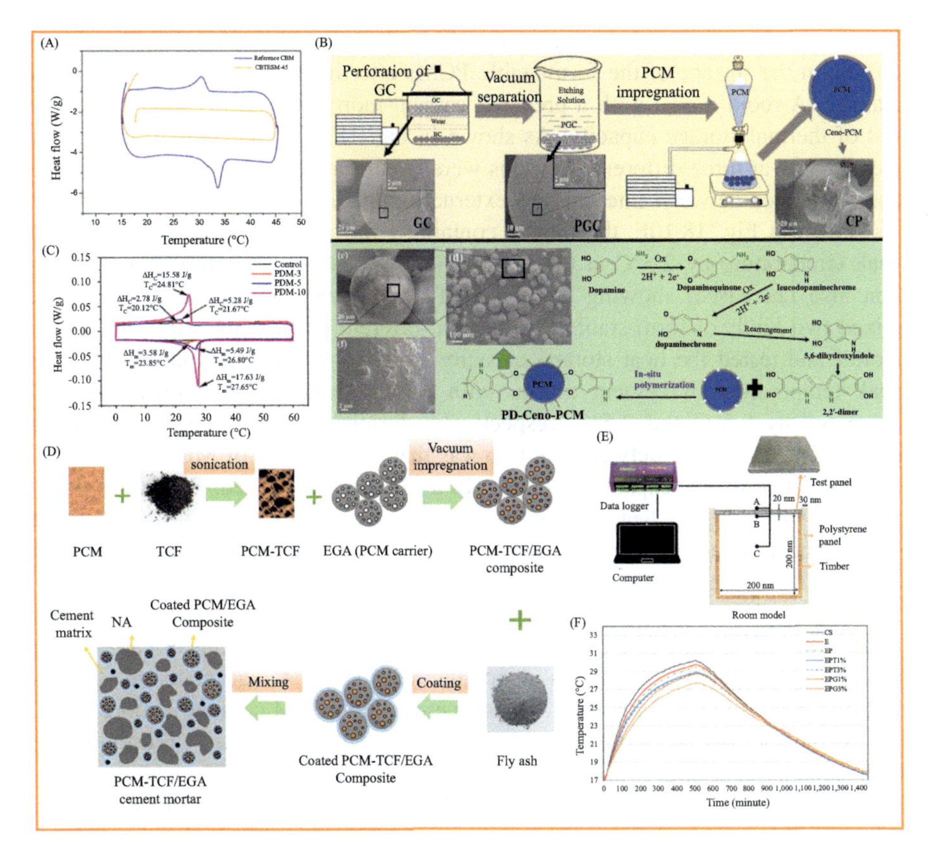

Figure 18.10 (A) Differential scanning calorimeter (DSC) curves of cement with and without form-stable phase change materials (PCMs) (Gencel et al., 2021), (B) scheme for the modification of form-stable PCMs, (C) DSC curves of cement paste with different form-stable PCMs (Halder et al., 2022), (D) conceptual illustration of the fabrication process, (E) room model designed for the test of thermal storage capacity, (F) temperature–time variation curve of mortar containing form-stable PCMs (Yousefi et al., 2022).

results, cement paste mixed with the form-stable PCMs was used in the DSC test. The DSC curves of cement paste with different form-stable PCM content were shown in Fig. 18.10C. It was observed that the cement mortar exhibited phase change behavior, with a melting enthalpy of 15.58 J/g and a PCM content of 10 wt.%.

However, some researchers mentioned that the results of concrete enthalpy obtained directly using the DSC test may not be accurate because of the extremely small sample quality in one DSC test. Therefore the thermal storage capacity of concrete can be indirectly characterized by testing the temperature–time variation curve during the temperature rise and fall process (Kumar, 2023). Yousefi et al. (2022) impregnated PCMs into expanded glass aggregates (EGA) to prepare form-stable PCMs. The conceptual illustration of the fabrication process was shown in Fig. 18.10D. The graphite powder and nano-TiO_2 were added to improve the thermal conductivity of the form-stable PCMs. Furthermore, a layer of fly ash was coated on the surface of the form-stable PCMs after impregnation in order to avoid leakage. A room model with a prepared cement mortar board was designed for the test of thermal storage capacity. As shown in Fig. 18.10E, temperature thermocouples located at three different positions were used to record the temperature change and difference between internal and external temperatures in the test room model. As shown in Fig. 18.10F, the mortar containing form-stable PCMs can reduce the peak temperature by 2.5°C at the highest, and the day–night temperature fluctuation also decreased, showing a good heat storage capacity. Wang et al. (2022) mixed two kinds of form-stable PCMs with different particle sizes into the cement mortar and tested the heat storage capacity in a way similar to Yousefi et al. The peak temperature of cement mortar with MPCMs and nanoencapsulated PCMs decreased by 5.5°C and 2.5°C, respectively, during the same temperature change period. The difference between the two was that the enthalpy of MPCMs was higher than that of nanoencapsulated PCMs. In all studies mentioned, PCMs can endow concrete with a higher heat storage capacity, which was like their influence on the specific heat capacity. The heat storage capacity of concrete was also positively correlated with the enthalpy and content of PCMs.

In conclusion, thermal conductivity of smart thermal storage energy concrete varies from different PCM types. However, both of the specific heat capacity and heat storage capacity for various smart thermal energy storage concretes increase with the increase in PCM content. Consequently, high PCM content is more beneficial to the thermal properties of smart thermal storage concrete. Unfortunately, a high content of PCMs would cause a significant loss of mechanical properties, which leads to the current content of PCMs in concrete has to be maintained at a low level (Marani et al., 2022). Innovative ways to balance the relationship between mechanical properties and thermal storage function are valuable research directions for the future.

18.4 Conclusion

Smart thermal energy storage concrete can spontaneously store or release thermal energy during the ambient temperature changing, which has broad application prospects in building thermal comfort regulation and frost damage prevention on roads in cold regions. In this chapter, we introduce the types and incorporation methods of PCMs used to prepare smart thermal storage concrete. The mechanical and thermal properties of the smart thermal storage concrete are also summarized. The main conclusions are as follows.

The methods for incorporation of PCMs into concrete include direct incorporation and indirect incorporation with form-stable PCMs. Indirect incorporation method is the mainstream method for the preparation of heat storage intelligent concrete. Physical absorption and core–shell encapsulation are commonly used strategies to prepare form-stable PCMs used in smart thermal energy storage concrete. The encapsulation and incorporation methods of PCMs are important factors affecting the performance of smart thermal storage concrete, so they should be selected carefully according to the application scenarios of the required smart thermal storage concrete.

The introduction of PCMs leads to a certain loss of the mechanical properties of smart thermal storage concrete. The decline in mechanical properties of the smart thermal storage concrete is the macroscopic manifestation of the weak interface between PCMs and the concrete matrix. The hydrophobic characteristics of PCMs hinder the hydration reaction of cement, which is also one of the main reasons for the strength loss. Modification of PCMs is an effective way to reduce the loss of mechanical properties of smart thermal storage concrete. In addition, adding fiber to concrete to strengthen the matrix strength can also effectively make up for the decline in mechanical properties caused by PCMs.

The latent heat of PCMs can endow concrete with good thermal properties. The temperature regulation function of thermal storage concrete is also closely related to its thermal conductivity, and different types and incorporate on methods of PCMs have different effects on the thermal conductivity of concrete. Therefore the demand for thermal conductivity in the application scenario should be considered when designing the mix proportion and selecting the preparation process for thermal storage concrete.

The main problem limiting the further development and application of smart thermal storage concrete is that it must sacrifice a certain strength while obtaining good thermal properties. Consequently, the strengthening of thermal and mechanical properties in coordination is a very important factor in promoting the development of smart thermal energy storage concrete, which is worth more consideration and research in the future.

Acknowledgments

The financial support from the National Natural Science Foundation of China (No. 51902068).

References

Abbasi, T., & Abbasi, S. A. (2011). Decarbonization of fossil fuels as a strategy to control global warming. *Renewable and Sustainable Energy Reviews*, *15*(4), 1828–1834. Available from https://doi.org/10.1016/j.rser.2010.11.049.

Akeiber, H., Nejat, P., Majid, M. Z. A., Wahid, M. A., Jomehzadeh, F., Zeynali Famileh, I., Calautit, J. K., Hughes, B. R., & Zaki, S. A, (2016). A review on phase change material (PCM) for sustainable passive cooling in building envelopes. *Renewable and Sustainable Energy Reviews*, *60*, 1470–1497. Available from https://doi.org/10.1016/j.rser.2016.03.036.

Al-Absi, Z.A., Hafizal, M.I. M., & Ismail, M.. (2023). Innovative PCM-incorporated foamed concrete panels for walls' exterior cladding: An experimental assessment in real-weather conditions. *Energy and Buildings, 288*, 113003.

Alehosseini, E., & Jafari, S. M. (2019). Micro/nano-encapsulated phase change materials (PCMs) as emerging materials for the food industry. *Trends in Food Science and Technology, 91*, 116–128. Available from https://doi.org/10.1016/j.tifs.2019.07.003. Available from: http://www.elsevier.com/wps/find/journaldescription.cws_home/601278/description#description.

Alva, Guruprasad, Lin, Yaxue, Liu, Lingkun, & Fang, Guiyin (2017). Synthesis, characterization and applications of microencapsulated phase change materials in thermal energy storage: A review. *Energy and Buildings, 144*, 276–294. Available from https://doi.org/10.1016/j.enbuild.2017.03.063.

Arabzadeh, Ali, Notani, Mohammad Ali, Kazemiyan Zadeh, Ayoub, Nahvi, Ali, Sassani, Alireza, & Ceylan, Halil (2019). Electrically conductive asphalt concrete: An alternative for automating the winter maintenance operations of transportation infrastructure. *Composites Part B: Engineering, 173*. Available from https://doi.org/10.1016/j.compositesb.2019.106985.

Arora, A., Sant, G., & Neithalath, N. (2017). Numerical simulations to quantify the influence of phase change materials (PCMs) on the early- and later-age thermal response of concrete pavements. *Cement and Concrete Composites, 81*, 11–24. Available from https://doi.org/10.1016/j.cemconcomp.2017.04.006. Available from: http://www.sciencedirect.com/science/journal/09589465.

Arshad, Adeel, Jabbal, Mark, Yan, Yuying, & Darkwa, Jo (2019). The micro-/nano-PCMs for thermal energy storage systems: A state of art review. *International Journal of Energy Research, 43*(11), 5572–5620. Available from https://doi.org/10.1002/er.4550.

Berardi, U., & Gallardo, A. A. (2019). Properties of concretes enhanced with phase change materials for building applications. *Energy and Buildings, 199*, 402–414. Available from https://doi.org/10.1016/j.enbuild.2019.07.014. Available from: https://www.journals.elsevier.com/energy-and-buildings.

Borhani, S. M., Hosseini, M. J., Pakrouh, R., Ranjbar, A. A., & Nourian, A. (2021). Performance enhancement of a thermoelectric harvester with a PCM/Metal foam composite. *Renewable Energy, 168*, 1122–1140. Available from https://doi.org/10.1016/j.renene.2021.01.020.

Cabeza, L. F., Castell, A., Barreneche, C., de Gracia, A., & Fernández, A. I. (2011). Materials used as PCM in thermal energy storage in buildings: A review. *Renewable and Sustainable Energy Reviews, 15*(3), 1675–1695. Available from https://doi.org/10.1016/j.rser.2010.11.018.

Cao, L., Tang, F., & Fang, G. (2014). Preparation and characteristics of microencapsulated palmitic acid with TiO_2 shell as shape-stabilized thermal energy storage materials. *Solar Energy Materials and Solar Cells, 123*, 183–188. Available from https://doi.org/10.1016/j.solmat.2014.01.023.

Cao, V. D., Pilehvar, S., Salas-Bringas, C., Szczotok, A. M., Rodriguez, J. F., Carmona, M., Al-Manasir, N., & Kjøniksen, A. L. (2017). Microencapsulated phase change materials for enhancing the thermal performance of Portland cement concrete and geopolymer concrete for passive building applications. *Energy Conversion and Management, 133*, 56–66. Available from https://doi.org/10.1016/j.enconman.2016.11.061.

Cao, V. D., Pilehvar, S., Salas-Bringas, C., Szczotok, A. M., Valentini, L., Carmona, M., Rodriguez, J. F., & Kjøniksen, A. L. (2018). Influence of microcapsule size and shell polarity on thermal and mechanical properties of thermoregulating geopolymer concrete for passive building applications. *Energy Conversion and Management, 164*, 198–209. Available from https://doi.org/10.1016/j.enconman.2018.02.076.

Cao, Z., Zhang, G., Liu, Y., & Zhao, X. (2022). Thermal performance analysis and assessment of PCM backfilled precast high-strength concrete energy pile under heating and cooling modes of building. *Applied Thermal Engineering*, 216, 119144.

Chen, Keping, Yu, Xuejiang, Tian, Chunrong, & Wang, Jianhua (2014). Preparation and characterization of form-stable paraffin/polyurethane composites as phase change materials for thermal energy storage. *Energy Conversion and Management*, *77*, 13–21. Available from https://doi.org/10.1016/j.enconman.2013.09.015.

Chen, Liangjie, Zou, Ruqiang, Xia, Wei, Liu, Zhenpu, Shang, Yuanyuan, Zhu, Jinlong, Wang, Yingxia, Lin, Jianhua, Xia, Dingguo, & Cao, Anyuan (2012). Electro- and photo-driven phase change composites based on wax-infiltrated carbon nanotube sponges. *ACS Nano*, *6*(12), 10884–10892. Available from https://doi.org/10.1021/nn304310n.

Chen, Lin, Sun, Ying-Ying, Lin, Jun, Du, Xiao-Ze, Wei, Gao-Sheng, He, Shao-Jian, & Nazarenko, Sergei (2015). Modeling and analysis of synergistic effect in thermal conductivity enhancement of polymer composites with hybrid filler. *International Journal of Heat and Mass Transfer*, *81*, 457–464. Available from https://doi.org/10.1016/j.ijheatmasstransfer.2014.10.051.

Chen, Lin, Wang, Ting, Zhao, Yan, & Zhang, Xin-Rong (2014). Characterization of thermal and hydrodynamic properties for microencapsulated phase change slurry (MPCS). *Energy Conversion and Management*, *79*, 317–333. Available from https://doi.org/10.1016/j.enconman.2013.12.026.

Chung, D. D. L. (2021). Self-sensing concrete: From resistance-based sensing to capacitance-based sensing. *International Journal of Smart and Nano Materials*, *12*(1), 1–19. Available from https://doi.org/10.1080/19475411.2020.1843560.

Cui, H., Zou, J., Gong, Z., Zheng, D., Bao, X., & Chen, X. (2022). Study on the thermal and mechanical properties of steel fibre reinforced PCM-HSB concrete for high performance in energy piles. *Construction and Building Materials, 350*, 128822.

Cui, Hongzhi, Liao, Wenyu, Mi, Xuming, Lo, Tommy Y., & Chen, Dazhu (2015). Study on functional and mechanical properties of cement mortar with graphite-modified microencapsulated phase-change materials. *Energy and Buildings*, *105*, 273–284. Available from https://doi.org/10.1016/j.enbuild.2015.07.043.

Cunha, S., Silva, M., & Aguiar, J. (2020). Behavior of cementitious mortars with direct incorporation of non-encapsulated phase change material after severe temperature exposure. *Construction and Building Materials, 230*, 117011.

D'Alessandro, Antonella, Pisello, Anna Laura, Fabiani, Claudia, Ubertini, Filippo, Cabeza, Luisa F., & Cotana, Franco (2018). Multifunctional smart concretes with novel phase change materials: Mechanical and thermo-energy investigation. *Applied Energy*, *212*, 1448–1461. Available from https://doi.org/10.1016/j.apenergy.2018.01.014.

Das, R., Siva Ranjani Gandhi, I., & Muthukumar, P. (2022). Use of agglomerated micro-encapsulated phase change material in cement mortar as thermal energy storage material for buildings. *Materials Today: Proceedings*, *65*, 808–814. Available from https://doi.org/10.1016/j.matpr.2022.03.316. Available from: https://www.sciencedirect.com/journal/materials-today-proceedings.

Deng, Yong, Li, Jinhong, Deng, Yanxi, Nian, Hongen, & Jiang, Hua (2018). Supercooling suppression and thermal conductivity enhancement of $Na_2 HPO_4 \cdot 12H_2O$/expanded vermiculite form-stable composite phase change materials with alumina for heat storage. *ACS Sustainable Chemistry & Engineering*, *6*(5), 6792–6801. Available from https://doi.org/10.1021/acssuschemeng.8b00631.

Ding, S. (2023). Self-heating ultra-high performance concrete with stainless steel wires for active deicing and snow-melting of transportation infrastructures. *Cement and Concrete Composites*, *138*, 105005.

Dora, S., Barta, R. B., & Mini, K. M. (2023). Study on foam concrete incorporated with expanded vermiculite/capric acid PCM – A novel thermal storage high-performance building material. *Construction and Building Materials*, *392*, 131903.

Eddhahak-Ouni, Anissa, Drissi, Sarra, Colin, Johan, Neji, Jamel, & Care, Sabine (2014). Experimental and multi-scale analysis of the thermal properties of Portland cement concretes embedded with microencapsulated Phase Change Materials (PCMs). *Applied Thermal Engineering*, *64*(1–2), 32–39. Available from https://doi.org/10.1016/j.applthermaleng.2013.11.050.

Feng, D.L., Zang, Y.Y., Li, P., Feng, Y.H., Yan, Y.Y., & Zhang, X.X. (2021). Polyethylene glycol phase change material embedded in a hierarchical porous carbon with superior thermal storage capacity and excellent stability. *Composites Science and Technology*, *210*, 108832.

Figueiredo, António, Lapa, José, Vicente, Romeu, & Cardoso, Claudino (2016). Mechanical and thermal characterization of concrete with incorporation of microencapsulated PCM for applications in thermally activated slabs. *Construction and Building Materials*, *112*, 639–647. Available from https://doi.org/10.1016/j.conbuildmat.2016.02.225.

Fu, Lulu, Wang, Qianhao, Ye, Rongda, Fang, Xiaoming, & Zhang, Zhengguo (2017). A calcium chloride hexahydrate/expanded perlite composite with good heat storage and insulation properties for building energy conservation. *Renewable Energy*, *114*, 733–743. Available from https://doi.org/10.1016/j.renene.2017.07.091.

Gao, D. C., Sun, Y., Fong, A. M., & Gu, X. (2022). Mineral-based form-stable phase change materials for thermal energy storage: A state-of-the art review. *Energy Storage Materials*, *46*, 100–128. Available from https://doi.org/10.1016/j.ensm.2022.01.003. Available from: http://www.journals.elsevier.com/energy-storage-materials/.

Gaumet, A. V., Ball, R. J., & Nogaret, A. (2021). Graphite-polydimethylsiloxane composite strain sensors for in-situ structural health monitoring. *Sensors and Actuators A: Physical*, *332*, 113139.

Gencel, O., Sarı, A., Ustaoglu, A., Hekimoglu, G., Erdogmus, E., Yaras, A., Sutcu, M., & Cay, V. V. (2021). Eco-friendly building materials containing micronized expanded vermiculite and phase change material for solar based thermo-regulation applications. *Construction and Building Materials*, *308*, 125062.

Gencel, Osman, Subasi, Serkan, Ustaoglu, Abid, Sarı, Ahmet, Marasli, Muhammed, Hekimoğlu, G. ökhan, & Kam, Erol (2022). Development, characterization and thermoregulative performance of microencapsulated phase change material included-glass fiber reinforced foam concrete as novel thermal energy effective-building material. *Energy*, *257*. Available from https://doi.org/10.1016/j.energy.2022.124786.

Gencel, Osman, Ustaoglu, Abid, Benli, Ahmet, Hekimoğlu, G. ökhan, Sarı, Ahmet, Erdogmus, Ertugrul, Sutcu, Mucahit, Kaplan, G. ökhan, & Yavuz Bayraktar, Oguzhan (2022). Investigation of physico-mechanical, thermal properties and solar thermoregulation performance of shape-stable attapulgite based composite phase change material in foam concrete. *Solar Energy*, *236*, 51–62. Available from https://doi.org/10.1016/j.solener.2022.02.042.

Guardia, Cynthia, Barluenga, Gonzalo, Palomar, Irene, & Diarce, Gonzalo (2019). Thermal enhanced cement-lime mortars with phase change materials (PCM), lightweight aggregate and cellulose fibers. *Construction and Building Materials*, *221*, 586–594. Available from https://doi.org/10.1016/j.conbuildmat.2019.06.098.

Halder, S., Wang, J., Fang, Y., Qian, X., & Imam, M.A. (2022). Cenosphere-based PCM microcapsules with bio-inspired coating for thermal energy storage in cementitious materials. Materials Chemistry and Physics, *291*, 126745

Han, Baoguo, Ding, Siqi, & Yu, Xun (2015). Intrinsic self-sensing concrete and structures: A review. *Measurement*, *59*, 110–128. Available from https://doi.org/10.1016/j.measurement.2014.09.048.

Hao, L., Xiao, J., Sun, J., Xia, B., & Cao, W. (2022). Thermal conductivity of 3D printed concrete with recycled fine aggregate composite phase change materials. *Journal of Cleaner Production*, *364*, 132598.

Hattan, H. A., Madhkhan, M., & Marani, A. (2021). Thermal and mechanical properties of building external walls plastered with cement mortar incorporating shape-stabilized phase change materials (SSPCMs). *Construction and Building Materials*, *270*, 121385.

Hawes, D. W., Banu, D., & Feldman, D. (1989). Latent heat storage in concrete. *Solar Energy Materials*, *19*(3–5), 335–348. Available from https://doi.org/10.1016/0165-1633(89)90014-2.

Hu, P., Feng, Y., Li, Q., Lin, C.-H., Ning, Y.-H., Li, Y.-T., Yu, L.-P., Cao, Z., & Zeng, J. (2022). Preparation and characterization of n-octadecane @ calcium fluoride microencapsulated phase change materials. *Solar Energy Materials and Solar Cells*, *237*, 111571.

Huang, J., Su, J., Weng, M., Xiong, L., Wang, P., Liu, Y., Lin, X., & Min, Y. (2022). An innovative phase change composite with high thermal conductivity and sensitive light response rate for thermal energy storage. *Solar Energy Materials and Solar Cells*, *245*, 111872.

Hunger, M., Entrop, A. G., Mandilaras, I., Brouwers, H. J. H., & Founti, M. (2009). The behavior of self-compacting concrete containing micro-encapsulated phase change materials. *Cement and Concrete Composites*, *31*(10), 731–743. Available from https://doi.org/10.1016/j.cemconcomp.2009.08.002.

Jayalath, Amitha, San Nicolas, Rackel, Sofi, Massoud, Shanks, Robert, Ngo, Tuan, Aye, Lu, & Mendis, Priyan (2016). Properties of cementitious mortar and concrete containing micro-encapsulated phase change materials. *Construction and Building Materials*, *120*, 408–417. Available from https://doi.org/10.1016/j.conbuildmat.2016.05.116.

Jiang, Y., Ding, E., & Li, G. (2001). Study on transition characteristics of PEG/CDA solid-solid phase change materials. *Polymer*, *43*(1), 117–122. Available from https://doi.org/10.1016/S0032-3861(01)00613-9. Available from: http://www.journals.elsevier.com/polymer/.

Kalombe., Sobhansarbandi, S., & Kevern, J. (2023). Low-cost phase change materials based concrete for reducing deicing needs. *Construction and Building Materials*, *363*129129.

Kashyap, S., Kabra, S., & Kandasubramanian, B. (2020). Graphene aerogel-based phase changing composites for thermal energy storage systems. *Journal of Materials Science*, *55*(10), 4127–4156. Available from https://doi.org/10.1007/s10853-019-04325-7. Available from, http://www.springer.com/journal/10853.

Kehli, K., Belhadj, B., & Ferhat, A. (2023). Development of a new lightweight gypsum composite: Effect of mixed treatment of barley straws with hot water and bio-based phase change material on the thermo-mechanical properties. *Construction and Building Materials*, *389*131597.

Kheradmand, M., Abdollahnejad, Z., & Pacheco-Torgal, F. (2020). Alkali-activated cement-based binder mortars containing phase change materials (PCMs): Mechanical properties and cost analysis. *European Journal of Environmental and Civil Engineering*, *24*(8), 1068–1090. Available from https://doi.org/10.1080/19648189.2018.1446362.

Kocyigit, F., Bayram, M., Hekimoglu, G., Cay, V. V., Gencel, O., Ustaoglu, A., Sari, A., Erdogmus, E., & Ozbakkaloglu, T. (2023). Thermal energy saving and physico-mechanical properties of foam concrete incorporating form-stabilized basalt powder/capric acid based composite phase change material. *Journal of Cleaner Production*, *414*, 137617.

Kousksou, T., Bruel, P., Jamil, A., El Rhafiki, T., & Zeraouli, Y. (2014). Energy storage: Applications and challenges. *Solar Energy Materials and Solar Cells, 120*, 59–80. Available from https://doi.org/10.1016/j.solmat.2013.08.015.

Kumar, D. (2023). Comparative analysis of form-stable phase change material integrated concrete panels for building envelopes. *Case Studies in Construction Materials.*

Lee, Jongki, Wi, Seunghwan, Yun, Beom Yeol, Chang, Seong Jin, & Kim, Sumin (2019). Thermal and characteristic analysis of shape-stabilization phase change materials by advanced vacuum impregnation method using carbon-based materials. *Journal of Industrial and Engineering Chemistry, 70*, 281–289. Available from https://doi.org/10.1016/j.jiec.2018.10.028.

Lee, T., Hawes, D. W., Banu, D., & Feldman, D. (2000). Control aspects of latent heat storage and recovery in concrete. *Solar Energy Materials and Solar Cells, 62*(3), 217–237. Available from https://doi.org/10.1016/s0927-0248(99)00128-2.

Liang, Q., Pan, D., & Zhang, X. (2023). Construction and application of biochar-based composite phase change materials. *Chemical Engineering Journal, 453*, 139441.

Ling, T. C., & Poon, C. S. (2013). Use of phase change materials for thermal energy storage in concrete: An overview. *Construction and Building Materials, 46*, 55–62. Available from https://doi.org/10.1016/j.conbuildmat.2013.04.031.

Liu, Z., Tang, B., & Zhang, S. (2020). Novel network structural PEG/PAA/SiO2 composite phase change materials with strong shape stability for storing thermal energy. *Solar Energy Materials and Solar Cells, 216*, 110678.

Liu, D., Wang, C., Gonzalez-Libreros, J., Guo, T., Cao, J., Tu, Y., Elfgren, L., & Sas, G. (2023). A review of concrete properties under the combined effect of fatigue and corrosion from a material perspective. *Construction and Building Materials, 369*, 130489.

Liu, L., Xu, J., Yin, T., Wang, Y., & Chu, H. (2021). Improved conductivity and piezoresistive properties of Ni-CNTs cement-based composites under magnetic field. *Cement and Concrete Composites, 121*(3), 104089.

Liu, M. Y. J., Alengaram, U. J., Jumaat, M. Z., & Mo, K. H. (2014). Evaluation of thermal conductivity, mechanical and transport properties of lightweight aggregate foamed geopolymer concrete. *Energy and Buildings, 72*, 238–245. Available from https://doi.org/10.1016/j.enbuild.2013.12.029. Available from: https://www.journals.elsevier.com/energy-and-buildings.

Liu, Y., & Yang, Y. (2017). Preparation and thermal properties of $Na_2CO_3 \cdot 10H_2O$-$Na_2HPO_4 \cdot 12H_2O$ eutectic hydrate salt as a novel phase change material for energy storage. *Applied Thermal Engineering, 112*, 606–609. Available from https://doi.org/10.1016/j.applthermaleng.2016.10.146. Available from: http://www.journals.elsevier.com/applied-thermal-engineering/.

Lopez-Arias, M., Francioso, V., & Velay-Lizancos, M. (2023). High thermal inertia mortars: New method to incorporate phase change materials (PCMs) while enhancing strength and thermal design models. *Construction and Building Materials, 370*, 130621.

Lutz, W., & Kc, S. (2010). Dimensions of global population projections: What do we know about future population trends and structures? *Philosophical Transactions of the Royal Society B: Biological Sciences, 365*(1554), 2779–2791. Available from https://doi.org/10.1098/rstb.2010.0133. Available from: http://rstb.royalsocietypublishing.org/content/365/1554/2779.full.pdf + html.

Madbouly, A. I., Mokhtar, M. M., & Morsy, M. S. (2020). Evaluating the performance of rGO/cement composites for SHM applications. *Construction and Building Materials, 250.*

Marani, A., Zhang, L. V., & Nehdi, M. L. (2022). Multiphysics study on cement-based composites incorporating green biobased shape-stabilized phase change materials for thermal energy storage. *Journal of Cleaner Production*, *372*, 133826.

Mehrali, Mohammad, Latibari, Sara Tahan, Mehrali, Mehdi, Indra Mahlia, Teuku Meurah, Cornelis Metselaar, Hendrik Simon, Naghavi, Mohammad Sajad, Sadeghinezhad, Emad, & Akhiani, Amir Reza (2013). Preparation and characterization of palmitic acid/graphene nanoplatelets composite with remarkable thermal conductivity as a novel shape-stabilized phase change material. *Applied Thermal Engineering*, *61*(2), 633–640. Available from https://doi.org/10.1016/j.applthermaleng.2013.08.035.

Mohammadi, B., Shekaari, H., & Zafarani-Moattar, M.T. (2022). Study of the nano-encapsulated vitamin D3 in the bio-based phase change material: Synthesis and characteristics. *Journal of Molecular Liquids, 350*, 118484.

Mondal, S. (2008). Phase change materials for smart textiles – An overview. *Applied Thermal Engineering*, *28*(11–12), 1536–1550. Available from https://doi.org/10.1016/j.applthermaleng.2007.08.009.

Muchtar, A.R., Hassam, C.L., Srinivasan, B., Berthebaud, D., Mori, T., Soelami, N., & Yuliarto, B. (2022). Shape-stabilized phase change materials: Performance of simple physical blending synthesis and the potential of coconut based materials. *Journal of Energy Storage, 52*, 104974.

Navarro, Lidia, de Gracia, Alvaro, Niall, Dervilla, Castell, Albert, Browne, Maria, McCormack, Sarah J., Griffiths, Philip, & Cabeza, Luisa F. (2016). Thermal energy storage in building integrated thermal systems: A review. Part 2. Integration as passive system. *Renewable Energy*, *85*, 1334–1356. Available from https://doi.org/10.1016/j.renene.2015.06.064.

Nomura, T., Okinaka, N., & Akiyama, T. (2009). Impregnation of porous material with phase change material for thermal energy storage. *Materials Chemistry and Physics*, *115* (2–3), 846–850. Available from https://doi.org/10.1016/j.matchemphys.2009.02.045.

Pan, Xiaoying, & Gencturk, Bora (2023). Self-healing efficiency of concrete containing engineered aggregates. *Cement and Concrete Composites*, *142*. Available from https://doi.org/10.1016/j.cemconcomp.2023.105175.

Park, Sangphil, Lee, Yeongmin, Kim, Yong Seok, Lee, Hyang Moo, Kim, Jung Hyun, Cheong, In. Woo, & Koh, Won-Gun (2014). Magnetic nanoparticle-embedded PCM nanocapsules based on paraffin core and polyurea shell. *Colloids and Surfaces A: Physicochemical and Engineering Aspects*, *450*(1), 46–51. Available from https://doi.org/10.1016/j.colsurfa.2014.03.005.

Pasupathy, A., Velraj, R., & Seeniraj, R. V. (2008). Phase change material-based building architecture for thermal management in residential and commercial establishments. *Renewable and Sustainable Energy Reviews*, *12*(1), 39–64. Available from https://doi.org/10.1016/j.rser.2006.05.010.

Peng, Suqing, Zhong, Weilin, Zhao, Hailun, Wang, Chao, Tian, Zhipeng, Shu, Riyang, & Chen, Ying (2022). Solar-driven multifunctional Au/TiO_2@PCM towards bio-glycerol photothermal reforming hydrogen production and thermal storage. *International Journal of Hydrogen Energy*, *47*(98), 41573–41586. Available from https://doi.org/10.1016/j.ijhydene.2022.03.273.

Pilehvar, Shima, Cao, Vinh Duy, Szczotok, Anna M., Valentini, Luca, Salvioni, Davide, Magistri, Matteo, Pamies, Ramón, & Kjøniksen, Anna-Lena (2017). Mechanical properties and microscale changes of geopolymer concrete and portland cement concrete containing micro-encapsulated phase change materials. *Cement and Concrete Research*, *100*, 341–349. Available from https://doi.org/10.1016/j.cemconres.2017.07.012.

Pilehvar, Shima, Szczotok, Anna M., Rodríguez, Juan Francisco, Valentini, Luca, Lanzón, Marcos, Pamies, Ramón, & Kjøniksen, Anna-Lena (2019). Effect of freeze-thaw cycles on the mechanical behavior of geopolymer concrete and portland cement concrete containing micro-encapsulated phase change materials. *Construction and Building Materials, 200*, 94–103. Available from https://doi.org/10.1016/j.conbuildmat.2018.12.057.

Qian, Tingting, Li, Jinhong, Min, Xin, Deng, Yong, Guan, Weimin, & Ning, Lei (2015). Diatomite: A promising natural candidate as carrier material for low, middle and high temperature phase change material. *Energy Conversion and Management*, *98*, 34–45. Available from https://doi.org/10.1016/j.enconman.2015.03.071.

Rebelo, F., Figueiredo, A., Vicente, R., & Ferreira, V.M. (2022). Study of a thermally enhanced mortar incorporating phase change materials for overheating reduction in buildings. *Journal of Energy Storage, 46*, 103876.

Ren, M., Xiaodong, W., Gao, X., & Liu, Y. (2021). Thermal and mechanical properties of ultra-high performance concrete incorporated with microencapsulated phase change material. *Construction and Building Materials, 273*(1), 121714.

Rong, H., Wang, C., Liu, X., Zhuang, Y., Zeng, Z., Wu, T., & Hu, Y. (2022). A novel elastomeric copolymer-based phase change material with thermally induced flexible and shape-stable performance for prismatic battery module. *International Journal of Thermal Sciences, 174*, 107435.

Ryms, M., Januszewicz, K., Haustein, E., Kazimierski, P., & Lewandowski, W.M. (2022). Thermal properties of a cement composite containing phase change materials (PCMs) with post-pyrolytic char obtained from spent tyres as a carrier. *Energy, 239*, 121936.

Samimi, Fereshteh, Babapoor, Aziz, Azizi, Mohammadmehdi, & Karimi, Gholamreza (2016). Thermal management analysis of a Li-ion battery cell using phase change material loaded with carbon fibers. *Energy*, *96*, 355–371. Available from https://doi.org/10.1016/j.energy.2015.12.064.

Sarı, A., Biçer, A., & Alkan, C. (2017). Thermal energy storage characteristics of poly(styrene-co-maleic anhydride)-graft-PEG as polymeric solid–solid phase change materials. *Solar Energy Materials and Solar Cells, 161*, 219–225.

Schackow, Adilson, Effting, Carmeane, Folgueras, Marilena V., Güths, Saulo, & Mendes, Gabriela A. (2014). Mechanical and thermal properties of lightweight concretes with vermiculite and EPS using air-entraining agent. *Construction and Building Materials, 57*, 190–197. Available from https://doi.org/10.1016/j.conbuildmat.2014.02.009.

Shao, R., Wu, C., & Li, J. (2022). A comprehensive review on dry concrete: Application, raw material, preparation, mechanical, smart and durability performance. *Journal of Building Engineering, 55*(1), 104676.

Sharma, Atul, Tyagi, V. V., Chen, C. R., & Buddhi, D. (2009). Review on thermal energy storage with phase change materials and applications. *Renewable and Sustainable Energy Reviews, 13*(2), 318–345. Available from https://doi.org/10.1016/j.rser.2007.10.005.

Sharshir, S.W., Joseph, A., Elsharkawy, M., Hamada, M.A., Kandeal, A.W., Elkadeem, M. R., Thakur, A.K., Ma, Y., Moustapha, M.E., Rashad, M., & Arıcı, M. (2023). Thermal energy storage using phase change materials in building applications: A review of the recent development. *Energy and Buildings, 285*, 112908.

Shen, Y. (2021). Experimental thermal study of a new PCM-concrete thermal storage block (PCM-CTSB). *Construction and Building Materials*, *293*123540.

Singh, A.K., Rathore, P.K. S., Sharma, R.K., Gupta, K., & Kumar, R. (2023). Experimental evaluation of composite concrete incorporated with thermal energy storage material for improved thermal behavior of buildings. *Energy, 263*, 125701.

Soares, N., Costa, J. J., Gaspar, A. R., & Santos, P. (2013). Review of passive PCM latent heat thermal energy storage systems towards buildings' energy efficiency. *Energy and Buildings*, *59*, 82–103. Available from https://doi.org/10.1016/j.enbuild.2012.12.042.
Su, J. C., & Liu, P. S. (2006). A novel solid-solid phase change heat storage material with polyurethane block copolymer structure. *Energy Conversion and Management*, *47* (18–19), 3185–3191. Available from https://doi.org/10.1016/j.enconman.2006.02.022.
Su, W., Darkwa, J., & Kokogiannakis, G. (2015). Review of solid-liquid phase change materials and their encapsulation technologies. *Renewable and Sustainable Energy Reviews*, *48*, 373–391. Available from https://doi.org/10.1016/j.rser.2015.04.044. Available from: https://www.journals.elsevier.com/renewable-and-sustainable-energy-reviews.
Sukontasukkul, Piti, Uthaichotirat, Pattra, Sangpet, Teerawat, Sisomphon, Kritsada, Newlands, Moray, Siripanichgorn, Anek, & Chindaprasirt, Prinya (2019). Thermal properties of lightweight concrete incorporating high contents of phase change materials. *Construction and Building Materials*, *207*, 431–439. Available from https://doi.org/10.1016/j.conbuildmat.2019.02.152.
Sun, Qinrong, Zhang, Nan, Zhang, Haiquan, Yu, Xiaoping, Ding, Yulong, & Yuan, Yanping (2020). Functional phase change composites with highly efficient electrical to thermal energy conversion. *Renewable Energy*, *145*, 2629–2636. Available from https://doi.org/10.1016/j.renene.2019.08.007.
Sun, X., Zhang, Y., Xie, K., &, Medina, M.A. (2022). A parametric study on the thermal response of a building wall with a phase change material (PCM) layer for passive space cooling. *Journal of Energy Storage, 47*, 103548.
Trigui, A., Karkri, M., & Krupa, I. (2014). Thermal conductivity and latent heat thermal energy storage properties of LDPE/wax as a shape-stabilized composite phase change material. *Energy Conversion and Management*, *77*, 586–596. Available from https://doi.org/10.1016/j.enconman.2013.09.034.
Tuncel, E. Y., & Pekmezci, B. Y. (2018). A sustainable cold bonded lightweight PCM aggregate production: Its effects on concrete properties. *Construction and Building Materials*, *181*, 199–216. Available from https://doi.org/10.1016/j.conbuildmat.2018.05.269.
Wang, H., Deng, Y., Wu, F., Dai, X., Wang, W., Mai, Y., Gu, Y., & Liu, Y. (2021). Effect of dopamine-modified expanded vermiculite on phase change behavior and heat storage characteristic of polyethylene glycol. *Chemical Engineering Journal, 415*, 128992.
Wang, J., Zhai, X., Zhong, Z., Zhang, X., & Peng, H. (2022). Nanoencapsulated n-tetradecane phase change materials with melamine–urea–formaldehyde–TiO_2 hybrid shell for cold energy storage. *Colloids and Surfaces A: Physicochemical and Engineering Aspects*, *636*, 128162.
Wang, Y., Li, Q., Miao, W., Su, Y., He, X., & Strnadel, B. (2022). The thermal performances of cement-based materials with different types of microencapsulated phase change materials. *Construction and Building Materials, 345*(2), 128388.
Wang, Yuanyuan, Bailey, Josh, Zhu, Yuan, Zhang, Yingrui, Boetcher, Sandra K. S., Li, Yongliang, & Wu, Chunfei (2022). Application of carbon nanotube prepared from waste plastic to phase change materials: The potential for battery thermal management. *Waste Management*, *154*, 96–104. Available from https://doi.org/10.1016/j.wasman.2022.10.003.
Wu, M. Q., Wu, S., Cai, Y. F., Wang, R. Z., & Li, T. X. (2021). Form-stable phase change composites: Preparation, performance, and applications for thermal energy conversion, storage and management. *Energy Storage Materials*, *42*, 380–417. Available from https://doi.org/10.1016/j.ensm.2021.07.019.

Xia, B., Ding, T., & Xiao, J. (2020). Life cycle assessment of concrete structures with reuse and recycling strategies: A novel framework and case study. *Waste Management*, *105*, 268–278. Available from https://doi.org/10.1016/j.wasman.2020.02.015. Available from, http://www.elsevier.com/locate/wasman.

Xiao, S., Hu, X., Jiang, L., Ma, Y., Che, Y., Zu, S., & Jiang, X. (2022). Nano-Ag modified bio-based loofah foam/polyethylene glycol composite phase change materials with higher photo-thermal conversion efficiency and thermal conductivity. *Journal of Energy Storage, 54*, 105238.

Xie, S., Ma, C., Ji, Z., Wu, Z., Si, T., Wang, Y., & Wang, J. (2023). Electromagnetic wave absorption and heat storage dual-functional cement composites incorporated with carbon nanotubes and phase change microcapsule. *Journal of Building Engineering, 67*, 105925.

Yan, K., Qiu, L., & Feng, Y. (2023). Erythritol/expanded graphite form-stable phase change materials with excellent thermophysical properties. *Journal of Energy Storage, 68*, 107667.

Ying, H., Wang, S., Lu, Z., Liu, B., Cui, L., Quan, X., Liu, K., & Zhao, N. (2023). Development and thermal response of concrete incorporated with multi-stage phase change materials-aggregates for application in seasonally frozen regions. *Journal of Building Engineering*, *71*(4), 106562.

Yousefi, A., Tang, W., Khavarian, M., & Fang, C. (2022). Effects of thermal conductive fillers on energy storage performance of Form-Stable phase change material integrated in cement-based composites. *Applied Thermal Engineering, 212*,118570.

Yu, K., Liu, Y., Jia, M., Wang, C., Yang, Y. (2022). Thermal energy storage cement mortar containing encapsulated hydrated salt/fly ash cenosphere phase change material: Thermo-mechanical properties and energy saving analysis. *Journal of Energy Storage, 51*, 104388.

Yu, K., Jia, M., Liu, Y., & Yang, Y. (2023). Binary decanoic acid/polyethylene glycol as a novel phase change material for thermal energy storage: Eutectic behaviors and energy conservation evaluation. *Journal of Energy Storage*, *68*(5997), 107663.

Yu, K., Liu, Y., & Yang, Y. (2021). Review on form-stable inorganic hydrated salt phase change materials: Preparation, characterization and effect on the thermophysical properties. Applied Energy, *292*, 116845.

Zhang, G. H., Bon, S. A. F., & Zhao, C. Y. (2012). Synthesis, characterization and thermal properties of novel nanoencapsulated phase change materials for thermal energy storage. *Solar Energy*, *86*(5), 1149–1154. Available from https://doi.org/10.1016/j.solener.2012.01.003.

Zhang, M., Li, M., Zhang, J., Liu, D., Hu, Y., Ren, Q., & Tian, D. (2020). Experimental study on electro-thermal and compaction properties of electrically conductive roller-compacted concrete overwintering layer in high RCC dams, *Construction and Building Materials, 263*, 120248.

Zhang, X., Zhang, Y., Li, H., &, Chen, Z. (2023). Enhanced thermal conductivity and photo-thermal effect of microencapsulated n-octadecane phase change material with calcium carbonate-polydopamine hierarchical shell for solar energy storage. *Solar Energy Materials and Solar Cells, 256*, 112336.

Zhang, Y., Jia, Z., Hai, A.M., Zhang, S., & Tang, B. (2022). Shape-stabilization micromechanisms of form-stable phase change materials-A review. *Composites Part A: Applied Science and Manufacturing, 160*, 107047.

Zhang, Y., Xiu, J., Tang, B., Lu, R., & Zhang, S. (2018). Novel semi-interpenetrating network structural phase change composites with high phase change enthalpy. *AIChE Journal*, *64*(2), 688–696. Available from https://doi.org/10.1002/aic.15956.

Zhu, L., Dang, F., Ding, W., Sang, G., Wang, Q., & Jiao, K. (2023). Thermo-physical properties of light-weight aggregate concrete integrated with micro-encapsulation phase change materials: Experimental investigation and theoretical model. *Journal of Building Engineering, 69*, 106309.

Zhu, L., Dang, F., Xue, Y., Jiao, K., & Ding, W. (2021). Multivariate analysis of effects of microencapsulated phase change materials on mechanical behaviors in light-weight aggregate concrete. *Journal of Building Engineering, 42*, 102783.

Smart sustainable concrete materials and structures

19

Antonella D'Alessandro
Department of Civil and Environmental Engineering, University of Perugia, Perugia, Italy

19.1 Introduction

This chapter aims to describe the most promising techniques and materials in the field of sustainable and multifunctional concrete. Several approaches highly promising for enhancing the performance and functionality of concrete structures have emerged.

One notable technique is the use of advanced additives and admixtures, such as nanomaterials, fibers, and polymers. These additives can improve the mechanical properties, durability, and crack resistance of concrete, while also enabling functionalities such as self-healing and self-cleaning.

Another area of focus is the development of high-performance and eco-friendly cement replacements, such as geopolymers and supplementary cementitious materials (SCMs). These alternatives reduce the carbon footprint associated with traditional cement production and offer enhanced durability characteristics.

In terms of structural design, researchers are exploring the development of self-sensing materials and systems within concrete elements to enable real-time performance assessment and structural health monitoring. This allows for proactive maintenance and timely interventions, ensuring the long-term safety and sustainability of structures.

In addition, ongoing research is investigating the use of recycled aggregates and industrial by-products as sustainable alternatives to traditional concrete ingredients. These efforts aim to reduce the environmental impact associated with the extraction and production of raw materials.

By highlighting these promising techniques and materials, this chapter focuses on multifunctional cement-based materials that contribute to the advancement of sustainable and smart concrete solutions, paving the way for more resilient and environmentally conscious construction practices.

In particular, the following sections consider advanced sustainable and smart concrete solutions for resilient and eco-friendly construction practices, starting from the description of the cementitious materials for the identification of possible and promising approaches for the development of less impactful composites, more resistant construction materials, innovative and enhanced solutions for the extension of the service life of constructions, and the increase of the safety of the users.

This chapter is organized as following: first, the peculiarities of concrete are briefly described, to identify possible ways of modification, enhancement of the material, and improvement of its weaknesses. The main possible approaches are then

Sustainable Concrete Materials and Structures. DOI: https://doi.org/10.1016/B978-0-443-15672-4.00019-X

described, with particular attention to: energy efficiency, filler for enhanced properties, three-dimensional (3D) printing, recycled components, carbonation, self-healing, and self-sensing.

Such approaches represent the advanced and promising research development of smart concretes with sustainable and smart properties, for the development of multifunctional building materials, with improved capabilities, and less environmental impact.

The intention of this chapter is to provide insights and avenues for investigation and development toward the reduction of the environmental impact of the construction field, with a special focus on building materials and specifically cementitious ones.

19.2 Why smart concretes?

Concrete is one of the most widely used materials in the construction industry. It possesses a complex chemical structure that influences its physical, chemical, and mechanical properties, due to the particular chemical reactions, which determine its strength, that can be enhanced of affected by the curing conditions. Moreover, the presence of nano- and microsized pores or mineral phases and components allows the eventuality of modification at different scales. In recent years, researchers have focused on the nanometric features within concrete, aiming to understand how these characteristics can be leveraged to bring about significant modifications to the material.

The main components of concretes are cement, sand, and coarse aggregates: the first modification action for enhancing the sustainability of the final product consists in the use of eco-friendlier materials, such as less impactful cements, alternative binders, and recycled or enhanced aggregates. A second possible action is in the choice or tailoring of specific additives with peculiar properties, able to modify the curing characteristics or change the internal chemical structure.

The modification of the nanometric and micrometric structures in concrete, by a resizing of the internal porosity, can bring to an improvement of the material's properties, such as durability and environmental resistance. For instance, the addition of specific additives with peculiar granulometric characteristics can influence the formation and distribution of porosity in concrete during the mixing and curing period, increasing the service life of the structures constituted by it, allowing a more enhanced management of the maintenance activity and the disposal times.

Likewise, nanometric structures in concrete have become the subject of intensive research, especially in the last years, because of the great expansion of the field of nanotechnology, in the development of novel or modified particles and fillers, in the construction of enhanced tools and instruments, in the progress of tailored techniques and procedures for the production, and characterization of cementitious materials. Several papers emphasize the importance of nanoparticles in enhancing concrete properties. The addition of nanoparticles can positively influence the strength and durability of concrete, owing to their high surface area and reinforcing properties. Fibers create a 3D network within the material, increasing its ability to absorb energy and withstand external stresses.

The choice of the peculiar approach for the specific binder, or application, potentially lead to an infinite set of combination of components and preparation process. The knowledge of the innovative and advanced products and techniques is a powerful possibility for engineers, designers, administrations, and users, for the increase in the awareness in the positive approaches for all the choice related to those activities in the construction sector which determine an impact to the environment. The next section describes the main available approaches for enhancing the sustainability of structures and infrastructures.

19.3 Sustainable approaches for construction materials

It is worth noting that the building industry is responsible for nearly 40% of solid waste generation, 40% of energy consumption, 12% of water depletion, and 46% of anthropogenic and greenhouse gas (GHG) emissions (Mindess, 2019). Therefore the sustainability of materials and construction processes is crucial, given the increasing industrialization and large-scale construction projects worldwide, whose impacts can be significant for the environment (Ma et al., 2022). In particular, concrete is responsible for 8%–9% of global anthropogenic GHG emissions and constitutes a significant constituent of demolition waste in the phases of construction maintenance and disposal. Part of the concrete greenhouse emissions is reabsorbed through carbonation in the service life of the new materials, a natural, but slow chemical process. As a matter of fact, concrete is a versatile and widely used construction material as it allows the design of complex architectures coupling easiness of production and application. In recent years, there has been increasing interest in developing intelligent and innovative concretes with enhanced properties for the improvement of its issues related to its durability, and its environmental impact (Miller et al., 2016). These advanced concretes offer improved performance, durability, sustainability, and even the ability to respond to environmental stimuli. Various approaches have been explored to achieve these intelligent characteristics in concrete: the main ones, described in the next sections, can be defined as following (Fig. 19.1):

1. *Incorporation of additives.* One approach is to incorporate additives into the concrete mix design. These additives can be in the form of fillers, or chemical admixtures, enhancing specific properties, such as strength and durability. For example, the addition of fibers, such as steel or polymeric fibers, can significantly improve the crack resistance and tensile strength of concrete. Other additives, like superplasticizers, can enhance workability and reduce water content, leading to denser and stronger concrete. In particular, the use of nanomaterials, such as nanoparticles or nanofibers, offers another way for developing smart concretes. The large surface area and high reactivity of nanoparticles facilitate a better cement hydration and can lead to denser microstructures.
2. *Use of high efficient materials, as phase change materials (PCMs).* The use of specific fillers could enhance the thermal properties of the cementitious composites to increase the efficiency of structures, and the comfort of internal users through the proper interaction with environmental conditions. For example, PCMs, substances that can store and release thermal energy during phase transitions, could allow a smart temperature regulation.

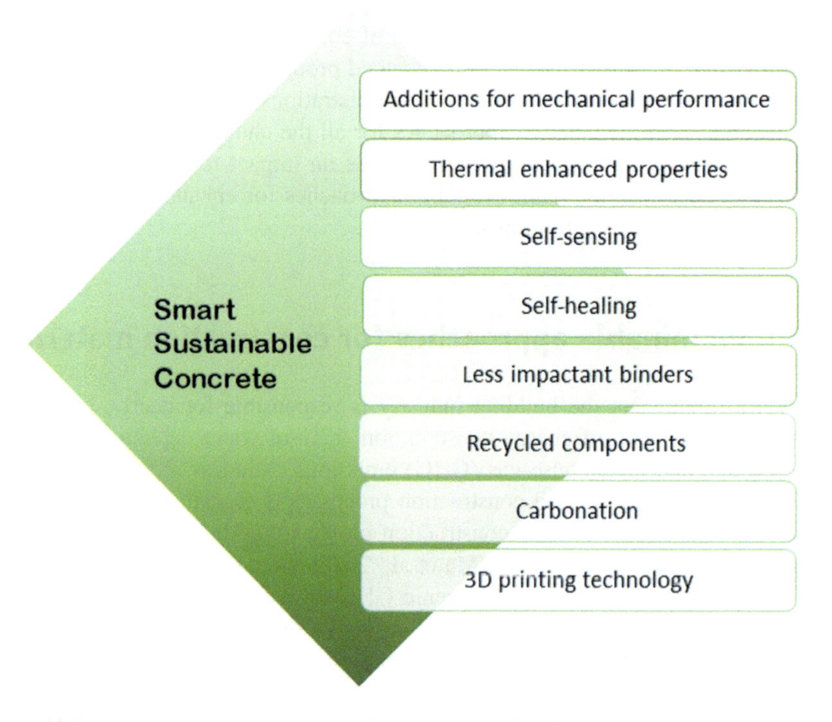

Figure 19.1 Main approaches for developing smart sustainable concrete.

3. *Implementation of sensing technologies.* Advancements in sensing technologies enable real-time monitoring of concrete's structural health and performance. Embedding sensors or smart concrete elements can provide valuable data on strain, stress, temperature, and moisture levels. This information helps in early detection of damage or structural changes, facilitating timely maintenance and proactive decision-making.
4. *Implementation of self-healing technologies.* The use of specific and advanced components allows the development of composites able to self-repair incipient damages and cracks, allowing the enhancement of the durability of the structures. This property could be originated by the use of specific fillers or biological organisms.
5. *Use of less impactful binders, as geopolymers and alkali-activated binders.* To diminish the environmental impact of the preparation process of the different types of cements, also in the energy consumption, novel binders, not cement-based, appear more sustainable anyway maintaining a mechanical resistance and performance suitable for construction applications.
6. *Use of recycled components.* As a composite material, concrete is suitable for the substitution of the various components with less impactful ones. The aim is to reduce the raw materials, for preserving natural resources, and utilize waste products from other anthropic and industrial activities, for solving disposal issues and consequent environmental damages.
7. *Enhancement of carbonation.* As a matter of fact, carbonation is a natural process. However, it is slow, and, in normal conditions, it is not able to reabsorb the high quantities of GHGs produced by several human activities and production processes.

8. *3D printing of concrete.* This advanced technique allows a faster and more performant construction of structures and infrastructures, due to the automated process that optimizes the building steps and avoid high production of waste.

The pursuit of intelligent and innovative concretes has opened up new possibilities in the construction industry. Through the incorporation of additives, nanomaterials, PCMs, and sensing technologies, concrete can possess enhanced properties, such as improved strength, durability, thermal regulation, and real-time monitoring capabilities. Researchers and engineers are continuously exploring and developing these approaches to create concrete materials that meet the evolving needs of modern construction. By incorporating these intelligent concretes, we can enhance sustainability, resilience, and performance in infrastructure projects, leading to safer, more efficient, and longer-lasting structures.

19.3.1 Thermal properties

One of the challenges in sustainable constructions is the enhancement of energy efficiency, in terms of decrease in emissions during the service life of the structures. Constructions are responsible for almost 40% of the worldwide energy demand (Berardi & Gallardo, 2019). Among all building materials, concrete is extensively used, both in the residential and commercial buildings, because it possesses many outstanding properties such as good fire-resistant properties, versatility, moldability, and good compressive strength. One of the drawbacks of such a material is its thermal performance, energy consumption, and emissions that can be reduced.

The GHG emissions associated with each individual concrete component and its production need to be carefully considered and refined. It is important to tailor the mix design and utilize SCMs that are readily available to reduce the embodied CO_2 emissions by up to 16% compared with general practices (Robati et al., 2016). In addition, the thermal conductivity of a concrete mix is influenced by altering the density of aggregates and the proportion of cementitious materials. Innovating the built environment of Europe to zero assumes a minimization of the energy consumption of buildings, eradication of the energy poverty, and mitigation of the urban heat island and the local climate change (Santamouris, 2016). To achieve a zero-emission built environment in Europe, it is crucial to minimize energy consumption in buildings, eliminate energy poverty, and mitigate the urban heat island effect and local climate change.

There has been a growing emphasis on the development and integration of energy-efficient materials and technologies in buildings to meet cooling energy requirements. In recent years, there has been a significant push toward modifying building envelopes to reduce energy consumption. One approach involves customizing construction materials, such as mortar, with heat storage materials. This modification enables better control of indoor temperature and ultimately leads to improved energy efficiency (Rao et al., 2018).

In recent years, there has been a multitude of approaches to enhance the thermal conductivity of concrete. The core concept involves incorporating highly insulating materials to reduce the thermal conductivity of the resulting composite. This can be accomplished through various methods, including the use of foams (Amran et al., 2015), lightweight aggregates (LWAs) like expanded clay and expanded perlite (Rashad, 2018), or expanded polystyrene (Sayadi et al., 2016; Wu et al., 2015). In addition, recent advancements have explored the utilization of cenospheres, glass aggregates (Zeng et al., 2018), and aerogels (Fickler et al., 2015) for this purpose. The LWAs reduced the overall density of the concrete, while the insulating fillers reduced heat transfer, leading to improved thermal insulation (Dixit et al., 2019). The nano- and microparticles modified the pore structure that could result in reduced thermal conductivity and improved insulation capabilities.

There has been a growing momentum in the development and integration of energy-efficient materials and technologies in buildings to meet cooling energy requirements. Recognizing that building envelopes can often contribute to higher energy consumption, efforts have been directed toward modifying them through the use of construction materials, such as mortar, combined with heat storage materials. This strategic modification enables effective regulation of indoor temperature, resulting in enhanced energy efficiency. By tailoring construction materials with heat storage capabilities, building envelopes can play a crucial role in optimizing energy usage and creating more sustainable indoor environments.

Extensive research has focused on optimizing the energy efficiency of concrete by incorporating PCMs, which can store heat in building constructions (Cabeza et al., 2011; D'Alessandro et al., 2018). The use of PCMs has shown promising results in improving the energy efficiency of buildings by reducing temperature fluctuations and decreasing thermal loads. PCMs are widely used in building applications as a thermal storage medium for passive thermal regulation and to enhance the efficiency of heating, ventilation and air conditioning systems. These applications have demonstrated significant potential in reducing energy demand and peak loads for both heating and cooling in buildings. In hot climates or structures with high thermal mass requirements, PCMs effectively absorb excess heat during the day and release it at night, resulting in reduced energy consumption and improved temperature stability (Fig. 19.2).

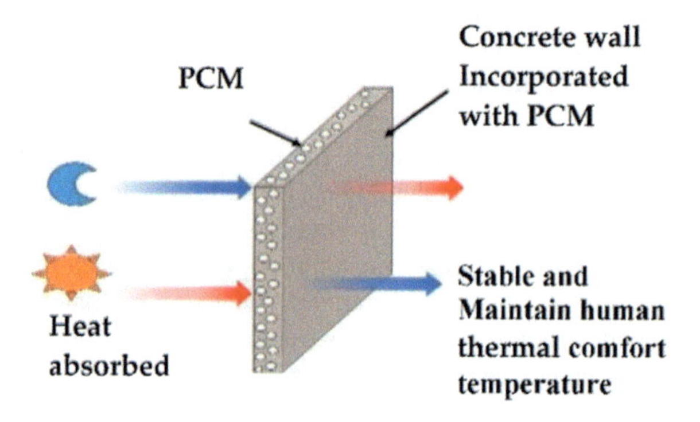

Figure 19.2 Structure and working principle of PCM (Bat-Erdene & Pareek, 2022).

PCMs can be categorized into three types: organic, inorganic (salt hydrates), and eutectic compositions. Within the organic PCM group, there are two distinct subtypes: paraffin and nonparaffin compounds. While inorganic PCMs generally outperform organic PCMs in terms of cost-effectiveness and thermal energy storage capacity, their usage is limited due to issues such as high-volume changes, corrosion, and potential subcooling. This led to an increased utilization of organic PCMs in conjunction with concrete (Asadi et al., 2022).

One significant challenge with PCM microcapsules is their susceptibility to shell breakage during mixing or loading processes, primarily due to their low intrinsic strength and polymeric composition (Alsaadawi et al., 2022). To address this, previous studies have recommended focusing on methods to minimize or eliminate the reduction in mechanical strength. An effective solution would involve the development of robust PCM microcapsules with enhanced outer shell materials that resist rupture under loading conditions. This would contribute to reducing the loss in mechanical strength, enabling a higher incorporation of PCM into concrete, thereby improving its thermal energy storage capacity.

By enhancing the strength and durability of PCM microcapsules, concrete can achieve better thermal performance and enhanced thermal comfort. This advancement has the potential to significantly increase the amount of PCM that can be effectively integrated into concrete, resulting in improved thermal energy storage capabilities.

19.3.2 Filler for enhanced concrete properties

Smart concretes, fortified with nano- and microfibers, have emerged as a revolutionary material in the construction industry. These fibers, when incorporated into concrete, can significantly enhance its properties, such as strength, durability, and sustainability. The development of mechanical properties plays a fundamental role in the design and construction of structures using cementitious materials. It is crucial for ensuring the structural integrity and performance of such structures. In particular, the early age cracking phenomenon is strongly influenced by the progress in stiffness and tensile properties. The strength development significantly impacts construction processes, such as determining the appropriate time for formwork removal and prestressing activities. As structures age, the continuous improvement of mechanical properties has a positive impact on their load-bearing capacity and overall stiffness. Therefore the ongoing enhancement of mechanical properties is essential for ensuring the long-term durability and structural reliability of cementitious structures (Habel et al., 2006). Numerous researchers have investigated the use of fibers to increase the strength of concrete. The findings of the studies demonstrate the considerable potential of fibers in augmenting the tensile and fracture strengths for achieving ultra-high performance concrete (UHPC) (Wen et al., 2022). The optimal volume fraction of fibers appears to be closely associated with the specific fiber type used. Interestingly, the aspect ratio of the fibers does not appear to have a discernible impact on the properties of UHPC. However, the effects of combining different types of fibers, known as hybrid fiber combinations, remain uncertain and contingent on the synergistic interactions between the fibers. It is important to note that variations in factors such as the water/binder ratio, mix proportion, and curing system can also influence the observed results. Hence, it is imperative to consider these variables in UHPC research and design to ensure accurate and reliable outcomes.

Fibers are commonly used to enhance various aspects of cement composites (Paul et al., 2020), including shrinkage cracking mitigation, flexural strength, toughness, and impact resistance (Park et al., 2022). In the case of mono fiber-reinforced cement mortar, the potential for synergistic effects of fiber reinforcement becomes limited within a certain range. However, the utilization of hybrid fiber-reinforced cement mortar, where two or more different fibers are thoughtfully incorporated, can lead to significant improvements in mechanical properties and impact resistance due to the complementary roles of different fiber types (Fig. 19.3). Micro- and macrofibers are the two commonly used types of fibers in the construction of concrete structures, distinguished by their size (length and diameter). Microfibers typically range from 5 to 10 mm in length and 7–30 μm in diameter. These microfibers can be carbon fiber, glass fiber, and basalt fiber based on their material composition (Ivorra et al., 2010). On the other hand, macrofibers possess a length of 25–60 mm and a diameter of 0.2–0.8 mm and include types such as steel and polymeric fibers.

The studies showed that the presence of fibers created a reinforcing network, effectively distributing stress and preventing crack propagation.

Enhancing the long-term durability of concrete has been a subject of considerable interest for researchers over the past decades. In their persistent pursuit, researchers have explored various methods to improve durability. One effective technique involves incorporating new materials into the concrete matrix (Balapour et al., 2018).

Several literature works focus on the modification effects of nanomaterials and the underlying mechanisms that influence the properties of cementitious materials (Shah et al., 2016). The specific features related to nanomaterials, such as the hydration seeding effect, filling effect, thixotropy-modifying effect, and chemical reactivity feature, were examined. The results suggested that the incorporation of nanomaterials can lead to the production of cementitious materials that are stronger,

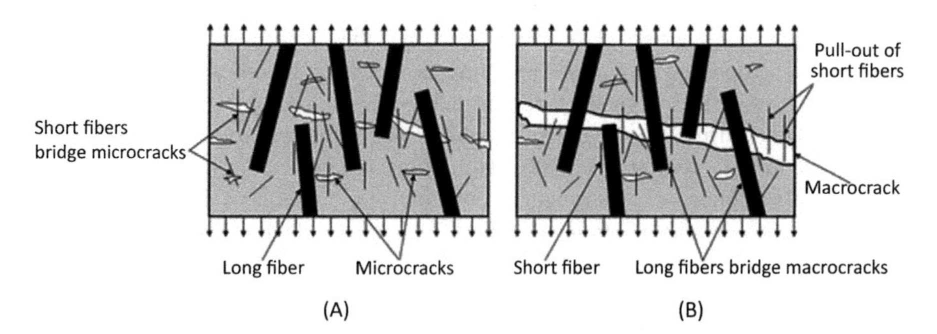

Figure 19.3 Bridging action of the hybrid with different size: (A) first phase of loading and (B) second phase of loading (Park et al., 2022).

more environmentally friendly, and exhibit improved durability, and nanomodification holds the potential to enhance the performance and sustainability of cementitious materials. As an example, adding appropriate carbon nanotubes (CNTs) into concrete can effectively improve its mechanical properties and durability (Khan & Siddique, 2011) because nucleation can promote cement hydration, which explains the compressive strength of concretes with CNTs is better than that of ordinary concrete, and CNTs can fill the internal pores, reducing the porosity and subsequently the infiltration of chloride and sulfate ions, enhancing durability.

Another diffuse and interesting filler for enhancing concrete performance is silica fume (Zhang et al., 2023). Silica fume is composed of ultrafine vitreous particles approximately 100 times smaller than the average cement particle. Due to its exceptional fineness and high silica content, silica fume exhibits excellent pozzolanic properties, making it a highly effective material. Researches have shown that the inclusion of silica fume improves various aspects of concrete performance. It enhances compressive strength, bond strength, and abrasion resistance, while also reducing permeability. By doing so, silica fume plays a crucial role in safeguarding reinforcing steel from corrosion.

Observations have provided evidence of the potential of concrete additives to enhance also the freeze–thaw durability of concrete by utilizing four mechanisms (Ebrahimi et al., 2018): (1) introducing air bubbles in concrete to create additional space for ice expansion; (2) reducing the porosity of concrete through the incorporation of pozzolans and fillers; (3) mitigating crack propagation by incorporating microfibers, nanotubes, and nanosheets; and (4) decreasing water absorption through the use of hydrophobic agents. By implementing these mechanisms, concrete can significantly enhance its ability to withstand freeze–thaw cycles, which can otherwise lead to deterioration and reduced durability.

19.3.3 Three-dimensional printing

The advent of 3D-printing technology has revolutionized various industries, including construction. 3D-printed concrete, also known as additive manufacturing of concrete, involves the layer-by-layer deposition of concrete material to create complex and customized structures (Rob et al., 2015). Understanding the concept of 3D-printed concrete, recent studies and the contributions of key researchers are essential acts to grasp the advancements in this innovative construction technique. 3D-printed concrete involves the use of specialized printers capable of depositing layers of concrete material according to a predetermined design. These printers utilize various techniques, such as extrusion- or powder-based processes, to create concrete structures with intricate geometries and enhanced design flexibility with efficiency in terms of time, material optimization, and waste reduction. Recent researches emphasize the importance of material selection, mix design, and postprocessing technique, together with the influence of printing parameters, such as layer height and printing speed, in achieving high-quality printed structures (Al-Noaimat et al., 2023; Mohan et al., 2021).

The concept of green buildings is being furthered by the implementation of 3D-printing technology, which offers reduced environmental impact and energy requirements (Prince, 2014). This innovative construction method brings about numerous benefits, including a significant reduction in construction waste by 30%–60%, labor costs by 50%–80%, and construction time by 50%–70% (Ma et al., 2022). However, the successful application of printable composites relies on the properties of the binder being tailored to achieve desired pumping, extruding, and printing characteristics. Therefore further research is needed to further investigate rheological properties, printability, setting time, and their interrelationships. In the development of this technology, it is crucial to consider hardened properties such as interlayer bonding, fracture performance, and shrinkage. As a result, several studies have explored the use of SCMs as an alternative to ordinary Portland cement in 3D printing of concrete (Brun et al., 2020). Moreover, incorporating various types of fibers has been found to enhance the mechanical properties and provide excellent crack resistance in printed structures (Zhou et al., 2023). Fillers like steel and carbon can improve rheological properties, while polymer-based fillers hold promise for enhancing bonding and shrinkage properties. The high level of automation and precision offered by 3D printing enables the development of structures with extended service life, utilizing the capabilities of specialized fiber-reinforced composites and sensors (Zhang et al., 2019). Continued research and innovation in 3D-printed concrete will pave the way for further advancements, ultimately transforming the construction industry with efficient and sustainable building practices (Bos et al., 2022; Kaszyńska et al., 2020).

3D printing has emerged as a disruptive and innovative technology that can transform digital design models into physical objects. The advantages of 3D printing in the construction sector are noteworthy. This technology enables optimal utilization of environmental and financial resources while allowing for the creation of esthetically and structurally complex architectural designs (Cao et al., 2022). Furthermore, 3D printing in building construction opens up new possibilities for adopting automation technologies at construction sites (Guamán-Rivera et al., 2022). Robotic platforms used in 3D printing offer solutions to challenges related to structure design and the size of printed elements. However, the use of robots on-site and off-site introduces additional challenges such as location constraints and interactions with unexpected events such as human entry or interactions with other robots. Integrating innovations into conventional construction techniques requires collaboration among various fields of expertise, including civil, systems, electronic, and mechanical engineering, graphic design, and architecture. Contributions from these disciplines have focused on solving challenges related to structural design, the development of concrete-based materials, algorithm implementation, optimization of 3D-printing models, and the establishment of robust systems that could permit the effective use of this technology for different applications and building typologies (Fig. 19.4).

19.3.4 Recycled components

Concrete production consumes a significant amount of natural resources, leading to environmental concerns. In recent years, the use of recycled components in concrete

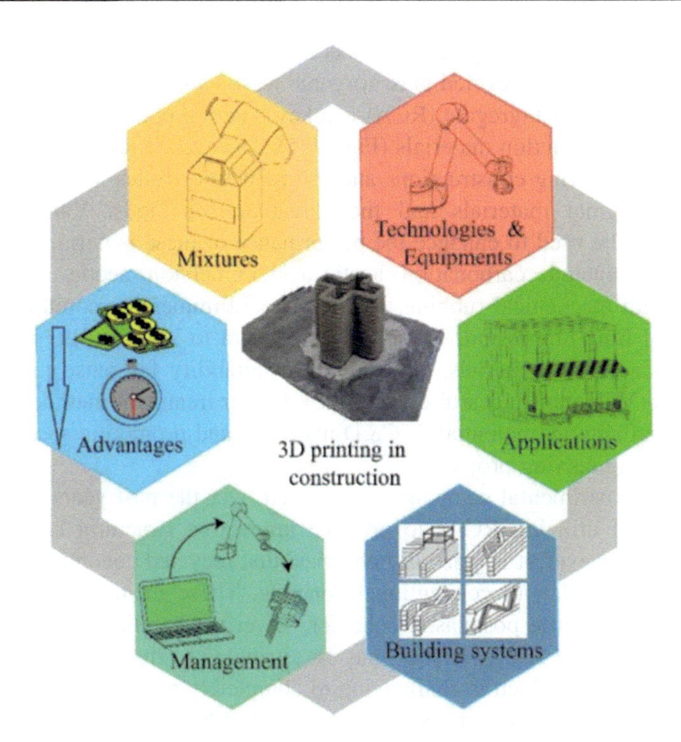

Figure 19.4 Peculiarities of building construction process using 3D-printing technology (Guamán-Rivera et al., 2022).

has gained attention as a sustainable alternative. This approach involves incorporating recycled materials, such as aggregates, cementitious binders, and SCMs, into concrete mixtures. Understanding the use of recycled components in concrete, the investigation of prevalent materials, recent studies, and the activity of key researchers are all crucial features in promoting sustainable construction practices (Tam et al., 2018).

Component recycling serves to reduce the consumption of raw materials during production and minimize waste generated from demolitions. Recycled materials can include aggregates sourced from construction & demolition (C&D) waste or recycled lightweight materials. In addition, binders such as kaolin or modified earth, as well as additives such as blast furnace slag, oils, or industrial waste, can be recycled. Steel reinforcements can also be recycled or a combination of these recycled components can be used. To enhance the performance of binders, more effective materials can be utilized and developed, while their environmental impact can be minimized by substituting them with more sustainable alternatives. Furthermore, the environmental footprint of construction processes can be reduced by utilizing local or recycled resources and implementing lower-impact procedures. Due to its composite nature (Mehta & Monteiro, 2014), concrete is suitable for the addition of various typologies of recycled substances: plastic waste, organic waste, and several other natural or by-products materials (Zamora-Castro et al., 2021).

Examples of possible recycled products that can be reused in concrete are C&D waste, exhausted tyres, recycled glass, organic waste, and industrial by-products.

Recycled concrete aggregate (RCA) is a key material obtained from crushed and processed old construction materials (Fig. 19.5). The successful reuse of CD materials in civil engineering constructions and infrastructures relies heavily on the quality of the original materials and their treatment methods. Various treatment techniques can be used to enhance the performance of these materials. Some potential techniques include carbonation, acetic acid immersion, acetic acid immersion combined with mechanical rubbing, and acetic acid immersion combined with carbonation and lime immersion. These treatments aim to improve the properties and characteristics of the materials, ensuring their suitability for reuse in construction applications (Villagrán-Zaccardi et al., 2022). Proper treatment methods play a vital role in maximizing the potential of C&D materials and promoting sustainable practices in civil engineering projects.

Extensive experimental campaigns conducted over the past years have explored the potential of utilizing RCAs to produce innovative structural concrete (Liotta et al., 2016). However, it is generally observed that recycled concrete exhibits lower properties compared with conventional concrete. These include a reduced elastic modulus, a more brittle postelastic behavior, lower workability, and higher shrinkage and creep. These characteristics may be attributed to the presence of mortar in recycled aggregates, which is challenging to completely remove. To ensure optimal performance of recycled concrete, it is crucial to use high-quality RCAs and adhere to specific design and production processes (RILEM Recommendation, 1994). By selecting appropriate RCAs and following meticulous procedures, it is possible to achieve satisfactory performance and overcome the inherent challenges associated with recycled concrete.

Figure 19.5 Natural aggregated (A) and recycled aggregates (B) from C&D (Kong et al., 2022).

There are several intriguing industrial by-products that have shown successful implementation in concrete technology (the so-called SCMs). Two notable examples include silica fume, a by-product of the semiconductor industry, and fly ash, a by-product of coal-fired electric power plants (Guo et al., 2020). Extensive literature showcases numerous papers documenting the benefits of silica fume, which serves as both a pozzolan and a filler material, recognized for its ability to enhance the strength of concrete, or as an additive to improve various properties of both fresh and hardened concrete. The inclusion of silica fume in concrete contributes to the production of high-strength concrete, demonstrating its significant potential in enhancing concrete performance and sustainability (ACI Committee 234, 2006). Fly ash is capable of provide beneficial effects in cementitious materials, such as the decrease in the heat of hydration, the improvement of the durability, and the enhancement of strength.

In addition, crushed recycled glass and rubber from tires can be incorporated as aggregate replacements in specialized applications. Rubber and cement matrices exhibit contrasting mechanical properties compared with traditional concrete composed solely of natural aggregates. The incorporation of rubber results in a significant decrease in compressive and tensile strengths, as well as stiffness, with potential strength losses of up to 80%. However, studies exploring the use of old tires as aggregate replacements in recycled concrete have shown promising results (Hernandez-Olivares et al., 2002). Such additions enhance toughness, sound insulation properties, and energy absorption capacity. Furthermore, replacing fine aggregates with recycled rubber can decrease the thermal conductivity of concrete, offering improved thermal insulation characteristics (Pacheco-Torgal et al., 2012).

The incorporation of crushed glass as aggregates in Portland cement concrete can have some detrimental effects on its mechanical properties. However, complete replacement of traditional aggregates with crushed glass is still feasible in practical applications (Shayan & Xu, 2004). The primary concerns associated with using crushed glass as aggregates are expansion and cracking caused by the glass particles. It has been found that replacing 20% of the cement with waste glass offers a suitable balance between performance, cost-effectiveness, and environmental impact reduction. Substituting up to 30% of the cement content in certain concrete mixes can lead to satisfactory strength development (Islam et al., 2017). While the use of waste glass as concrete aggregates may slightly affect workability, strength, and resistance to freezing–thawing cycles, the drying shrinkage of recycled concrete with glass powder appears to be adequate.

Despite the introduction of degradable plastics in packaging and disposable items, nonbiodegradable plastics still dominate the market. Polyethylene terephthalate (PET) is one of the most prevalent types of polymer waste found in solid urban waste. Literature investigations have confirmed the potential of PET waste as a substitute for traditional aggregates in concrete (Jo et al., 2008). The inclusion of PET waste aggregates has been shown to enhance the workability of concrete. In addition, researchers have demonstrated that incorporating recycled monofilament PET fibers increases the toughness of cementitious composites, and composites made from PET and polycarbonate waste exhibit excellent energy-absorbing performance.

Numerous other polymeric wastes have also been studied in the literature for their potential as recycled aggregates in cementitious composites: it should be noted that some of these materials may undergo chemical degradation when exposed to the alkaline environment of cementitious materials (Siddique et al., 2008). Further research and exploration are needed to fully understand the performance and long-term behavior of these polymeric waste-based composites in cementitious applications. However, the investigation of various polymeric wastes as potential recycled aggregates highlights the potential for reducing plastic waste and incorporating sustainable materials into concrete production.

As concerns the organic waste, the residues generated from thermochemical processes used for treating organic municipal solid waste in urban areas can give rise to solid waste ash that can be incorporated into construction materials. In fact, literature suggests that introducing these materials into the cement industry as replacements for binders is a promising option for effectively utilizing large quantities of waste materials (Joseph et al., 2018). Continued research and exploration of solid waste ash utilization in construction materials are necessary to ensure optimal performance, compatibility, and adherence to relevant standards.

19.3.5 Carbonation

Carbonation is a significant phenomenon related to the performance and service life of concrete structures. It is a chemical process wherein carbon dioxide from the surrounding environment reacts with calcium hydroxide in concrete to form calcium carbonate. In nature, carbonation takes place over incredibly long periods, contributing to the formation of limestone ($CaCO_3$) and dolomite ($MgCO_3$). However, recent advancements have focused on enhancing and expediting this reaction for a range of applications, primarily aimed at carbon dioxide sequestration. Sequestration of CO_2 in cement-based materials is widely recognized as an effective and energy-efficient alternative. This approach facilitates the conversion of CO_2 into stable carbonates, requiring minimal additional energy consumption (Meng et al., 2023). The concrete's capacity to absorb CO_2 holds important environmental significance. This unique attribute provides a promising avenue for reducing the carbon footprint of the construction industry.

Some notable applications of accelerated carbonation include:

1. *Manufacture of artificial aggregates* (Ren et al., 2021). Artificial carbonates can be produced by exposing natural minerals or industrial/construction waste materials to concentrated CO_2 and optimal reaction conditions, such as high temperature and pressure. These carbonates can then be utilized as aggregates in the construction industry. While these applications are currently limited in scale, the use of carbonation in producing artificial building materials holds immense potential as an effective approach for carbon capture and utilization. However, its success relies on the availability of sufficient quantities of feedstock and CO_2, as well as a market for the resulting products.
2. *Concrete curing* (Zhang et al. 2017). In recent years, there has been a growing focus on carbonation as a curing method for cement-based materials. This heightened attention is primarily driven by the increasing awareness and initiatives surrounding carbon emissions.

Carbonation for curing is being recognized as a viable and environmentally friendly approach. The ongoing sequestration of carbon dioxide in building products aligns with the concrete industry's commitment to sustainability, fulfilling the requirements for a reduced carbon footprint. Nevertheless, it offers a cost-effective solution for carbon sequestration and has the potential to recover a significant portion of the CO_2 generated during cement production.

3. *Carbonation of recycled concrete* (Silva et al., 2015). By subjecting crushed concrete to a controlled carbonation process, the carbon dioxide reacts with the alkaline compounds within the concrete matrix, resulting in the formation of calcium carbonate. This carbonation of recycled concrete not only contributes to carbon dioxide sequestration but also enhances the mechanical properties and long-term stability of the material. The properties of concrete significantly influence its carbon absorption capacity (Andersson et al., 2019). Studies have indicated that higher water-to-cement ratios in the initial concrete mix positively influence the rate of carbonation. Additional factors that impact carbonation include humidity levels, porosity, temperature, binder content, particle size, partial pressure of CO_2, as well as the presence and type of SCMs.

19.3.6 Self-healing

Concrete is susceptible to cracking, caused by load or deformation-induced stresses throughout its service life: this is a well-recognized challenge and a significant obstacle to the long-term durability of concrete structures. Cracks serve as direct pathways for the entry of detrimental substances into concrete. Despite the ability of modern high-performance technologies to produce concrete with low porosity in its bulk matrix, the presence of cracks undermines its overall durability. This issue could be solved by the enhancement of self-healing properties of binders. Self-healing refers to the ability of the material to repair cracks and restore its structural integrity without the need for external intervention. This is achieved through the incorporation of self-activating mechanisms, such as encapsulated healing agents or intrinsic healing materials, which are released when damage occurs. The aforementioned capability, which relies on factors such as the age of cracking, crack width, and the substantial presence of water, can be considered a valuable advantage. In specific types of structures such as tanks and reservoirs, where exposure conditions involving the presence of water are experienced, self-healing can effectively mitigate the disadvantages of early-age shrinkage cracking (Ferrara et al., 2018). Recent advancements in self-healing concrete concern the autogenous or intrinsic healing of conventional concrete, as well as the stimulated autogenous healing achieved through the incorporation of mineral additives, crystalline admixtures, or superabsorbent polymers. The autonomous self-healing mechanisms could involve the application of micro-, macro-, or vascular-encapsulated polymers, minerals, or bacteria (De Belie, 2018). Autogenous healing of cementitious materials is a fundamental phenomenon that contributes to the partial or complete self-closure of cracks. As a result, it enables a partial restoration of the initial durability and physical–mechanical properties of the composites (Fig. 19.6).

There are many possible causes for the self-healing phenomena, where the main mechanism is believed to be attributed to the crystallization of calcium carbonate

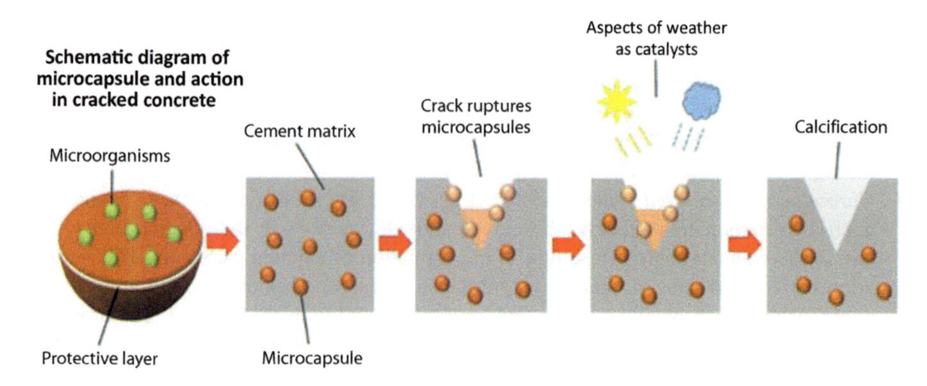

Figure 19.6 Example of healing of cracks in concrete (Roque et al., 2023).

(Wu et al., 2012): (1) formation of calcium carbonate or calcium hydroxide; (2) blockage of cracks by impurities in the water and loose concrete particles caused by crack spalling; (3) continued hydration of unreacted cement or cementitious materials; and (4) expansion of the hydrated cementitious matrix in the crack flanks.

In the context of self-healing in cementitious materials, the key concept is to supply the essential components that can effectively fill the cracks when damage occurs: fibers (Dry, 1994), internal encapsulation of healing adhesives (Li & Lim, 1998), microencapsulation (Roque et al., 2023), use of expansive agents and mineral admixtures (Ahn & Kishi, 2010), bacteria (Tittelboom et al., 2010), and shape memory materials (Zhang et al., 2020).

The integration of new technologies, such as nanotechnology, geopolymer technology, 3D printing/digital production technology, biotechnology, self-assembly technology, and organic–inorganic copolymerization technology, may significantly contribute to the development of self-repairing concrete. Further research is required to investigate the long-term reliability of self-healing behavior, particularly considering the limited shelf life of healing materials (Meraz et al., 2023).

19.3.7 Self-sensing

Among the different strategies used to meet technical and sustainable criteria, an interesting and innovative approach is the use of smart self-sensing concretes for structural health monitoring (Nalon et al., 2022). Such cementitious materials, through a specific production, tailoring, and setup, could provide multifunctional properties, coupling desirable mechanical resistance and sensing capabilities for strain and damage detection (Han et al., 2015).

Self-sensing is a technique that enables the detection of structural effects produced by external factors, such as changes in loading conditions, environmental variations, and temperature fluctuations, by converting them into electrical output properties (Fig. 19.7). This conversion can be achieved through the addition of conductive fillers to the matrix material, the application of an electric field or the connection of the

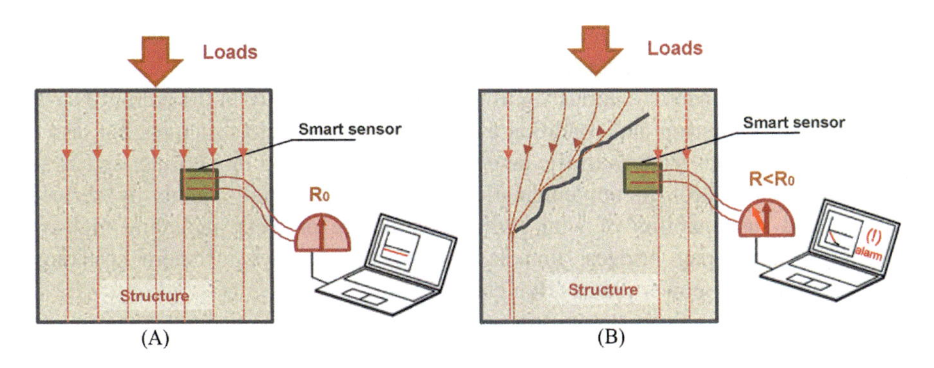

Figure 19.7 Concept of a self-sensing material (D'Alessandro, Birgin, Cerni, & Ubertini, 2022).

composite with equivalent circuit models. Although the initial cost of research and implementing, self-sensing technology may be higher compared with traditional sensors, the subsequent maintenance and repair costs are significantly reduced, making it a cost-effective choice. Various characteristics of the material, including electromechanical, electromagnetic, electrochemical, dielectric, magnetic, and optical properties, can influence sensing capabilities for stress or strain measurements (Ramachandran et al., 2022). Furthermore, smart and sustainable concretes offer enhanced operational durability and exceptional structural bonding capabilities due to their ability to be embedded within structures or even form complete structural elements (D'Alessandro, Birgin, & Ubertini, 2022). This enables the structure itself to sense its strain state and detect changes in its behavior and integrity. The self-monitoring functionality is achieved by correlating variations in strains or stresses with changes in electrical properties (D'Alessandro, Birgin, Cerni, & Ubertini, 2022).

To achieve electrical conductivity, the building material should incorporate conductive fillers into its matrix. Typically, a voltage is applied to conductive electrodes connected to the material, and the resulting variation in electrical current or voltage is measured. The applied signal can be direct current, alternating current, or biphasic. The sensitivity of the smart construction composite depends on various factors, including the piezoresistive properties of the fillers, the electrical characteristics of the matrix, and the contact resistance between the fillers, the matrix, and the electrodes (Galao et al., 2017).

Steel- or carbon-based fillers are commonly utilized as suitable options for such applications. Possible suitable fillers are carbon nano- and microfibers, CNTs, graphene, graphene nanoplatelets, graphite, and carbon black (Taheri et al., 2022). Achieving adequate dispersion of fillers is crucial to ensure the material's homogeneity and isotropy, and it can be accomplished through the utilization of dispersants, mechanical stirring, sonication, or component functionalization (Konsta-Gdoutos et al., 2010).

Smart materials, obtained through nanomodification, hold immense potential and could be utilized in the near future within the concrete industry for practical applications. They offer the ability to monitor the stress levels of reinforced concrete elements exposed to static, dynamic, and impact loads. Specifically, the data on actual stress experienced under dynamic and impact loads could enhance design procedures for protective structures (Coppola et al., 2011).

Self-sensing concrete exhibits significant potential not only in the realm of structural health monitoring and condition assessment for concrete structures but also in diverse applications such as traffic detection, rebar corrosion monitoring, and structural vibration control. Its implementation guarantees the safety, durability, serviceability, and sustainability of vital civil infrastructure projects, including high-rise buildings, large-span bridges, tunnels, high-speed railways, offshore structures, dams, and nuclear power plants (Han et al., 2014).

19.4 Conclusions

Smart and sustainable concretes are paving the way for a more environmentally friendly approach to construction materials. Several approaches can be followed to reduce the environmental footprint of concrete. These include recycling a portion of the components, utilizing nonimpactful materials, minimizing harmful emissions during production, maintenance, and disposal phases, as well as enhancing or expanding the properties of concrete to produce multifunctional composites. In addition to possessing suitable mechanical properties for structural applications, these innovative concretes demonstrate new potential for increasing building energy efficiency, improving durability and structural safety, enabling continuous monitoring of building performance, and extending the lifespan while facilitating easier maintenance and control. The market and scientific literature offer numerous possibilities within the various growing approaches that can be tailored for specific applications.

This chapter aims to synthesize current and ongoing research in the scientific landscape, describing key innovations and future developments in the field of smart and sustainable concretes.

Acknowledgments

The author gratefully acknowledges the support of the Italian Ministry of University and Research within the "Vitality" ecosystem of innovation funded through the Italian National Recovery and Resilience Plan (PNRR).

References

ACI Committee 234. (2006). Guide for the use of silica fume in concrete. *American Concrete Institute Report 234R-06.*

Ahn, T. H., & Kishi, T. (2010). Crack self-healing behavior of cementitious composites incorporating various miner admixtures. *Journal of Advanced Concrete Technology, 8* (2), 171–186.

Al-Noaimat, Y. A., Ghaffar, S. H., Chougan, M., & Al-Kheetan, M. J. (2023). A review of 3D printing low-carbon concrete with one-part geopolymer: Engineering, environmental and economic feasibility. *Case Studies in Construction Materials*, *18*. Available from https://doi.org/10.1016/j.cscm.2022.e01818.

Alsaadawi, Mostafa M., Amin, Mohamed, & Tahwia, Ahmed M. (2022). Thermal, mechanical and microstructural properties of sustainable concrete incorporating Phase change materials. *Construction and Building Materials*, *356*129300. Available from https://doi.org/10.1016/j.conbuildmat.2022.129300.

Amran, H. M., Farzadnia, N., & Ali, A. A. A. (2015). Properties and applications of foamed concrete: A review. *Construction and Building Materials*, *101*, 990–1005. Available from https://doi.org/10.1016/j.conbuildmat.2015.10.112.

Andersson, R., Stripple, H., Gustafsson, T., & Ljungkrantz, C. (2019). Carbonation as a method to improve climate performance for cement based material. *Cement and Concrete Research*, *124*105819. Available from https://doi.org/10.1016/j.cemconres.2019.105819.

Asadi, I., Baghban, Mohamed H., Hashemi, Mohamed, Izadyar, N., & Sajadi, B. (2022). Phase change materials incorporated into geopolymer concrete for enhancing energy efficiency and sustainability of buildings: A review. *Case Studies in Construction Materials*, *17*e01162. Available from https://doi.org/10.1016/j.cscm.2022.e01162.

Balapour, Mohammad, Joshaghani, A., & Althoey, F. (2018). Nano-SiO_2 contribution to mechanical, durability, fresh and microstructural characteristics of concrete: A review. *Construction and Building Materials*, *181*, 27–41. Available from https://doi.org/10.1016/j.conbuildmat.2018.05.266.

Bat-Erdene, P.-E., & Pareek, S. (2022). Experimental study on the development of fly ash foam concrete containing phase change materials (PCMs). *Materials*, *15*, 8428. Available from https://doi.org/10.3390/ma15238428.

Berardi, U., & Gallardo, A. A. (2019). Properties of concretes enhanced with phase change materials for building applications. *Energy and Buildings*, *199*, 402–414. Available from https://doi.org/10.1016/j.enbuild.2019.07.014.

Bos, F. P., Menna, C., Pradena, M., Kreiger, E., da Silva, W. R. L., Rehman, A. U., Weger, D., Wolfs, R. J. M., Zhang, Y., Ferrara, L., & Mechtcherine, V. (2022). The realities of additively manufactured concrete structures in practice. *Cement and Concrete Research.*, *156*. Available from https://doi.org/10.1016/j.cemconres.2022.106746.

Brun, F., Gaspar, F., Mateus, A., Vitorino, J., & Diz, F. (2020). Experimental study on 3D printing of concrete with overhangs. In F. Bos, S. Lucas, R. Wolfs, & T. Salet (Eds.), *RILEM international conference on concrete and digital fabrication* (pp. 778–789). Springer. Available from https://doi.org/10.1007/978-3-030-49916-7_77.

Cabeza, L. F., Castell, A., Barreneche, C., De Gracia, A., & Fernández, A. I. (2011). Materials used as PCM in thermal energy storage in buildings: A review. *Renewable and Sustainable Energy Reviews*, *15*(3), 1675–1695.

Cao, X., Yu, S., Cui, H., & Li, Z. (2022). 3D printing devices and reinforcing techniques for extruded cement-based materials: A review. *Buildings*, *12*, 453. Available from https://doi.org/10.3390/buildings12040453.

Coppola, L., Buoso, A., & Corazza, F. (2011). Electrical properties of carbon nanotubes cement composites for monitoring stress conditions in concrete structures. *Applied Mechanics and Materials.*, *82*, 118–123.

De Belie, N., Gruyaert, E., Al-Tabbaa, A., Antonaci, P., Baera, C., Bajare, D., Darquennes, A., Davies, R., Ferrara, L., Jefferson, T., Litina, C., Miljevic, B., Otlewska, A., Ranogajec, J., Roig-Flores, M., Paine, K., Lukowski, P., Serna, P., Tulliani, J.-M., &

Jonkers, H. M. (2018). Review of self-healing concrete for damage management of structures. *Advanced Materials Interfaces*, *5*(17). Available from https://doi.org/10.1002/admi.201800074.

Dixit, A., Pang, S. D., Kang, S.-H. , & Moon, J. (2019). Lightweight structural cement composites with expanded polystyrene (EPS) for enhanced thermal insulation. *Cement and Concrete Composites*, *102*, 185–197. Available from https://doi.org/10.1016/j.cemconcomp.2019.04.023.

Dry, C. (1994). Matrix cracking repair and filling using active and passive modes for smart timed release of chemicals from fibers into cement matrices. *Smart Mater Struct*, *3*(2), 118–123.

D'Alessandro, A., Pisello, A. L., Fabiani, C., Ubertini, F., Cabeza, L. F., & Cotana, F. (2018). Multifunctional smart concretes with novel phase change materials: mechanical and thermo-energy investigation. *Applied Energy*, *212*(January), 1448–1461.

D'Alessandro, A., Birgin, H. B., & Ubertini, F. (2022). Carbon microfiber-doped smart concrete sensors for strain monitoring in reinforced concrete structures: An experimental study at various scales. *Sensors*, *22*(16), 6083. Available from https://doi.org/10.3390/s22166083.

D'Alessandro, A., Birgin, H. B., Cerni, G., & Ubertini, F. (2022). Smart infrastructure monitoring through self-sensing composite sensors and systems: A study on smart concrete sensors with varying carbon-based filler. *Infrastructures*, *7*, 48. Available from https://doi.org/10.3390/infrastructures7040048.

Ebrahimi, K., Daiezadeh, Mohammad J., Zakertabrizi, Mohammad, Zahmatkesh, F., & Habibnejad Korayem, A. (2018). A review of the impact of micro- and nanoparticles on freeze-thaw durability of hardened concrete: Mechanism perspective. *Construction and Building Materials*, *186*, 1105–1113. Available from https://doi.org/10.1016/j.conbuildmat.2018.08.029.

Ferrara, L., Van Mullem, T., Alonso, M. C., Antonaci, P., Borg, R. P., Cuenca, E., Jefferson, A., Ng, P.-L. , Peled, A., R.-F., M., Sanchez, M., Schroefl, C., Serna, P., Snoeck, D., Tulliani, J. M., & De Belie, N. (2018). Experimental characterization of the self-healing capacity of cement based materials and its effects on the material performance: A state of the art report by COST Action SARCOS WG2. *Construction and Building Materials*, *167*, 115–142. Available from https://doi.org/10.1016/j.conbuildmat.2018.01.143.

Fickler, S., Milow, B., Ratke, L., Schnellenbach-held, M., & Welsch, T. (2015). Development of high performance aerogel concrete. *Energy Procedia*, 6–11.

Galao, O., Baeza, F. J., Zornoza, E., & Garcés, P. (2017). Carbon nanofiber cement sensors to detect strain and damage of concrete specimens under compression. *Nanomaterials*, *7*, 413.

Guamán-Rivera, R., Martínez-Rocamora, A., García-Alvarado, R., Muñoz-Sanguinetti, C., González-Böhme, L. F., & Auat-Cheein, F. (2022). Recent developments and challenges of 3D-printed construction: A review of research fronts. *Buildings*, *12*, 229. Available from https://doi.org/10.3390/buildings12020229.

Guo, Z., Jiang, T., Zhang, J., Kong, X., Chen, C., & Lehman, D. E. (2020). Mechanical and durability properties of sustainable self-compacting concrete with recycled concrete aggregate and fly ash, slag and silica fume. *Construction and Building Materials*, *231*, 117115.

Habel, K., Viviani, M., Denarié, E., & Brühwiler, E. (2006). Development of the mechanical properties of an ultra-high performance fiber reinforced concrete (UHPFRC). *Cement and Concrete Research*, *36*(7), 1362–1370.

Han, B., Yu, X., & Ou, J. (2014). *Self-sensing concrete in smart structures*. Butterworth-Heinemann.

Han, B., Ding, S., & Yu, X. (2015). Intrinsic self-sensing concrete and structures: A review. *Measurement*, *159*, 110–128. Available from https://doi.org/10.1016/j.measurement.2014.09.048.

Hernandez-Olivares, F., Barluenga, G., Bollati, M., & Witoszek, B. (2002). Static and dynamic behaviour of recycled tyre rubber-filled concrete. *Cement and Concrete Research*, *32*(10), 1587–1596.

Islam, G. M. S., Rahman, M. H., & Kazy, N. (2017). Waste glass powder as partial replacement of cement for sustainable concrete practice. *International Journal of Sustainable Built Environment*, *6*, 37–44.

Ivorra, S., Garcés, P., Catalá, G., Andión, L. G., & Zornoza, E. (2010). Effect of silica fume particle size on mechanical properties of short carbon fiber reinforced concrete. *Materials & Design*, *31*, 1553–1558.

Jo, B., Park, S., & Park, J. (2008). Mechanical properties of polymer concrete made with recycled PET and recycled concrete aggregates. *Construction and Building Materials*, *22*, 2281–2291.

Joseph, A. M., Snellings, R., Van den Heede, P., Matthys, S., & De Belie, N. (2018). The use of municipal solid waste incineration ash in various building materials: A Belgian point of view. *Materials (Basel, Switzerland)*, *11*(1), 141. Available from https://doi.org/10.3390/ma11010141.

Kaszyńska, M., Skibicki, S., & Hoffmann, M. (2020). 3D concrete printing for sustainable construction. *Energies*, *13*, 6351. Available from https://doi.org/10.3390/en13236351.

Khan, M. I., & Siddique, R. (2011). Utilization of silica fume in concrete: Review of durability properties. *Resources, Conservation and Recycling*, *57*, 30–35.

Kong, X., Yao, Y., Wu, B., Zhang, W., He, W., & Fu, Y. (2022). The impact resistance and mechanical properties of recycled aggregate concrete with hooked-end and crimped steel fiber. *Materials*, *15*(19), 7029. Available from https://doi.org/10.3390/ma15197029.

Konsta-Gdoutos, M. S., Metaxa, Z. S., & Shah, S. P. (2010). Highly dispersed carbon nanotube reinforced cement based materials. *Cement and Concrete Research.*, *40*, 1052–1059.

Li, V. C., Lim, Y. M., & Chan, Y. W. (1998). Feasibility of a passive smart self-healing cementitious composite. *Composites Part B: Engineering*, *29B*, 819–827.

Liotta, M. A., Viviani, M., & Rodriquez, C. (2016). Structural concrete with recycled aggregate: Advances in mechanical properties, durability and sustainability. *Applied Mechanics and Materials*, *847*, 553–558.

Ma, L., Zhang, Q., Lombois-Burger, H., Jia, Z., Zhang, Z., Niu, G., & Zhang, Y. (2022). Pore structure, internal relative humidity, and fiber orientation of 3D printed concrete with polypropylene fiber and their relation with shrinkage. *Journal of Building Engineering*, *61*. Available from https://doi.org/10.1016/j.jobe.2022.105250.

Mehta, P. K., & Monteiro, P. J. M. (2014). *Concrete: Microstructure, properties, and materials*. McGraw Hill, 704 pp..

Meng, D., Unluer, C., Yang, E.-H. , & Qian, S. (2023). Recent advances in magnesium-based materials: CO_2 sequestration and utilization, mechanical properties and environmental impact. *Cement and Concrete Composites*, *138*104983. Available from https://doi.org/10.1016/j.cemconcomp.2023.104983.

Meraz, M. M., Mim, N. J., Mehedi, M. T., Bhattacharya, B., Aftab, M. R., Billah, M. M., & Meraz, M. M. (2023). Self-healing concrete: Fabrication, advancement, and effectiveness for long-term integrity of concrete infrastructures. *Alexandria Engineering Journal*, *73*, 665–694.

Miller, S. A., Horvath, A., & Monteiro, P.-J. M. (2016). Readily implementable techniques can cut annual CO_2 emissions from the production of concrete by over 20%. *Environmental Research Letters*, *11*, 074029.

Mindess, S. (2019). *Sustainability of concrete*. Elsevier Ltd. Available from https://doi.org/10.1016/B978-0-08-102616-8.00001-0.

Mohan, M. K., Rahul, A. V., De Schutter, G., & Van Tittelboom, K. (2021). Extrusion-based concrete 3D printing from a material perspective: a state-of-the-art review. *Cement and Concrete Composites, 115*103855. Available from https://doi.org/10.1016/j.cemconcomp.2020.103855.

Nalon, G. H., Santos, R. F., Soares de Lima, G. E., Roch Andrade, I. K., Pedroti, L. G., Lope Ribeiro, J. C., & Franco de Carvalho, J. M. (2022). Recycling waste materials to produce self-sensing concretes for smart and sustainable structures: A review. *Construction and Building Materials, 1325*, 126658. Available from https://doi.org/10.1016/j.conbuildmat.2022.126658.

Pacheco-Torgal, F., Ding, Yining, & Jalali, Said (2012). Properties and durability of concrete containing polymeric wastes (tyre rubber and polyethylene terephthalate bottles): An overview. *Construction and Building Materials, 30*, 714–724.

Park, J.-G., Seo, D.-J., & Heo, G.-H. (2022). Impact resistance and flexural performance properties of hybrid fiber-reinforced cement mortar containing steel and carbon fibers. *Applied Sciences., 12*, 9439. Available from https://doi.org/10.3390/app12199439.

Paul, S. C., van Zijl, G. P. A. G., & Šavija, B. (2020). Effect of fibers on durability of concrete: A practical review. *Materials, 13*, 4562. Available from https://doi.org/10.3390/ma13204562.

Prince, J. D. (2014). 3D printing: An industrial revolution. *Journal of Electronic Resources in Medical Libraries, 11*, 39–45. Available from https://doi.org/10.1080/15424065.2014.877247.

RILEM Recommendation. (1994). Specifications for concrete with recycled aggregates. *Materials & Structures., 27*, 557–559.

Ramachandran, K., Vijayan, P., Murali, G., & Vatin, N. I. (2022). A review on principles, theories and materials for self sensing concrete for structural applications. *Materials, 15* (11), 3831. Available from https://doi.org/10.3390/ma15113831.

Rao, V. V., Parameshwaran, R., & Ram, V. V. (2018). PCM-mortar based construction materials for energy efficient buildings: A review on research trends. *Energy and Buildings, 158*, 95–122. Available from https://doi.org/10.1016/j.enbuild.2017.09.098.

Rashad, A. M. (2018). Lightweight expanded clay aggregate as a building material – an overview. *Construction and Building Materials, 170*, 757–775.

Ren, P., Ling, T.-C. , & Mo, K. H. (2021). Recent advances in artificial aggregate production. *Journal of Cleaner Production, 291*125215. Available from https://doi.org/10.1016/j.jclepro.2020.125215.

Robati, M., McCarthy, T. J., & Kokogiannakis, G. (2016). Incorporating environmental evaluation and thermal properties of concrete mix designs. *Construction and Building Materials, 128*, 422–435.

Roque, B. A. C., Brasileiro, P. P. F., Brandão, Y. B., Casazza, A. A., Converti, A., Benachour, M., & Sarubbo, L. A. (2023). Self-healing concrete: Concepts, energy saving and sustainability. *Energies, 16*, 1650. Available from https://doi.org/10.3390/en16041650.

Santamouris, M. (2016). Innovating to zero the building sector in Europe: minimising the energy consumption, eradication of the energy poverty and mitigating the local climate change. *Solar Energy, 128*, 61–94.

Sayadi, A., Tapia, J. V., Neitzert, T. R., & Clifton, G. C. (2016). Effects of expanded polystyrene (EPS) particles on fire resistance, thermal conductivity and compressive strength of foamed concrete. *Construction and Building Materials, 112*, 716–724. Available from https://doi.org/10.1016/j.conbuildmat.2016.02.218.

Shah, S. P., Hou, Pengkun, & Konsta-Gdoutos, M. S. (2016). Nano-modification of cementitious material: toward a stronger and durable concrete. *Journal of Sustainable Cement-Based Materials*, *5*(1-2), 1–22. Available from https://doi.org/10.1080/21650373.2015.1086286.

Shayan, A., & Xu, A. (2004). Value-added utilisation of waste glass in concrete. *Cement and Concrete Research*, *34*(1), 81–89. Available from https://doi.org/10.1016/S0008-8846(03)00251-5.

Siddique, R., Khatib, J., & Kaur, I. (2008). Use of recycled plastic in concrete: A review. *Waste Management*, *28*(10), 1835–1852. Available from https://doi.org/10.1016/j.wasman.2007.09.011.

Silva, R. V., Neves, R., de Brito, J., & Dhir, R. K. (2015). Carbonation behaviour of recycled aggregate concrete. *Cement and Concrete Composites*, *62*, 22–32. Available from https://doi.org/10.1016/j.cemconcomp.2015.04.017.

Taheri, S., Georgaklis, J., Ams, M., et al. (2022). Smart self-sensing concrete: The use of multi-scale carbon fillers. *Journal of Materials Science*, *57*, 2667–2682, https://doi.org/10.1007.

Tam, V. W. Y., Soomro, M., & Evangelista, A. C. J. (2018). A review of recycled aggregate in concrete applications (2000–2017). *Construction and Building Materials*, *172*, 272–292.

Tittelboom, K. V., Belie, N. D., Muynck, W. D., & Verstraete, W. (2010). Use of bacteria to repair cracks in concrete. *Cement and Concrete Research*, *40*, 157–166.

Villagrán-Zaccardi, Y. A., Marsh, A. T. M., Sosa, M. E., Zega, C. J., De Belie, N., & Bernal, S. A. (2022). Complete re-utilization of waste concretes – valorisation pathways and research needs. *Resources, Conservation & Recycling*, *177*, 105955.

Wen, C., Zhang, P., Wang, J., & Hu, S. (2022). Influence of fibers on the mechanical properties and durability of ultra-high-performance concrete: A review. *Journal of Building Engineering*, *52*104370. Available from https://doi.org/10.1016/j.jobe.2022.104370.

Rob, W., Salet, T., & Hendriks, B. (2015). 3D printing of sustainable concrete structures. In: *Proceedings of IASS annual symposia. International Association for Shell and Spatial Structures (IASS)*, Vol. 2015, no. 2.

Wu, M., Johannesson, B., & Geiker, M. (2012). A review: Self-healing in cementitious materials and engineered cementitious composite as a self-healing material. *Construction and Building Materials*, *28*(1), 571–583. Available from https://doi.org/10.1016/j.conbuildmat.2011.08.086.

Wu, Y., Wang, J., Monteiro, P. J. M., & Zhang, M. (2015). Development of ultra-lightweight cement composites with low thermal conductivity and high specific strength for energy efficient buildings. *Construction and Building Materials*, *87*, 100–112. Available from https://doi.org/10.1016/j.conbuildmat.2015.04.004.

Zamora-Castro, S. A., Salgado-Estrada, R., Sandoval-Herazo, L. Cs, Melendez-Armenta, R. A., Manzano-Huerta, E., Yelmi-Carrillo, E., & Herrera-May, A. L. (2021). Sustainable development of concrete through aggregates and innovative materials: A review. *Applied Sciences.*, *11*, 629.

Zeng, Q., Mao, T., Li, H., & Peng, Y. (2018). Thermally insulating lightweight cement-based composites incorporating glass beads and nano-silica aerogels for sustainably energy-saving buildings. *Energy and Buildings*, *174*, 97–110. Available from https://doi.org/10.1016/j.enbuild.2018.06.031.

Zhang, D., Ghouleh, Z., & Shao, Y. (2017). Review on carbonation curing of cement-based materials. *Journal of CO_2 Utilization*, *21*, 119–131. Available from https://doi.org/10.1016/j.jcou.2017.07.003.

Zhang, J., Wang, J., Dong, S., Yu, X., & Han, B. (2019). A review of the current progress and application of 3D printed concrete. *Composites Part A: Applied Science and Manufacturing*, *125*105533. Available from https://doi.org/10.1016/j.compositesa.2019.105533.

Zhang, P., Su, J., Guo, J., & Hu, S. (2023). Influence of carbon nanotube on properties of concrete: A review. *Construction and Building Materials, 369*130388. Available from https://doi.org/10.1016/j.conbuildmat.2023.130388.

Zhang, W., Zheng, Q., Ashour, A., & Han, B. (2020). Self-healing cement concrete composites for resilient infrastructures: A review. *Composites Part B: Engineering, 189*107892. Available from https://doi.org/10.1016/j.compositesb.2020.107892.

Zhou, Y., Jiang, D., Sharma, R., Xie, Y. M., & Singh, A. (2023). Enhancement of 3D printed cementitious composite by short fibers: A review. *Construction and Building Materials, 362*. Available from https://doi.org/10.1016/j.conbuildmat.2022.129763.

Detachable connections for future reuse of structural concrete elements

Wei Zhou
School of Civil Engineering, Harbin Institute of Technology, Harbin, P.R. China

20.1 Introduction

The construction industry provides infrastructure and buildings to society by consuming a large amount of nonrenewable resources, which will lead to a large amount of high greenhouse gas CO_2 emissions (Huang et al., 2018). At the same time, a lot of concrete demolition waste (CDW) will be generated at the life cycle end of the building (Gastaldi et al., 2015). According to statistics, more than 11 billion tons of CWD globally each year, of which concrete waste accounts for 50%–70% (Kim & Kim, 2007; Tam, 2008), and the disposal of these solid wastes in landfills and incinerators is potentially hazardous to the residents in the surrounding areas (Hogland et al., 2004; Mattiello et al., 2013). Promoting the development of low-carbon building materials, energy-efficient use of building machinery, and using renewable energy in the building sector are three key opportunities to reduce carbon emissions in the building (Huang et al., 2018). In the building industry, the "3R" waste principle, that is, reducing, reusing, and recycling, is generally followed to minimize the negative impacts of buildings on the environment, society, and the economy (Ding et al., 2018). Industrialized buildings rely on quality growth, environmental integration, sound design, planning, and optimization to achieve sustainable buildings (Zabihi & Mirsaeedie, 2013).

Reducing strategy means avoiding over-design of buildings by optimizing design and material solutions (Andrew, 2019). This strategy focuses on the energy efficiency of the whole cycle of building construction, use, and demolition and reduces the carbon emissions of this cycle through optimization (Tingley & Davison, 2011). In addition to reducing carbon emissions by optimizing structural design solutions, reducing carbon emissions during cement production is also an essential part of the reducing strategy. Researchers have tried to propose new formulations to reduce carbon emissions during cement production (Meyer, 2009), such as alkali-activated concretes (Swamy, 1998; Zakka et al., 2021; Zhe & Zhou, 2023), which efficiently utilizes fly ash (Kurtoğlu et al., 2018) and slag (Cong et al., 2020). In addition, the development and application of improved concrete materials such as self-compacting concrete (Scrivener & Kirkpatrick, 2008) and recycled concrete (Zhang et al., 2023) have also reduced energy consumption and carbon emissions. However, applying these improved concrete materials may need further development considering technical and financial considerations.

Sustainable Concrete Materials and Structures. DOI: https://doi.org/10.1016/B978-0-443-15672-4.00020-6

By designing concrete buildings in a demountable form, the reuse strategy allows reusable parts to remain after demolition. This type of removable concrete building form makes it more economical to maintain and repair throughout its life cycle. Demolishable building forms require the development of demountable building components. This fact gave rise to the design for deconstruction (DFD), a concept popularized in mechanical engineering, which is very important for demountable buildings because DFD components not only reduce the resource consumption and carbon emissions associated with the production of new components but also reduce the maintenance costs of a demountable building. Utilizing as few resources as possible to achieve more functionality is the key to green and sustainable development in almost every field. For wood and steel construction, years of assembly building research have made it very easy to realize demountable building solutions. Concrete buildings have always been predominantly poured on-site, meaning their integrity is affected by traditional concrete construction processes. Demolition of concrete buildings has generally had to rely on methods such as bulldozers and explosives. The program offered by the DFD suggests the reuse of dismantled concrete elements, which means that research on concrete specimens needs to focus on three aspects. The first aspect is demolition friendly, the second aspect is to improve the assembly efficiency of the product, and the third is to improve the finishing performance. The application of these strategies in concrete structures is very challenging. Strict policies, DFD processes and capabilities, material recycling, material reuse, and flexibility in building design are five critical factors for DFD success (O.O. Akinade et al., 2017). Engineering methods such as life cycle assessment (Densley Tingley & Davison, 2012) and building information modeling (O. Akinade et al., 2020) are gradually being used to assess the economic and environmental effects of DFD. However, the reliability of these DFD schemes still needs a research basis, which limits the application and diffusion of some compelling and concise constructions.

For DFD programs for concrete buildings, the connection is the most important, which determines accountability, constructability, and seismic performance. Connection form influences all aspects of concrete building performance from construction to the entire life cycle. Currently, some concrete connection forms (Ding et al., 2018; Figueira et al., 2021; Xiao et al., 2017) (dry/semidry connections) with low site casting requirements are being developed and promoted for concrete buildings in DFD schemes. In this chapter, we summarize the current research on dry/semidry connections for precast reinforced concrete and some typical construction forms proposed. Furthermore, we provide our insights into the challenges and future directions of DFD schemes and precast concrete connection forms.

20.2 Design for deconstruction program with dry connection

Casting is a necessary process in concrete construction. The tricky part of the assembled concrete building program is the connections. The reliability of the connections directly influences the behavior of the concrete building in some extreme situations. On the one hand, we want the concrete building to be demountable, and on the other hand, we hope the concrete building can withstand extreme conditions, such as earthquakes. DFD is still a novel concept for concrete buildings, and the existing solutions are still far from ready for large-scale implementation. In addition, the detailed conceptual definitions of DFD schemes are subject to some controversy in engineering, which is inevitable in a maturing discipline. Concrete connections can be categorized as dry/semidry, with the concrete discipline referring to connections that require grouting or postcasting as semidry and those that are dry. In general, the application of dry connections in concrete construction can be challenging, thus leading to less research on this type of construction. Several very typical dry connections are described in detail below.

20.2.1 Steel connectors-bolt program

Dry connections in DFD for concrete buildings are constructed through bolts and steel plates. With the tests conducted by Sun et al. (2016), the concrete slabs and bases were connected through steel connectors and bolts, as shown in Fig. 20.1. Such a DFD program provides reliable connections and seismic performance, as the steel connections provide sufficient deformation capacity, allowing the concrete building to maintain good seismic loading and energy dissipation capacity.

However, the reason that such a connection program does not require grout or postcast concrete is that the slab is so thin that bolts can run through both ends of the slab. With the steel connectors restraining the concrete slab and the base, a very reliable connection can be formed. Furthermore, because of the steel connectors, there are virtually no stress concentrations near the open concrete slab. Overall, this is a very reliable solution for connecting the base to the slab.

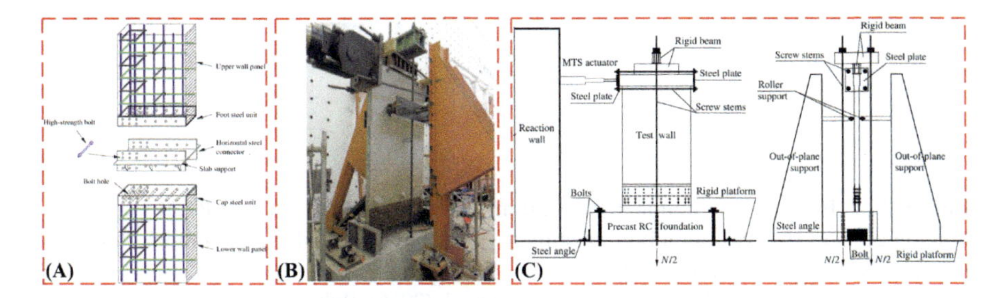

Figure 20.1 Steel connectors—bolt program (Sun et al., 2016). (A) Schematic of the system concept; (B) photograph; and (C) loading system elevation.
Source: From Sun, J., Qiu, H., & Lu, Y. (2016). Experimental study and associated numerical simulation of horizontally connected precast shear wall assembly. *The Structural Design of Tall and Special Buildings*, *25*(13), 659–678. https://doi.org/10.1002/tal.1277.

20.2.2 Bolted shear connectors program

The second type of dry connection in the DFD program also has to rely on bolts to provide shear capacity, such as the concrete-steel beam connection proposed by Lawan et al. (2016), which uses bolts to provide shear capacity. Fig. 20.2 shows the construction of this connection program, which is in line with the first program, which also requires that the concrete members not be too thick. The distribution of bolts must be manageable, which means that the program needs to consider the utilization of both steel and concrete materials when considering the distribution of bolts, making full use of the mechanical properties of both materials, which means we need to consider the utilization of both steel and concrete materials when considering the distribution of bolts, to give full play to the mechanical properties of the two materials.

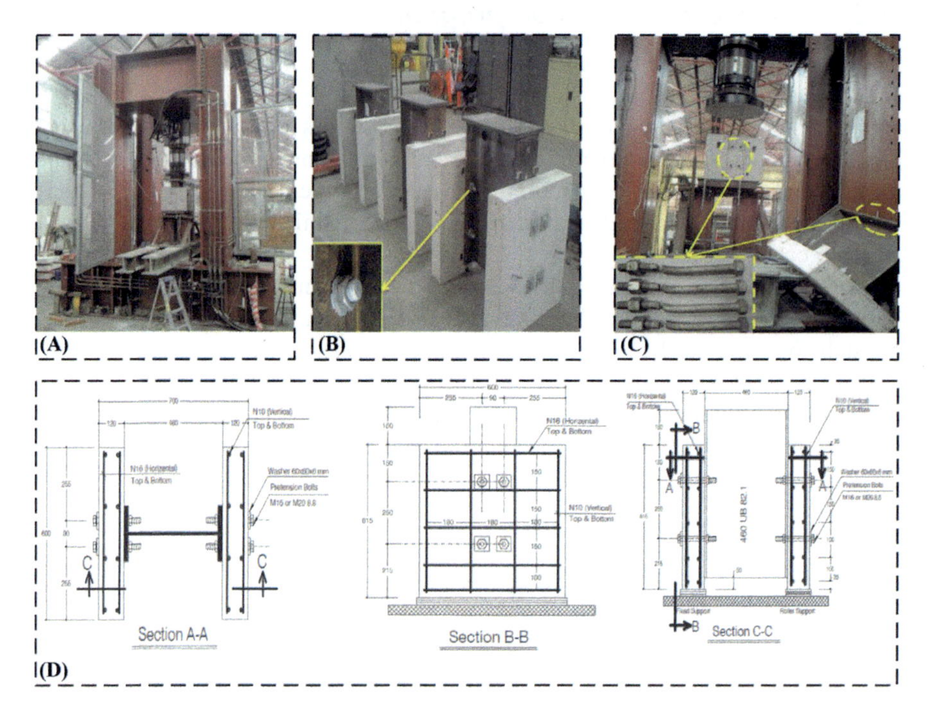

Figure 20.2 Bolted shear connectors program (Lawan et al., 2016). (A) Photograph of loading device; (B) photograph of specimens; (C) photograph of failure mode; and (D) details of specimen construction.

Source: From Lawan, M. M., Tahir, M. M., & Mirza, J. (2016). Bolted shear connectors performance in self-compacting concrete integrated with cold-formed steel section. *Latin American Journal of Solids and Structures*, *13*(4), 731–749. https://doi.org/10.1590/1679-78252004.

20.2.3 Demountable angle/tube connections program

The third form of connection we show is a DFD program for beam–column connections. Concrete beams and columns are challenging to make skinny, but it is still possible to apply dry connections such as the configuration proposed by Aninthaneni and Dhakal (2019). Aninthaneni utilizes bolts to anchor the angles or steel tubes to hold the concrete beams and columns, and this configuration is very similar to some typical steel joint configurations. From the failure mode shown in Fig. 20.3, the cracking of the concrete members is mainly concentrated at the beam ends, which implies that this connection also meets the seismic design requirements for concrete buildings.

20.2.4 Bolt-concrete program

In addition to the beam–column and base–slab connections, some particular concrete building forms also produce unique slab–slab connections, as shown in Fig. 20.4. Such slab–slab connection programs (Cai et al., 2019) may produce localized damage in the first place, and this phenomenon also tells us that the integrity of DFD concrete construction can also be used as a reference criterion.

20.2.5 Steel plate-bolt-concrete program

Unlike the bolt-steel plate scheme, which is well suited for connecting concrete slabs, as presented in steel connectors-bolt program, Almahmood et al. (2022)

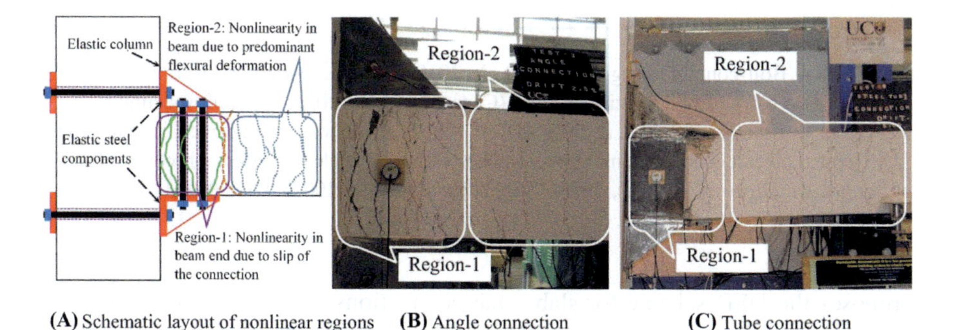

(A) Schematic layout of nonlinear regions **(B)** Angle connection **(C)** Tube connection

Figure 20.3 Demountable angle/tube connections program. (A) Schematic layout of nonlinear regions; (B) angle connection; and (C) tube connection.
Source: From Aninthaneni, P. K., & Dhakal, R. P. (2019). Analytical and numerical investigation of "dry" jointed precast concrete frame sub-assemblies with steel angle and tube connections. *Bulletin of Earthquake Engineering*, *17*(9), 4961–4985. https://doi.org/10.1007/s10518-019-00663-8.

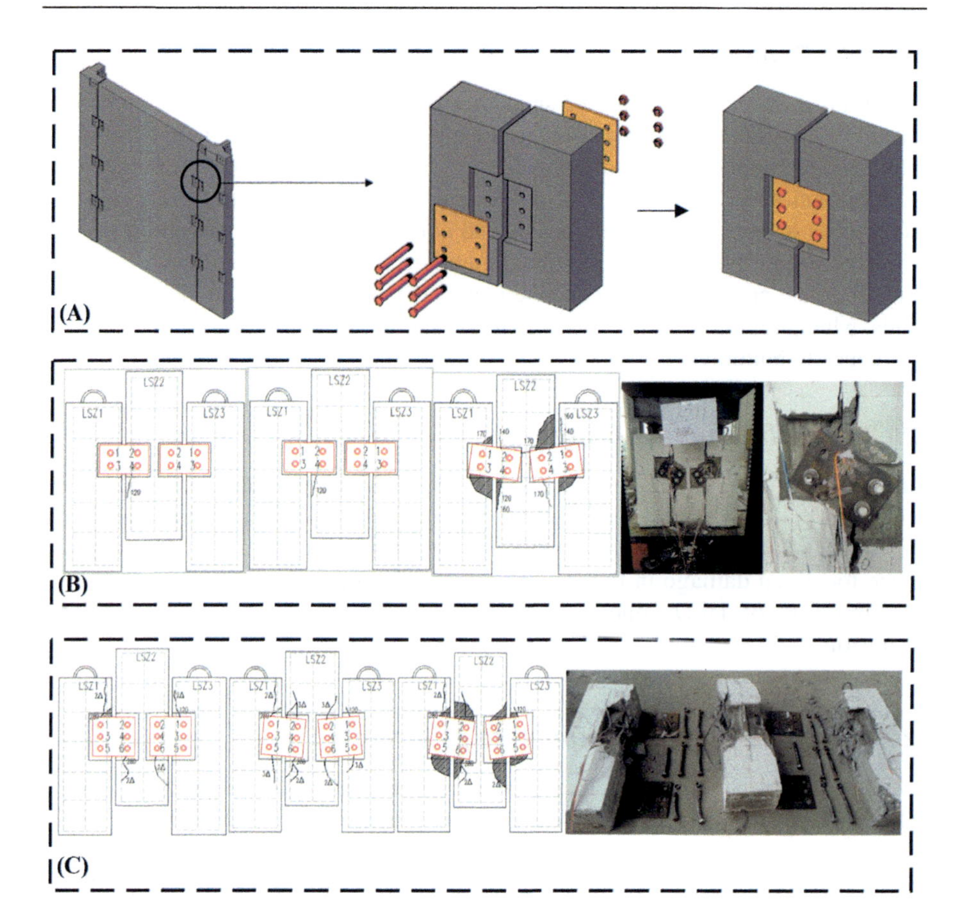

Figure 20.4 Bolt-concrete program (Cai et al., 2019). (A) The bolted connection; (B) test results of two rows of bolt specimens; and (C) test results of three-row bolt specimens. *Source*: From Aninthaneni, P. K., & Dhakal, R. P. (2019). Analytical and numerical investigation of "dry" jointed precast concrete frame sub-assemblies with steel angle and tube connections. *Bulletin of Earthquake Engineering*, *17*(9), 4961–4985. https://doi.org/10.1007/s10518-019-00663-8.

proposes the DFD scheme for slab–slab connections. By embedding ribbed steel blocks or new joint configurations, concrete realizes a reliable slab–slab connection (Fig. 20.5).

20.3 Design for deconstruction program with semidry connection

Semidry connections in concrete buildings with DFD programs are more mature than dry connections. This phenomenon is because the friction between the bolts or

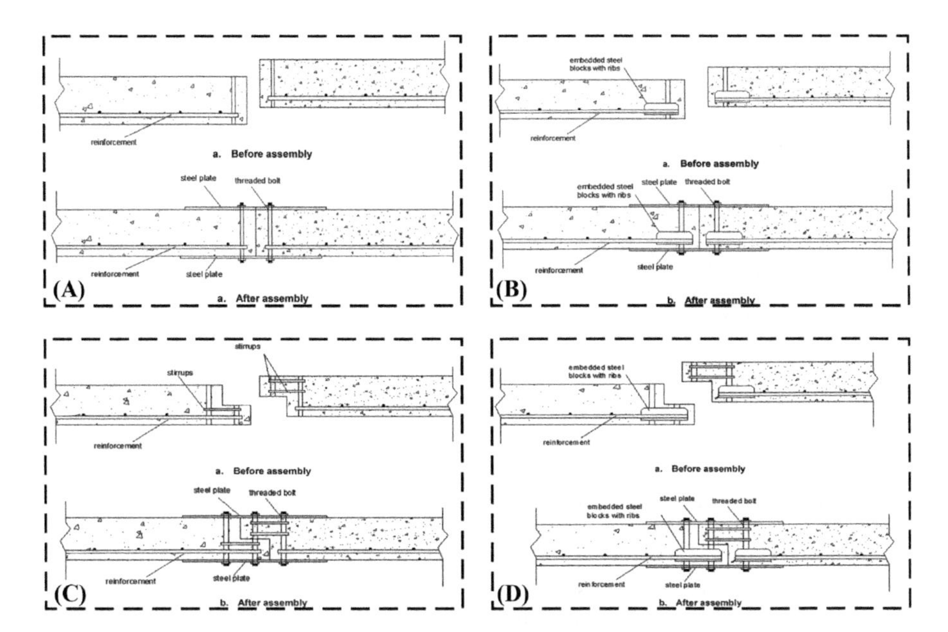

Figure 20.5 Steel plate-bolt-concrete program. (A) Connection details with no key nor embedded block; (B) connection details of the slab with embedded block but no key; (C) connection details of slabs connected with a key and no embedded block; and (D) connection details of the slabs connected with a key and an embedded block.
Source: From Almahmood, H., Ashour, A., Figueira, D., Yıldırım, G., Aldemir, A., & Sahmaran, M. (2022). Tests of demountable reinforced concrete slabs. *Structures*, *46*, 1084–1104. https://doi.org/10.1016/j.istruc.2022.10.097.

reinforcement and the grouted concrete can provide good pullout resistance for the joint. In this chapter, we introduce readers to several DFD program joints for concrete buildings that require concrete grout.

20.3.1 Reinforcement with bushes program

The friction between the reinforcement and the concrete grout can significantly improve the reliability of the connection, so adding grout to the sleeve would be an outstanding construction for semidry connections, as proposed by Metelli et al. (2012). As shown in Fig. 20.6, this sleeve + grout connection scheme is simple and efficient and shows good seismic and energy dissipation capacity.

20.3.2 Postcast concrete program

In contrast to the sleeve-grout solution, the solution with a backing strip at the joints also fits well with the nature of the concrete building, such as the construction proposed by Xiao et al. (2017) (Fig. 20.7). The program of applying a backing strip to connect concrete beams and columns has been

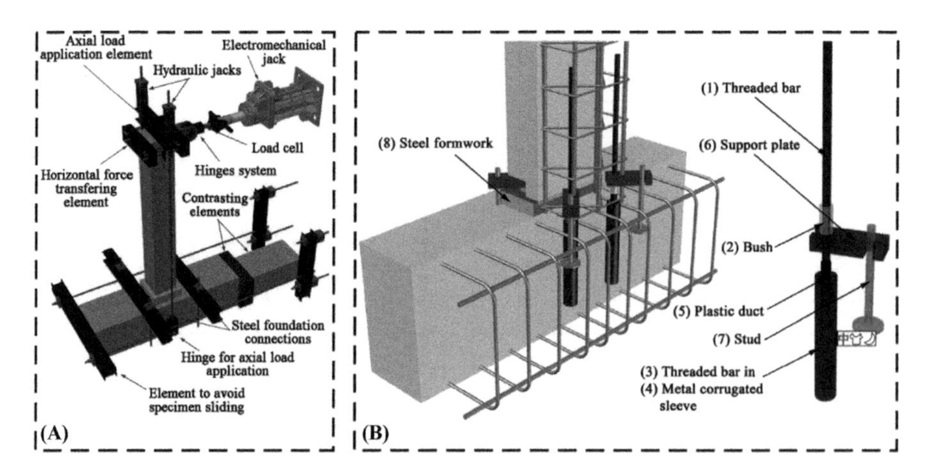

Figure 20.6 Reinforcement with bushes program. (A) Test set up and (B) the connection system for column-to-foundation joints.
Source: From Metelli, G., Beschi, C., & Riva, P. (2012). Cyclic behaviour of a column to foundation joint for concrete precast structures. *European Journal of Environmental and Civil Engineering*, *15*(9), 1297–1318. https://doi.org/10.1080/19648189.2011.9714856.

used in many projects, and the positioning of the concrete elements in this type of program relies on the occlusion of the joints. As a result, such post-formed concrete joints tend to exhibit a mortise and tenon shape at both ends.

20.3.3 Composite connection program

The construction proposed by Aktepe et al. (2023) utilizes both sleeves, bolts, and steel plates, which are commonly used in assembled concrete structures, as shown in Fig. 20.8. This design allows the friction between the concrete and the bolt as well as the preload provided by the bolts and the steel plates to contribute to the connection of the concrete members. Complex construction does not imply redundancy, and the coordinated and efficient utilization of various types of construction is an essential reference for studying concrete buildings under the DFD program.

20.4 Demountable precast concrete–frame building system

The dry/semidry connections in Sections 20.2 and 20.3 of this chapter only describe the construction of localized connections such as beam–column, base–slab, and slab–slab. Concrete building is a whole, and different connection forms serve to enhance the overall reliability of the building. Fig. 20.9 shows the overall building scheme in Aninthaneni's study (Aninthaneni & Dhakal, 2017), where different parts need to be designed with different types of connection forms to maximize overall reliability and cost-effectiveness. The study of concrete joint construction from the overall performance of assembled concrete buildings is an issue that researchers must focus on.

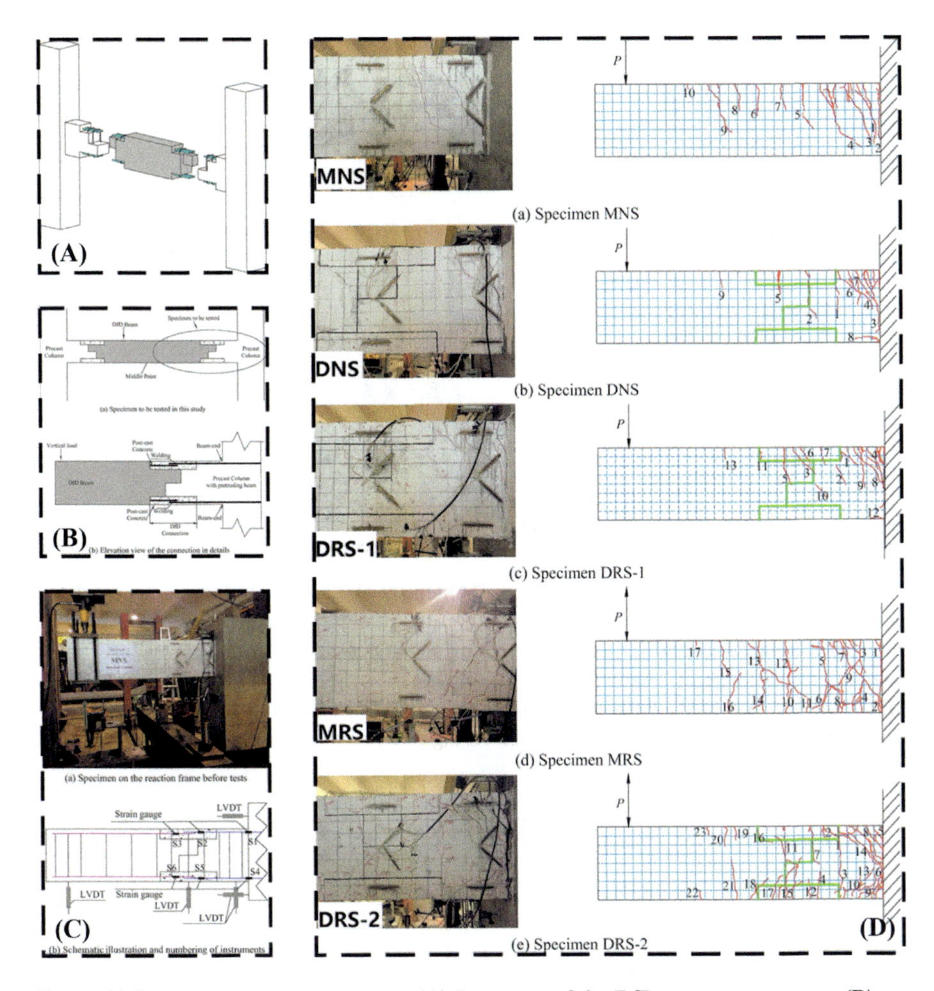

Figure 20.7 Postcast concrete program. (A) Prototype of the DfD concrete structure; (B) elevation view of the connection in details; (C) schematic illustration and numbering of instruments; and (D) crack patterns of specimens.
Source: From Xiao, J., Ding, T., & Zhang, Q. (2017). Structural behavior of a new moment-resisting DfD concrete connection. *Engineering Structures*, *132*, 1–13. https://doi.org/10.1016/j.engstruct.2016.11.019.

20.5 Challenges and prospects

Table 20.1 summarizes some typical constructions for dry/semidry concrete connections. Except for the postcasted program, the use of bolts is unavoidable in all of these constructions. For concrete buildings in the DFD program, bolt-based connections relying on preload alone are insufficient because the distribution of bolts must be limited, and the tensile and shearing capacity provided

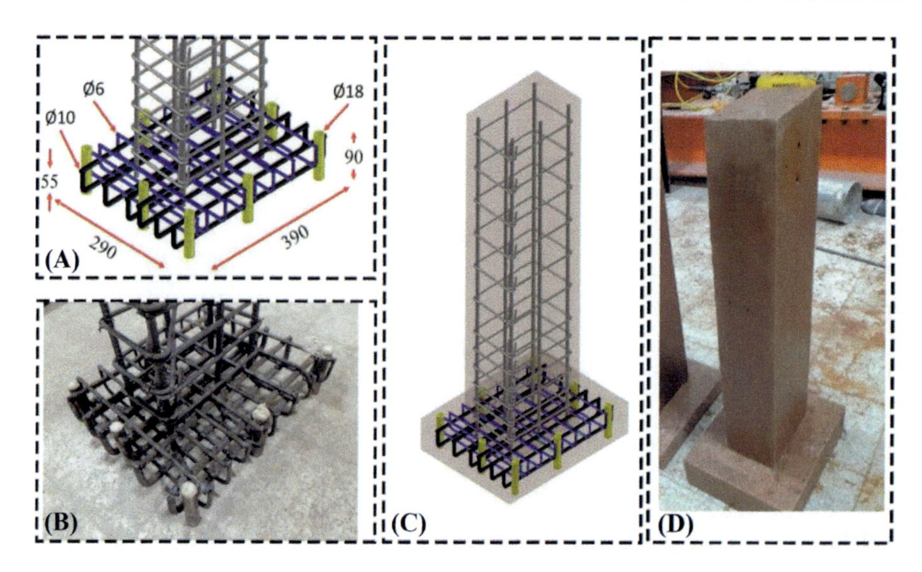

Figure 20.8 Composite connection program (Aktepe et al., 2023). (A) Reinforcement details in connection region; (B) steel reinforcement construction; (C) general views of the composite connection; and (D) composite connector object picture.
Source: From Aktepe, R., Akduman, S., Aldemir, A., Ozcelikci, E., Yildirim, G., Sahmaran, M., & Ashour, A. (2023). Fully demountable column base connections for reinforced CDW-based geopolymer concrete members. *Engineering Structures*, *290*. https://doi.org/10.1016/j.engstruct.2023.116366.

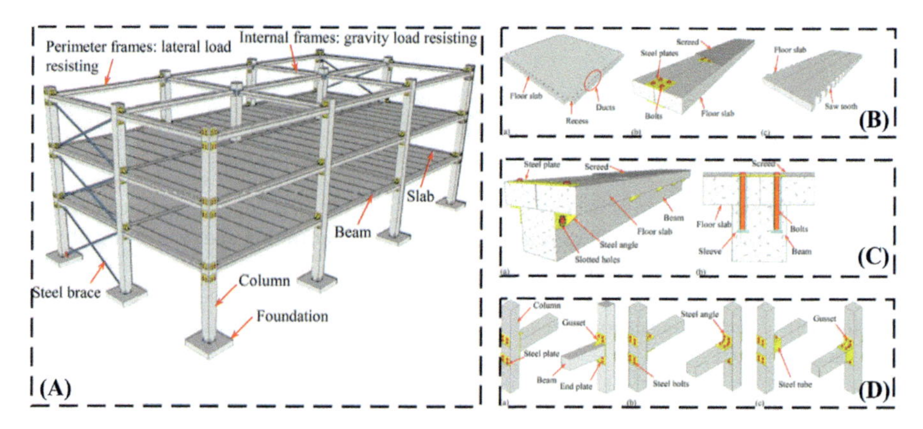

Figure 20.9 Demountable precast concrete–frame building system (Aninthaneni & Dhakal, 2017). (A) Layout of the proposed demountable precast concrete frame–building system; (B) layout of possible types of removable floor-to-floor connections; (C) layout of possible types of removable floor-to-beam connections; and (D) layout of possible types of removable rigid beam–column connections.
Source: From Aninthaneni, P. K., & Dhakal, R. P. (2017). Demountable precast concrete frame–building system for seismic regions: Conceptual development. *Journal of Architectural Engineering*, *23*(4). https://doi.org/10.1061/(asce)ae.1943-5568.0000275.

Table 20.1 Summary of dry DFD connections.

Researcher	Dry connection	Constructional form	Researcher	Semidry connection	Constructional form
Sun et al. (2016)	From Sun, J., Qiu, H., & Lu, Y. (2016). Experimental study and associated numerical simulation of horizontally connected precast shear wall assembly. *The Structural Design of Tall and Special Buildings*, *25*(13), 659–678. https://doi.org/10.1002/tal.1277	Steel connectors-bolt program	Metelli et al. (2012)	From Metelli, G., Beschi, C., & Riva, P. (2012). Cyclic behaviour of a column to foundation joint for concrete precast structures. *European Journal of Environmental and Civil Engineering*, *15*(9), 1297–1318. https://doi.org/10.1080/19648189.2011.9714856	Reinforcement with bushes program
Lawan et al. (2016)	From Lawan, M. M., Tahir, M. M., & Mirza, J. (2016). Bolted shear connectors performance in self-compacting concrete integrated with cold-formed steel section. *Latin American Journal of Solids and Structures*, *13*(4), 731–749. https://doi.org/10.1590/1679-78252004	Bolted shear connectors program	Xiao et al. (2017)	From Xiao, J., Ding, T. & Zhang, Q. (2017). Structural behavior of a new moment-resisting DfD concrete connection. *Engineering Structures*, 132, 1–13. https://doi.org/10.1016/j.engstruct.2016.11.019	Postcast concrete program
Aninthaneni and Dhakal (2019)	From Aninthaneni, P. K., & Dhakal, R. P. (2019). Analytical and numerical investigation of “dry” jointed precast concrete frame sub-assemblies with steel angle and tube connections. *Bulletin of Earthquake Engineering*, *17*(9), 4961–4985. https://doi.org/10.1007/s10518-019-00663-8	Demountable angle/tube connections program	Pul and Senturk (2017)	From Pul, S., & Senturk, M. (2017). A bolted moment connection model for precast column-beam joint. In *World congress on civil, structural, and environmental engineering*. Avestia Publishing. https://doi.org/10.11159/icsenm17.129.	Precast concrete-bolt program

(*Continued*)

Table 20.1 (Continued)

Researcher	Dry connection	Constructional form	Researcher	Semidry connection	Constructional form
Cai et al. (2019)	From Cai, G., Xiong, F., Xu, Y., Si Larbi, A., Lu, Y., & Yoshizawa, M. (2019). A demountable connection for low-rise precast concrete structures with dfd for construction sustainability: A preliminary test under cyclic loads. *Sustainability*, *11*(13). https://doi.org/10.3390/su11133696	Bolt-concrete program	Ataei et al. (2014)	From Ataei, A., Bradford, M., & Liu, X. (2014). Sustainable composite beams and joints with deconstructable bolted shear connectors. In *Proceedings of the 23rd Australasian conference on the mechanics of structures and materials (ACMSM23)* (pp. 9–12). (Original work published 2014.)	Bolted shear connectors program
Almahmood et al. (2022)	From Almahmood, H., Ashour, A., Figueira, D., Yıldırım, G., Aldemir, A., & Sahmaran, M. (2022). Tests of demountable reinforced concrete slabs. *Structures*, *46*, 1084–1104. https://doi.org/10.1016/j.istruc.2022.10.097	Steel plate-bolt-concrete program	Aktepe et al. (2023)	From Aktepe, R., Akduman, S., Aldemir, A., Ozcelikci, E., Yildirim, G., Sahmaran, M., & Ashour, A. (2023). Fully demountable column base connections for reinforced CDW-based geopolymer concrete members. *Engineering Structures*, *290*. https://doi.org/10.1016/j.engstruct.2023.116366	Composite connection program

by a small number of bolts are difficult to work with concrete members with large cross-sections. Consequently, dry connections are almost always applied to slab–slab connections in concrete buildings because slabs are more accessible to open compared to beams and columns. The shear and tension forces provided by the bolts are of the same order of magnitude as the shear and bending moments that the concrete can resist on its own, which is the basis of a reliable connection.

Semidry connections have friction between the bolts and grout, which provides more excellent bending moments in the concrete connection, and this is why the DFD program for concrete buildings favors semidry connections more. More importantly, the semidry connection effectively improves the integrity of the concrete building that is why it possesses good seismic performance. For concrete buildings with the DFD program, the authors believe that researchers need to develop connection forms that improve the overall performance of the concrete building to improve the performance of the concrete building under extreme conditions. This phenomenon means that the optimal concrete connection form must be different for different building parts, and a good combination will be an essential reference for developing demountable concrete buildings.

20.6 Summary and conclusions

This chapter describes the need to develop demountable concrete buildings from an environmental perspective and presents the history and development of DFD solutions for concrete buildings. Furthermore, some typical dry/semidry concrete connection configurations are selected, and their respective characteristics are presented. Dry connections are suitable for concrete slab connections because of the larger opening space, and the tensile and shear capacities provided by the bolted sections are of the same order of magnitude as those provided by the concrete specimen sections. Semidry joints can effectively improve the integrity of concrete buildings for DFD schemes. In addition, this chapter concludes that the future development of DFD schemes or new concrete connections should be based on the reference of improving the overall performance of concrete buildings to maximize the performance of each component of the assembled building and to make each part work together.

References

Akinade, O., Oyedele, L., Oyedele, A., Davila Delgado, J. M., Bilal, M., Akanbi, L., Ajayi, A., & Owolabi, H. (2020). Design for deconstruction using a circular economy approach: Barriers and strategies for improvement. *Production Planning and Control*, *31*(10), 829–840. Available from https://doi.org/10.1080/09537287.2019.1695006.

Akinade, O. O., Oyedele, L. O., Ajayi, S. O., Bilal, M., Alaka, H. A., Owolabi, H. A., Bello, S. A., Jaiyeoba, B. E., & Kadiri, K. O. (2017). Design for deconstruction (DfD): Critical success factors for diverting end-of-life waste from landfills. *Waste Management*, *60*, 3–13. Available from https://doi.org/10.1016/j.wasman.2016.08.017.

Aktepe, R., Akduman, S., Aldemir, A., Ozcelikci, E., Yildirim, G., Sahmaran, M., & Ashour, A. (2023). Fully demountable column base connections for reinforced CDW-based geopolymer concrete members. *Engineering Structures, 290*. Available from https://doi.org/10.1016/j.engstruct.2023.116366.

Almahmood, H., Ashour, A., Figueira, D., Yıldırım, G., Aldemir, A., & Sahmaran, M. (2022). Tests of demountable reinforced concrete slabs. *Structures*, *46*, 1084–1104. Available from https://doi.org/10.1016/j.istruc.2022.10.097.

Andrew, R. M. (2019). Global CO_2 emissions from cement production, 1928–2018. *Earth System Science Data*, *11*(4), 1675–1710. Available from https://doi.org/10.5194/essd-11-1675-2019.

Aninthaneni, P. K., & Dhakal, R. P. (2017). Demountable precast concrete frame–building system for seismic regions: Conceptual development. *Journal of Architectural Engineering*, *23* (4). Available from https://doi.org/10.1061/(asce)ae.1943-5568.0000275.

Aninthaneni, P. K., & Dhakal, R. P. (2019). Analytical and numerical investigation of "dry" jointed precast concrete frame sub-assemblies with steel angle and tube connections. *Bulletin of Earthquake Engineering*, *17*(9), 4961–4985. Available from https://doi.org/10.1007/s10518-019-00663-8.

Ataei, A., Bradford, M., & Liu, X. (2014). Sustainable composite beams and joints with deconstructable bolted shear connectors. In *Proceedings of the 23rd Australasian conference on the mechanics of structures and materials (ACMSM23)*, pp. 9–12.

Cai, Gaochuang, Xiong, Feng, Xu, Yong, Si Larbi, Amir, Lu, Yang, & Yoshizawa, Mikio (2019). A demountable connection for low-rise precast concrete structures with DfD for construction sustainability—A preliminary test under cyclic loads. *Sustainability*, *11* (13). Available from https://doi.org/10.3390/su11133696.

Cong, X., Zhou, W., & Elchalakani, M. (2020). Experimental study on the engineering properties of alkali-activated GGBFS/FA concrete and constitutive models for performance prediction. *Construction and Building Materials*, *240*. Available from https://doi.org/10.1016/j.conbuildmat.2019.117977.

Densley Tingley, D., & Davison, B. (2012). Developing an LCA methodology to account for the environmental benefits of design for deconstruction. *Building and Environment*, *57*, 387–395. Available from https://doi.org/10.1016/j.buildenv.2012.06.005.

Ding, T., Xiao, J., Zhang, Q., & Akbarnezhad, A. (2018). Experimental and numerical studies on design for deconstruction concrete connections: An overview. *Advances in Structural Engineering*, *21*(14), 2198–2214. Available from https://doi.org/10.1177/1369433218768000.

Figueira, D., Ashour, A., Yıldırım, G., Aldemir, A., & Şahmaran, M. (2021). Demountable connections of reinforced concrete structures: Review and future developments. *Structures*, *34*, 3028–3039. Available from https://doi.org/10.1016/j.istruc.2021.09.053.

Gastaldi, D., Canonico, F., Capelli, L., Buzzi, L., Boccaleri, E., & Irico, S. (2015). An investigation on the recycling of hydrated cement from concrete demolition waste. *Cement and Concrete Composites*, *61*, 29–35. Available from https://doi.org/10.1016/j.cemconcomp.2015.04.010.

Hogland, W., Marques, M., & Nimmermark, S. (2004). Landfill mining and waste characterization: A strategy for remediation of contaminated areas. *Journal of Material Cycles and Waste Management*, *6*(2). Available from https://doi.org/10.1007/s10163-003-0110-x.

Huang, L., Krigsvoll, G., Johansen, F., Liu, Y., & Zhang, X. (2018). Carbon emission of global construction sector. *Renewable and Sustainable Energy Reviews*, *81*, 1906–1916. Available from https://doi.org/10.1016/j.rser.2017.06.001.

Kim, G. D., & Kim, T. B. (2007). Development of recycling technology from waste aggregate and dust from waste concrete. *Journal of Ceramic Processing Research, 8* (1), 82–86.

Kurtoğlu, A. E., Alzeebaree, R., Aljumaili, O., Niş, A., Gülşan, M. E., Humur, G., & Çevik, A. (2018). Mechanical and durability properties of fly ash and slag based geopolymer concrete. *Advances in Concrete Construction, 6*(4), 345–362. Available from https://doi.org/10.12989/acc.2018.6.4.345.

Lawan, M. M., Tahir, M. M., & Mirza, J. (2016). Bolted shear connectors performance in self-compacting concrete integrated with cold-formed steel section. *Latin American Journal of Solids and Structures, 13*(4), 731–749. Available from https://doi.org/10.1590/1679-78252004.

Mattiello, A., Chiodini, P., Bianco, E., Forgione, N., Flammia, I., Gallo, C., Pizzuti, R., & Panico, S. (2013). Health effects associated with the disposal of solid waste in landfills and incinerators in populations living in surrounding areas: A systematic review. *International Journal of Public Health, 58*(5), 725–735. Available from https://doi.org/10.1007/s00038-013-0496-8.

Metelli, G., Beschi, C., & Riva, P. (2012). Cyclic behaviour of a column to foundation joint for concrete precast structures. *European Journal of Environmental and Civil Engineering, 15*(9), 1297–1318. Available from https://doi.org/10.1080/19648189.2011.9714856.

Meyer, C. (2009). The greening of the concrete industry. *Cement and Concrete Composites, 31*(8), 601–605. Available from https://doi.org/10.1016/j.cemconcomp.2008.12.010.

Pul, S., & Senturk, M. (2017). A bolted moment connection model for precast column-beam joint. In *World congress on civil, structural, and environmental engineering*. Avestia Publishing. Available from https://doi.org/10.11159/icsenm17.129.

Scrivener, K. L., & Kirkpatrick, R. J. (2008). Innovation in use and research on cementitious material. *Cement and Concrete Research, 38*(2), 128–136. Available from https://doi.org/10.1016/j.cemconres.2007.09.025.

Sun, J., Hongxing, Q., & Lu, Y. (2016). Experimental study and associated numerical simulation of horizontally connected precast shear wall assembly. *The Structural Design of Tall and Special Buildings, 25*(13), 659–678. Available from https://doi.org/10.1002/tal.1277.

Swamy, R. N. (1998). Designing concrete and concrete structures for sustainable development. Sustainable development of the cement and concrete industry. In *Proceedings of CANMET/ACI international symposium*, Ottawa, Canada. pp. 245–255.

Tam, V. W. Y. (2008). Economic comparison of concrete recycling: A case study approach. *Resources, Conservation and Recycling, 52*(5), 821–828. Available from https://doi.org/10.1016/j.resconrec.2007.12.001.

Tingley, D. D., & Davison, B. (2011). Design for deconstruction and material reuse. *Proceedings of Institution of Civil Engineers: Energy, 164*(4), 195–204. Available from https://doi.org/10.1680/ener.2011.164.4.195.

Xiao, J., Ding, T., & Zhang, Q. (2017). Structural behavior of a new moment-resisting DfD concrete connection. *Engineering Structures, 132*, 1–13. Available from https://doi.org/10.1016/j.engstruct.2016.11.019.

Zabihi, H., & Mirsaeedie, L. (2013). Towards green building: Sustainability approach in building industrialization. *International Journal of Architecture and Urban Development, 3*(3), 49–56.

Zakka, W. P., Abdul Shukor Lim, N. H., & Chau Khun, M. (2021). A scientometric review of geopolymer concrete. *Journal of Cleaner Production, 280*. Available from https://doi.org/10.1016/j.jclepro.2020.124353.

Zhang, T., Chen, M., Wang, Y., & Zhang, M. (2023). Roles of carbonated recycled fines and aggregates in hydration, microstructure and mechanical properties of concrete: A critical review. *Cement and Concrete Composites, 138*. Available from https://doi.org/10.1016/j.cemconcomp.2023.104994.

Zhe, R., & Zhou, W. (2023). Study of dynamic mechanical properties of UHPC-AAC composites based on SHPB test. *Journal of Building Engineering, 78*. Available from https://doi.org/10.1016/j.jobe.2023.107668.

Integration of new technologies with sustainable concrete materials and structures

21

Payam Hosseini
Genex Systems, Turner-Fairbank Highway Research Center, McLean, VA, United States

21.1 Introduction

The global urban population is projected to increase by 2.5 billion by 2050, leading to a growing demand for new structures and infrastructure to accommodate the expanding population (Habert et al., 2020; Swilling et al., 2018). Concrete, being a widely favored construction material due to its numerous advantages, including cost-effectiveness, fire resistance, high durability, recyclability, lower carbon footprint, and the local availability of its ingredients, is the most extensively used engineered material on a global scale (Miller et al., 2016; Petek Gursel et al., 2014). Consequently, meeting the high demand for concrete construction necessitates large-scale production of concrete materials to address the challenges posed by population growth. In this context, the surge in concrete consumption is anticipated to reach 12%–23% higher than its 2014 consumption levels by the year 2050 (Lim et al., 2019).

Furthermore, annual global concrete consumption stands at 25 billion tonnes, equivalent to over 3.8 tonnes per capita, underscoring its immense significance among various construction materials (Lim et al., 2019; Petek Gursel et al., 2014).

Despite concrete's lower carbon footprint in comparison to other construction materials (Coffetti et al., 2022), its substantial volume consumption contributes significantly to global warming. Studies indicate that concrete production is accountable for 9% of total greenhouse gas emissions, with cement production representing 85%–95% of this share (Coffetti et al., 2022; Global Material Resources Outlook to 2060, 2018; Miller et al., 2016).

Despite heightened commitments and objectives to address climate change, prevailing policies continue to chart a trajectory toward approximately 2.7°C of global warming by the end of the century, surpassing the ambitious goal of the Paris Agreement to limit global warming to 1.5°C above preindustrial levels (Lenton et al., 2023; Meinshausen et al., 2022). However, to prevent catastrophic economic, social, and environmental consequences, the temperature increase must be restricted to 1.5°C (Coffetti et al., 2022; The Intergovernmental Panel on Climate Change IPCC, 2018). Achieving this goal necessitates the adoption of new techniques and technologies to reduce the carbon footprint of concrete materials and structures.

Sustainable Concrete Materials and Structures. DOI: https://doi.org/10.1016/B978-0-443-15672-4.00021-8

In recent years, various approaches have been implemented to neutralize the carbon footprint of concrete products while simultaneously extending the service life of concrete structures and diminishing their impact on global warming throughout their life cycle. This chapter aims to introduce and briefly describe different innovative techniques employed to develop sustainable concrete materials and structures.

21.2 New technologies for sustainable concrete materials

Current approaches for developing sustainable concrete materials encompass the utilization of supplementary cementitious materials (SCMs), recycled aggregates, wastewater, seawater, and waste fibers in the manufacturing of concrete products (Hassan et al., 2000; Juenger et al., 2019; Kumaresan et al., 2022; Shi et al., 2016; Su et al., 2002; Younis et al., 2018). However, significant attention has been directed toward two key areas: reducing the environmental impact of cement clinker production and minimizing its usage in cement manufacturing, as cement clinker possesses the highest carbon footprint among all concrete ingredients (Hassan et al., 2000; Juenger et al., 2019; Khozin et al., 2020; Kumaresan et al., 2022; Shi et al., 2016; Su et al., 2002; Younis et al., 2018).

In addition to the existing techniques, several novel technologies have emerged to further mitigate the environmental impact of concrete materials, while simultaneously enhancing their mechanical strength and durability. The following sections provide a comprehensive review and discussion of the most prominent technologies employed to develop more sustainable concrete products.

21.2.1 Modern composite cements

While there are additional strategies implemented in cement plants to reduce the carbon footprint of cement, such as enhancing energy efficiency in the production process or utilizing fuels with a lower carbon footprint (Gonnon & Lootens, 2023; Khozin et al., 2020), the most effective and straightforward approach to decreasing the carbon footprint of concrete materials is by reducing the quantity of clinker in the final product. This can be achieved through two parallel methods: reducing clinker in cement and decreasing the cement content in the concrete mixture.

Regarding the first strategy, blended cements were among the initial products developed to decrease the clinker factor in cement. Various types of SCMs, including blast-furnace slag (BFS), natural pozzolan, fly ash, and silica fume, have been successfully employed to produce blended cement and reduce the clinker content (Standard Specification for Blended Hydraulic Cements, n.d.). However, the current supply of conventional SCMs, such as BFS and fly ash, falls short in meeting the demand for reactive SCMs. This is due to the increasing demand for concrete materials, efforts to reduce production waste in the steel-making industry, and the transition from coal to renewable energy sources, resulting in a decline in the availability of conventional SCMs for blended cements (EUROFER, 2023; Snellings, 2016).

Consequently, more attention has been focused on identifying SCMs from sources that are abundantly available worldwide, such as limestone and clay (Juenger et al., 2019; Snellings et al., 2023).

Portland-limestone cement (PLC) has been available for many years, but the decreasing availability of conventional SCMs used in blended cements has prompted cement plants to adopt PLC. PLC offers the advantage of reducing clinker quantity at a low cost, making it a greener option for the concrete industry. The limestone content in PLC typically ranges between 5% and 15% (Standard Specification for Blended Hydraulic Cements, n.d.). However, higher replacement of limestone can lead to a decline in the mechanical strength and durability of concrete materials. This is because the reactivity of limestone particles within an ordinary portland cement (OPC)-based matrix is limited, and further reduction in cement content results in a loss of the main binder (Marzouki & Lecomte, 2017; Ramezanianpour et al., 2009).

Nevertheless, by optimizing the particle size distribution of limestone fines, it is possible to achieve up to a 50% reduction in the carbon footprint of cement mortar and concrete while maintaining similar mechanical properties compared to control mixtures without limestone. This is attributed to improved particle packing, which reduces the water content required to achieve a specific workability (Gonnon & Lootens, 2023). Modified PLC (M-PLC) presents a novel pathway for developing concrete materials with enhanced sustainability.

Direct application of limestone powder or clay does not typically result in high mechanical strength and durability, as they are not reactive in their natural state. However, low-purity clay can be calcined to produce a reactive SCM. Combining calcined clay with limestone further enhances the properties of concrete materials, as the calcium carbonate ($CaCO_3$) in limestone can react with aluminate phases in the calcined clay, forming carboaluminates (Zunino et al., 2021). This reaction also occurs when blending cement with limestone, but the quantity of reaction products is limited due to the lower quantity of aluminate phases in cement clinker. Nevertheless, blending cement with limestone and calcined clay can result in a denser matrix, leading to higher mechanical strength and durability due to the pozzolanic reactivity of calcined clay, the formation of carboaluminates, and the micro-filling effect of limestone particles (Juenger et al., 2019; Zunino et al., 2021). This development has led to the introduction of limestone calcined clay cement (LC^3), which significantly reduces the carbon footprint of OPC. Studies have shown that LC^3-50 can reduce the clinker factor to 50% while offering similar strength compared to OPC (Zunino et al., 2021).

21.2.2 Nanotechnology

Nanotechnology has made significant contributions to various fields, including the construction industry (Papadaki et al., 2018). One of its major applications lies in the production of concrete materials by incorporating nanomaterials such as nanoparticles, nanofibers, and nanoplatelets (Hosseini et al., 2014; Sanchez & Sobolev, 2010). The inclusion of nanomaterials imparts superior performance to concrete materials, including high early strength, improved early-age and long-term durability, self-sensing capabilities, self-cleaning properties, self-healing behavior, and high tensile strength (Dimov et al., 2018; Dong et al., 2021; Hosseini et al., 2010; Khaloo et al., 2016; Ruan et al., 2021; Shen et al., 2015; Shi et al., 2022; Sujitha et al., 2023). These functionalities can be utilized to develop functionalized sustainable concrete materials or enhance the performance of existing materials.

Nanomaterials are introduced during the mixing process to enhance the strength and durability of concrete materials, especially when incorporating waste materials such as upcycled SCMs like BFS, fly ash, sewage sludge ash, as well as recycled aggregates and waste fibers (Bahadori & Hosseini, 2018; Hosseini et al., 2018; Hosseini, Mohamad, et al., 2011; Hosseinpourpia et al., 2012; Said et al., 2021). This is particularly important for concrete mixtures that contain a large volume of waste materials or low-reactivity SCMs, as nanomaterials can further promote the development of sustainable concrete materials (Hosseini et al., 2020; Varghese et al., 2019). Nanomaterials facilitate the acceleration of cement hydration by providing nucleation sites for cement hydrates. Some nanomaterials, such as nano-SiO_2 particles, exhibit fast pozzolanic reactions, resulting in a compact microstructure even at the early stages of cement hydration (Hosseini et al., 2020; Khaloo et al., 2016).

Nanoparticles, including nano-SiO_2 particles, nano-$Ca(OH)_2$ particles, and nano graphite platelets, have been employed to enhance the mechanical strength and durability properties of concrete materials containing recycled aggregates (Ahmad et al., 2022; Hosseini, Booshehrian, et al., 2011; Zhang et al., 2022). These nanomaterials can be directly added to the concrete matrix during the mixing process or coated onto the surface of waste aggregates. This application improves the characteristics of recycled aggregate concrete by accelerating cement hydration, providing a micro-filling effect, and enhancing the bonding between the aggregate and matrix. Consequently, the pore structure within the sustainable concrete matrix becomes more refined (Ahmad et al., 2022; Chen & Jiao, 2022; Hosseini, Booshehrian, et al., 2011; Sahu et al., 2022; Zhang et al., 2022).

In addition to improving the mechanical and durability performance of sustainable concrete materials, reactive nanoparticles like nano-SiO_2 particles have been utilized to reduce the cement content in normal strength concrete (Bahadori & Hosseini, 2012). Moreover, apart from synthetic nanomaterials manufactured through a bottom-up approach (Sanchez & Sobolev, 2010), nanoparticles derived from waste materials such as glass, ceramic, and rice husk ash have been produced via milling (top-down method). These waste-derived nanoparticles have been successfully utilized to enhance the compressive strength and reduce the corrosion rate of rebars in concrete materials (Fahmy et al., 2022; Faried et al., 2021).

While the use of synthetic nanomaterials offers significant improvements in the mechanical, durability, and self-sensing characteristics of sustainable concrete materials due to their controlled production process and wide range of properties, it is important to note that these materials often have a high carbon footprint (Roes et al., 2010). As a result, researchers have been exploring new synthesis routes to reduce the carbon footprint of nanomaterials that can be utilized in the production of sustainable concrete materials (Gao & Yu, 2019; Lazaro et al., 2016).

Reactive nanoparticles can also be employed to enhance the properties of blended and composite cements, such as PLC. In a recent study (Liu et al., 2023), nano-SiO_2 particles were loaded onto calcite particles using a low-cost and environmentally friendly method, which shows promising potential for developing high-performance, low-carbon footprint cements.

Furthermore, nano-TiO_2 particles have been successfully incorporated into concrete mixtures to provide self-cleaning and air purification capabilities (Shen et al., 2015). Nano-TiO_2 particles exhibit photocatalytic activity when activated by UV light, leading to the mitigation of color change caused by air pollution and the degradation of pollutants such as methylene blue, soot, and NO_x. However, it's important to note that only nanoparticles available on the concrete surface or in close proximity to the surface are involved in air purification/self-cleaning, while the remaining nano-TiO_2 particles within the matrix contribute to the enhancement of mechanical strength and durability by providing nucleation sites for the growth of calcium-silicate-hydrate (C-S-H) and exerting a micro-filling effect.

Various types of concrete, such as self-compacting concrete and ultra-high-performance concrete (UHPC), require the use of different admixtures. Currently, there is a growing focus on controlling the rheological (flow) behavior of concrete materials to ensure the quality of concrete on the jobsite remains intact during transportation and handling processes. Some admixtures may experience performance degradation after prolonged transportation, while certain emerging SCMs can have variable effects on the rheological properties of concrete, making the control of flow behavior increasingly significant.

The concept of active rheology control deals with real-time regulation of concrete mixture flow or stiffening. Magnetic nanoparticles like Fe_3O_4 have shown promise in effectively controlling the rheological properties of concrete, finding application in reducing or preventing formwork leakage under pressure (Chibulu et al., 2023). This approach is particularly crucial in concrete three-dimensional (3D) printing, where the fresh mixture must meet pumpability, extrudability, and buildability requirements to achieve the desired characteristics in the final product. In this regard, nano-Fe_3O_4 particles have been utilized to accelerate structural build-up after printing, demonstrating that the magneto-rheology control method can enhance the buildability of 3D-printed concrete (Jiao et al., 2022).

21.2.3 Artificial intelligence and machine learning

The advent of artificial intelligence (AI) has brought various capabilities to the fields of engineering and science. One significant application of AI in concrete science is predicting the performance of materials and developing materials with specific characteristics (Hu et al., 2021; Li et al., 2022). Different machine learning (ML) methods have been developed and utilized to optimize concrete mixtures and predict their behavior in both fresh and hardened states (Hu et al., 2021; Li et al., 2022).

ML techniques, such as artificial neural networks, support vector machines, and random forest, have been employed to predict the rheological (e.g., slump, viscosity, yield stress), mechanical (e.g., compressive strength, flexural strength, modulus of elasticity), and durability (e.g., permeability, corrosion rate) properties of sustainable concrete mixtures (Han et al., 2020; Hu et al., 2021; Li et al., 2022; Rehman et al., 2022). This enables researchers to optimize sustainable mixtures by reducing costs, lowering environmental impact, and improving mechanical strength and durability (Naseri et al., 2020, 2023; Shobeiri et al., 2023). Concrete mixture design is optimized by adjusting the mix proportions to achieve the optimal performance in the final product (Naseri et al., 2020, 2023; Shobeiri et al., 2023). For example, maximizing mechanical strength while minimizing cost, environmental impact, and permeability is desired for sustainable concrete mixtures. AI and ML methods can be specifically employed to further reduce cement content while increasing the utilization of waste SCMs, leading to a significant reduction in the carbon footprint of sustainable concrete (Naseri et al., 2023; Shobeiri et al., 2023). In this context, utilizing coyote optimization programming, Naseri et al. (2023) demonstrated that when using SCMs, a specific SCM could lead to the lowest global warming potential (GWP) at each compressive strength level of normal strength concrete. For instance, concrete mixtures containing fly ash exhibited the lowest GWP at the designed compressive strength levels of 40, 50, and 60 MPa, while concrete mixtures with ground granulated blast furnace slag (GGBFS) had the lowest GWP at the 30 MPa compressive strength level. Consequently, the utilization of GGBFS and fly ash resulted in a reduction of GWP by 54.6%, 50.9%, 44.2%, and 35.5% for concrete mixtures with designed compressive strengths of 30, 40, 50, and 60 MPa, respectively, compared to the control concrete mixture (i.e., made without SCMs).

21.2.4 Carbon capture, utilization, and storage

Given the substantial global production volume of concrete materials, reducing their carbon footprint is of utmost importance in current efforts to mitigate their impact on global warming (Habert et al., 2020; Ostovari et al., 2021). Utilizing waste materials as SCMs and aggregates presents a cost-effective and straightforward approach to lowering the carbon footprint of concrete. Nanomaterials can also contribute to improving the mechanical strength and durability of concrete that incorporates waste materials. However, the high carbon footprint, cost, and challenges with dispersion make the practical use of high quantities of synthetic nanomaterials less feasible at present.

AI and ML methods can be employed to optimize mixture designs in a way that reduces the carbon footprint of concrete materials. However, achieving a significant reduction in the environmental impact of concrete through high-volume replacement of cement, which has the highest carbon footprint among concrete ingredients, is not feasible due to the loss of the main binder in cement-based materials. This loss would result in lower mechanical strength and durability. Taking all these factors into consideration, the mineralization of CO_2 captured from the atmosphere or flue gas from air-polluting industries, such as cement plants, emerges as a novel method to substantially reduce the carbon footprint of sustainable concrete materials. This avenue opens a new door toward achieving carbon-neutral concrete products (Winnefeld et al., 2022).

CO_2 can be directly mineralized in fresh concrete mixtures as an admixture added during the mixing process (Monkman et al., 2023). When CO_2 dissolves in mixing water, it forms carbonic acid, which reacts with calcium ions from the dissolution of calcium silicate phases (C_2S and C_3S) to generate C-S-H and $CaCO_3$ (Fig. 21.1). The nanoscale $CaCO_3$ acts as seeding agents for the further growth of cement hydration products, particularly C-S-H (Monkman et al., 2023).

Carbon mineralization can also occur indirectly in fresh concrete mixtures through the utilization of carbonated SCMs, carbonated aggregates, and carbonated mixing water (Chen et al., 2021; Jiang et al., 2022; Kusin et al., 2020; Liu, Shen, et al., 2022; Liu et al., 2021; Monkman, Hanmore, et al., 2022; Ren et al., 2022; Shen et al., 2022; Skevi et al., 2022). Various types of waste powders, such as steel slag, concrete fines, biomass bottom ash, municipal solid waste incineration fly ash, sintering red mud, and sedimentary mine waste, have been subjected to wet or dry carbonation processes. These materials are then used to prepare low-carbon cement-based materials with improved mechanical strength and durability (Chen et al., 2021; Kusin et al., 2020; Liu, Shen, et al., 2022; Liu et al., 2021; Ren et al., 2022; Shen et al., 2022; Skevi et al., 2022).

CO_2 can be mineralized on recycled concrete aggregates (RCAs), enhancing their bonding capacity within the concrete matrix (Jiang et al., 2022). Unhydrated calcium silicates and calcium hydroxide present on the surface of RCAs undergo carbonation through CO_2 treatment, leading to the formation of $CaCO_3$ crystals and a silica-rich layer. Additionally, wastewater from washing central concrete mixers and concrete trucks can be treated with CO_2 and subsequently utilized in concrete mixtures to reduce their carbon footprint. The calcium ions in the wash water react with dissolved CO_2 (as carbonic acid), resulting in the formation of fine calcite particles in the solution. Studies have shown that the use of CO_2-treated wash water increases the compressive strength of concrete while having minimal impact on its durability (Monkman, Hanmore et al., 2022). Note that concrete elements can also undergo curing under a CO_2 atmosphere, which will be further discussed in Section 21.3.3.

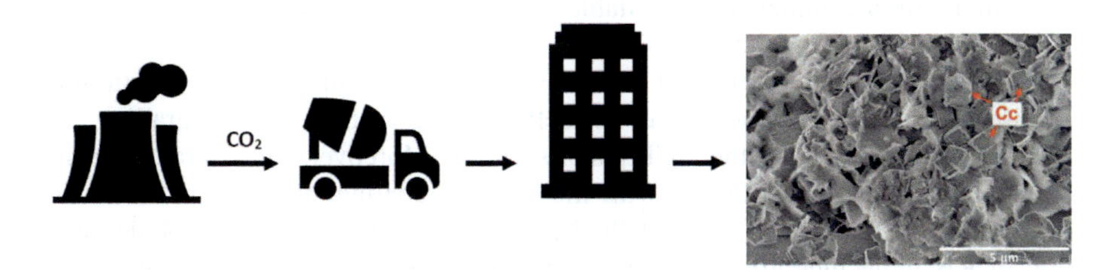

Figure 21.1 CO_2 storage in fresh concrete mixture. From CO_2 capture to precipitation of calcium carbonate microcrystals (Cc) within the concrete matrix (Monkman, Sargam, et al., 2022).
Source: From Monkman, S., Sargam, Y., Naboka, O., & Lothenbach, B. (2022). Early age impacts of CO_2 activation on the tricalcium silicate and cement systems. *Journal of CO_2 Utilization, 65*. https://doi.org/10.1016/j.jcou.2022.102254.

21.3 New technologies for sustainable concrete structures

Sustainable concrete structures are currently being constructed using waste materials such as SCMs like BFS and fly ash, as well as recycled aggregates, to replace a portion of cement and aggregates. However, the primary focus has been on partially substituting waste SCMs for cement, given that cement has the highest carbon footprint among concrete ingredients. Additionally, there is variability in the characteristics of recycled aggregates.

To further promote the construction of sustainable concrete structures, new methods and strategies need to be employed. These techniques encompass the application of nanotechnology, AI, carbon utilization, additive manufacturing (AM), and self-healing methods. The following sections provide a review and discussion of these new technologies that can be utilized in the design and construction of concrete structures to enhance their sustainability.

21.3.1 Nanotechnology

The contribution of nanotechnology to the advancement of sustainable concrete structures can be categorized into four main categories: self-cleaning, self-curing, self-healing, and self-sensing capabilities (Dinesh et al., 2022; Ding et al., 2019; Khannyra et al., 2022; Roopa et al., 2020; Sujitha et al., 2023; Wang et al., 2020).

Nanotechnology provides new possibilities to enhance the sustainability of concrete structures by incorporating smart features. Coatings containing nano-TiO_2 particles can be applied to the exposed surfaces of concrete structures, enabling self-cleaning and air purification capabilities (Khannyra et al., 2022; Wang et al., 2020). In a study conducted by Wang et al. (2020), a TiO_2 hydrosol coating was synthesized and applied to the surface of hydrated cement paste (HCP) panels to evaluate its self-cleaning efficacy. The photocatalytic and self-cleaning properties of the TiO_2 hydrosol coating were assessed by measuring the color change rate of Rhodamine B (RB) and the NO_x conversion under UV irradiation. Three different levels of TiO_2 hydrosol coating—0.77, 1.54, and 3.08 g/m^2—were examined and compared with a commercially available nano-TiO_2 product under identical experimental conditions (Wang et al., 2020).

The findings revealed that all three levels of TiO_2 hydrosol coating exhibited significantly superior photocatalytic self-cleaning abilities in degrading the organic dye RB films on the HCP panel surfaces. Moreover, the degradation rate of RB increased proportionally with the level of coating. Even with the lowest coating level (0.77 g/m^2), the TiO_2 hydrosol coating achieved over 85% degradation of RB molecules after 24 hours of UV irradiation, whereas the commercial nano-TiO_2 coating, applied at a level of 1.54 g/m^2, only degraded approximately 64.9% of RB molecules. Furthermore, as depicted in Fig. 21.2, at the intermediate level of TiO_2 hydrosol coating (1.54 g/m^2), both NO and NO_x conversion rates surpassed 90%.

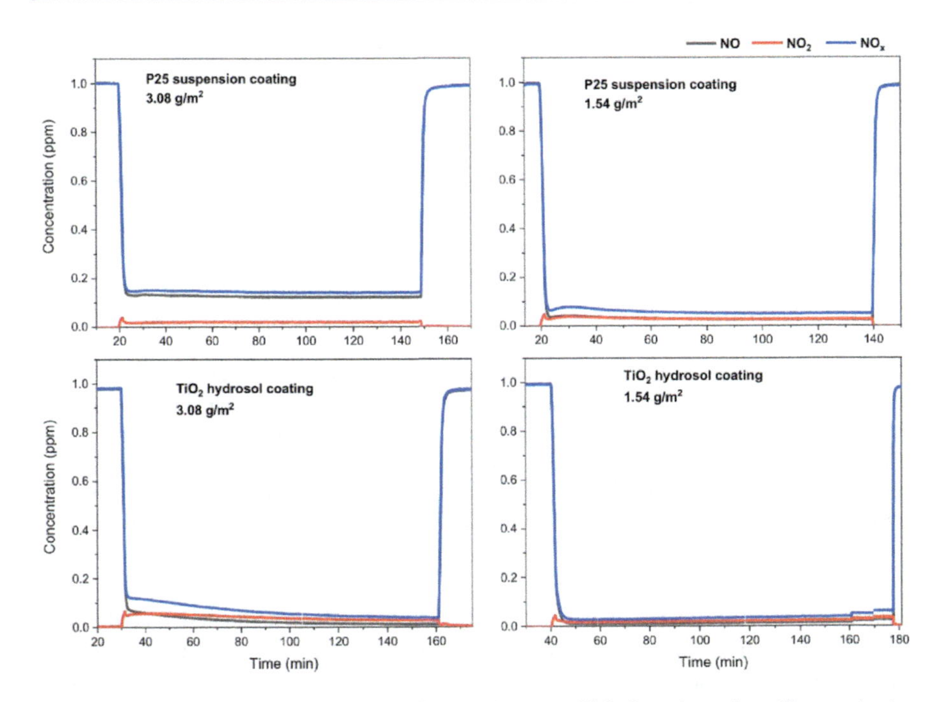

Figure 21.2 Efficient removal of air pollutants by nano-TiO_2-based coating. Changes in the concentration of NO, NO_2, and NO_x at two levels of TiO_2 hydrosol coating versus commercial TiO_2 (P25) (Wang et al., 2020).
Source: From Wang, Z., Gauvin, F., Feng, P., Brouwers, H. J. H., & Yu, Q. (2020). Self-cleaning and air purification performance of portland cement paste with low dosages of nanodispersed TiO_2 coatings. *Construction and Building Materials, 263*, 120558. https://doi.org/10.1016/j.conbuildmat.2020.120558.

Additionally, nanocomposites can be utilized to achieve self-curing and self-healing behaviors in sustainable concrete structures (Pourjavadi et al., 2013; Sujitha et al., 2023). For instance, silane-functionalized superabsorbent polymers (SAPs)/nanoalumina composites have been developed and utilized to create self-cured cement-based mortar for plastering applications (Sujitha et al., 2023).

Moreover, concrete structures can exhibit self-sensing capabilities through the use of fibers with nano/micro dimensions, such as carbon nanotubes (CNTs) and carbon nanofibers (CNFs), as well as nanocomposites like CNTs/nano carbon black (NCB) composite fillers and CNT/CNF/graphene composites (Dinesh et al., 2022; Ding et al., 2019; Roopa et al., 2020). This capability is valuable for structural health monitoring of sustainable concrete structures. Ding et al. (2019) utilized cementitious sensors created with electrostatically self-assembled CNTs/NCB composite fillers to advance the development of smart structural elements. These smart concrete columns were assessed for their self-sensing capabilities under cyclic and monotonic loading by measuring the fractional change in resistivity (FCR) of the

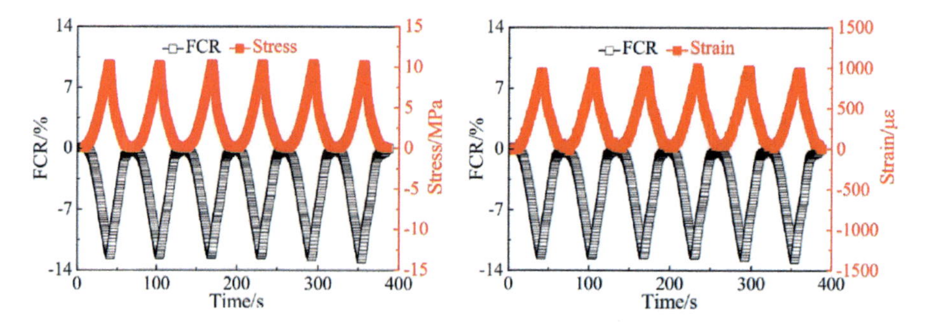

Figure 21.3 Changes in the properties of the cementitious sensor during loading. Relationships between fractional change in resistivity (FCR) and compressive stress/strain of sensor under cyclic compressive loading (Ding et al., 2019).
Source: From Ding, S., Ruan, Y., Yu, X., Han, B., & Ni, Y. Q. (2019). Self-monitoring of smart concrete column incorporating CNT/NCB composite fillers modified cementitious sensors. *Construction and Building Materials, 201*, 127–137. https://doi.org/10.1016/j.conbuildmat.2018.12.203.

embedded cementitious sensor (Fig. 21.3). Their findings indicated that the presence of the sensor did not impact the bearing capacity of the concrete column. Moreover, the piezoresistivity exhibited high stability and repeatability under cyclic loading within the elastic range, suggesting that the developed cementitious sensor is well suited for embedding into prefabricated components for real-time structural monitoring.

Changes in the conductivity of the cementitious matrix in the presence of cracks and during crack propagation can provide opportunities for early detection of damage to concrete structures and prompt repair actions. This can significantly extend the service life of structures, thereby reducing the environmental impact over their operational lifespan.

21.3.2 Artificial intelligence and machine learning

In the design of sustainable concrete structures, life-cycle assessment (LCA) and environmental impact assessment (EIA) have recently been integrated (Danatzko & Sezen, 2011; Rezaei et al., 2019). The main objectives during the design process are to minimize construction costs, extend the service life, and reduce the environmental impact of concrete structures by minimizing energy and raw material usage while maximizing the utilization of waste materials. LCA-informed design has been employed to address the need for sustainability improvement in concrete structures. However, to optimize the design process, other methods need to be employed.

AI-based techniques can play a crucial role in optimizing the design process and advancing sustainability within the constraints of available resources. These methods can be utilized to consider all available options for sustainability improvements, such as incorporating waste materials into concrete mixtures while achieving the required design strength and durability, predicting the behavior of structures when using sustainable concrete materials, and more.

Another application of AI-based techniques in developing sustainable concrete structures is in structural health monitoring. These programs can enable autonomous early detection of damage through image processing techniques, leading to prompt repair of damaged sections and increasing the service life of sustainable concrete structures (Adel et al., 2021; Thai, 2022; Wu et al., 2022).

Furthermore, ML methods have been employed to predict concrete pouring time and optimize performance by suggesting changes to the mixture design (Fahim et al., 2021). This AI-based technology utilizes data collected by internet of things (IoT) sensors from numerous concrete mixtures across various projects and countries to improve the future performance of concrete in the field. Additionally, the prediction of concrete performance can provide practitioners with strategies to save cement while achieving the required performance through modifications in the mix design (Fahim et al., 2021).

21.3.3 Carbon capture, utilization, and storage

Curing structural members, particularly precast concrete elements, in a CO_2 environment can enhance their mechanical strength and durability. This curing process is known as "CO_2 curing" or "carbonation curing." CO_2 gas penetrates the concrete from its surface and reacts with unhydrated calcium silicates and calcium hydroxide crystals, forming C-S-H and $CaCO_3$. This reaction leads to micro-filling, resulting in higher strength and durability. Typically, a concrete element is placed in an enclosed system at a specific relative humidity (RH) under elevated CO_2 pressure, facilitating the gas penetration into the concrete matrix (Monkman & Shao, 2010a, 2010b).

However, due to practical limitations, curing structural elements in a pressurized vessel for an extended period is not feasible. Consequently, during short-time carbonation curing, the penetration of CO_2 is limited to the concrete surface and does not reach significant depths. Thus short-time carbonation curing primarily affects the permeability of the concrete, leading to higher durability of concrete elements.

Surface treatment of concrete with carbonation densifies the microstructure, resulting in a significant reduction in permeability against harmful gases and liquids (Wang et al., 2021). Building upon this concept, researchers have developed a carbon capture (CC) coating that can be sprayed onto the surface of concrete elements to create a dense coating upon carbonation treatment (Peng et al., 2023). Such coatings typically consist of materials that exhibit high reactivity with CO_2 gas, such as dicalcium silicate or belitic (C_2S) binder. Peng et al. (2023) demonstrated that a CC coating, developed using CO_2-activated dicalcium silicate (with a C_2S cement:distilled water:superplasticizer weight ratio of 1:0.29:0.06), achieved a notable carbonation degree of 23.3% within just 30 minutes. The resultant CC coating, at a thickness of about 200 μm after this short duration of carbonation, effectively reduced the capillary water absorption of concrete samples by 20.5%. Additionally, it exhibited a satisfactory bonding strength of 2.74 MPa, signifying the potential for the CC coating to significantly extend the service life of concrete structures while concurrently curbing carbon emissions. Fig. 21.4

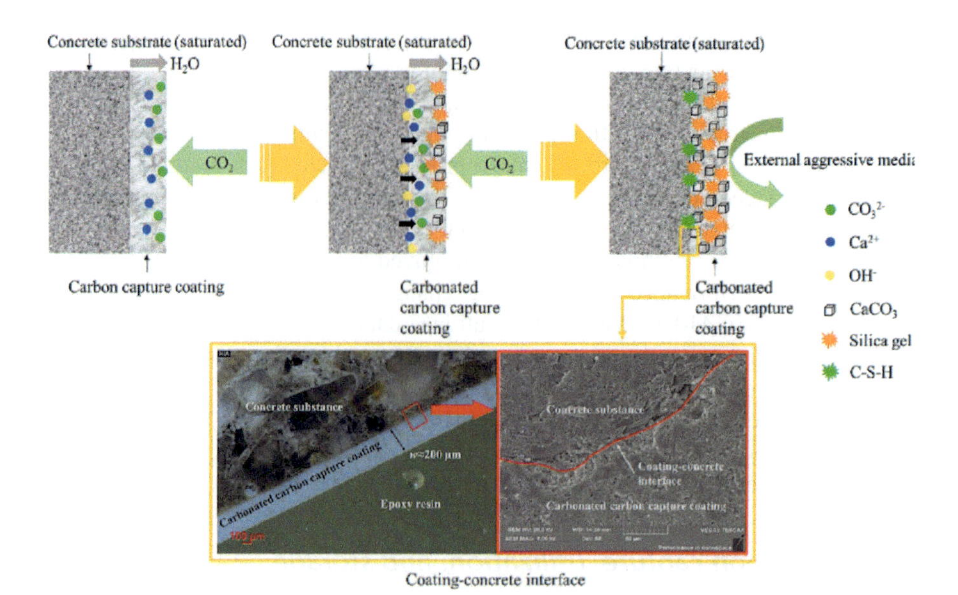

Figure 21.4 Development of carbon capture coating-concrete substrate interface . Mechanism of interface enhancement via using carbon capture coating (Peng et al., 2023). *Source*: From Peng, L., Shen, P., Poon, C.-S., Zhao, Y., & Wang, F. (2023). Development of carbon capture coating to improve the durability of concrete structures. *Cement and Concrete Research, 168*, 107154. https://doi.org/10.1016/j.cemconres.2023.107154.

illustrates the mechanism behind how the CC coating enhances the durability of the concrete substrate, as well as delineates the characteristics of the interface between the CC coating and the concrete.

Furthermore, carbonation treatment can be employed to rapidly repair damaged structures caused by fire. Recurring the concrete elements through water–CO_2 cyclic curing has shown promising results. Even at high–temperature exposures (e.g., 1000°C), compressive strength recovery has been observed due to the formation of carbonation products that partially fill microcracks within the matrix (Li et al., 2023).

21.3.4 Additive manufacturing

AM or 3D printing has gained significant attention in the field of concrete construction due to its potential for reducing construction waste, lowering costs, and ensuring high safety and health standards on construction sites. The digital fabrication technology used in 3D printing allows for the creation of complex shapes without the need for traditional formworks, enabling rapid construction and a significant reduction in construction duration. Additionally, carefully designed concrete mixtures in 3D printing can help minimize construction waste and develop mixtures with a low carbon footprint (Batikha et al., 2022; Flatt & Wangler, 2022). Within this context, Batikha et al. (2022) conducted a comparative analysis involving 3D concrete printing (3DCP) for designing a two-story building, juxtaposed with several common construction methods: in-situ reinforced concrete (RC) construction, hot rolled steel (HRS), cold-formed steel (CFS), and prefabricated concrete construction (PCC). As depicted in Fig. 21.5, except for prefabricated modular concrete, 3DCP significantly reduces construction time by approximately 95% compared to the conventional cast-in-situ RC method. Furthermore, 3DCP exhibited the smallest overall carbon footprint among the evaluated construction techniques. While the total cost of the RC construction method might be lower compared to the 3DCP technique, the reduced carbon footprint and shorter construction time of 3DCP can compensate, especially in light of new global policies imposing carbon taxes on various products, including buildings.

To further advance sustainability in 3D-printed concrete structures, other technologies can be employed, such as the utilization of industrial waste materials, carbonation curing, and nanotechnology (Dey et al., 2022; Liu, Jiang, et al., 2022; Wang et al., 2023). Precast structural concrete members can be 3D-printed in a factory setting and subsequently cured with CO_2 to enhance their durability (Fig. 21.6). Wang et al. (2023) highlighted that carbonated cement-based 3D-printed structures exhibit significant CC potential. For example, a 24-hour carbonation process of 3D-printed mortar strips measuring 10 mm in width yielded a carbonation percentage of 24%.

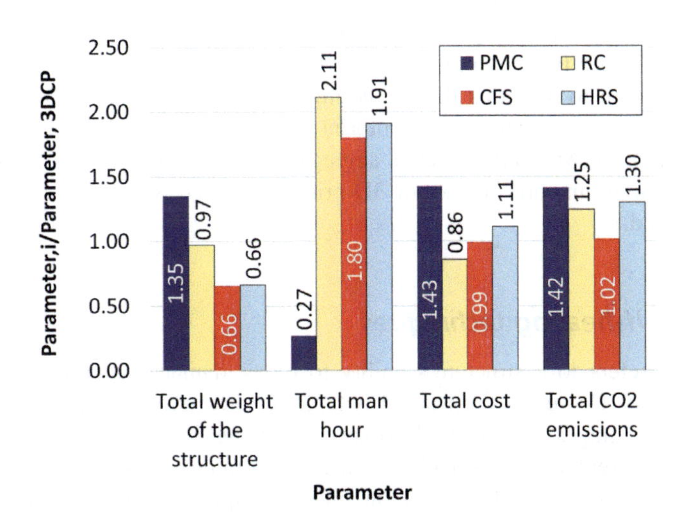

Figure 21.5 Comparative changes in the factors affecting sustainable construction of a two-story building for different construction methods. Relative construction, cost, and environmental factors for various construction methods compared to three-dimensional concrete printing (Batikha et al., 2022).
Source: From Batikha, M., Jotangia, R., Baaj, M. Y., & Mousleh, I. (2022). 3D concrete printing for sustainable and economical construction: A comparative study. *Automation in Construction, 134*, 104087. https://doi.org/10.1016/j.autcon.2021.104087.

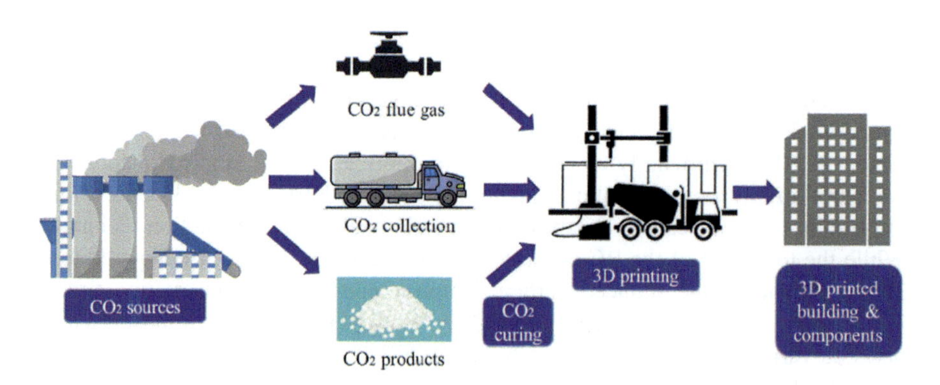

Figure 21.6 From capturing CO_2 **at the plant to utilizing it for curing three-dimensional (3D)-printed concrete products**. Incorporation of CO_2 curing technologies into 3D-printed concrete construction (Wang et al., 2023).
Source: From Wang, D., Xiao, J., Sun, B., Zhang, S., & Poon, C. S. (2023). Mechanical properties of 3D printed mortar cured by CO_2. *Cement and Concrete Composites, 139*, 105009. https://doi.org/10.1016/j.cemconcomp.2023.105009.

Wang et al. (2023) demonstrated that CO_2 curing positively influenced the compressive strength and splitting tensile strength of 3D-printed cement mortars, particularly noticeable in the early stage (3 days). Additionally, CO_2 curing enhanced the interlayer bonding strength of 3D-printed mortar. The enhancement in bonding strength between adjacent layers surpassed that between stacked layers, indicating a more pronounced efficiency in interlayer cohesion.

It is important to note that due to the higher binder content in 3D printing concrete mixtures, their initial carbon footprint tends to be higher. However, by incorporating waste SCMs and recycled aggregates in combination with carbonation curing, the environmental impact of 3D-printed concrete structures can be significantly reduced.

21.3.5 Self-healing techniques

Self-healing refers to a structure's ability to repair damages caused by internal or external factors. However, self-healing is typically effective for small-scale damages or at the initial stages of damage. Various self-healing techniques have shown effectiveness in concrete structures, including the use of microcapsules containing healing agents, microbially induced calcium carbonate precipitation (MICP), shape memory polymers (SMPs), SAP nanocomposites, and vascular flow networks for delivering healing agents (Davies et al., 2018; Jiang et al., 2020; Sujitha et al., 2023).

In a Jiang et al. (2020) investigation, *Bacillus cohnii* (*B. cohnii*) was utilized as the bacterial agent in formulating concrete mixtures, and the study focused on the mineralization activity of spores postgermination, along with the crack-healing capability of the concrete. The findings revealed that *B. cohnii* spores exhibit mineralizing properties after germination, producing pure calcite crystals through MICP. The research indicated a gradual reduction in crack width for bacterial concrete specimens with increasing healing time. When the initial crack width was small (0.43 mm), complete repair occurred after 3 days of healing, whereas a larger crack width (1.23 mm) required 28 days for full restoration. Significantly, control concrete specimens lacking bacteria exhibited unrepaired cracks, underscoring the pivotal role of bacteria in the crack-healing process observed in bacterial concrete.

Davies et al. (2018) investigated the effectiveness of various self-healing technologies in promoting crack recovery in five in-situ cast concrete panels. The study employed four distinct self-healing techniques, both individually and in combination: (1) the utilization of microcapsules containing mineral healing agents, (2) bacterial healing, (3) a system based on SMP for crack closure, and (4) the delivery of a mineral healing agent through a vascular flow network. The fifth panel was devoid of any self-healing technology and functioned as a control. Table 21.1 provides a summary of the self-healing technologies applied in the panels, while Fig. 21.7 illustrates the construction details of the panels under investigation.

Based on field data (Davies et al., 2018), the average crack width measured 6 months following the initial loading/unloading experiment, with all self-healing technologies fully engaged, were documented as 0.10, 0.07, 0.07, 0.12, and 0.23 mm for panels A through E, respectively. Fig. 21.8 displays images depicting cracks immediately after the initial loading/unloading and their condition 6 months later, with the full implementation of self-healing techniques. These findings highlight the effective use of suitable self-healing techniques in the large-scale applications.

Concrete structures equipped with self-healing capabilities offer extended service life, indirectly contributing to sustainability by reducing their carbon footprint over their lifespan. The ability to autonomously repair and mitigate damage ensures that the structure remains functional and durable for a longer period, thus reducing the need for frequent repairs or replacements and minimizing the overall environmental impact and cost.

Table 21.1 Panel label and the utilized self-healing technology (Davies et al., 2018).

Panel	Self-healing technology
A	Microcapsules containing sodium silicate cargo
B	Shape memory polymers and flow networks
C	Bacteria-infused perlite, nutrient-infused perlite and flow networks
D	Control panel (C40/50 reinforced concrete)
E	Control with flow networks

Source: From Davies, R., Teall, O., Pilegis, M., Kanellopoulos, A., Sharma, T., Jefferson, A., Gardner, D., Al-Tabbaa, A., Paine, K., & Lark, R. (2018). Large scale application of self-healing concrete: Design, construction, and testing. *Frontiers in Materials, 5*. https://doi.org/10.3389/fmats.2018.00051.

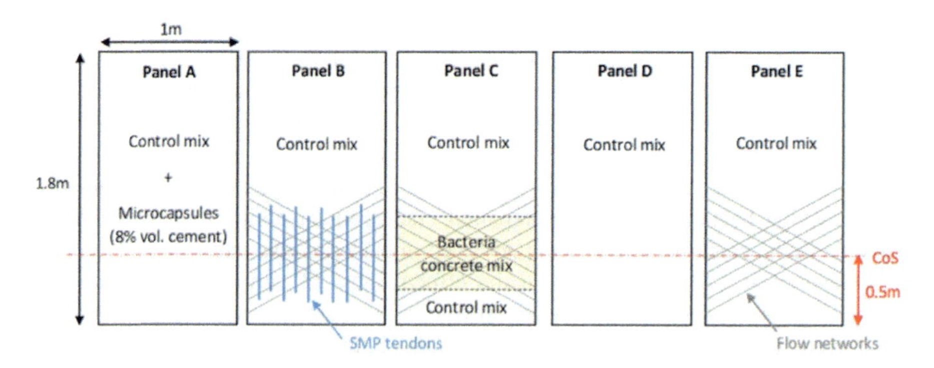

Figure 21.7 Configuration of self-healing techniques used in the concrete panels. Self-healing technology used in the panels for field study (Davies et al., 2018).
Source: From Davies, R., Teall, O., Pilegis, M., Kanellopoulos, A., Sharma, T., Jefferson, A., Gardner, D., Al-Tabbaa, A., Paine, K., & Lark, R. (2018). Large scale application of self-healing concrete: Design, construction, and testing. *Frontiers in Materials, 5*. https://doi.org/10.3389/fmats.2018.00051.

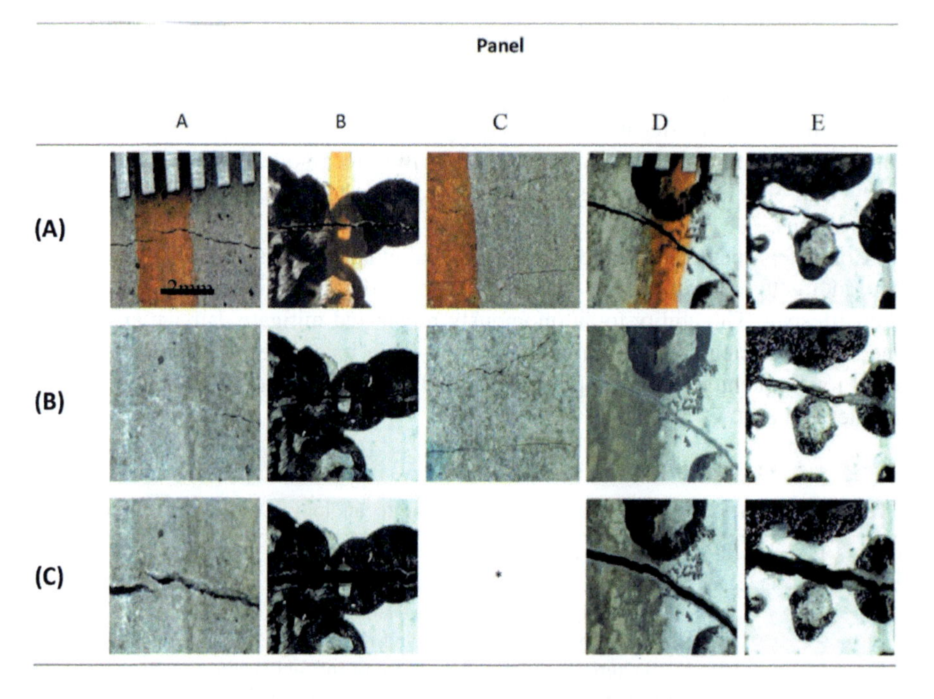

Figure 21.8 Microscopic images depict the evolution of cracks on concrete panels from the initial loading to the final loading. Crack condition assessment for various panels with and without self-healing technology after (A) initial loading, (B) 6 months, and (C) final loading (Davies et al., 2018).
Source: From Davies, R., Teall, O., Pilegis, M., Kanellopoulos, A., Sharma, T., Jefferson, A., Gardner, D., Al-Tabbaa, A., Paine, K., & Lark, R. (2018). Large scale application of self-healing concrete: Design, construction, and testing. *Frontiers in Materials, 5*. https://doi.org/10.3389/fmats.2018.00051.

21.4 Conclusions

To date, several technologies have been introduced to enhance the mechanical and durability properties of concrete materials and structures. Among these are nanotechnology, AI, carbon capture, utilization, and storage (CCUS), AM, and self-healing techniques, which can be utilized to improve the sustainability of concrete materials and structures. This extends to materials and structures that are currently designed and constructed with sustainability in mind. In this chapter, various novel technologies were introduced, and their application in developing sustainable concrete materials and structures was reviewed. The development of modern composite cements, such as M-PLC and LC^3, can reduce the clinker factor in cement while providing mechanical and durability performance similar to or better than OPC.

Nanomaterials, whether synthetic or derived from waste materials, can be harnessed to enhance the mechanical and durability properties of cement-based materials while simultaneously bolstering sustainability. This is achieved by substituting a portion of cement in the matrix and/or introducing novel characteristics like air purification capabilities and active rheology control in concrete materials. Furthermore, nanotechnology-enabled cement-based materials have found applications in the development of sustainable structures with diverse functionalities, including self-sensing and self-healing capabilities. These advancements enable the detection of cracks and swift repairs to prevent immediate failures, affording time for assessing damage conditions and implementing comprehensive repairs. Another noteworthy concept is self-curing, which enhances the dimensional stability of concrete structures constructed with materials containing a high proportion of fine constituents, such as UHPC. This is particularly valuable in mitigating the pronounced early-age shrinkage issues that typically lead to durability concerns and a reduced service life for concrete structures.

In addition to nanomaterials, various other approaches can be employed to imbue sustainable concrete materials and structures with self-healing capabilities. These include the use of bacterial strains, microcapsules containing expansive agents, SAPs modified with nanomaterials, SMPs, and vascular-based technology.

AI and ML techniques can be employed to optimize both the mix design of concrete materials and the structural design of concrete structures, thereby enhancing their sustainability. This optimization entails the judicious utilization of SCMs to attain superior mechanical strength and durability while simultaneously minimizing material consumption and cost. Additionally, AI and ML can contribute to improved energy efficiency in construction through structural design optimization.

To further diminish the carbon footprint of sustainable concrete materials and structures, diverse technologies have been adopted for the utilization of captured carbon, whether sourced from the air or a point source. Key applications of captured CO_2 in reducing the carbon footprint of concrete structures include curing precast concrete elements within a CO_2 environment, employing carbonated SCMs, and injecting CO_2 into the mixture during the mixing process.

AM stands as another innovative technology for the development of sustainable structures by minimizing construction time, enhancing energy efficiency, and reducing costs. While 3D printable concrete materials typically necessitate a high binder content, their carbon footprint can be significantly reduced through the incorporation of waste materials and the application of carbonation curing methods.

Figs. 21.9 and 21.10 provide a summary of the novel technologies reviewed in this chapter that are utilized to advance the sustainability of concrete materials and structures, respectively.

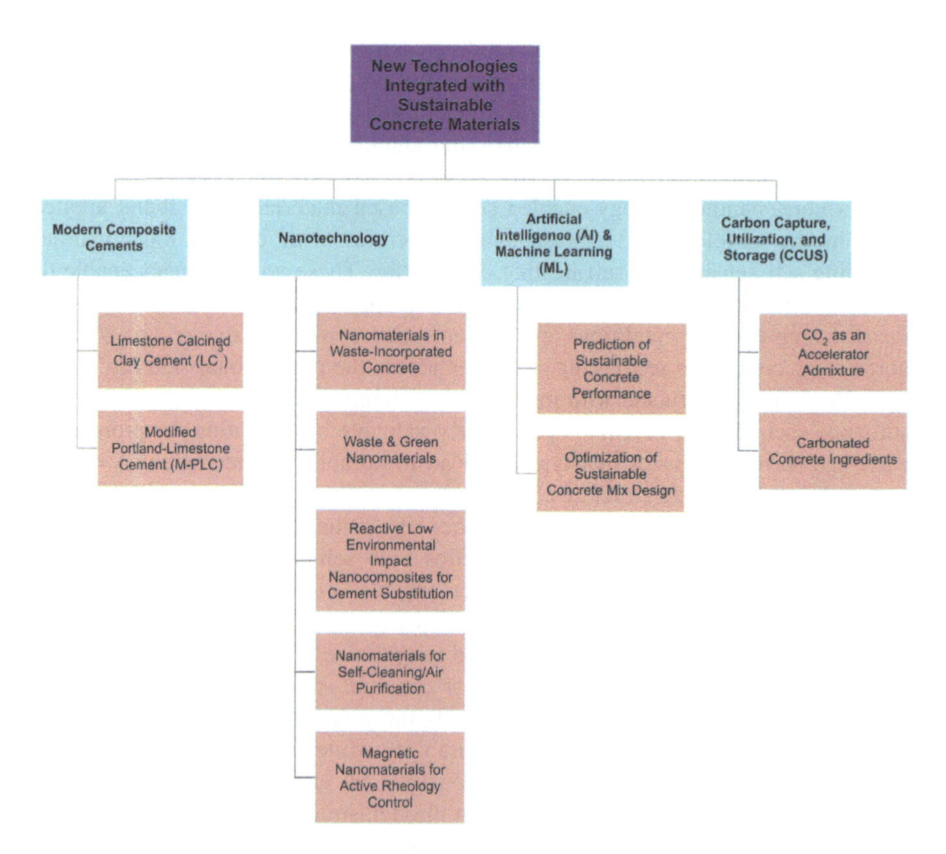

Figure 21.9 New technologies to develop sustainable concrete materials. New technologies for the advancement of sustainable concrete materials.

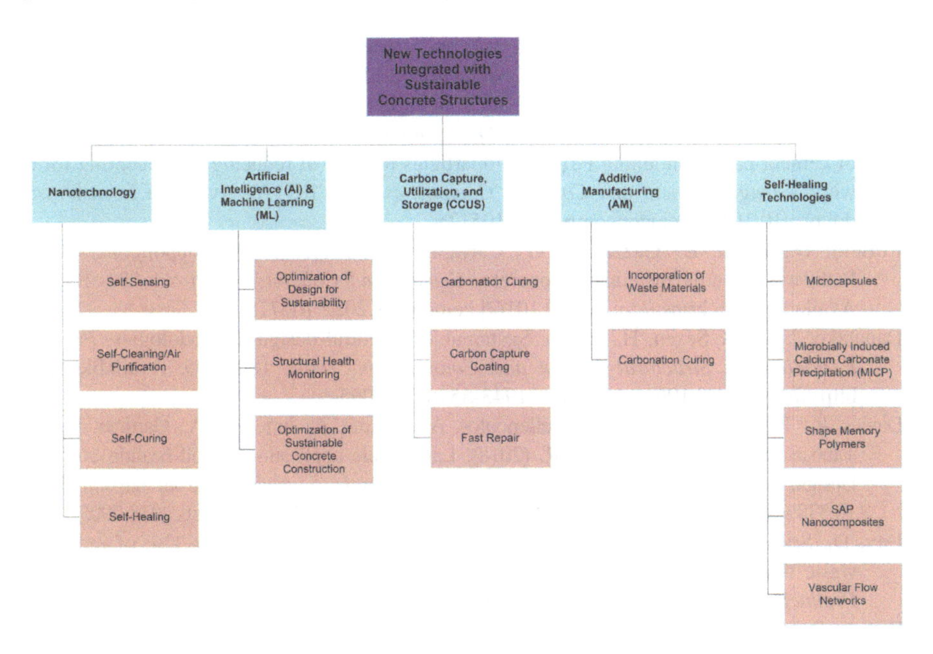

Figure 21.10 New technologies to develop sustainable concrete structures. New technologies for enhancement of sustainability in concrete structures.

References

Adel, M., Yokoyama, H., Tatsuta, H., Nomura, T., Ando, Y., Nakamura, T., Masuya, H., & Nagai, K. (2021). Early damage detection of fatigue failure for RC deck slabs under wheel load moving test using image analysis with artificial intelligence. *Engineering Structures*, *246*, 113050. Available from https://doi.org/10.1016/j.engstruct.2021.113050.

Ahmad, F., Qureshi, M. I., & Ahmad, Z. (2022). Influence of nano graphite platelets on the behavior of concrete with E-waste plastic coarse aggregates. *Construction and Building Materials*, *316*, 125980. Available from https://doi.org/10.1016/j.conbuildmat.2021.125980.

Bahadori, H., & Hosseini, P. (2012). Reduction of cement consumption by the aid of silica nanoparticles (investigation on concrete properties). *Journal of Civil Engineering and Management*, *18*(3), 416–425. Available from https://doi.org/10.3846/13923730.2012.698912.

Bahadori, H., & Hosseini, P. (2018). Developing green cement paste using binary and ternary cementitious blends of low pozzolanic sewage sludge ash and colloidal nanosilica (short-term properties). *Asian Journal of Civil Engineering*, *19*(4), 501–511. Available from https://doi.org/10.1007/s42107-018-0037-0, https://rd.springer.com/journal/42107.

Batikha, M., Jotangia, R., Baaj, M. Y., & Mousleh, I. (2022). 3D concrete printing for sustainable and economical construction: A comparative study. *Automation in Construction*, *134*, 104087. Available from https://doi.org/10.1016/j.autcon.2021.104087.

Chen, X. F., & Jiao, C. J. (2022). Microstructure and physical properties of concrete containing recycled aggregates pre-treated by a nano-silica soaking method. *Journal of Building Engineering*, *51*. Available from https://doi.org/10.1016/j.jobe.2022.104363, http://www.journals.elsevier.com/journal-of-building-engineering/.

Chen, Z., Li, R., & Liu, J. (2021). Preparation and properties of carbonated steel slag used in cement cementitious materials. *Construction and Building Materials*, *283*, 122667. Available from https://doi.org/10.1016/j.conbuildmat.2021.122667.

Chibulu, C., Jiao, D., Yardimci, M. Y., & De Schutter, G. (2023). Magneto-rheology control of cement paste containing Fe_3O_4 nanoparticles in view of reducing or preventing formwork leakage. *Cement and Concrete Composites*, *142*, 105176. Available from https://doi.org/10.1016/j.cemconcomp.2023.105176.

Coffetti, D., Crotti, E., Gazzaniga, G., Carrara, M., Pastore, T., & Coppola, L. (2022). Pathways towards sustainable concrete. *Cement and Concrete Research*, *154*, 106718. Available from https://doi.org/10.1016/j.cemconres.2022.106718.

Danatzko, J. M., & Sezen, H. (2011). Sustainable structural design methodologies. *Practice Periodical on Structural Design and Construction*, *16*(4), 186–190. Available from https://doi.org/10.1061/(ASCE)SC.1943-5576.0000095.

Davies, R., Teall, O., Pilegis, M., Kanellopoulos, A., Sharma, T., Jefferson, A., Gardner, D., Al-Tabbaa, A., Paine, K., & Lark, R. (2018). Large scale application of self-healing concrete: Design, construction, and testing. *Frontiers in Materials*, *5*. Available from https://doi.org/10.3389/fmats.2018.00051, https://www.frontiersin.org/articles/10.3389/fmats.2018.00051/pdf.

Dey, D., Srinivas, D., Panda, B., Suraneni, P., & Sitharam, T. G. (2022). Use of industrial waste materials for 3D printing of sustainable concrete: A review. *Journal of Cleaner Production*, *340*, 130749. Available from https://doi.org/10.1016/j.jclepro.2022.130749.

Dimov, D., Amit, I., Gorrie, O., Barnes, M. D., Townsend, N. J., Neves, A. I. S., Withers, F., Russo, S., & Craciun, M. F. (2018). Ultrahigh performance nanoengineered graphene–concrete composites for multifunctional applications. *Advanced Functional Materials*, *28*(23). Available from https://doi.org/10.1002/adfm.201705183, http://onlinelibrary.wiley.com/journal/10.1002/(ISSN)1616-3028.

Dinesh, A., Durgadevi, S., Veeraraghavan, S., & Janani Praveena, S. (2022). Carbon-based nanomaterial embedded self-sensing cement composite for structural health monitoring of concrete beams-A extensive review. *Materials Research Proceedings*, *23*, 217–230. Available from https://doi.org/10.21741/9781644901953-25, http://www.mrforum.com/materials-research-proceedings-volumes/.

Ding, S., Ruan, Y., Yu, X., Han, B., & Ni, Y. Q. (2019). Self-monitoring of smart concrete column incorporating CNT/NCB composite fillers modified cementitious sensors. *Construction and Building Materials*, *201*, 127–137. Available from https://doi.org/10.1016/j.conbuildmat.2018.12.203.

Dong, S., Li, L., Ashour, A., Dong, X., & Han, B. (2021). Self-assembled 0D/2D nano carbon materials engineered smart and multifunctional cement-based composites. *Construction and Building Materials*, *272*, 121632. Available from https://doi.org/10.1016/j.conbuildmat.2020.121632.

EUROFER, Low Carbon Roadmap 2019 15. (2023). 6 Brussels Unpublished content Pathways to a CO2-neutral European steel industry https://www.eurofer.eu/assets/Uploads/EUROFER-Low-Carbon-Roadmap-Pathways-to-a-CO2-neutral-European-Steel-Industry.pdf.

Fahim, A., Mehdi, T., Taheri, A., Ghods, P., Alizadeh, A., De Carufel, S. (2021). The use of machine learning algorithms and IoT sensor data for concrete performance testing and analysis. American Concrete Institute. ACI Special Publication. SP-350 9781641951623, 140–150 American Concrete Institute Canada. https://www.concrete.org/topicsinconcrete/topicdetail/special%20publication.

Fahmy, N. G., Hussien, R. M., el-Hafez, L. M. A., Mohamed, R. A. S., & Faried, A. S. (2022). Comparative study on fresh, mechanical, microstructures properties and corrosion resistance of self compacted concrete incorporating nanoparticles extracted from industrial wastes under various curing conditions. *Journal of Building Engineering*, *57*. Available from https://doi.org/10.1016/j.jobe.2022.104874, http://www.journals.elsevier.com/journal-of-building-engineering/.

Faried, A. S., Mostafa, S. A., Tayeh, B. A., & Tawfik, T. A. (2021). Mechanical and durability properties of ultra-high performance concrete incorporated with various nano waste materials under different curing conditions. *Journal of Building Engineering*, *43*. Available from https://doi.org/10.1016/j.jobe.2021.102569, http://www.journals.elsevier.com/journal-of-building-engineering/.

Flatt, R. J., & Wangler, T. (2022). On sustainability and digital fabrication with concrete. *Cement and Concrete Research*, *158*, 106837. Available from https://doi.org/10.1016/j.cemconres.2022.106837.

Gao, X., & Yu, Q. L. (2019). Effects of an eco-silica source based activator on functional alkali activated lightweight composites. *Construction and Building Materials*, *215*, 686–695. Available from https://doi.org/10.1016/j.conbuildmat.2019.04.251.

Global Material Resources Outlook to 2060. Organization for Economic and Co-operation and Development (OECD). (2018). Unpublished content. https://www.oecd.org/environment/waste/highlights-global-material-resources-outlook-to-2060.pdf.

Gonnon, P., & Lootens, D. (2023). Toward net zero carbon for concrete and mortar: Clinker substitution with ground calcium carbonate. *Cement and Concrete Composites*, *142*, 105190. Available from https://doi.org/10.1016/j.cemconcomp.2023.105190.

Habert, G., Miller, S. A., John, V. M., Provis, J. L., Favier, A., Horvath, A., & Scrivener, K. L. (2020). Environmental impacts and decarbonization strategies in the cement and concrete industries. *Nature Reviews Earth and Environment*, *1*(11), 559–573. Available from https://doi.org/10.1038/s43017-020-0093-3, http://nature.com/natrevearthenviron/.

Han, T., Siddique, A., Khayat, K., Huang, J., & Kumar, A. (2020). An ensemble machine learning approach for prediction and optimization of modulus of elasticity of recycled aggregate concrete. *Construction and Building Materials*, *244*, 118271. Available from https://doi.org/10.1016/j.conbuildmat.2020.118271.

Hassan, K. E., Brooks, J. J., & Erdman, M. (2000). The use of reclaimed asphalt pavement (RAP) aggregates in concrete. *Waste Management Series*, *1*(C), 121–128. Available from https://doi.org/10.1016/S0713-2743(00)80024-0.

Hosseini, P., Abolhasani, M., Mirzaei, F., Kouhi Anbaran, M. R., Khaksari, Y., & Famili, H. (2018). Influence of two types of nanosilica hydrosols on short-term properties of sustainable white portland cement mortar. *Journal of Materials in Civil Engineering*, *30*(2). Available from https://doi.org/10.1061/(ASCE)MT.1943-5533.0002152, http://ascelibrary.org/mto/resource/1/jmcee7/.

Hosseini, P., Booshehrian, A., & Farshchi, S. (2010). Influence of nano-SiO_2 addition on microstructure and mechanical properties of cement mortars for ferrocement. *Transportation Research Record: Journal of the Transportation Research Board*, *2141* (1), 15–20. Available from https://doi.org/10.3141/2141-04.

Hosseini, P., Booshehrian, A., & Madari, A. (2011). Developing concrete recycling strategies by utilization of nano-SiO_2 particles. *Waste and Biomass Valorization*, *2*(3), 347–355. Available from https://doi.org/10.1007/s12649-011-9071-9.

Hosseini, P., Mohamad, M. I., Nekooie, M. A., Taherkhani, R., & Booshehrian, A. (2011). Toward green revolution in concrete industry: The role of nanotechnology (a review). *Australian Journal of Basic and Applied Sciences*, *5*(12), 2768–2782. Available from http://www.insipub.net/ajbas/2011/December-2011/2768-2782.pdf.

Hosseini, P., Hosseinpourpia, R., Pajum, A., Khodavirdi, M. M., Izadi, H., & Vaezi, A. (2014). Effect of nano-particles and aminosilane interaction on the performances of cement-based composites: An experimental study. *Construction and Building Materials*, *66*, 113–124. Available from https://doi.org/10.1016/j.conbuildmat.2014.05.047.

Hosseini, P., Yavari, A., Lotfi, H., Kouhi Anbaran, M. R., Khaksari, Y., & Famili, H. (2020). Properties of nanosilica-reinforced green architectural cement composites incorporating ground granulated blast furnace slag with low activity. *European Journal of Environmental and Civil Engineering*, *24*(12), 1901–1920. Available from https://doi.org/10.1080/19648189.2018.1492973, http://www.tandfonline.com/loi/tece20.

Hosseinpourpia, R., Varshoee, A., Soltani, M., Hosseini, P., & Ziaei Tabari, H. (2012). Production of waste bio-fiber cement-based composites reinforced with nano-SiO_2 particles as a substitute for asbestos cement composites. *Construction and Building Materials*, *31*, 105–111. Available from https://doi.org/10.1016/j.conbuildmat.2011.12.102.

Hu, X., Li, B., Mo, Y., & Alselwi, O. (2021). Progress in artificial intelligence-based prediction of concrete performance. *Journal of Advanced Concrete Technology*, *19*(8), 924–936. Available from https://doi.org/10.3151/jact.19.924, https://www.jstage.jst.go.jp/article/jact/19/8/19_924/_pdf/-char/en.

Jiang, L., Jia, G., Wang, Y., & Li, Z. (2020). Optimization of sporulation and germination conditions of functional bacteria for concrete crack-healing and evaluation of their repair capacity. *ACS Applied Materials and Interfaces*, *12*(9), 10938–10948. Available from https://doi.org/10.1021/acsami.9b21465, http://pubs.acs.org/journal/aamick.

Jiang, Y., Shen, P., & Poon, C. S. (2022). Improving the bonding capacity of recycled concrete aggregate by creating a reactive shell with aqueous carbonation. *Construction and Building Materials*, *315*, 125733. Available from https://doi.org/10.1016/j.conbuildmat.2021.125733.

Jiao, D., Shi, C., & De Schutter, G. (2022). Magneto-rheology control in 3D concrete printing: A rheological attempt. *Materials Letters*, *309*, 131374. Available from https://doi.org/10.1016/j.matlet.2021.131374.

Juenger, M. C. G., Snellings, R., & Bernal, S. A. (2019). Supplementary cementitious materials: New sources, characterization, and performance insights. *Cement and Concrete Research*, *122*, 257–273. Available from https://doi.org/10.1016/j.cemconres.2019.05.008, http://www.sciencedirect.com/science/journal/00088846.

Khaloo, A., Mobini, M. H., & Hosseini, P. (2016). Influence of different types of nano-SiO_2 particles on properties of high-performance concrete. *Construction and Building Materials*, *113*, 188–201. Available from https://doi.org/10.1016/j.conbuildmat.2016.03.041.

Khannyra, S., Luna, M., Almoraima Gil, M. L., Addou, M., & Mosquera, M. J. (2022). Self-cleaning durability assessment of TiO_2/SiO_2 photocatalysts coated concrete: Effect of indoor and outdoor conditions on the photocatalytic activity. *Building and Environment*, *211*, 108743. Available from https://doi.org/10.1016/j.buildenv.2021.108743.

Khozin, V., Khokhryakov, O., & Nizamov, R. (2020). A «carbon footprint» of low water demand cements and cement-based concrete. *IOP Conference Series: Materials Science and Engineering*, *890*(1), 012105. Available from https://doi.org/10.1088/1757-899x/890/1/012105.

Kumaresan, M., Nachiar., & Anandh, S. (2022). Implementation of waste recycled fibers in concrete: A review. *Materials Today: Proceedings.*, *68*, 1988–1994. Available from https://doi.org/10.1016/j.matpr.2022.08.228.

Kusin, F. M., Hasan, S. N. M. S., Hassim, M. A., & Molahid, V. L. M. (2020). Mineral carbonation of sedimentary mine waste for carbon sequestration and potential reutilization as cementitious material. *Environmental Science and Pollution Research*, *27*(11), 12767–12780. Available from https://doi.org/10.1007/s11356-020-07877-3, https://link.springer.com/journal/11356.

Lazaro, A., Yu, Q. L., & Brouwers, H. J. H. (2016). *Nanotechnologies for sustainable construction* (pp. 55–78). Elsevier BV. Available from 10.1016/b978-0-08-100370-1.00004-4.

Lenton, T. M., Xu, C., Abrams, J. F., Ghadiali, A., Loriani, S., Sakschewski, B., Zimm, C., Ebi, K. L., Dunn, R. R., Svenning, J. C., & Scheffer, M. (2023). Quantifying the human cost of global warming. *Nature Sustainability*, *6*(10), 1237–1247. Available from https://doi.org/10.1038/s41893-023-01132-6, https://www.nature.com/natsustain/.

Lim, T., Ellis, B. R., & Skerlos, S. J. (2019). Mitigating CO_2 emissions of concrete manufacturing through CO_2-enabled binder reduction. *Environmental Research Letters*, *14*(11). Available from https://doi.org/10.1088/1748-9326/ab466e, https://iopscience.iop.org/article/10.1088/1748-9326/ab466e.

Liu, Q., Hosseini, P., Ragipani, R., & Wang, B. (2023). Wet coating of calcite with silica nanoparticles in CO_2 environment. *Journal of Coatings Technology and Research*. Available from https://doi.org/10.1007/s11998-023-00812-4.

Liu, Q., Jiang, Q., Huang, M., Xin, J., & Chen, P. (2022). The fresh and hardened properties of 3D printing cement-base materials with self-cleaning nano-TiO_2:An exploratory study. *Journal of Cleaner Production*, *379*, 134804. Available from https://doi.org/10.1016/j.jclepro.2022.134804.

Liu, S., Shen, Y., Wang, Y., Shen, P., Xuan, D., Guan, X., & Shi, C. (2022). Upcycling sintering red mud waste for novel superfine composite mineral admixture and CO_2 sequestration. *Cement and Concrete Composites*, *129*, 104497. Available from https://doi.org/10.1016/j.cemconcomp.2022.104497.

Liu, G., Schollbach, K., Li, P., & Brouwers, H. J. H. (2021). Valorization of converter steel slag into eco-friendly ultra-high performance concrete by ambient CO_2 pre-treatment. *Construction and Building Materials*, *280*, 122580. Available from https://doi.org/10.1016/j.conbuildmat.2021.122580.

Li, Y., Wang, H., Shi, C., Zou, D., Zhou, A., & Liu, T. (2023). Effect of post-fire lime-saturated water and water–CO_2 cyclic curing on strength recovery of thermally damaged high-performance concrete with different silica contents. *Cement and Concrete Research*, *164*, 107050. Available from https://doi.org/10.1016/j.cemconres.2022.107050.

Li, Z., Yoon, J., Zhang, R., Rajabipour, F., Srubar, W. V., Dabo, I., & Radlińska, A. (2022). Machine learning in concrete science: Applications, challenges, and best practices. *npj Computational Materials*, *8*(1). Available from https://doi.org/10.1038/s41524-022-00810-x, https://www.nature.com/npjcompumats/.

Marzouki, A., & Lecomte, A. (2017). Durability of cementitious composites mixed with various portland limestone cement-cements. *ACI Materials Journal*, *114*(5), 763–773. Available from https://doi.org/10.14359/51700798, https://www.concrete.org/publications/internationalconcreteabstractsportal.aspx?m = details&ID = 51700798.

Meinshausen, M., Lewis, J., McGlade, C., Gütschow, J., Nicholls, Z., Burdon, R., Cozzi, L., & Hackmann, B. (2022). Realization of Paris Agreement pledges may limit warming just below 2°C. *Nature*, *604*(7905), 304–309. Available from https://doi.org/10.1038/s41586-022-04553-z, http://www.nature.com/nature/index.html.

Miller, S. A., Horvath, A., & Monteiro, P. J. M. (2016). Readily implementable techniques can cut annual CO_2 emissions from the production of concrete by over 20%. *Environmental Research Letters*, *11*(7). Available from https://doi.org/10.1088/1748-9326/11/7/074029, http://iopscience.iop.org/article/10.1088/1748-9326/11/7/074029/pdf.

Monkman, S., Cialdella, R., & Pacheco, J. (2023). Performance, durability, and life cycle impacts of concrete produced with CO_2 as admixture. *ACI Structural Journal*, *120*(1), 53–62. Available from https://doi.org/10.14359/51734732, https://www.concrete.org/publications/internationalconcreteabstractsportal.aspx?m = details&id = 51734732.

Monkman, S., Hanmore, A., & Thomas, M. (2022). Sustainability and durability of concrete produced with CO_2 beneficiated reclaimed water. *Materials and Structures/Materiaux et Constructions*, *55*(7). Available from https://doi.org/10.1617/s11527-022-02012-9, https://link.springer.com/journal/11527.

Monkman, S., Sargam, Y., Naboka, O., & Lothenbach, B. (2022). Early age impacts of CO_2 activation on the tricalcium silicate and cement systems. *Journal of CO_2 Utilization*, *65*. Available from https://doi.org/10.1016/j.jcou.2022.102254, https://doi.org/10.1016/j.jcou.2022.102254.

Monkman, S., & Shao, Y. (2010a). Integration of carbon sequestration into curing process of precast concrete. *Canadian Journal of Civil Engineering*, *37*(2), 302–310. Available from https://doi.org/10.1139/L09-140Canada, http://article.pubs.nrc-cnrc.gc.ca/RPAS/RPViewDoc?_handler_ = HandleInitialGet&calyLang = eng&journal = cjce&volume = 37&articleFile = l09-140.pdf.

Monkman, S., & Shao, Y. (2010b). Carbonation curing of slag-cement concrete for binding CO_2 and improving performance. *Journal of Materials in Civil Engineering*, *22*(4), 296–304. Available from https://doi.org/10.1061/(ASCE)MT.1943-5533.0000018.

Naseri, H., Hosseini, P., Jahanbakhsh, H., Hosseini, P., & Gandomi, A. H. (2023). A novel evolutionary learning to prepare sustainable concrete mixtures with supplementary cementitious materials. *Environment, Development and Sustainability*, *25*(7), 5831–5865. Available from https://doi.org/10.1007/s10668-022-02283-w, https://www.springer.com/journal/10668.

Naseri, H., Jahanbakhsh, H., Hosseini, P., & Moghadas Nejad, F. (2020). Designing sustainable concrete mixture by developing a new machine learning technique. *Journal of Cleaner Production*, *258*, 120578. Available from https://doi.org/10.1016/j.jclepro.2020.120578.

Ostovari, H., Müller, L., Skocek, J., & Bardow, A. (2021). From unavoidable CO_2 source to CO_2 sink? A cement industry based on CO_2 mineralization. *Environmental Science and Technology*, *55*(8), 5212–5223. Available from https://doi.org/10.1021/acs.est.0c07599, http://pubs.acs.org/journal/esthag.

Papadaki, D., Kiriakidis, G., & Tsoutsos, T. (2018). *Applications of nanotechnology in construction industry. Fundamentals of nanoparticles: Classifications, synthesis methods, properties and characterization* (pp. 343–370). Greece: Elsevier. Available from http://www.sciencedirect.com/science/book/9780323512558, 10.1016/B978-0-323-51255-8.00011-2.

Peng, L., Shen, P., Poon, C.-S., Zhao, Y., & Wang, F. (2023). Development of carbon capture coating to improve the durability of concrete structures. *Cement and Concrete Research*, *168*, 107154. Available from https://doi.org/10.1016/j.cemconres.2023.107154.

Petek Gursel, A., Masanet, E., Horvath, A., & Stadel, A. (2014). Life-cycle inventory analysis of concrete production: A critical review. *Cement and Concrete Composites*, *51*, 38–48. Available from https://doi.org/10.1016/j.cemconcomp.2014.03.005, http://www.sciencedirect.com/science/journal/09589465.

Pourjavadi, A., Fakoorpoor, S. M., Hosseini, P., & Khaloo, A. (2013). Interactions between superabsorbent polymers and cement-based composites incorporating colloidal silica nanoparticles. *Cement and Concrete Composites*, *37*(1), 196–204. Available from https://doi.org/10.1016/j.cemconcomp.2012.10.005.

Ramezanianpour, A. A., Ghiasvand, E., Nickseresht, I., Mahdikhani, M., & Moodi, F. (2009). Influence of various amounts of limestone powder on performance of portland limestone cement concretes. *Cement and Concrete Composites*, *31*(10), 715–720. Available from https://doi.org/10.1016/j.cemconcomp.2009.08.003.

Rehman, F., Khokhar, S. A., & Khushnood, R. A. (2022). ANN based predictive mimicker for mechanical and rheological properties of eco-friendly geopolymer concrete. *Case Studies in Construction Materials, 17*. Available from https://doi.org/10.1016/j.cscm.2022.e01536, http://www.journals.elsevier.com/case-studies-in-construction-materials/.

Ren, P., Ling, T.-C., & Mo, K. H. (2022). CO_2 pretreatment of municipal solid waste incineration fly ash and its feasible use as supplementary cementitious material. *Journal of Hazardous Materials, 424*, 127457. Available from https://doi.org/10.1016/j.jhazmat.2021.127457.

Rezaei, F., Bulle, C., & Lesage, P. (2019). Integrating building information modeling and life cycle assessment in the early and detailed building design stages. *Building and Environment, 153*, 158–167. Available from https://doi.org/10.1016/j.buildenv.2019.01.034, http://www.elsevier.com/inca/publications/store/2/9/6/index.htt.

Roes, A. L., Tabak, L. B., Shen, L., Nieuwlaar, E., & Patel, M. K. (2010). Influence of using nanoobjects as filler on functionality-based energy use of nanocomposites. *Journal of Nanoparticle Research, 12*(6), 2011–2028. Available from https://doi.org/10.1007/s11051-009-9819-3.

Roopa, A. K., Hunashyal, A. M., Venkaraddiyavar, P., & Ganachari, S. V. (2020). Smart hybrid nano composite concrete embedded sensors for structural health monitoring. *Materials Today: Proceedings, 27*, 603–609. Available from https://doi.org/10.1016/j.matpr.2019.12.071, https://www.sciencedirect.com/journal/materials-today-proceedings.

Ruan, C., Lin, J., Chen, S., Sagoe-Crentsil, K., & Duan, W. (2021). Effect of graphene oxide on the pore structure of cement paste: Implications for performance enhancement. *ACS Applied Nano Materials, 4*(10), 10623–10633. Available from https://doi.org/10.1021/acsanm.1c02090, https://pubs.acs.org/journal/aanmf6.

Sahu, A., Chakraborty, S., & Dey, T. (2022). Performance evaluation of sustainable recycled aggregate concrete with colloidal nano-silica. *European Journal of Environmental and Civil Engineering, 26*(15), 7878–7898. Available from https://doi.org/10.1080/19648189.2021.2012263, http://www.tandfonline.com/loi/tece20.

Said, A. M., Islam, M. S., Zeidan, M. S., & Mahgoub, M. (2021). Effect of nano-silica on the properties of concrete and its interaction with slag. *Transportation Research Record, 2675*. Available from https://doi.org/10.1177/0361198120943196, http://journals.sagepub.com/home/trr.

Sanchez, F., & Sobolev, K. (2010). Nanotechnology in concrete - A review. *Construction and Building Materials, 24*(11), 2060–2071. Available from https://doi.org/10.1016/j.conbuildmat.2010.03.014.

Shen, P., Zhang, Y., Jiang, Y., Zhan, B., Lu, J., Zhang, S., Xuan, D., & Poon, C. S. (2022). Phase assemblance evolution during wet carbonation of recycled concrete fines. *Cement and Concrete Research, 154*, 106733. Available from https://doi.org/10.1016/j.cemconres.2022.106733.

Shen, W., Zhang, C., Li, Q., Zhang, W., Cao, L., & Ye, J. (2015). Preparation of titanium dioxide nano particle modified photocatalytic self-cleaning concrete. *Journal of Cleaner Production, 87*(1), 762–765. Available from https://doi.org/10.1016/j.jclepro.2014.09.014, https://www.journals.elsevier.com/journal-of-cleaner-production.

Shi, T., Liu, Y., Hu, Z., Cen, M., Zeng, C., Xu, J., & Zhao, Z. (2022). Deformation performance and fracture toughness of carbon nanofiber-modified cement-based materials. *ACI Materials Journal, 119*(5), 119–128. Available from https://doi.org/10.14359/51735976, https://www.concrete.org/publications/internationalconcreteabstractsportal.aspx?m = details&ID = 51735976.

Shi, C., Li, Y., Zhang, J., Li, W., Chong, L., & Xie, Z. (2016). Performance enhancement of recycled concrete aggregate - A review. *Journal of Cleaner Production, 112*, 466–472. Available from https://doi.org/10.1016/j.jclepro.2015.08.057.

Shobeiri, V., Bennett, B., Xie, T., & Visintin, P. (2023). Mix design optimization of concrete containing fly ash and slag for global warming potential and cost reduction. *Case Studies in Construction Materials, 18*, e01832. Available from https://doi.org/10.1016/j.cscm.2023.e01832.

Skevi, L., Baki, V. A., Feng, Y., Valderrabano, M., & Ke, X. (2022). Biomass bottom ash as supplementary cementitious material: The effect of mechanochemical pre-treatment and mineral carbonation. *Materials, 15*(23), 19961944. Available from https://doi.org/10.3390/ma15238357, http://www.mdpi.com/journal/materials.

Snellings, R., Suraneni, P., & Skibsted, J. (2023). Future and emerging supplementary cementitious materials. *Cement and Concrete Research, 171*, 107199. Available from https://doi.org/10.1016/j.cemconres.2023.107199.

Snellings, R. (2016). Assessing, understanding and unlocking supplementary cementitious materials. *RILEM Technical Letters, 1*, 50–55. Available from https://doi.org/10.21809/rilemtechlett.2016.12, http://letters.rilem.net/index.php/rilem/article/download/12/15.

Standard Specification for Blended Hydraulic Cements (n.d.). ASTM C595/C595M – 21. ASTM International, ASTM International.

Sujitha, V. S., Ramesh, B., & Xavier, J. R. (2023). Effects of silane-functionalized nanocomposites in superabsorbent polymer and its reinforcing effects in cementitious materials. *Polymer Bulletin, 14362449*. Available from https://doi.org/10.1007/s00289-023-04888-1, https://www.springer.com/journal/289.

Su, N., Miao, B., & Liu, F. S. (2002). Effect of wash water and underground water on properties of concrete. *Cement and Concrete Research, 32*(5), 777–782. Available from https://doi.org/10.1016/S0008-8846(01)00762-1.

Swilling, M., Hajer, M., Baynes, T., Bergesen, J., Labbé, F., Musango, J.K., Ramaswami, A., Robinson, B., Salat., S., Suh, S., Currie, P., Fang, A., Hanson, A., Kruit, K., Reiner, M., Smit, S., & Tabory, S. (2018). United Nations Environment Programme Nairobi, Kenya Unpublished content. The Weight of Cities: Resource Requirements of Future Urbanization.

Thai, H. T. (2022). Machine learning for structural engineering: A state-of-the-art review. *Structures, 38*, 448–491. Available from https://doi.org/10.1016/j.istruc.2022.02.003, https://www.journals.elsevier.com/structures.

The Intergovernmental Panel on Climate Change (IPCC). (2018). United Nations Unpublished content. Global Warming of 1.5°C. https://www.ipcc.ch/sr15/.

Varghese, L., Rao, V. V. L. K., & Parameswaran, L. (2019). Study on high-volume unprocessed fly ash concrete with colloidal nanosilica under chloride exposure. *ACI Materials Journal, 116*(4), 61–68. Available from https://doi.org/10.14359/51716711, https://www.concrete.org/publications/internationalconcreteabstractsportal.aspx?m = details&ID = 51716711.

Wang, Z., Gauvin, F., Feng, P., Brouwers, H. J. H., & Yu, Q. (2020). Self-cleaning and air purification performance of portland cement paste with low dosages of nanodispersed TiO_2 coatings. *Construction and Building Materials, 263*, 120558. Available from https://doi.org/10.1016/j.conbuildmat.2020.120558.

Wang, Y., Lu, B., Hu, X., Liu, J., Zhang, Z., Pan, X., Xie, Z., Chang, J., Zhang, T., Nehdi, M. L., & Shi, C. (2021). Effect of CO_2 surface treatment on penetrability and microstructure of cement-fly ash–slag ternary concrete. *Cement and Concrete Composites, 123*, 104194. Available from https://doi.org/10.1016/j.cemconcomp.2021.104194.

Wang, D., Xiao, J., Sun, B., Zhang, S., & Sun Poon, C. (2023). Mechanical properties of 3D printed mortar cured by CO_2. *Cement and Concrete Composites*, *139*, 105009. Available from https://doi.org/10.1016/j.cemconcomp.2023.105009.

Winnefeld, F., Leemann, A., German, A., & Lothenbach, B. (2022). CO_2 storage in cement and concrete by mineral carbonation. *Current Opinion in Green and Sustainable Chemistry*, *38*, 100672. Available from https://doi.org/10.1016/j.cogsc.2022.100672.

Wu, P., Liu, A., Fu, J., Ye, X., & Zhao, Y. (2022). Autonomous surface crack identification of concrete structures based on an improved one-stage object detection algorithm. *Engineering Structures*, *272*, 114962. Available from https://doi.org/10.1016/j.engstruct.2022.114962.

Younis, A., Ebead, U., Suraneni, P., & Nanni, A. (2018). Fresh and hardened properties of seawater-mixed concrete. *Construction and Building Materials*, *190*, 276–286. Available from https://doi.org/10.1016/j.conbuildmat.2018.09.126.

Zhang, B., Ji, T., Ma, Y., Yang, Y., & Hu, Y. (2022). Workability and strength of ultra-high-performance concrete containing nano-$Ca(OH)_2$-modified recycled fine aggregate. *ACI Materials Journal*, *119*(5), 165–174. Available from https://doi.org/10.14359/51735953, https://www.concrete.org/publications/internationalconcreteabstractsportal.aspx?m = details&ID = 51735953.

Zunino, F., Martirena, F., & Scrivener, K. (2021). Limestone calcined clay cements (lc3). *ACI Materials Journal*, *118*(3), 49–60. Available from https://doi.org/10.14359/51730422, https://www.concrete.org/publications/internationalconcreteabstractsportal.aspx?m = details&id = 51730422.

CO_2 capture and storage for sustainable concrete production

22

Musab Alhawat[1,2], *Ashraf Ashour*[1] *and Gurkan Yildirim*[1,3]
[1]Faculty of Engineering and Digital Technologies, University of Bradford, Bradford, United Kingdom, [2]Departement of Civil Engineering, Elmergib University, Al Khums, Libya, [3]Department of Civil Engineering, Hacettepe University, Ankara, Turkey

22.1 Introduction

The global construction industry faces significant challenges in reaching sustainability goals due to its enormous influence on resource consumption and greenhouse gas emissions. Concrete is one of the most widely used construction materials, contributing significantly to greenhouse gas emissions, primarily through cement manufacturing. In recent years, there has been an increasing focus on sustainable construction practices, with a particular emphasis on lowering the carbon footprint of building materials (Islam et al., 2019). One viable way to achieve this aim is to use recycled aggregate (RA) coupled with the innovative technique of rapid carbonation.

Recycled aggregate concrete provides a sustainable alternative to typical building materials by decreasing the consumption of virgin resources and the energy required during production, and reducing waste generation. RA is obtained from the crushing and processing of construction and demolition waste (CDW), such as bricks, tiles, and concrete (Tam et al., 2020). However, there is a restricted usage of CDW in concrete industry, which is mainly associated with the presence of old adhered mortar and old interfacial transition zones (ITZs), making RA more absorbing to water (2.3–4.6 times), and more crushable than conventional aggregates (33%–45%), in addition to the issues related to durability (Nayana & Kavitha, 2017; Sultan et al., 2022). These concerns may be considerably alleviated by applying appropriate techniques, which can improve the quality of CDW.

Although various methods have been proposed to enhance the properties of RA (e.g., mechanical grinding, ultrasonic cleaning, polymer emulsion, and nanomaterials), carbon dioxide capture is considered one of the most effective techniques that can improve the quality of aggregate, while simultaneously reducing the amount of greenhouse gases from the atmosphere (Araba et al., 2021; Kaliyavaradhan & Ling, 2017). Accelerated carbonation has demonstrated a high potential to improve the characteristics of RA and concrete by lowering the porosity and water absorption of recycled materials through the filling influence of calcium carbonate, while also sequestering carbon dioxide (CO_2) from the environment (Li et al., 2020). The capture and storage of carbon dioxide are employed to accelerate the carbonation reactions between CO_2

Sustainable Concrete Materials and Structures. DOI: https://doi.org/10.1016/B978-0-443-15672-4.00022-X

and calcium-bearing components. The procedure includes exposing concrete buildings to a high concentration of CO_2 underregulated settings (Bosque et al., 2020/2020) which causes a chemical reaction between the gas and the calcium hydroxide (CH) found in concrete (Bosque et al., 2020/2020; Sultan et al., 2022). This reaction produces calcium carbonate ($CaCO_3$), a mineral molecule that successfully binds CO_2 and permanently stores it inside the concrete matrix (Wang et al., 2020). Therefore the success of applying this technique would enable RA to replace conventional aggregate in a more environmentally friendly way.

The concept of carbonation in concrete is not totally novel, as natural carbonation develops gradually over time. However, the traditional carbonation process is very sluggish and may take several decades to attain its full potential (Zhan et al., 2016). Accelerated carbonation, on the other hand, speeds up the process by varying factors such as CO_2 concentration, temperature, humidity, and exposure duration. This technology allows for the attainment of optimum carbonation levels in a matter of weeks or months, greatly decreasing the overall carbon footprint of concrete structures (Tam et al., 2020). Therefore it is feasible to gain large environmental benefits and improve the overall performance of concrete structures by integrating recycled materials into the concrete mix and exposing RA and concrete to rapid carbonation. As research and development in this area continue to advance, it is believed that the capture and storage of carbon dioxide is anticipated to gain further recognition and adoption for generating sustainable built environments.

This chapter provides a comprehensive review of the latest research related to the accelerated carbonation technique applied for producing green concrete, including the most recent knowledge about the reaction processes and the affecting parameters on accelerated carbonation. Moreover, the characteristics of carbonated recycled aggregates were discussed, and the mechanical and durability performance of carbonated concrete were also analyzed. The environmental and economic viability of this technology were also discussed along with identifying industrial challenges and research gaps for future work. This chapter could be regarded as a cutting-edge review of the use of accelerated carbonation technology in concrete industry.

22.2 Carbon dioxide capture methodologies

Accelerated carbonation has been applied to concrete industry in two distinct ways (1) either coarse or fine recycled aggregates prior to mixing (Cuenca-Moyano et al., 2019; Li et al., 2020; Wang et al., 2020) and (2) concrete prepared with recycled and conventional aggregate (Bosque et al., 2020/2020; Shi et al., 2018; Zhan et al., 2016). In terms of carbonated concrete, CO_2 curing works by permeating CO_2 into concrete through pores, making the microstructure denser, which also reduces permeability, speeds up the curing process, and enhances mechanical characteristics. In the case of carbonated aggregate, applying CO_2 curing can decrease water absorption, porosity, and crushing value of aggregate. Despite the similarity of carbonation mechanisms in the two processes (recycled aggregate and concrete), more challenges arise with applying carbonation to concrete since it is limited to precast elements. In contrast, concrete can be produced in situ by using carbonated recycled aggregate (Bosque et al., 2020/2020; Kaliyavaradhan & Ling, 2017). Moreover, carbonating concrete requires reaching CO_2 through the whole concrete mass through pores, which seems hard to achieve. Thus a special mix design might be needed to allow carbonation of the entire concrete. In reinforced concrete, corrosion might develop due to carbonation when the pH of carbonated concrete decreases (Ekolu, 2016; Shi et al., 2018; Xuan et al., 2018). This issue might be avoided in the case of carbonating recycled aggregate prior to mixing. On the other hand, carbonating concrete is the ideal choice for shortening setting time and enhancing early strength (Tam et al., 2020). In addition, the carbonation of concrete seems more beneficial in terms of sequestering more carbon dioxide, compared to the carbonation of RA under the same carbonation conditions, as stated in a recent study (Tang et al., 2023). The carbon capture of industrial byproducts like ground granulated blast furnace slag (GGBS) and fly ash has also exhibited a high potential for reducing carbon dioxide emissions and enhancing the performance of these materials. GGBS and fly ash are waste products generated in the steel and power industries, respectively, and are rich in calcium, magnesium, and silicon compounds. Efforts have been made to capture CO_2 from these materials by exposing them to various carbonation techniques. Accelerated carbonation and wet carbonation are two commonly employed methods. Accelerated carbonation involves subjecting GGBS and fly ash to high-pressure CO_2 to form stable carbonates like calcium and magnesium carbonates, effectively sequestering CO_2 within the materials (Wang et al., 2024). Wet carbonation, on the other hand, involves a reaction with CO_2 in a water-based environment to form carbonates (Valencia-Saavedra et al., 2022). Recent research has shown promising results in terms of CO_2 capture efficiency and the transformation of waste materials into valuable resources. Carbonated GGBS and fly ash exhibit enhanced mechanical properties, reduced leachability of harmful elements, and increased chemical stability (Kurda et al., 2019; Li, Farzadnia, et al., 2017). Moreover, the carbonation process may enhance the pozzolanic reactivity of these materials, making them even more attractive for use in cementitious applications (Suescum-Morales et al., 2022). In the case of carbonating aggregates, CO_2 interacts with the old mortar adhered to RA surface to improve the old paste and ITZ of produced concrete, while CO_2 react with old and new mortar in the case of carbonation after mixing to densify cement paste, enhancing the old paste as well as old and new ITZ, as presented in Fig. 22.1.

22.3 Existing accelerated carbonation methods

Carbonation is a very slow process in nature, and the concentration of CO_2 in the atmosphere is about 0.03% (Li et al., 2023). Thus different approaches have been developed to accelerate the carbonation process of RA and concrete, as shown in Fig. 22.2.

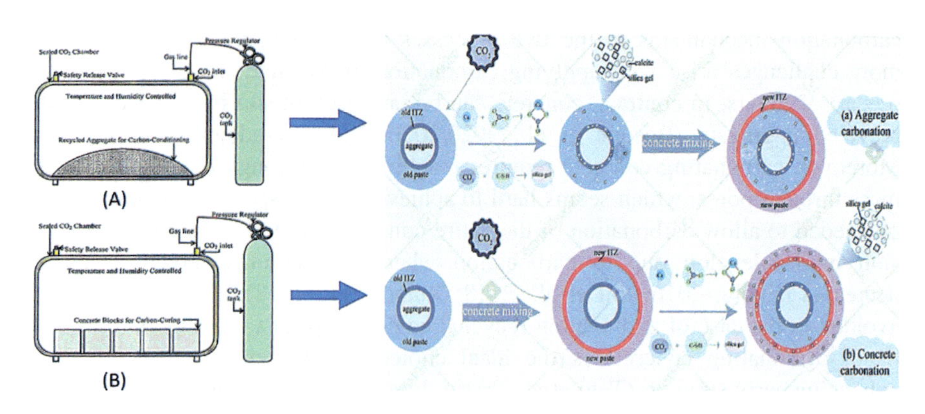

Figure 22.1 Carbonation mechanism of (A) recycled aggregates and (B) concrete blocks (Tang et al., 2023).
Source: From Tang, B., Fan, M., Yang, Z., Sun, Y., & Yuan, L. (2023). A comparison study of aggregate carbonation and concrete carbonation for the enhancement of recycled aggregate pervious concrete. *Construction and Building Materials, 371*. https://doi.org/10.1016/j.conbuildmat.2023.130797.

Standard carbonation is a common method applied in concrete research investigations, which is based on the Chinese standard GB50082–2009. According to the standard, the following conditions are recommended; CO_2 concentration (20% $\pm$ 3%), relative humidity (RH) (70% $\pm$ 5%), and temperature (20°C $\pm$ 2°C). However, long exposure times (over 1 week) might be needed to enhance the relatively low efficiency of this approach. The second approach applied is pressurized carbonation, in which RH (50% $\pm$ 5%) and temperature (25°C $\pm$ 2°C), are suggested, while gas pressure can be ranged from 0.1 to 4 atm (Kou et al., 2014; Monkman & Shao, 2010). In this approach, the pressure inside the carbonation chamber is vacuumed until it reaches about −0.5 bar, and then aggregate is placed inside the chamber and a certain level of CO_2 pressure (0.1–5.0 bar) is applied. The third approach is called "flow-through CO_2 curing," where a mixture of air and CO_2 is applied inside a carbonation chamber from one side and expelled from the other, with a suggested optimum of temperature, CO_2 content, CO_2 flow rate, and RH of 23°C, 10%, 5 L/min and 50% $\pm$ 5%, respectively (Fang et al., 2021; Suescum-Morales et al., 2021), exhibiting rapid carbonation efficiency. The concept of the third approach "water-CO_2 cooperative curing approach" is based on injecting a combination of CO_2, N_2, and air into the water placed inside the carbonation chamber. The technology of accelerated carbonation through aqueous environments seems more successful in increasing the rate of carbonation because the reactions between the mixture of CO_2 and water, and hydrated products can be rapidly facilitated (Li et al., 2019; Zhang et al., 2022). However, the efficiency of the accelerated carbonation approach has been mainly proven with fine recycled particles (Fang et al., 2021; Liu et al., 2021). Some pretreatment steps have also been proposed before placing RA in the chamber, which may effectively promote the carbonation process such as immersing RA in calcium nitrate water or limewater solution and spraying reclaimed wastewater (Fang et al., 2020; Zhan et al., 2018).

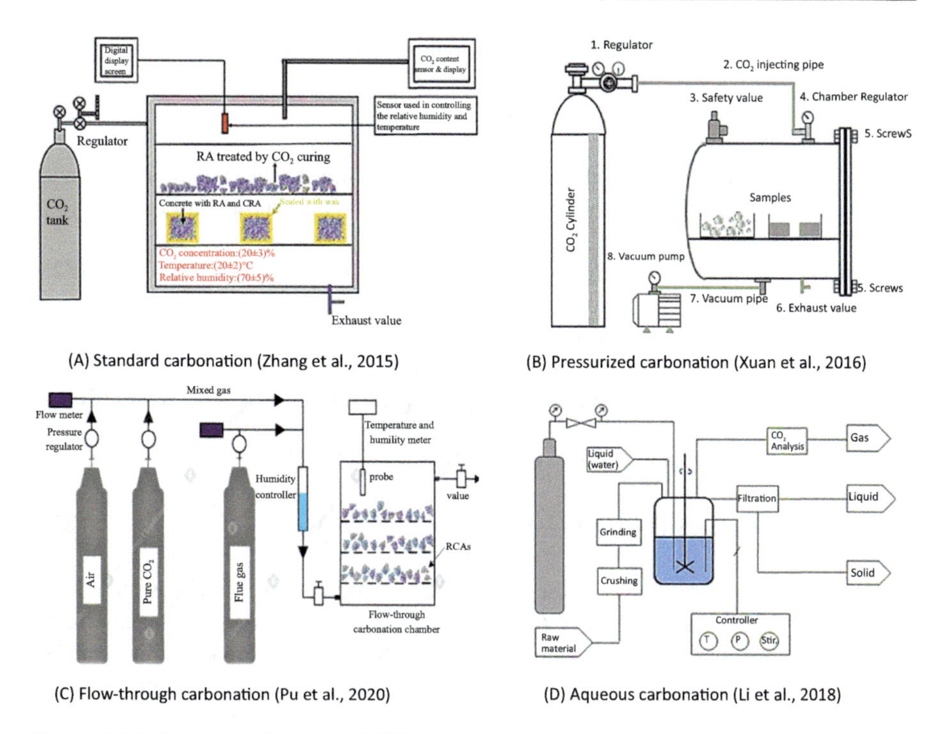

(A) Standard carbonation (Zhang et al., 2015)

(B) Pressurized carbonation (Xuan et al., 2016)

(C) Flow-through carbonation (Pu et al., 2020)

(D) Aqueous carbonation (Li et al., 2018)

Figure 22.2 Schematic diagram of different existing carbonation methods.

22.4 Mechanisms of accelerated carbonation process

Carbonation is an extremely slow chemical process, naturally occurring in concrete exposed to carbon dioxide from the atmosphere through the reaction with the main hydration products [e.g., calcium hydroxide $Ca(OH)_2$ and calcium silicate hydrate (C-S-H)], affecting mostly the areas around the surface of concrete. Therefore to accelerate this process, all surrounding conditions need to be enhanced and optimized (Kou et al., 2014). It is important to note that carbonation mechanisms can be highly influenced by the carbonation approach used and conditions/pretreatments applied as well as recycled material characteristics. In the case of carbonating concrete, the phenomenon occurs when carbon dioxide gas penetrates concrete through microcracks and pores, and then dissolves in pore water, leading to the formation of carbonic oxide. The latter reacts with calcium ions that are dissolved from calcium hydroxide and calcium silicate hydrate to eventually generate calcium carbonate crystal ($CaCO_3$), as described in the equations below (Liang et al., 2019; Zhan et al., 2016).

$$CO_2 + H_2O \rightarrow H_2CO_3 \tag{22.1}$$

$$H_2CO_3 + Ca(OH)_2 \rightarrow CaCO_3 + 2H_2O \tag{22.2}$$

Carbonation of C-S-H occurs when the majority of calcium hydroxide is depleted.

$$CO_2 + \text{C-S-H} \rightarrow CaCO_3 + SiO_2 \cdot H_2O \quad (22.3)$$

Carbon dioxide may also interact with calcium aluminate hydrate to produce additional calcium carbonate (Kaliyavaradhan & Ling, 2017), as shown in Eq. (22.4).

$$4CO_2 + 4CaO \cdot Al_2O_3 \cdot 13H_2O \rightarrow 4CaCO_3 + 10H_2O + 2Al(OH)_3 \quad (22.4)$$

In most cases, carbonation takes place before the completion of hydration process, and thus, unhydrated cement products like belite (C_2S) and alite (C_3S) may also be involved in the carbonation process, as presented in Eqs. (22.5) and (22.6).

$$2CO_2 + 2CaO \cdot SiO_2 + H_2O \rightarrow 2CaCO_3 + SiO_2 \cdot H_2O \quad (22.5)$$

$$3CO_2 + 3CaO \cdot SiO + H_2O \rightarrow SiO_2 \cdot H_2O + 3CaCO_3 \quad (22.6)$$

The carbonation reactions in the case of RA are quite similar to those that occur in concrete to eventually leading to the formation of calcium carbonate due to the reactions between carbon dioxide and calcium hydroxide and calcium silicate hydrate found on the old mortar-attached aggregate. However, silica might be added to the atmosphere to contribute to forming additional calcium silicate hydrate, which, in turn, can react with CO_2, in addition to releasing extra silica, as illustrated in Eq. (22.7).

$$3Ca(OH)_2 + 2SiO_2 + 2CO_2 \rightarrow Ca_3SiO \cdot 5CO_3 + 3H_2O \quad (22.7)$$

Calcium carbonate, which is the main product formed during this process, eventually participates in the cracks and pores of concrete as calcite, vragonite, and vaterite, resulting in filing up the pores in the microstructure, and thus enhancing concrete permeability, making it more durable (Nayana & Kavitha, 2017; Pu, Li, Wang, Shi, Fu, et al., 2021).

22.5 Parameters affecting carbonation capture performance

The accelerated carbonation process of RA and concrete can be highly affected by a number of parameters, including carbonation conditions such as gas pressure applied, CO_2 concentration, RH, temperature and exposure period, in addition to those related to the characteristics of the original recycled materials such as particle size, water content, and old mortar. Therefore controlling and optimizing these factors can significantly assist in enhancing the performance of the carbonated product.

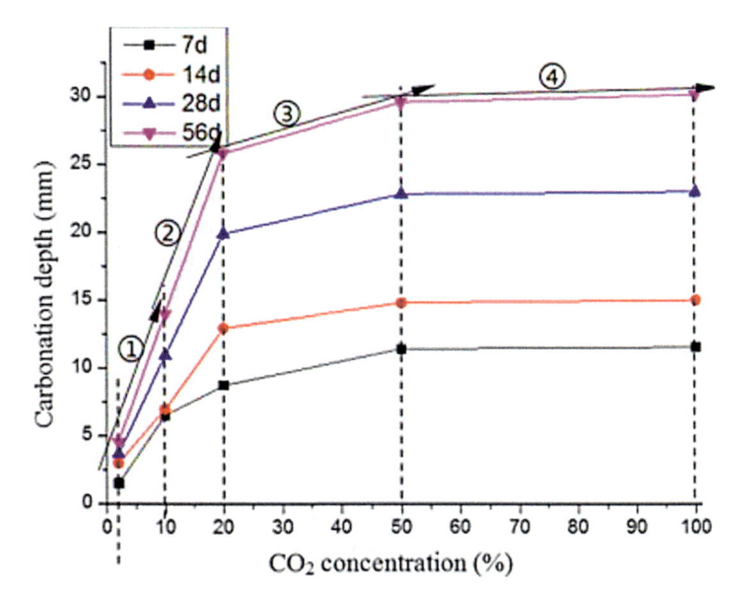

Figure 22.3 CO_2 concentration effect on concrete the carbonation depth (Cui et al., 2015). *Source:* From Cui, H., Tang, W., Liu, W., Dong, Z., & Xing, F. (2015). Experimental study on effects of CO_2 concentrations on concrete carbonation and diffusion mechanisms. *Construction and Building Materials, 93*, 522–527. https://doi.org/10.1016/j.conbuildmat.2015.06.007.

22.5.1 Carbon dioxide concentration

The efficiency of carbonation process can be highly enhanced by increasing the concentration applied up to a certain limit (Fang et al., 2020), and a limited increase to a steady level or even negative efficiency might then occur (Zhang et al., 2020). Higher CO_2 concentrations could speed up carbonation reactions and fill up the pore solution with more stable carbon dioxide content, especially in early periods. Different optimum concentration values have been suggested in the literature. For instance, Cui et al. (2015) investigated the carbonation speed under various CO_2 concentrations (i.e., 2%, 10%, 20%, 50%, and 100%) and for different curing periods (i.e., 7, 14, 28, and 56 days). The findings presented in Fig. 22.3 indicate that the fastest carbonation rate occurs with increasing CO_2 concentration from 2% to 20%, whereas limited progress takes place when the concentration applied is between 20% and 50%, and almost no diffusion improvement once the concentration exceeds 50% CO_2. A similar conclusion was drawn in some other studies (Fang et al., 2020; Kashef-Haghighi et al., 2015; Pu, Li, Wang, Shi, Fu, et al., 2021), indicating the carbonation diffusion rate is dramatically increased during the first day with increasing CO_2 concertation up to about 20%, leading to a significant decrease in porosity, while minor changes might be observed with higher levels. Fast carbonation rates were observed even with higher CO_2

concentrations (70%–73%) in different studies conducted on recycled fine and coarse aggregate (Pan et al., 2017; Zhang et al., 2022). However, the excessive levels of CO_2 could also result in the formation of undesirable carbonates (e.g., magnesium and iron carbonates), resulting in a high decalcification of C-S-H gel, and increasing the possibility of carbonation shrinkage (Liu et al., 2019).

The above results show that the optimum CO_2 concentration for carbonation capture is not conclusive, possibly due to the number of parameters controlling the process, including curing circumstances as well as recycled material properties. Therefore to optimize CO_2 concentrations and maximize the rapid carbonation, it is critical to consider the relevant circumstances for each application accordingly.

22.5.2 Gas pressure

The carbonation efficiency is strongly influenced by the pressure of CO_2 gas, and its influence on the rapid carbonation of concrete and RA has been evaluated in several studies (Fang et al., 2021; Liu et al., 2021; Xuan et al., 2018). Higher gas pressure can enhance the carbonation process due to the quick permeating and dissolving of carbon dioxide, resulting in faster reaction rates, however, such a relationship is not necessarily to be linear (Zhan et al., 2016). For instance, the water absorption measured by Xuan et al. (2016) clearly dropped as CO_2 pressure climbed from 0 to 0.5 MPa, but higher pressures led to higher absorptivity. Similarly, it is noted that the water absorption of recycled coarse aggregate (RCA) obviously decreased by applying 0.1 MPa of CO_2 pressure, while less effect was observed with higher gas pressure (0.5 MPa) (Fang et al., 2020). Bukowski and Berger (1979) reported the highest concrete compressive strength was achieved at a pressure of 0.31 MPa, while applying higher pressure (5.6 MPa) led to a slight strength decrease. The presence of increased pressures could potentially account for the promotion of additional calcium carbonate precipitation, impeding the penetration of carbon dioxide. Consequently, the inner region of RCA might not undergo complete carbonation (Pu, Li, Wang, Shi, Fu, et al., 2021). Applying excessive gas pressure may cause an adverse impact on RCA quality such as the formation of cracks, as reported by Pan et al. (2017), who found the water absorption of RCA can reduce up to 58% at 0.5 MPa gas pressure, while higher pressure levels (1.0 and 1.5 MPa) led to lowering the absorption decrease rate to 46% and 36%, respectively. Furthermore, large and loosely crystal calcium carbonates are typically generated under high pressure, while those created under low pressure are mostly tiny and firmly linked to each other (Pu, Li, Wang, Shi, Fu, et al., 2021; Tang et al., 2023). It can be said that the ideal gas pressure can be in the range of 0.1–1.0 MPa, depending on the material type and other conditions applied.

22.5.3 Temperature

Temperature has an obvious impact on the efficiency of carbon capture, and carbonation occurs faster with higher temperatures (Liu et al., 2019). This is due to the fact that higher temperatures hasten chemical reactions by participating in the diffusivity of carbon dioxide, leaching calcium ions, and the formation of metastable carbonate calcium (Drouet et al., 2019). It is noted that the water absorption of carbonated aggregates reduced from 2.5% to 1.45% by increasing the temperature from 20°C to 75°C (Zhan et al., 2018). Nevertheless, this effect is restricted by the fact that high temperatures can decrease CO_2 solubility and result in a slower rate of carbonation dissolution accompanied by concrete cracking, which eventually result in the disintegration of carbonates over time. High temperatures can also contribute to leaching Ca^2+ ions from silicates. The data collected from previous studies (Chen et al., 2018; Jae-Dong et al., 1990; Liu et al., 2001; Papadakis et al., 1991; Wang et al., 2019; Xuan et al., 2016), and presented in Fig. 22.4, indicated that the optimal temperature ranges between 20°C and 80°C, while others limited the temperature to be under 30°C (Li, Farzadnia, et al., 2017; Pu, Li, Wang, Shi, Fu, et al., 2021; Wang et al., 2020). The effect of temperature might be varied depending on other related factors.

22.5.4 Relative humidity

The speed of carbonation can also be affected by the RH applied. As the solubility of carbon dioxide can be slowly progressing under low RH, the ingress of CO_2 would be inhabited under high RH. In accordance with the model suggested by Fang et al. (2017), the rate of CO_2 uptake might be enhanced up to about 200% at the ideal RH. According to Wang et al. (2022), the reaction speed is determined by the water content of RA rather than CO_2 diffusion when the RH is less than 50%.

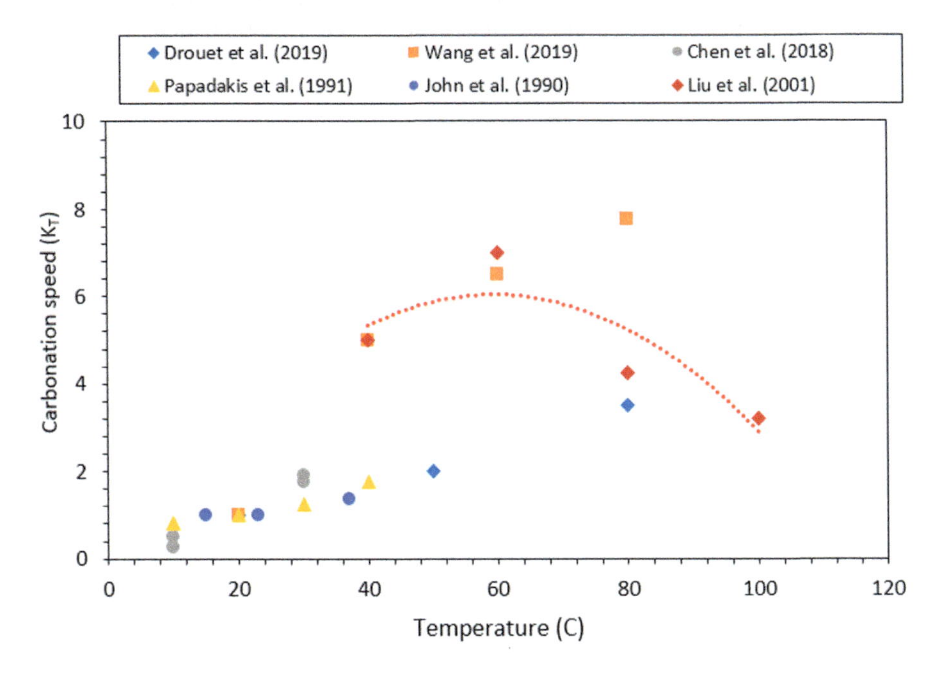

Figure 22.4 Effect of temperature on the efficiency of carbonation process.

The findings from previous studies suggested that the range of 40%–70% RH is the ideal RH for most carbonation cases (Fang et al., 2020; Liang et al., 2019; Zhan et al., 2014; Zhan et al., 2016). Similarly, Elsalamawy et al. (2019) reported that the highest depth of carbonation can be achieved at RH of 65%. Jae-Dong et al. (1990) highly linked the ideal RH with the w/b ratio; a higher w/b requires a higher RH. Based on these previous studies, the best carbonation could be obtained when RH ranges from 50% to 70%, depending on other affected parameters, as shown in Fig. 22.5. However, lower or higher RH levels could be more suitable for other samples since RH is influenced by the samples' pore structure. Porosity is strongly related to the w/b ratio, and the samples with higher w/b and more porous can rapidly be carbonated at higher RH. Jae-Dong et al. (1990) studied the influence of RH on the carbonation depth of mortar using varying w/b ratios, and found that the optimum RH levels decreased from 51% to 45%, and then to 40% when the w/c ratio reduced from 70% to 60%, and 55%, respectively.

22.5.5 Exposure period

The time of CO_2 exposure directly influences the level of carbonation; the capture of carbon typically rises by increasing the period until full carbonation is achieved. However, the relationship between exposure time and carbonation rate is not always linear. In the early phases of carbonation, the reactions are typically fast, and the absorbing rate of CO_2 is relatively high. The reaction rate, then, tends to slow down over time until it eventually becomes insignificant, or even no more carbonation occurs. This is mostly due to the fact that a slower rate of carbonation is produced

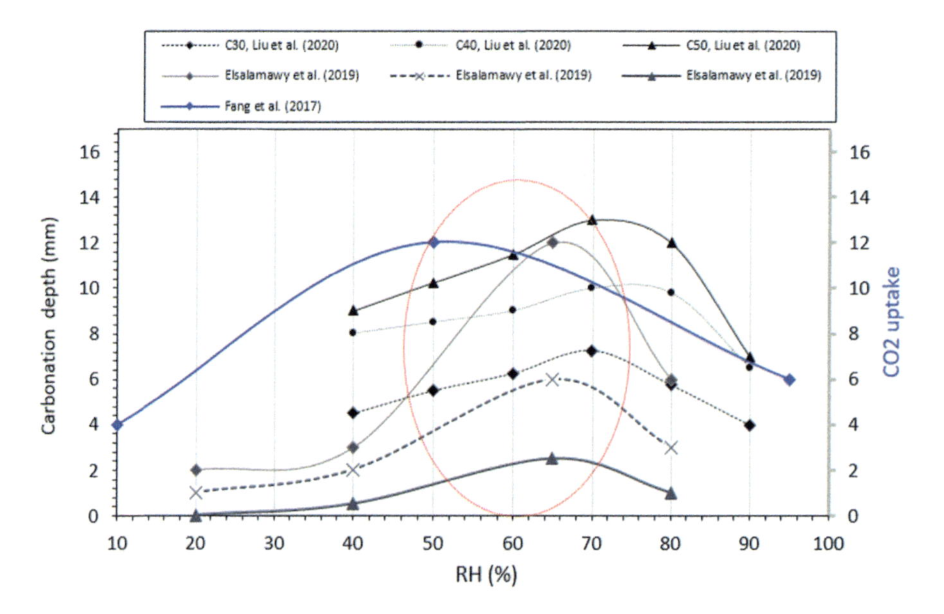

Figure 22.5 Effect of relative humidity on the performance of accelerated carbonation.

as the carbonation reaction advances and CO_2 penetrates deeper, increasing the distance over which CO_2 can disperse and react. The formation and precipitation of calcium carbonate in concrete pores during carbonation process can act as a barrier, hindering further penetration of CO_2. The efficiency of carbonation time is also strongly related to other parameters such as CO_2 concentration, temperature, and RH applied, in addition to the characteristics of original material. According to Pu, Li, Wang, Shi, Luan, et al. (2021), the process of carbonation can be divided into three phases: rapid phase, slow phase, and stable phase. Zhan et al. (2014) stated that the carbonation rate of RA substantially rose within the first 2 hours, while the speed of carbonation enhanced by only 13% after 4 hours. According to Wang et al. (2019), carbon dioxide diffusion mostly occurred in the first 6 hours of carbonation curing, whereas the subsequent exposure time was not highly significant due to the production of calcium carbonate, which blocks CO_2 diffusion. Other studies (Kou et al., 2014; Luo et al., 2018) established a correlation of the efficiency of carbonation to the initial 24-hour period, regardless of the quality of RA. A similar conclusion was made in another study (Luo et al., 2018), highlighting the increase in calcium carbonate content by about 95% and a reduction in calcium hydroxide by approximately 42% within the initial 24 hours.

22.5.6 Recycled aggregate characteristics

The water content of RA has a substantial impact on the carbonation process since CO_2 needs to be dissolved in water solution prior to the occurrence of carbonation. Carbonation would not occur in general without the catalysis of free water, while those that did not contain water content showed negligible carbonation, as confirmed by Wang et al. (2020). Water can contribute to decomposing calcium ions in addition to taking part in diffusing CO_2 into the mineral matrix, increasing the overall reaction rate. A low-water content indicates very dry conditions that are less conducive to the dissolution of carbon dioxide. However, an extremely high-water content might also have a negative impact on the carbonation process. This is because water can act as a barrier to CO_2 diffusion, lowering the amount of CO_2 that can enter the mineral matrix and participate in the process. Furthermore, extra water can dilute the quantity of CO_2, reducing the driving force for the reaction to occur. The water content is directly linked to the content of adhered mortar of RA as well as the RH used.

Due to its effect on the surface area and reactivity of RA, its particle size can also influence the effectiveness of accelerated carbonation. The impact of RA particle size on the acceleration of carbonation has been examined in several previous works. For instance, Wu et al. (2022) studied the influence of particle size on the carbon capture efficiency of RA sourced from concrete waste, and found that smaller aggregate recycled particles (0.5 mm) demonstrated higher carbon capture efficiencies than larger particles. Zhan et al. (2019) found that the carbonation degree for RA with 5–10 mm recorded at 65%, while those for larger sizes (14–20 mm) reached only 37%, suggesting that smaller RCAs have a higher capacity to absorb CO_2. This can be explained by the fact that smaller particles have higher surface areas that would be exposed to CO_2, compared to those with larger sizes (Bai et al., 2020). In addition, smaller RA particles contain higher amounts of attached mortar, which have high calcium contents and a higher capacity to absorb oxides like CO_2, as illustrated in Fig. 22.6.

The relationship between recycled fine aggregate particle size and the carbonation kinetics during aqueous carbonation was thoroughly examined in a recent work by Jiang et al. (2022). This relationship presented in Fig. 22.7 indicates that carbonation of aggregates with coarse particles (0.6–2.36 mm) was developed mainly by enhancing their surface characteristics through the creation of reactive shells, in addition to densifying their microstructure. These effects were attributed to an initial carbonation process occurring in the bulk solution, and progressed to internal carbonation. On the other hand, the finer particles (less than 0.15 mm) experienced total disintegration and transformed into calcium carbonate with silica crystals, while those between 0.15 and 0.6 mm were partially decomposed and transformed into calcium carbonate (Mehdizadeh et al., 2021). It is important to note that the effect of aggregate size on accelerated carbonation is influenced by various other factors, such as water-to-cement ratio, type of cement used, and curing conditions. These factors interact with aggregate size to determine the overall carbonation rate and depth in concrete.

22.6 Physical characteristics of carbonated aggregates

22.6.1 Water absorption

Water absorption is one of the fundamental characteristics used to assess the quality of aggregates. Typically, RA possesses a higher water absorption capacity (3.0%–7.0%), compared to that found in conventional aggregates (1.0%–2.5%), depending on the attached mortar content and aggregate size (Alhawat & Ashour, 2019). As illustrated in Fig. 22.8, the water absorption of RA could be clearly decreased by 7%–57% after rapid carbonation, based on the characteristics of virgin aggregate as well as the

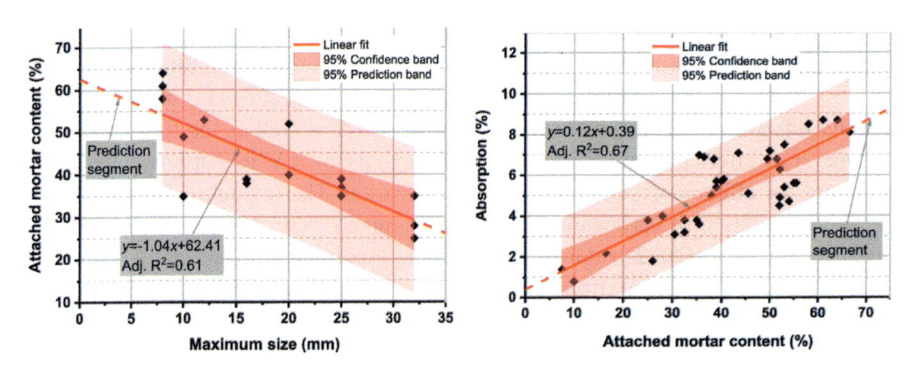

Figure 22.6 Correlation between aggregate size, attached mortar, and absorption. (Bai et al., 2020). *Source:* From Bai, G., Zhu, C., Liu, C., & Liu, B. (2020). An evaluation of the recycled aggregate characteristics and the recycled aggregate concrete mechanical properties. *Construction and Building Materials, 240*, 117978. https://doi.org/10.1016/j.conbuildmat.2019.117978.

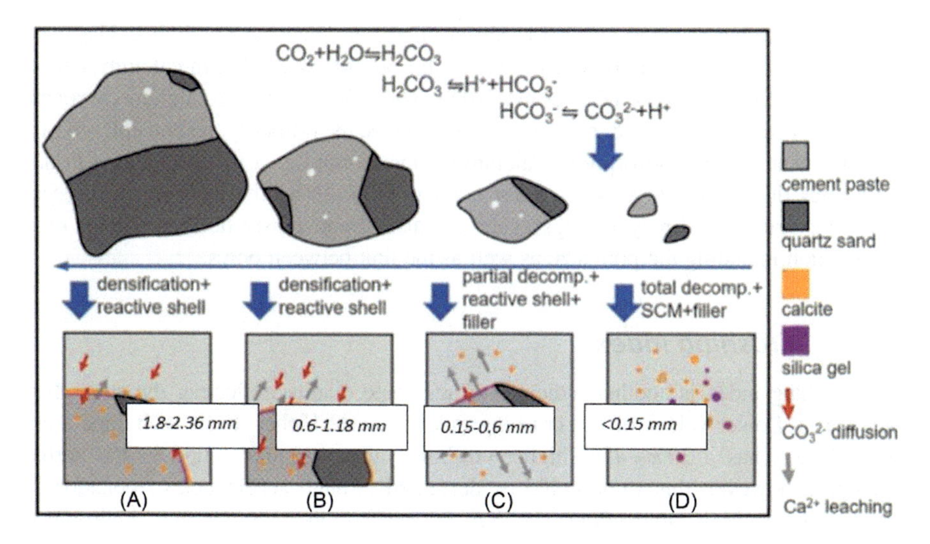

Figure 22.7 Carbonation mechanism of different recycled fine aggregates (Jiang et al., 2022). *Source:* Jiang, Y., Li, L., Lu, J. X., Shen, P., Ling, T. C., & Poon, C. S. (2022). Mechanism of carbonating recycled concrete fines in aqueous environment: The particle size effect. *Cement and Concrete Composites, 133*, 104655.

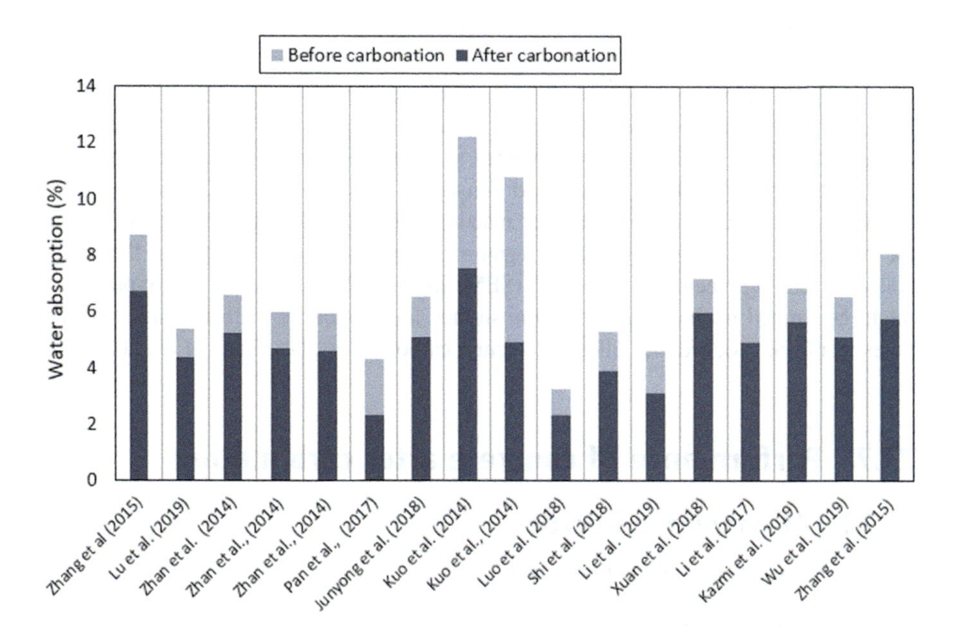

Figure 22.8 Change in the water absorption of recycled aggregate/concrete after accelerated carbonation.

conditions applied (Kou et al., 2014; Shi et al., 2018; Wang et al., 2020). The water absorption rate might decrease even lower in the case of presoaking in calcium hydroxide solution by up to 70% (Pan et al., 2018). The filling effect of calcium carbonate, which reduces porosity and refines pore structures, is primarily responsible for the improvement in water absorption. Calcium carbonate can be easily precipitated in narrow pores due to the increase in water condensation during the carbonation process. The pore structure of recycled aggregate can also be obviously decreased after rapid carbonation including the pore size as well as the link between pores.

22.6.2 Crushing index

The crushing index is another indicator of aggregate quality, reflecting the strength of aggregates. Lower values of the crushing index indicate higher levels of hardness and strength. RA has a 30%–45% higher crushing value than that found in the normal aggregate, however, this value could be decreased through accelerated carbonation by up to about 31%, depending on RA properties and the curing environment (Li, Farzadnia, et al., 2017; Luo et al., 2018; Xuan & Poon, 2018). A further decrease might be obtained (up to 44%) in the case of presoaking treatment (Zhan et al., 2017). The precipitation of crystal calcium carbonate inside aggregate pores through the carbonation process results in minimizing the pore volume of old ITZ and attached mortar.

22.6.3 Apparent density

RA generally has a lower density than normal aggregate, which is mainly attributed to the old mortar adhering to the aggregate surface, resulting in higher porosity (Silva et al., 2019). Most previous studies documented that the density of RA slightly improved (1.0%–7.0%) after being carbonated (Luo et al., 2018; Wang et al., 2020; Zhan et al., 2014), and it might be extended to 12% in some cases (Li, Farzadnia, et al., 2017). The primary cause of increased density is the conversion of some of calcium hydroxide into calcium carbonate after reacting with CO_2. This process leads to filling the pores and cracks in recycled aggregate microstructure, as illustrated in the images collected scaning electron microscope (SEM) in Fig. 22.9. However, this improvement might be limited to the area around the surface layer since the path to reach the inner pores would be blocked and filled by calcium carbonate crystals (Fang et al., 2020).

22.7 Performance of concrete after carbonation

22.7.1 Mechanical properties

22.7.1.1 Compressive strength

The use of fully RA can cause a clear decrease in concrete compressive strength by 10%–30%, owing to the effect of porous old mortar and weak old ITZ (Kazmi et al., 2019). However, this negative impact on the compressive strength

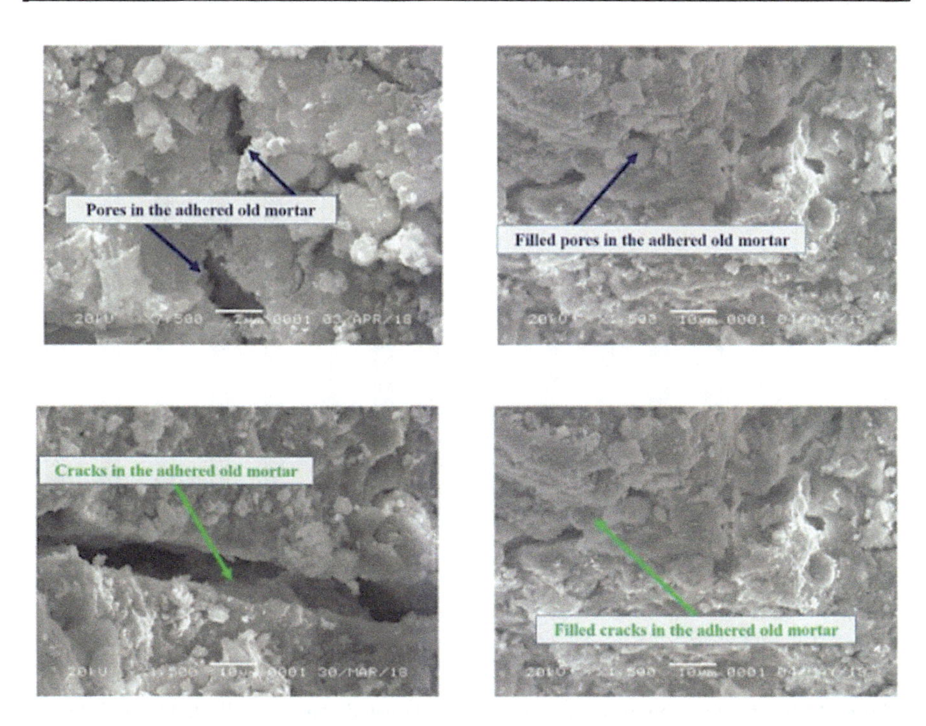

Figure 22.9 SEM images of recycled aggregate before and after carbonation (Liang et al., 2019). *Source:* From Liang, C., Ma, H., Pan, Y., Ma, Z., Duan, Z., & He, Z. (2019). Chloride permeability and the caused steel corrosion in the concrete with carbonated recycled aggregate. *Construction and Building Materials, 218*, 506–518. https://doi.org/10.1016/j.conbuildmat.2019.05.136.

of concrete mixtures can be mitigated through the carbonation treatment of RA, and the improvement rate can be increased with higher contents of carbonated aggregates. For example, Xuan et al. (2018) investigated the influence of the inclusion of RA collected from demolished concrete structures before and after carbonation at various substitution levels (i.e., 20%, 40%, 60%, 80%, and 100%). The findings showed almost a 26% drop in the compressive strength of concrete made with 100% RCA, whereas the use of 30% and 60% led to a negligible reduction. Interestingly, the compressive strength increased by roughly 22.5% when 100% carbonated aggregates were used to be comparable to the normal ones. The increased density and lower crush index of carbonated aggregates over noncarbonated ones contribute to enhancing compressive strength. The decreased water absorption of carbonated aggregates also leads to requiring less water for mixing concrete, and then the actual water-to-binder ratio would be minimized and compressive strength would be improved (Pu, Li, Wang, Shi, Fu, et al., 2021). Fig. 22.10 exhibits the relative compressive strength of carbonated concrete (the ratio between the compressive strength of carbonated concrete to unncarbonated ones) collected from different sources from the literature

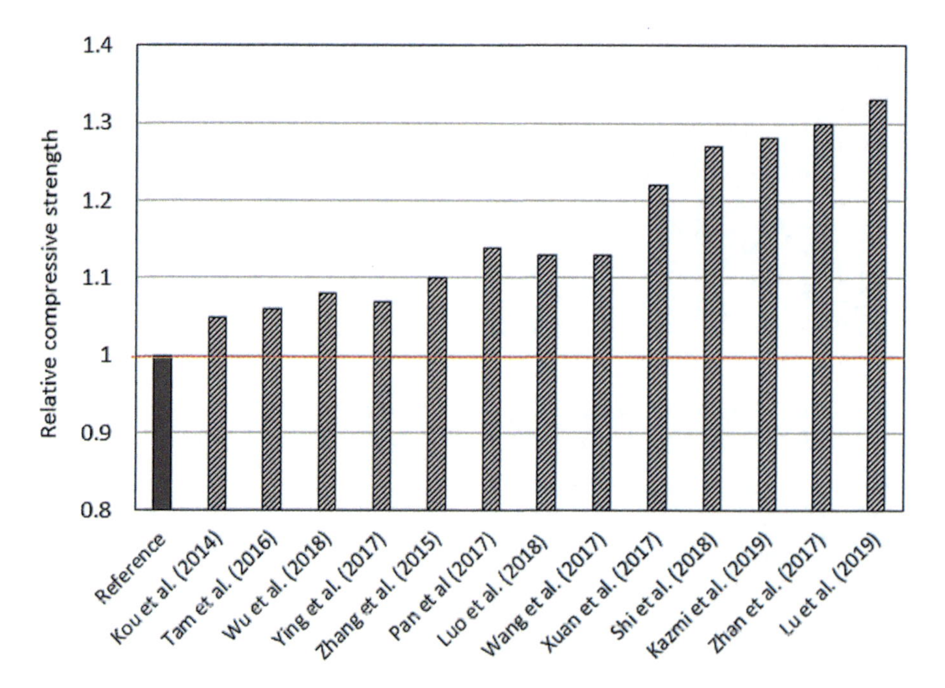

Figure 22.10 Relative compressive strength of carbonated concrete collected from different sources.

(Kazmi et al., 2019; Luo et al., 2018). It can be noted from Fig. 22.9 that the strength enhancement of carbonated concrete ranges from 5% to 33%, in comparison to uncarbonated concrete (reference mixture), depending on the properties of virgin aggregate and conditions applied.

22.7.1.2 Tensile and flexural strengths

The splitting tensile strength of recycled aggregate concrete is generally slightly inferior to that made of normal natural aggregate (less than a 10% decrease), but it could be recovered through accelerated carbonation to be comparable or even higher to that of conventional concrete (Kazmi et al., 2019; Kou et al., 2014; Wu et al., 2018). Tensile strength is primarily affected by the effective water-to-binder ratio and density of concrete (Tam et al., 2016). Thus improving these properties through accelerated carbonation can directly reflect on the tensile performance of produced concrete. The increase in tensile strength may become more noticeable by extending the curing period. Kou et al. (2014) found the tensile strength of concrete made with carbonated recycle aggregate was inferior to conventional concrete at 28 days, but a higher enhancement ratio was reported at 90 days, surpassing the ones of normal concrete by (5%–10%). However, a negative impact could be found in the scenario of using the same total water content in the mixes made with noncarbonated aggregate, since less water absorption is expected for carbonated aggregate, and thus the effective w/b ratio would be higher (Tam et al., 2016).

Similar to tensile strength, the flexural performance trend is governed by the old attached mortar content found in RA. In general, concrete made with carbonated recycled aggregates exhibited higher flexural strength (5%–28%) in the most available investigation, compared to noncarbonated ones (Kazmi et al., 2019; Xuan et al., 2018), reaching up to about 50% in some cases (Tam et al., 2016).

22.7.1.3 Modulus of elasticity

The use of carbonated recycled aggregates showed little influence on concrete elastic modulus (1%–6%) (Luo et al., 2018), and it might reach 21%, as reported in some studies (Kazmi et al., 2019; Xuan & Poon, 2018). Soaking RA in lime solution prior to applying accelerated carbonation may further promote the elastic modulus of concrete up to 27%, since additional calcium compounds would be formed. However, this enhancement could be enough to compensate for the loss that occurred due to RA. The residual elastic modulus of carbonated aggregate concrete can also be improved after being exposed to elevated temperatures, as demonstrated by Xuan et al. (2018).

22.7.2 Durability properties

22.7.2.1 Corrosion resistance

The incorporation of RA in concrete offers more pathways for harmful agents to penetrate concrete due to the higher porous medium found in the matrix compared to normal concrete. The concrete pore solution must be maintained at a high alkaline condition (pH over 12) to prevent the oxide layer on the steel surface from corrosion. Corrosion occurs mainly through two possible ways, chloride penetration, which is responsible for more than 80% of concrete structure failure, and carbonation (Alhawat et al., 2020; El-Reedy, 2017). Although reactions between $Ca(OH)_2$ and CO_2 would decrease the alkalinity of concrete and increase the depassivation potential of steel reinforcement through carbonation treatment, the pore structure of concrete could be improved, resulting in better corrosion resistance. According to Cui and Cahyadi (2001), liquid and gas can pass through pores with 10–100 μm, while it is hard to move through pores smaller than 10 μm. Liang et al. (2019) found that the corrosion current density of reinforced concrete produced with carbonated aggregate could decrease by around 41%, compared with noncarbonated ones due to the reduction in chloride permeability. Corrosion resistance is expected to be more obvious with low-quality concretes (Zhang et al., 2015). Furthermore, carbonated reinforced concrete exhibited less corrosion-induced cracking compared to that made with noncarbonated ones, as illustrated by Zhao et al. (2021), in Fig. 22.11, owing to the excellent dimensional stability and crystallization of calcium carbonate.

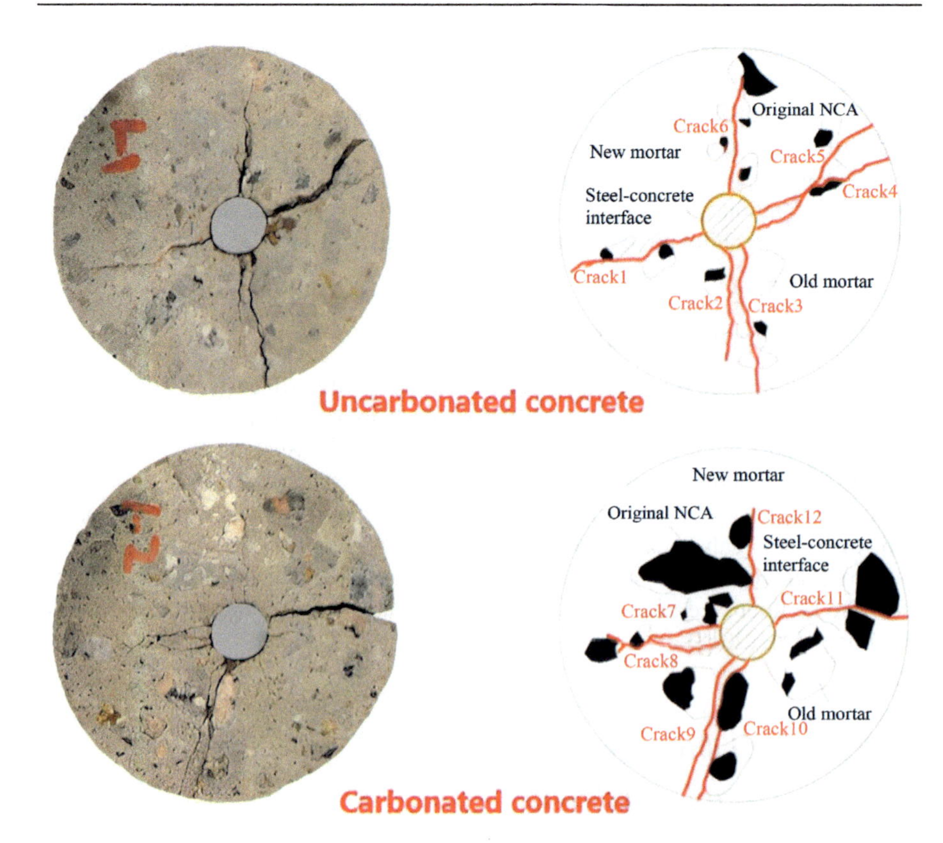

Figure 22.11 Comparison of crack due to corrosion for carbonated and uncarbonated concrete (Zhao et al., 2021).
Source: From Zhang, D., Liu, T., & Shao, Y. (2020). Weathering carbonation behavior of concrete subject to early-age carbonation curing. *Journal of Materials in Civil Engineering, 32*(4). https://doi.org/10.1061/(ASCE)MT.1943−5533.0003087.

Zhan et al. (2019) found less steel reinforcement corrosion in the samples prepared with RA than that made with carbonated aggregate during the first 10 weeks of dry and wet cycles. However, a clear drop occurred, then, only in noncarbonated samples in week 11, as shown in Fig. 22.12. This might be attributed to the increased alkalinity due to the tendency of old mortar, while the alkalinity of carbonated aggregate tends to decrease due to the less availability of free alkali metal after carbonation (Zhan et al., 2019). Moreover, concrete prepared with carbonated aggregates can decrease the carbonation depth of recycled aggregate concrete by up to about 28% (Shi et al., 2018). However, the carbonation depth level is still high, in comparison with normal concrete.

Several studies also reported a decrease in autogenous shrinkage in concrete prepared with carbonated aggregates (Kou et al., 2014; Xuan et al., 2017). However, an opposite effect might occur since higher water absorption can be found in noncarbonated aggregates, which may provide further internal curing, and thus minimizing autogenous shrinkage, as stated by Zhang et al. (2015).

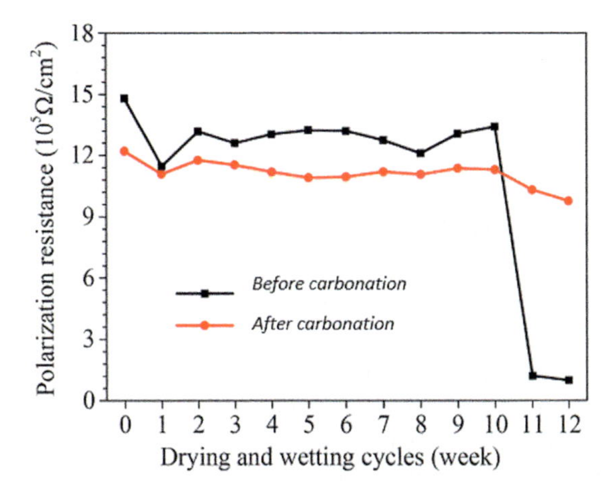

Figure 22.12 Polarization corrosion resistance of uncarbonated/carbonated samples (Zhan et al., 2019).
Source: From Zhan, B. J., Xuan, D. X., Zeng, W., & Poon, C. S. (2019). Carbonation treatment of recycled concrete aggregate: Effect on transport properties and steel corrosion of recycled aggregate concrete. *Cement and Concrete Composites, 104*. https://doi.org/10.1016/j.cemconcomp.2019.103360.

22.7.2.2 High-temperature resistance

The mechanical characteristics of concrete dramatically decrease after exposure to elevated temperatures owing to the formation of water vapor pressure after the decomposition of aggregate and mortar, causing thermal incompatibility among other elements in concrete. However, studies found that the use of carbonated aggregate can help to alleviate the detrimental effects of high temperatures, which can be explained by the late decomposition of calcium carbonate (at 550°C–950°C), formed during carbonation process, compared to the rapid breakdown of calcium hydroxide at 450°C–550°C (Morandeau et al., 2014; Xuan et al., 2018).

22.7.2.3 Freeze-thaw resistance

Concrete made with RA typically has weaker resistance against freeze-thaw exposure compared to normal concrete since it has higher porosity. However, carbonated aggregate concrete showed a higher relative dynamic elastic modulus accompanied by less mass loss after exposure to 50 freeze-thaw cycles, compared to that made with noncarbonated ones (Liang et al., 2019). Similar resistance improvement was reported by Li, Farzadnia, et al. (2017) after exposure to 300 cycles, and the behavior was found to be highly affected by w/b ratio used in old concrete since the relative dynamic elastic modulus of carbonated aggregate concrete can increase from 11.9% to 79% when the w/b ratio of old mortar rises from 0.3 to 0.5. The fundamental reason for this improvement is related to the formation of calcium carbonate at filling, which lowers the pore structure in the matrix. Table 22.1 illustrates the enhancement that occurred in durability properties collected from several studies in the literature.

22.8 Environmental and cost impact

It is stated that producing one cubic meter of concrete made with fully RA can save about 44% of mineral resources compared to that prepared with normal aggregate (Ding et al., 2016). According to some statistics (Guo et al., 2018; Li et al., 2020), RA after carbonation treatment have the potential to absorb up to 20 kg of CO_2 per each cubic meter of concrete, while others limit the absorption of recycled aggregate to 8–11 kg of CO_2 (Kikuchi & Kuroda, 2011; Xuan et al., 2018). According to research conducted by Chisalita et al. (2019), rapid carbonation can save CO_2 emissions from concrete manufacturing by up to 50%. The findings obtained from a recent study (Xiao et al., 2022) showed that by replacing normal aggregate with carbonated recycled aggregate, the total CO_2 footprints would be reduced by 7.1%–13.3%. A study conducted by Iacovidou et al. (2021) and Huang et al., 2019) indicated that the global warming potential (GWP) of one cubic meter of accelerated carbonation-cured concrete ranged from 292 to 454 kg CO_2-eq GWP, based on the conditions applied. This represents a significant reduction when compared to traditional steam curing, which recorded a GWP of 419 kg CO_2-eq, leading to up to 30% of CO_2 emissions reduction. The quantitative analysis done by Zhang et al. (2020) illustrated that around 457.7 Mt of CO_2 could be potentially sequestered from concrete waste by 2035 under the optimal capturing conditions, compared with only 74.9 Mt of CO_2 sequestration potential in 2000. The Scenarios showed the tremendous potential of CO_2 reduction.

In a recent study, life cycle assessment and costing were carried out by Zhang et al. (2022) to assess the potential emission mitigation and economic efficiency of employing carbonation technology in the CDW industry for 14 global regions. The outcomes of the analysis demonstrated notable viriations in the average unit net CO_2 and economic benefits across multiple nations, reflecting the diverse strategies in industrialising carbonated recycled concrete aggregate technology. While CO_2 benefits vary from 0.7 tons of CO_2 emissions per ton of CO_2 uptake in Brazil to 2.6 tons in Pakistan, economic advantages range from 18.5 dollars per ton of concrete debris in the United States to −5.6 in Pakistan. It is also suggested that few nations are able to simultaneously produce net negative CO_2 emissions and positive economic values, emphasizing the necessity for further production process optimization.

The method used and the energy needed for CO_2 injection are the two key factors affecting the energy consumption related to accelerated carbonation in concrete. To evaluate the total environmental effect, it is critical to take the energy input for the carbonation process into account. However, there are little particular statistical data on energy usage in this setting in the literature that is currently

Table 22.1 Durability performance of carbonated concrete.

References	Durability properties	Test	Test unit	Carbonation conditions					Enhancement rate (%)
				C (%)	T (°C)	RH (%)	P (psi)	D (days)	
Liang et al. (2019)	Corrosion	Current density	mA/cm^2	20	20	70	–	–	41.46
		Corrosion potential	(MV)	20	20	70	–	–	5.50
Shi et al. (2018)	Carbonation	Carbonation depth	mm	20	20	70	–	28	28.68
Liang et al. (2019)		Carbonation depth	mm	20	20	70	–	10	22.32
Zhang et al. (2020)		Carbonation depth 12 weeks; w/c = 0.40	mm	99.8	12	–	0.5	–	5.0
Zhang et al. (2020)		Carbonation depth 12 weeks; w/c = 0.65	mm	99.8	12	–	0.5	–	28.1
Zhang et al. (2020)		Carbonation coefficient w/c = 0.40	%	99.8	12	–	0.5	–	8.3
Zhang et al. (2020)		Carbonation coefficient w/c = 0.65	%	99.8	12	–	0.5	–	61.5
Liang et al. (2019)	Freeze-thaw	Mass loss after 50 cycles	%	20	20	70	–	10	75.0
Li, Poon, et al. (2017)		Mass loss after 300 cycles	%	20	20	70	–	10	48.39
Liang et al. (2019)		Relative dynamic elastic modulus after 50 cycles	%	20	20	70	–	10	79.56
Li, Poon, et al. (2017)		Relative dynamic elastic modulus after 300 cycles	%	95	20	70	10	7	9.12
		Mass loss after 150 cycles	%	99.5	12	–	0.5	–	95.0

(Continued)

Table 22.1 (Continued)

References	Durability properties	Test	Test unit	Carbonation conditions					Enhancement rate (%)
				C (%)	T (°C)	RH (%)	P (psi)	D (days)	
Xuan et al. (2018)	High temperature	Residual elastic modulus at 400°C	%	100	25	50	500	1	14.29
Xuan et al. (2018)		Residual compressive strength at 600°C	%	100	25	50	500	1	70.00
Xuan et al. (2018)		Residual strain at 800°C	%	100	25	50	500	1	21.10
Park et al. (2018)		Residual compressive strength at 600°C and 800°C	%	10	20	60	0	672	56.0 11.5
Meng et al. (2019)		Residual compressive strength at 600°C and 800°C	%	10	20	60	0	672	24.8 1.6
Meng et al. (2019)		Sorptivity at 600°C	mm/s	20	20	70	0.1	672	21.20

C, CO_2 concentration; *T*, temperature; *RH*, relative humidity (%); *D*, duration (days); *P*, pressure.

accessible. It is important to note that the environment of rapid carbonation might vary based on local conditions such as CO_2 source availability, waste products, and regulatory support.

While there are certain upfront expenses associated with deploying accelerated carbonation technology, the long-term benefits, such as lower carbon taxes and possible income streams from selling carbonated concrete products, may make it economically feasible. It is predicted to cost between $25 and $56 USD to sequester 1 ton of CO_2 within cement-based products (Jang & Lee, 2016). The location of the materials used for gas streams can limit the use of current carbonation methods, which makes it difficult to deploy the technology effectively in large-scale applications. In the foreseeable future, the manufacturing of carbonated aggregates is not expected to be cost-effective, in comparison with the quarrying sector. As a result, in the absence of appropriate incentives, carbonated concrete may be more marketable than carbonated aggregates. A recent assessment conducted by Zhang et al. (2022) showed that the economic and environmental benefits of carbonation technology are currently not high as they have been claimed in the literature. This is because the operability of this technology differs from one region to another. Moreover, the market demand for carbonated products in the construction sector is critical to the economic viability of accelerated carbonation. The availability of suitable waste materials and the competitiveness of the products in comparison to conventional alternatives are critical concerns.

Accelerated carbonation in concrete has the potential to offer environmental and economic benefits, however, significant related aspects such as waste management, energy consumption, and durability must all be considered.

22.9 Accelerated carbonation applications in concrete industry

The capture and storage of CO_2 emissions through accelerated carbonation can offer promising opportunities to enhance the performance of concrete. The application of carbon capture technology in concrete production can be mainly classified into four main groups.

- The carbonation of cement wastes (e.g., cement kiln dust) and concrete slurry waste can be employed to extract pure calcium carbonate and sequester carbon dioxide, which eventually can be used as raw materials for producing cement. This is mainly because these materials are rich in calcium oxides. Cement kiln dust is a fine-grained solid waste formed during cement production, and contains unreacted blended raw feed, while concrete slurry waste is made of almost 70% of hydration products. It should be noted that the calcium oxide content in concrete powder may be varied based on the quality of original concrete. According to Katsuyama et al. (2005), the projected cost of extracting calcium carbonate from cement powder waste through carbonation process is around 136 USD per metric ton, which is substantially lower than the cost of extracting calcium carbonates from natural limestone (200–350 USD/metric ton).

- A wide range of industrial byproducts and wastes such as slag, fly ash, and mine tailings can be treated and stabilized using accelerated carbonation. This process can reduce waste materials' environmental impact while simultaneously allowing for their beneficial reuse in construction applications (Aguayo et al., 2020). For example, carbonated slag and fly ash can be used as cementitious ingredients in concrete production, enhancing their strength and durability performance, while carbonated bottom ash can be utilized as a partial replacement for cement and aggregate in concrete production (Roessler et al., 2016; Schnabel et al., 2021).
- Accelerated carbonation of recycled aggregate can assist in neutralizing alkalinity and improving the quality of recycled aggregates. Thus recycled aggregates subjected to accelerated carbonation can be used in the manufacture of concrete as a substitute for natural aggregates, offering an ecofriendly substitute for conventional building materials, and making it appropriate for both structural and nonstructural uses.
- Accelerated carbonation has also the potential to be used as a concrete curing procedure. When concrete is exposed to high levels of CO_2, CO_2 reacts with calcium hydroxide to produce calcium carbonate, which in turn precipitates within the concrete matrix, enhancing its strength and durability. The carbonation reaction can accelerate the hardening of concrete products and enhance the production efficiency by lowering curing time (Shi et al., 2018). The most common applications are precast concrete products such as blocks, panels, and pavers.

Apart from concrete production, CO_2 capture and storage might be employed in some construction applications such as improving soil characteristics for geotechnical applications, particularly those with high clay content. The injection of CO_2 into soil could induce mineral reactions that can improve the stiffness and strength of unstable and weak soils, in addition to reducing their susceptibility to corrosion. Carbonation-based soil stabilization can be very beneficial for construction sites, slope stabilization and road construction projects (Cruz et al., 2023; Deneele et al., 2021).

22.10 Current challenges

Despite the benefits that can be obtained from applying accelerated carbonation technology, there are still several challenges that need to be addressed.

- Limited applicability: structures with high porosity, like precast concrete or concrete blocks, are best suited for accelerated carbonation technology. It may be less effective in denser concrete structures, restricting its use in some construction projects.
- High-energy consumption: the rapid carbonation process requires the use of high energy, which is normally produced by the combustion of fossil fuels. As a result, this energy-intensive approach can lead to increased greenhouse gas emissions, which can potentially counterbalance the environmental benefits of carbonation.
- Carbon dioxide availability and transportation: carbon dioxide availability and transportation can present challenges, particularly in certain rural areas or regions without convenient access to CO_2 sources. These circumstances can result in increased costs and logistical complexities when implementing rapid carbonation technology.
- Structural integrity concerns: while accelerated carbonation can increase the compressive strength of concrete, it may also impact other qualities. Carbonation can reduce the

alkalinity of concrete, potentially leading to reinforcing steel corrosion if not managed effectively. Over time, such corrosion can compromise the structural integrity of structures.
- Quality control: quality management and monitoring are essential when implementing accelerated carbonation technology to achieve consistent and desirable results. To obtain the appropriate carbonation levels while maintaining structural performance, factors such as CO_2 concentration, exposure time, and ambient conditions must be precisely controlled and monitored throughout the process.
- Long-term performance uncertainties: the long-term performance of concrete structures treated with accelerated carbonation technique remains inadequately studied.
- Cost considerations: The implementation of accelerated carbonation technology can involve additional costs, such as CO_2 purchase and shipping, specialized equipment, and monitoring systems. These supplementary expenses can render certain construction projects, particularly those of smaller scale, economically less feasible.

22.11 Conclusions and recommendations

Accelerated carbonation in recycled aggregate and concrete represents a promising approach toward sustainable construction. The chapter reviewed various CO_2 capture technologies, and evaluated their feasibility to be applied in concrete industry, in addition to discussing the affecting parameters. Moreover, the characteristics of carbonated recycled aggregates were discussed, and the mechanical and durability performance of carbonated concrete were also analyzed.

CO_2 capture and storage technique generally exhibited promising results in terms of enhancing the properties of aggregate and concrete. However, there appears to be a considerable difference in the findings collected from the literature.

- Accelerated carbonation has been performed to improve the quality of concrete through two different methodologies either carbonating recycled aggregates prior to mixing, or carbonating concrete blocks. Despite the similarity between the two approaches, more CO_2 sequestration can be achieved by carbonating concrete under exposure to the same curing conditions with recycled aggregates; however, more challenges may arise with carbonation concrete in situ.
- Some intercorrelated factoring parameters, such as CO_2 concentration, gas pressure, RH, temperature, and water content, need to be precisely evaluated in order to effectively enhance carbonation process.
- In most cases, the carbonation reaction starts out quickly before slowing down, while recycled aggregate with smaller particles can be carbonated more easily than those with bigger sizes.
- CDW can benefit from CO_2 capture and storage technology; the porosity, pore structure, and microcracks at ITZ can be reduced. As a result, the characteristics of carbonated recycled aggregates (e.g., crush index and water absorption) could be obviously decreased accompanied by denser microstructure, while others, like apparent density, may be slightly improved.
- Although the majority of concrete properties made with carbonated aggregates are improved, in comparison with uncarbonated ones, these properties do not always reach those containing only natural aggregates, especially in terms of durability.

- Some strategies can significantly enhance the efficiency of carbonation technology (e.g., presoaking in calcium hydroxide solution); however, these techniques are mostly costly or complicated, limiting their wider applicability and adoption.

Based on the current work, several future investigations on the application of accelerated CO_2 technology are suggested to address the issues raised in this review chapter:

- Despite CO_2 curing can improve the characteristics of recycled aggregates and concrete, investigations on steel-reinforced concrete elements are still susceptible because of impaired pH and corrosion concerns of reinforcing steel. Further, the durability behavior of carbonated aggregate concrete under long-term exposure needs to be urgently studied.
- Although rapid carbonation can improve the characteristics of recycled aggregates and produced concrete, more energy consumption is expected, especially in the case of high temperatures. Therefore a detailed life cycle and economic assessment should be conducted to accurately assess these aspects.
- The feasibility of CO_2 technology in the construction industry has been proved in terms of lab-scale systems; however, investigations on large-scale samples are rarely conducted. Thus further research should be performed to enhance the industrialization of carbon capture technology, especially in terms of the structural performance of concrete made with carbonated aggregate.
- The feasibility of applying CO_2 from direct sources needs to be modified, while the quality of recycled aggregate sourced from various areas needs to be determined in order to standardize this approach.

References

Aguayo, Federico, Torres, Anthony, Kim, Yoo-Jae, & Thombare, Omkar (2020). Accelerated carbonation assessment of high-volume fly ash voncrete. *Journal of Materials Science and Chemical Engineering*, *08*(03), 23–38. Available from https://doi.org/10.4236/msce.2020.83002.

Alhawat, Musab, & Ashour, Ashraf (2019). Bond strength between corroded steel reinforcement and recycled aggregate concrete. *Structures*, *19*, 369–385. Available from https://doi.org/10.1016/j.istruc.2019.02.001, https://linkinghub.elsevier.com/retrieve/pii/S2352012419300256.

Alhawat, Musab, Khan, Amir, & Ashour, Ashraf (2020). Evaluation of steel vorrosion in concrete structures using impact-echo method. *Advanced Materials Research*, *1158*(1), 147–164. Available from https://doi.org/10.4028/http://www.scientific.net/AMR.1158.147, https://www.scientific.net/AMR.1158.147.

Araba, Almahdi Mohamed, Memon, Zubair Ahmed, Alhawat, Musab, Ali, Mumtaz, & Milad, Abdalrhman (2021). Estimation at completion in civil engineering projects: Review of regression and soft computing models. *Knowledge-Based Engineering and Sciences*, *2*(2), 1–12. Available from https://doi.org/10.51526/kbes.2021.2.2.1-12.

Bai, Guoliang, Zhu, Chao, Liu, Chao, & Liu, Biao (2020). An evaluation of the recycled aggregate characteristics and the recycled aggregate concrete mechanical properties. *Construction and Building Materials*, *240*, 117978. Available from https://doi.org/10.1016/j.conbuildmat.2019.117978.

Bosque., Heede., de Belie, N., Rojas., & Medina, C. (2020). Carbonation of concrete with construction and demolition waste based recycled aggregates and cement with recycled content. *Construction and Building Materials, 234.*

Bukowski, J. M., & Berger, R. L. (1979). Reactivity and strength development of CO_2 activated non-hydraulic calcium silicates. *Cement and Concrete Research, 9*(1), 57–68. Available from https://doi.org/10.1016/0008-8846(79)90095-4, https://linkinghub.elsevier.com/retrieve/pii/0008884679900954.

Chen, Ying, Liu, Peng, & Yu, Zhiwu (2018). Effects of environmental factors on concrete carbonation depth and compressive strength. *Materials, 11*(11), 2167. Available from https://doi.org/10.3390/ma11112167.

Chisalita, D. A., Petrescu, L., Cobden, P., van Dijk, H. E., Cormos, A. M., & Cormos, C. C. (2019). Assessing the environmental impact of an integrated steel mill with post-combustion CO2 capture and storage using the LCA methodology. *Journal of Cleaner Production, 211*, 1015–1025.

Cruz, N., Ruivo, L., Avellan, A., Römkens, P. F. A. M., Tarelho, L. A. C., & Rodrigues, S. M. (2023). Stabilization of biomass ash granules using accelerated carbonation to optimize the preparation of soil improvers. *Waste Management, 156*, 297–306. Available from https://doi.org/10.1016/j.wasman.2022.11.011.

Cuenca-Moyano, G. M., Martín-Morales, M., Bonoli, A., & Valverde-Palacios, I. (2019). Environmental assessment of masonry mortars made with natural and recycled aggregates. *International Journal of Life Cycle Assessment, 24*(2), 191–210. Available from https://doi.org/10.1007/s11367-018-1518-9, http://www.springerlink.com/content/0948-3349.

Cui, H., Tang, W., Liu, W., Dong, Z., & Xing, F. (2015). Experimental study on effects of CO_2 concentrations on concrete carbonation and diffusion mechanisms. *Construction and Building Materials, 93*, 522–527. Available from https://doi.org/10.1016/j.conbuildmat.2015.06.007, https://www.journals.elsevier.com/construction-and-building-materials.

Cui, L., & Cahyadi, J. H. (2001). Permeability and pore structure of OPC paste. *Cement and Concrete Research, 31*(2), 277–282. Available from https://doi.org/10.1016/S0008-8846(00)00474-9, http://www.sciencedirect.com/science/journal/00088846.

Deneele, Dimitri, Dony, Anne, Colin, Johan, Herrier, Gontran, & Lesueur, Didier (2021). The carbonation of a lime-treated soil: Experimental approach. *Materials and Structures, 54*(1). Available from https://doi.org/10.1617/s11527-021-01617-w.

Ding, T., Xiao, J., & Tam, V. W. Y. (2016). A closed-loop life cycle assessment of recycled aggregate concrete utilization in China. *Waste Management, 56*, 367–375. Available from https://doi.org/10.1016/j.wasman.2016.05.031, http://www.elsevier.com/locate/wasman.

Drouet, E., Poyet, S., Le Bescop, P., Torrenti, J. M., & Bourbon, X. (2019). Carbonation of hardened cement pastes: Influence of temperature. *Cement and Concrete Research, 115*, 445–459. Available from https://doi.org/10.1016/j.cemconres.2018.09.019, http://www.sciencedirect.com/science/journal/00088846.

Ekolu, S. O. (2016). A review on effects of curing, sheltering, and CO_2 concentration upon natural carbonation of concrete. *Construction and Building Materials, 127*, 306–320. Available from https://doi.org/10.1016/j.conbuildmat.2016.09.056.

El-Reedy, M. A. (2017). *Steel-reinforced concrete structures: Assessment and repair of corrosion, Second edition*, 1–208. Available from https://doi.org/10.1201/b22237, http://www.tandfebooks.com/doi/book/10.1201/b22237.

Elsalamawy, M., Mohamed, A. R., & Kamal, E. M. (2019). The role of relative humidity and cement type on carbonation resistance of concrete. *Alexandria Engineering Journal*, *58*(4), 1257–1264. Available from https://doi.org/10.1016/j.aej.2019.10.008, http://www.elsevier.com/wps/find/journaldescription.cws_home/724292/description#description.

Fang, X., Xuan, D., & Poon, C. S. (2017). Empirical modelling of CO_2 uptake by recycled concrete aggregates under accelerated carbonation conditions. *Materials and Structures*, *50*, 1–13.

Fang, Xiaoliang, Zhan, Baojian, & Poon, Chi Sun (2020). Enhancing the accelerated carbonation of recycled concrete aggregates by using reclaimed wastewater from concrete batching plants. *Construction and Building Materials*, *239*, 117810. Available from https://doi.org/10.1016/j.conbuildmat.2019.117810, https://linkinghub.elsevier.com/retrieve/pii/S0950061819332635.

Fang, Xiaoliang, Zhan, Baojian, & Poon, Chi Sun (2021). Enhancement of recycled aggregates and concrete by combined treatment of spraying Ca^{2+} rich wastewater and flow-through carbonation. *Construction and Building Materials*, *277*, 122202. Available from https://doi.org/10.1016/j.conbuildmat.2020.122202, https://linkinghub.elsevier.com/retrieve/pii/S0950061820342057.

Guo, H., Shi, C., Guan, X., Zhu, J., Ding, Y., Ling, T. C., Zhang, H., & Wang, Y. (2018). Durability of recycled aggregate concrete – A review. *Cement and Concrete Composites*, *89*, 251–259. Available from https://doi.org/10.1016/j.cemconcomp.2018.03.008, http://www.sciencedirect.com/science/journal/09589465.

Huang, H., Guo, R., Wang, T., Hu, X., Garcia, S., Fang, M., Luo, Z., & Maroto-Valer, M. M. (2019). Carbonation curing for wollastonite-Portland cementitious materials: CO_2 sequestration potential and feasibility assessment. *Journal of Cleaner Production*, *211*, 830–841.

Iacovidou, E., Gerassimidou, S., & Martin, O. V. (2021). Environmental Life Cycle Impacts of Inert/Less Reactive Waste Materials: A report prepared for Defra. Department for Environment, Food and Rural Affairs.

Islam, R., Nazifa, T. H., Yuniarto, A., Shanawaz Uddin, A. S. M., Salmiati, S., & Shahid, S. (2019). An empirical study of construction and demolition waste generation and implication of recycling. *Waste Management*, *95*, 10–21. Available from https://doi.org/10.1016/j.wasman.2019.05.049, http://www.elsevier.com/locate/wasman.

Jae-Dong, John, Hirai, Kazunobu, & Mihashi, Hirozo (1990). Influence of environmental moisture and temperature on carbonation of mortar. *Concrete Research and Technology*, *1*(1), 85–94. Available from https://doi.org/10.3151/crt1990.1.1_85, https://www.jstage.jst.go.jp/article/crt1990/1/1/1_85/_article/-char/ja/.

Jang, J. G., & Lee, H. K. (2016). Microstructural densification and CO_2 uptake promoted by the carbonation curing of belite-rich Portland cement. *Cement and Concrete Research*, *82*, 50–57. Available from https://doi.org/10.1016/j.cemconres.2016.01.001, http://www.sciencedirect.com/science/journal/00088846.

Jiang, Y., Li, L., Lu, J. X., Shen, P., Ling, T. C., & Poon, C. S. (2022). Mechanism of carbonating recycled concrete fines in aqueous environment: The particle size effect. *Cement and Concrete Composites*, *133*. Available from https://doi.org/10.1016/j.cemconcomp.2022.104655, http://www.sciencedirect.com/science/journal/09589465.

Kaliyavaradhan, S. K., & Ling, T. C. (2017). Potential of CO_2 sequestration through construction and demolition (C&D) waste - An overview. *Journal of CO_2 Utilization*, *20*, 234–242. Available from https://doi.org/10.1016/j.jcou.2017.05.014, http://www.journals.elsevier.com/journal-of-co2-utilization/.

Kashef-Haghighi, S., Shao, Y., & Ghoshal, S. (2015). Mathematical modeling of CO_2 uptake by concrete during accelerated carbonation curing. *Cement and Concrete Research*, *67*, 1–10. Available from https://doi.org/10.1016/j.cemconres.2014.07.020, http://www.sciencedirect.com/science/journal/00088846.

Katsuyama, Y., Yamasaki, A., Iizuka, A., Fujii, M., Kumagai, K., & Yanagisawa, Y. (2005). Development of a process for producing high-purity calcium carbonate ($CaCO_3$) from waste cement using pressurized CO_2. *Environmental Progress*, *24*(2), 162–170. Available from https://doi.org/10.1002/ep.10080.

Kazmi, S. M. S., Munir, M. J., Wu, Y. F., Patnaikuni, I., Zhou, Y., & Xing, F. (2019). Influence of different treatment methods on the mechanical behavior of recycled aggregate concrete: A comparative study. *Cement and Concrete Composites*, *104*. Available from https://doi.org/10.1016/j.cemconcomp.2019.103398, http://www.sciencedirect.com/science/journal/09589465.

Kikuchi, T., & Kuroda, Y. (2011). Carbon dioxide uptake in demolished and crushed concrete. *Journal of Advanced Concrete Technology*, *9*(1), 115–124. Available from https://doi.org/10.3151/jact.9.115Japan, http://gcs.jstage.jst.go.jp/article/jact/9/1/9_115/_article?from = Scopus.

Kou, Shi-Cong, Zhan, Bao-jian, & Poon, Chi-Sun (2014). Use of a CO_2 curing step to improve the properties of concrete prepared with recycled aggregates. *Cement and Concrete Composites*, *45*, 22–28. Available from https://doi.org/10.1016/j.cemconcomp.2013.09.008, https://linkinghub.elsevier.com/retrieve/pii/S0958946513001364.

Kurda, R., De Brito, J., & Silvestre, J. D. (2019). Carbonation of concrete made with high amount of fly ash and recycled concrete aggregates for utilization of CO_2. *Journal of CO_2 Utilization*, *29*, 12–19. Available from https://doi.org/10.1016/j.jcou.2018.11.004, http://www.journals.elsevier.com/journal-of-co2-utilization/.

Li, L., Poon, C. S., Xiao, J., & Xuan, D. (2017). Effect of carbonated recycled coarse aggregate on the dynamic compressive behavior of recycled aggregate concrete. *Construction and Building Materials*, *151*, 52–62. Available from https://doi.org/10.1016/j.conbuildmat.2017.06.043.

Li, N., Farzadnia, N., & Shi, C. (2017). Microstructural changes in alkali-activated slag mortars induced by accelerated carbonation. *Cement and Concrete Research*, *100*, 214–226. Available from https://doi.org/10.1016/j.cemconres.2017.07.008, http://www.sciencedirect.com/science/journal/00088846.

Li, Y., Liu, W., Xing, F., Wang, S., Tang, L., Lin, S., & Dong, Z. (2020). Carbonation of the synthetic calcium silicate hydrate (C-S-H) under different concentrations of CO_2: Chemical phases analysis and kinetics. *Journal of CO2 Utilization*, *35*, 303–313. Available from https://doi.org/10.1016/j.jcou.2019.10.001, http://www.journals.elsevier.com/journal-of-co2-utilization/.

Li, Y., Zhang, S., Wang, R., Zhao, Y., & Men, C. (2019). Effects of carbonation treatment on the crushing characteristics of recycled coarse aggregates. *Construction and Building Materials*, *201*, 408–420. Available from https://doi.org/10.1016/j.conbuildmat.2018.12.158.

Li, L., Ziyabek, N., Jiang, Y., Xiao, J., & Poon, C. S. (2023). Effect of carbonation duration on properties of recycled aggregate concrete. *Case Studies in Construction Materials*, *19*, e02640.

Liang, C., Ma, H., Pan, Y., Ma, Z., Duan, Z., & He, Z. (2019). Chloride permeability and the caused steel corrosion in the concrete with carbonated recycled aggregate. *Construction and Building Materials*, *218*, 506–518. Available from https://doi.org/10.1016/j.conbuildmat.2019.05.136.

Liu, Kaiwei, Xu, Wanyu, Sun, Daosheng, Tang, Jinhui, Wang, Aiguo, & Chen, Dong (2021). Carbonation of recycled aggregate and its effect on properties of recycled aggregate concrete: A review. *Materials Express*, *11*(9), 1439–1452. Available from https://doi.org/10.1166/mex.2021.2045, https://www.ingentaconnect.com/content/10.1166/mex.2021.2045.

Liu, L., Ha, J., Hashida, T., & Teramura, S. (2001). Development of a CO_2 solidification method for recycling autoclaved lightweight concrete waste. *Journal of materials science letters*, *20*(19), 1791–1794. Available from https://doi.org/10.1023/A:1012591318077.

Liu, Peng, Chen, Ying, Yu, Zhiwu, & Zhang, Rongling (2019). Effect of temperature on concrete carbonation performance. *Advances in Materials Science and Engineering*, *2019*, 1–6. Available from https://doi.org/10.1155/2019/9204570.

Luo, Surong, Wu, Wenda, & Wu, Kaiyun (2018). Effect of recycled coarse aggregates enhanced by CO_2 on the mechanical properties of recycled aggregate concrete. *IOP Conference Series: Materials Science and Engineering*, *431*(10), 102006. Available from https://doi.org/10.1088/1757-899x/431/10/102006.

Mehdizadeh, H., Ling, T. C., Cheng, X., & Mo, K. H. (2021). Effect of particle size and CO_2 treatment of waste cement powder on properties of cement paste. *Canadian Journal of Civil Engineering*, *48*(5), 522–531. Available from https://doi.org/10.1139/cjce-2019-0574, http://www.nrcresearchpress.com/loi/cjce.

Meng, Y., Ling, T. C., Mo, K. H., & Tian, W. (2019). Enhancement of high temperature performance of cement blocks via CO_2 curing. *Science of the Total Environment*, *671*, 827–837. Available from https://doi.org/10.1016/j.scitotenv.2019.03.411, http://www.elsevier.com/locate/scitotenv.

Monkman, Sean, & Shao, Yixin (2010). Integration of carbon sequestration into curing process of precast concrete. *Canadian Journal of Civil Engineering*, *37*(2), 302–310. Available from https://doi.org/10.1139/L09-140, http://www.nrcresearchpress.com/doi/10.1139/L09-140.

Morandeau, A., Thiéry, M., & Dangla, P. (2014). Investigation of the carbonation mechanism of CH and C-S-H in terms of kinetics, microstructure changes and moisture properties. *Cement and Concrete Research*, *56*, 153–170. Available from https://doi.org/10.1016/j.cemconres.2013.11.015.

Nayana, A. Y., & Kavitha, S. (2017). Evaluation of CO_2 emissions for green concrete with high volume slag, recycled aggregate, recycled water to build eco environment. *International Journal of Civil Engineering and Technology*, *8*(5), 703–708. Available from http://www.iaeme.com/ijciet/index.asp.

Pan, G., Zhan, M., Fu, M., Wang, Y., & Lu, X. (2017). Effect of CO_2 curing on demolition recycled fine aggregates enhanced by calcium hydroxide pre-soaking. *Construction and Building Materials*, *154*, 810–818. Available from https://doi.org/10.1016/j.conbuildmat.2017.07.079.

Pan, Ganghua, Shen, Qizhen, & Li, Jiao (2018). Microstructure of cement paste at different carbon dioxide concentrations. *Magazine of Concrete Research*, *70*(3), 154–162. Available from https://doi.org/10.1680/jmacr.17.00106, https://www.icevirtuallibrary.com/doi/10.1680/jmacr.17.00106.

Papadakis, V. G., Vayenas, C. G., & Fardis, M. N. (1991). Fundamental modeling and experimental investigation of concrete carbonation. *ACI Materials Journal*, *88*(4), 363–373.

Park, S. M., Seo, J. H., & Lee, H. K. (2018). Thermal evolution of hydrates in carbonation-cured Portland cement. *Materials and Structures*, *51*(1), 7. Available from https://doi.org/10.1617/s11527-017-1114-7, http://link.springer.com/10.1617/s11527-017-1114-7.

Pu, Y., Li, L., Wang, Q., Shi, X., Fu, L., Zhang, G., Luan, C., & Abomohra, A. E. F. (2021). Accelerated carbonation treatment of recycled concrete aggregates using flue gas: A comparative study towards performance improvement. *Journal of CO2 Utilization, 43*. Available from https://doi.org/10.1016/j.jcou.2020.101362, http://www.journals.elsevier.com/journal-of-co2-utilization/.

Pu, Yunhui, Li, Lang, Wang, Qingyuan, Shi, Xiaoshuang, Luan, Chenchen, Zhang, Guomin, Fu, Ling, & El-Fatah Abomohra, Abd (2021). Accelerated carbonation technology for enhanced treatment of recycled concrete aggregates: A state-of-the-art review. *Construction and Building Materials, 282*, 122671. Available from https://doi.org/10.1016/j.conbuildmat.2021.122671, https://linkinghub.elsevier.com/retrieve/pii/S0950061821004311.

Roessler, J., Paris, J., Ferraro, C. C., Watts, B., & Townsend, T. (2016). Use of waste to energy bottom ash as an aggregate in portland cement concrete: Impacts of size fractionation and carbonation. *Waste and Biomass Valorization, 7*(6), 1521–1530. Available from https://doi.org/10.1007/s12649-016-9545-x, http://www.springer.com/engineering/journal/12649.

Schnabel, K., Brück, F., Mansfeldt, T., & Weigand, H. (2021). *Full-scale accelerated carbonation of waste incinerator bottom ash under continuous-feed conditions, . Waste Management* (125, pp. 40–48). Germany: Elsevier Ltd.. Available from http://www.elsevier.com/locate/wasman, 10.1016/j.wasman.2021.02.027.

Shi, C., Wu, Z., Cao, Z., Ling, T. C., & Zheng, J. (2018). Performance of mortar prepared with recycled concrete aggregate enhanced by CO_2 and pozzolan slurry. *Cement and Concrete Composites, 86*, 130–138. Available from https://doi.org/10.1016/j.cemconcomp.2017.10.013, http://www.sciencedirect.com/science/journal/09589465.

Silva, R. V., de Brito, J., & Dhir, R. K. (2019). Use of recycled aggregates arising from construction and demolition waste in new construction applications. *Journal of Cleaner Production, 236*117629. Available from https://doi.org/10.1016/j.jclepro.2019.117629.

Suescum-Morales, D., Bravo, M., Silva, R. V., Jiménez, J. R., Fernandez-Rodriguez, J. M., & de Brito, J. (2022). Effect of reactive magnesium oxide in alkali-activated fly ash mortars exposed to accelerated CO_2 curing. *Construction and Building Materials, 342*. Available from https://doi.org/10.1016/j.conbuildmat.2022.127999, https://www.journals.elsevier.com/construction-and-building-materials.

Suescum-Morales, David, Kalinowska-Wichrowska, Katarzyna, Fernández, José María, & Jiménez, José Ramón (2021). Accelerated carbonation of fresh cement-based products containing recycled masonry aggregates for CO_2 sequestration. *Journal of CO2 Utilization, 46*. Available from https://doi.org/10.1016/j.jcou.2021.101461.

Sultan, H. K., Zinkaah, O. H., Rasheed, A. A., Alridha, Z., & Alhawat, M. (2022). Producing sustainable modified reactive powder concrete using locally available materials. *Innovative Infrastructure Solutions, 7*(6), 23644184. Available from https://doi.org/10.1007/s41062-022-00948-z, https://www.springer.com/journal/41062.

Tam, Vivian W. Y., Butera, Anthony, & Le, Khoa N. (2016). Carbon-conditioned recycled aggregate in concrete production. *Journal of Cleaner Production, 133*, 672–680. Available from https://doi.org/10.1016/j.jclepro.2016.06.007, https://linkinghub.elsevier.com/retrieve/pii/S0959652616306758.

Tam, Vivian W. Y., Butera, Anthony, Le, Khoa N., & Li, Wengui (2020). Utilising CO_2 technologies for recycled aggregate concrete: A critical review. *Construction and Building Materials, 250*, 118903. Available from https://doi.org/10.1016/j.conbuildmat.2020.118903, https://linkinghub.elsevier.com/retrieve/pii/S0950061820309089.

Tang, Bowen, Fan, Meng, Yang, Zhengquan, Sun, Yongshuai, & Yuan, Linjuan (2023). A comparison study of aggregate carbonation and concrete carbonation for the enhancement of recycled aggregate pervious concrete. *Construction and Building Materials*, *371*. Available from https://doi.org/10.1016/j.conbuildmat.2023.130797.

Valencia-Saavedra, W. G., Aguirre-Guerrero, A. M., & Mejía de Gutiérrez, R. (2022). Alkali-activated concretes based on high unburned carbon content fly ash: carbonation and corrosion performance. *European Journal of Environmental and Civil Engineering*, *26*(8), 3292–3312. Available from https://doi.org/10.1080/19648189.2020.1785948, http://www.tandfonline.com/loi/tece20.

Wang, Dianchao, Xiao, Jianzhuang, & Duan, Zhenhua (2022). Strategies to accelerate CO_2 sequestration of cement-based materials and their application prospects. *Construction and Building Materials*, *314*, 125646. Available from https://doi.org/10.1016/j.conbuildmat.2021.125646.

Wang, Hui, Zhang, Ailian, Zhang, Linchun, Liu, Junzhe, Han, Yan, & Wang, Jianmin (2020). Research on the influence of carbonation on the content and state of chloride ions and the following corrosion resistance of steel bars in cement paste. *Coatings*, *10* (11). Available from https://doi.org/10.3390/coatings10111071.

Wang, J., Xu, H., Xu, D., Du, P., Zhou, Z., Yuan, L., & Cheng, X. (2019). Accelerated carbonation of hardened cement pastes: Influence of porosity. *Construction and Building Materials*, *225*, 159–169. Available from https://doi.org/10.1016/j.conbuildmat.2019.07.088, https://www.journals.elsevier.com/construction-and-building-materials.

Wang, Z., Cui, L., Liu, Y., Hou, J., Li, H., Zou, L., & Zhu, F. (2024). High-efficiency CO_2 sequestration through direct aqueous carbonation of carbide slag: determination of carbonation reaction and optimization of operation parameters. *Frontiers of Environmental Science and Engineering*, *18*(1). Available from https://doi.org/10.1007/s11783-024-1772-y, https://www.springer.com/journal/11783.

Wu, Y., Mehdizadeh, H., Mo, K. H., & Ling, T. C. (2022). High-temperature CO2 for accelerating the carbonation of recycled concrete fines. *Journal of Building Engineering*, *52*, 104526.

Wu, H. L., Zhang, D., Ellis, B. R., & Li, V. C. (2018). Development of reactive MgO-based engineered cementitious composite (ECC) through accelerated carbonation curing. *Construction and Building Materials*, *191*, 23–31. Available from https://doi.org/10.1016/j.conbuildmat.2018.09.196.

Xiao, Jianzhuang, Zhang, Hanghua, Tang, Yuxiang, Deng, Qi, Wang, Dianchao, & Poon, Chi-sun (2022). Fully utilizing carbonated recycled aggregates in concrete: Strength, drying shrinkage and carbon emissions analysis. *Journal of Cleaner Production*, *377*, 134520. Available from https://doi.org/10.1016/j.jclepro.2022.134520, https://linkinghub.elsevier.com/retrieve/pii/S0959652622040926.

Xuan, D., & Poon, C. S. (2018). Sequestration of carbon dioxide by RCAs and enhancement of properties of RAC by accelerated carbonation. *New Trends in Eco-efficient and Recycled Concrete*, 477–497. Available from https://doi.org/10.1016/B978-0-08-102480-5.00016-6, http://www.sciencedirect.com/science/book/9780081024805.

Xuan, D., Zhan, B., & Poon, C. S. (2016). Development of a new generation of eco-friendly concrete blocks by accelerated mineral carbonation. *Journal of Cleaner Production*, *133*, 1235–1241. Available from https://doi.org/10.1016/j.jclepro.2016.06.062.

Xuan, D., Zhan, B., & Poon, C. S. (2017). Durability of recycled aggregate concrete prepared with carbonated recycled concrete aggregates. *Cement and Concrete Composites*, *84*, 214–221. Available from https://doi.org/10.1016/j.cemconcomp.2017.09.015, http://www.sciencedirect.com/science/journal/09589465.

Xuan, Dongxing, Zhan, Baojian, & Poon, Chi Sun (2018). Thermal and residual mechanical profile of recycled aggregate concrete prepared with carbonated concrete aggregates after exposure to elevated temperatures. *Fire and Materials, 42*(1), 134–142. Available from https://doi.org/10.1002/fam.2465, https://onlinelibrary.wiley.com/doi/10.1002/fam.2465.

Zhan, B., Poon, C. S., Liu, Q., Kou, S., & Shi, C. (2014). Experimental study on CO_2 curing for enhancement of recycled aggregate properties. *Construction and Building Materials, 67*, 3–7. Available from https://doi.org/10.1016/j.conbuildmat.2013.09.008.

Zhan, Bao Jian, Xuan, Dong Xing, & Poon, Chi Sun (2018). Enhancement of recycled aggregate properties by accelerated CO_2 curing coupled with limewater soaking process. *Cement and Concrete Composites, 89*, 230–237. Available from https://doi.org/10.1016/j.cemconcomp.2018.03.011, https://linkinghub.elsevier.com/retrieve/pii/S0958946517311733.

Zhan, B. J., Xuan, D. X., Poon, C. S., & Shi, C. J. (2016). Effect of curing parameters on CO_2 curing of concrete blocks containing recycled aggregates. *Cement and Concrete Composites, 71*, 122–130. Available from https://doi.org/10.1016/j.cemconcomp.2016.05.002, http://www.sciencedirect.com/science/journal/09589465.

Zhan, B. J., Xuan, D. X., Zeng, W., & Poon, C. S. (2019). Carbonation treatment of recycled concrete aggregate: Effect on transport properties and steel corrosion of recycled aggregate concrete. *Cement and Concrete Composites, 104*. Available from https://doi.org/10.1016/j.cemconcomp.2019.103360, http://www.sciencedirect.com/science/journal/09589465.

Zhan, M., Pan, G., Wang, Y., Fu, M., & Lu, X. (2017). Effect of presoak-accelerated carbonation factors on enhancing recycled aggregate mortars. *Magazine of ConcreteResearch, 69*(16), 838–849. Available from https://doi.org/10.1680/jmacr.16.00468, http://www.icevirtuallibrary.com/content/serial/macr.

Zhang, Duo, Liu, Tianlu, & Shao, Yixin (2020). Weathering carbonation behavior of concrete subject to early-age carbonation curing. *Journal of Materials in Civil Engineering, 32*(4), 04020038. Available from https://doi.org/10.1061/(ASCE)MT.1943-5533.0003087, https://ascelibrary.org/doi/10.1061/%28ASCE%29MT.1943-5533.0003087.

Zhang, J., Shi, C., Li, Y., Pan, X., Poon, C. S., & Xie, Z. (2015). Influence of carbonated recycled concrete aggregate on properties of cement mortar. *Construction and Building Materials, 98*, 1–7. Available from https://doi.org/10.1016/j.conbuildmat.2015.08.087.

Zhang, Ning, Zhang, Duo, Zuo, Jian, Miller, Travis R., Duan, Huabo, & Schiller, Georg (2022). Potential for CO_2 mitigation and economic benefits from accelerated carbonation of construction and demolition waste. *Renewable and Sustainable Energy Reviews, 169*. Available from https://doi.org/10.1016/j.rser.2022.112920.

Zhao, Yuxi, Peng, Ligang, Zeng, Weilai, Poon, Chi sun, & Lu, Zhenmei (2021). Improvement in properties of concrete with modified RCA by microbial induced carbonate precipitation. *Cement and Concrete Composites, 124*, 104251. Available from https://doi.org/10.1016/j.cemconcomp.2021.104251.

Conclusions and recommendations

23

Ashraf Ashour[1], *Xinyue Wang*[2] *and Baoguo Han*[3]
[1]Faculty of Engineering and Digital Technologies, University of Bradford, Bradford, United Kingdom, [2]School of Civil Engineering, Tianjin University, Tianjin, P.R. China, [3]School of Civil Engineering, Dalian University of Technology, Dalian, Liaoning, P.R. China

Concrete possesses outstanding mechanical properties, notably exceptional resistance to both compressive and cyclic loads. Moreover, it exhibits waterproof characteristics, rendering it corrosion-resistant, and is inherently fire-resistant. Concrete structures are highly safe and durable, and require little maintenance. When considering its affordability, ease of production, and widespread availability, concrete structures have indeed exerted a transformative impact on the world and will continue in shaping the future (Alhawat et al., 2022; Ding et al., 2023; Elhacham et al., 2020). The extensive application of concrete in infrastructure, however, results in a considerable impact on resources, energy, and the environment. Hence, the adoption of sustainable concrete materials and structures becomes an inevitable trend to mitigate the substantial environmental and energy challenges posed by conventional concrete. For this purpose, various forms of sustainable concrete materials and structures have been developed, involving several technical routes, such as modification of cement performance, reduction of cement/concrete consumption, development of new sustainable binders, utilization of supplementary cementitious materials, use of industrial wastes, recycle materials, and regional natural materials, adoption of advanced manufacturing/construction technology, enhancement of performance, and development of functional/smart performance. The detailed investigations and technologies of these development have been well presented in Chapters 2–22 of this book, thus encouraging the construction industry to adopt and invest in the latest innovative sustainable concrete materials and structures to reap significant environmental and economic benefits.

Achieving net-zero emissions in the concrete industry is a challenging task but essential goal for mitigating climate change (Han et al., 2017, 2019; Wang et al., 2024). Some key strategies and considerations for achieving net-zero emissions in the concrete industry have been presented in various chapters in this book, for example, reducing CO_2 emissions from cement production (Chapter 2), developing less intensive carbon emitters binders (Chapters 4, 5, 10, and 21), carbon capture and storage technologies (Chapter 22), use of recycled aggregates and locally sourced materials to reduce transportation-related emissions (Chapters 6, 7, 8, 9, 15, and 16), improving the energy efficiency of concrete production processes, such as kiln operations and grinding, to reduce energy consumption and emissions

Sustainable Concrete Materials and Structures. DOI: https://doi.org/10.1016/B978-0-443-15672-4.00023-1

(Chapter 2), optimizing the use of concrete by adopting innovative technologies, for example, 3D printing technology (Chapters 11–14 and 19), use of supplementary cementitious materials (Chapter 3), development of demountable connections, allowing future reuse of structural concrete elements (Chapter 20) and enhancing concrete performance for less cement use (Chapter 17) or thermal energy storage of concrete (Chapters 18 and 19). On the other hand, at the structural level, PAS 2080 proposed the hierarchy of carbon reduction potential presented in Fig. 23.1 (The British Standards Institution, 2023).

Achieving net-zero emissions in the concrete industry will require a concerted effort from all stakeholders, including researchers, engineers, manufacturers, and policymakers. Continuous research and development, along with a commitment to sustainable practices, will be crucial in reaching this ambitious goal.

While significant progress has been achieved, the sustainable development of concrete materials and structures continues to face challenges before widespread implementation (Figueira et al., 2021; Han et al., 2014, 2017; Ozcelikci et al., 2023). Primarily, the adoption of novel sustainable concrete materials and structures could potentially introduce new environmental and energy-related challenges. For example, geopolymer concrete usually needs strong alkali or strong acid activations. The massive use of bases and acids, on the one hand, requires new scarce raw materials, for example, sodium silicate; on the other hand, their production consumes a large amount of energy and has potential environmental harm due to their leaking risk. Similarly, for other forms of sustainable concrete materials and structures, all their potential environmental impact should also be considered throughout. Secondly, it is imperative to prioritize the development of scalable and cost-effective sustainable concrete manufacturing technologies, given that current efforts are predominantly conducted on a laboratory scale. Recent 3D concrete printing technologies remain prohibitively expensive for widespread practical use, and their suitability for large-scale concrete structures necessitates additional validation and encounters several challenges, such as addressing longitudinal reinforcement and producing sizable flat slabs. Thirdly, more potential forms of sustainable concrete materials and structures, for example, special concrete materials and structures with nondispersible underwater, antistatic, radiation shielding, and bactericidal capacities, need further exploration and verification. Finally, the life-cycle assessment and design standard of sustainable concrete materials and structures face obstacles due

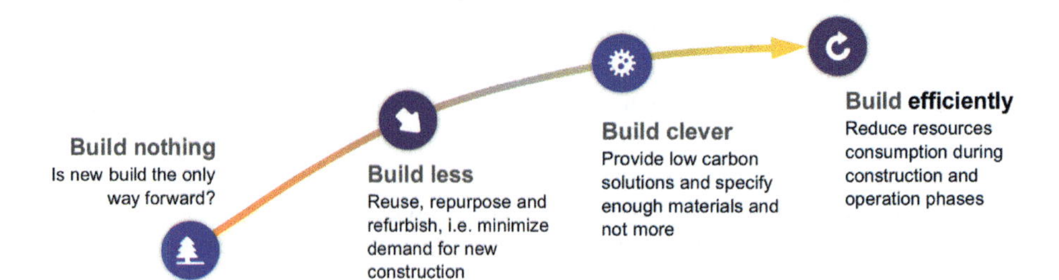

Figure 23.1 Hierarchy of carbon reduction potential.

to the lack of long-term durability data and insufficient mechanical experimental data. Although many researchers have noticed the importance of sustainability of concrete materials and structures, the lack of a coordinating mechanism leads to duplication of research, resulting in a huge amount of effort only generating a limited data and knowledge. More effective and fast-track channels should be developed to entrench smooth communication for the updated findings and applications, considerably boosting the development of sustainable concrete materials and structures.

In conclusion, sustainable concrete materials and structures offer promise for the future, yet there is still a significant journey ahead to fully realize their potential. As shown in Fig. 23.2, it is necessary not only to underpin the relationship between the composition, structures, process, and properties of developed sustainable concrete materials and structures, but also to assess their life-cycle costs in terms of economic, social, and environmental aspects. To achieve these goals, research clusters comprising diverse institutions should cooperate to uncover the basic principles, integrate advanced technologies such as nanotechnology, biotechnology, materials genome engineering, metamaterial, artificial intelligence, digital manufacturing, digital twin, intelligent construction, and establish guidelines and standards for the

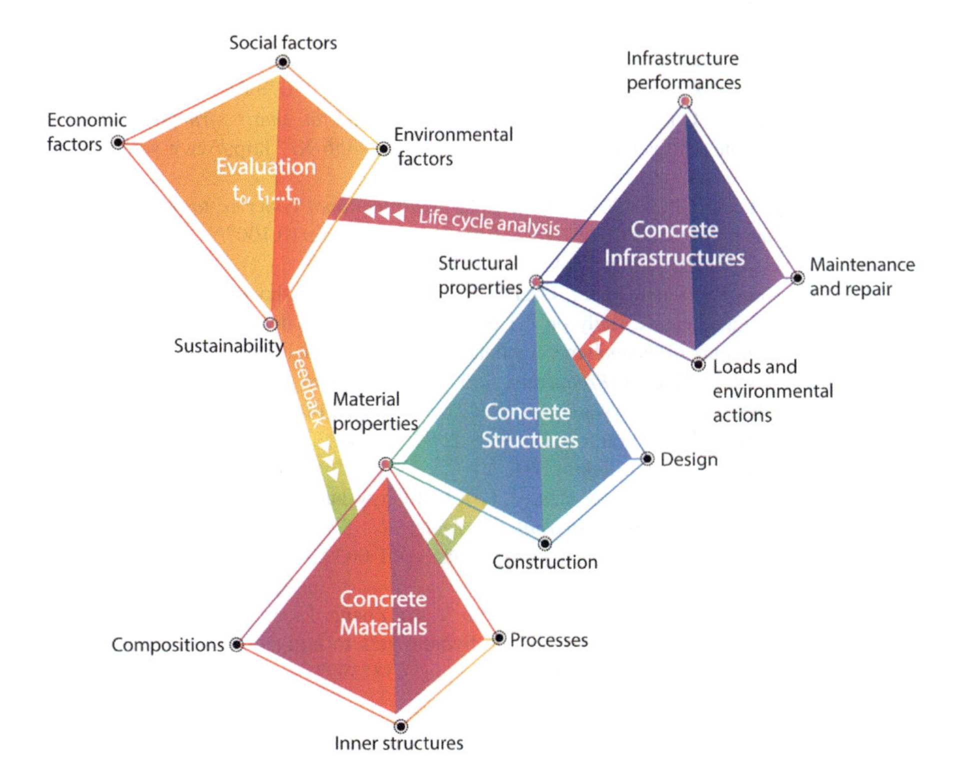

Figure 23.2 Research and development roadmap for sustainable concrete materials and structures (in which $t_1, t_2, \ldots, t_n$ mean the different influencing factors).

advancement of sustainable concrete materials and structures (Ding et al., 2023; Figueira et al., 2021; Han et al., 2017; Wang et al., 2021).

References

Alhawat, Musab, Ashour, Ashraf, Yildirim, Gurkan, Aldemir, Alper, & Sahmaran, Mustafa (2022). Properties of geopolymers sourced from construction and demolition waste: A review. *Journal of Building Engineering*, *50*104104. Available from https://doi.org/10.1016/j.jobe.2022.104104.

Ding, S., Wang, X., & Han, B. (2023). *New-generation cement-based nanocomposites*. Springer. Available from https://doi.org/10.1007/978-981-99-2306-9, https://link.springer.com/book/10.1007/978-981-99-2306-9.

Elhacham, E., Ben-Uri, L., Grozovski, J., Bar-On, Y. M., & Milo, R. (2020). Global human-made mass exceeds all living biomass. *Nature Research*, *588*(7838), 442–444. Available from https://doi.org/10.1038/s41586-020-3010-5.

Figueira, D., Ashour, A., Yıldırım, G., Aldemir, A., & Şahmaran, M. (2021). Demountable connections of reinforced concrete structures: Review and future developments. *Structures*, *34*, 3028–3039. Available from https://doi.org/10.1016/j.istruc.2021.09.053.

Han, B., Ding, S., Wang, J., & Ou, J. (2019). *Nano-engineered cementitious composites: Principles and practices*. Springer. Available from https://doi.org/10.1007/978-981-13-7078-6, https://link.springer.com/book/10.1007/978-981-13-7078-6.

Han, B., Yu, X., & Ou, J. (2014). *Self-sensing concrete in smart structures*. Elsevier. Available from https://doi.org/10.1016/C2013-0-14456-X, http://www.sciencedirect.com/science/book/9780128005170.

Han, B., Zhang, L., & Ou, J. (2017). *Smart and multifunctional concrete toward sustainable infrastructures*. Springer. Available from https://doi.org/10.1007/978-981-10-4349-9, https://www.springer.com/in/book/9789811043482.

Ozcelikci, E., Kul, A., Gunal, M. F., Ozel, B. F., Yildirim, G., Ashour, A., & Sahmaran, M. (2023). A comprehensive study on the compressive strength, durability-related parameters and microstructure of geopolymer mortars based on mixed construction and demolition waste. *Journal of Cleaner Production*, *396*. Available from https://doi.org/10.1016/j.jclepro.2023.136522, https://www.journals.elsevier.com/journal-of-cleaner-production.

The British Standards Institution. (2023). *PAS 2080: Carbon management in buildings and infrastructure guidance*. London: BSI.

Wang, X., Dong, S., Ashour, A., & Han, B. (2021). Energy-harvesting concrete for smart and sustainable infrastructures. *Journal of Materials Science*, *56*(29), 16243–16277. Available from https://doi.org/10.1007/s10853-021-06322-1, http://www.springer.com/journal/10853.

Wang, X., Ding, S., Ashour, A., Ye, H., Thakur, V. K., Zhang, L., & Han, B. (2024). Back to basics: Nanomodulating calcium silicate hydrate gels to mitigate CO_2 footprint of concrete industry. *Journal of Cleaner Production*, *434*139921. Available from https://doi.org/10.1016/j.jclepro.2023.139921.

Index

Note: Page numbers followed by "*f*" and "*t*" refer to figures and tables, respectively.

D

E

F

Q

R

9780443156724